XIV International Conference on Small-Angle Scattering 2009

AA008604

(SAS 2009)

Journal of Physics: Conference Series
Volume 247

Oxford, United Kingdom
13-18 September 2009

Editors:

Goran Ungar and Team of Scientific Editors

Graham Douglas
Johanna Schwarz

ISBN: 978-1-61782-076-2

Printed from e-media with permission by:

Curran Associates, Inc.
57 Morehouse Lane
Red Hook, NY 12571

CURRAN ASSOCIATES INC. proceedings.com

Some format issues inherent in the e-media version may also appear in this print version.

Printed by Curran Associates, Inc. (2010)

Institute of Physics
Dirac House, Temple Back
Bristol BS1 6BE UK

Phone: 44 1 17 929 7481
Fax: 44 1 17 920 0979

techtracking@iop.org

Additional copies of this publication are available from:

Curran Associates, Inc.
57 Morehouse Lane
Red Hook, NY 12571 USA
Phone: 845-758-0400
Fax: 845-758-2634
Email: curran@proceedings.com
Web: www.proceedings.com

TABLE OF CONTENTS

INVITED PAPER AND REVIEWS

BEAM LINES AND INSTRUMENTATION

THEORY, DATA PROCESSING AND MODELLING

BIOLOGICAL SYSTEMS AND MEMBRANES

CERAMICS, GLASSES AND POROUS MATERIALS

COLLOIDS AND SOLUTIONS

HIERARCHICAL STRUCTURES AND FIBRES

METALLIC AND MAGNETIC SYSTEMS

POLYMERS

TIME-RESOLVED DIFFRACTION, KINETIC AND DYNAMICAL STUDIES

PREFACE

The XIV International Conference on Small-Angle Scattering, SAS-2009, was held in Oxford UK, 13-18 September 2009, and was jointly organised under the auspices of the *International Union of Crystallography Commission on SAS* by a team from the Diamond Light Source and the ISIS Pulsed Neutron Source – their first such joint venture - with help from the UK Science & Technology Facilities Council. It was the first time that this long running and successful series of conferences on the application, science and technology of small-angle scattering techniques had been staged in the UK.

The UK has a proud heritage in small-angle scattering: as home to one of the world's first SANS instruments (at AERE Harwell), as the site of the world's first 2nd generation X-ray Synchrotron (the SRS at Daresbury with its suite of SAXS beamlines), and latterly as the location of the world's most successful pulsed source SANS instrument. Indeed, 2009 also marked the 25th Anniversary of neutron operations at ISIS and the opening of a Second Target Station. Whilst the SRS ceased operations in 2008, its mantle has been inherited by the Diamond synchrotron. Many delegates took the opportunity to visit both Diamond and ISIS during a conference excursion.

Despite the prevailing global economic downturn, we were delighted that 434 delegates from 32 different countries were able to attend SAS-2009; two-thirds were drawn from the UK, Germany, Japan, the USA and France, but there were also sizeable contingents from Australia, Korea, Taiwan and South America. In many ways this geographical spread reflects the present and emerging distribution, respectively, of 3rd generation X-ray synchrotrons and high-flux neutron sources, although the scope of the conference was not solely limited to these probes.

Financial support from the IUCr enabled us to grant bursaries to attend SAS-2009 to 12 delegates from emerging countries (Algeria, Argentina, Brazil, India, Nepal, Romania, Russia and the Ukraine).

The scientific heart of the conference comprised 10 plenary sessions, interspersed by 39 'themed' parallel sessions, 2 poster sessions, an afternoon tour of Diamond and ISIS, and a week-long exhibition. There were 144 contributed oral presentations and 308 poster presentations across a total of 21 themes. Over half of all presentations fell under 6 themes: biological systems, colloids and solutions, instrumentation, kinetic and time-resolved measurements, polymers, and surfaces and interfaces. The importance of SAS techniques to the study of biology, materials science and soft matter/nanoscience is clear.

The plenary presentations, which covered topics as diverse as advanced analysis techniques, biology, green chemistry, materials science and surfaces, were delivered by Frank Bates, *Minnesota, USA,* Peter Fratzl, *MPI Golm, Germany,* Buxing Han, *Bejing, China,* Julia Kornfield, *CIT, USA,* Jan Skov Pedersen, *Aarhus, Denmark,* Moonhor Ree, *Pohang, Korea,* Mitsuhiro Shibayama, *Tokyo, Japan,* Robert Thomas, *Oxford, UK,* Jill Trewhella, *Sydney, Australia,* and Thomas Zemb, *ICSM Bagnols, France.*

Instigated by representatives of the Belgian and Dutch SAS communities one parallel session was dedicated to a tribute for Michel Koch, the pioneer of so many novel applications of SAXS, who retired after 30 years at the EMBL Hamburg in late 2006. With a supporting cast that included Wim Bras, *ESRF, France,* Tony Ryan, *Sheffield, UK* and Joe Zaccai, *ILL,France,* and watched by former colleague André Gabriel, Michel treated the audience to a fascinating - and at times light-hearted - retrospective of the evolution of synchrotron SAXS.

Another parallel session was devoted to the work of the *canSAS* (Collective Action for Nomadic Small-Angle Scatterers) network of large-facility representatives and instrument scientists in areas such as data file formats, intensity calibration and software development. For further information see http://www.smallangles.net/wgwiki/index.php/canSAS_Working_Groups.

A total of nine awards were presented at the conference. The *Lifetime Achievement*, or *'Andre Guinier', Award*, given to those who have made a sustained and recognised contribution to the development or application of Small-Angle Scattering, went to Vittorio Luzzati, Emeritus Research Scientist at the Centre de Génétique Moléculaire du CNRS, France. Dr Luzzati has had a long and distinguished career in X-ray scattering publishing over 170 research papers - 10 in *Nature* – which have so far accumulated over 3500 citations.

The award for 'Excellence in SAS Technical/Instrumental Development' went to J Polte, *BAM, Germany,* for *'New insights into nucleation and growth processes of gold nanoparticles derived via coupled in-situ methods'.* That for 'Excellence in the Theoretical Development of SAS' went to C Gommes, *Liege, Belgium,* for *'SAXS Data Analysis of Ordered and Disordered Morphologies with Gaussian Random Field Models'.* B Pauw, *Technical University, Denmark,* received the award for 'Excellence in the Application of SAS' for work on *'Strain-induced Internal Fibrillation of Aramid Filaments'.* And the award for 'Excellence in the Communication of SAS Science' went to J G Grossmann, *Liverpool, UK,* for his talk on *'Probing the Structure of Biological Macromolecules in the Gas Phase'* .

A Hexemer, *LBNL, USA,* won the prize for the 'Best Poster in Technical/Instrumental Development' for *'SAXS/WAXS using a Multilayer Monochromator'.* The prize for 'Best Poster in Theoretical Development' went to S Haas, *Helmholtz Centre Berlin, Germany,* for *'Simultaneous structure and chemical nano-analysis of an efficient frequency upconversion glass-ceramic by ASAXS'.* And in a remarkable 'double', the prizes for 'Best Poster for Application in Life Sciences' and 'Best Poster for Application in Physical Sciences' went to A Maerten and J Prass, respectively, both from *MPI Golm, Germany,* for their work on *'SAXS studies of human tooth dentine: analysis of a spatially inhomogeneous and varying bio-material'* and *'Analysis of sorption strains in ordered mesoporous materials by in-situ small-angle x-ray diffraction'.*

The conference could not have been staged without the support and commitment of a large number of people and organizations, many of whom are listed separately within these Proceedings, and it is only right that we acknowledge their contribution. The generous financial support from our sponsors was particularly welcome given the economic climate. To all of them we offer our heartfelt and grateful thanks.

We would also like to thank our Guest Editor, Goran Ungar, his team of Section Editors, and the several dozen anonymous referees, for so admirably managing the scientific review of the manuscripts in these Proceedings, and Graham Douglas and Johanna Schwarz at IoP Publishing for their assistance in bringing these Proceedings to fruition.

Stephen King
Nicholas Terrill
SAS-2009 Co-Chairmen & Local Organisers

SAS-2012 will be held in Sydney, Australia, 18-23 November 2012.

INTERNATIONAL ADVISORY COMMITTEE

I Hamley, *Reading, UK*, E Gilbert, *OPAL, AUSTRALIA,* T Hanley, *ANSTO, AUSTRALIA,* G Warr, *Sydney, AUSTRALIA,* O Glatter, *Graz, AUSTRIA,* B Goderis, *Leuven, BELGIUM,* I Torriani, *Campinas, BRAZIL,* B Han, *Bejing, CHINA,* C Han, *Academy of Sciences, CHINA,* P Štěpánek, *IMC, Prague, CZECH REPUBLIC,* J S Pedersen, *Aarhus, DENMARK,* J-F Berret, *Paris 7, FRANCE,* W Bras, *ESRF, Grenoble, FRANCE,* T Forsyth, *ILL, Grenoble, FRANCE,* T Narayanan, *ESRF, Grenoble, FRANCE,* T Zemb, *CEA, Saclay, FRANCE,* P Fratzl, *MPI Golm, GERMANY,* D Richter, *Julich, GERMANY,* D Svergun, *DESY, Hamburg, GERMANY,* S Mazumder, *BARC, INDIA,* A Ikram, *NSL Batan, Serpong, INDONESIA,* J Holmes, *Cork, IRELAND,* Y Cohen, *Technion, ISRAEL,* F Carsughi, *Ancona, ITALY,* E Fratini, *Firenze, ITALY,* M Furusaka, *KEK, JAPAN,* T Hashimoto, *Kyoto, JAPAN,* M Shibayama, *Tokyo, JAPAN,* S Sakurai, *Kyoto Institute of Technology, JAPAN,* N Yagi, *Spring-8, JAPAN,* A Petukhov, *Utrecht, NETHERLANDS,* A Ozerin, *ISPM, Moscow, RUSSIA,* D Uhríková, *Bratislava, SLOVAKIA,* A Stradner, Freibourg, *SWITZERLAND,* A Allen, *NIST, USA,* N Balsara, *Berkeley, USA,* P Butler, *NIST, USA,* T Irving, *APS, USA,* T Lodge, *Minnesota, USA,* P Thiyagarajan, *IPNS, USA,* L Walker, *Carnegie Mellon, USA.*

SCIENTIFIC PROGRAMME COMMITTEE

A Ryan, *Sheffield, UK,* T Wess, *Cardiff, UK,* J Eastoe, *Bristol, UK,* K Edler, *Bath, UK,* S Eichorn, *Manchester, UK,* M J Lawrence, *King's College London, UK,* S Lee, *St Andrews, UK,* J Penfold, *ISIS & Oxford, UK,* G Sankar, *Royal Institution London, UK,* J Seddon, *Imperial College London, UK,* R Thomas, *Oxford, UK,* E Valsami Jones, *Natural History Museum, London, UK,* J van Duijneveldt, *Bristol, UK,* R Winter, *Aberystwyth, UK.*

LOCAL ORGANISING COMMITTEE

S King, *ISIS,* N Terrill, *Diamond,* T Arnold, *Diamond,* I Boscaro-Clarke, *Diamond,* J Comfort, *Diamond* , P Elcock, *ISIS,* S Fletcher, *Diamond,* R Heenan, *ISIS,* J Hiller, *Diamond,* S Hills, *STFC,* K Inoue, *Diamond,* M Malfois, *Diamond,* C Moores, *Diamond,* C Nicklin, *Diamond,* C Pizzey, *Diamond,* D Price, *Diamond,* J L Rawle, *Diamond,* S Rogers, *ISIS,* P Taylor, *ISIS,* A Terry, *ISIS,* M Wells, *STFC.*

SPONSORS AND EXHIBITORS

Anton-Paar, Advance Design Consulting, Advent Research Materials, ANSTO, Australian Synchrotron, Bruker-AXS, ESRF, ILL, Hecus, Helmholtz-Zentrum Berlin, Kurt J Lesker, Mirrortron, *Neutron News*, Rayonix, *Synchrotron Radiation News* and *Soft Matter*. There were also exhibits by Diamond, ISIS and the IUCr.

GUEST EDITORS

G Ungar, *Sheffield, UK*, R Heenan, *ISIS, UK,* C Neylon, *ISIS, UK,* R Winter, *Aberystwyth, UK,* P Griffiths, *Cardiff, UK,* C Knupp, *Cardiff, UK,* C Rau, *Diamond, UK,* W Bouwmann, *Delft, The Netherlands,* E Valsami Jones, *Natural History Museum, London, UK,* S Eichhorn, *Manchester, UK,* I Robinson, *University College London, UK,* N Rhodes, *ISIS, UK,* E Gilbert, *ANSTO, Australia,* A Hoell, *Helmholtz-Institute, Berlin, Germany,* S Lee, *St Andrews, UK,* P Fairclough, *Sheffield, UK,* K Edler, *Bath, UK,* J S Pedersen, *Aarhus, Denmark,* J Cabral, *Imperial College London, UK,* W Bras, *ESRF, France,* S Holt, *ANSTO, Australia,* F Fernandez-Alonso, *ISIS, UK.*

Editorial

Proceedings of the XIV International Conference on Small-Angle Scattering, SAS-2009

There are 52 papers in these Proceedings. The papers are divided into 10 thematic sections and a section for invited papers and reviews. The sections and the respective section editors are given below.

Section	*Editor(s)*
Invited Papers and Reviews	Peter Griffiths, Wim Bras, Rudolf Winter
Beamlines and Instrumentation	Elliot Gilbert, Wim Bras, Nigel Rhodes
Theory, Data processing and Modelling	Jan Skov Pedersen, Carlo Knupp
Biological Systems and Membranes	Richard Heenan, Cameron Neylon
Ceramics, Glasses and Porous Materials	Rudolf Winter
Colloids and Solutions	Peter Griffiths
Hierarchical Structures and Fibres	Steve Eichhorn, Karen Edler
Metallic and Magnetic Systems	Armin Hoell
Polymers	Patrick Fairclough
Time resolved Diffraction, Kinetic and Dynamical Studies	João Cabral, Christoph Rau

We are grateful to all section editors and the many anonymous referees for their invaluable effort which made the publication of the Proceedings possible. The refereeing process was strict and thorough, some papers were rejected and most were improved. The resulting compendium gives a good overview of recent developments in small-angle X-ray and neutron scattering theory, application, methods of analysis and instrumentation. Thus it should be a useful source of reference for a number of years to come. The papers are a good reflection of the material presented at the meeting. Because of the general high quality of the articles, it was difficult to decide which to highlight and be fair to all contributors. The following in particular have caught the attention of the editors.

Highlighted papers

A statistical survey of **publications** reporting the application of SAXS and SANS by Aldo Craievich (paper 012003) is recommended reading for anyone needing convincing about the vibrancy of this scientific field and the ever expanding use of these techniques. Two aspects of **coherent X-ray scattering**, made available by the advent of the 3rd generation synchrotron sources, are discussed in the papers by Felisa Berenguer and the Diamond/UCL team (012004) and by Birgit Fischer and the DESY/Rostock team (012026). The former describe their effort to reconstruct the image of wet collagen tissue from the speckled pattern of a narrow transmitted beam and complement the results of imaging by scanning SAXS. Fischer et al. utilize the coherence of the X-ray beam in the photon correlation spectroscopy mode to determine the dynamic structure factor, from which the q-dependence of the relaxation times was determined for solutions of charged colloidal particles. The nature of particle motion was thus determined, i.e. ballistic at lower concentrations vs. restricted by "caging" at higher concentrations. The use of scanning **microbeam**

diffraction was also described in the paper by the group of Peter Fratzl (012031), who combined it with tilting the sample in order to obtain what is referred to as 3D SAXS. In this way the orientation distribution of crystallites in callus tissue during bone fracture healing can be explored. Naoto Yagi et al. (012024) used microbeam SAXS and WAXS to study dental enamel crystals in a caries lesion.

Anomalous, or resonant, X-ray scattering at low angles (ASAXS) was employed by Guenter Goerigk and Norbert Mattern (Jülich / Dresden) (012022) to study phase separation in the ternary alloy Ni-Nb-Y. Experiments were performed near the K-absorption edges of the three elements and the authors describe how they determined quantitatively the chemical composition of the different separate phases as well as short-range concentration fluctuations during spinodal decomposition. For organic materials, where heavy atoms are not present, typically sulphur would be replaced by Se, so ASAXS could be performed near the Se K-edge. However, Masashi Handa et al. from the University of Tokyo report in paper 012006 a feasibility study of using sulphur itself as the label. They deal with the challenging task of using low-energy X-rays at the sulphur K-edge.

Alexandra Vasilieva and teams from St Petersburg and Moscow describe in paper 012029 a study of a FCC/HCP photonic crystal with a unit cell of 650 nm using the DUBBLE beamline at ESRF, where the incident beam is focused with a compound Berilium lens to give divergence in the **microradian** range. Defects, correlation lengths, mosaic spread and twinning could be studied in a thin film. The use of grazing incidence small- and intermediate-angle X-ray scattering (**GISAXS**) was illustrated on several examples of complex liquid crystal phases and nanoparticle systems by Xiangbing Zeng and the team from Sheffield in paper 012032. Apart from the intrinsic interest in the structure of thin films of these materials, the close proximity of the surface imposes usually a very high degree of preferred orientation, facilitating structural analysis.

From the papers describing the use of neutron scattering, we highlight the study by Léon van Heijkamp and the Delft group, paper 012016, which describes proof-of-principle experiments using **spin-echo SANS** (SESANS) for the study of D_2O-labeled liposomes and their destruction with a view of providing a nondestructive technique for monitoring targeted drug delivery for the destruction of tumor tissue. This technique is based on monitoring the decay of spin polarization of scattered neutrons, and it can probe correlation lengths from nanometres to tens of microns.

In paper 012002 Thomas Zemb discusses the various ways in which **SAXS and SANS** data can be combined to investigate microemulsions of surfactants lacking long-range order. Thus, for example, it is described how form and structure factors can be separated, or how the degree of connectivity can be determined between rod-like entities forming “molten cubic” or “sponge” phases.

In the **Theory, Data Processing and Modelling** section we highlight two papers. Paper 0122012, by Carlos Cabrillo and co-workers from Madrid, presents a real-space model for powder samples of relatively highly ordered colloidal particles based on the radial pair distance distribution function. The model takes account of different types of disorder, i.e. of both 1st (“thermal”) and 2nd (“paracrystalline”) kind, as well as finite size, stacking and orientational disorder. In paper 012014 Salvino Ciccariello works out the 3D correlation function of plane objects, specifically a triangle of a general shape. This paves the way to analysing morphologies that can be approximated by cylinders of different cross-sections.

Papers 012051 (Polte et al.) and 012047 report the use of SAS in combination with **X-ray spectroscopy**. Ristic et al. describe a real-time study of **crystal nucleation** induced by ultrasound (012049), while Marianne Imperor-Clerc et al. present an investigation of lyotropic **liquid crystals under shear** (012052). In the **Polymers** section Geoff Mitchell describes modelling of SANS data on electrospun fibres (012042), while Cabral et al. describes a study of nanoparticle aggregation in bulk and thin polymer films (012046). Work on **phospholipid membranes** was reported by Spinozzi et al. and by Onai et al. in papers 012019 and 012018, respectively. The latter deals with the effect of osmotic pressure on model "lipid rafts", these being considered as having an important role in the function of the mammalian cell membrane. Readers interested in **biological** systems will find a number of other interesting papers describing the use of either SAXS or SANS in the **Biological Systems and Membranes** section. A number of descriptions of recent designs of SAXS (synchrotron) and SANS beamlines, as well as detector developments can be found in the section on **Beamlines and Instrumentation**.

Goran Ungar, Editor-in-Chief
Richard Heenan, Deputy Editor-in-Chief

SAXS Instrumentation for Synchrotron Radiation then and now

M.H.J. Koch

Müllenhoffweg 33a, D-22607 Hamburg, Germany

koch@embl-hamburg.de

Abstract. A brief overview is given of the development of instruments and methods for synchrotron radiation small angle scattering and diffraction on biological systems and synthetic polymers. It is shown that many of the developments were largely accidental and although there were a number of innovations, in most cases other fields provided the sources of novelty.

1. The road to the tipping point

The present article aims at giving an – inevitably biased - overview of the development of instruments for small angle X-ray scattering (SAXS) and diffraction (SAXD), in particular for applications in biology and synthetic polymers with an emphasis on time resolved measurements. It does not cover important areas where I have no experience like inorganic applications, anomalous scattering, grazing incidence scattering or coherent applications. Since many things happened in parallel the story is not linear and frequent backtracking is unavoidable.

As noted by Patricia Fara in her recent book on the history of science, during the 20th century, (big) science was driven by the 5M (military, money, manpower, machines, media) [1].This is perhaps not surprising as the Greek philosopher Heraclitus (600BC) had noted more generally that war is the father of all things and somewhat later Thucydides (400BC) had diagnosed very human characteristics, honour, fear and interest, as the causes of war. Is this really what drives our activity?

The nearly simultaneous development of synchrotron radiation in Europe, the USA and the Soviet Union (for a short history see http://xdb.lbl.gov/xdb.pdf) apparently – but only apparently - followed the 5M pattern. The first synchrotron was built in 1946 by a group of the British Admiralty [2] and the second one at the General Electric labs, where work on betatrons as potential X-ray sources had been going on for some years (reviewed in [3, 4]). Both machines were intended to produce X-rays by sending the accelerated electrons to a target like in conventional sources. Although early synchrotrons were built from surplus radar sets and capacitors the aim was actually to develop non-weapons facilities after world war II [5]. Synchrotron radiation was only observed later and its properties as light source were only fully realized in the 1960s (see [4]). Further accelerator developments took place in high energy physics (HEP) laboratories, among others at CERN. The change of mood and funding patterns in those years is best illustrated by the fact that the decision to join CERN by the British government, for example, was based on the assumption that no work of military importance would be done there [6]. With insertion devices there is a slightly different story. The first undulator built in 1953 [7], and the first wigglers used as radiation sources around 1980 were based on electromagnets and soon after on strong permanent magnets, which had become available in the early 1970s, and on superconducting devices [8, 9]. Wigglers and undulators thus existed well before funds for large devices like those used in free electron lasers later started to flow in the frame of the Strategic Defence Initiative (1983-1992) [10]. SSRL's first wiggler has now been put to the ultimate peaceful use: it decorates the garden at the SLAC National Accelerator Laboratory.

High energy physicists like to point out that synchrotron radiation was a by-product of their experiments and although this is largely true, it is perhaps more accurate to say that storage rings are the descendents of (commercially) failed X-ray sources.

The development of detectors follows a pattern similar to that for the sources. Multiwire proportional detectors with wires connected in parallel and a readout based on valves had already been used as particle detectors in the Manhattan project (see e.g [11]) and delay lines had been developed for early radar and computer applications [5]. These developments were, of course, only described in the late 1940s but were a source of inspiration for high energy physicists trying to replace bubble chambers and photographic methods. After spark chambers, which had very low count rates, various forms of multiwire chamber readouts, including some involving delay lines, were proposed in the 1960s. Electromagnetic delay lines, which were capacitively coupled to the wires, were introduced by V. Perez-Mendez and his colleagues at Berkeley [12]. Two events which had an influence on detectors occurred in 1968 – but remember that events mark the end of a process, not its beginning: G. Charpak and coworkers using newly available solid state amplifiers developed the readout of individual wires [13], which was very successful in high energy physics and for a while also in protein crystallography, whereas C. Borkowski and M. Kopp developed the RC delay line readout for various types of detectors [14]. The propagation of signals along a delay line (i.e. Kelvin's 1855 model and Heaviside's modern form of the telegrapher's equation), the electrostatic field around a wire or a grid and the properties of RC and LC delay lines (ladder networks) were all standard textbook knowledge at that time [15]. It was essentially progress in electronics, made for entirely different purposes, which enabled an accurate readout of the position of the events. This led to the success of multiwire chambers in HEP and in X-ray detection. The mere existence of such devices was, however, not sufficient, they had to be brought in a new context. For SAXS this occurred a few years later when the experiments on lipids made in V. Luzzatti's laboratory with an RC delay line detector built by A. Gabriel [16], following a suggestion of M. Chabre, attracted the attention of biophysicists to these devices. Due to this transfer delay, the first experiments with SR were still done with film, but as from the mid-1970s they were made with gas detectors (see e.g. [17]). What made delay line detectors a success for SAXS with synchrotron radiation was the introduction of LC readout with inductances directly connected to the cathode strips by A. Gabriel [18]. The replacement of the resistive wire of the RC detectors by metal wires gave the new detectors a much higher radiation resistance (For an introduction to these detectors see [19]).

The proceedings of the 1977 synchrotron radiation instrumentation conference in Orsay [9] and of the 1980 conference on X-ray detectors for synchrotron radiation in Hamburg [20] give a good overview of the detector work going on at that time at the different facilities worldwide. What is striking with hindsight is that there was only a brief mention of the potential of CCD detectors and that imaging plates and pixel detectors, the disruptive innovations of the next decade, do not even appear. It is also useful to realize, when reading such documents, that many more devices got proposed, or even built, than were actually used afterwards for any meaningful measurements.

Efforts were also underway in the late 1970's to develop integrating detectors. These were based on vidicons, and although they were useful for strong scatterers like muscle or polymers [21, 22], their high background made them unsuitable for weak scatterers. Fortunately, the development of gas detectors survived a bout of exuberant but fruitless enthusiasm for TV-based detectors at the IPPC (Instrumentation policy planning committee) and the instrumentation division of the European Molecular Biology Laboratory (EMBL) in Heidelberg.

Apart from a source and a detector a SAXS instrument is conceptually very simple as it consists of some optics, slits and vacuum pipes. These all existed and the instruments at synchrotron radiation sources, not unexpectedly, naturally evolved from instruments used on conventional sources. The history of some of the early devices and their designs have been recently described [23-25].

The story is thus very complicated and essentially illustrates that scientists quite flexibly and, forced by circumstances, sometimes somewhat deviously tap the flows of money in society. No wonder thus that even historians may get confused about the way things work.
Sociologists [26] and evolutionists [27] therefore often prefer diffusion models to explain how effectively uncorrelated events are associated at another level of complexity with apparently ordered patterns in a system. Sociologists also distinguish between *invention* (i.e. the process by which a new idea is discovered or created) and *innovation* (i.e. an idea, practice, or object that is perceived as new by an individual or other unit of adoption, regardless of its objective novelty). Diffusion of ideas and practices depends on the environment in which it takes place and because of the majority's aversion towards novelty, resistance or friction is highest just before things start moving. Diffusion models of successful projects lead to sigmoid growth curves as illustrated in Figure 1. The two most important points in this curve are the tipping point, where the process of growth or acceptance becomes self-sustained, and the point of diminishing returns. The best time to move on for a scientist is somewhere between these two points since as Rutherford observed "when the man in the street starts believing it, the man in the lab knows that it is finished". In other words, when the hill is taken, the commandos are replaced by the infantry. Of course, most projects are unsuccessful and peter out even before they reach the tipping point. This well-known Christmas turkey effect [28] is also illustrated in Figure 1. For how long a given type of instrument (or species) survives depends on so many largely uncorrelated factors that selection can be considered a random process. The only constant is the need to work. This view also better reflects my personal experience and I have come to believe that at the more practical level science is driven by the five F's: Funds, Freedom (there is none without funds), Fun (there is none without freedom), Fortune and Fame. For completion, at least in the case of experimenters, one should perhaps add one more F on which one rarely insists in more romantic presentations. As pointed out by J. Gleick: "The experimenter's lovers sweat, complain, and fart" [30].

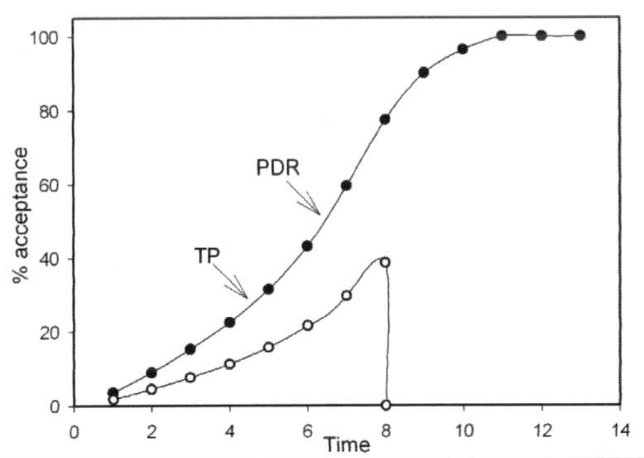

Figure 1. Growth of acceptance of a new device or practice in a group. Beyond the tipping point (TP) the process becomes self-sustained. PDR: point of diminishing returns. Most projects fail as a result of unexpected events like a discontinuity in funding or the discovery of negative side effects. Performance curves of devices [29] or investments [28] against time have the same shape.

I joined the EMBL outstation on the site of the Deutsches Elektronen Synchrotron (DESY) in Hamburg in 1977, soon after H. Stuhrmann, with whom I had collaborated for several years on neutron small angle scattering, had become the head of outstation. Part of the attraction was the stated purpose of the EMBL, which was (and still is!) to "promote cooperation among European States in fundamental research, in the development of advanced instrumentation .....in molecular biology...".
Clearly, by then most of the equipment and concepts to do small angle diffraction and scattering with synchrotron radiation had already been available in some form or other in different places. Developments in SAXS/SAXD were also under way at LURE (Orsay) [31], the Stanford synchrotron radiation laboratory (SSRL) [32], Novosibirsk [33, 34], Daresbury, where H. Huxley had installed a camera [35], and among other SAXS devices, a white beam camera using an energy dispersive

detector had been built [36], and in Japan where soft X-ray SAXS had been done [37]. Of course, as a newcomer in the field I – perhaps fortunately - did not yet know this.

The closure of NINA at Daresbury in 1977 not only brought J. Bordas, now director of the ALBA synchrotron source in Barcelona, and somewhat later P. Clout, now president of Vista Control Systems in Los Alamos, who had, respectively, experience in synchrotron radiation instrumentation and computer control and data acquisition, to Hamburg, but also a steady stream of visitors and some equipment from Britain. The storage ring DORIS at DESY was starting to work for colliding beam experiments and a tunable wavelength instrument for fibre diffraction (X11) designed by G. Rosenbaum already existed in the experimental area of the EMBL in Bunker IV, but it had not been foreseen to have other instruments in this hall or to operate a multipurpose facility. Installation of additional instruments and shielding in the hall thus had to be planned so that X11 could remain in place [38] as illustrated in Figure 2. To complicate matters, and give the engineer, the late H. Ludwig, quite a few headaches, all of this also required the *a posteriori*, installation of an overhead crane. As the HEP program at DORIS was somewhat unpredictable, one of the main practical problems was that as the energy increased from 3 to 5.6 GeV and the current in the ring from 10 to over 100 mA, the shielding, which consisted of large barite concrete blocks, had to be regularly modified and reinforced to cope with the higher radiation levels.

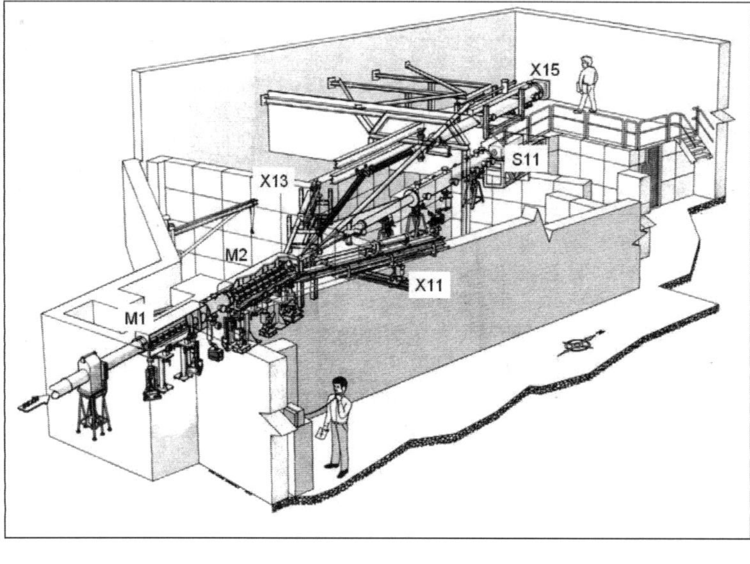

Figure 2. Layout of the instruments in Bunker IV at the EMBL outstation in Hamburg at the end of the 1970s. X15: anomalous scattering camera. S11: EXAFS spectrometer, X11: tunable wavelength mirror-monochromator camera. X13: fixed wavelength mirror-monochromator camera- M1 vacuum box: M2 vacuum mirror - monochromator box for X11 and X13

For a while, the tunnel on the ground floor was shared between the EXAFS spectrometer and H. Huxley's camera which had been brought from Daresbury. Later this camera was replaced by the dance floor instrument of the Risoe group, who could test their instruments while HASYLAB V was under construction. On the roof of the tunnel there was the anomalous scattering camera X15 [39] and a Mössbauer experiment.

The experimental area of X13, a fixed wavelength double focusing mirror monochromator camera [40], was only accessible over the tunnel, but this had the advantage that it was a rather quiet place. At the front end the mirrors and monochromator were located in the original X11 mirror box (the second one in Figure 2) which was a low budget design. The optical system was based on geometrical considerations as ray tracing programs were not available. Mirrors were not readily available and for the first set, made by Astron, a subsidiary of Ferranti, the special tools required to make them also had to be paid for. These two companies no longer exist and subsequent mirrors were made by Zeiss. X13 had a simpler design than X11. The DC motors were controlled with a multiplexer rather than with individual potentiometers for each motor as in the original design of X11. Note that the monochromators were located after the mirror which consisted of eight flat quartz segments in a bicycle chain arrangement. No attempt was made at bending the segments. The initial X11 design used a four point bender for the monochromator which gave some problems and could only be used with short crystals. The main source of progress is often imitation with modification. Hence, as the classical mechanical solution for bending implemented in the triangular monochromator at LURE gave satisfactory results [41] we also used it, but replaced the micrometer screw of the bender by an eccentric shaft to avoid breaking the rather expensive 18 cm long asymmetrically cut triangular Germanium monochromator. A similar monochromator was later also installed at the Photon Factory [42]. The plexiglass tubes filled with helium were soon replaced by metal vacuum tubes with mica front windows instead of Kapton, which much improved the background. The advantage of instruments with sideways deflection rapidly became clear. Since they were outside the cone of the radiation shower which might have occurred in case of electron beam loss in the hall, with an absorber in the beam all adjustments in the hutch could be made manually. The fact that this is no longer allowed only proves that safety regulations are not based on accident statistics.

X-ray monochromators have a narrow bandpass, and especially those who had done neutron scattering dreamed of larger bandpasses. All sorts of techniques were tried out ranging from ion implantation to soap films (for a review of early proposals see [43]). It took a while for some to understand that although bent monochromators gave a wavelength spread this did not result from a higher bandpass but from the fact that each volume element reflected a slightly different wavelength with the same narrow bandpass as the flat crystal. There was also a proposal for a white beam camera at the EMBL, which was, however, not implemented [44]. Pink beam cameras were only implemented much later at the APS (Advanced Photon Source, Argonne) and the ESRF (European Synchrotron Radiation Facility, Grenoble), where they are used for ultrafast time resolved measurements. There are hitherto too few applications [45, 46] to judge how generally useful such devices will be for work on biological systems.

Multilayer monochromators, with a much wider bandpass, had been patented in 1974 by D. Caspar and B. Schoenborn, based on their experiments on nerve fibers [47] (but see earlier references in [48]). It was, however, only in 1998 that H. Tsuruta and his colleagues at the SSRL installed a routinely usable multilayer monochromator for SAXS [49].

The main thing that remained to be done after the successful feasibility and demonstration studies in the late 1970s was to show that SAXS with synchrotron radiation could really contribute something useful to a number of scientific questions of the time. The crucial factor in making use of the available intensity was to increase the readout rate of the data acquisition system. With the first system, based on a time to pulse height converter (TPHC) and a programmable multichannel analyzer (Intertechnique IN90), the total count rate was limited to about 10 kHz. The data were stored on tape and had to be transferred to a PDP 11/45 (248kB) computer for data analysis. The IN90 was, however, sufficient to do experiments on unstriated molluscan muscle which required a time resolution of 0.5 s and where sufficient statistics could be accumulated from about ten contractions. The results of these experiments were, however, only published much later [50]. To speed up the appraisal and interpretation of what was for the time a rather large amount of data some software was also needed as none of the modern packages existed and most users were not as familiar with computers as later generations. This was the origin of the OTOKO program [51, 52], which was quite popular for a while, and still exists in a Windows version.

To make best use of the parasitic time which corresponded to single bunch operation of the storage ring J. Hendrix and the late H. Fuerst built a drift chamber [53] which was, however, never put into operation because soon after the high energy physicists settled for multibunch operation, which rendered the device unusable. This was obviously an example of an interesting idea, rendered obsolete by an unpredicted change of circumstances.

A new data acquisition system based on CAMAC, a standard developed in the early 1970s, soon replaced the IN90. The CAMAC serial highway [54] connected to the PDP 11/45 served three experiments. The data acquisition system consisted of commercial modules for the amplifiers, the

Journal of Physics: Conference Series **247** (2010) 012001 doi:10.1088/1742-6596/247/1/012001

TPHC, the analog to digital converter (ADC) and the 64K histogramming memory. The decisive jump in performance was made by replacing the TPHC/ADC by a time to digital converter (TDC) which directly digitized the time differences between the arrival of the start and stop pulses from the delay line in the detector. The first TDC was a module designed at CERN [55] which had been adapted by C. Boulin to the readout of delay line detectors [56]. The timing of the experiments was controlled by a time frame generator (TFG) and ancillary data like temperature, force etc. were recorded as analog signals, digitized with voltage to frequency converters and stored in a calibration channel unit (CCU). The first prototypes of these modules were built by E. Dorrington in Hamburg [57] and a second version, which could also be distributed, was designed by C. Boulin at the EMBL-Heidelberg [58]. The system was programmed in CATY [59], a BASIC-like language for CAMAC control and data acquisition by F. Golding and D. Dainton.

The CAMAC system was used for more than 20 years with various upgrades. The local intelligence provided initially by the LSI 11 (32K and later 64K), later by more powerful auxiliary controllers (CES, Starburst) was replaced by an IBM PC (AT) in 1988. This was not necessarily a popular development and even the EMBL council briefly took interest. As PCs became more powerful the serial highway could be replaced by other connections and the data acquisition programs were rewritten in C^{++} by F. Golding. These developments, which also made it possible to improve display and data appraisal were also passed on to other experiments at the EMBL and to other labs using this system.

When the CAMAC system became difficult to maintain at the turn of the century it was replaced by National Instruments cards [60] and the N110 TDC developed at the ESRF [61]. Although this system is still in use in a number of places, its maximum count rate, which is limited by the data transfer to the PC, and its flexibility in terms of timing and control of the experiment are lower than that of the CAMAC system, a clear case of regression.

As soon as X13 was equipped with the new acquisition system the stage was set to make the first millisecond time resolved measurements on muscle, which had been a Holy Grail for so many years, possible [21]. In London, the Times even reported the event (March 14, 1980). These and similar experiments have been extensively reviewed [23, 35]. The data acquisition system was very flexible and could be used not only with the delay line detectors but also with the parallel readout detector built by J. Hendrix [62]. This detector had a lower spatial resolution (1 mm) than the delay line detectors (0.2 mm) but could cope with very high count rates which made sub-ms time resolutions possible with muscle [63] or lipids [64]. Its lower spatial resolution was not necessarily a handicap because methods were available to increase the resolution of a scattering pattern by small horizontal displacements of the detector [65]. Unfortunately, with time this device could no longer be maintained.

Experiments on frog muscle necessitated the accumulation of the results of many contractions and thus required long periods of stable beam, which were only available during a limited number of main user shifts. The parasitic time, which was available during colliding beam experiments, was very unpredictable, which made most biological experiments impossible. Nevertheless it seemed a waste of our most important asset not to use it. A very effective use for it was found through collaboration with the group of the late G. Zachmann at the University of Hamburg, who worked on synthetic polymers. With these systems useful time resolved experiments could usually be done in single shots [66, 67] and unlike biological samples, polymers are stable and usually available in large quantities. In the following years a number of people, among them C. Riekel, R. Gehrke, P. Boesecke and S. Seifert, who would later contribute to the development of SAXS at DESY, the ESRF and the APS acquired experience in this group.

Not everything had been perfect with X13. The effect of heat load and radiation damage on various components showed up when the energy and current of DORIS were increased. The mirrors, which were fortunately not coated, were exposed to white radiation in a poor vacuum which resulted in cracks. There were also deposits of cracked oil on their surface which degraded their performance and they regularly had to be dipped in aqua regia to clean them. The insulation of electric cables was damaged and finally even the lead slits melted (see Figure 3). The heatload was in some instances also a problem in the ring and one day we chased the beam for hours until one had to admit that there was an obstruction in the beampath inside the ring. Indeed a piece of stainless steel had melted! This all resulted in quite a lot of maintenance and a steep learning curve.

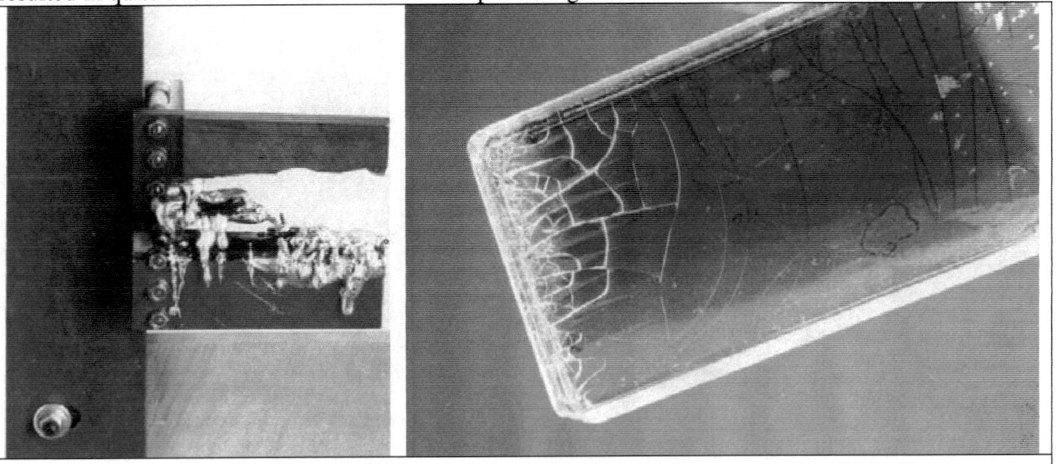

Figure 3. Left: Lead slits exposed to white radiation in vacuum melted away. Removal of heat with copper straps in contact with the mirror box cured the problem. Right: Mirrors exposed to white radiation in poor vacuum rapidly developed cracks and were coated with cracked oil deposits.

Fortunately, we soon could – or rather had to - build a new instrument (X33) in HASYLAB V because dedicated beamtime was only going to be available there. One great advantage of the new construction was that Al/Pb sandwich plates replaced concrete blocks for shielding. Given the previous experience, when X33 was built the relative position of the mirror and monochromator were inverted. Stepping motors were used which facilitated computer control and the mirror boxes and other vacuum components were designed with more care. Airpaths were reduced to a minimum while retaining the flexibility required to handle very different sample environments. Some things were somewhat more difficult than now. As no stepping motors for vacuum nor affordable rotary feedthroughs were available, for example, the motors had to be disassembled to replace all lubricants and re-magnetized in a dipole magnet that was made available for this purpose by DESY. We could also not afford reliable position encoders, which would in any case not have been very useful because of regular changes in the position of the beam when one switched from HEP operation to dedicated beam.
The real difficulties were, however, of a political or managerial nature with the delegations of some countries very directly interfering in purchasing procedures. Moreover, due to staffing limitations at the outstation, imposed as a condition for the participation of Italy to the EMBL, some of the designs had to be made in the instrumentation division in Heidelberg, which did not simplify matters. Despite their enormous good will those involved never really got used to the deadlines and strict schedules in Hamburg, but they still remember this as their most interesting project (H. Wittman, private communication, 2006). As illustrated in Figure 4, X33 was a sturdy instrument which served in its original version for over twenty years.

Besides muscle there were several other scientific projects mostly with external collaborations, which dealt with the assembly of fibers (microtubules, actin, collagen) and in-house research on the superstructure of chromatin with L. Perez-Grau. Promising results had also been obtained with a USAXS camera. The use of magnetic orientation had been tested by A. Fowler at the EMBL in Heidelberg, but the purchase of a strong in-line magnet was turned down. Such experiments were later done on tubulin in Daresbury [68]. Recent experiments on actin indicate that strong magnetic fields give a better orientation than previous methods [69]. Right- or left-circular orientation of tobacco mosaic virus was also observed in time resolved measurements using periodic magnetic fields and gradients [70]. This all suggests that a somewhat more systematic investigation of these methods may be worthwhile.

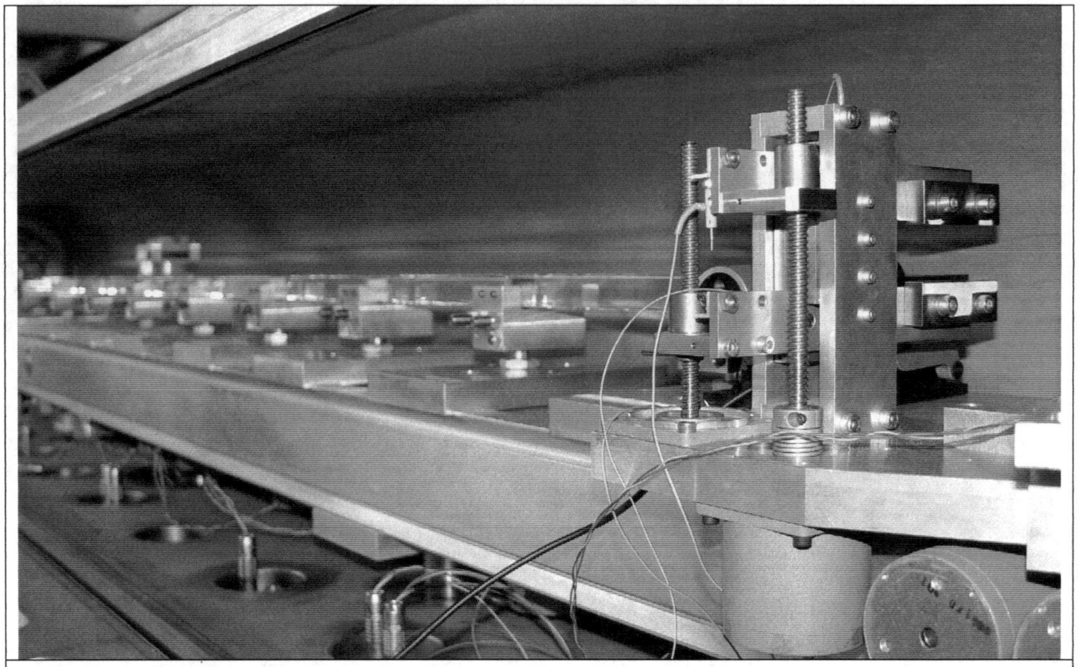

Figure 4. The segmented mirror of X33 monochromator-mirror camera was still in good condition after more than 20 years of use.

The first T-jump used in the assembly studies was an extremely simple device which allowed jumps up or down in the appropriate temperature range – usually 4 to 40°C [71]. It used four water baths [very hot, hot, cold, very cold] and a switching system based in the final version on washing machine valves. By appropriately adjusting the switching points it was possible to avoid over- or undershoots. A device based on Peltier elements with rapid current reversal, which would nowadays be the preferred solution, was used at LURE for studying phase transitions in lipids [72].

The development of stopped flow techniques illustrates how imitation with improvements rapidly led to better devices and how technology diffused between facilities. On X13 the size of the beam at the sample was too large to make use of commercial stopped flow devices or those adapted for work with lasers and the first attempts therefore did not lead to anything useful. The first useful experiments were done with the group of M. Moody from the EMBL Heidelberg using a purpose-built device [73, 74]. The postdoc doing the hard work was P. Vachette. Some of these experiments used a circular delay line detector, the predecessor of the quadrant detectors. The experiments on assembly of tubulin and other proteins were done with a slower mixer based on motor driven syringes [71]. A new mixing device was designed a few years later by the late C. Berthet at the EMBL-Grenoble [75] for studies on the assembly of viral capsids [76]. After some very preliminary experiments [77] in Hamburg H. Kihara went back to Japan and developed a system [78] which was first used at the Photon Factory to study protein folding and the allosteric transition of aspartate transcarbamoylase (ATC-ase) [79] and later at the SSRL for, among others, iconic studies on virus assembly (see e.g. [80]). At third generation sources commercial devices made for optical measurements can be used without difficulty. We therefore also discontinued this type of work when beamlines like ID02 at the ESRF became available. Very fast mixing (<150 μs) using microfluidic mixing devices has been achieved on third generation sources [45, 81]. It is, however, unclear from these applications whether such devices are generally usable.

The end of the beginning

With all the ongoing activity at the EMBL in the early 1980s it thus looked for a brief while as if further progress might be made rather rapidly. All this changed abruptly and following a number of reorganisations the EMBL decided – like everybody else- that it would put the emphasis on protein crystallography. This also meant that, perhaps more than elsewhere, support for further development in SAXS instrumentation was halted and the outstation was reorganized into independent groups. Many years later I found a quotation which George Hampton, a former administrative director of CERN, circulated to his colleagues after a discussion of a reorganisation in that institution (CERN Courier, May 2004, p4): "It seemed that every time we were beginning to form up into teams, we would be disbanded. I was to learn later in life that we tend to meet any new situation by reorganizing; and a wonderful method it can be for creating the illusion of progress while producing confusion, inefficiency and demoralization. Petronius Arbiter 57 AD." I did not check the authenticity of the quotation but it certainly very well describes what happened at regular intervals at the EMBL outstation in subsequent years.

H. Stuhrmann went to the GKSS (Geesthacht) where he pursued his work on resonant and soft X-ray scattering in HASYLAB. P. Laggner went back to Graz and later built the AustroSAXS beamline at ELETTRA [82] and J. Bordas returned to the SRS at Daresbury to set up the Biology Support Laboratory (BSL) [83]. The rivalry between the EMBL and the SRS, which some in the media would have liked to see [83], never occurred – one should not always believe the media - and good collaborations including transfer of equipment and software and common projects continued until the BSL became fully operational.

In 1984, Zehra Sayers, who had been a regular user for many years, joined my group. Together with an enthusiastic young biochemistry technician, A-M Michon and later P. Brouillon, and at times a postdoc or PhD student we concentrated on biophysical projects. By that time the polymer community worldwide had realized the potential of synchrotron radiation for SAXS and new approaches were rapidly being developed. As an example, the first experiments combining results of differential scanning calorimetry (DSC) and time resolved SAXS at the SSRL [85] and in Hamburg [86], which were going to be followed by many others, were published within months of each other.

As intensities increased, the ability to collect short time frames even for projects which did not strictly require time resolution became an important asset in detecting the onset of radiation damage to which nucleic acids are even more sensitive than proteins. Radiation damage of synthetic polymers or even inorganic systems [87], although also often observed, remains much less well documented.

Over the following years, the EMBL detectors and data acquisition systems were made available to LURE, the SRS, SSRL, the X27C beamline at the NSLS and DUBBLE at the ESRF. More recently D2L, the company which A. Gabriel created after retiring from the EMBL, also delivered linear and area detectors to the NSRRC [88] and other facilities. The need to produce more detectors for X-rays and neutrons prompted A. Gabriel and F. Dauvergne in Grenoble to turn to more industrial means of production. Initially, the delay lines had been handmade. As this limited the production, delay lines with surface mounted technology (SMT) were designed as illustrated in Figure 5. Not every novelty is an improvement and although the SMT delay lines can be produced rapidly and very reproducibly, they have significantly higher resistance (8 Ohms) than the hand wound lines (3 Ohms). This was going to take its revenge ten years later when the demand for simultaneous SAXS/WAXS experiments on lipids and polymers arose, as we shall see later.

In the process of our work and that at Daresbury the detectors and data acquisition systems were further developed, especially large quadrant detectors which were more adapted to solution scattering experiments. The first TDC was replaced by the combination of modified Lecroy TDCs and a purpose-built readout module depending on the type of detector used (linear or area) [89]. When Lecroy proposed to commercialize a data acquisition system for LC delay line detectors in 1982, the UC Berkeley administration informed them that they thought that this might infringe patents taken by V. Perez-Mendez in 1973, which they owned (U.S. patents 3,842,373 and 3,772,521). The matter was not pursued as the coupling methods used in the two types of devices were sufficiently different, but with humour A. Gabriel still regards this as the highest recognition he received during his career. The project also never really took off because in the mid 1980s the firms involved in fast electronics for data acquisition lost interest in such developments as they moved to the much more lucrative telecommunications market.

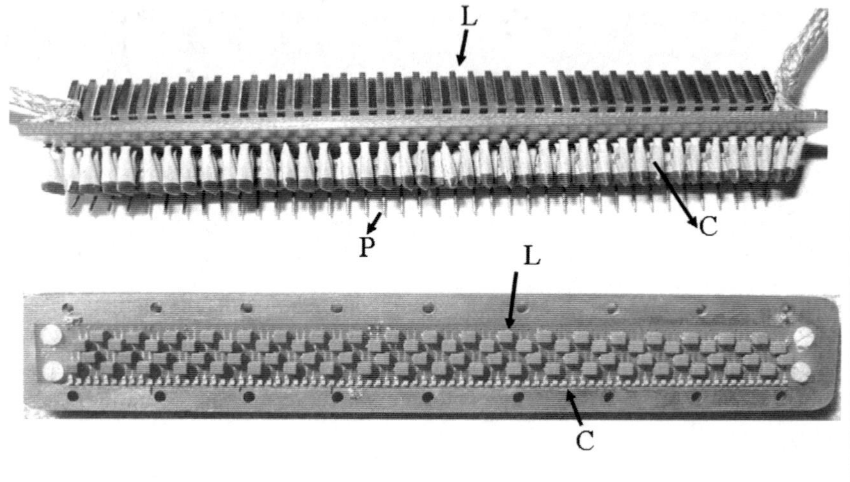

Figure 5. Top: Handmade LC delay line. L: inductances, C: capacitors. The pins (P) connect the delay line to a separate circuit with the cathode strips. Bottom: a SMT LC delay line integrated with the cathode circuit. The cathode strips are on the other side of the circuit.

The area detectors had less impact in time resolved measurement than initially believed. Several factors contributed to this. Rapid accumulation of frames required large histogramming memories, which were not readily available. A proposal in 1982 to have an external company build such a large memory had been rejected partly due to the opposition of the instrumentation groups.

We fortunately later had access to such a memory built in Daresbury [83] which allowed us to carry out some projects on Limulus and frog or fish muscle (only the Limulus work could be published [90]). Although some of the data looked quite acceptable, as illustrated in Figure 6, it was claimed that the resolution and efficiency of the detectors were too low. Later improvements in these two parameters, when other types of detectors became available, also did not bring the breakthroughs that some had promised in fibre diffraction, suggesting that the problem was at least partly lying elsewhere.

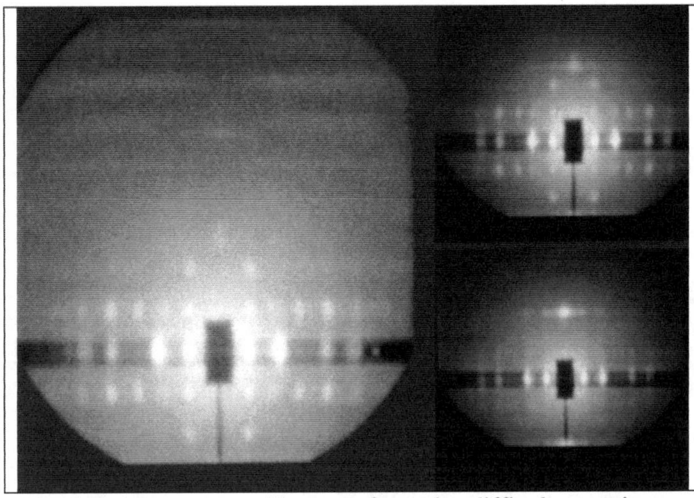

Figure 6: Left: Diffraction pattern of a resting fish muscle recorded in 1986 with a gas detector with LC delay line readout. The strong equatorial reflections were covered with an Aluminium strip to avoid space charge effects. The layer lines are clearly visible in the outer part of the pattern.
Right: central part of the pattern in the resting state (top) and during contraction (bottom).

The detectors were usable even for quite difficult experiments like electric field scattering, where tobacco mosaic virus or DNA could be fully oriented in solution using a short electric pulse [91]. This experiment was only possible because we were lucky enough to be able to borrow a high voltage pulse generator from one of the DESY groups.

When large CAMAC memories [92] based on new and much less expensive technology and new TFG [93] and CCU [94] modules with extended capabilities, based on field programmable gate arrays (FPGA), became available at the EMBL ten years later, they were very useful but the emphasis had shifted to applications which did not require area detectors. The EMBL area detectors were much more extensively used elsewhere, e.g. on the JUSIFA beamline [95], the soft X-ray scattering instrument [96] or BW4 [97] in HASYLAB and later at the DUBBLE beam line at the ESRF [98].

An unexpected event changed the course of things in 1986. Y. Maeda, whose group worked on muscle, brought back from Japan some imaging plates developed by Fujifilm for medical applications, which were in use at the Photon Factory. Initially these plates could only be scanned at a hospital in Hannover, a 2hr car drive from Hamburg. The results convinced J. Hendrix and his group to abandon their work on gas detectors and to build a scanner, which two years later became a very successful device for protein crystallography. This success also marked the end of all detector development in Hamburg.

Imaging plates did not have a significant impact on solution scattering from biological macromolecules. Their slow readout and the difficulty to guarantee full erasure of the previous pattern contributed to an uncertainty on the background to be subtracted, especially in the case of solutions. They were, however, briefly used a few years ago at the EMBL [99] but rapidly replaced by other detectors and are, apparently also still in use for solution scattering on BL08B2 and BL45XU at Spring8. (www.spring8.org.jp).

X13 was rebuilt by G. Rapp during an eighteen months shutdown in 1987-88. This time the segments of the mirror were halved in length and placed on a continuous Aluminium bender with the aim of improving the vertical focusing [100]. The gains compared with the previous system were not spectacular but it was a very good instrument. Soon after, it was decided that DORIS would become a dedicated source, which finally happened in 1993. This also meant that instruments like the new X13, which were now pointing in the wrong direction, had to be dismounted, mothballed for a year, and turned around following a period of extensive remodelling of the buildings. By that time several good instruments had been built at different facilities, as illustrated by the fact that whereas in 1988 about ten SAXS instruments were operational worldwide [101], ten years later the number was already three times larger [102] and it continues to grow (see e.g. [103, 104]). It is therefore not possible in a limited space to give a fair account of all contributions and I shall therefore concentrate on the ones I followed most closely.

A lot of good work was done with the new X13 including some on muscle, lipids and bacteriorhodopsin as well as some of the first SAXS pressure experiments on lipids [105] and solutions. The first pressure experiment on solutions was actually done at SPring8 [106], where the conditions were, of course, much better and in Europe too this type of work soon moved to third generation sources.

Although formally belonging to different groups G. Rapp and I had several productive collaborations and during the time that X13 was unavailable we also shared beamtime on X33.

One of G. Rapp's strengths was in the construction of UV/visible flash lamps for time resolved studies based on photolysis of caged compounds and laser infrared light sources. Such a flash lamp was used to study the structural changes during the photocycle of bacteriorhodopsin [107] which had previously been detected by neutron scattering on frozen samples [108]. One of his erbium infrared laser was used to follow structural changes upon temperature jumps of 10K in 1-2 ms with sub-ms time resolution in phospholipid phase transitions [64]. Such studies in the non-equilibrium non-linear regime still offer many opportunities, especially in materials science. Previous studies on lipid phase transitions using microwaves had achieved jumps of $30Ks^{-1}$ [109].

By the mid-1990s research on muscle had migrated to 3rd generation sources and some of us briefly regretted the late night frog legs with garlic or occasional lobster. Measurements on overexpressed proteins have definitely less to offer in this respect! The next reorganisation at the EMBL was, however, around the corner and the new X13 was dismantled again and converted into another protein crystallography instrument. Its crew was dispersed. Later G. Rapp set up a very successful company (www.rapp-opto.com), S. Funari is now responsible for HASYLAB's most productive SAXS beamline (A2) and M. Rappolt joined the AustroSAXS beamline at ELETTRA.

At about the same time a similar fate struck the X12B SAXS beamline at the NSLS where over the years the groups collaborating with M. Capel had produced many interesting results using gas detectors with delay line readout built by G. Smith [110].

It was also clear by the mid-1990s that more demanding work on synthetic polymers could better be pursued at specialized beamlines such as those at the ESRF or the NSLS, where B. Hsiao, also a frequent visitor in Hamburg, and his colleagues had started a very successful program and several advances in instrumentation were made [111]. New detectors like the cooled CCD combined with an image intensifier developed by Y. Amemiya and his group [112] were also becoming realistic alternatives to proportional area detectors for time resolved measurements, even if they were less straightforward to use and required quite a few corrections [113].

With hindsight some of the early beam lines at the ESRF look somewhat like "missing links" in as much as a double monochromator was followed by a horizontally focusing bent triangular monochromator and gas detectors with delay line readout were used beside imaging plates [114]. In contrast, the new beamlines at 3rd generation sources usually have a tunable double monochromator, sometimes with sagital focussing which deliver much more flux at the sample than those of the previous generation. They were fortunately also built with budgets that one could only have dreamed of even a few years earlier and are equipped with solid state detectors which cost orders of magnitude more than delay line detectors. Recent visits at some of the newer facilities have, however, sometimes reminded me not only of X13 (see Figure 7) but also of R. Feynman's comparison between the MIT and Princeton cyclotrons [115] The first one was a new, beautifully engineered, machine, the other one was a setup run by scientists, which was so chaotic that it finally burned down, but it had produced lots of interesting results. Fortunately, nothing as drastic happened to X13.

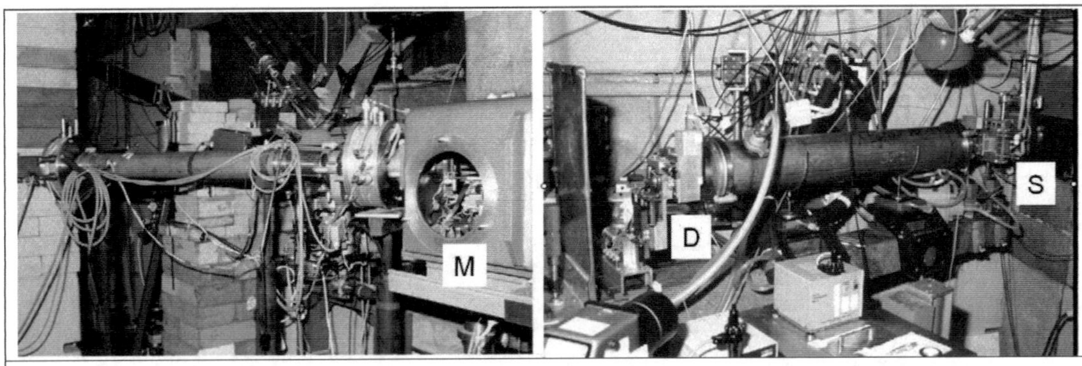

Figure 7. Left: Front end of X13. The monochromator in the mirror box (M) was followed by a slit and a vacuum pipe leading to the experimental area. Note the makeshift reinforcements of the shielding with lead bricks at the back. Right: Experimental area of X13 (the „monkey cage"). The sample was placed after the slit S and the pattern was recorded with a linear gas detector (D), with motorized XY adjustments.

The most complex and expensive instruments are not always the most useful which is also what gives recent instruments on some of the smaller sources [116] a fair chance to compete, especially for static measurements. The best example of this was probably D24 at LURE, which for many years produced very high quality data mainly on protein solutions. The device was very compact. Its optical system was limited to a curved triangular monochromator and it was equipped with a gas detector. D24 was probably the first beamline specialized for solution scattering of biological macromolecules which was equipped with a sample cell in vacuum [117]. New experiments suggest that a further reduction in the background may be possible by working with microdrops [118] or even in the gas phase [119].
In contrast with D24, some of the other early beamlines at a number of facilities have recently received a new lease of life. X33 was upgraded [99], after my departure from the EMBL in 2006, largely automated for solution scattering [120] and equipped with a PILATUS pixel detector [121]. Around the same time, BL4-2 at the SSRL was also upgraded and equipped with an additional USAXS device [122]. High throughput systems aiming at obtaining low resolution models of proteins are also being installed at other facilities [123]. The future will show to what extent this flurry of description will help to understand biological systems.

2. Delay line readout

The reconstruction of DORIS in the early 1990s was accompanied by a long period of beam instability until the ring was switched to positrons. Shutdowns and periods when the beam was unusable were always some of the most productive periods in terms of instrumentation as they gave the opportunity to explore some new ideas.

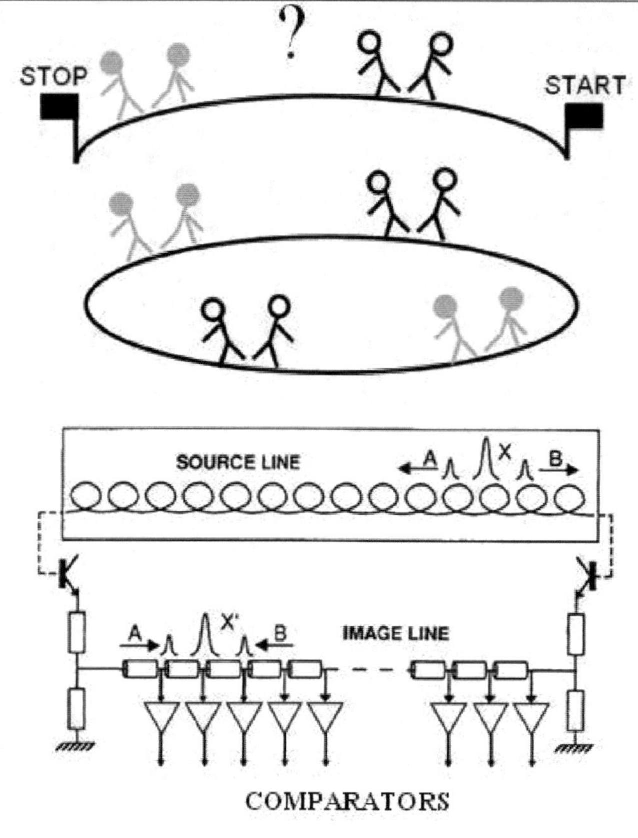

Figure 8. Top: The avalanches in a gas detector induce two pulses travelling in opposite directions on a delay line with delay δ. The spatial position of a single avalanche (e.g. the starting position of the pair of black runners) is readily obtained from the time difference between the arrivals of the two pulses at the ends of the delay line. The position of simultaneous avalanches can, however, not be unambiguously determined with such a standard time to digital converter, because pulses from different avalanches are used as start and stop signals. Middle: With space-time-space conversion the circuit is closed with a second delay line of the same total delay (δ). If the clock is started by the prompt anode pulse and a snapshot is taken after a time δ, the pairs of pulses correctly recombine on the lower part of the track. Bottom: In practice, the circuit is implemented by triggering a set of comparators at a time δ after the arrival of the anode pulse.

During one of these periods A. Gabriel and I visited K. Kuroda at CERN to talk about multianode photomultipliers but came out of there with an idea for a system with two back to back delay lines which could be used for the readout of delay line detectors [124]. Over a period of about ten years this was going to lead step by step to the solution, at least for linear detectors, of a problem which had been considered a major drawback for delay readout i.e. its inability to deal with multiple events as illustrated in Figure 8.

The feasibility was first shown with the help of F. Cipriani [125], who had been hired for the project, but soon after was asked to built a microdiffractometer for protein crystallography. The next step, which involved the production of application specific integrated circuits (ASICs), was only made possible by a collaboration with Smart Silicon Systems in Lausanne. Tests with a prototype data acquisition system based on a PC card [126] proved the good performance of the ASICs which were then used in two NIM data acquisition modules. These devices performed very well and were used for a number of experimental projects (e.g. [127]). They also simplified the read out of multianode detectors, without loss in performance, by connecting groups of eight wires to a short delay line segments connected to one ASIC as illustrated in Figure 9.

A different type of readout was tested with the prototype of a multihit time-stamp TDC, developed for the COMPASS experiment at CERN [128]. The results were very promising as these devices made it possible to detect double events and fluorescence events and, using the signal of the bunch clock, to observe the time structure of the beam [129]. This feature may be useful for measurements on pulsed sources.

The possibility of using gas detectors for measurements where energy resolution matters was also briefly explored. For this, the beam must enter through the side of the detector rather than through the

front and one measures the number of photons absorbed as a function of depth in the detector. This is a topic which may be worth investigating in more detail for special applications (e.g. harmonic detection, or SWAX with two wavelengths) [130].
Some may see these developments as a case of undue persistence but delay lines have had so many applications in data acquisition that it would be astonishing if some of this technology would not also find other applications.

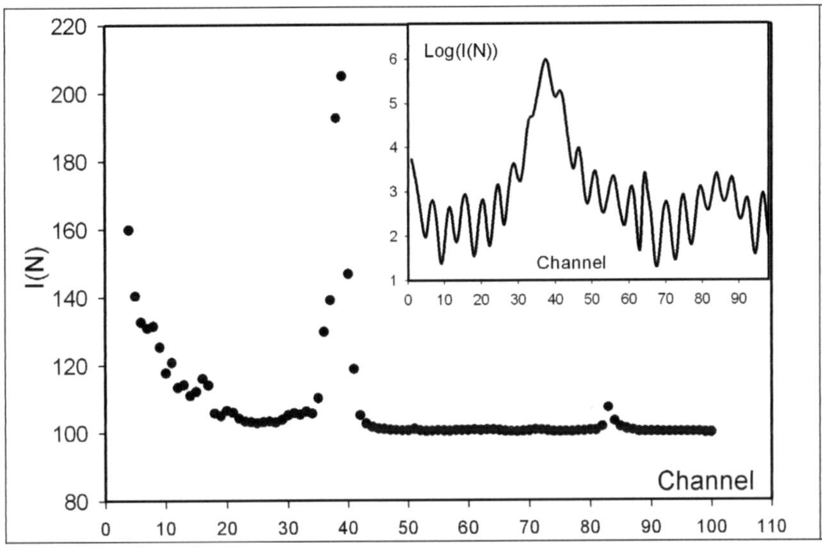

Figure 9. Readout of a parallel anode (wire per wire) detector with a time to space conversion module. The insert illustrates the wire structure in a pattern before correction for the detector response.

Gas detectors may in some cases still have an advantage over integrating detectors for fast (sub-ms) measurements, especially, if like the RAPID detector [131] or other recent detectors [132], they can cope with high count rates. With these detectors the readout is sufficiently fast to collect n successive frames without significant deadtime. Repeating the sequence m times to achieve statistics uses a total sample volume of mV, where V is the volume used for each shot.
In contrast, the relatively long readout times of integrating detectors (about 60-200 ms for the FReLoN detector at the ESRF [133] and 2.5s for a full image or a few ms in frameshift mode for the latest MarCCD (SX165) (www.mar-usa.com)) has hitherto imposed to collect frames at each time point and hence uses a volume nmV of sample. In the first case the sample is, of course, irradiated for a time nt, where t is the frame length, whereas in the second case it is only irradiated for a time t. The best choice between the two approaches will thus depend on the radiation sensitivity of the sample and on the relative efficiency of the detectors. Time resolutions of the order of 2.5 ms exposure and 2.5 ms readout have recently been achieved in continuous stopped flow experiments with a PILATUS detector [121] at the SSRL (H. Tsuruta, private communication). The readout performance of integrating and pixel area detectors for SAXS could perhaps still be improved by taking into account that the number of photons in a single shot does not require the large number of bits used in present devices.

3. Simultaneous small and wide angle scattering (SAXS/WAXS)

The development of simultaneous SAXS/WAXS or SWAX which was perhaps less accidental than some other projects, clearly illustrates that the most successful ideas in terms of acceptance are not necessarily the cleverest but often the simplest. Around 1990 a number of time resolved projects on lipids and polymers required to measure SAXS and WAXS patterns. Rather than measuring those separately it was obviously more efficient to measure them simultaneously. One of the first projects to use SWAX was that of the late K. Westesen from the institute of pharmaceutical technology in

Braunschweig investigating lipid nanoparticles as potential drug carriers [134]. The project, which yielded interesting physical chemistry beside practical applications, implied making a large number of measurements to obtain e.g. phase diagrams. Sample cells with several compartments were moved in the beam at the end of each measurements and the temperature of the cells was remotely controlled, a very modest first step towards automation. Over the years, many such mechanical and electronic improvements were made by the technical staff of the outstation, and in particular R. Klaering and B. Rohbrahn, to meet the requirement of different types of projects. These included various sample cells and holders for ancillary equipment, the use of a photodiode in the beamstop to measure absorption, which was especially useful with polymers samples where the thickness can change (e.g. during heating or upon stretching), mini-shutters to protect samples from unnecessary irradiation, various control modules and other useful gadgets. They were sometimes mentioned in the methods section of some papers but never independently published.
Various approaches to the problem of simultaneous SAXS and WAXS recordings had been proposed in different laboratories around that time (see [101]) but delay line detectors certainly offered by far the least expensive solution [135] and also solved the problem of exact synchronization between the SAXS and WAXS patterns. This was, however, not a novelty as delay line detectors in series had already been used more than ten years earlier for a fluctuation X-ray scattering experiment on frozen particles [136]. To circumvent the problem of the resistance of the new SMT delay lines the signals from the new detectors had to be delayed as illustrated in Figure 10, rather than by simply connecting the detectors in series.
SWAXS became very popular especially among polymer scientists (e.g. [137]), so much so, that under the influence of W. Bras at the SRS a beamline was dedicated to these applications [138]. One of the important results of SWAX was the proof that under appropriate circumstances one can circumvent Babinet's principle in SAXS and unambiguously decide whether the crystallinity of a polymer is above or below 50% [139].
In the meantime, SAXS/WAXS with two area detectors is available at the ID02 at the ESRF [140] and at the BioCAT beamline at the APS (T. Irving, private communication).

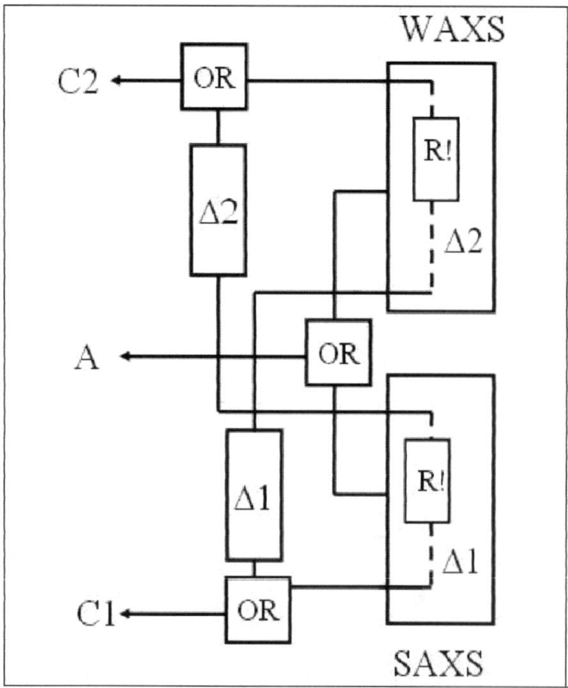

Figure 10. Alternative SAXS/WAXS connection of linear detectors. The signals at the output of the detector preamplifiers are first input to a constant fraction discriminator (not shown). One of the signals from each detector is delayed by a time corresponding to the delay line of the other detector to produce a circuit which is equivalent to two detectors connected in series. This avoids the damping of the signals due to the higher resistance of SMT delay lines (R!) as well as the gap due to impedance mismatch (see [17]) when delay line detectors are simply connected in series with cables.

4. Data analysis

Unforeseen events in the east of Europe at the end of the 1980s, which resulted in the collapse of the Soviet Union and the reunification of Germany, a bit later led to the development of the ATSAS suite of programs [141]. In 1988 I started a collaboration with the group of A. Schellenberger from the University of Halle [142], which was at the time still in the German Democratic Republic. Two years later this had changed and D. Svergun from the Institute of Crystallography from the (by then) Russian Academy of Sciences had joined the group. The projects with the Halle group served as test cases in the further development of shape determination methods using spherical harmonics, which describe models in terms of a few parameters, proposed by Stuhrmann [143]. The huge progress in computers and display facilities made over the previous twenty years certainly also contributed to the success of this project. Starting from there the programs of the ATSAS suite were successively developed. This process will not be described in detail here as D. Svergun will certainly some time in future provide a better historical overview. Spherical harmonics are now only used to speed up the calculations but they provided an indispensable stepping stone towards the current methods. Significant progress was made when Chacon et al. introduced *ab initio* bead modelling based on a genetic algorithm [144] and this prompted D. Svergun to write the program DAMMIN [145]. Bead modelling had previously been extensively used by several authors but not for *ab initio* modelling, whereas the simulated annealing techniques [146] used in the ATSAS programs had also been used in previous work on phasing of crystal structures [147]. A survey of the literature suggests that, unlike the early 1970s when H. Stuhrmann first proposed shape determination against strong resistance, times were ripe for such developments, as they took place in different laboratories over a few years (see [148, 149]) (i.e. a sufficient number of people believed that one could restore a three-dimensional model from a one-dimensional scattering curve).

The success of modelling methods (see e.g. [150]) and the ease with which higher resolution data can be collected from protein solutions with appropriate protocols at 3rd generation sources [151] have recently led to interesting empirical [152] and theoretical [153] approaches to the interpretation of high angle scattering using also molecular dynamics calculations with explicit solvent models [154].

The properties of protein solutions can be strongly affected by crowding [155, 156] or cosolutes like urea or trimethylamine-N-oxide (TMAO) [157] and it may be difficult to separate the effects due to changes in solvent structure and in the protein structure. For such problems isotopic substitution neutron scattering methods and the EPSR (Empirical Potential Simulation Refinement) method [158], which also makes extensive use of molecular dynamics calculations, may provide additional information at least on the structure of the solvent. This technique recently gave the explanation for the antagonistic effects of urea and TMAO [159].

The effects of crowding on the properties of biological macromolecules, which is probably one of the most important questions at this stage, could probably usefully be studied by neutron scattering using a deuterated protein in high concentrations of a protonated protein. Alternatively, heavy labels, as used in some recent studies on the structure and dynamics of DNA [160], at the C- and N-terminals of an unstructured protein may be useful to monitor structural changes in a crowded solution by X-rays.

The combination of molecular dynamics simulations and high angle X-ray scattering data also plays a major role in the analysis of fast reactions triggered by a laser pulse (see e.g. [161]). It thus seems that this is an area where further progress could be made, especially as scattering patterns contain information about longer range interactions than molecular dynamics can accurately reproduce.

The success of a program suite like ATSAS is certainly also due to the fact that it was made easily available, maintained and improved over the years. This is where a problem may lie in future, since as J. Monnet once said in a different context, "nothing gets done without the individuals, but nothing remains without the institutions", but where are the institutions where useful documentation about hardware like detectors or software are being conserved and made available to those interested?

Unfortunately, no suite of programs comparable to ATSAS has yet been developed for time resolved measurements on polymers or colloidal systems, although bits and pieces of software exist in many groups. The result is that large amounts of data remain un-interpreted. This situation certainly provides an opportunity waiting to be taken.

5. Questions for the near future

The drive towards high throughput at most facilities has put a strong emphasis on routine static measurements, at the expense of more experimental work and instruments have become more specialized. This will sooner or later raise the question whether the use of synchrotron radiation from high brilliance sources, is the most cost-effective way of performing these measurements, especially as considerable progress has also been made with instruments on conventional sources like the optimized NanoSTAR [162] from Bruker (www.bruker-axs.com), the S3-Micropix system of HECUS (www.hecus.at) or the SAXSess of Anton-Paar (www.anton-paar.com).

For an analytical technique like SAXS to be useful for longer than academic fashions last, it must yield information which has operational value. This is where time resolved measurements during the processing of materials in close to industrial conditions offer many opportunities (see e.g. [163]).

High brilliance sources will remain indispensable for microfocus applications [164] and ultrafast kinetics [165] where much progress in instrumentation has recently been made. For these methods to become generally useful ways will have to be found to deal with radiation damage. In the case of solutions of biological macromolecules it may help to move the sample in the beam [166] but it is also increasingly important to check the samples (e.g. by in-line chromatography) as introduced recently on a number of beamlines [167-169]. This may also help separating species which are in slow equilibrium in solution. It is more difficult but also useful to check the samples after irradiation.

For synthetic polymers other means like differential scanning calorimetry must be used to check, before and after exposure to X-rays, whether samples undergoing the same processes in the absence of beam behave identically.

Ultrafast kinetic experiments with X-ray scattering are hitherto usually performed on small molecules where total sample volumes are not limiting (see e.g. [157]). This may not be the case for biological macromolecules, which are also easily damaged by the laser trigger pulses [46].

6. Looking back into the future

Looking back, many things, even the collective fears (nowadays: climate, pandemics, energy…) have changed during the last 30 years. In instrumentation, computers and software, sensors, modularization and miniaturization have simplified many things. In the environment there were some very positive changes, like the opening of borders in and around Europe and much easier travel and communication which facilitate the diffusion of ideas and practices. There were, of course, also less positive developments like the increasing bureaucratization of science and changes in attitudes towards intellectual property in public research. One may regret that, as a consequence, facilities are perhaps less open and prone to share their developments with others than in the past. This slows down imitation with amelioration and hence also progress. Most facilities have also evolved towards a service model, including "Fedex" operation rather than a scientifically much more efficient and personally more satisfying collaborative model, and emphasize metric-driven rather than curiosity-driven research. One can only hope that the driving force for developments in instrumentation will remain curiosity and that the deeper motivations for automation will not become boredom and laziness.

We have had the good fortune at the EMBL outstation in Hamburg to have been a small group of enthusiastic scientists, engineers and technicians with access to one of the best storage rings and to the best detectors at a time when synchrotron radiation was taking off. There were lots of innovations to be made, perhaps because we knew so little. Being embedded in an established large organization like DESY, without being part of it, gave us some freedom, at least for a few years, and the fun we had made up for the relative lack of funds. Nowadays, there are more funds and many opportunities, which are admittedly somewhat more demanding than in the early days, but unfortunately also less freedom. If any lessons can be drawn from history the recent evolution of some HEP laboratories into photon science laboratories suggests that the best way to prepare the future is to set up some niches where things slightly beyond ones remit, and in which the majority does not yet believe, can be freely tried out. Instrumentation alone will, however, not guarantee success because, ever since Roger Bacon was walking down the streets of Oxford, the main question when building instruments has remained "What is the question?"

Acknowledgments
I am deeply indebted to my former colleagues and coworkers and to all external collaborators without whose contributions SAXS could not have been developed at the EMBL outstation in Hamburg. I thank W. Bras, A. Gabriel, Z. Sayers, D. Svergun, H. Tsuruta and P. Vachette for their useful comments on the manuscript. Any errors or omissions are however mine.

References

[1] Fara P 2009 Science A four thousand year history. (New York: Oxford University Press) p310
[2] Goward F K and Barnes D E 1946 *Nature* **158** 413
[3] Blewett J P 1988 *Nucl. Instrum. and Methods A* **266** 1-9
[4] Blewett J P 1998 *J. Synchrotron Rad.* **5** 135-9
[5] Galison P 1997 Image and logic. (Chicago:The University of Chicago Press) pp 498-515
[6] Krige J 1990 The research system in transition ed S E Cozzens, P Healey, A Rip and J Ziman (Dordrecht: Kluwer Academic Publishers) pp 179-97
[7] Motz H 1951 *J. Appl. Phys.* **22** 527-35
[8] Winick H and Brown G 1982 *Nucl. Instrum. and Methods* **195** 347-8
[9] Wuilleumier F and Farge Y ed. 1978 *Proc. Int. Conf.on Synchrotron Radiation Instrumentation and New Developments Orsay September 12-14 1977* (Amsterdam: North-Holland)
[10] Board of Physics and Astronomy 2009 *Scientific Assessment of High Power Free electron Laser Technology* (Washington DC: The National Academies Press)
[11] Rossi B B and Staub H H 1949 *Ionization chambers and counters. Experimental Techniques* [New York: McGraw-Hill] pp 97-98
[12] Grove R, Lee K, Perez-Mendez V and Sperinde J 1970 *Nucl. Instrum. and Method* **89** 257-62
[13] Charpak G, Bouclier R, Bressani T, Favier J and Zupancic C 1968 *Nucl. Instrum. and Methods* ***62*** 262-68
[14] Borkowski C J and Kopp MK 1968 *Rev. Sci. Instrum.* **39** 1515-22
[15] Feynman R P, Leighton R B and Sands M 1967 *The Feynman Lectures on Physics* (Reading: Addison Wesley)
[16] Dupont Y, Gabriel A, Chabre M, Gulik-Krzywicki T and Schechter, E 1972 *Nature* **238** 331-3
[17] Goody R S, Barrington Leigh J, Mannherz H G, Tregear R T and Rosenbaum G 1976 *Nature* **262** 613-5
[18] Gabriel A 1977 *Rev. Sci. Instrum.* **48** 1303-5.
[19] Petrascu A-M, Koch M H J and Gabriel A 1998 *Journ. Macromolec. Sci. Physics. B* **37** 463-83
[20] Bordas J, Fourme R and Koch M H J eds. 1980. *Proc. Int. Conf. on X-ray detectors for synchrotron radiation Hamburg 17-21 November Nuclear Instrum. and Methods in Physics Research* **201**
[21] Huxley H E, Faruqi A R, Bordas J, Koch M H J and Milch, J. 1980 *Nature* **284** 140-3
[22] Prieske W, Riekel C, Koch M H J and Zachmann H G 1983 *Nucl. Instrum. and Methods* **208** 435-8
[23] Huxley H E and Holmes K C 1997 *J. Synchrotron Rad.* **4** 366-79
[24] Holmes K C 1989 *Topics in Current Chemistry* **151** 1-7
[25] Holmes K C and Rosenbaum G 1998 *J. Synchrotron Rad.* **5** 147-53

[26] Rogers E M 2005 *Diffusion of Innovations* (New York: Free Press)
[27] Gould S J 1996 *Full House. The spread of excellence from Plato to Darwin* (New York: Harmony Books) pp 149-51
[28] Taleb N N 2007 *The Black Swan* (New York: Random House) p 41
[29] Martino J P 1969 *Air University Review, March-April* pp 11-20
[30] Gleick J 1987 *Chaos, making a new science* (New York:Viking Penguin) p 125
[31] Tchoubar D, Rousseaux F, Pons C-H and Lemonnier M 1978 *Nucl. Instrum. and Methods* **152** 301-5
[32] Webb N G, Samson S, Stroud R M, Gamble R C and Baldeschwieler J D 1977 *J. Appl. Cryst.* **10** 104-10
[33] Vazina A A 1979 *Sov. Phys. Usp.* **22**:385-6
[34] Vazina A A 2007 Brilliant Light in Life and Material Sciences ed V Tsakanov and H Wiedemann (Heidelberg: Springer) pp 121-31
[35] Huxley H E and Faruqi A R 1983 *Ann. Rev. Biophys. Bioeng.* **12** 381-417
[36] Bordas J, Munro I H and Glazer A M 1976 *Nature* **262** 541-5
[37] Wakabayashi K, Kakizaki A, Siota Y, Namba K, Kurita K, Yokata M, Tagawa H, Inoko Y, Mitsui T, Wada E, Ueki T, Nagakura I and Matsukawa T 1978 *J. Phys. Soc. Jpn.* **44** 1314-22
[38] Stuhrmann H 1978 *Nucl. Instrum. and Methods* **152** 297-300
[39] Stuhrmann H and Gabriel A 1983 *J. Appl. Crystallogr.* **16** 563-71
[40] Hendrix J, Koch M H J and Bordas J 1979 *J. Appl. Crystallogr.* **12** 467-72
[41] Lemonnier M, Fourme R, Rousseaux F and Kahn R 1978 *Nucl. Instrum. and Methods* **152** 173-7
[42] Amemiya Y, Wakabayashi K, Hamanaka T, Wakabayashi T, Matsushita and Hashizume, H. 1983 *Nucl. Instrum. and Methods* **208** 471-7
[43] Bilderback D 1982 *Nucl. Instrum. and Methods* **195** 67-72
[44] Glatter O and Laggner P 1983 *J. Appl. Crystallogr.* **16** 42-6
[45] Pollack L et al. 2001 *Phys. Rev. Lett.* **86** 4962-5
[46] Cammarata M, Levantino M, Schotte F, Anfinrud P A, Ewald F, Choi J, Cupane A, Wulff M, Ihee H. 2008 *Nature Methods* **5** 881-6
[47] Schoenborn B P, Caspar D L D and Kammerer O F 1974 *J. Appl. Crystallogr.* **7** 508-10
[48] Russell T P 1991 *Handbook on Synchrotron Radiation* ed G S Brown and D E Moncton (Amsterdam: North Holland) pp 381-469
[49] Tsuruta H, Brennan S, Rek Z U, Irving T C, Tompkins W H and Hodgson K O 1998 *J. Appl. Crystallogr.* **31** 672-82
[50] Lowy J and Poulsen F R 1982 *Nature* **299** 308-12
[51] Koch M H J and Bendall P J 1981 *Proc. of the Digital Equipment Corporation Users Society DECUS UK* pp 13-16
[52] Boulin C, Kempf R, Koch M H J and Mc Laughlin S M 1986 *Nucl. Instruments and Methods A* **249** 399-407
[53] Hendrix J and Fuerst H 1980 *IEEE Trans. Nucl. Sci. NS-* **27** 145-9
[54] Clout P N 1982 *Nucl. Instrum. and Methods* **201** 225-30
[55] Engster C, Van Koningsveld L and Verwey H 1975 CERN NP/ 1975-0620
[56] Boulin C 1982 *Nucl. Instrum. and Methods* **201** 237-9
[57] Dorrington E 1982 *Nucl. Instrum. and Methods* **201** 247-50
[58] Boulin C 1982 *Nucl. Instrum. and Methods* **201** 251-9
[59] Golding F R 1982 *Nucl. Instrum. and Methods* **201** 231-5
[60] Shang W, Robrahn B, Golding F and Koch M H J 2004 *Nucl. Instrum. and Methods A* **530** 513 -20
[61] Hervé C and Torki K 2001 *Nucl. Instrum. and Methods A* **481** 566-74
[62] Hendrix J, Fuerst H, Hartfiel B and Dainton D 1982 *Nucl. Instrum. and Methods* **201** 139-44
[63] Kress M, Huxley H E, Faruqi A R and Hendrix J 1986 *J. Mol. Biol.* **188** 325-42
[64] Laggner P, Kriechbaum M and Rapp G 1991 *J. Appl. Crystallogr.* **24** 836-42

[65] Chamayou J M F 1980 *Computer Phys. Comm.* ***21*** 145-61
[66] Koch M J H, Bordas J, Schöla E and Broecker H C 1979 *Polymer Bull.* **1** 709-14
[67] Elsner G, Koch M H J, Bordas J and Zachmann H G 1981 *Makromol. Chem.* **182** 1263-69
[68] Bras W, Diakun G P, Diaz J F, Maret G, Kramer H, Bordas J and Medrano F J 1998 *Biophys. J* **74** 1509-21
[69] Lednev V V, Popp D and Jahn W 1987 *J. Mol. Biol.* **197** 679-84
[70] Hirai M, Koizumi M, Han R, Hayakawa T and Sano Y 2003 *J. Appl. Crystallogr.* **36** 520-24
[71] Renner W, Mandelkow E M, Mandelkow E and Bordas J 1983 *Nucl. Instrum. and Methods A* **208** 535-40
[72] Ranck J L, Letellier Y, Shechter E, Krop B, Pernot P, Tardieu A 1984 *Biochemistry* **23** 4955-61
[73] Moody M F, Vachette P, Foote A M, Tardieu A M, Koch M H J and Bordas J. 1980 Proc. Natl. Acad. Sci. USA **77** 4040-43
[74] Fowler A G, Foote A M, Moody M F, Vachette P, Provencher S W, Gabriel A, Bordas J and Koch M H J 1983 *Journ. Biochem. Biophys. Meth.* **7** 387-92
[75] Berthet-Colominas C, Bois J M, Cuillel M, Sedita J and Vachette P 1984 *Rev. Phys. Appl.* **19:**769-72
[76] Berthet-Colominas C, Cuillel M, Koch M H J, Vachette P and Jacrot B 1987 *Europ. Journ. Biophys.* **15** 159-68
[77] Inoko Y, Kihara H and Koch M H J 1983 *Biophys. Chem.* **17** 171-4
[78] Tsuruta H, Nagamura T, Kimura K, Igarashi Y, Kajita A, Wang Z X, Wakabayashi K, Amemiya Y and Kihara H 1989 *Rev. Sci. Instrum.* **60** 2356-8
[79] Tsuruta H, Sano T, Vachette P, Tauc P, Moody M F, Wakabayashi K, Amemiya Y, Kimura K, Kihara H 1990 **263** 66-8
[80] Lee K K, Gan L, Tsuruta H, Moyer C, Conway J F, Duda R L, Hendrix R W, Steven A C, Johnson J E 2008 *Structure* **16** 1491-1502
[81] Akiyama S T S, Kimura T, Ishimori K, Morishima I, Nishikawa Y and Fujisawa T. 2002 *Proc. Natl. Acad. Sci. USA* **99**:1329-34
[82] Amenitsch H, Bernstorff S and Laggner P 1995 *Rev. Sci. Instrum.* **66** 1624-6.
[83] Towns-Andrews E, Berry A, Bordas J, Mant G R, Murray P K, Roberts K, Sumner I, Worgan J S, Lewis R and Gabriel A 1989 *Rev. Sci. Instrum.* **60** 2346-9
[84] Newmark, P. 1983 *Nature* **301** 459
[85] Russel T P, Koberstein J T 1985 *J. Polym.Sci., Polym. Phys. Ed.* **23** 1109-15
[86] Schouterden P, Riekel C, Koch M, Groeninckx G and Reynaers H 1985 *Polymer Bull.* **13** 533-9
[87] Gommes C J, Goderis B and Pirard J P 2007 *J. Phys. Chem. C* **111** 11150-6.
[88] Lai Y-H et al. 2006 *J. Appl. Crystallogr.* **39** 871-7
[89] Boulin C J, Kempf R, Gabriel A and Koch M H J 1988 *Nucl. Instrum. and Methods A* **269** 312 -20
[90] Maeda Y, Boulin C, Gabriel A, Koch M H J and Sumner I 1986 *Biophys. J.* **50** 1035-42
[91] Koch M H J, Dorrington E, Klaering R, Michon A M, Sayers Z, Marquet R and Houssier C 1988 *Science* **240** 194-6
[92] Epstein A and Boulin C 1996 *IEEE Trans. Nucl. Sci.* **43** 159
[93] Khazaie J B and Boulin C 1996 *IEEE Trans. Nucl. Sci.* **43** 197-201
[94] Khazaie J B and Boulin C 1996 *IEEE Trans. Nucl. Sci.* **43** 193-6
[95] Haubold H-G et al. 1989 *Rev. Sci. Instrum.* **60** 1943-6
[96] Stuhrmann H. 1985 Advances in Polymer Science **67** 123
[97] Gehrke R, Bark M, Lewin D and Cunis S 1995 *Rev. Sci. Instrum.* **66** 1354-6
[98] Borsboom M. et al. 1998 *J. Synchrotron Rad.* **5** 518-520.
[99] Roessle M W et al. 2007 *J. Appl. Crystallogr.* **40** s190-4
[100] Rapp G 1992 *Acta Phys. Polonica A* **82** 103-9
[101] Koch M H J 1988 *Makromolekulare Chem., Macromol. Symp.* **15** 79-90
[102] Bras W and Ryan A J 1998 *Advances in Colloid and Interface Science* **75** 1-43

[103] Liu D G et al. U.S. 2009 *J. Synchrotron Radiat.* **16** 97-104
[104] Korneev V N, Shlektarev V A, Zabelin A.V, Medvedev B I, Sharafutdinov M R, Tolochko V P, Sheromov M A, Kondratyev V I, Umnov A I and Vazina A A 2007 *Nucl. Instrum. and Methods* **575** 134-9
[105] Czeslik C, Winter R, Rapp G. and Bartels K 1995 *Biophys. J.* **68** 1423-9
[106] Kato M and Fujisawa T 1998 *J. Synchrotron Rad.* **5** 1282-6
[107] Koch M H J, Dencher N A, Oesterhelt D, Ploehn H-J, Rapp G and Bueldt G 1991 *EMBO J.* **10** 521-6
[108] Dencher N A, Dresselhaus D and Zaccai G 1989 *Proc. Natl. Acad. Sci. USA* **86** 7876 9
[109] Caffrey M, Fanger G, Magin R L and Zhang J 1990 *Biophys. J.* **58** 677-88
[110] Capel M S, Smith G C and Yu B 1995 *Rev Sci Instrum* **66** 2295-9
[111] Chu B and Hsiao B S 2001 *Chem. Rev.* **101** 1727-61
[112] Amemiya Y, Ito K, Yagi N, Asano Y, Wakabayashi K, Ueki T and Endo T 1995 *Rev. Sci. Instrum.* **66** 2290-4
[113] Fujisawa T, Inoko Y and Yagi N 1999 *J. Synchrotron Rad.* **6** 1106-14
[114] Riekel C, Boesecke P, Diat O, Lorenzen M, Sanchez del Rio M and Snigireva I 1995 *Rev. Sci Instrum.* **66** 987-94
[115] Feynman R P 1985 *Surely You're Joking Mr. Feynman!* (Toronto: Bantam) pp 49-51
[116] Knaapila M et al. 2009 *J. Synchrotron Rad.* **16** 498-504
[117] Dubuisson J-M, Decamps T and Vachette P. 1997 *J. Appl. Crystallogr.* **30** 49-54
[118] Graceffa R, Burghammer M, Davies R J, Ponchut C and Riekel C 2009. *Appl. Phys.Lett.* **94** 062902
[119] McCullough B J, Entwistle A, Konishi X, Buffey S, Hasnain S S, Brancia F L, Grossmann, J G and Gaskell S J 2009 *Anal. Chem.* **81** 3392-7
[120] Round A R, Franke, D, Moritz S, Huchler R, Fritsche M, Malthan D, Klaering R, Svergun D I and Roessle M 2008 *J. Appl. Crystallogr.* **41** 913-7
[121] Kraft P et al. 2009 *J. Synchrotron Rad.* **16** 368-75
[122] Smolsky I L, Liu P, Niebuhr M, Ito K, Weiss T M and Tsuruta H 2007 *J. Appl. Crystallogr.* **40**:s453-8
[123] Hura G L et al. 2009 *Nat. Methods* **6** 606-12
[124] Gabriel A and Koch M H J 1992 *Nuclear Instrum. and Methods A* **313** 549-54
[125] Cipriani F, Gabriel A and Koch M H J 1994 *Nuclear Instrum. and Methods A* **346** 286-91
[126] Mio H, Chemloul M, Laggner P, de Raad Iseli C, Krummenacher F, Gabriel A and Koch M H J 1998 *Nuclear Instrum. and Methods A* **419** 715-20
[128] Unruh T, Bunjes H, Westesen K, Bösecke P, Lindner P and Koch M H J 2002 *Langmuir* **18** 1796-1800
[129] Fischer H et al. 2002 *IEEE Trans. Nucl. Sci.* **NS49** 443-7
[129] Niebuhr M, Gabriel A and Koch M H J 2003 *Nuclear Instrum. and Methods A* **510** 309-17
[130] Gabriel A, Dauvergne F, Nolting H F and Koch M H J 1994 *Nuclear Instrum. and Methods A* **349** 461-5
[131] Lewis R A, Berry A, Hall C J, Helsby W I and Parker B T 2000 *Nucl. Instrum. and Methods A* **454** 165-72
[132] Kocsis M et al. 2006 *Nucl. Instrum. and Methods A* **563** 172-6
[133] Coan P, Peterzol A, Fiedler S, Ponchut C, Labiche J C and Bravina A 2006 *J Synchotron Radiat.* **13** 260-70
[134] Westesen K, Siekmann B and Koch M H J 1993 *Int. Journ. Pharmaceutics* **93** 189-99
[135] Rapp G, Gabriel A, Dosiere M and Koch M H J 1995 *Nuclear Instrum. and Methods A* **357** 178-82
[136] Kam Z, Koch M H J and Bordas J. 1981 *Proc. Natl. Acad. Sci. USA* **78** 3559-62
[137] Fougnies C, Damman P, Villers D, Dosière M and Koch M H J 1997 *Macromolecules* **30** 1385 -91

[138] Bras W, Derbyshire G , Ryan A J, Mant G R, Felton A, Lewis R A, Hall C J and Greaves G N 1993 *Nuclear Instrum. and Methods A* **326** 587-91
[139] Goderis B, Reynaers H, Scherrenberg R, Mathot V B F and Koch M H J 2001 *Macromolecules* **34** 1779-87
[140] Urban V, Panine P, Ponchut C, Boesecke P and Narayanan T 2003 *J. Appl. Crystallogr.* **36** 809 -11
[141] Konarev P V, Petoukhov M V, Volkov V V and Svergun D I 2006 *J. Appl. Crystallogr.* **39** 277 -86
[142] Huebner G, Koenig S, Schellenberger A and Koch M H J 1990 *FEBS Lett.* **266** 17-20
[143] Stuhrmann H B 1970 *Acta Crystallogr. A* **26** 297-306
[144] Chacon P, Moran F, Diaz J F, Pantos E and Andreu J M 1998 *Biophys. J.* **74** 2760-75
[145] Svergun D I 1999 *Biophys. J.* **76** 2879-86
[146] Kirckpatrick S, Gelatt C D Jr and Vecci M P 1983 Science **220** 671-80
[147] Subbiah S 1991 *Science* **252** 129-33
[148] Koch M H J, Vachette P and Svergun D I 2003 *Quart. Rev. Biophys. 36* 147-227
[149] Förster F, Webb B, Krukenberg K A, Tsuruta H, Agard D A and Sali A 2008 *J. Mol.Biol.* **382** 1089-1106
[150] Petoukhov M V and Svergun D I 2007 *Curr. Opin. Struct. Biol.* **17** 562-71
[151] Fischetti R F, Rodi D J, Mirza A, Irving T C, Kondrashkina E and Makowski L 2003 *J. Synchrotron Rad.* **10** 398-404
[152] Makowski L, Rodi D J, Mandava S, Devarapalli S and Fischetti R F 2008 *J. Mol. Biol* **383** 731-44
[153] Shneerson V L and Saldin D K 2009 *Acta Crystallogr. A* **65** 128-34
[154] Park S, Bardhan J P, Roux B and Makowski L 2009 *J. Chem. Phys.* **130** 134114
[155] Makowski L, Rodi D J, Mandava S, Minh D, Gore D B and Fischetti R F 2008 *J. Mol. Biol.* **375** 529-46
[156] Homouz D, Perham M, Samiotakis A, Cheung M S and Wittung-Stafshede P 2008 *Proc. Natl Acad. Sci. USA* 105 11754–9
[157] Niebuhr M and Koch M H J 2005 *Biophys. J:* **89** 1978-84
[158] Soper A K 2005 *Phys. Rev. B* **72** 104204
[159] Meersman F, Bowron D, Soper A K and Koch M H J 2009 *Biophys. J.* in press.
[160] Mathew-Fenn R S, Das R and Harbury P A B 2008 *Science* **322** 446-9
[161] Kong Q, Lee J H, Plech A, Wulff M, Ihee H and Koch M H J 2008 *Angew. Chem, Int. Ed. Engl.* **47** 5550-3
[162] Pedersen J S 2004 *J. Appl. Crystallogr.* **37** 369-80
[163] Mitchell J B A, LeGarrec J L, Sztucki M, Narayanan T, Dikhtyar V and Jerby E 2008 *Phys. Rev. Lett.* **100** 065001
[164] Riekel C, Burghammer R, Davies R, Gebhardt R and Popov D 2009 *Applications of Synchrotron Radiation Light to Scattering and Diffraction in Materials and Life Sciences* ed T A Ezquerra, M Garcia-Gutierrez, A. Nogales and M. Gomez (Heidelberg: Springer) pp 91 -104
[165] Plech A, Wulff M, Bratos S, Mirloup F, Vuilleumier R, Schotte F and Anfinrud P A 2004 *Phys. Rev. Lett.* **92** 125505
[166] Hong X and Hao Q 2009 *Rev. Sci. Instrum.* **80** 014303
[167] Mathew E,Mirza A and Menhart N 2004 *J. Synchrotron Rad.* **11** 314-8
[168] David G and Pérez J 2009 *J. Appl. Crystallogr.* **42** (in press)
[169] Watanabe Y and Inoko Y 2009 *J. Chromatogr. A* (in press)

What can we learn from combined SAXS and SANS measurements of the same sample containing surfactants?

Thomas Zemb and Olivier Diat

Institut de Chimie séparative de Marcoule, UMR5257 CEA/CNRS/UM2/ENSCM, F30207 Bagnols sur Cèze (France)

thomas.zemb@icsm.fr

Abstract. We review the information gained when studying the same sample containing surfactants by SANS and SAXS. Specific information can be obtained if intensities I(q) are determined and used on absolute scale as well as over a large q-window, thus extending to the « Porod limit » asymptotic range. Comparing SAXS and SANS peak positions (when present) allows unambiguous identification of structure factors separated from underlying form factors. Absolute scale refers to scattering cross-section while resolution relates to the q_{max}/q_{min} ratio. Taking into account "external" knowledge of molecular volumes in constrained fitting from explicit models gives better results on aggregation numbers as well as on interfacial thickness than considering only Patterson functions.
We assemble in this review results on adsorption isotherms on surfactant films as well as equations of state, quantifying colloidal interactions. Identification of the topological origin of swelling behaviour are made possible by using combined SAXS and SANS on the largest possible q-range. Best results using this general methodology up to now were obtained by considering data separated from background up to q_{max}= 0.6 Å^{-1} - 0.8 Å^{-1}.

1. Structure factors and dilution plots

In the seventies, "vesicles" and "micelles" made from amphiphilic molecules were not distinguished because of the similarities and interplay between core-shell form factors dominating the SAXS signal and the structure factor more apparent in SANS. Broad peaks observed typically in the range q= 0.01 Å^{-1} to q= 0.1 Å^{-1} arise either from the structure factor S(q) or from the from factor P(q) due to the high electron density present in polar head-groups. Comparing SAXS and SANS spectra from the same sample allowed for the first time direct separation between form and structure factors. As a direct extrapolation of this separation procedure, it has been demonstrated that decomposition in form and structure factors – initially introduced for discrete globular particles by Hayter and Penfold [1], remains valid for connected structures. In this case, space has to be tessellated in Voronoï cells of density n. Average distance between cell centres is, in that case, the origin of the maximum of structure factors [2]. The structure factor S(q) gives information on the network of cell centres, which is Fourier transformed to obtain the pair correlation function $\gamma(\vec{r})$. On this lattice, any set of connected structures is geometrically convoluted by a local microstructure: this can be either a droplet, a connected cylinder or a connected bilayer at first approximation. The product of the underlying S(q), the distribution of seed points, by the local shape characterized by P(q) corresponds still to the experimental I(q) at a very good approximation. This has been numerically verified by Welberry [2] and Barnes [3]. This applies to most micelles and microemulsions in the bicontinuous [4], including those with a stiff interfaces, where bending energy overcomes entropy effects. [1, 5].

The implicit decomposition of a random network into local structure convoluted on a set of seed points is schematised on figure 1: on the left part, a typical tesselation of space in adjacent cells is shown. Each of these Voronoi cells may be decorated with a droplet at centre, or connected cylinders perpendicular to cells faces or swollen bilayers decorating the faces. The 2D (or 3D) structure can further be described by a connectivity Z or a symmetry ψ, characterizing the dissymetry of the two spaces on both sides of the interface. Figure 1 to the left shows local microstructures as connected cylinders with Z=2, alias a living polymer or fluid giant micelle (top) or a connected cylinder microstructure with connectivity 4 (bottom), which can be understood as a molten cubic phase with local diamond structure. If the maximum in SAXS and SANS coincides, this can be identified safely as a measure of the most probable distance between centres of Voronoï cells.

From the plot of the scaled peak position D*. Σ versus volume fraction ϕ, such as the one shown on figure 1 (central part), different types of microemulsions can be discriminated, since the swelling depends on the dimensionality of the aggregates. Reduced units must be used in so-called "swelling plots", which links the dimensionless product of the peak position by the specific water/oil interface versus the specific volume Σ, in units of area per unit volume (reciprocal length). Power-law description of swelling law versus volume fraction is simple in the case of droplets (slope: -1/3), cylinders (-1/2) or smectics (-1). However, all intermediary values exist: swelling plots can be used to identify local dimensionality of the microstructure [2].

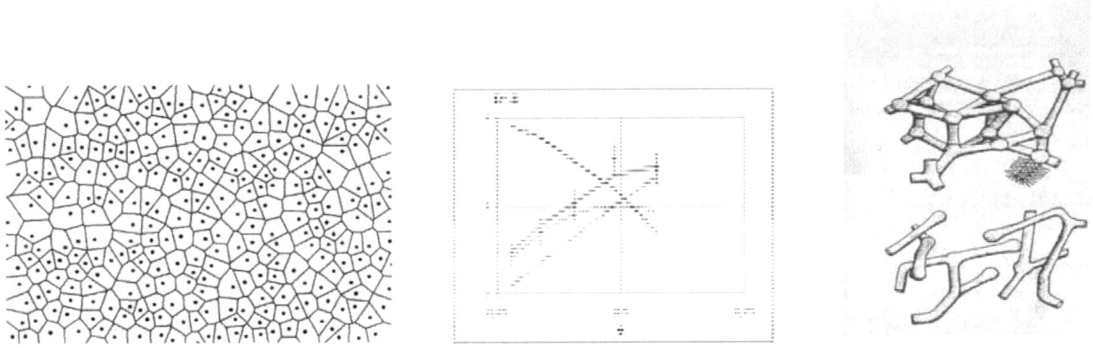

Figure 1. Left: 2D-model of a Voronoï tesselation of space: partition of each Voronoï cell in a « polar » and « non-polar » part separated by a film allows to build a complete family of connected structures of known volume fraction ϕ, specific surface and structure factor. Middle: Comparison of an experimental swelling curve, linking scattering peak position D*. Σ (-S/V) to volume fraction ϕ, to the expected behaviours expected from different types of microemulsions. Continuous line: DOC locally lamellar model; double-dotted increasing line: water-in oil spheres; triple-dotted decreasing line: biliquid o/w foam, thin dotted line: standard flexible model of microemulsions using De Gennes-Taupin prediction. Right: scaled sketches of randomly connected locally cylindrical w/o microemulsions. Top: a "molten" cubic phase; bottom: a living polymer, with connection points and end-caps [reproduced from 5 and 17].

In the case of small droplet micelles formed by surfactants, it has been shown by Hayter and Penfold [1] that self-consistent fitting on absolute scale has not to rely on adjustment of radii, but only of aggregation and hydration numbers. This allows to define uniquely the intermediary values with a physical meaning such as radii and scattering length densities [6]. There are two ways to generate simulated I(q) for micellar systems: one method is to choose arbitrarily "best" radii for SAXS and SANS, these are different because excess electron relevant in SAXS density profiles are not proportional to scattering length density profiles relevant for SANS. The other method is to use number of molecules –i.e. surfactant, co-surfactant, "bound" water molecules per aggregate (or area per head-group). Using known molecular volumes, one calculates then the radii of the "core" and the "shell" of micelle as a function of aggregation numbers, hydration numbers, and fraction of bound counter-ions only. Typically radii involved are of the order of nanometre; therefore in the classical q-range up to 0, 6 $Å^{-1}$, one gains no information on attempting to separate micelle in a third layer, such as Palisade or Stern layer. Such details are beyond resolution of SAXS/SANS if the q-range considered is not extended above 0.6 $Å^{-1}$. Since chemical formulae of molecules involved are known, and when proton exchange or counter-ion distributions are taken into account, the scattering length densities are calculated by simple addition and are not fitting parameters. Hence, SAXS and SANS can be calculated together –using the same S(q) of course- for one unique choice of aggregation and hydration numbers [7]. Simultaneous fit on absolute scale and over the whole q-range ensures consistency of the result and good determination of micellar structure. Any procedure allowing arbitrary fitting on radii and scattering length density may produce inconsistent values of radii and masses. The result coming out may be perfect reproduction of scattering curves, but with values inconsistent with known molecular volumes taken from density or area par head-group, specially if SANS or SAXS data are fitted "alone".

On figure 2, we show an example for a typical surfactant micellar system (20g/l), the broad scattering peak observed by SAXS and SANS are of different origin. However, the same S(q) comes in the form factor-structure factor iterative separation procedure and the two spectra are well fitted by the same ensemble of hydration numbers, aggregation numbers and effective charge parameters [7,8, 9]. In the same approach, effective charge has to be uniquely defined as the un-compensated charge due to all counter-ions not included in the polar part of the micelle. Thus, fitting simultaneously SAXS and SANS on absolute scale with the same basic quantities, i.e. numbers of molecules and effective charge allows unique determination of osmotic compressibility, charge and mass of micelles.

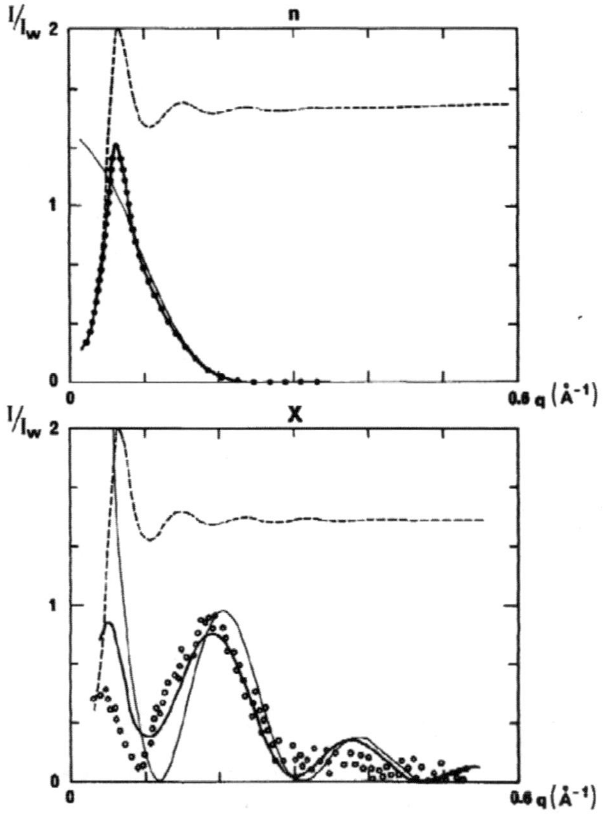

Figure 2. Using the scattering of pure water as a reference as well for SAXS than for SANS, this figure shows that broad peaks observed for some micelles made from single-chain ionic surfactants arise from the structure factor (SANS, upper graph) or from the form factor of a "hollow shell" of head-groups (SAXS, lower graph). Figure shows decomposition in structure (dashed line) and form factors (dotted line), obtained from the same values of fitting parameters: Aggregation number, hydration number and effective charge [extracted from ref. 7].

2. High-resolution combined SAXS and SANS extended to high-q region

On a wide q-range, i.e. for a q_{max}/q_{min} of at least two decades, micellar aggregates can be studied at so-called "high resolution". Figure 3 shows in the right plot, a "Porod plot", the product $q^4I(q)$ versus q, for diluted ionic micelles versus q. Measurement has to be extended at least up to q= 0.6 $Å^{-1}$, i.e. the spatial resolution to $2\pi/q_{max}$ of 10 Å. Patterns arising from intra-micellar form factor P(q) clearly differ, as shown on figure 3 (left). Combined with high-resolution SAXS experiments made on the same q-range, data as those shown on figure 3 (left) can be compared to predictions arising from different models of packing of chains in the micellar core.

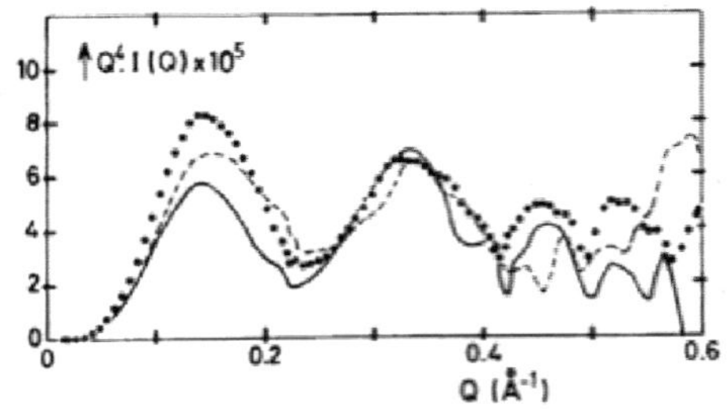

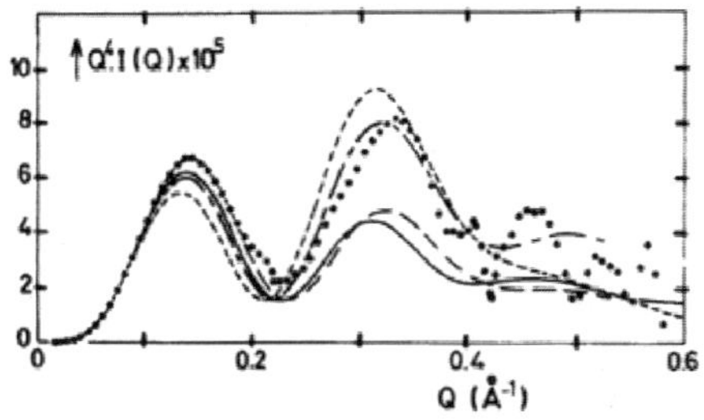

Figure 3. Left: Wide angle neutron scattering (WANS) observed up to large q-window for SDS micelles with specifically labelled groups. The "Porod plot" q4.I(q) versus q enhancing the high-resolution part of the spectrum demonstrates differences in the localisation of the deuterated groups. Right: Comparison between statistical models of packing of chains in the core of micelles illustrates the high degree of disorder in the chains. Dots: full contrast, dashed line: deuterated in position 3; continuous line: deuterated at the end of chain [from ref. 8].

This type of experiments demonstrated that chains in the micellar core are disordered and liquid-like. With a probability of a "gauche" conformation 1/3 along SDS chains, the methylated end-groups are frequently present at the oil-water interface: this explains the "anomalous reactivity" of functional groups attached close to the end-groups of hydrophobic chains in micelles. Thus, for ionic micelles, frequent location of methyl-terminal chains could be ascertained by comparing SAXS and SANS data from specifically deuterated surfactants in D_2O solutions up to 0.6 $Å^{-1}$ [8].

In this early work, the thickness of the polar layer made by the head-groups has been determined to be 4 Å in the case of sodium as a counter-ion. In the case of caesium, this may be different. Fitting on absolute scale on the largest available q-range with the SAME input parameters the scattering of a micelle has been performed again by Vass and co-workers using anionic micelles with caesium as a counter-ion: these authors have shown that provided the q-range is extended up to 0.6 $Å^{-1}$, the layer of caesium as counter-ion can be modelled as a diffuse layer within approximation in standard Poisson-Boltzmann distributions [9]. However, four radial layers where needed to model perfectly the results: the classical core, shell and cation diffuse cloud, but also a fourth layer probably consistent with the Hofmeister effect related to the polarisable caesium ion [10].

Another way to produce more cohesive and diffuse polar layer of micelles is to make catanionics. Salt-free catanionics are produced by using H^+ and OH^- counter-ions only while five component catanionic samples are produced by simple mixing of two ionic surfactants. Using a q-range in SANS and SAXS extending up to q= 0.6 $Å^{-1}$, it has been possible to determine the "thickness" of the layer made by charged polar heads in the case of catanionic micelles. The "diffusivity" of mixtures of anionic and cationic surfactants, i.e. catanionic bilayers above chain melting temperature, has been studied by using a similar combined SAXS-SANS methodology, but without the constraint in fitting given by known molecular volumes, and therefore calculating normalized Patterson functions. Kaler and co-workers have found that head-groups form a diffuse boundary when 5% salt is added, i.e. that the layer formed by head-groups has an extension similar to the Debye screening length, a value compatible with expected cohesive energies of surfactants [11]. It is important to notice that close to

Journal of Physics: Conference Series **247** (2010) 012002 doi:10.1088/1742-6596/247/1/012002

equimolarity, the thickness of the "polar" layer sometimes called "Palisade layer" is typically double for catanionics than for anionic or cationic micelles including quaternary ammoniums or sulfate head-groups only, going from 5Å to 9Å. To our knowledge, these polar layer thicknesses have not yet been determined with sufficiently high resolution - i.e. extending the q-range up to 1 $Å^{-1}$ - in the case of surfactants with variable charge depending on pH, such as phosphates or phosphonates, or when variable chain aliphatic alcohols are added.

What happens wit the polar layer thickness and the curvature when the micelle is "stabilized" by polymerisation of its constituent? To answer this question, polymerisable catanionic mixtures were studied comparing absolute scale of the scattering differing in SAXS and SANS. Unfortunately, the q-range of measurement only extended to 0.3 $Å^{-1}$, so the associated spatial resolution is 20 Å. However, it could be confirmed that whatever composition, structure of micelles and microemulsions where conserved upon partial polymerisation of the surfactants [**12**].

The so-called Brij 700 is an important "giant surfactant", with C18 chains counter-balanced by hundred oxyethylen groups. Each micelle is stabilized with a nonionic brush. SANS-SAXS experiments were performed on absolute scale and successful models could only be applied after taking into consideration the crossed terms in form factor as well as the contribution of corona on the average "background solvent" cross-section term. It was shown that corona properties diverge only close to phase inversion temperature (PIT), i.e. within 5°C, associated to dehydration of ethoxy groups. This observation questions the usual assumption about the linearity of spontaneous dependence of interfacial curvature which is usually assumed for non ionic microemulsions [**13**]. In continuity with the study of protruding EO chains, Arleth and co-workers have studied by SAXS/SANS experiments the interfacial thickness of hairy mixed micelles containing PEGylated lipids, proposed to be used as "stealth" liposomes in emerging drug delivery technology. Using Patterson functions, it was found that spheres or cylinders can be observed by varying the mole ratio between the two components. Whatever the shape of the aggregate, the stabilizing protruding chains where interpenetrating and remain Gaussian chains with constant gyration radius even in concentrated solutions. These micelles are stabilized as osmotic brushes [**14**].

In the microemulsion regime, adding Brij700 is a well known method used in industry to increase the « efficiency » of a surfactant mixture: it is supposed intuitively that protruding ethoxy brushes act as stabilizers for droplets. Due to the difficulty of making osmotic measurement with systems containing $C_{12}E_5$ nonionic surfactants, only combined SAXS and SANS can directly proven the mechanism, since in light scattering, interaction and structure are intrinsically mixed in the q-range available. They used a q-range up to 0.3Å-1 for microemulsions of 50 Å minimum curvature radius, i.e. a packing parameter of the order of 0.8. With Brij load up to 5% in mass, a rod to sphere transition as well as an increase of repulsion between corona-stabilized micelles could be demonstrated. Quantitative determination of corona extension as well as height of repulsive potential resulting would require higher q-range measurements [**15**].

Extending this methodology, SAXS and SANS experiments have both be performed up to the high-resolution limit. Both intensity curves can then be divided ($I_{SANS}(q)/I_{SAXS}(q)$) and the resulting ratio is exempt from S(q). This ratio can be compared to the "predicted" from any assumed microstructural molecular packing model, moreover independent from colloidal scale interactions governing the structure factor. Figure 4 shows the case of introducing a small amount of water (typically 10%) in a pure liquid non-ionic surfactant, which is liquid at room temperature and miscible to water at any proportion in a given temperature range. The data obtained have been compared to all proposed structural and thermodynamic models of water-surfactant miscibility, at least in water-poor or solvent-poor domains of ternary phase diagrams. As can be seen in figure 4, all current models up to now models fail to predict the observed ratio $I_{SANS}(q)/I_{SAXS}(q)$ [16]. We believe that this division method on absolute scale and at high resolution remains one of the most powerful methods to investigate concentrated isotropic solutions, involving short surfactant or hydrotrope species, including water-alcohol mixtures.

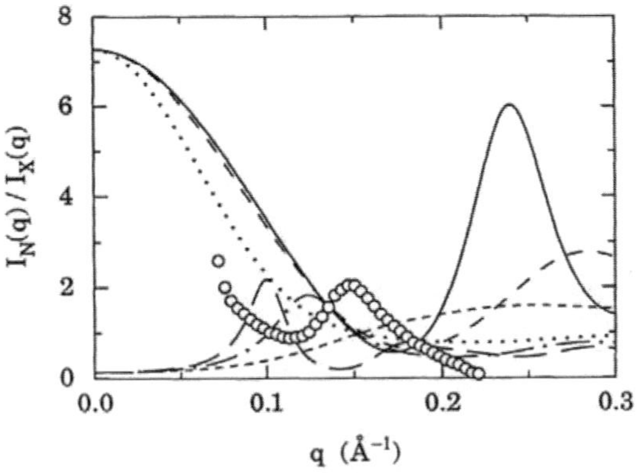

Figure 4. Dividing the SANS by the SAXS spectrum obtained in the same q-range gives a signal depending on P(q) only: this allows a critical test and challenges the models of microstructure in concentrated solutions. The figure shows the failure of currently accepted models of packing to predict the features of IN(q)/Ix(q) in the case of 20% of water added to pure non-ionic surfactant [from ref. 16].

3. Usage of Porod plots to determine thermodynamic quantity: area per surfactant.

Absolute scaled calculations of expected intensities derived from moles to be tested and comparing data on logarithmic scale allows determination of the so-called Porod limit. If investigated aggregate has typical radius R, the asymptotic Porod domain starts for qR above 6 to 10. In some cases, these values obtain by SAXS or SANS differ in quantities as soon as curvatures are large (packing parameters differing from one). From differential geometry, the difference between the values is a direct way to measure curvature and hence determination the packing parameter of the surfactant film. The quantities that have to be known to derive safely specific area per volume of sample (S/V), or per surfactant are either:

- the electron densities (or scattering length) difference between both sides of the water-oil interface
- the "invariant" of the full spectrum, i.e. the second moment of the scattering spectrum I(q) which is known from composition [17].

Measuring a sample in the Porod regime requires good statistics for subtracting the background. Often, experiment give more than the expected value. It is always worth to investigate this difference: corrections due to precise determination of solvent scattering length densities for example evidence initially unexpected monomer-aggregate equilibrium. An example along a dilution line in a ternary system involving a cationic synthetic lipid, water and oil is shown on figure 5. As can be seen in log-log plots, the Porod asymptote occurs in over three decades in intensity. The peak positions can be determined by SAXS and SANS, since they occur at similar positions in q-space and may even be refined considering the small correction by taking the maximum of I(q) divided by a good approximation of form factor. This correction rarely exceeds a few percent. It should be mentioned that fitting of data in q-space versus radii only is often made on linear scale, plotting the fit and data on logarithmic scale shows inconsistency: intensity can be easily misevaluated in the high-q part of the region. This is because surfactant assembly has to be characterized with three parameters: one linked to volume, one to curvature and one to available area in the most general case [18].

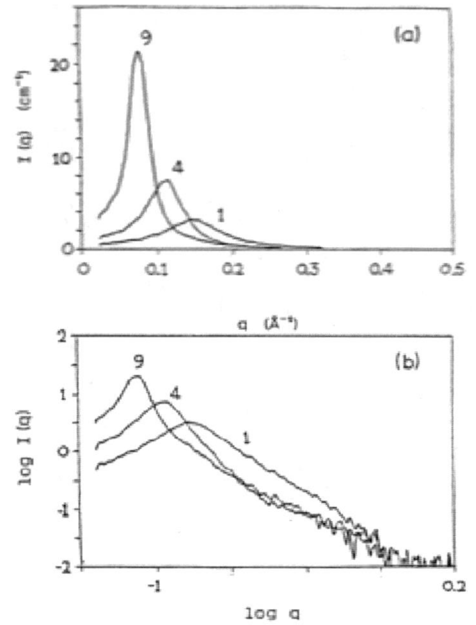

Figure 5. Linear and logarithmic plot of the Intensity I(q) in cm-1, along a dilution line in a ternary or quaternary system allows to determine independently specific surface and "lattice constant" of the underlying Voronoï (or generalized Wigner) cells. Product of peak position by specific surface gives a dimensionless number, crucial for evaluating the swelling behaviour of microstructures [from ref. 18].

The amount of surfactant participating in the surfactant film has in a first step to be separated from the surfactant present as monomers in the "solvent" phase. Once this is taken into account, determination by SAXS and SANS of the basic thermodynamic quantity, i.e. the area of interface per surfactant (or lipid) head participating to the continuous film, can be determined safely with a 10% precision. When compared to values determined on a "flat" -i.e. macroscopic- interface as derived from water-oil surface tension measurements, combined SAXS and SANS experiments performed in the high-q regime allows the determination of the position of the neutral plane of bending. This gives access to the molecular origin of the bending mechanism of surfactant films.

Moreover, the total water-oil contact area per unit sample volume, i.e. the available area (S/V) can be scaled to the peak position, a dimensionless number is then obtained. When microemulsions are made from flexible films, De Gennes and Taupin have demonstrated that this numerical value cannot exceed three [18] (see figure 6).

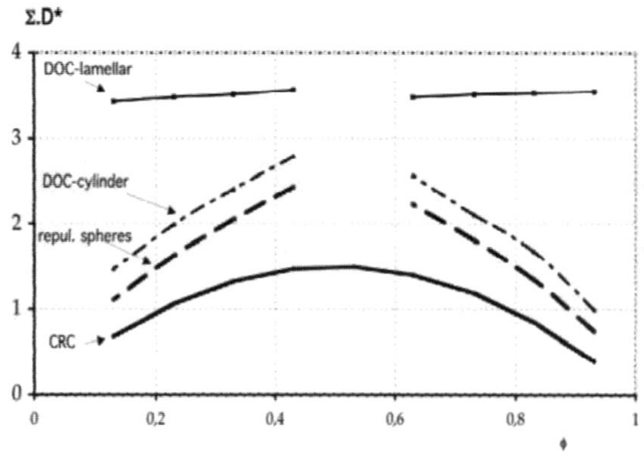

Figure 6. Swelling of the most popular models of microemulsions: the cubic random cell model (CRC, alias De Gennes and Taupin, based on thermodynamics and valid for flexible films; repulsive spherical droplets, connected DOC-cylinders or DOC-lamellar (alias "sponge" phases). The more connected the structure, the higher the value of the dimensionless product .D* [from ref. 17].

More complex structures, which can be understood as molten liquid crystals, "consume" more contact area per unit volume. Hence, the corresponding dilution plot lies above the De Gennes-Taupin relation. De Gennes and Taupin have given a simple analytic expression. Validity is implicitly assumed in most microemulsion studies, such as those implying single-chain non-ionic surfactants [19-20]. However, a large number of results in the literature correspond to other behaviour of swelling plots, meaning that the microemulsion is not made only from a flexible film: curvature effects have to be taken into account. To our knowledge, realistic space representations of random as well as connected structures, thermodynamically equilibrated, and made outside of the degenerated symmetric volume fraction of 50% polar/apolar volume have only be performed once [21].

Establishing experimentally and calculating from random wavelet models dilution plots in scaled units involving only polar/apolar volume fraction and the product of specific area to S(q) peak position translated from reciprocal to real space, has for example allowed to identify the geometrical packing constraints involved in progressive structural transition form globules to connected cylinders and bilayers, a reminiscence of the universal series of liquid crystal structures encountered with co-polymers. Moreover, the procedure introduced by Marcelja allows to represent in real space equilibrium microstructures corresponding to an imposed peak position (using random wavelets) as well as volumes and surfaces for any value of volume fraction, stiffness and spontaneous curvature [21].

For a peculiar class of micelles, known as water-poor w/o microemulsions, used industrially in liquid-liquid extraction processes, simultaneous fitting of I(q) with a large dynamic range in q-space (i.e. $q_{max}/q_{min} > 30$) with the same parameters on an absolute scale allows, on small reverse micelles, unprecedented analytic results. For instance, the effective packing parameter of the highly curved surfactant film can be derived from simultaneous SAXS and SANS modelling, using aggregation numbers as parameters instead of the usual core-shell radii and scattering length densities. The classically considered equation of state relies on osmotic pressure, temperature and the average density between colloidal aggregates. A more delicate thermodynamic relation is the "lateral" equation of state, linking osmotic pressure, temperature and the area per molecule. We have shown that SAXS/SANS comparison allows for the first time, to our knowledge, derivation of not only "perpendicular", but also the "lateral" equations of state governing reverse micelle stability [4].

4. Adsorption isotherms in complex as deduced from combined SAXS and SANS contrast variation

Contrast variation in small angle scattering dates back to sugar/water mixtures used in the investigation of bilayers by Vittorio Luzzatti and co-workers. Serendipously, it was found at that occasion that equilibrium distances between bilayers were strongly increased by the difference in dielectric constants in the van der Waals forces as well as the hydration force between lipid aggregates. Contrast variation, using H/D contrast and specific labelling came into a major technique in physical chemistry and structure determination: from the simple mixture of surfactants [22] to complex structures such as the ribosome [23].

Adsorption of any solute on a pre-existing micelle or microemulsion can be considered with the same concepts than adsorbtion of molecules from a gas or liquid phase on a solid. Combined SANS and SAXS experiments provide an extremely, general and powerful method for direct determination of the adsorption isotherm on any interface made from a surfactant film.

The principle is as follows: Consider for example a lamellar phase in equilibrium with excess water. Then add a solute. The solute may be depleted or adsorbed in the multi-layered vesicle (i.e. the lamellar phase) in equilibrium with excess water. This can be quantified by dosage of the rest of the solute in the excess phase. In parallel, the periodicity shifts to lower values if the solute penetrates in the film and acts as a co-surfactant while periodicity increase if the solute only weakly adsorbs on the surfactant. Accurate SAXS recording of the low-q or high-q peak position shift induced by addition of a solute allows the determination of a possible insertion of the solute in the bilayer (or cylindrical core in the case of an hexagonal phase). For example, partial penetration of hydrophobically modified cyclodextrins were obtained by this method [24]. But how is it possible to determine the adsorption isotherm, i.e. weak adsorption of the solute onto a bilayer?

This has been shown to be possible if one takes into account following observation: if a solute is mixed to the solvent inside a bilayer, it participates to the average scattering length density of the solvent. On the contrary, any solute "bound", i.e. with an interaction energy larger than k_BT towards the interface, the scattering length density of the solute participates to the scattering length density of the periodic bilayer. Using for example H_2O/D_2O mixtures and carbohydrates, or deuterated pentanol, points of zero contrast can vary typically from 5% to 20% in deuterium content [25]. The exact value of zero contrast has to be taken from the square root of intensity at first Bragg peak, and not from the limit at zero-q as usually done in polymer mixtures. Figure 7 shows a contrast plot of this type used to determine adsorption of small sugars onto neutral lipids [26-27]. As an example, depletion from a non-amphiphilic carbohydrate from the interior of a multi-layered liposome could be understood as a simple exclusion from the water in the first shell of hydration of the phospholipids [28]. It should be noted here that "thicknesses" of a heterogeneous bilayer, made from "interior" and "polar head-group" layer, differ naturally if determined by SAXS and SANS: using only the area per head-group and hydration number allows to simulate SAXS and SANS spectra from consistent density profiles. This requires care in using precise definition of the membrane "thickness", since basic variable is area brought per molecule. From this moments of scattering density profiles can be calculated and the same assumption of molecular packing must get to best fitting of scattering data in the high-q regime, as has been shown by Cantu and co-workers [29]. The constraints on acceptable values for the structure of a bilayer, using extra information such as the incompressibility of a fluid with known density are similar in the case of bilayers than the case of micelles as pioneered by Hayter and Luzzatti thirty years ago. This SAXS/SANS method of contrast variation has been extended to investigate hybrid titania/zirconia nanoparticles [30] as well as competitive adsorption of mixed solvents on calcium porous nanoparticles [31].

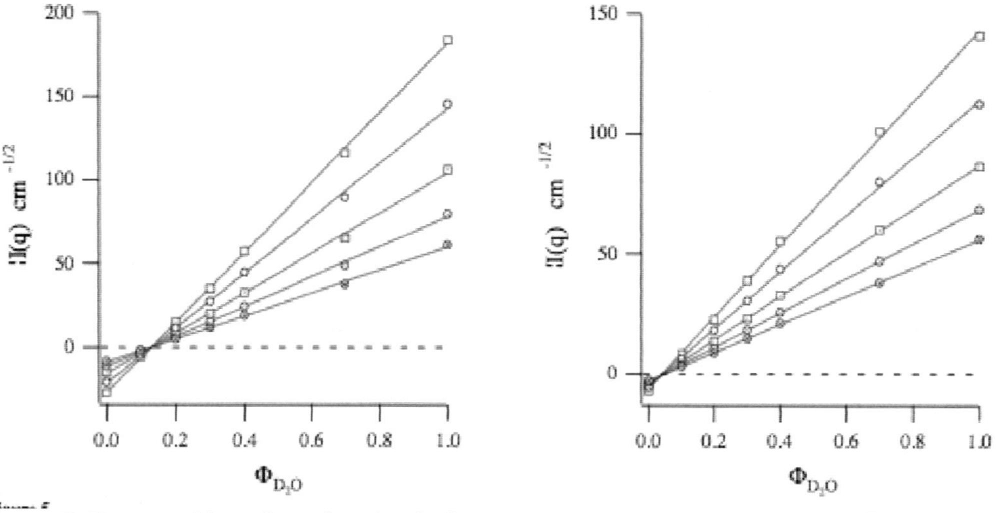

Figure 7. Exact position of match-point during a contrast variation in an organized sample containing a solute depends on the adsorption isotherm of the solute on the water-oil interface. Case of small sugars in phospholipids is shown. The position of the peak as observed by SAXS complements the exact information about periodicity, and detects possible area variations associated to the presence of the solute [from ref. 25].

Journal of Physics: Conference Series **247** (2010) 012002 doi:10.1088/1742-6596/247/1/012002

5. SAXS and SANS experiments to obtain lateral equations of state of surfactants.

The lateral equation of state of surfactant is the relationship between the area per head-group and the osmotic pressure for different temperatures. Only very few complete lateral equations of state are available (see ref. 4 and references cited wherein). An important subset of the lateral equation of state is the relation between molecular area and salinity in the solvent phase. For example in the case of a triple chain lipophilic surfactant, area per molecule and its variation depending on the solute to be extracted in a solvent could be determined using simultaneous SANS and SAXS fitting on an absolute scale. Aggregation numbers, as well as area per head-group tacking into account the volume of the ions and molecules present in the reverse micellar core, could be determined [32-33]. Area per head-group at the water-oil interface, as well as the packing parameter, unambiguously determined from the radii of core and shell of this core-shell structures, have been determined [34]. Figure 8 shows the SAXS and SANS spectra fitted on absolute scale. In this case the scattering from monomers present at high concentration with micelles as well as the crossed term to be considered in the modelisation of scattering to be compared to the experimental results.

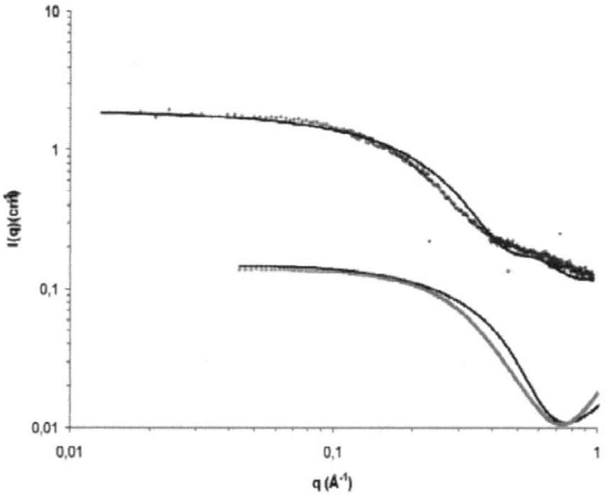

Figure 8. Scattered intensity in log-log and absolute scale in SANS (top) and SAXS (bottom) spectrum of the same reverse micelle, as obtained with a triple chain amphiphilic molecule, compared to models using simultaneous fitting with only one parameter: average aggregation number of the small w/o reverse micelles coexisting with monomers. Solvent is deuterated heptane. Co-solubilized water contained in the core is determined separately by titration [from ref. 34].

The form factor to consider here is the square of the amplitude. The amplitude arising from monomers is not negligible when weak surfactants are considered: typically, one third of the molecules are present in monomers, in coexistence with aggregated pentamers. In this case, form factors of monomers and aggregates, not neglecting the crossed term, has been taken into account in order to obtain agreement of SAXS and SANS on absolute scale. Here again, only aggregation numbers and sticky hard sphere attraction potential compensating sterical repulsions are used as constrained fitting parameters [35]. The sticky hard sphere parameter used in Baxter model can also be determined independently from the phase diagram, knowing the location of the critical point on the liquid-gas transition line of the complex fluid made from reverse micelles [36]. The lateral variation of polar/apolar contact area per molecule and the film curvature (expressed in terms of the dimensionless packing parameter p) are shown in figure 9. Here again, comparison between the "hole in deuterated solvent" as seen by SANS and the volume of the polar core with high electron density as measured by SAXS gives the clue: area per molecule are different at the polar/apolar interface as seen with SAXS, while SANS measures the area of contact of the complete reverse micelle with the solvent [33, 35].

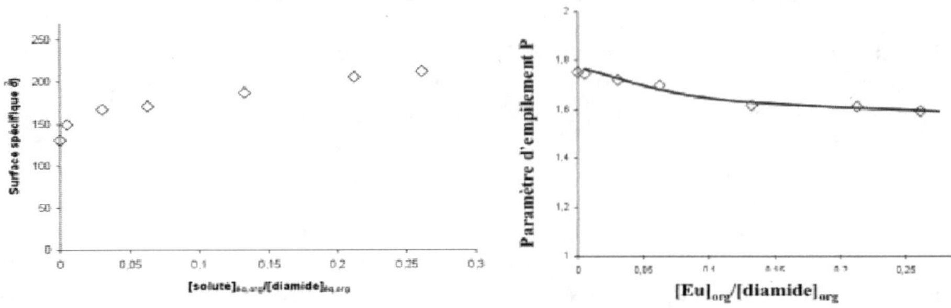

Figure 9. Partial equation of state, related to area and curvature versus ionic strength obtained from simultaneous SANS and SAXS fitting on absolute scale. First direct measure of the effect of ions extracted on the area per head-group versus mole ratio of extracted ions per surfactant as well as interfacial curvature, expressed in terms of dimensionless packing parameter [from ref. 34].

6. SAXS and SANS experiments in frozen surfactant films to determine local order

It has been recently shown that in the so-called "true" catanionic systems, i.e. mixtures of anionic surfactants in acidic form and cationic surfactants presents hydroxide, micelles of different shape are stable as poorly charged colloids, even in the "frozen" state of the surfactant chains. These micelles, prepared by reversible cycling temperatures or soft dialysis are thermodynamically stable. They are in a situation described as gel and sub-gel phases for lipids. Shapes of the micelles depend on the composition and are consistently understood in ternary phase diagrams [37]. Using mixtures of deuterated anionic components and hydrogenated cationic surfactants (or the reverse), fundamental experiments can be done by comparing the Bragg peaks and diffuse scattering. It turns out that SAXS and SANS spectra are complex: interaction peaks, form factors at several scales are seen in SAXS and SANS. Hoverer, extending the q-range of observation to 2.0 $Å^{-1}$ allows observation of Bragg peak by WANS and WAXS: the first order of packing corresponds to hexagonal order of hydrocarbon chains, around 1.6 $Å^{-1}$. Due to labeling of one of the components, a broad diffuse band appears with a maximum around 0.8 $Å^{-1}$. This diffuse band is due to local correlation between anionic and cationic components in the plane of head-groups. This diffuse band can be simulated using Monte-Carlo. This broad diffuse band due to the in-plane order of the catanionic bilayer does not exist in the case of random distribution of anions and cations in the plane and sharpens if there is a two-dimensional crystal when mole ratio of component is 2 to 1, analogue to anti-ferromagnetic order. Extreme cases of this shape of the diffuse band seen by WANS are shown in figure 10 [38]. In the present case of local order in frozen chains, it has been shown by combined WANS and WAXS that an in-plane order propagates till the third neighbor.

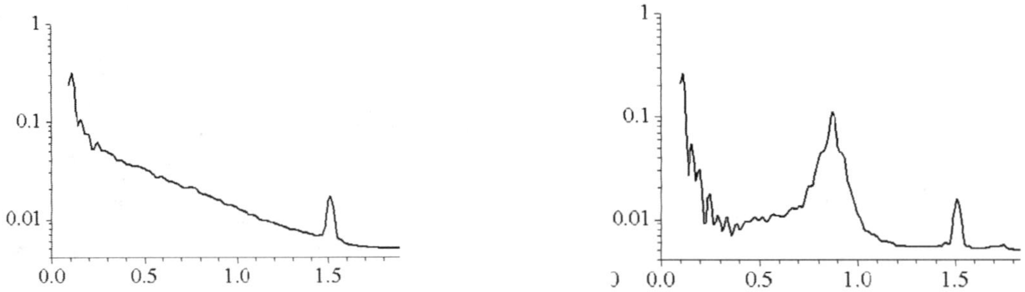

Figure 10. Simulation of diffuse scattering observable in frozen catanionic bilayers at roughly half of the first order in the WANS region due to chain-chain distance. At this q-range, no detection of diffuse scattering is seen in WAXS. Left: limit of random in-plane distribution of anionic and cationic component in a bilayer. Right: with strong correlation due to anionic-cationic component interactions. Comparing simulations to observations of the diffuse band specific of SANS allowed the first determination of the local order in frozen bilayers [from ref. 38].

7. Conclusion

We have shown here that combining SAXS and SANS at high resolution, i.e. up to large q-values and using number of molecules of known chemical composition and molar volume (to be determined by densitometry) gives a very precise description of molecular assembly, including equations of state when osmotic pressure is known.

Using contrast variation, adsorption isotherms on surfactant films as well as location of the solute "in" or "on" the film can be determined. Even with the simplest approach, identification of the maximum of S(q) associated to dilution experiments, can give a direct clue on progressive shape transformations of colloidal aggregates. Combining SAXS and SANS remains one of the most powerful methods for overcoming the mathematical difficulty of getting back to real space from reciprocal space measurements. This is especially relevant if measurements are made on absolute scale, i.e. if scattering cross-sections are experimentally determined using well-defined standards, such as thin layers of pure water at a known temperature, a convenient reference for swollen systems and using SAXS and SANS techniques.

A recent example of this concerns the study maltosides [**39**], single chain amphiphiles. Due to temperature independent hydration of the sugar head-group, packing parameters do not depend on temperature or salinity, making them useful detergents. Using a q-range limited to 0.5Å^{-1}, it has been shown that these molecules form cylindrical micelles with spherical end-caps in water, since the volume of the polar layer is incompressible [40]. What happens when maltosides are mixed with more classical linear nonionic ethoxylates has also been studied by combined SANS-SAXS, taking into account molecular volumes and absolute scale measurements. Surprisingly, instead of the micelle of "average" curvature expected from dynamical equilibrium in micelles, the coexistence of spherical micelles poor in maltoside with cylindrical micelles rich in maltosides, probably enriched at end-caps in one of the components, was recently proposed. This is possible if there are competing attraction/repulsions within the same micelle of the two components; free energy can be minimized by coexistence between A-rich and B-rich micelles of different curvatures in the same sample, still minimizing free energy expressed in the term of packing frustration. Extension to higher q-values and usage of contrast-variation in SANS will probably soon open the door to the first determinations of interaction parameters derived from the structure of micelles: those have to be compared with the one determine from CMC equilibria.

Still a large number of questions in this field are open. For instance, determination of molecular interactions within a given micelle/microemulsions controlling spontaneous packing and coexistence of different micellar shapes –and corresponding solubilisation properties are still ahead!

The review here concerns only surfactants measured by SAXS and SANS: papers are available describing the same SANS/SANS WAXS/WANS methodology used for studying aerogels [**41**], fullerenes [**42**], off-equilibrium coarse emulsions [**43**], rocks [**44**], starch when processed [**45**], and elastomers made from polymers and fillers [**46**]. As can be judged from the variety of colloidal samples investigated, the experimental possibilities opened in the eighties by SANS-contrast variation/SAXS comparison on absolute scale is an emerging methodology which can be used to investigate multi-scale colloidal structures as well as the energetics of the colloidal interaction.

References

[1] Hayter J B and Penfold R. 1981 *J. Chem. Soc. Faraday Soc.* I **77** 1851
[2] Welberry T R and Zemb T 1988 *J. Colloid Interface Science* **123** 413 ; Zemb T 1997 *Colloids Surf. A* **129-130**, 435
[3] Barnes I S and Zemb T 1988 *J. Appl. Crystallogr.* **21** 373
[4] Chevalier Y and Zemb T 1990 *Reports in Progress of Physics* **53** 279
[5] Zemb T 1997 *Colloids and Surfaces A* 1997 **129-130** 342
[6] Hayter J B, Penfold J and Zemb T 1982 *Chem. Phys. Lett.,* **93** 91

[7] Charpin P, and Zemb T 1985 *J. Physique Paris* **46** 249
[8] Cabane B, Duplessix R and Zemb T 1985 *J. Physique Paris* **46** 2161
[9] Vass Sz. *J. Phys. Chem. B* 2001, ***105***, 455. ; Vass Sz, Plestil J, Laggner P, Gilanyi T, Borbely S, Kriechbaum M, Jakli G, Decsy Z ,Abuja P M. 2003 *J. Phys. Chem. B* **107**, 12752
[10] Kunz W , in « Specific ion effects », World scientific publ. Singapore (2010)
[11] Iampietro D J, Brasher L L and Kaler E W, Stradner A and Glatter O, *J. Phys. Chem. B* 1998**,** *102,* 3105
[12] Summers M, Eastoe J, Davis S, Du Zhipig, Richardson R M, Heean R K and Steytler D, 2001, *Langmuir* **17**, 5388
[13] Sommer C, Pedersen J S and Garamus V M, 2005 *Langmuir* **21**, 2137
[14] Arleth L, Ashok B, Onyuksel H, Thiyagarajan P, Jacob J and Hjelm R P, 2005 *Langmuir* **21**, 3279
[15] Sommer C, Deen G R, Struntz O., Pedersen J S and Garamus V M, 2007 *Langmuir* **23**, 6544
[16] Barnes I S , Corti M, Degiorgio V and Zemb T 1993 Progr. Colloid Polym. Sci. **93** 205
[17] Spalla O 1991 in *"Neutron, X-Rays and Light Scattering"*, ed. P. Lindner and T. Zemb, publ. Elsevier 177
[18] Barnes I S, Derian P J, Hyde S T, Ninham B W and Zemb T 1990 J. Phys. France **51** 2605
[19] Zemb T 2009 *Comptes Rendus Chimie* **12** 218
[20] Welberry T R and Zemb T 1993 *Progr. Colloid Polym. Sci.* **271** 124
[21] Arleth L, Marcelja S and Zemb T 2001 *J. Phys. Chem.* **8** 3923
[22] Hayter J B, Chevalier Y and Zemb T 1984 *Colloid & Polymer Sci* **262** 798
[23] May R 1991 in *"Neutron, X-Rays and Light Scattering"*, ed. P. Lindner and T. Zemb, publ. Elsevier 214
[24] Auzely-Velty R, Djedaini-Pilard F, Taché O, Zemb T, Perly B and Dalbiez J P 1999 *Carbohydrate Research* **318** 82
[25] Demé B , Dubois M, Zemb T and Cabane B 1996 *J. Phys. Chem.* **100** 3828
[26] Demé B and Zemb T 2000 *J. Appl. Crystallogr.* **33** 569
[27] F. Ricoul , Dubois M and Zemb Th 1997 *J. Phys. II France* 7, 69
[28] Demé B et al. 1997 *Colloids Surfaces A*, **121** 135
[29] Cantu L, Corti M, Del Favero E, Dubois M and Zemb T 1998 *J. Phys. Chem. B* **102** 5737
[30] Bartlett J, Gazeau D, Woolfrey J and Zemb T 1998, *Langmuir* **14** 3538
[31] Fratini E, Page M G, Giorgi R, Coelfen H, Baglioni P, Demé B and Zemb T, 2007 *Langmuir* **23** 2330.
[32] Coulombeau H, Testard F, Zemb T and Larpent Ch. 2004 *Langmuir* **12**, 4840
[33] Erlinger C , Gazeau D, Zemb T, Madic C, Lefrançois L, Hebrant M and Tondre C 1998 *Solvent Ext. Ion Exch.* **16** 707
[34] Meridiano Y , phD thesis, University of Orsay, Paris XI (France) , 2009
[35] Nave S, Mandin C, Martinet L, Berthon L, Testard F, Madic C and Zemb T 2003 Physical Chemistry Chemical Physics **6 79**9
[36] Berthon L, Martinet L, Testard F, Zemb T and Madic C 2007 *Solvent Extraction and Ion Exchange* **25** 545
[37] Zemb T and Dubois M 2003 *Austr. Australian Journal of Chemistry* **56** 971
[38] Carrière D, Belloni L, Demé B, Dubois M, Vautrin C and Zemb T 2009 *Small* **5** 1816
[39] Bäverback P., Oliveira C L P, Garamus v M, Varga I, Claesson P M and Pedersen J S, 2009 *Langmuir* **25**, 7296
[40] Cecutti C, Focher B, Perly B and Zemb T, 1991 *Langmuir* **7**, 2580
[41] Ballauff M, 2001 *Curr. Op. in Coll. Interf.* **6**, 132
[42] Jeng U, Lin T L, Liu W J, Tsao S, Canteenwala T, Chiyang L I, Sung L P, Han C C, 2002 *Physica A*, **304**, 191
[43] Radlinski A P, mastalerz M, Hinde A L, Hainbuchner M, Rauch M, Baron M, Lin J S, Fan P and Thiyagarajan , 2004, *International journal of coal geology* **59**, 245

[44] Lesieur P, Lindner P, Desforge C, Lambard J and Zemb T , 1992 *Physica B*, **180**, 564
[45] Jenkins P J and Donald A M, 1998 *Carbohydrate research* **308**, 133
[46] Botti A, Pyckhout-Hintzen D, Richter D and Straube E, 2002, *Physica A*, **304**, 230

Quantitative analysis and relevant features of the scientific literature related to SAXS and SANS

Aldo F. Craievich[1] and Hannes Fischer[2]

[1]Institute of Physics, University of São Paulo, São Paulo-SP, Brazil
[2]Jacto Máquinas Agrícolas SA, Pompéia-SP, Brazil,

E-mail: craievich@if.usp.br, hannes@jacto.com.br

Abstract. We present and discuss here numerical information derived from a systematic searching of scientific papers related to SAXS and SANS published in indexed journals - from 1945 until nowadays - recorded by the Web of Science Data Bank (WoS). We have detected interesting features regarding the time dependence of the number of papers/year, $N(t)$, indicating the existence of three well-defined periods of historical evolution with rather well-defined boundaries. All three periods exhibit a positive and approximately linear variation of $N(t)$ but, at the two transitions between periods, the rate of growth exhibits clear and strong increases. Differences of the historical evolutions in the numbers of papers/year related to SAXS and to SANS were established. The different behaviours regarding the numbers of papers/year related to SAXS and to SANS and the existence of three different and well defined periods for $N(t)$ can be qualitatively understood as a consequence of the progressive and increasing availability along the last three decades of very brilliant synchrotrons, last generation commercial X-ray sources, new neutron facilities, powerful computers and novel theoretical approaches for SAS data analysis. The rates of growth in the number of papers/year published by authors from a set of different countries are approximately constant along the last two decades. For other countries we have detected a slowing down effect in the number of papers/year while a clear acceleration could be noticed for the production of SAS papers by authors from several emerging countries. These opposite trends compensate in such a way that the number of SAS (SAXS+SAXS) articles published per year all around the world maintained a vigorous linear growth - during more than 20 years - at a constant rate of 60 papers/year, without any indication of eventual saturation. The observed distribution of articles among different journals indicates that a very high fraction of the volume of SAS research is focused to the structure of soft matter.

1. Introduction

The rate of published articles in the scientific literature in most research areas is nowadays extremely high and is still increasing at a very high pace. This is also the case for the literature related to small-angle X-ray scattering (SAXS) and small-angle neutron scattering (SANS) techniques. We present here numerical results of a systematic searching of published papers related to SAXS and SANS in the Web of Science Data Bank (WoS) from Thomson Reuters. In order to establish eventually relevant features of the historical evolution, the number of published papers/year, $N(t)$, was recorded as a function of the year of publication. The papers published from 1991 until now days recorded by us are those that contain a predefined set of words. The search query used at WoS in the topic field was: *((ultra*small-angle same scattering) or (gi*sans or u*sans or su*sans or gi*saxs or u*saxs or su*saxs) or ((small or low) same angle same scattering or saxs or sans))*. In order to be retrieved an article published from 1991 until now days should contain these words in their title, key words or abstract and those published between 1945 and 1990 in their title or key words. The retrieval date was 9th April 2009. The final and total number of articles retrieved was 25791. The bibliographic data of these articles were then downloaded and analyzed using a data-mining software named VantagePoint 6.0 from Search Technology, Inc (www.thevantagepoint.com) [1]. The WoS Data Bank does not include the abstracts of papers published before 1991 so as the plot of the number of (SAXS+SANS) articles/year exhibits a sharp increase between 1990 and 1991. We have suppressed this apparent discontinuity in the annual rate $N(t)$ by multiplying the numbers of articles/year before 1991 by a factor 2.2.

The completeness of our searching procedure was estimated by checking the whole texts of all papers published by one of the authors (AFC) in indexed journals. We have verified in this particular case that 95% of the total number of published papers related to SAXS and SANS were actually recorded by the retrieval procedure applied here. This finding suggests that our retrieval procedure recorded most of the published papers related to SAXS and SANS, the fraction of articles that were missed being only a few percent.

This article aims at determining the number of papers published each year related to SAXS and SANS and their growth rate as functions of time, establishing the distributions of the number of papers/year among different countries and among different scientific journals.

2. Historical evolution of the annual number of articles related to SAXS and SANS

The number of articles related to SAS (either SAXS or SANS) published each year is displayed as a function of time, $N(t)$ in figure 1. We can clearly notice three different periods in the plot shown in figure 1, all of them with an essentially linear time dependence but with clearly different slopes, $\Delta N / \Delta t$, that indicate different growth rates. The first period ending by 1965 is featured by a low number of articles published per year and by a rather slow growth in the number of papers/year. The second period - starting by 1965 and ending by 1990 - is characterized by a moderate growth in the number of papers/year, the function $N(t)$ increasing from about 20/year in 1965 up to 300/year in 1990. Over the third period - starting in 1988 and persisting until now days - $N(t)$ exhibits an also linear time dependence but a much faster growth rate. The number of published articles increases in this period at a pace of 60 articles/year circa, from 300 articles/year in 1990 up to 1500/year in 2008.

The first period mentioned above was featured by a very low number of articles/year so as it can be considered a sort of "incubation" period for the current SAS literature. During the second period starting in 1965 - coincidently the year during which the first International SAS Conference was held - the rate of growth, $\Delta N / \Delta t$, became rather high. Finally during the last two decades the number of papers/year was characterized by a much stronger growth rate, which persists until nowadays without any indication of eventual saturation.

The strong growth in the number of papers/year during the third period starting by 1990 (Figure 1) can be explained by the increasing number and wider availability of (i) new synchrotron and neutron SAS setups, (ii) powerful personal computers, and (iii) newly developed theoretical approaches and fitting procedures applied to SAS data. An additional and relevant reason for the observed strong growth in the number of paper/year is the vigorous increase in the interest of materials scientists for the study of nanomaterials, which started to be widely produced during the last two decades. The growing needs of researchers for the structural characterization of newly developed nanostructured materials induced the correlated growth of the literature related to SAXS [2] and to SANS [3] and also to the more recently developed GISAXS and GISANS techniques. Another important contribution to the very high rate of growth in the number of SAS papers observed during the last two decades was the strong and increasing interest in the study of biological macromolecules in solution (proteins and others), this contribution being enhanced by the availability of new developments of powerful simulation tools [4].

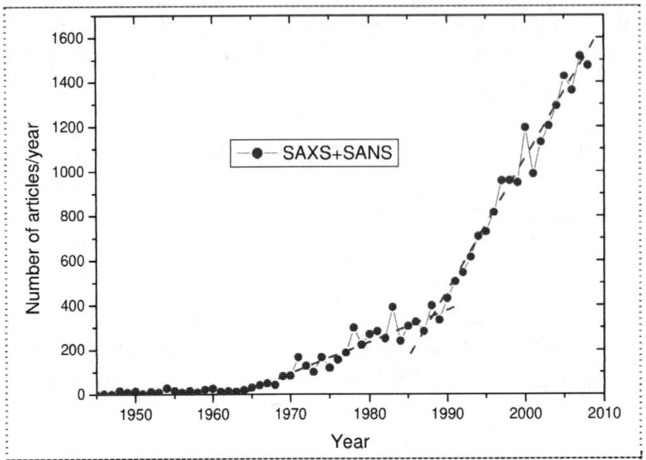

Figure 1. Number of articles/year related to SAS measurements (SAXS or SANS).

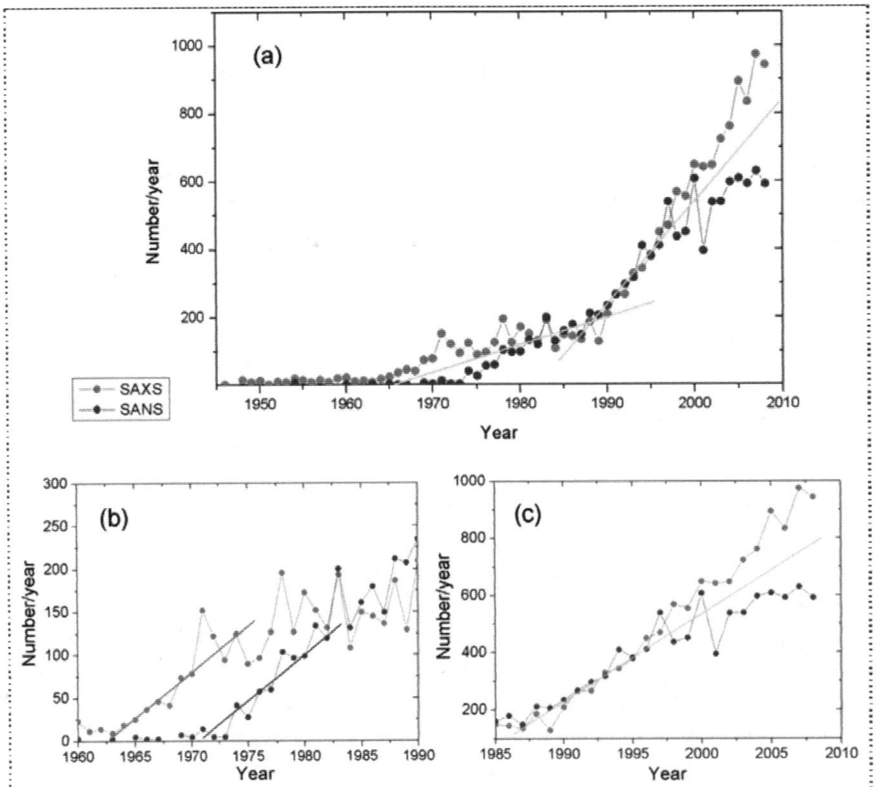

Figure 2. (a) Number of paper published per year containing results referring to SAXS and SANS separately. Details corresponding to 1960/1990 and 1985/2008 periods are displayed in (b) and (c), respectively.

Figure 2a exhibits separately the number of papers/year related to SAXS and those associated to SANS. Details of these curves corresponding to the periods 1960/1990 and 1990/2008 are plotted in Figure 2b and 2c, respectively. Articles published until 1973 were almost exclusively related to the SAXS technique. This happened because of the absence of a significant number of neutron facilities with SANS setups in operation until the end of the sixtieth. In 1973 the number of papers/year associated to SANS started to grow and became soon, by 1978, similar to the number related to SAXS. The equivalence of the $N(t)$ functions related to SAXS and to SANS persisted from 1978 until 2000. Nevertheless, by 2000, the number of articles/year based on SANS results exhibits a clear tendency to saturation and becomes nowadays nearly constant (about $N = 600$ papers/year). This saturation in the number of published papers related to SANS is being accompanied by a parallel acceleration in the growth rate of the number of articles of papers/year related to SAXS, reaching now days a number $N = 1000$ circa. i. e. ~70% above those related to SANS. These positive (SAXS) and negative (SANS) deviations of $N(t)$ from a linear time dependence compensate so as the total number of (SAXS+SANS) papers/year persists in its linear growth until nowadays, as it can be seen in Figure 1.

3. Distribution among countries

Since many papers are authored by scientists from different countries, we have assigned each paper to the country in which the first author works. This seems reasonable because the order of authors is expected to be geographically distributed at random. The histogram displayed in Figure 3 exhibits the total number of papers related to SAS published from 1945 until now by first authors from different countries. The countries with the highest total numbers of published articles are USA, five European countries (France, Germany, Russia, Italy and Netherlands), and three Asiatic countries (Japan, China and India).

The numbers of articles published per year by first authors from different countries, $N(t)$, are displayed in figures 4a, b and c. These figures also show solid lines corresponding to second order polynomials, $N(t) = a_0 + a_1(t - t_0) + a_2(t - t_0)^2$, where $t_0 = 1990$, fitted to the retrieved $N(t)$ function for every country, from 1991 until 2008. The smoothed (best fitted) $N(t)$ functions were extrapolated a few years, these extrapolations expecting to yield a reasonable prediction for the near future, provided the individual trends during the last two decades are preserved.

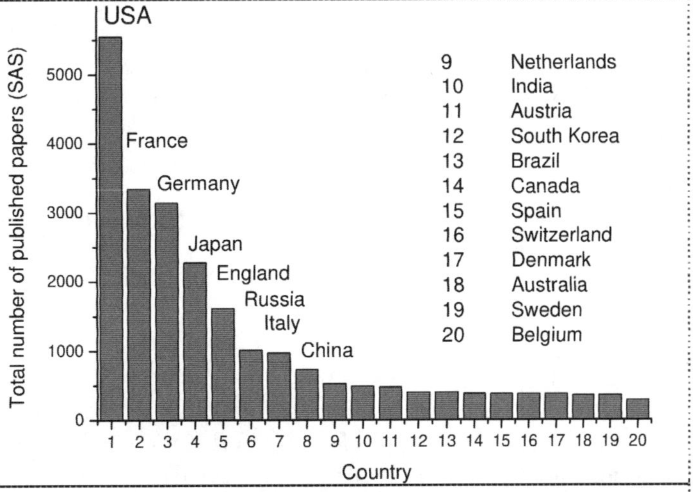

Figure 3. Geographical distribution of the total number of papers related to SAS (SAXS or SANS).

The numbers of papers/year related to SAS (SAS+SANS) published from 1991 exhibit different overall features for three different groups of countries, as it can be seen in Figure 4a, b and c. For the first group composed of France, Russia, Italy and Austria (Figure 4a) the number of papers/year grows linearly, i. e. $N(t) = a_0 + a_1(t - t_0)$. For all these countries, the second order coefficient of the fitted polynomials, a_2 , is smaller than the corresponding statistical error bar, so as the number of papers from these countries exhibit the same linear and growing behavior as that observed for the total number of SAS papers/year published by authors from all countries (Figure 1). On the other hand, USA, Germany, Japan, England, Netherlands, Denmark, Sweden and Belgium (Figure 4b) belong to a second group of countries for which the second order polynomial coefficient is negative, $a_2 < 0$, so indicating a slowing down in the growth rate of the number of articles/year. On the contrary, for a third group of countries, composed of China, India, South Korea, Brazil, Spain, Australia, Switzerland and Canada, the second order polynomial coefficient is positive, $a_2 > 0$, thus evidencing an acceleration in the growth rate of the number of papers/year (Figure 4c). The acceleration noticed for the third group of countries compensates the slowing down observed for countries of the second group, so as the total number of paper published per year by authors from all countries preserves its linear growth over the whole 20 year period, from 1988 until now days, as previously verified in the plot displayed in Figure 1.

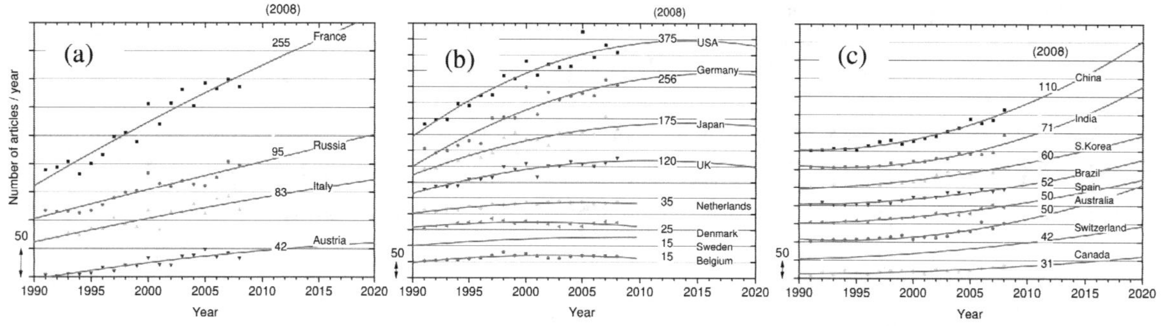

Figure 4. Annual numbers of papers published by first authors from different countries. The curves were vertically shifted for clarity. (a) Countries with an annual number of published papers exhibiting a linear growth over the whole period. (b) Countries that exhibit a slowing down in the growth rate of the number of papers/year. (c) Countries that exhibit a positive acceleration in the growth rate. The distances between horizontal lines correspond to 50 articles. The numbers of papers published in 2008 corresponding to all countries are indicated.

Not surprisingly, all countries that exhibit the highest numbers of published papers (USA, France, Germany, Japan, UK, Italy and Russia) and also all of those that showed a positive acceleration in the growth rate of published papers (China, India, South Korea, Brazil, Switzerland, Australia, Canada, Switzerland and Spain) posses one or several synchrotron radiation sources in operation and/or under construction.

The histogram displayed in figure 5 indicates the number of paper/year published 10 years ago (red) and nowadays (green), and the expected values within 10 year provided the parabolic trend observed during the past 20 years is preserved along the forthcoming decade (blue). We can notice that the distribution among countries within 10 years is expected to be more homogeneous than now days, i. e. the differences between the number of papers published per year by developed countries such as USA, Germany and France, and those published by emerging countries tend to decrease. Emerging countries such as China and India exhibit a particularly strong increase in the growth rate of the number of published papers/year so as one can guess that in the forthcoming years and provided the trends prevailing during the past 20 years persist, these countries will reach a number of papers/year comparable to most of the developed countries.

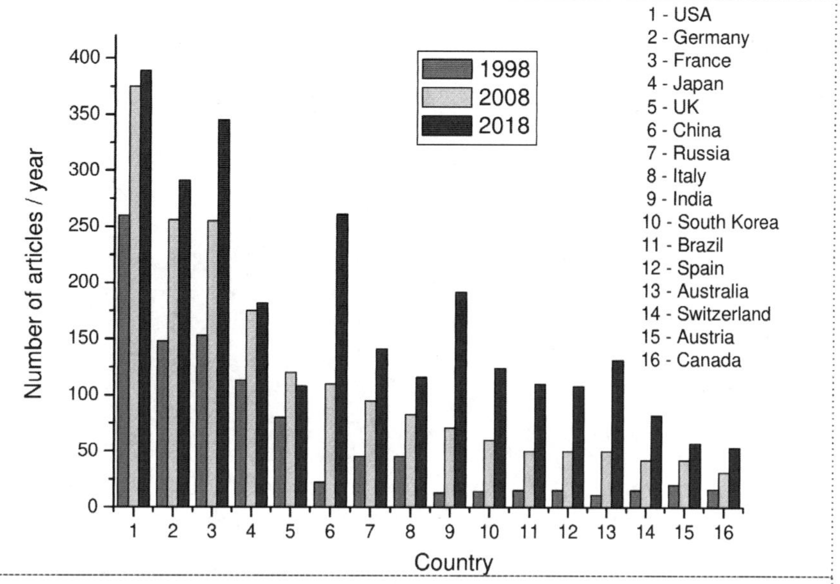

Figure 5. Histogram of the number of articles published per year in 1998 (red) and 2008 (green) by authors from different countries. The blue bars correspond to the expected annual numbers under the assumption that the trends observed during the last 20 years will be preserved during the forthcoming decade.

4. Distribution among journals

The number of articles/year based on SAXS and SAXS techniques in the eight journals that have published until now the highest total number of articles related to SAXS and SANS are displayed in figure 6a, in which the different curves were vertically displaced for clarity. The numbers of SAXS or SANS papers published per year in each journal exhibit many oscillations over time but an approximately linear overall trend prevails. The numbers of articles/year published by the Journal of Applied Crystallography and Physica B exhibit particularly strong oscillations and nearly periodic and narrow peaks. The peaks corresponding to Journal of Applied Crystallography occur at years during which the Proceedings of the International SAS Conferences appeared [5]. Similarly, the peaks associated to Physica B occur at years during which Proceedings of International SANS Conferences were published.

Figure 6b exhibits the straight lines, without vertical displacements, that best fitted to the data plotted in Figure 6a. Three journals with the highest number of papers/year in 2008 (Macromolecules, Langmuir and Polymers) publish results of investigations dealing with proteins, thin organic films, polymeric materials and other "soft" materials. On the other hand, the next journals in the ranking – Journal of Physical Chemistry B, Journal of Applied Crystallography and Physica B - publish papers focusing to all types of materials. This obviously indicates that SAXS and SANS techniques are more widely applied to the investigation of soft matter, the highest number of papers being published by Macromolecules, followed by Langmuir and Polymers. The results plotted in Figure 6b qualitatively suggest that the observed difference between the number of articles/year related to structural characterization of soft matter (organic films, polymers and protein in solution) and those related to inorganic materials still nowadays tends to increase.

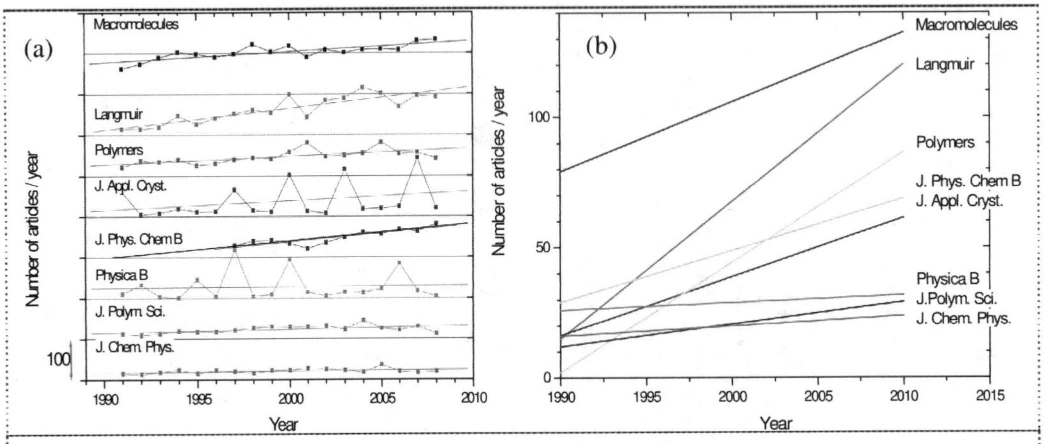

Figure 6. (a) Number of paper/year (SAXS and SANS) published by different journals. The curves are vertically shifted for clarity. (b) Comparative trends for different journals described by the straight lines (without vertical displacement) that best fit the data plotted in (a).

5. Most cited articles

We have also retrieved from WoS Data Bank the articles most cited in the literature related to SAXS and SANS. Ten articles received more than 400 citations and two of them, both related to SAXS, received more than 1000 citations. The references corresponding to the two most cited articles are:

-"CdSe/ZnS core-shell quantum dots: Synthesis and characterization of a size series of highly luminescent nanocrystallites" [6] (1206 citations).

-"Buildup of ultra thin multilayer films by a self-assembly process. Consecutively alternating adsorption of anionic and cationic polyelectrolytes on charged surfaces" [7] (1136 citations).

We have tried to establish the common features of highly cited articles related to SAXS and SANS. In most of these articles, several experimental techniques for materials characterization are applied, so as attracting the attention of a spectrum of readers wider than in case of the application of a single technique. Another common feature of many highly cited articles is that they include results of SAXS or SANS - and eventually of other techniques – for structural characterization, this structural information being also applied to understand different properties of the studied materials.

6. Contribution of SAS to the general scientific literature

We have already pointed out that the rate of growth of the scientific literature during the last decades is very high. The same can be said about the volume of the literature related to SAS (SAXS+SANS), which increased from about 20 published papers per year in the sixties of last century up to 1500 papers/year circa in 2008. We have determined the fraction of the annual number of papers related to SAS with respect to the total annual number of papers published in the literature, and established the trend of this fraction as a function of time. In order to perform this analysis, eight journals were selected, in which papers related to SAS are more commonly published. These journals were those also selected in section 4 for the plots displayed in figure 6. We have then determined the annual number of articles related to SAS, N_{SAS}, and the annual number of all articles, N_{Total}, published in these journals, from 1980 until 2008. The fraction of the number of papers related to SAS with respect to the total number of published articles, N_{SAS}/N_{Total}, is displayed as a function of time in figure 7.

This function exhibits a set of nearly periodic peaks that, again, are clear effects coming from the contribution of the periodic proceedings of the International SAS Conference series.

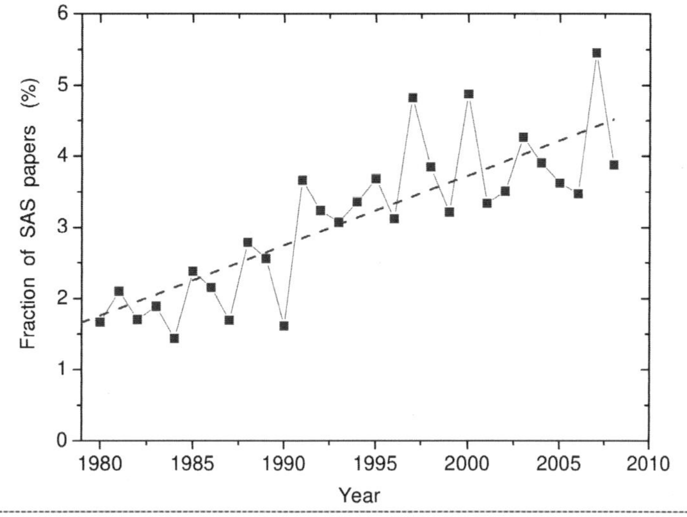

Figure 7. Number fraction of articles related to SAS with respect to the total number of articles published in eight selected journals. The dashed straight line is a guide for the eyes.

The results plotted in figure 7 demonstrate that the fraction of the number of papers related to SAS with respect to those of the whole literature published in the selected journals, exhibits a significant increase, from less than 2.0% in 1980 up to 4.5% circa in 2008. This trend indicates that beside the strongly increasing volume of published research related to SAXS and SANS during the last three decades, these experimental techniques also exhibit a clearly increasing relevance with respect to the general scientific literature.

7. Main conclusions

The presented results of our searching over the SAXS/SANS literature obviously do not yield an evaluation of the quality of the research performed but they allowed us to infer some interesting insights into historical trends, geographical distribution and other features of the published articles related to these experimental techniques.

The time dependence of the number of papers published per year from 1945 until 2008 clearly indicate the existence of three periods of the historical evolution of the SAXS/SANS literature (1945/1965, 1965/1990 and 1990/2008), each of them with different and successively increasing growth rates. The relevant features of these three periods of growing activity seem to be well correlated to the parallel and increasing availability of brighter (synchrotron and commercial) X-ray sources, new neutron facilities and more powerful personal computers and also to the development of new theoretical approaches for SAS analysis. Another relevant reasons for the observed strong growth in the number of SAXS/SANS paper/year during the last two decades are (i) the vigorous increase in the interest of materials scientists for new nanomaterial development and characterization and (ii) the significant increase in the number of biology oriented researchers doing SAS studies of macromolecules in solution taking benefit from the availability of new powerful simulation tools for data analysis.

The results regarding the distribution of papers among authors from various countries during the last two decades showed three different groups of countries with different historical trends. A first group exhibits a linear growth in the number of papers/year, a second group composed of developed countries shows a slowing down in the growth rate and, inversely, a third group of emerging countries exhibit a clear positive acceleration. The opposite trends of the second and third groups tend to

Journal of Physics: Conference Series **247** (2010) 012003 doi:10.1088/1742-6596/247/1/012003

compensate in such a way that the linear growth and high slope in the number of SAS (SAXS+SAXS) articles/year produced all around the world are preserved over two decades, starting in 1990 and up to now without any apparent saturation effect.

References

[1] Zhu D and Porter A L 2002 *Technological Forecasting and Social Change* **69** 495
[2] Peterlik H and Fratzl P 2006 *Monatshifte fur Chemie* **137** 529-543
[3] Meinichenko Y B and Wignall G D 2007 *J. Appl Phys* **102** 02101
[4] Petoukhov M V and Svergun D I 2007 *Curr. Opin Struct. Biol.* **17** 562
[5] Craievich A 2003 *J. Appl. Cryst. Proc. of the XII International SAS Conference. Foreword.*
[6] Dabbousi B O, Rodriguez-Viejo J et al. 1997 J Phys Chem B **101** 9463-9475
[7] Decher G, Hong J D et al. 1992 *Thin Solid Films* **210** 831-835

Journal of Physics: Conference Series **247** (2010) 012004 doi:10.1088/1742-6596/247/1/012004

Collagen imaged by Coherent X-ray Diffraction: towards a complementary tool to conventional scanning SAXS

Felisa Berenguer de la Cuesta[1,*], Richard J Bean[1], Catriona McCallion[2], Kris Wallace[2], Laurent Bozec[1, 3], Jen C Hiller[4], Nicholas J Terrill[4], Ian K Robinson[1,4]

[1] London Centre for Nanotechnology (LCN), University College London (UCL), London WC1H 0AH, United Kingdom

[2] Department of Physics and Astronomy, University College London (UCL), London WC1E 6BT, United Kingdom

[3] Biomaterials and Tissue Engineering, Eastman Dental Institute, UCL, London WC1X 8LD, United Kingdom

[4] Diamond Light Source, Harwell Science and Innovation Campus, Didcot, Oxfordshire OX11 0DE, United Kingdom

Corresponding author e-mail address: f.berenguer@ucl.ac.uk

Abstract. Third generation x-ray sources offer unique possibilities for exploiting coherence in the study of materials. New insights in the structure and dynamics of soft condensed matter and biological samples can be obtained by coherent x-ray diffraction (CXD). However, the experimental procedures for applying these methods to collagen tissues are still under development. We present here an investigation for the optimal procedure in order to obtain high quality CXD data from collagen tissues. Sample handling and preparation and adequate coherence defining apertures are among the more relevant factors to take into account. The impact of the results is also discussed, in particular in comparison with the information that can be extracted from conventional scanning small angle x-ray scattering (SAXS). Images of collagen tissues obtained by CXD reconstructions will give additional information about the local structure with higher resolution and will complement scanning SAXS images.

1. Introduction

Small Angle X-ray Scattering (SAXS) has been traditionally applied to the study of a large range of biomaterials, from protein solutions to fibres and composites, as well as biological tissues such as muscle and bone. SAXS provides valuable information about the mesoscale of these systems, in particular over the range of a few tens of nanometers. However, the structural information in the sample is averaged over an area of typically 200-300 microns due to the size of the beam used. While some studies do not require a higher spatial resolution, in the case of biological tissues (that are highly hierarchical structures) the possibility of reducing the spot size together with the resolution is very appealing.

The introduction of microbeam facilities in SAXS beamlines [1] was a first step towards exploiting SAXS as an imaging technique for biomaterials. The use of microbeams in scanning and/or tomographic setups has opened a new range of possibilities in the study of tissues and fibres, with beam spot sizes of typically 3-50 microns [2]. Currently this technique is widely used in the study of biomaterials, for instance, in bone structure and micromechanics [3], soft tissue and cornea [4], natural polymers structure [5], or parchment and ancient materials [6]. With the advent of nanobeams in some facilities, such as ID13 at the European Synchrotron Radiation Facility (ESRF), we can expect new possibilities in the study of local structure, although the effects of the radiation damage induced by the highly intense focused nanobeams will be a relevant issue in soft matter samples.

The natural step further is to fully exploit the coherence of 3rd generation sources, together with micro or nano beam sizes, in order to extract as much information as possible from each exposure of the sample, therefore limiting the onset of radiation damage. When using a fully coherent beam, the scattered amplitudes of different areas of the sample interfere coherently, and the resulting diffracting pattern is proportional to the square of the sum of these amplitudes. On the other hand, if the incident beam is incoherent, the resulting diffraction pattern will be proportional to the sum of the intensities of the scattered waves from the different regions of the sample. Therefore, it is clear that a diffraction pattern from a coherent beam contains additional information about the sample nano and microstructure which is masked otherwise if the incident beam is incoherent. The extra information appears as speckles, or modulations of the intensity distribution and can be used by phase retrieval algorithms [7-10] in order to recover the amplitude and phase transmission functions of the sample with a resolution (ideally) only limited by the wavelength. This information can be combined with the results from scanning microbeam SAXS and/or tomography SAXS to get a more complete picture of the sample structure.

We are particularly interested in the applications of Coherent X-ray Diffraction (CXD) to collagen tissues in the small angle regime, in particular tendon and bone. In the case of bone, current non-coherent scanning SAXS can give relevant information about the location and orientation of the mineral phase [3], in some cases with 1 micron spatial resolution. However, in order to study the osteogenesis and mineralization processes it would be very interesting to obtain similar information, with improved resolution, about the disposition and orientation of the organic matrix of collagen fibres. Some recent studies using scanning SAXS in tendon samples show promising results [4] although the spatial resolution is still unsatisfactory (20 microns). In the present paper, we present CXD methods, in particular ptychography, as a potential and powerful complementary method that may be used for obtained dark field images of collagen tissues with a resolution of some tens of nanometers.

2. Coherent X-ray Diffraction

Coherent X-ray Diffraction imaging is based on the inversion of speckled far-field diffraction patterns in order to obtain the complex-valued (amplitude and phase) transmission function of the sample. This so-called inversion process involves, as a first step, the retrieval of the phases on the detector plane of the diffracted waves using iterative phasing algorithms. This is achieved using the intensities recorded in the diffraction patterns together with some constraints imposed from the sample transmission function, the illuminating wavefront, or both. Once the amplitude and phase of the diffracted wave has been determined on the detector plane the next step is a propagation of the wave back to the sample exit plane (done usually by calculating the wave inverse Fourier transform) to obtain the sample complex-valued transmission function. CXD inversion algorithms require the patterns to be oversampled [11], so the maximum spatial frequency in the pattern is at least two times bigger than the detector pixel size. In most cases, this is achieved by illuminating a small isolated sample under a fully coherent beam, so that all the density of the object is contained in a delimited region of the beam footprint, whereas the sample density outside this area is 0. This fact can be translated onto a mathematical real-space constraint in most part of the CXD phasing algorithms, as one requires that the density of the reconstructed sample is contained in a portion (the so-called "support") of the total area covered by the beam. The use of this real-space constraint in the inversion process has proven to be a successful approach in a range of problems both in material science [12] and biology [13]. However, the requirement for a small isolated object being smaller than the beam footprint constitutes a limitation for a wide range of scientific cases. In the past few years several new approaches have proposed tackling the inversion problem from alternative perspectives; namely ptychography, "keyhole" diffractive imaging and wavefront modulation-based imaging.

Ptychography is based on a different approach to the inversion process of a series of diffraction patterns, each one recorded from different illuminated areas in the sample [8]. A given amount of overlap (ideally a minimum of ~60% [14]) is required between different illuminated areas. Here the redundant information in successive patterns gives a strong constraint to the inversion algorithm. An efficient first ptychography algorithm was presented by Rodenburg and Faulkner [8] and demonstrated for x-rays shortly afterwards [15]. However, the first algorithms required an extremely good knowledge of the probe wavefront profile in order to obtain a good convergence. This limitation has been overcome recently by Thibault and colleagues [9], where they demonstrated that both the sample transmission function (amplitude and phase) and the probe wavefront profile can be obtained from the same ptychography dataset. Other recent methods [10] also reconstruct the illumination function. This set of ptychography algorithms has proved to be powerful in the reconstruction of some relevant samples not only with visible light [10] but also in the case of hard x-rays [16] and electrons [17]. In all these cases, the data were obtained in a forward scattering geometry; ptychography has not yet been proved to work under Bragg diffraction. Although the use of the latest algorithms that reconstruct also the illuminating wavefield has eased the inversion problem, the technique is still quite sensitive to errors in the probe position, effects due to partial coherence in the beam and poor signal-to-noise ratio.

The phase retrieval process in both conventional "support"-based methods and ptychography evolves by Fourier transformations of the scattered waves between real and reciprocal spaces in two planes: the one of the sample and the one of the detector, where the diffraction patterns are recorded. When using the information from a third plane, additional constraints are added to the iterative process and the reconstruction of the sample transmission function is in principle improved. Both keyhole coherent diffraction and wavefront modulation-based methods are rooted in this idea. Keyhole coherent diffractive imaging has been demonstrated recently [18]. It uses a curved illumination and solves initially for the illumination wavefront and then, in a second step, the sample transmission function [19]. The method is powerful although it is sensitive to errors in the optical elements distances determination.

Another interesting possibility is to introduce a known phase modulator downstream of the sample [20]. Here a set of diffraction patterns are recorded for different transverse positions of the phase modulator, illuminating the same area of the sample. The method has been demonstrated for visible light [21] and hard x-rays [22] and nicely overcomes some of the requirements in the probe characteristics and signal-to-noise ratio needed in other CXD methods such as ptychography and keyhole imaging. However, the fact that the same area of the sample is illuminated during the modulator scan might induce the onset of radiation damage in the case of some biological samples, which could be delayed when scanning the sample instead (as done in ptychography).

In the present paper, we focus on the possibilities of Coherent X-ray Diffraction (in particular ptychography) for the imaging of biological specimens as a complementary tool to conventional scanning SAXS, providing extra information that could not be accessible otherwise. The technique has already been successful in obtaining images of biosamples such as bacteria [16] and blood cells [23]. However, we are interested in extending ptychography to the Bragg diffraction geometry. In such a way, the technique could be used for obtaining dark field images of biological extended samples, in particular collagen [24] or cellulose-based tissues. Here dark field is used to describe the diffraction contrast of the different components of the sample which can be exploited in order to obtain images where only the distribution of one single component (the one from which the Bragg reflections have been used for the CXD reconstruction process) is visible, whereas all the other components are filtered. This is attractive, for example, in the study of composite materials or biological tissues, as we can obtain images with the distribution of each one of the components separately. When imaging materials where the different components show small differences in terms of absorption and phase contrast, the ability to exploit the diffraction contrast provided by dark field ptychography can be useful. Our motivation here was to study the experimental requirements needed in Bragg diffraction ptychography of collagen tissues taking into account the more or less value of the output information that may be obtained from a given sample reconstruction.

3. Materials and methods

Samples consisted of collagen fibres carefully extracted from rat tail tendons [25]. The tails, from young adult healthy rats, were kept frozen until dissection. The dissection involved a careful extraction of the four tendons of the tail after skin removal. This operation is difficult as tendons are strongly attached to the tiny bones of the tail and needs to be done by means of short, quick and precise stroke movements of the dissection tweezers; otherwise the superficial part of the tendon is damaged. We carefully removed the blood vessels and connective tissue attached to the tendons. Collagen fibres of ca. 10 mm long and about 0.3-0.5 mm diameter were extracted simply by pulling them from the central part of the tendon in saline solution and mounted slightly stretched in a suitable cell filled with phosphate buffer solution (7.4 pH). The cell is a brass cell specially designed to clamp the fibres keeping them parallel without introducing additional stretching. It has a reservoir for buffer solution and two mica windows (each 25 μm thick) with a total x-ray path length of 2 mm. The scatter from the solution volume was found to be relevant close to the beamstop, but had less affect in the Q-range of interest for this experiment (around 0.009 A^{-1}). In this way the samples were kept wet during the data acquisition. This increased the useful lifetime of the samples exposed to the beam, as the liquid reduced the heat load on the collagen fibres and hence the radiation damage produced by thermal denaturalization of collagen. Although the presence of liquid accelerates the other mechanisms of radiation damage in collagen, such as hydrogen bonds dissociation, which are firstly noticeable as a lost of intensity in the highest orders of diffraction, here we were concerned only about the first meridional reflection preservation. This peak is rapidly affected by the process of thermal denaturalization, so the presence of buffer solution greatly helps us to record diffraction patterns with 120 sec exposure time before the effects of radiation damage were noticeable.

Data were collected in three runs at beamline I22, Diamond Light Source (UK). The beamline is fed by an IV-25 undulator. The double crystal fixed exit Si (111) monochromator offers an energy resolution of 10^{-4} and the beamline uses a pair of Kirkpatrick-Baez mirrors for focusing and harmonics rejection. The beam energy was set to 8.5 keV, corresponding to a wavelength λ of about 0.1459 nm. An optically-coupled CCD camera (PI-SCX 1300 from Princeton Instruments, with 1340 x 1300 pixels, pixel size 20 microns, 16-bites dynamic range) was mounted at a distance of 5.2-5.7 m (depending on the experiment) from the sample in order to cover the Q-range of interest (0.006 to 0.011 A^{-1}). The distance sample-to-detector matters in a CXD experiment in the small angle regime as it is linked to the resolution of the reconstruction given by the algorithm as well as to the defining coherence aperture-to-detector distance. The former defines the size of the speckles observed on the detector and needs to be adjusted so the speckles are oversampled (at least 2 CCD pixels per speckle) [11] and hence make possible the use of iterative phase retrieval algorithms for reconstructing the sample.

We used two different configurations for the coherence defining aperture. In a first setup, the aperture was a square of 25 x 25 μm^2 defined by a pair of slits at 1.4 m upstream the sample. The last set of slits in the beamline, located at 0.4 m in front of the sample, was used as guard slits. Because of partial coherence effects observed (see Results section) we used a set of 2 pinholes and modified distances between the elements for the most recent run. The first pinhole (20 μm) was located at 11 cm upstream from the sample, followed by a second 100 μm pinhole at 3 cm for removing parasitic scattering. Each sample was measured in a set of overlapping positions spaced 2 μm, 3μm or 4μm apart, depending on the experiment.

4. Results

Figure 1 shows a typical diffraction pattern from the collagen fibres in one position of the ptychography scan. Because of the long distance between the sample and the CCD detector, only the first meridional orders of collagen (corresponding to a spacing of approximately 1/67 nm^{-1}, Q = 0.009 A^{-1}) are recorded. Based on experience of macroscopic rat-tail collagen samples, we would expect these reflections to form an arc with an angular size related to the degree of axial disorder of the fibrils within our sample [26]. However, in our data the reflections can be seen to form straight lines, which indicate that the fibrils are almost parallel, with a high degree of axial order in the tissue (compare figures 1a and 1b). Not all samples showed this phenomenon, but it was our impression that it represented our "best" samples, those which had undergone the least mechanical handling during preparation. The reflections are now confined in a smaller area of the diffraction pattern, hence reducing the intensity spread and increasing the signal-to-noise ratio of the Bragg peaks of interest. This is particularly relevant for us as the phase retrieval algorithms usually require strong measured intensities, and it was achieved thanks to an improved sample preparation protocol. In our previous work [24] the signal-to-noise ratio of the obtained diffraction patterns was low; hence phase retrieval algorithms could not be applied straightforwardly. It was evident that an investigation of the requirements for the data acquisition process was needed, in particular regarding the sample preparation protocols and the maximization of the beams natural coherence.

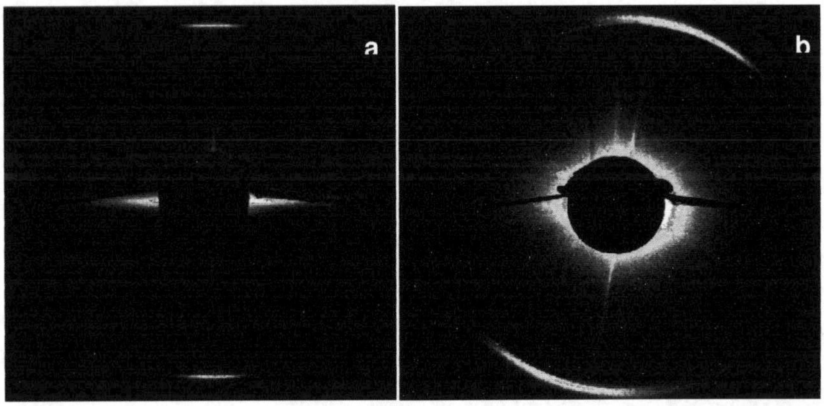

Figure 1. Diffraction patterns from different sections of the same rat tail tendon. For figure a) an accurate protocol for the sample preparation and handling was followed. The protocol included keeping the sample immersed in phosphate buffer solution at all times (see text). For b) this protocol was not followed and the handling process introduced a high degree of axial disorder in the fibrils distribution of the sample, as appreciated by the high azimuthal angle of the first meridional orders of collagen.

Another interesting fact is that the FWHM of the entire envelope of the Bragg peak along a raster line is ca. 100-200 pixels (with some sample-to-sample variation). This corresponds to a 0.3-0.6 μm width in real space of the diffracting motif, in this case, the fibrils within our sample. This value nicely fits with the typical diameter size of the fibrils expected in rat tail tendon. Obviously this approach is only valid in the case where the motifs (fibrils) can be modeled as an array of cylinders with almost identical diameter and negligible axial disorder. We are confident that we can apply this model for data sets as the one in figure 1a as the reflections are almost straight lines.

The peaks show a rich speckled structure, easily noticeable in the zooms in figure 2. This is due to the spatial coherence in the illuminating beam, as we reported previously [24]. In the data set where a set of slits was used as coherence defining aperture, the first order reflections contain around 20-30 speckles of ca. 45-50 μm size (around 2.5 pixels in the CCD), corresponding to a square aperture of 25 x 25 μm^2 at a 7 m distance from the detector, as derived from the optical "grating" formula:

$$s = \frac{\lambda \cdot D}{d} \qquad \text{(eq. 1)}$$

where s is the speckle size, D is the aperture-detector distance, d the aperture size and λ the wavelength. Speckles of similar size could be observed when the defining aperture was a 20 μm pinhole at a distance of 7 m from the detector. This simple experiment shows that equation 1 is a good approximation for calculating the expected speckle size in coherent diffraction. Note that a good oversampling is achieved: the speckles are bigger than 2 CCD pixels, which is considered the minimum required in order to retrieve the phases by iterative algorithms [11].

A good estimate of the degree of coherence in the beam is the speckle visibility (in 1D) or speckle contrast (in 2D). The visibility can be calculated as the ratio (from the type of Young's double slit experiment described in [27]):

$$V = \frac{I_{max} - I_{min}}{I_{max} + I_{min}} \qquad \text{(eq. 2)}$$

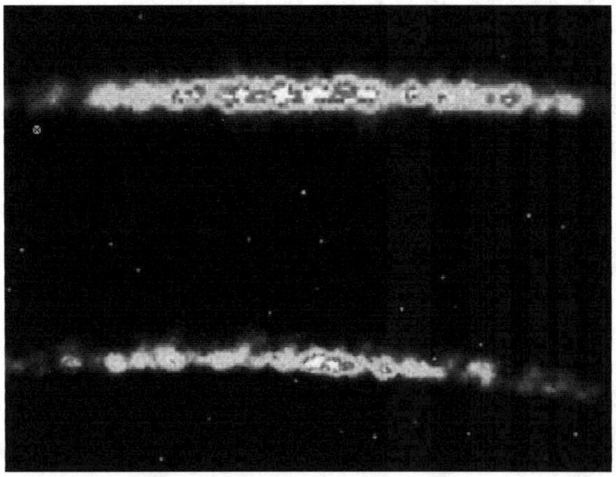

Figure 2. Zoom of the first meridional reflection of rat tail tendon fibres. The upper panel was obtained with the first setup, using a 25 x 25 μm^2 slit aperture. The bottom panel corresponds to the most recent setup with a 20 μm pinhole as the coherence-defining aperture. The contrast of the speckles is slightly better in the second setup (see text) although the small difference is not clearly appreciable in the 2D data. The upper pattern is more "one-dimensional" in the sense that it has less up-down variation in the speckle positions.

Figures 3a and 3b show the intensity of the central raster line through the first Bragg peak of two patterns for the 25 μm^2 slit aperture and the 20 μm pinhole, respectively. The visibility, as defined by equation 2, for the first case varies between 40% and 50% from sample to sample and with the total exposure time (this is an indicator of radiation damage starting to appear). However, for the more recent setup with the pinhole, the visibility increases to ~50-57% depending on the sample under investigation. There are several factors contributing to this difference: firstly, both measurements were made with samples from different specimens, so we should expect a small variation in the visibility due to this fact. Also, there is some loss of contrast inherent to the sample: simple simulations of the sample as a 3D array of cylinders with a distribution of different diameters predict a minimum visibility of at least 20%. Finally, the buffer solution contributes to the incoherent scattered intensity in each peak, thus increasing the I_{min} recorded and effectively reducing the visibility.

There are also other factors than the ones directly related to the sample nature and characteristics, in particular the degree of coherence in the beam notably affects the visibility of the speckles in CXD studies. If there is any incoherent part of the beam, it will give rise to the smooth part of the Bragg peak underneath the speckles. In the limiting case where the beam is totally incoherent, the speckles will not appear and the Bragg peak will be constituted simply by the smooth intensity distribution. In the opposite case, a fully coherent beam will give (ideally) a speckled pattern with up to 100% contrast, that is, which means an intensity value reaching zero between speckles. In our experiments, the dimensions of the pinhole (20 μm) are within the transverse coherence length values expected for the beamline. The 25 μm^2 slit aperture is close to the limit, so a great loss of visibility and speckle contrast should be expected for this case, as shown in figure 3.

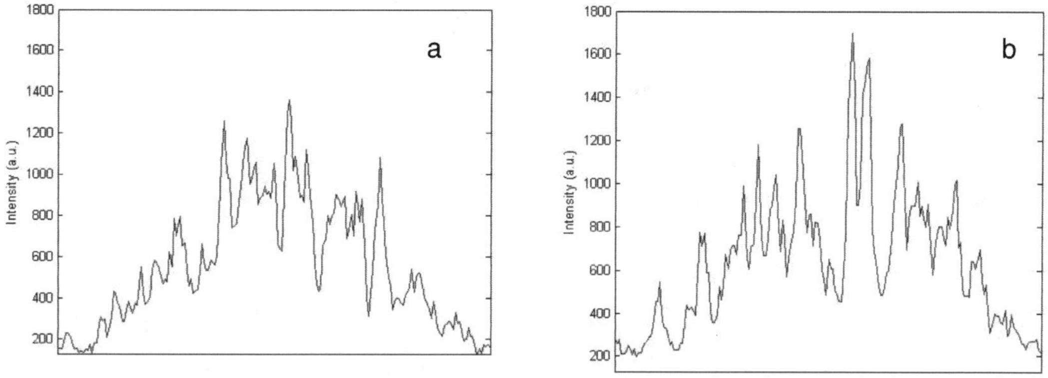

Figure 3. One dimensional profiles of the first order Bragg peak of rat tail tendon for 2 different defining coherent apertures. In a), we used 25 x 25 μm^2 slits; in b) a 20 μm pinhole. The visibility is greater in the second case. Note that some variation of the visibility due to the sample is also expected as both figures were measured with different fibres (see text).

5. Discussion

CXD inversion algorithms require good quality data in order to converge appropriately. First, the signal-to-noise ratio must be optimized, with intensities at least 3 orders of magnitude bigger than the background counts. Achieving this condition with biological samples is difficult as they scatter weakly. Collagen Bragg peaks can be more than 10^6 times weaker than the transmitted beam, and this fact together with the limited dynamic range of the available detectors compromises the signal-to-noise ratio, therefore the ability to obtain a dark field image of the sample from an inversion algorithm. A way to improve the signal-to-noise ratio is to search for more ordered areas of the sample: if the measured volume is the same, the integrated diffracted intensity is conserved, but it is now distributed within a smaller Bragg peak. Collagen fibres within native tail tendon have a naturally high degree of axial order (with deviations of approximately 5°), although sample preparation and handling usually disturb the natural order of the fibres. For instance, Bragg peaks from our first coherent diffraction experiment on tail tendon samples [24] were 40° wide arcs. Since then we have improved our sample preparation protocol. Fibres are now kept wet during all the steps of the sample mounting and data acquisition processes. This has also increased the samples resistance to radiation damage by a factor of 5 approximately, as the liquid reduces the heat load on the fibers and delays the onset of the thermal denaturalization in collagen. It is also crucial to minimize, as much as possible, the time between tendon dissection and data recording, as the natural cross linkage of the fibres

Journal of Physics: Conference Series **247** (2010) 012004 doi:10.1088/1742-6596/247/1/012004

degrades quickly within a few hours. Because of the small size of the fibres used, handling must be extremely careful in order to avoid any damage induced in the surface of the sample by the dissection tweezers during the mounting process. All these factors are particularly relevant in CXD experiments as both the sample and the beam used are much smaller than the ones in conventional SAXS, therefore the data are much more sensitive to small disorders induced in the sample. This new protocol was successfully applied in the experiments presented here, thus obtaining much narrower first order reflections from the collagen samples. These peaks are much less noisy that the ones we obtained in our first experiment [24], which will aid the phase retrieval process. We are developing also a new liquid cell for the samples with a design that minimizes the x-ray path within the buffer solution in order to decrease the incoherent scattering from the liquid.

Another requirement for easy convergence of the algorithms is a good degree of coherence in the beam; otherwise the speckle contrast is spoiled. The setup with a smaller coherence aperture (20 μm pinhole) at a shorter distance from the sample is more convenient for this aim. We would expect a better value of the speckle contrast if we had used even smaller apertures, or removed the mirrors of the beamline from the optical path, as slope errors and surface defects on the mirrors spoil the beam natural spatial coherence. It is also important to minimize the distance between the defining coherence aperture and the sample as well as to use a guard aperture at a suitable distance in order to remove the undesired parasitic scatter from the first aperture, hence "cleaning" the background counts. We found that the use of the second aperture increased the contrast of the patterns as we could exploit the dynamic range of the detector to its maximum.

We have shown that an undulator SAXS beamline can easily be modified for this type of study by the incorporation of two apertures for defining the coherence length. By this means, CXD data can be recorded and extra information from the sample can be extracted. Scanning SAXS gives the average orientation of fibres or crystallites within a matrix as well as the sample mean absorption with a resolution equivalent to the beam size. All this information is extracted from the magnitude of the scattered intensity. Iterative algorithms in CXD recover the transmission function of the sample both in amplitude and phase. This means that the reconstructed image would include extra information of the sample. For instance, the short range disorder in a fibre or crystallite matrix is encoded in the phase differences of the speckles within a Bragg peak. As a consequence, the reconstruction of the phase of the scattered wave from the sample contains information about the displacements from the ideal positions of the diffraction units, independently of the cause of these displacements. With this information it is possible to create strain or dislocation maps of the sample with extremely high resolution, as it has been already done in CXD for crystals [24]. This is particularly appealing for collagen tissues, as information concerning the mechanical properties of the tissue in the nanoscale could be directly obtained. Some scanning SAXS studies of these tissues give the average value of the orientation of the crystallites or fibres within the tissue with some tens of micron resolution. From these measurements an indirect estimate of some simple mechanical properties can be obtained. CXD could complement these studies with much higher resolution images. Moreover, when reconstructing coherent diffraction data from isolated Bragg peaks, the result will be a so-called dark field image of the sample, where the collagen matrix would be visible with contrast enhancement. These images will be very useful in bone studies when combined with conventional scanning SAXS data from where the distribution of apatite crystals can be extracted, to give a complete picture of the tissue.

6. Summary and outlook

We have discussed the improvements we have made in our sample preparation and experimental procedure in order to obtain high quality coherent diffraction data from collagen tissues. This is particularly relevant as only with high quality data, CXD reconstructions can be envisaged, hence all factors affecting the quality of data must be considered. Accurate sample handling protocols and the use of adequate coherence defining apertures are two relevant factors for increasing the signal-to-noise ratio of the measured Bragg peaks as well as their speckle contrast. Inversion of the obtained patterns has still not been achieved, but is now more likely to be possible as CXD algorithms require high quality data. Eventually this will give dark field images of the tissues, with both amplitude and phase information from the sample. The reconstructed phase will be related to local disorder of the fibre positions on the short range, therefore will complement the information obtained by a conventional scanning and/or tomographic SAXS setup where the average orientation of the fibres within the tissue could be obtained. This will have consequences for micro and nanomechanics studies of collagen tissues, as high resolution information can be obtained without averaging large areas of the tissue.

The development of CXD methods applied to biological tissues is still at its initial stages; although we expect a rapid development of the techniques with the increasing availability of coherent sources (including the x-ray free electron lasers). This is a step towards the full imaging of biological tissues at the nanoscale.

Acknowledgments

We gratefully acknowledge the use of Diamond Light Source and the assistance of M. Malfois and C. Pizzey during the experiments. We also thank J. Rodenburg from University of Sheffield for fruitful discussions and T. Wess from University of Cardiff for providing some samples. The authors wish to acknowledge the support of EPSRC for funding this work through a Basic Technology Grant (EP/E034055/1) "Ultimate Microscopy: wavelength limited resolution without high quality lenses" and a Next Generation Facility development grant (EP/F020767/1) "Development of X-ray Ptychography at the Diamond Light Source.

References

[1] Roth S V, Burghammer M, Riekel C, Müller-Buschbaum P, Diethert A, Panagiotou P and Walter H 2003 *Appl. Phys. Lett* **82** 1935

[2] Schroer C G, Kuhlmann M, Roth S V, Gehrke R, Stribeck N, Almandarez Camarillo A and Lengeler B 2006 *Appl. Phys. Lett* **88** 164102

[3] Zizac I, Paris O, Roschger P, Bernstoff S, Amenitsch H, Klaushofer K and Fratzl P 2000 *J. Appl. Cryst.* **33** 820-823

Gupta H S, Seto J, Wagermaier W, Zaslansky P, Boesecke P and Fratzl P 2006 *Proc. Nat. Acad. Sci. USA* **47** 17741-17746

Gourrier A, Wagermaier W, Burghammer M, Lammie D, Gupta H S, Fratzl P, Riekel C, Wess T J and Paris P 2007 *J. Appl. Cryst.* **40** s78-82

Burger C, Zhou H, Wang H, Sics I, Hsiao B, Chu B, Graham L and Glimcher M J 2008 **95(4)** 1985-92

[4] Bunk O, Bech M, Jensen T H, Feidenhans'l R, Binderup T, Menzel A and Pfeiffer F 2009 *New Journal of Physics* **11** 123016

Boote C, Dennis S, Newton R H, Puri H and Meek K M 2003 *Invest. Ophthalmol. Vis. Sci.* **44** 2941-2948

Quantock A J, Boote C, Young R D, Hayes S, Tanioka H, Kawasaki S, Ohta N, Iida T, Yagi N, Kinoshita S and Meek K M 2007 *J. Appl. Cryst.* **40** s335-s340

[5] Riekel C, Madsen B, Knight D and Vollrath F 2000 *Biomacromolecules* **1(4)** 622-626

[6] Kennedy C J, Hiller J C, Lammie D, Drakopoulos M, Vest M, Cooper M, Adderley W P and Wess T J 2004 *NanoLetters* **8** 1373-1380

Hiller J C and Wess T J 2006 *J. Archeol. Sc.***33** 560-572

[7] Gerchberg R W and Saxton W O 1972 *Optik* **35** 237-246

Fienup J R 1982 *Appl. Opt.* **21(15)** 2758-2769

Elser V 2003 *J. Opt. Soc. Am. A* **20(1)** 40-55

Guizar-Sicairos M and Fienup J R 2008 *Opt.Express* **16(10)** 7264-7278

[8] Rodenburg J M and Faulkner H M L 2004 *Appl. Phys. Lett.* **85 4795-4797**
[9] Thibault P, Dierolf M, Menzel A, Bunk O, David C, Pfeiffer F 2008 *Science* **321** 379-382
[10] Maiden A M and Rodenburg J M 2009 *Ultramicroscopy* **109(10)** 1256-1262
[11] Sayre D 1952 *Acta Crystallogr.* **5(6)** 843
[12] Miao J, Charalambous P, Kirz J and Sayre D 1999 *Nature* **400** 342:344
Zuo J M, Vartanyants I, Gao M, Zhang R and Nagahara L A 2003 *Science* **300** 1419–1421
Williams G J, Pfeifer M A, Vartanyants I A and Robinson I K 2003 *Phys. Rev. Lett.* **90** 175501
Pfeifer M A, Williams G J, Vartanyants I A, Harder R and Robinson I K 2006 *Nature* **442** 63:66
[13] Miao J, Hodgson K O, Ishikawa T, Larabell C A, LeGros M A and Nishino Y 2003 *Proc. Natl. Acad. Sci. USA* **100** 110:112.
Shapiro D, Thibault P, Beetz T, Elser V, Howells M R, Jacobsen C, Kirz J, Lima E, Miao H, Nieman A M and Sayre D 2005 *Proc. Natl Acad. Sci. USA* **102** 15343:15346
Jiang H, *et al.* (2008) Nanoscale Imaging of Mineral Crystals inside Biological Composite Materials Using X-Ray Diffraction Microscopy *Phys. Rev. Lett.* 100, 38103
Song C, Jiang H, Mancuso A, Amirbekian B, Peng L, Sun R, Shah S S, Hong Zhou Z, Ishikawa T and Miao J 2008 *Phys. Rev. Lett.* **101** 158101
Nishino Y, Takahashi Y, Imamoto N, Ishikawa T and Maeshima K 2009 *Phys. Rev. Lett.* **102** 018101
[14] Bunk O, Dierolf M, Kynde S, Johnson I, Marti O and Pfeiffer F 2008 *Ultramicroscopy* **108(5)** 481-487
[15] Rodenburg J M, Hurst A C, Dobson B R, Pfeiffer F, Bunk O, David C, Jefimovs K and Johnson I 2007 *Phys. Rev. Lett.* **98** 034801
[16] Giewekemeyer K, Thibault P, Kalbfleisch S, Beerlink A, Kewish C M, Dierolf M, Pfeiffer F and Salditt T 2010 *Proc. Natl Acad. Sci. USA* **107** 529-534
[17] Hüe F, Rodenburg J M, Maiden J M and Midgley P A 2010 *submitted*
[18] Williams G J, Quiney H M, Dhal B B, Tran C Q, Nugent K A, Peele A G, Patterson D and De Jonge M D 2006 *Phys. Rev. Lett.* **97** 025506
[19] Abbey B, Nugent K A, Williams G J, Clark J, Peele A G, Pfeifer M A, De Jonge M and McNulty I 2008 *Nature Phys.* **4** 394-398
[20] Zhang F, Pedrini G and Olsen W 2006 *Opt. Lett.* **31(11)** 1633
[21] Zhang F, Pedrini G and Olsen W 2007 *Phys. Rev. A* **75** 043805
[22] Johnson I, Jefimovs K, Bunk O, David C, Dierolf M, Gray J, Renker D and Pfeiffer F 2008 *Phys. Rev. Lett.* **100** 155503
[23] Menzel A, Kewish C M, Dierolf M, Thibault P, Kraft P, Bunk O, Jefimovs K, David C and Pfeiffer F 2009 *J. of Phys: Conf Series* **186** 012054
[24] Berenguer de la Cuesta F, Wenger M P E, Bean R J, Bozec L, Horton M A and Robinson I K 2009 *Proc. Natl Acad. Sci. USA* **106** 15297-15301
[25] Rowe R W D 1985 *Conn. Tiss. Res.* **14(1)** 9-20
[26] Hulmes D J, Wess T J, Prockop D J and Fratzl P 1995 *Biophys. J.* **68** 1661:1670.
[27] Thompson B J and Wolf E 1957 *J. Opt. Soc. Am.* **47** 895

The Absolute Calibration of a Small-Angle Scattering Instrument with a Laboratory X-ray Source

Lixin Fan[1], Mike Degen[1], Scott Bendle[1], Nick Grupido[1] and Jan Ilavsky[2]
[1]Rigaku Innovative Technologies Inc., Auburn Hills, MI 48326, USA
[2]Advanced Photon Source, Argonne National Laboratory, Argonne, IL 60439, USA

E-mail: lixin.fan@rigaku.com

Abstract. Absolute calibration of small-angle scattering data (in units of differential cross-section per unit sample volume per unit solid angle) is necessary for the determination of molecular weights, the number density of particles, the scattering-length densities of phases in multiphased systems, volume fraction, the specific surface area of the scatterers and to restrict the parameters of a given model to the set which reproduces the observed intensity. It is also a useful means for the detection of artefacts in SAS experiments. Absolute intensities from the same sample also allows intercalibration among different instruments. This work details the absolute calibration procedure of a small-angle X-ray scattering instrument, the Rigaku S-Max3000. Absolute calibration is achieved by using two standards: homogeneous and stable glassy carbon and water. The scattering intensity of glassy carbon is calibrated by comparison with the absolute-calibration measurements taken on the USAXS instrument located at the 32ID beamline of the Advanced Photon Source in Argonne National Laboratory. This instrument has primary calibration capability. Scattering from water is angle-independent and only depends on the physical property of isothermal compressibility. Absolute calibrations using two standards were compared. The agreement of scale factors obtained using two standards suggests that precalibrated glassy carbon can serve as a convenient standard for all types of materials under study.

1. Introduction

Small-angle X-ray scattering (SAXS) is a well established technique that allows us to obtain structural information about inhomogeneities in materials with a characteristic length from 1nm to 100nm. SAXS is thus an appropriate technique to characterize a very large variety of samples, ranging from colloids of all types, metals, cements, nanoparticles, oils, polymers, plastics, proteins, surfactants, foods and pharmaceuticals. Absolute calibration relates the measured (arbitrary) intensity $\boldsymbol{I}$ to the differential scattering cross section per unit volume, $\partial\Sigma/\partial\Omega$, in units of cm^{-1}. While small-angle neutron scattering data is commonly converted to an absolute scale taking advantage of strong incoherent scattering of water [1-3], a significant fraction of X-ray data is still published in arbitrary units. Particularly, absolute calibration is less of a routine procedure for SAXS instruments with laboratory X-ray sources. Several reasons may exist for this lack of calibration. The use of suitable calibration techniques is by no mean universal and straightforward. The first method, the so-called direct method or premier method, attenuates the beam mechanically by using a rotating disc [4] or a moving slit [5]. This method is not well suited for use with position-sensitive detectors [6]. Detailed analysis of the attenuator transmission measurements is often nontrivial as well [7]. The second method compares the scattering intensity from the sample with that from a calibration standard sample of theoretically known scattering cross-section, such as water, or pre-calibrated standards. Given the lower flux, when compared to synchrotrons, laboratory SAXS measurements of water can be time consuming. Obtaining stable pre-calibrated secondary standards is also difficult. Some bench-top SAXS cameras may also face the practical complexity and difficulty in routine processes such as removing and realigning the beamstop [8] and data smearing issues. The Rigaku S-Max3000 SAXS camera uses pinhole collimation and a beamstop with an embedded photodiode to record direct and transmitted beam intensities. The problems of data smearing and removing and realigning the beamstop are thus avoided. With improved flux using a MicroMax 002+ X-ray source and focusing confocal optics, the Rigaku S-Max3000 enhances the routine use of water as a premier standard.

The absolute scale is not necessary for the measurement of the interference functions [9]. For example, determination of radius of gyration, phase transitions and the derivation of the correlation function can be done using measured arbitrary intensity. However, placing the data on an absolute scale is necessary for the determination of molecular weights [10], the number density of particles [11], the scattering-length densities of phases in multiphase systems[12], volume fraction, the specific surface area of the scatterers and to restrict the parameters of a given model to the set which reproduces the observed intensity[13,14,1,2]. An absolute scale is also useful for the detection of artefacts in small angle scattering (SAS) experiments [9]. The absolute intensity is independent of sample thickness, SAXS camera geometry and measurement time. Therefore, absolute intensities from the same sample also allow inter-calibration among different instruments [9].

Due to the aforementioned difficulties of the direct method, the use of calibration standards is more common for absolute calibration in SAXS. Water and pure liquids are typically used as primary standards [15-17] since their differential scattering cross sections depend only on the temperature and isothermal compressibility and therefore can be calculated. However these standards are not suitable for a laboratory SAXS camera with low-flux sources because the scattering is very weak and measurement time is long. In addition, these standards are more suitable for biosciences; however, their X-ray scattering cross sections are too low to serve as a standard in the areas of materials science, polymer science, ceramics, chemistry, etc. Alternatively, monodisperse colloidal suspensions have been used as primary intensity calibration standards [18], for which the differential cross section can be obtained from Porod's law as the interfacial region is well defined. However, the metastable nature of these colloidal systems leaves their long-term usage in question. As secondary standards, vitreous or glassy carbon [19], aluminium-silver alloys [20], silica suspensions [21, 18], Lupolen [22-24] and calibrated foils have been used [25-27].

Among the many intensity calibration standard materials, glassy carbon is a strong scatterer and has a good temperature tolerance [19]. It has been informally used as a calibration standard. As with other secondary standards, its differential cross section needs to be measured. Traditionally, this measurement has been done either at a neutron facility or through calibration, using X-ray data of primary standards. Now there is the possibility to measure it directly at a Synchrotron. The versatile ultra-small-angle X-ray scattering (USAXS) instrument at the Advanced Photon Source (APS) at Argonne National Laboratory (ANL) offers quantitative primary calibration of X-ray scattering cross sections without the need for beam attenuation or primary standards of any kind [28]. Designed after the Bonse-Hart double-crystal configuration [29-31], the APS USAXS instrument takes advantage of a silicon photodiode detector that is linear over 9 decades of scattering intensity. This broad linearity enables the instrument to scan through the full incident beam without loss of linearity in measurement of the beam intensity. This instrument is not limited to ultra small angles. A wide scattering vector range from 0.0001 to1 $Å^{-1}$ can be covered.

This work details the absolute calibration procedure of a small-angle X-ray scattering instrument, the Rigaku S-Max3000. One should be note that this procedure is applicable to any type of SAXS camera with a laboratory source. We first give a short review of the principle and theory for absolute calibration, followed by an investigation of two standards: water and glassy carbon. The intensity of glassy carbon in absolute scale is obtained by measuring the sample on the USAXS instrument located at the 32ID beamline of APS at Argonne National Laboratory. We compare scale factors obtained using these two standards. We discuss the suitability of two standards and suggest glassy carbon as a general standard material for absolute intensity calibration.

2. Theory

Detailed theory and relevant expressions for absolute calibrations can be found elsewhere [32, 33, 34]. Here we restate the formulas used for extracting the calibration scale factor. The intensity scattered from a sample can be described as

$$I_s(q) = I_0(\lambda)A\Delta\Omega\eta(\lambda)T_s(\lambda)d_s(\frac{\partial\Sigma}{\partial\Omega})_s(q) + BG_s \ , \qquad (1)$$

where $q = 4\pi\sin(\theta)/\lambda$ is the momentum transfer; θ is half of the scattering angle; λ is the wavelength; $I_s(q)$ is measured intensity of the sample; $I_0(\lambda)$ is the incident flux (in units of photons/s^{-1}/cm^{-2}); $\Delta\Omega$ is the solid angle element defined by the size of a detector pixel; A is the area illuminated by the beam; η is the detector efficiency; $T_s(\lambda)$ is the transmission of the sample; d_s is the thickness of the sample; BG_s is the scattering background; $(\frac{\partial\Sigma}{\partial\Omega})_s$ is the differential scattering cross section(cm^{-1}) of the sample and is the quantity obtained from the absolute calibration of the measured intensity.

The absolute intensity of the sample can be obtained by using standards with known differential scattering cross section $(\frac{\partial\Sigma}{\partial\Omega})_{st}(q)$ and thickness d_{st}:

$$(\frac{\partial\Sigma}{\partial\Omega})_s(q) = (\frac{\partial\Sigma}{\partial\Omega})_{st}(q)\frac{(I_s(q) - BG_s)}{d_sT_{s+cell}} \Big/ \frac{(I_{st}(q) - BG_{st})}{d_{st}T_{st+cell}} \ , \qquad (2)$$

where $T_{s+cell}(\lambda)$ and $T_{st+cell}(\lambda)$ are the transmissions of the sample and standard in the containers (cells), respectively. $I_{st}(q)$ is the measured intensity of the standard. Therefore, the intensity of the sample in absolute scale can be obtained by multiplying the measured intensity by the scale factor. Please note that the measured intensity of the sample should first be normalized by transmission and thickness of the sample.

$$ScaleFactor = (\frac{\partial\Sigma}{\partial\Omega})_{st} \Big/ \frac{(I_{st}(q) - BG_{st})}{d_{st}T_{st+cell}} \qquad (3)$$

3. Experimental Methods

3.1. Materials

Distilled water and glassy carbon were studied for the calibration of SAXS experiments. We used commercially available glassy carbon plate from Alpha Aesar. We chose type 2 (stock number: #38021), which has stronger scattering than type 1 (stock number: #38024). The glassy carbon plates have dimensions of 50 mm × 50 mm × 1mm and ≈ 3.55 g/ 50 x 50 mm. The glassy carbon was cut to a smaller piece for X-ray measurements. Please note that scattering from glassy carbon may vary from plate to plate even if the plates have the same stock number. Therefore, the absolute intensity data for a particular piece of glassy carbon may not be suitable for calibration of another glassy carbon plate.

Journal of Physics: Conference Series **247** (2010) 012005 doi:10.1088/1742-6596/247/1/012005

For accuracy, the exact same sample should be used for the calibration measurements at the synchrotron and home labs.

3.2. Laboratory-based small angle X-ray scattering measurements

The measurements were performed on a Rigaku three-pinhole SAXS camera S-Max3000 (Figure 1). This camera is a dedicated SAXS instrument suitable for measuring both isotropic and anisotropic systems. The instrument uses Cu Kα radiation (λ = 1.54 Å). The combination of sealed tube microfocus source MicroMax 002+ and focusing confocal optics provides a small beam spot and high flux on the sample. The high brilliance microfocus source MicroMax 002+ requires low maintenance and can operate at low power consumption, typically at a current of 0.88 mA and a potential difference of 45 kV. The scattering data were collected by a fully integrated 2D multiwire proportional counting gas-filled detector that has low background. The diameter of the detector was 120 mm. A photodiode embedded in the beamstop was used allowing continuous intensity monitoring and direct measurement of the transmission. The two sample chambers provide flexibility in covering a large q range without the need to move the detector and subsequent alignment. The water sample was placed in a sealed capillary with a diameter of 0.15 cm. The glassy carbon was directly mounted into the beam path without a container. The measurements were performed in vacuum and at room temperature. Samples were placed in two sample chambers and measured at two sample-detector distances (1.5 m and 0.44 m) to cover total q range from 0.0067 Å^{-1} to 0.55 Å^{-1}. The empty capillary, empty beam, dark current and transmissions of the samples were also measured for data correction. The data were reduced and analyzed using Rigaku's SAXSGUI and Nanosolver software packages.

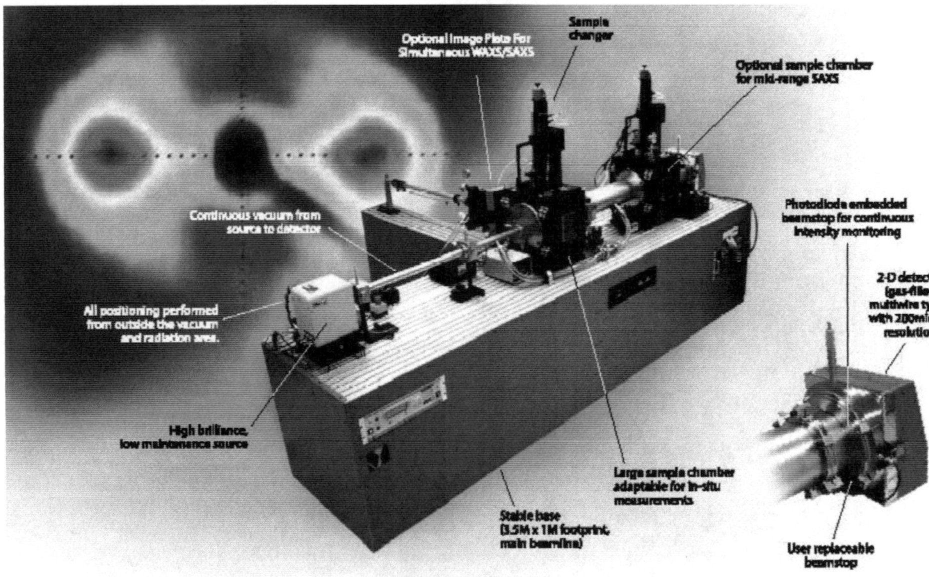

Figure1. Rigaku three-pinhole SAXS camera S-Max3000

3.3. Ultra Small angle X-ray scattering measurements at synchrotron

The absolute intensity of glassy carbon was measured on a USAXS instrument at beamline 32-ID of the APS, ANL. This USAXS instrument offers quantitative primary calibration of X-ray scattering cross sections, a scattering vector range from 0.0001 to 1 Å^{-1}, and intensity range of up to 9 decades [28]. The data were collected using monochromatic 12 keV radiation, beam size 2 mm x 1 mm (width x height), with exposure of 1 second at each point, resulting in total data collection time about 10 minutes for the measurement. USAXS data were reduced, calibrated, and corrected for all instrumental effects (such as slit smearing) using the Indra and Irena software [35].

.

4. Results and discussions

Figure 2 shows the scattering curves (with error bars) of water in the capillary, empty capillary and pure water after subtraction of these two. Data were collected using the Rigaku SAXS camera S-Max3000. Each curve is a combination of two data sets collected in two sample chambers e.g. at two different sample-detector distances. Water is a very weak scatterer hence the dark current is significant and can't be neglected. The raw data were corrected for dark current and transmission following standard procedures during the data reduction using SAXSGUI software, but not yet normalized with the thickness of the sample.

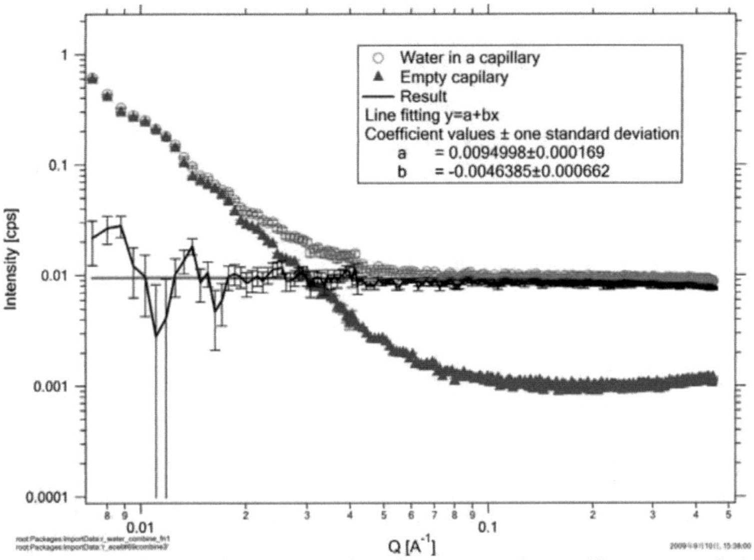

Figure 2. Scattering profiles: scattering from water in capillary are red circles; empty capillary- blue triangles; pure water-dark; horizontal linear fitting in solid red line.

The data exhibits a straight line after $q = 0.02$ Å^{-1}. The linear fitting extrapolates to intensity at zero angle of $I_{H2O}(0) = 9.5\times10^{-3}\,s^{-1} \pm 0.2\times10^{-3}\,s^{-1}$. The scattering of water only depends on temperature and the physical property of the isothermal compressibility according to

$$(\frac{\partial\Sigma}{\partial\Omega})_{H2O}(0) = \rho^2 KT\chi_T\,, \tag{4}$$

where ρ is the scattering-length density and χ_T is the isothermal compressibility. The differential scattering cross section is $1.65\times10^{-2}\,(cm^{-1})$ at 293 K and 10^5 Pa with the isothermal compressibility of $4.591\times10^{-10}\,Pa^{-1}$ [36].

As we mentioned above, the data in Figure 2 was corrected with the transmission factor, but not the thickness of the sample. The diameter of the capillary is 0.15 cm. Therefore, the scale factor should be:

$$SF = \frac{(\frac{\partial\Sigma}{\partial\Omega})_{H2O}(0)}{I_{H2O}(0)} \times d_{H2O} = 1.65\times10^{-2}\,\frac{0.15}{9.5\times10^{-3}} = 0.261$$

To scale to absolute intensity, the intensity (after correction of background, transmission and thickness) measured on the lab-based SAXS should be multiplied by this scale factor. Taking into account the error of 0.2×10^{-3} s^{-1} for $I_{H2O}(0)$, the error for the scale factor from water is estimated to be 0.003(0.3%).

Please note in the calculation above we used the thickness of capillary, provided by the manufacturer, as 0.15 cm and the error in the thickness was not taken into account. Since the X-ray beam spot is not a point, the actual beam path length through the capillary is not the diameter of the capillary and it is not trivial to define. It is better to use a flat quartz cell for the measurements. More accurate estimation of the thickness (averaged beam path length) of water in the capillary can be done by measuring the transmissions of water in the capillary and empty capillary and using the known mass attenuation coefficient of water. The transmission of a sample can be described as

$$T = I_t / I_0 = e^{-(\mu/\rho)X} = e^{-\sum_i w_i(\mu/\rho)_i X},$$

where I_o and I_t are the intensities of incident beam and transmitted beam, respectively; x is the mass thickness obtained by multiplying the thickness t by the density ρ i.e., $x = \rho t$. For mixtures and compounds, the values of the mass attenuation coefficient, μ/ρ, can be obtained by simple additivity: $\sum_i w_i(\mu/\rho)_i$,where w_i is the fraction by weight of the ith atomic constituent. Therefore, the transmission of water in a capillary can be described as

$$T_{water+cell} = T_{cell} \times e^{-(\mu/\rho)_{water}\rho_{water}t}$$

The mass attenuation coefficient for water, $(\mu/\rho)_{water}$, at a photon energy of 8 Kev is 10.37 (cm^2/g) according to http://physics.nist.gov/PhysRefData/XrayMassCoef/ComTab/water.html. For the particular quartz capillary used for our measurements, $T_{water+cell}$ =0.1458 and T_{cell} =0.6098. The density of water is 1g/cm^3. Then the thickness of water can be calculated as 0.1384 cm. Comparing to the nominal diameter of the capillary of 0.15 cm, the error is ~7.7%. Therefore the error in the water thickness can dominate the error in the scale factors. It needs to be carefully defined. Taking into account the errors from sample thickness and intensity at zero angle, the total error for the scale factor from water was estimated to be 0.02. The scale factor with new sample thickness of 0.1384 cm can be recalculated as SF = 0.24 ± 0.02.

For the above described Rigaku instrument setup, 2 hours exposure time can produce scattering data from water with reasonably good statistics. To improve the statistics, longer measuring time is needed. Our measurements were repeated 4 times with an exposure time of 2 hours each. Four data sets were summed for data reduction and analysis. The data showed in Figure 2 have a total of 8 hours exposure time for each sample. For other low flux SAXS cameras, even longer measuring time would be required. The Rigaku S-Max3000 provides a second optional sample chamber closer to detector that allows exploration of a middle q range without moving the detector or realigning the beam. This setup also makes water calibration much less time consuming. The integrated counts for water in the second chamber are 1.92×10^6 per 4 hours, where as 7.56×10^5 per 8 hours for water in first chamber. The count rate for water in second chamber is about 4 times for water in first chamber that means for same statistics only about 1/4 measuring time is needed. Particularly, the flat region of scattering profile of water used for the calibration is in larger q area. The second chamber covers this q area. Therefore, practically, data from water in second chamber alone are enough for intensity calibration. Measuring water in first chamber is optional.

For intensity calibration, accurate transmission is required. Another advantage of the Rigaku S-Max3000 system is that it uses a photodiode embedded beamstop. It allows continuous monitoring of the beam intensity yielding a direct measurement of the sample transmission. A seriesl of transmission measurements can be easily measured and then averaged. The obtained transmission is pretty accurate. The standard deviation is normally within 0.005.

Traditionally, the intensity in absolute scale for glassy carbon has been measured at a neutron facility. Using neutron data for X-ray intensity calibration has the following disadvantages: first, the scattering contrasts for neutrons and X-rays are different as determined by the nuclear scattering length and by the electron density difference, respectively. Therefore, the differential cross section obtained with neutrons needs to be converted to that for X-rays as follows:

$$\left(\frac{\partial\Sigma}{\partial\Omega}\right)_{x-ray}(q) = \left(\frac{\partial\Sigma}{\partial\Omega}\right)_{neutron}(q)\frac{(\rho_{matrix} - \rho_{particle})^2_{x-ray}}{(\rho_{matrix} - \rho_{particle})^2_{neutron}} \quad (5)$$

Second, establishing the linear region of the scattering profile from glassy carbon with any certainty is very difficult. Therefore, a non-negligible error is obtained when the differential cross section at $q=0$ is extrapolated. Using absolute intensity data measured at the synchrotron X-ray instrument allows one to scale the data collected at lab-based SAXS directly into absolute units.

The same glassy carbon with a thickness of 1mm was measured on the USAXS instrument at APS of ANL and on a Rigaku S-Max3000 SAXS. The exposure time was 10 and 5 minutes, respectively. The scattering profiles are shown in Figure 3. The data measured on the Rigaku S-Max3000 is scaled to USAXS data. We can see that two data sets coincide well within a common q range.

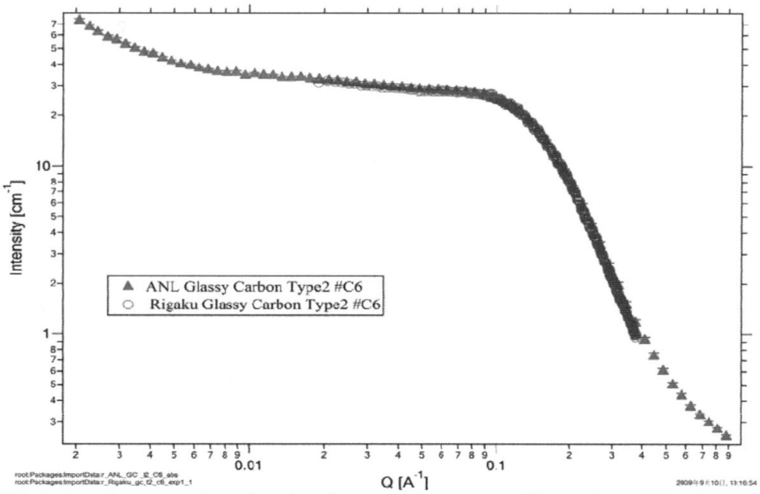

Figure 3. SAXS from glassy carbon in absolute units. Data collected on USAXS at APS, Argonne National Lab are in red; Rigaku S-Max3000 data are in blue.

We selected several q positions to evaluate the scale factors. For data obtained on Rigaku S-MAX3000, the transmission factor of the glassy carbon was corrected during data reduction, but the thickness is not yet normalized, Therefore the scale factor is calculated as $SF = (I_{ANL}/I_{Rigaku}) \times d_{GC}$, where I_{ANL} and I_{Rigaku} are measured intensities on USAXS at Advanced Photon Source of Argonne National Laboratory and on Rigaku S-Max3000 with laboratory X-ray source, respectively. The glassy carbon is in a form of a flat plate. Its thickness can be easily measured. d_{GC} is 0.1 ±0.8% cm. The q positions and obtained I_{ANL}/I_{Rigaku} are listed in Table 1.

Journal of Physics: Conference Series **247** (2010) 012005 doi:10.1088/1742-6596/247/1/012005

Table 1. I_{ANL}/I_{Rigaku} at different q positions

q (Å^{-1})	I_{ANL}/I_{Rigaku}
0.0525	2.64
0.0753	2.64
0.1606	2.62
Average	2.63
Standard deviation	0.01

Using $I_{ANL}/I_{Rigaku}= 2.63 \pm 0.01$ and d_{GC},= 0.1 ±0.8% cm, we estimated the scale factor from glassy carbon as SF = 0.26 ± 0.02. Once determined, the scaling factor can be used to convert other SAXS data, collected on the same SAXS instrument configuration, to absolute units by a simple multiplication.

The scale factor obtained with the glassy carbon standard is 0.26 ± 0.02 which is in an agreement within the error bars with the value obtained using water, 0.24 ± 0.02. The measuring time for glassy carbon is just 5 minutes on the Rigaku S-Max3000, compared to 8 hrs for water plus 8 hrs for capillary making glassy carbon a much more convenient standard for a lab-based SAXS instrument. Over the years our lab at Rigaku Innovative Technologies, Inc., has regularly used glassy carbon from Alpha Aesar to check the performance of the SAXS camera. We found that the scattering profiles are repeatable within experimental errors and they are considered stable with no radiation damage. The temporal stability of glassy carbon was systematically studied [37], and it was proven that the glassy carbon microstructure is stable over time and is not affected by atmosphere or radiation exposure. During the inter-laboratory calibration, although the same block of glassy carbon is measured, different sample area may be probed due to the difference in beam sizes and sample location. Therefore, the spatial homogeneity of glassy carbon is concern for the calibration. Recently, the spatial homogeneity of one type of commercially available glassy carbon was studied [37]. It was demonstrated that glassy carbon is sufficiently homogeneous. These factors suggest that pre-calibrated glassy carbon is capable of serving as a general intensity calibration standard.

The high scattering power, temporal stability and sufficient homogeneoity make glassy carbon a convenient standard. One should note that such a strong scatterer might cause a saturation problem on some detectors both at synchrotrons and in home laboratories [38]. For a CCD detector, the photon counts can be reduced by measuring for shorter times or using a filter. For lab-based gas proportional counters, the count rate is the key factor for evaluating the saturation. For many gas proportional counters, the count rate from the sample should be kept below 10^5 cps. There are many types of glassy carbons. One should be carefully chosen to avoid saturation effects. In our studies, the count rates were 2.17×10^4 and 3.57×10^4 cps for glassy carbon located in first and second chamber, respectively.

One should note that water still remains a useful calibration standard for dilute solutions of weakly scattering particles (proteins, micelles etc). The exposure time for the samples is similar to that of water and therefore the measurement of water does not represent a large 'overhead'. Particularly, when a capillary is used as the solution container, it is not trivial to estimate the effective sample thickness due to the beam spot size. Using water as a standard and using the same capillary for the measurements of the solution samples, the thicknesses for water and samples are cancelled out. Therefore, solution sample thickness is no longer needs to be determined.

5. Conclusions

This work has detailed the absolute intensity calibration procedures for lab-based SAXS instruments using two standards: water and homogeneous/stable glassy carbon. Water has a known differential cross section and no pre-calibration is required; however, the measurement is time consuming. Moreover, the beam path length through the capillary is not trivial to define due to beam spot size. We described a method to evaluate this value by measuring transmissions of the water in the capillary and empty capillary. The absolute calibration method using glassy carbon was based on a primary calibration X-ray instrument (the USAXS instrument at the APS at ANL). In this method the data collected on the lab-based camera can be scaled directly to data in absolute units measured on USAXS. This procedure avoids the non-negligible error produced by converting the differential cross section obtained with neutrons to X-rays. We compared scale factors obtained using water and glassy carbon. The good agreement of the scale factors suggest that pre-calibrated glassy carbon can serve as a general intensity calibration standard for all types of materials. Particularly, in the fields of materials science, polymer science, ceramics, and chemistry, when samples are more likely to have strong scattering, glassy carbon should be the standard. Water still remains a suitable standard for diluted solutions and serves bioscience well.

Acknowledgement
We thank Chris Breese, Bart Ruc, John Yezback and Moe Farhat for practical help and support.

References

[1] Zemb T and Charpin P 1985 *J. Phys. (Paris)* **46** 249-256
[2] Wignall G D and Bates, F. S. 1987 *J. Appl. Cryst.* **20** 28-40
[3] Ghosh R E, Egelhaaf S U and Rennie A R 1998 *A Computing Guide for Small-Angle Scattering Experiments* (ILL Grenoble)
[4] Kratky O and Wawra H. 1963 *Monatsh. Chem.* **94** 981-87
[5] Stabinger H and Kratky O 1978 *Makromol. Chem.* **179** 1655-59
[6] Russell T P 1983. *J. Appl. Cryst.* **16** 473-78
[7] Orthaber D 1999 *PhD thesis* (University of Graz, Austria).
[8] Strunz P, Saroun J, Keiderling U, Wiedenmann A and Przenioslo R 1999 *In XIth International Conference on Small-Angle Scattering* (Long Isl City, New York) 829-33.
[9] Russell T P 1988 *J. Appl. Cryst.* **21** 629-38
[10] Kratky O, Porod G and Kahovec 1951 *Z. Elektrochem.* **55** 53-59
[11] Hendricks R W, Schelten J and Schmatz W 1974 *Philos. Mag.* **30** 819-37.
[12] Alexander L E 1969 *X-ray Diffraction Methods in Polymer Science* (New York: Wiley)
[13] Hayter J I and Penfold J 1983 *Colloid Polym. Sci.* **261** 1022-30.
[14] Cabane B, Duplessix R and Zemb T J 1985 *J. Phys.* (Paris) **46** 2161-78
[15] Orthaber D, Bergmann A and Glatter O 2000 *J. Appl. Cryst.* **33** 218-25
[16] Dingenouts N, Bolze J, Potschke D and Ballauff M 1999. *Adv. Polym. Sci.* **144** 1–47
[17] Zemb T, Tache′ O, Ne′ F and Spalla O 2003 *J. Appl. Cryst.* **36** 800–805
[18] Russell T P *J. Appl. Cryst.*1983 **16** 473-78
[19] Perret R and Ruland W O 1972 *J. Appl. Cryst.* **5** 116-119
[20] Walker, C B and Guinier A 1955 *Acta Metall.* **1** 568-577
[21] Patel I S and Scitmidt P W 1971 *J. Appl. Cryst.* **4** 50-55.
[22] Kratky O 1964. *Z. Anal. Chem.* **201** 161-174.
[23] Kratky O, Pilz I and Schmitz P J 1966 *J. Colloid Interface Sci.* **21** 24-34.
[24] Pilz I 1969 *J. Colloid Interface Sci.* **30** 140-144.
[25] Hendricks R W 1972 *J. Appl. Cryst.* **5** 315-324.
[26] Shaffer L B and Hendricks R W 1974 *J. Appl. Cryst.* **7** 159-163
[27] Russell T P, Lin J S, Spooner S and Wignall G D 1988 *J. Appl. Cryst.* **21** 629-638.

[28] Ilavsky J, Jemian P R, Allen A J, Zhang F, Levine L E and Long G G **2009** *J. Appl. Cryst.* **42** 469-479
[29] Long G G, Jemian P R, Weertman J R, Black D.R, Burdette H E and Spal R 1991 *J Appl Cryst,* **24** 30-37
[30] Sztucki M and Narayanan T 2007 *J. Appl. Cryst.* **40** *s459*-s62
[31] Jemian P R and Long G G *1990 J Appl. Cryst.* **23** 430-32
[32] King S M 1999 *Modern Techniques for Polymer Characterization* edited by Pethrick R A and Dawkins J V (New York: John Wiley) 171–232.
[33] Lindner P 2002 *Neutrons, X-rays and Light: Scattering Methods Applied to Soft Condensed Matter* edited by Lindner P
[34] Dreiss C A, Jackb K S and Parkerc A P 2006 *J. Appl Cryst.* **39** 32–38
[35] Ilavsky J and Jemian P R 2009 *J. Appl. Cryst.* **42** 347-353
[36] CRC 1983 *Handbook of Chemistry and Physics* (Boca Raton, Florida: CRC Press)
[37] Zhang F, Ilavsky J, Long G, Allen A J and Jemian P R 2009 *Metallurgical and Materials Transactions A* 1073-5623 (Print) 1543-1940 (Online) 10.1007/s11661-009-9950-x
[38] Pedersen J. S., 2004 *J. Appl. Cryst.* **37** 369-380

Feasibility Study on Anomalous Small-Angle X-ray Scattering near Sulphur *K*-edge

Masashi Handa[1], Yuya Shinohara[1], Hiroyuki Kishimoto[2], Yusuke Tamenori[3], and Yoshiyuki Amemiya[1]

[1]Department of Advanced Materials Science, Graduate School of Frontier Sciences, The University of Tokyo, 5-1-5 Kashiwanoha, Kashiwa, Chiba, Japan
[2]Sumitomo Rubber Industries Co. Ltd., Kobe, Hyogo 651-0071, Japan
[3]JASRI/SPring-8, 1-1-1 Kouto, Sayo 679-5198, Japan

E-mail: yuya@k.u-tokyo.ac.jp

Abstract. Anomalous small-angle X-ray scattering (ASAXS) is expected to be a powerful and vital tool for the structural study of soft matter. We investigated feasibility of ASAXS near sulphur *K* absorption edge at SPring-8. Scattering pattern was successfully obtained and the dependence of scattering intensity on the energy of X-ray well agreed with the calculated one. This study can be the basis of structural study of soft matter using ASAXS, especially non-crystalline/amorphous materials.

1. Introduction

Small-angle Scattering (SAS) is a powerful tool to investigate the structure of soft matter, especially for non-crystalline and inhomogeneous structures (for example [1]). By using SAS, one can obtain the structural information of such non-crystalline/inhomogeneous structure. In the case of SAXS (Small-Angle X-ray Scattering) in which an X-ray is used as a probe, the scattering is caused by inhomogeneities of the electron scattering length density which depends on the density distribution of the atomic form factors inside the sample. The scattering contrast can be varied near the absorption edge of a specific element due to its anomalous dispersion. The dispersion of the atomic form factor f is described as

$$f(E) = f_0 + f'(E) + \mathrm{i}f''(E), \tag{1}$$

where f', and f'' represent the real and imaginary parts of the dispersion corrections, respectively. By using this dispersion, one can perform not only the scattering measurement with a higher contrast but also element-specific structural studies. SAXS using this anomalous dispersion is called as anomalous SAXS (ASAXS) [2, 3]. Though ASAXS has been used in the fields of ceramics [4], catalysts [5], quasicrystals [6] alloys [7, 8, 9], ionomers[10], polyelectrolytes[11], nanoparticles [12], and nanocomposites containing high-Z elements [13, 14, 15], the application of ASAXS to soft matter has been restricted.

Soft matter is usually made of low-Z elements, such as carbon, nitrogen, oxygen, phosphorus, and sulphur. The difference of electron density between these elements is relatively small. This sometimes leads to the lack of contrast between elements which results in insufficient SAXS intensity for structural analysis. Thus it is effective to enhance the contrast of electron density by using ASAXS in the structural analysis of soft matter. ASAXS of soft matter, however, includes some technical challenges at the same time. First, SAXS has to be performed at lower X-ray energy than that used with conventional SAXS. Soft matter has its absorption edges at lower energy region: the K-absorption edge of low-Z elements such as sulphur ($Z = 16$) and phosphorus ($Z = 15$) is below 2.5 keV. This energy is much lower than the energy range of conventional SAXS, 8 - 15 keV, which is available at "hard X-ray" beamline where samples can be placed in ambient atmosphere due to beryllium windows. Though ASAXS study has been performed in many fields as is described above, there are few ASAXS studies performed near the K-edges of sulphur and phosphorus [16]. Thus we have to develop SAXS technique at the energy region, $E < 2.5$ keV, which is available at "soft X-ray" beamline where samples have to be kept in vacuum. Some attempts of sulphur ASAXS have been performed at hard X-ray beamlines by using the tail of anomalous dispersion. It is, however, obvious that ASAXS at the absorption edge is preferable from the viewpoint of signal-to-noise ratio than that at the tail of absorption edge. Thus ASAXS at soft X-ray beamlines should be developed.

In addition to the requirements to use soft X-rays, ASAXS of soft matter needs to be performed with an extremely high precision. In macromolecular crystallography, the importance of X-ray diffraction measurement at this energy region has been recognized for a long time and the diffraction studies using X-ray near the K-edge of sulphur and phosphorus were performed [17, 18]. Most of soft matter such as non-crystalline and inhomogeneous materials, however, does not show distinctive intensity profiles like Bragg peaks, which generally requires measurements with a higher precision than in macromolecular crystallography.

The goal of our research is to apply sulphur ASAXS to non-crystalline/amorphous materials. In the present study, we aimed at the measurement of accurate SAXS images near sulphur K-edge. As a first step, we used a model sample that consisted of silica particles and polymer matrix containing a large amount of sulphur. This is because only the scattering intensity, which is determined by the electron density contrast between particles and matrix, does depend on the energy of X-rays while the shape of scattering pattern does not. Because no soft X-ray beamline is designed for SAXS measurement, we have developed SAXS measurement at a soft X-ray beamline, BL27SU, SPring-8. In this report, SAXS experiment near the K-edge of sulphur is first described and then the change in X-ray scattering intensity near the sulphur K-edge is favorably compared with the calculated one.

2. Experimental Section

2.1. Beamline layout

ASAXS experiments was performed at BL27SU, SPring-8 (Hyogo, Japan). This beamline was originally designed for photochemical processes in inner-shell excitation. The experimental setup of our ASAXS measurement is described in Fig. 1. In this layout figure-8 undulator firstly generates X-rays in the range from 0.1 to 5.0 keV [19]. Then, two pre-mirrors (M0 and M1) focus the X-ray beam vertically on the entrance slit and horizontally on the sample, respectively. These two mirrors also play an important role in diverting the X-ray beam into this experimental arrangement. The X-rays with a specific energy is selected with a varied-line-spacing plane grating monochromator (VLS-PGM, [20, 21]), where M2 focuses the X-ray beam vertically on the exit slit. An upper available energy of X-ray is limited to 2.8 keV using the VLS-PGM. This monochromator has an energy resolution of $E/\Delta E \simeq 2000$ in 2.0 - 2.5 keV. Post-mirror (M3) focuses the monochromatized beam vertically on the sample. These mirrors hardly reflect higher harmonics.

A Remarkable feature of this beamline is the use of a differential pumping system [22] which allows windowless experiments under normal atmospheric conditions in soft X-ray region. This system is composed of three sections, which are separated with walls having a hole to let the X-ray beam go through. These three sections are evacuated with respective vacuum pumps, and

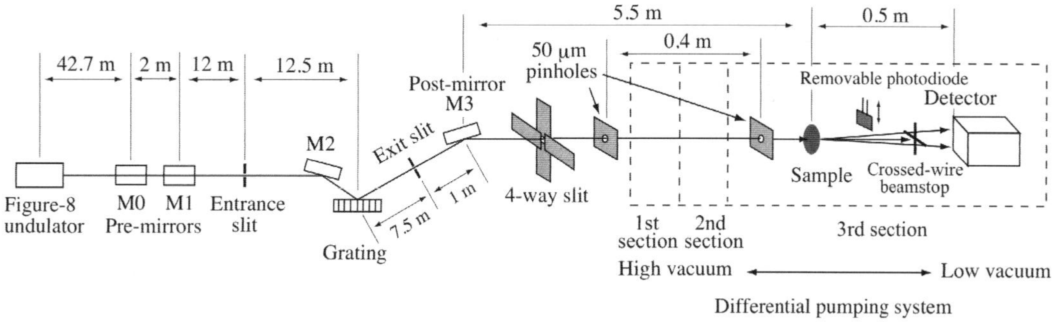

Figure 1. Schematic view of the experimental setup. X-ray beam is monochromatized with the combination of grating, mirrors, and slits. We adopt a differential pumping system which enables us to perform experiments without any vacuum window.

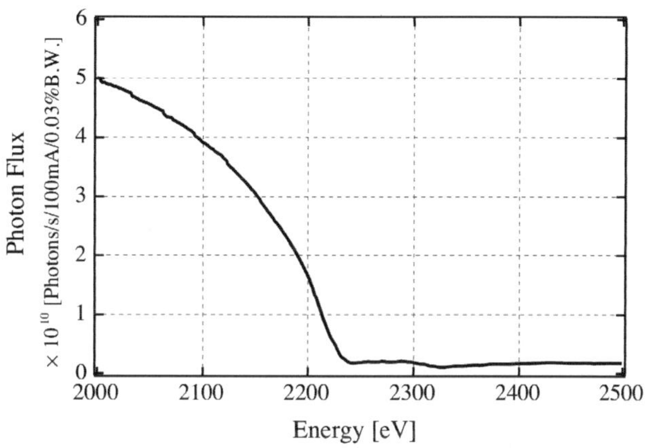

Figure 2. Photon flux at BL27SU, which is normalized to a ring current of 100 mA and $E/\Delta E = 3000$. This spectrum was measured with the photodiode placed in the 3rd section of the differential pumping system. It is clear that the flux sharply decreases in 2.00 – 2.25 keV. This decrease is caused by the absorption by gold that is coated on the mirrors and grating, and the low reflectivity of M2.

the degrees of vacuum are 10^{-5}, 10^{-1}, and 10^{1} Pa in the 1st, 2nd, and 3rd section, respectively.

We set a 4-way slit and a pinhole at the upstream of the differential pumping system, and another pinhole in the 3rd section in order to eliminate the parasitic scattering. These pinholes have the same diameter of 50 μm. A sample and a detector were also arranged in the 3rd section and the distance between them was about 50 cm. SAXS measurement was performed in a transmission mode. We put a removable photodiode (AXUVPSV, International Radiation Detectors Inc.) between the sample and the detector so that we can monitor the incident beam intensity during scattering measurement. Just downstream of the photodiode, a crossed-wire made of 200 μm tantalumn was placed and used as a beamstop.

Although a high flux is achieved in this beamline below 2.0 keV, it steeply decreases above 2.0 keV [23], as is shown in Fig. 2. The decrease of the flux above 2.0 keV is due to the X-ray absorption by all of the mirrors (M0, M1, M2 and M3) and grating, which are coated with gold having its M-edges at 2.21 and 2.29 keV. As a result, the flux decreases to

Table 1. Principal characteristics of the CCD detector

Format	1024 $\times$1024
Pixel Size	13 μm $\times$13 μm
Effective Area	13.3 mm $\times$ 13.3 mm
Frame Rate (slow scan)	0.28 frame/s
Readout Noise (catalogue)	6 electrons (r.m.s.)
Cooling Temperature	- 60 °C
Full Well	80,000 electrons

10^9 photons/s/100mA/0.03 %B.W. above 2.0 keV. Angles of incidence at these mirrors (M0, M1, M2 and M3) are 89.4, 89.4, 88.0 and 89.0 degrees, respectively. It should be noted that the critical angle is 2.3 − 1.8 degrees for 2.0 − 2.5 keV, which means that regarding M2, the angle of incidence does not meet the requirement for the total reflection in the higher region of 2.0 − 2.5 keV. This fact also cause the decrease of the photon flux.

2.2. Detector

We adopted an indirectly-illuminated X-ray area detector. This detector consists of a phosphor screen, an optical relay lens, and a CCD detector. In this system, an X-ray image was first converted to a visible light image by the phosphor screen, then it was focused onto the CCD detector by the optical relay lens. The CCD can be easily exchanged by the adoption of mechanical attachment that is called a Phillips mount. In the present study a full-frame back-illuminated CCD detector (C4742-98-KAG, Hamamatsu Photonics Co. Ltd.) was used, the characteristics of which being shown in Tab. 1.

Two types of phosphor screen available were tested: one was a thin plate of LSO (Lu_2SiO_5:Ce) single crystal and the other was P-43 (Gd_2O_2S:Tb) powders. By using the LSO screen, measurement with a higher-spatial resolution was available. When we used the LSO screen, however, stray light from upper-stream transmitted the screen and was incident onto CCD. This severely limited the signal-to-noise ratio of SAXS measurement. Furthermore the quantum detective efficiency with P-43 was much higher than that with LSO due to the high emission efficiency of P-43. Therefore the P-43 screen was used in the present study. The thickness of P-43 screen was 10 μm, which was thin enough to perform measurements with a high spatial resolution. In this whole system, one incident X-ray photon upon the P-43 screen produces about one electron in the CCD, considering the luminous efficiency of P-43, the transmission efficiency of the lens and the quantum efficiency of the CCD detector.

2.3. Materials

As a model sample, we used nanocomposite which contains a large amount of sulphur in its matrix part. The matrix is thiokol (LP-2, Toray Fine Chemicals Co. Ltd.), the chemical formula of which is -C_2H_4-O-CH_2-O-C_2H_4-S-S-. Monodispersive spherical silica particles (Seahoster, Nippon Shokubai Co. Ltd.) were added to thiokol. The volume fraction of particles was 20 %. After the particles were added to thiokol, it was carefully stirred and then kept at 120 °C for 48 hours so that it stiffens. Then the sample was sliced into a piece in a thickness of 30 μm by using a microtome.

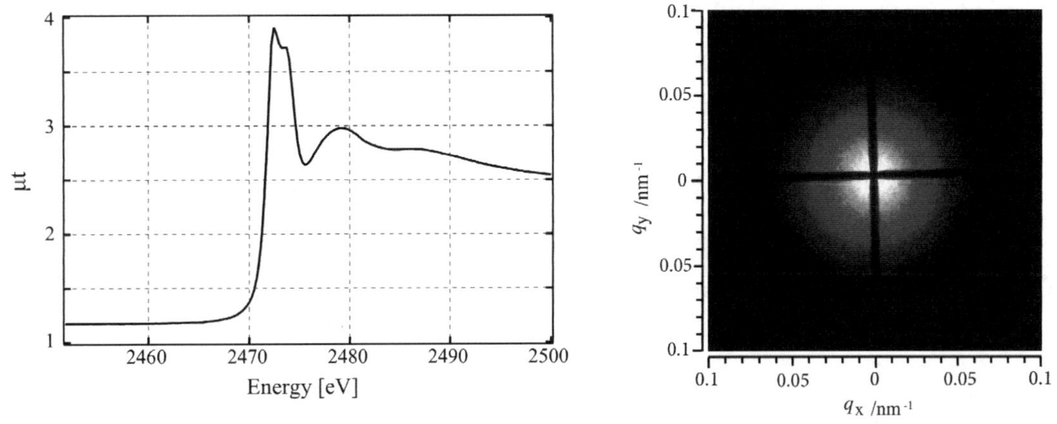

Figure 3. (Left) An X-ray absorption spectrum of thiokol near Sulphur K-edge. The intensity was measured with a 0.25 eV step and $E/\Delta E = 2000$ of the monochromator. (Right) SAXS image of silica particles filled in thiokol. The incident X-ray energy was 2100 eV.

3. Results and Discussion

Before measuring SAXS images, we measured the absorption spectrum of thiokol and the result is shown in Fig. 3 (left), which is calculated by

$$\mu t = -\ln I/I_0, \tag{2}$$

where μ, t, I, and I_0 represent the linear absorption coefficient, the thickness of sample, the intensity of transmitted X-ray, and that of incident X-ray, respectively. I and I_0 were measured with and without the sample, using the photodiode placed in the 3rd section. Fig. 3 (left) shows that the K-edge of sulphur in thiokol is hardly shifted from the edge of isolated sulphur atom, 2472 eV. We selected six energies, 2100, 2300, 2400, 2446, 2461 and 2466 eV to perform SAXS experiment.

An image taken at 2100 eV is shown in Fig. 3 (right). The exposure time was 300 sec. We processed the data in each quadrant to avoid the cross-shaped shadow due to the beamstop. From the obtained images we calculated scattering profiles by circularly averaging the intensities of corresponding pixels and normalized them with the effect of absorption by the sample and with the intensity of incident X-ray. The errors in the normalization process is estimated about 0.2 %. It should be noted that the energy dependence of the detector is ignored because we did not find any significant effect of energy on the sensitivity in the present study. When a measurement with an even higher precision is required, this factor needs to be carefully corrected.

The obtained profiles are shown in Fig. 4 (left). The scattering intensity profiles show a characteristic scattering pattern of isotropic monodispersive particles. The average diameter of contained silica particles is found to be 130 ± 11 nm, which is consistent with the SAXS measurement with a higher X-ray energy, 8 keV. It is clear that the scattered X-ray intensity increases as the energy approaches to the edge. This can be explained as follows: near the absorption edge, the atomic scattering factor of sulphur greatly depends on the energy of incident X-rays and is described as Eq. (1). In this expression, the real part of dispersion correction, $f'(E)$, sharply decreases as the energy approaches to the edge, and the imaginary part $f''(E)$ is nearly zero at the energy below the absorption edge. In SAXS experiments of nanocomposites, the intensity of scattered X-ray is determined by the difference of electron density between filled materials and matrix. This difference can be calculated by using an atomic scattering factor

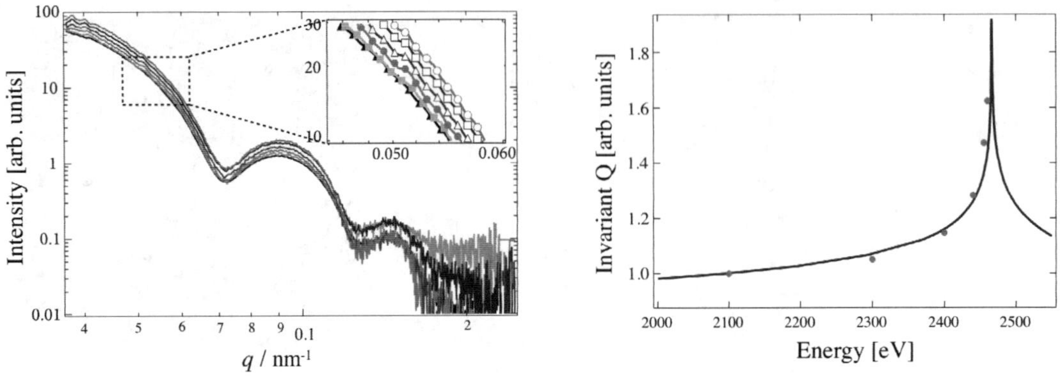

Figure 4. (Left) Scattering profile of thiokol at following energies, 2100 eV (closed triangles), 2300 eV (closed squares), 2400 eV (closed circles), 2446 eV (open triangles), 2461 eV (open squares) and 2466 eV (open circles). (Right) Energy dependence of invariant Q. Blue line shows theoretical calculation and red symbols show the measured ones.

mentioned above. In the present case, the intensity is proportional to

$$|\Delta\rho(E)|^2 = |\rho_{\mathrm{SiO_2}} - \rho_{\mathrm{thio}}(E)|^2, \tag{3}$$

where $\rho_{\mathrm{SiO_2}}$ and ρ_{thio} are the electron densities of silica particle and thiokol, respectively. It should be noted that the energy dependence of the X-ray intensity originates from only thiokol. In this equation, a decrease of real part of $f(E)$ enlarges the difference of electron density between silica and thiokol, which results in the increase of X-ray scattering intensity near the sulphur K-edge. The detailed value of $f(E)$ appears in some articles [24, 25], and we calculated $|\Delta\rho(E)|^2$ as shown in Fig. 4 (right, blue). Here we took it into consideration that the scattering intensity profiles at different energies needed to be convolved due to the finite resolution of the monochromator.

To compare the experimental results and the calculation, we used the invariant Q. It is defined in Eq. (4) and can be calculated for a two-phase system with Eq. (5):

$$\text{Invariant Q} = \int_0^\infty I(q)q^2 dq \tag{4}$$

$$= 2\pi^2\phi(1-\phi)(\Delta\rho)^2, \tag{5}$$

where ϕ is the volume fraction of one of the two phases. The experimental and expected invariants are compared in Fig. (4) (right side). The experimental values have been calculated with Eq. (4) and are shown as red dots. The expected energy dependence has been calculated with Eq. (5) and is shown as a blue line. The mean RMS deviation of the experimental values from the expected values is 7.5%. This is a fairly good value in consideration of the low flux of incident X-rays at the sample position in this energy region as shown in Fig. 2.

We used the sample which contains sulphur as much as 32 wt. %. Most non-crystalline/amorphous materials, however, contain less sulphur. For example, rubbery materials contain sulphur of only 1 – 2 wt. % which gives intensity variation of about 1 % near sulphur K-edge, assuming a similar two-phase system. In such ASAXS experiments the relative errors of less than 1 % are required. In order to reduce the relative errors, we need to improve the detector and the optical system in BL27SU.

4. Summary and Outlook

SAXS experiments near sulphur K-edge were performed and we obtained scattering intensity profiles of nanocomposites containing sulphur with a high accuracy. By changing the energy of incident X-rays near sulphur K-edge, we successfully measured the change of scattering intensity due to the sulphur component in the sample. The change of scattering intensity was well described by the energy dependence of anomalous dispersion of sulphur. This study can be the basis of the structural study of non-crystalline/amorphous materials by using ASAXS near the sulphur K-edge.

To apply ASAXS to non-crystalline/amorphous materials of interest, slight changes in the shape of scattering intensity profiles needs to be analyzed. To accomplish the analyses, the accuracy of measurement needs to be improved. We consider that the decrease of the photon flux above 2.0 keV and the low energy resolution near sulphur K-edge deteriorate the data quality, being limited by the use of the present optical system. Therefore we are planning to build a new experimental station on BL27SU, where we will install a silicon double-crystal monochromator to achieve an increased flux above 2.0 keV with an improved energy resolution. In addition, a new detector with higher quantum efficiency is being developed, which will be dedicated to SAXS experiments in soft X-ray region.

Acknowledgments

The authors acknowledge Drs. N. Ohta, Y. Suzuki, K. Uesugi, and N. Yagi (JASRI) for their kind support on the experiment. Dr. T. Itou and Y. Kajiura (Kao ltd.) supported us to perform ASAXS experiment. The ASAXS measurement was performed under the approval of JASRI program committee (Proposal No. 2008B1151 and 2009A1897).

References

[1] Lindner P and Zemb T (eds) 2002 *Neutron, X-ray and Light Scattering: Soft Condensed Matter* (Elsevier)
[2] Stuhrmann H B 1981 *Q. Rev. Biophys.* **14** 433–460
[3] Stuhrmann H B 2008 *Acta Cryst. A* **64** 181–191
[4] Barnardo T, Hoydalsvik K, Winter R, Martin C and Clark G 2009 *J. Phys. Chem. C* **113** 10021–10028
[5] Polizzi S, Riello P, Goerigk G and Benedetti A 2002 *J. Synchrotron Rad.* **9** 65–70
[6] Okuda H, Fukumoto T, Saida J, Ochiai S, Sasaki S and Masunaga H 2008 *J. Appl. Cryst.* **41** 675–9
[7] Lyon O, Hoyt J, Pro R, Davis B E C, Clark B, de Fontaine D and Simon J P 1985 *J. Appl. Cryst.* **18** 480–486
[8] Rice M, Wakatsuki S and Bienenstock A 1991 *J. Appl. Cryst.* **24** 598–602
[9] Yu C, Koh S, Leisch J, F T M and Strasser P 2008 *Faraday Discus.* **140** 283–96
[10] Chu B, Wu D, Lundberg R D and MacKnight W J 1993 *Macromolecules* **26** 994–999
[11] Guilleaume B, Blaul J, Ballauff M, Wittemann M, Rehahn M and Goerigk G 2002 *Euro. Phys. J. E* **8** 299–309
[12] Svergun D, Koin M B, Konarev P V, Shtykova E V, Volkov V V, Chernyshov D M, Valetsky P M and Bronstein L M 2000 *Cehm. Mater.* **12** 3552–3560
[13] Lee B, Lo C T, Seifert S, Rago N L D, Winans R E and Thiyagarayan P 2007 *Macromolecules* **40** 4235–43
[14] Vainio U, Pirllakainen K, Kisko K, Goerigk G, Kotelnikova N E and Serimaa R 2007 *Euro. Phys. J. D* **42** 93–101
[15] Morfin I, Ehrburger-Dolle, F Grillo I, Livet F and Bley F 2006 *J. Synchrotron Rad.* **13** 445–452
[16] Kühnholz O 1991 *J. Appl. Cryst.* **24** 811–814
[17] Scholl G, Dauvergne F, Gabriel A, Hutsch M, Marmotti M, Sayers Z, Stuhrmann S, Thomas J, Trame C and Stuhrmann H 1995 *Nucl. Instrum. Meth. B* **97** 303–7
[18] Carugo K, Helliwell J R, Stuhrmann H and Weiss M S 2005 *J. Synchrotron Rad.* **12** 410–9
[19] Tanaka T, Hara T, Oura M, Ohashi H, Kimura H, Goto S, Suzuki Y and Kitamura H 1999 *Rec. Sci. Instrum.* **70** 4153–60
[20] Hettrick M C, Underwood J H, Batson P J and Eckart M J 1988 *Appl. Optics* **27** 200–2
[21] Ohashi H, Ishiguro E, Tamenori Y, Okumura H, Hiraya A, Yoshida H, Senba Y, Okada K, Saito N, Suzuki I, Ueda K, Ibuki T, nagaoka S, Koyano I and Ishikawa T 2001 *Nucl. Instrum. Meth. A* **467-8** 533–6
[22] Tamenori Y 2010 *J. Synchrotron Rad.* **17** 243–9
[23] Tamenori Y, Ohashi H, Ishiguro E and Ishikawa T 2002 *Rec. Sci. Instrum.* **73** 1588–90

[24] Sasaki S 1989 *KEK Report* **88-14**
[25] Henke B L, Gullikson E M and Davis J C 1993 *Atom. Data Nucl. Data* **54** 181–342

A SAXS/WAXS/GISAXS Beamline with Multilayer Monochromator

Alexander Hexemer[1], Wim Bras[2], James Glossinger[1], Eric Schaible[1], Eliot Gann[1], Rick Kirian[1], Alastair MacDowell[1], Matthew Church[1], Bruce Rude[1], Howard Padmore[1]

[1]Lawrence Berkeley National Laboratory, 1 Cyclotron Road, Berkeley, CA 94720, United States

[2]Netherlands Organisation for Scientific Research (NWO), DUBBLE@ESRF, BP220, F38043 Grenoble Cedex, France

Abstract. We discuss the construction of a new SAXS/WAXS beamline at the Advanced Light Source at Lawrence Berkeley Laboratory. The beamline is equipped with a multilayer monochromator in order to obtain a high X-ray flux. The detrimental effects that the increased bandwidth transmitted by this monochromator could have on the data quality of the SAXS and WAXS patterns is shown to be negligible for the experimental program intended to be operated on this beamline.

1. Introduction

Demand for synchrotron-based SAXS remains high, despite recent improvements in generator-based SAXS equipment. Several reasons for this can be given. Synchrotron beamlines still produce at least an order of magnitude more flux, have better collimation, and can more easily accommodate the larger sample handling equipment associated with time-resolved experiments. Similarly, these beamlines are better suited for combining SAXS with other experimental techniques, either X-ray based or spectroscopic (e.g., Raman spectroscopy).

The combination of SAXS and WAXS is an obvious example of such technique combinations since it provides structural information over a large length scale range. The pioneering work in this was done in Hamburg[1]. After the first dedicated SAXS/WAXS beamline was constructed[2] it has become common practice to design SAXS beamlines on which a material science program is conducted with WAXS options as well[3].

In recent years the use of grazing incidence SAXS (GISAXS) has also gained in popularity. The reason for this is that the newer generation of synchrotron sources offers the X-ray beam stability that is required for this technique. At the same time, materials scientists are creating a wealth of new thin film molecular architectures with structural length scales that can be observed by SAXS methods. Since the experimental requirements for both SAXS and GISAXS are reasonably compatible we have implemented these techniques on the same beamline.

There are some minimum requirements regarding flux, beam size, low angle resolution etc. which have to be met in order for a beamline to be able to handle a competitive research program. Below we discuss some of these points with the understanding that it is possible to find exceptions to the general rules that are given.

The minimum X-ray flux required for a SAXS/WAXS beamline to be competitive is currently around 10^{11}-10^{12} photons/second. The specific number that is required depends on several factors, but the foremost considerations are the speed at which one wishes to do time-resolved experiments, and the statistical data quality one wishes to obtain. For a large variety of time-resolved experiments one can safely say that the lower limit is 0.1 sec/frame. Obviously there are experiments that require a higher time resolution but for most commonly used sample environments it will be difficult to perturb the samples faster and still retain a homogeneous sample (e.g., with regard to temperature or chemical state) over the entire X-ray beam size. The statistical data quality is obviously not only a matter of the intensity of the direct beam but is also dependent on the scattering contrast and the efficiency of the detectors used. With the recent introduction of the Si pixel detectors[4], which combine the high efficiency of CCD detectors with the single photon counting capacity of the traditional gas wire chambers, it is easier to obtain an acceptable data quality with less photons. It should be pointed out that the data quality issues are mainly limited to the wider angle region where one probes the matrix-particle interface and the scattered intensity is intrinsically low. In the lower scattering vector range the intensity is in general so high that even with old fashioned X-ray film one can still obtain decent results.

Another factor that should be considered is the beam size. There is a tendency to develop optics that minimize the beam size as much as possible but one should keep in mind that one still requires a certain number of photon-matter interactions in order to generate a scattering pattern. If all these interactions are taking place in a smaller volume there is an increased chance that the samples will suffer from radiation damage. This is less of an issue when one uses a micro sized beam for spatial scanning of a sample, but when one performs time-resolved experiments, and the sample is exposed to the beam for the entire period over which the sample evolution takes place, this can become a problem.

Although on some beamlines it is possible to observe q-values as low as $q = 0.012\ \text{nm}^{-1}$ one should keep in mind that, when using the conventional photon energy range of 8-12 keV, the entire scattering range of $0.01 < q < 0.06\ \text{nm}^{-1}$ will be mapped upon only a small number of pixels close to the shadow of the beam stop unless one uses unrealistic large sample-detector distances. It is questionable how high the information content of the data in this case is. However, to be able to observe a lower scattering angle limit of $q = 0.06\ \text{nm}^{-1}$ is desirable.

At the Advanced Light Source (ALS) in Berkeley, which is primarily designed as a third generation soft X-ray source, it requires some ingenuity in the optical design in order to be able to satisfy the above described boundary conditions for a competitive SAXS/WAXS and GISAXS beamline. In the energy range of 8 -12 keV the superbend magnets produce sufficient flux but there is only a limited amount of X-ray ports available on these magnets. Still, the presence on the Berkeley Laboratory site of both the National Center for Electron Microscopy and the Molecular Foundry creates strong local demand for a SAXS/WAXS/GISAXS beamline. The complimentarity between electron microscopy and SAXS is undisputed and there is a large synergy by having the option to locally characterize the nano structures created in the Molecular Foundry.

In the following we describe the design of and results from a new X-ray scattering beamline constructed on a conventional bending port of the ALS. This beamline is equipped with a high flux multilayer monochromator and a toroidal focusing mirror. The optical engineering solutions that we have implemented can be usefully applied at other bending magnet beamlines.

2. Materials and methods

2.1. Optics

The Advanced Light Source (ALS) at Lawrence Berkeley National Laboratory (LBNL) operates with a nominal energy of 1.9 GeV. In the recently introduced top-off mode the storage ring can operate with a constant beam current of 500 mA. One of the main benefits of this constant current are that the heat load on the optical elements is constant, which allows for stable operation of the beamlines.

The SAXS/WAXS beamline 7.3.3 receives the radiation from a bending magnet operated at 1.27 T, which means that the critical energy of the bend magnet is 3 keV. The source size is 0.1 x 0.01 mm^2 (H x V). The divergence of the beamline is 2.78 mrad. The beamline vacuum is separated from the storage ring vacuum by two 125 μm thick beryllium windows placed 13.7 m from the source. The beamline is designed to be operated at 10 keV, i.e. rather far above the critical energy and thus with a relatively low source intensity. In order to still retain a useful X-ray flux it was decided to utilize a multilayer monochromator build by RIGAKU. The monochromator consists of two multilayer mirrors of silicon <100> covered with 250 layers of alternating Mo and B_4C. The d-spacing was kept constant at 2.0nm +/- 0.1nm. The bandwidth of the multilayer was calculated and measured to be around 97eV and at 10keV this results in an $E/\Delta E$ ratio of 100. This monochromator has a roughly 10 times higher throughput compared to the often used Si(111) double crystal monochromator at the expense of transmitting a higher bandwidth (97 eV). For SAXS applications this poses no serious problems[5] and we will show later in the text that for WAXS applications the data quality remains similarly high.

To focus the beam we use a toroidal mirror (M1). The M1 mirror is located 2m downstream from the multilayer monochromator. The 500 A of Platinum coated toroid has a 1:1 focusing at 5.4 mrad incoming grazing angle with an rms roughness of 0.5 nm.

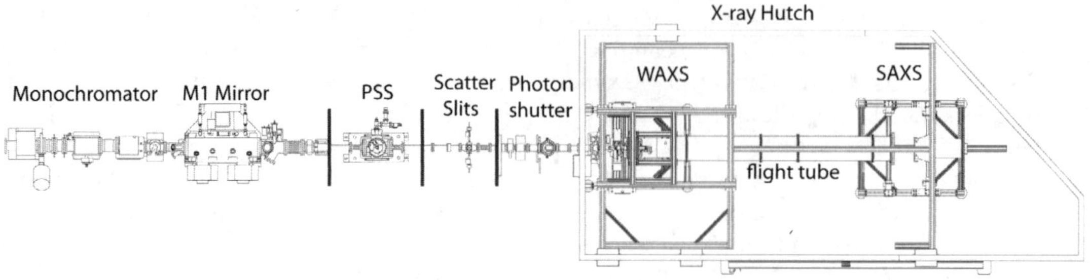

Figure 1. The schematic lay-out of the 7.3.3 SAXS/WAXS beamline.

The multilayer monochromator is placed at 14 m from the source point. The toroidal mirror is 16 m from the source. The sample position is at 32 m and the sample detector distance for the SAXS experiment can be varied between 1.3 m and 4.3 m. The beamline exit window material is a 100 μm thick muscovite mica.

At present the beamline is equipped with conventional knife edge slits but it is foreseen that newly designed scatterless slits will be installed in the near future[6]. These will allow the beam size to be tailored to the desired size without introducing parasitic scatter.

2.2. Detectors

There are several detectors available that are suitable for different aspects of the experimental program operated on the beamline. At present the workhorse detectors that are being used in the SAXS/WAXS mode are two ADSC Quantum 4r CCD detectors. The main advantage of these detectors is their large active area (188 x 188 mm^2). The drawback of these detectors is their limited dynamic range and the relatively slow read out speeds, which makes them less useful for time-resolved experiments.

For faster experiments a Pilatus 100k detector is available[4a]. This is a two dimensional single-photon counting detector with a high dynamics range of 20 bits and an active area of 83.8 x 33.5 mm^2. The Pilatus 100k is primarily installed for time resolved measurements due to its very fast readout time of 2.7 ms.

We have also ordered a Pilatus 1M detector, which will become available in the near future. This detector has a larger active area of 169 x 179 mm^2 and a maximum frame rate of 30Hz but for the rest has the same characteristics as the Pilatus 100k detector.

2.3. Controls and data acquisition software

The beamline control and data acquisition (DAQ) software has been written in LabVIEW. This allows for rapid integration of commercial off-the-shelf (COTS) devices and rapid development of user friendly Graphical User Interfaces (GUIs).

The beamline and experimental control is split between two PC's. The first PC controls low-level beamline hardware such as optical components, sample positioning, detector positioning, and the X-ray safety shutter. It also reads experimental data from ion chambers, thermocouples, and other sensors. The beamline control software can operate in a standalone mode but it also contains a TCP server so that the SAXS/WAXS DAQ software can communicate with it and retrieve the required experimental parameters.

The DAQ software controls the beamline control software in a master-slave configuration. It therefore effectively controls and coordinates the motors and sample environments. It also communicates with the independent control PC that is required for the different detectors. The interface that is available to the users is a GUI that presents a limited but flexible set of options. The user can take an image, find sample positions, or set sample temperatures with the push of a button. In addition, the user can graphically create extended command sequences, so as to automate more complex experimental protocols. These automated sequences are useful, for example, if one wishes to control temperature-annealing protocols or take SAXS/WAXS images at specified intervals. These sequences can be saved as a file and retrieved or modified for later use.

At each stage, the GUI provides feedback to the user. As images are taken, progress bars show their status and estimate the total time to complete all imaging. Temperature changes are displayed in real time. When automated sequences are run, the GUI displays them as a sequential queue in which the current step is highlighted.

The data reduction is performed using the NIKA and IRENA software[7] which has been adapted to be able to read the different proprietary data formats that are generated by the different detectors in use on the beamline. In order to correct for detector sensitivity we are using the flat-field correction provided by ADSC. Upstream of the sample as well as downstream we have ion chambers to correct for sample transmission. Typical samples for q-range calibrations are Silver behenate for the small-angle and Aluminum for the wide-angle range.

2.4. Sample handling

In the design of the beamline great care has been taken to provide as much flexibility and space around the sample position. This was done in order to be able to accommodate as large a variety of sample environments as possible. For a multi-user facility this, and the availability of additional experimental infrastructure, is a crucial requirement. A motorized 200 mm Newport translation stage as well as a motorized 76 mm Siskiyou vertical stage form the base of the sample setup.

Several dedicated sample cells have been developed, including a sample cell for solution scattering for proteins and a sample cell for high-throughput imaging of large numbers of capillary tubes. A conventional Linkam DSC[8] has been incorporated in the beamline and in the experimental control software. In addition a next generation heat stage is being developed to handle multiple samples with different annealing schedules.

For GISAXS experiments a Huber 5203.10 goniometer is used together with a Newport URS100PP rotational stage. An annealing stage for the GISAXS setup is currently under development.

3. Results

In this section we will show some representative results that have been obtained on beamline 7.3.3 in the different configurations that are available. It should be mentioned that the beamline is in full operation and that numerous publications have appeared [9]

The flux that is measured at the sample position is 1.3 10^{12}photon/sec at 10 keV photon energy. The lowest angle resolution that so far has been reached is $q = 2\pi/d= 0.002$ Å^{-1}. It is expected that this number will decrease to even lower Q values when the new scatterless guard slits are installed.

3.1. Wide Angle spatial resolution

The 2D WAXS detector is suspended on a gantry, which allows the placement of the detector in an arbitrary orientation and sample-detector distance. The q-vector range is in general calibrated using the (200) and (111) of an Aluminum foil. Figure 2 shows a cropped WAXS image of the (111) at 2.69 Å^{-1} and the (111) Aluminum (200) 3.1 Å^{-1} peak.

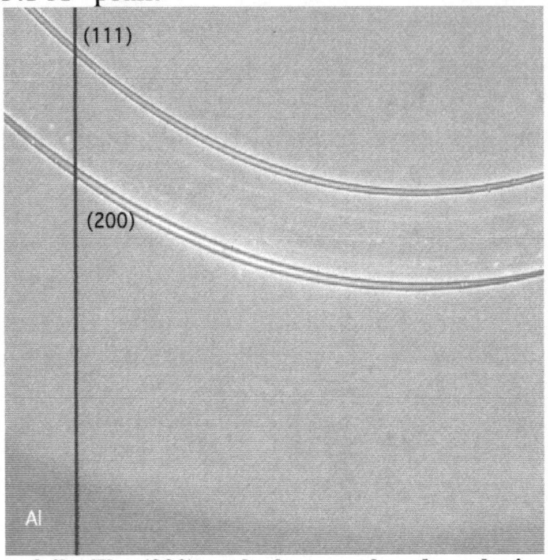

Figure 2. WAXS of rolled Aluminum foil. The (200) peak shows a clear broadening as a result of anisotropic grain orientation intrinsic to rolled Aluminum. The blue line on the left hand side is an instrumentation effect due to the gap between the CCD modules.

The Al (200) shows a clear peak broadening characteristic of Aluminum foil. The rolling of the foil during production creates texture and hence the peak broadening. It proves that the instrumental resolution is adequate to detect features such as line broadening in the WAXS regime and that the increased wavelength bandwidth does not pose a problem for the experiments for which the beamline is designed.

3.2. Simultaneous SAXS/WAXS experiments

Block copolymers (BCP) are composed of two or more monomer units covalently bound together. The variety in monomer units as well as length of the units creates a rich and diverse phase diagram with correspondingly diverse properties. In order to maximize the efficiency of such materials, complete control of the morphology has to be achieved. Here we show a rod-coil polymer system poly- (2,5-di(2′-ethylhexyloxy)-1,4-phenylenevinylene) (DEH-PPV) as a model system for materials with a potential use as organic electronics.[10] Questions regarding the morphology of this system, for example if the rods align perpendicular to the lamellae formed by the BCP or at a slight angle, can be answered with a combined SAXS/WAXS experiment. The question could be answered by aligning the entire sample in a magnetic **B** field and the expected scattering pattern for the case that the rods are perpendicular to the lamellae direction is shown in figure Figure 3. The lamellar spacing is around 20 nm, which is in the SAXS regime. The rod-rod distance is around 1Å and in the WAXS regime.

Journal of Physics: Conference Series **247** (2010) 012007 doi:10.1088/1742-6596/247/1/012007

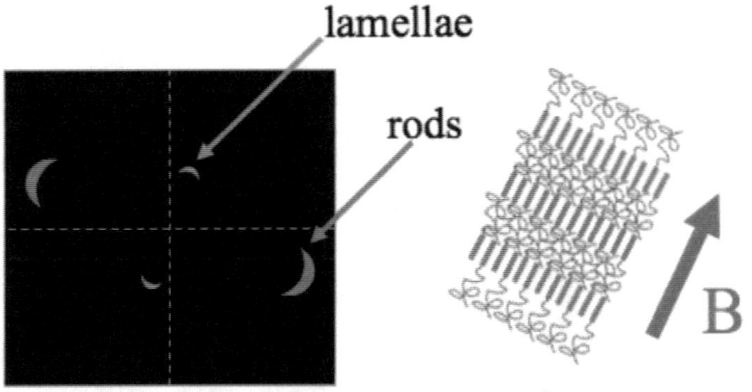

Figure 3. Expected Scattering pattering of a rod-coil block copolymer aligned in a magnetic field in which the polymer rods are at right angles with respect to the larger scale lamellae.

In the case that the angle between the rods and the lamellae would deviate from being perpendicular one would expect the WAXS 110 peak to split up into two peaks on either side of the equator.

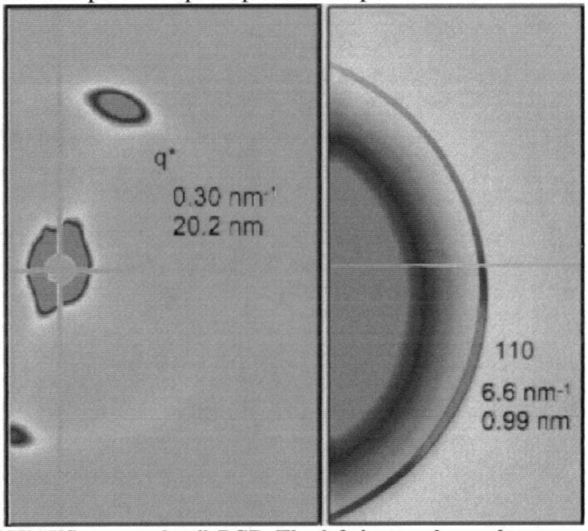

Figure 4. Simultaneous SAXS/WAXS on a rod-coil BCP. The left image shows the measured SAXS pattern, on the right the corresponding WAXS pattern.

In Figure 4 the SAXS pattern is shown on the left and the corresponding WAXS pattern on the right. The SAXS pattern shows the peak of the BCP lamellae at 20nm. Simultaneously we measured the WAXS pattern and the 110 peak of the tetragonal unit cell of the rods. The angle between the lamellae peak and the rod-rod peak was measured to virtually 90°, which proves that the rods are aligned perpendicular to the lamellae.

The X-ray beam was focused at 2.75m to a spot size of 250μm x 125 μm. The data quality in both the SAXS as well as the WAXS pattern is of high quality, and thus we can conclude that the over/under-focus, i.e. the depth of field, on both detectors does not cause a severe loss of resolution. This is not completely self evident when using a broad bandpass monochromator like the one used for this work.

3.3. Grazing incidence X-ray Scattering

Grazing Incidence X-ray Scattering (GISAXS) has become a vital tool to study surfaces and thin films. In particular in the field of polymer thin film GISAXS has received interest in the last few years. [9c, 9h, 11]

Thin block copolymer films show great potential in a wide variety of fields. Their potential applications include inexpensive printable photovoltaics, organic light emitting diodes, organic fuel cell membranes, and next generation hard drives. These new technologies are based on polymer self-assembly processes that happen on the nanometer length scale in films 10s to 100s of nanometers thick. These films are too thin to be measured in transmission geometry but can be studied by grazing incidence methods.

Figure 5 (a) shows a scanning force micrograph of the top surface of a thin poly(styrene)-poly(ethylene oxide) (M_w = 32k-11k) block copolymer (BCP) film.[9h] The BCP is spincast on the surface of a faceted sapphire substrate and solvent-annealed with THF and water. Figure 3 (b) shows the corresponding GIAXS data. The horizontal axis corresponds to the in-plane vector $\mathbf{q}_x$ and the vertical axis to the out-of-plane momentum transfer vector $\mathbf{q}_z$. In Figure 5 (b) we show the scattering pattern where the X-ray beam was aligned to coincide with the (10) plane of the hexagonally arranged standing up cylinders.

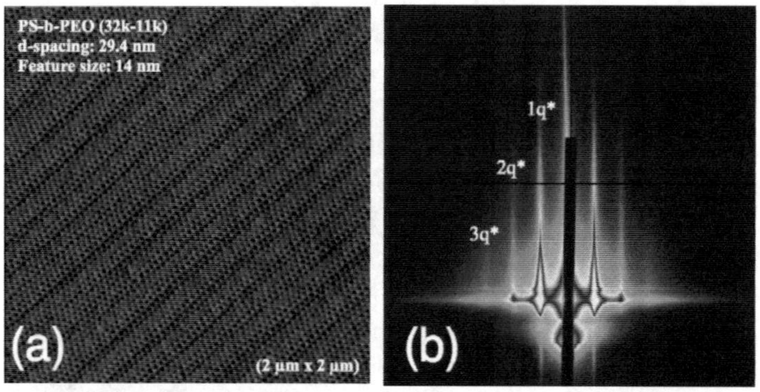

Figure 5 (a) Scanning force microscopy micrograph of a PS-PEO thin block copolymer film (b) corresponding GISAXS data along the (10) lattice axis

The observed peaks along the $\mathbf{q}_x$ direction correspond to the (10), (20) and the (30) plane. No other peaks are observed, which confirms that the standing up cylinders exhibit a single crystal orientation over the entire probed area as shown in ref 9h. Along the $\mathbf{q}_z$ direction streaks are observed corresponding to the form factor of the standing up cylinders.

3.4. Beamline operation

The stability of a beamline can be influenced by many factors, including the radiation source itself. Also the behavior of the X-ray optics under the thermal load of the white X-ray beam and the stability of the detector systems over an extended period is of importance. With the recent introduction of top-off mode at the ALS (cf. introduction) and the use of a Pilatus pixel detector we have been able to follow the development of density fluctuations in a temperature quenched glass that was annealed at elevated temperatures[12] (Figure 6).

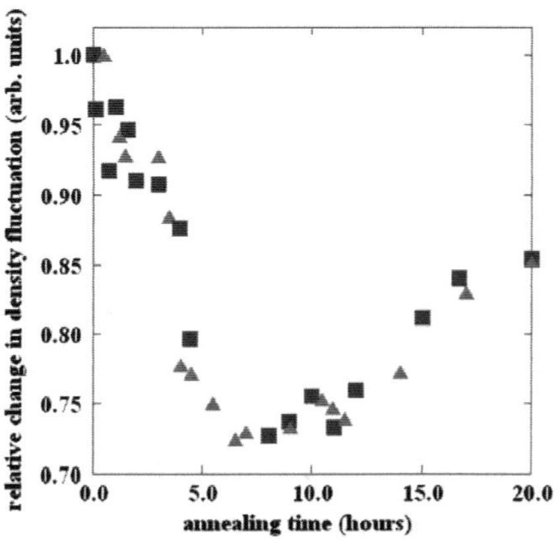

Figure 6 Scattered intensity due to density fluctuations in a temperature quenched and subsequently reheated glass sample. The development was followed over a period of 20 hours.

Density fluctuations in glass are in general very difficult to detect, since they can easily be swamped by only the smallest variations in beamline behavior. In fact, before the top-off operation mode of the ALS was introduced, after every machine refill there was a period of around 30 minutes in which the optics had to recover from the change in thermal load, and as a consequence the acquisition of this kind of data would not have been possible. The fact that we were able to follow the development for periods up to 24 hours is rather striking.

At present the time-resolution that can be achieved in combined SAXS/WAXS experiments is limited by the read out time of the CCD detectors and not by the available X-ray flux. The present limit is 10 sec/frame. The introduction of the new generation of pixel detectors, which have a read out time of only 0.03 seconds, will allow faster experiments to be carried out and will also improve the duty cycle (read out time vs. real data acquisition time). This is useful in materials science experiments where time resolutions of 0.1 sec/frame or longer are desired. However, it is also foreseen that time-resolved protein solution scattering experiments can be performed.

4. Conclusions

In combined SAXS/WAXS synchrotron radiation based experiments where independent detectors are used to collect the SAXS and WAXS patterns these detectors are often placed at considerable distance from each other. In theory this can be detrimental to the data quality since optimum data quality can only be achieved when the X-ray beam is focused onto a detector. When using a multilayer monochromator, with a transmission bandwidth that is larger than the commonly used Si-based monochromators, one aggravates this situation. However, we have shown that with the small X-ray beam divergence of the ALS and a sensible limit on the separation distance between the SAXS and WAXS detectors it is feasible to construct a beamline that has more than adequate operational performance in terms of X-ray flux, angle resolution, and diffraction resolution, even when using a multilayer monochromator.

The introduction of the ALS top-off mode and the use of a new generation of pixel detectors in combination with stable multilayer optics has allowed us to create a extremely versatile multi-user facility at which an extended materials science and biological research program can be carried out.

5. Acknowledgements

The Advanced Light Source is supported by the Director, Office of Science, Office of Basic Energy Sciences, Materials Sciences Division, of the U.S. Department of Energy under Contract No. DE-AC02-05CH11231 at Lawrence Berkeley National Laboratory and University of California, Berkeley, California. Dr. Jan Ilavsky is thanked for incorporating the ADSC format in the input routine for the Nika software, which was developed by him. We thank Prof. Balsara and S. Mullin for his support in developing the next generation heat stage. Special thanks also to Prof. Russell, Dr. Mauro and Prof. Segalman for providing the data presented here.

References

[1] M. Bark, H. G. Zachmann, R. Alamo, L. Mandelkern, *Makromolekulare Chemie-Macromolecular Chemistry And Physics* **1992**, *193*, 2363.
[2] W. Bras, G. E. Derbyshire, A. J. Ryan, G. R. Mant, A. Felton, R. A. Lewis, C. J. Hall, G. N. Greaves, *Nuclear Instruments & Methods in Physics Research Section a-Accelerators Spectrometers Detectors and Associated Equipment* **1993**, *326*, 587.
[3] a)W. Bras, I. P. Dolbnya, D. Detollenaere, R. van Tol, M. Malfois, G. N. Greaves, A. J. Ryan, E. Heeley, *Journal of Applied Crystallography* **2003**, *36*, 791; b)T. Narayanan, O. Diat, P. Bosecke, *Nuclear Instruments & Methods In Physics Research Section A-Accelerators Spectrometers Detectors And Associated Equipment* **2001**, *467*, 1005; c)R. J. Cernik, P. Barnes, G. Bushnell-Wye, A. J. Dent, G. P. Diakun, J. V. Flaherty, G. N. Greaves, E. L. Heeley, W. Helsby, S. D. M. Jacques, J. Kay, T. Rayment, A. Ryan, C. C. Tang, N. J. Terrill, *Journal Of Synchrotron Radiation* **2004**, *11*, 163.
[4] a)C. Broennimann, E. F. Eikenberry, B. Henrich, R. Horisberger, G. Huelsen, E. Pohl, B. Schmitt, C. Schulze-Briese, M. Suzuki, T. Tomizaki, H. Toyokawa, A. Wagner, *Journal Of Synchrotron Radiation* **2006**, *13*, 120; b)B. Schmitt, C. Bronnimann, E. F. Eikenberry, G. Hulsen, H. Toyokawa, R. Horisberger, F. Gozzo, B. Patterson, C. Schulze-Briese, T. Tomizaki, *Nuclear Instruments & Methods in Physics Research Section a-Accelerators Spectrometers Detectors and Associated Equipment* **2004**, *518*, 436.
[5] H. Tsuruta, S. Brennan, Z. U. Rek, T. C. Irving, W. H. Tompkins, K. O. Hodgson, *Journal of Applied Crystallography* **1998**, *31*, 672.
[6] Y. L. Li, R. Beck, T. Huang, M. C. Choi, M. Divinagracia, *Journal of Applied Crystallography* **2008**, *41*, 1134.
[7] J. Ilavsky, P. R. Jemian, *Journal of Applied Crystallography* **2009**, *42*, 347.
[8] W. Bras, G. E. Derbyshire, A. Devine, S. M. Clark, J. Cooke, B. E. Komanschek, A. J. Ryan, *Journal of Applied Crystallography* **1995**, *28*, 26.
[9] a)R. Beck, J. Deek, J. Jones, C. Safinya, in *Nat Mater*, **2009**; b)E. D. Gomez, A. Panday, E. H. Feng, V. Chen, G. M. Stone, A. M. Minor, C. Kisielowski, K. H. Downing, O. Borodin, G. D. Smith, N. P. Balsara, *Nano Letters* **2009**, *9*, 1212; c)S. S. Lee, C. S. Kim, E. D. Gomez, B. Purushothaman, M. F. Toney, C. Wang, A. Hexemer, J. E. Anthony, Y.-L. Loo, *Adv Mater* **2009**, *21*, 3605; d)A. Mastroianni, T. Schilling, J. Luther, in *Nat Mater*, **2009**; e)J. C. Mauro, S. S. Uzun, W. Bras, S. Sen, *Phys. Rev. Lett.* **2009**, *102*, 155506; f)A. Panday, S. Mullin, E. D. Gomez, N. Wanakule, V. L. Chen, A. Hexemer, J. Pople, N. P. Balsara, *Macromolecules* **2009**, *42*, 4632; g)M. J. Park, S. Kim, A. M. Minor, A. Hexemer, N. Balsara, *Adv Mater* **2009**, *21*, 203; h)S. Park, D. H. Lee, J. Xu, B. Kim, S. W. Hong, U. Jeong, T. Xu, T. P. Russell, *Science* **2009**, *323*, 1030; i)J. M. Virgili, A. Hexemer, J. A. Pople, N. P. Balsara, R. A. Segalman, *Macromolecules* **2009**, *42*, 4604; j)N. S. Wanakule, A. Panday, S. A. Mullin, E. Gann, A. Hexemer, N. P. Balsara, *Macromolecules* **2009**, *42*, 5642.
[10] B. D. Olsen, D. Alcazar, V. Krikorian, M. F. Toney, E. L. Thomas, R. A. Segalman, *Macromolecules* **2008**, *41*, 58.

[11] a)G. Renaud, R. Lazzari, F. Leroy, *Surface Science Reports* **2009**, *64*, 255; b)D. Smilgies, *Review of Scientific Instruments* **2002**, *73*, 1706; c)G. E. Stein, E. J. Kramer, X. Li, J. Wang, *Phys. Rev. Lett.* **2007**, *98*, 086101.

[12] J. C. Mauro, S. S. Uzun, W. Bras, S. Sen, *Physical Review Letters* **2009**, *102*.

Pixel detector system development at Diamond Light Source

J Marchal[1], **I Horswell, E N Gimenez and N Tartoni**

Diamond Light Source, Harwell Science and Innovation Campus, Oxfordshire, OX11 ODE UK
E-mail: julien.marchal@diamond.ac.uk

Abstract. Hybrid pixel detectors consisting of an array of silicon photodiodes bump-bonded to CMOS read-out chips provide high signal-to-noise ratio and high dynamic range compared to CCD-based detectors and Image Plates. These detector features are important for SAXS experiments where a wide range of intensities are present in the images. For time resolved SAXS experiments, high frame rates are compulsory. The latest CMOS read-out chip developed by the MEDIPIX collaboration provides high frame rate and continuous acquisition mode. A read-out system for an array of MEDIPIX3 sensors is under development at Diamond Light Source. This system will support a full resolution frame rate of 1 kHz at a pixel counter depth of 12-bit and a frame rate of 30 kHz at a counter depth of 1 bit. Details concerning system design and MEDIPIX sensors characterization are presented.

1. Introduction

SAXS requires fast X-ray area detectors with high spatial resolution over a large active area. Fast-readout can be obtained with multi-wire gas detectors working in coincidence mode such as the RAPID detector in use on the I22 beamline at the Diamond Light Source (DLS) [1]. A drawback of this gas detector detector technology is a low maximum global count rate due to the coincidence read-out mode and cumbersome ancillary equipment required to flush gas in the detector chamber. For less demanding experiments, standard CCD+scintillator detectors offer a straightforward solution. Hybrid pixel detectors are an alternative detector technology which fills the gap between CCD-based detectors and fast gas detectors. Hybrid pixel detectors provide excellent performance in terms of signal-to-noise ratio, dynamic range, maximum global count rate and frame rate. Hybrid pixel detectors consist of a pixellated read-out chip electrically connected to a pixellated semiconductor sensor (e.g. silicon) with small solder or indium bumps. X-rays interacting in the silicon sensor are converted into an electrical charge which is amplified and filtered locally in each pixel of the read-out chip. X-ray induced pulses which pass above a preselected threshold increment a counter implemented in each pixel. The counters are read out after the exposure time. Hybrid pixel detectors with 172 μm x 172 μm pixels capable of covering active areas up to 43 cm x 45 cm are commercially available (PILATUS detectors by DECTRIS Ltd. [2]) and are in use in a number of synchrotron beamlines around the world. A hybrid pixel detector with a pixel size of 130 μm (XPAD [3]) was also developed for X-ray scattering and diffraction experiments at SOLEIL, the French national synchrotron facility. MEDIPIX2 developed at CERN [4] is a hybrid pixel detector with pixels of 55 μm size. MEDIPIX2 chips are the core component of the MAXIPIX detector system developed by ESRF for synchrotron

[1] To whom any correspondence should be addressed.

experiments [5]. Single-chip MEDIPIX2 detectors have also been used for X-ray diffraction and imaging experiments on DLS synchrotron beamlines [6].

2. Charge-sharing in hybrid pixel detectors with small pixels

The main issue arising when reducing the pixel size of hybrid detectors is an increased proportion of X-ray hits generating a charge which is shared between two or more neighboring pixels. Due to the single-photon counting approach enabled by the presence of a signal discriminator in each pixel, charge-sharing between pixels will result in a strong dependence of the detector response on discriminator threshold and X-ray energy. When detecting monochromatic X-rays, a detection threshold set at 50% of the X-ray energy minimizes count loss or double counts at pixel edges. This nonetheless reduces the detection efficiency since X-rays interacting at pixel corners where charge is shared between more than two pixels are not detected. Count loss at pixel corners is visible in figure 1.a showing the response of a MEDIPIX2 detector when a 15 keV X-ray beam focused down to a 6 µm x 4 µm spot size was scanned across a 55 µm sided pixel. This data was acquired on B16 beamline at the Diamond Light Source for a MEDIPIX2 threshold set to 7.5 keV, i.e. 50% of the X-ray energy [7].

Detection efficiency loss is reduced even further when setting the detection threshold at a value higher than 50%, as is sometimes required in order to separate sample fluorescence background from the diffracted X-ray signal. Detection efficiency loss as a function of threshold is shown by the slope of the first part of the S-curve in figure 1.b, obtained by scanning the detection threshold of a MEDIPIX2 sensor illuminated with monochromatic X-rays.

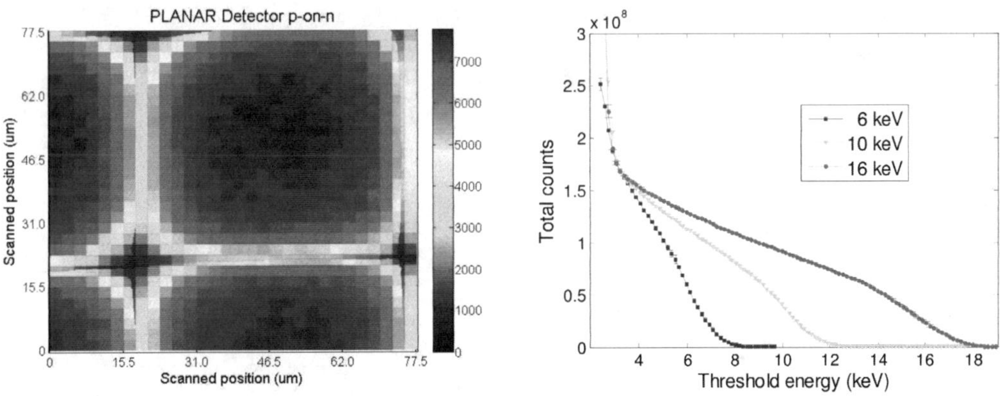

(a) (b)

Figure 1. (a) Mean detected count rate as a function of beam position when a 15 keV X-ray beam is scanned across a pixel of a MEDIPIX2. (b) S-curve obtained by plotting the number of X-ray counts as a function of detector threshold while illuminating the whole detector with X-rays of energy 6, 10 and 16 keV.

Discriminator thresholds are not uniform across the read-out chip of a hybrid pixel detector due to variations in the parameters of the CMOS components across the chip. Threshold trimming is usually performed to reduce threshold dispersion across the detector. Charge-sharing results in a higher sensitivity of the detector response to variations in the effective discriminator threshold set in each pixel and makes threshold trimming more difficult.

One of the aims of the new pixel architecture implemented in the read-out chip recently developed by the MEDIPIX3 collaboration is to suppress the effect of charge-sharing on image quality, as described in the following section.

3. MEDIPIX3 read-out chip

The MEDIPIX3 chip was developed by an international collaboration led by CERN, including several synchrotron radiation facilities [8]. It consists of an Application Specific Integrated Circuit (ASIC) designed in 0.13 μm CMOS technology which can be bump-bonded to a pixellated semiconductor sensor. A bare chip (without a sensor) is shown in figure 2.

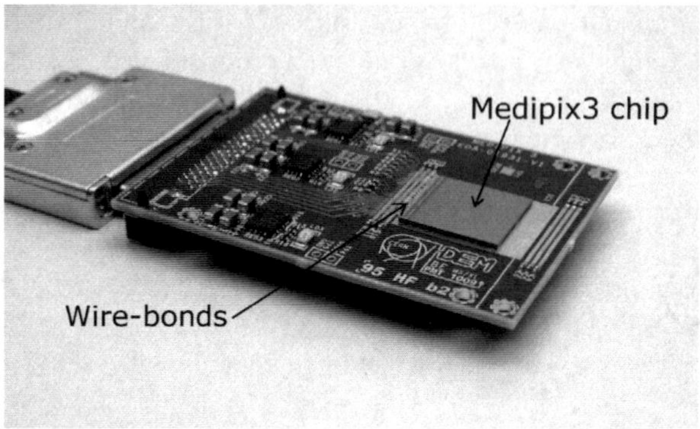

Figure 2. MEDIPIX3 read-out chip mounted on a carrier board developed by CERN

The chip is mounted on the carrier board and wire-bonds are used to connect the input/output lines of the chip to the carrier board in order to interface the chip with control and data acquisition electronics. Wire-bond connections are located on one side only of the chip so that MEDIPIX3 could be 3-side buttable, allowing several chips to be tiled together to form large detection areas with minimum dead-space. The main characteristics of the MEDIPIX3 chip are listed in table 1. Additional features of the MEDIPIX3 chip which are of interest for SAXS experiments on synchrotron beamlines are listed in the following sections.

Table 1 Basic specifications of a MEDIPIX3 read-out chip

Pixel size	55 μm x 55 μm
Format	256 x 256 pixels
Active area	1.41 x 1.41 cm^2
Count rate per pixel	$5x10^5$ counts/s @ 20% dead time
Count rate per mm^2	1.65 x 10^8 count/s @ 20% dead time
Counter depth	Configurable 1,4,12,24
Noise level	70 e- rms (single pixel mode) 144 e- rms (charge summing mode)
Power consumption	<1.5 W

1.1. Charge-sharing effect suppression

The MEDIPIX3 chip can be operated in the so-called "Charge Summing Mode" which prevents interacting X-rays from being detected more than once or from not being detected. This is performed by introducing interconnection between pixels so that analogue signals are summed in clusters of four pixels before in-pixel arbitration logic allocates a single hit to the interacting X-ray. This scheme maintains the spatial resolution of 55 μm. Since the signal from four read-out channels is summed before performing signal discrimination, rms noise which adds up quadratically is twice the rms noise present when operating the detector in Single Pixel Mode, where pixels work independently from each others. Charge Summing Mode therefore results in an increased noise level which leads to higher minimum detection levels compared to Single Pixel Mode.

1.2. Colour-mode operation

Each 55 µm x 55 µm pixel of the MEDIPIX3 chip contains 2 discriminators and 2 counters. In Colour Mode, pixels are grouped in 2x2 clusters to form 110 µm x 110 µm detecting elements each containing 8 counters associated with 8 discriminators. Since these 8 discriminators can be programmed with different discrimination thresholds, the detector can perform energy-sensitive (or "colour") X-ray imaging. This Colour Mode with 8 energy bins can be implemented in Single Pixel Operation where pixels operate independently as well as in Charge Summing Mode. Charge Summing Mode will suppress the effect of charge sharing on energy resolution while maintaining a spatial resolution of 110 µm x 110 µm corresponding to the size of the 2x2 pixel clusters.

1.3. Continuous read-write mode

A particular feature of MEDIPIX3 pixel architecture is the presence of two configurable 12-bit counters per pixels. These two counters can be operated separately and associated with a single discriminator. This allows continuous image acquisition without dead-time between frames since one counter can be read-out when the other is counting. In this mode of operation, a dead-time free frame rate of at least 2 kHz can be achieved with a 12-bit pixel depth. Counters can be configured to operate with one bit depth, allowing a dead-time free frame rate of 30 kHz. These two counters can also be combined to provide a pixel bit depth of 24 bits.

1.4. Option for edgeless tiling

MEDIPIX3 I/O architecture includes Though Silicon Via (TSV) pads in addition to wire-bond pads. TSV is a recently developed interconnection technique where signals are transferred from one side of a silicon chip to the other by drilling and metal plating via connections passing through the silicon thickness. The wire-bond pad section of the MEDIPIX3 chip can be diced away and TSV pads used to connect I/O, ground and sensor bias. This feature allows MEDIPIX3 to be four-side buttable and reduces the percentage of inactive area in a multi-chip detector assembly.

4. MEDIPIX3 data acquisition system based on a FPGA development kit

The MEDIPIX3 read-out chip must be interfaced with read-out electronics via LVDS I/O communications lines. A simple read-out set-up, shown in figure 3, is under design at the Diamond Light Source, based on a National Instruments PXI-7952R FlexRIO FPGA development kit including a Virtex-5 FPGA and 128 MB of onboard DDR2 DRAM. This set-up will allow image acquisition from a single MEDIPIX3 detector assembly at full frame rate and testing of the various modes of operation of the chip. It will also allow the development and test of calibration procedures such as equalization of discriminator thresholds across the chip. The chip carrier board is connected to an interface supplying chip bias voltages and performing LVDS to LVTTL conversion of I/O lines to allow the interfacing of the chip with the FPGA card. The FPGA and RAM module is connected to a PCI bus capable of 60 Mbytes/s data transfer rates.

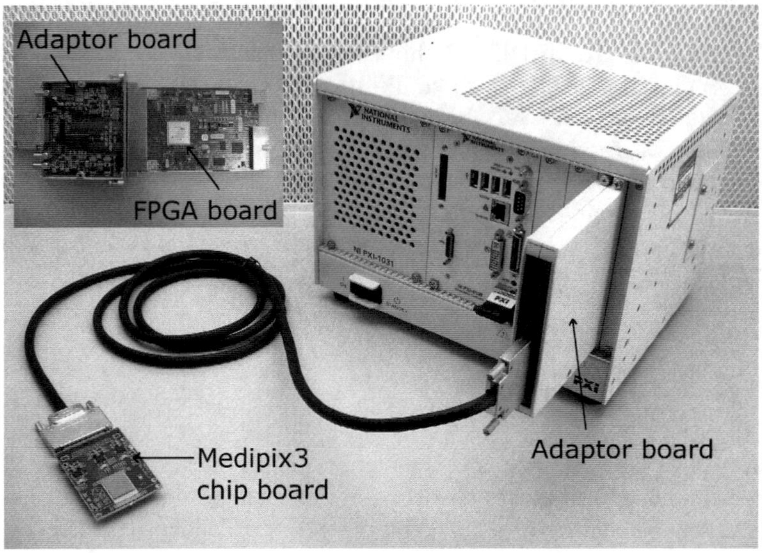

Figure 3. MEDIPIX3 read-out set up based on FPGA development kit

5. Conclusion

The MEDIPIX3 chip developed by the CERN based international collaboration offers a number of innovative design features such as small pixel size with charge-sharing effect suppression, potential for edgeless tiling, high frame rate without dead-time between frames and colour mode operation. Read-out electronics are under development at Diamond Light Source to operate a single-chip detector. The long-term goal of such project is to exploit the features of MEDIPIX3 by developing multi-chip modules covering a large active area. Synchrotron applications such as coherent diffraction could benefit from the pixel size and speed of such a detector. A modular design could also be scaled to fulfil the detector requirements of various applications including SAXS where a fast frame rate photon counting detector would overcome the limitations of CCD-based detectors and allow time-resolved experiments currently performed with gas detectors.

6. References

[1] Berry A, Helsby W, Parker B, Hall C, Buksh P, Hill A, Clague N, Hillon M, Corbett G, Clifford P et al 2003 *Nucl. Instr. Meth. Phys. Res.* A **513** 260-3

[2] http://www.dectris.com

[3] Basolo S, Bérar J-F, Boudet N, Breugnon P, Chantepie B, Clémens J-C, Delpierre P, Dinkespiler B, Hustache S, Medjoubi K et al 2008 *Nucl. Instr. Meth. Phys. Res.* A **589** 268-74

[4] http://medipix.web.cern.ch/MEDIPIX/Medipix2/indexMPIX2.html

[5] Ponchut C, Clément J, Rigal J-M, Papillon E, Vallerga J, LaMarra D and Mikulec B 2007 *Nucl. Instr. Meth. Phys. Res.* A **576** 109-12

[6] Pennicard D 2009 *3D detectors for synchrotron applications* PhD thesis, University of Glasgow

[7] E N Gimenez, Maneuski D, Mac Raighne A, Parkes C, Bates R, O'Shea V, Fleta C, Pellegrini G, Lozano M, Alianelli L et al 2010 *Proc. of 11th International Workshop On Radiation Imaging Detectors (Prague, Czech Republic, 28 June – 02 July 2009)*

[8] Ballabriga R, Campbell M, Heijne E H M, Llopart X and Tlustos L 2007 *IEEE Trans. Nucl. Sci.* **54** 1824-29

New beamline dedicated to solution scattering from biological macromolecules at the ESRF

P Pernot[1], P Theveneau, T Giraud, R Nogueira Fernandes, D Nurizzo, D. Spruce, J Surr and S McSweeney

ESRF, BP 220, Grenoble, France

E-mail: rejma@esrf.fr

A Round[1], F Felisaz, L Foedinger, A Gobbo, J Huet, C Villard and F Cipriani

EMBL Grenoble, BP 181, Grenoble, France

E-mail: around@embl.fr

Abstract. The new bio-SAXS beamline (ID14-3 at the ESRF, Grenoble, France) is dedicated exclusively to small-angle scattering experiments of biological macromolecules in solution and has been in user operation since November 2008. Originally a protein crystallography beamline, ID14-3 was refurbished, still as a part of the ESRF Structural Biology group, with the main aim to provide a facility with 'quick and easy' access to satisfy rapidly growing demands from crystallographers, biochemists and structural biologists. The beamline allows manual and automatic sample loading/unloading, data collection, processing (conversion of a 2D image to a normalized 1D X-ray scattering profile) and analysis. The users obtain on-line standard data concerning the size (radius of gyration, maximum dimension and volume) and molecular weight of samples which allow on-the fly ab-inito shape reconstruction in order to provide feedback enabling the data collection strategies to be optimized. Automation of sample loading is incorporated on the beamline using a device constructed in collaboration between the EMBL (Grenoble and Hamburg outstations) and the ESRF. Semi/automated data analysis is implemented following the model of the SAXS facility at X33, EMBL Hamburg. This paper describes the bio-SAXS beamline and set-up characteristics together with the examples of user data obtained.

1. Introduction

In January 2008 work began on the conversion of ID14-3 from a beamline dedicated to single-wavelength macromolecular crystallography (MX) experiments to one dedicated to small angle X-ray scattering of solutions of biological macromolecules (bio-SAXS). The new facility is intended to complement already existing SAXS facilities at the ESRF which are highly oversubscribed for a wide variety of experiments in the fields of Biology, Chemistry, Material Science etc. With a rapidly growing interest in bio-SAXS from the ESRF's MX user community beam-time on the existing resources was set to become even more sought after. Creating a dedicated to

[1] These authors contributed equally to this work.

Journal of Physics: Conference Series **247** (2010) 012009 doi:10.1088/1742-6596/247/1/012009

bio-SAXS facility alleviates the high rate of over-subscription, improve the throughput of such experiments by removing the need for the station to be optimized for every new user group, which results in more beam-time on the ESRF's existing SAXS end-stations being available for experiments in other fields. An additional benefit of building this new facility at the ESRF is the creation of collaboration with the already established small angle neutron scattering (SANS) facility at the ILL. Our intention is to allow users, via a single proposal, to access both facilities for appropriate experiments. This will allow the complementary information provided by neutron scattering (i.e. contrast variation) and X-ray scattering to be obtained in a single visit to the Grenoble site.

The conversion of ID14-3 has been undertaken in stages. The first of these allowed testing of the proposed design and proof-of-principle before the major renovation of the experimental hutch undertaken during the summer of 2008. The beamline has been in user operation since November 2008. Automation of sample loading has since been implemented on the beam-line using a device constructed in collaboration between the EMBL (Grenoble and Hamburg outstations) and the ESRF.

Access to the experimental facilities offered at the ESRF bio-SAXS beamline are exclusively through the ESRF MX rolling access system allowing data collection within 6 to 8 weeks of the experiments proposal submission [1]. Feedback from users is positive and the number of proposals for experiments and publications resulting from experiments at ID14-3 beamline are increasing.

2. Beamline and set-up characteristics

Beamline ID14 of the ESRF is an undulator beamline on a high-β section housing four independent stations allowing four different user groups to collect data simultaneously. Three undulators, provide high brilliance at 13.3 keV for ID14-1/2/3 fix energy end-stations, as well as limited tunability on ID14-4. The RMS source size is 402 μm (H) × 7.9 μm (V) and divergence 10.7 μrad (H) × 3.2 μrad (V). The beam size at the sample and at the detector during normal operation are 0.7 mm (H) and (V). ID14-3 was initially dedicated to monochromatic macromolecular crystallography, until it's conversion to a fixed energy bio-SAXS end-station. The U23 undulator is a single line undulator and provides the flux for the three fixed-energy stations. The other two undulators, U24.4 and U35 are used only by the tunable end station ID14-4. During the procedure to change energy at station ID14-4 the X-ray flux observed at the other end stations may change, for this reason during movement of the undulators data collection at ID14-3 will pause (at an appropriate point between measurements) and will continue once the undulators have reached the desired position. This procedure is foreseen to be completely automatic so as not to cause disruptions to any of the ID14 stations even during automated or remote access data collection.

The higher energy of this station compared to other fixed energy SAXS stations gives a smaller cone of scatter for the same size of scatterers, meaning the station can record a wider range of momentum transfer ($s = 4\pi \mathrm{Sin}\theta/\lambda$) with a shorter camera length and smaller detector. However, this same effect makes observing the very small angles more difficult requiring the small diameter beamstop to overcome this. ID14-3 has been optimized with the aim of undertaking experiments for macromolecules in solution only. The demand for variable energy solution scattering is low and the hope is that using the standard setup more than 95% of the solution SAXS experiments for the ESRF can be undertaken without modifying the experimental configuration. Much of the remaining 5% could be accommodated with some modifications and it is for this reason the experimental equipment is easily removable to allow users own devices to be installed if needed. However, when appropriate the experiment could be transferred to another of the ESRF SAXS stations if it was agreed it would be more suitable.

2.1. Optical elements

The optical elements of ID14-3 (diamond beam-splitter, Ge (220) monochromator, toroidal mirror) were kept the same as in the original set-up [2]. However, as SAXS experiments require parasitic scattering to be as low as possible new beam conditioning hardware was required. Redesigned attenuators have been installed upstream of the focusing optics in order that any X-rays scattered by the attenuators are rejected by the optics and slits downstream. New beam defining slits just after the mirror (see new elements in green on figure 1) as well as extra guard slits were installed. All hardware for all the beamline is now controlled by ESRF standard controllers (ICEPAPs).

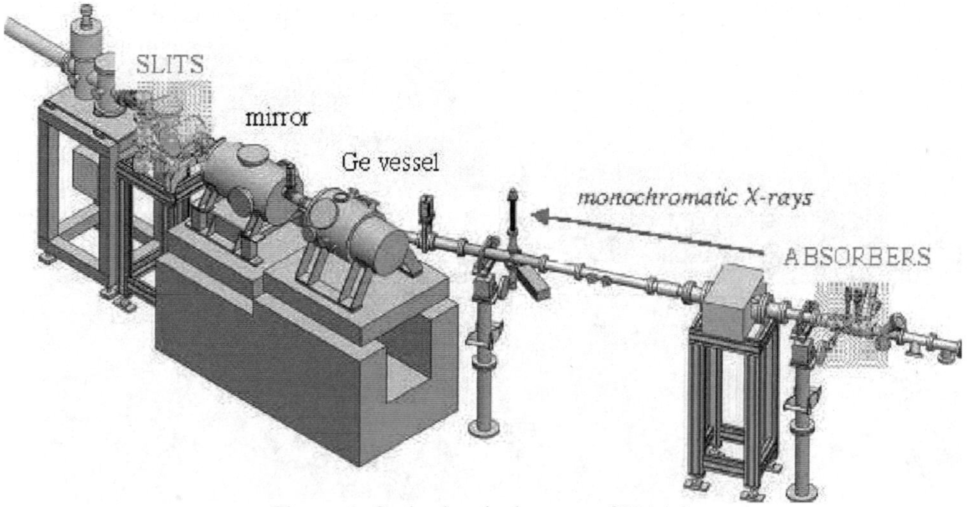

Figure 1. Optics hutch elements of ID14-3.

In order to maximize the available space for a long flight path between sample and detector in the experimental hutch, a “mini” hutch (the first of its type at the ESRF) has been built in the ID14 mezzanine, its photo is shown in figure 2, and houses new beam defining slits, experimental shutter and beam monitoring devices. The ‘mini’ hutch is situated about 9 m downstream from the Optics hutch and about 3 m upstream from the sample position in the Experimental hutch (figure 3).

Figure 2. ‘Mini’ lead hutch in the ID14 mezzanine.

2.2. Set-up description

The experimental hutch of ID14-3 has been completely refitted and now contains a 4 m long marble table which supports the new experimental equipment (figure 3). Two pairs of guard slits are mounted in a slit box prior to the sample exposure unit, which contains the 2 mm diameter quartz capillary (mounted in vacuum) in which the sample is held for measurement. The evacuated flight tube houses a beam-stop with separate (y, z) motor control sitting immediately before the 120 micron thick kapton window behind which is the 2D detector (Pilatus 1M, Dectris) mounted on an independent (y, z) motor stage. The flight tube itself is a modular design for variable length (0.5 to 2.5 m) and is mounted on motorised supports (y-lateral and z-vertical). The sample to detector dinstance for normal operation is 2.5 m giving a momentum transfer range, s ($4\pi \mathrm{Sin}\theta/\lambda$) from 0.05 to 5.8 nm^{-1}. The beamline is equipped with fluorescent screens with integral diodes after all beam conditioning elements to aid alignment. However, to prevent parasitic scattering during the data collections all the beam monitoring devices with the exception of the diode mounted in the beamstop (used for normalisation of X-ray intensity) are normally out of the beam and are only put into the monitoring position when required (using pneumatic actuators).

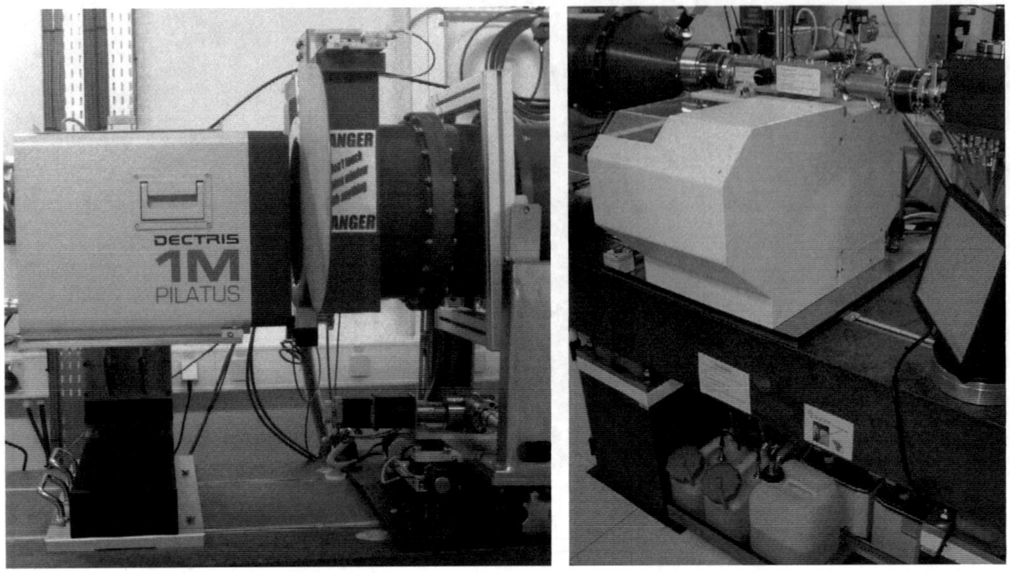

Figure 3. Experimental hutch for bio-SAXS at ID14-3.
Left: Flight tube exit (with safety screen) and Pilatus 1M detector.
Right: Sample changer (with touch screen control) and exposure unit mounted at the flight tube entrance

2.3. Data acquisition software

A new system named BsxCuBE (BioSAXS Customized Beamline Environment) has been developed to control the experiments at beamline ID14-3, integrating the disparate tools a user requires into a single interface (figure 4). The system has been in operation since summer 2009, with continuing improvements based on feedback from users. This new control system allows good quality data to be obtained even by users with little experience of the technique and has shown promising results: it is capable of a successful control of the data acquisition procedure and users feel comfortable using it. BsxCuBE can display, in near real time 2D scattering images being acquired as well as the processed 1D curves. At this stage, post-processing of 1D curves is also possible, such as averaging, cropping, subtracting or filtering according to the radiation damage. Further post-processing operations are planned for integration into the system.

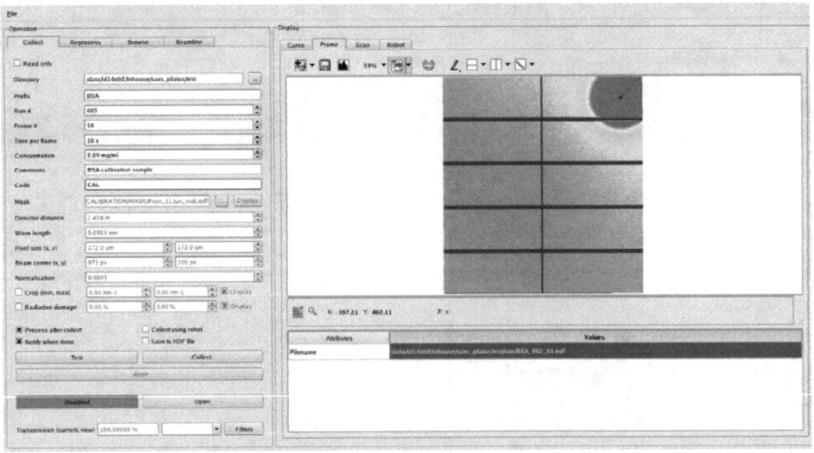

A

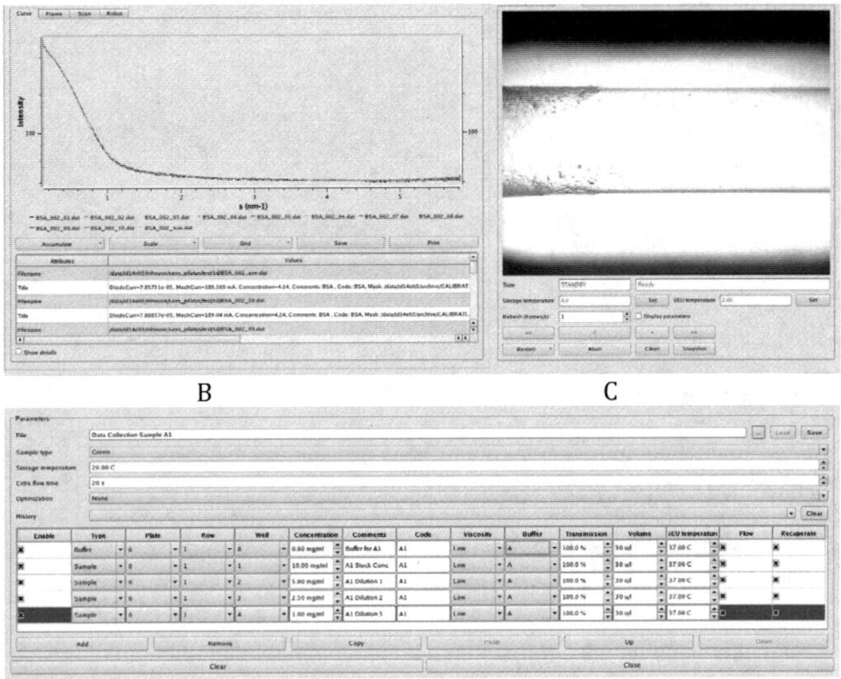

B C D

Figure 4. Snapshots of BxsCuBE software allowing data acquisition, and visualization (using tabs) of 2D data (A) radially integrated 1D curves (B) and control of the sample changer semi (C) and fully automated (D).

BsxCuBE was built using an in-house framework (Bliss Framework 4) which allows creation of sub-windows (Bricks) which are arranged in a main window to form a complete application. The communication with devices is separated into another layer called the control objects which connects to SPEC, ESRF experiment sequencer, and TACO and TANGO, the device server layers. BsxCuBE allows control of different components of the beamline, such as shutters or filters. Ongoing developments of this control software are: i) connection to the Information System for Protein Crystallography Beamlines (ISPyB) to provide automatic logging and sample tracking; ii) automatic batch experiments using the sample changer (basic functionality already available but currently requires sample information to be entered manually); iii) support of a new data storage paradigm called Hierarchical Data Format, following the NeXus convention; iv) integration to display the results of analysis of 1D curves developed by EMBL Hamburg i.e. automatic processing pipeline from 1D data to *ab-initio* models.

2.4. Sample changer

The sample changer is developed in collaboration between the EMBL (Grenoble and Hamburg outstations) and the ESRF. The robot allows automatic sample loading/unloading, cell cleaning: washing with detergent, rinsing with water and drying with air. The system provides and automated 'flow' option, i.e. data collection while flowing the available sample volume through the X-ray beam, thereby mitigating the effects of radiation damage on the data collected. There are two independent thermal controls of sample cell and of sample storage respectively, allowing temperatures in 2-60°C range. Sample storage can contain be optimized with a number of interchangeable configurations. 4 rows of 8 samples in standard PCR tubes (0.2 mL volume) with space for 3 buffers (per row) in eppendorf tubes (1.5 mL volume) or 96 well plates with either shallow or deep wells. The sample changer has place for 3 blocks giving a maximum of 288 separate aliquots for measurement.

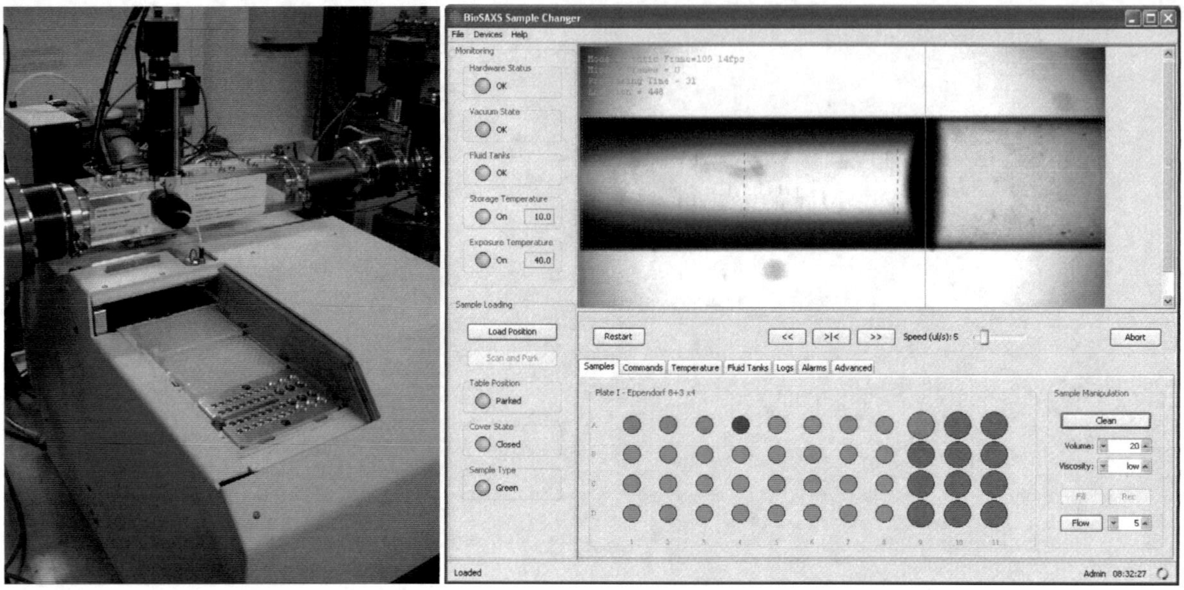

Figure 5: Left: Sample changer with storage tray in the loading position. Right: Sample Changer software interface.

On 12th June 2009 the first official users profited from the new sample changer set-up. From this date the robot has been in a constant use and highly appreciated by all visitors, external or in-house. It works very efficiently, reducing the time needed for cleaning and loading new samples from an average of 5 minutes manually to less than 30 seconds increasing throughput. The automated cleaning also improves reliability as the cleaning is more thorough reducing the risk of cross-contamination between samples.

2.5. Sample preparation laboratory

The half of the available bench space in the ID14 sample preparation room was re-equipped for the use of the bio-SAXS users (photo shown in figure 6). A cooled centrifuge is available for use as is a nanodrop spectrophotometer to enable measurements of the sample concentration immediately prior to the experiment. Final sample preparation and characterisation (dilutions, addition of ligands and or altering the concentration of additives) can be done immediately prior to the measurement and new samples prepared following feedback from data analysis.

Figure 6. Bio-SAXS sample preparation bench with the nanodrop spectrophotometer ND1000 and the centrifuge.

3. Data examples

As of the end of November 2009 more than 50 external user groups as well as a number of in-house researchers have visited ID14-3 for data collection. Papers presenting data collected at ID14-3 have already been published, the first (accepted 3 June 2009) was "Unusual bipartite mode of interaction between the nonsense-mediated decay factors, UPF1 and UPF2" [2]. This paper compared solution scattering data collected At ID14-3 with the theoretical scattering curve calculated from the crystal structure and presented the *ab initio* envelopes calculated from the data showing an elongated shape that accommodates the crystal structures well (see figure 7).

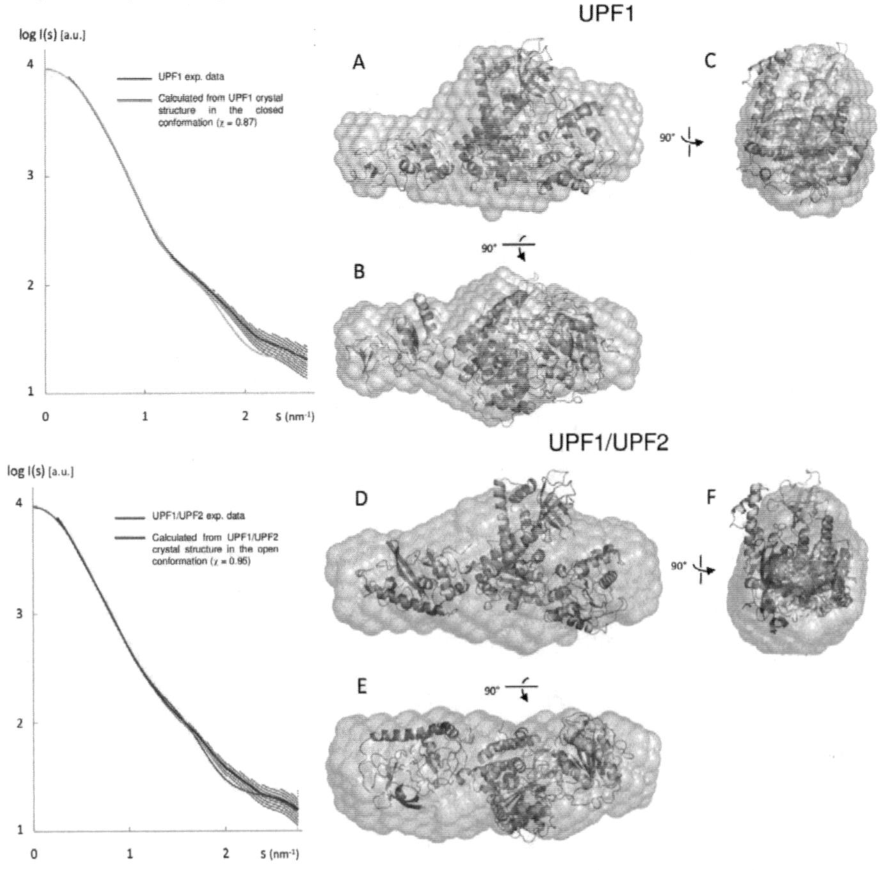

Figure 7. Left: Data for free UPF1(115-914) and UPF1 in complex with UPF2(1105-1207) (top and bottom respectively) showing experimental Scattering data (experimental data points were removed for clarity) with error bars overlaid with fits from theoretical crystal structure. Right: Ab initio envelopes derived from the small angle X-ray scattering data for UPF1(115-914) (A), (B) and (C) and UPF1(115-914)/UPF2(11051207) (D), (E) and (F) with the correspondent crystal structures showing good superimpositions.

Perhaps the clearest example of complimentary SAXS data for structural biology is structure validation, determining the physiologically relevant form of the protein in solution from a number of potential structures obtained from other techniques. An example of this was "The abscisic acid receptor PYR1 in complex with abscisic acid" [4]. The structure of PYR1 was solved using data collected at ID14-4 of the ESRF. However, the PYR1 crystallographic asymmetric unit contained four PYR1 monomers and all biochemistry data from PYR1 suggested that the protein was a dimer and thus the tetrameric form in the asymmetric unit was thought to be an artefact of crystal packing. In order to enable interpretation of the structure with regard to function under physiological conditions SAXS was used to determine which of the possible conformations the protein actually formed in solution. The theoretical scattering curves calculated for possible dimeric ensembles were calculated from the crystal structure and compared to the experimental scattering data (see figure 8). The curve for ensembles A-B (and C-D which is symmetrically identical) produced a good fit to the experimental data demonstrating that ensemble A-B/C-D corresponds to the dimeric form found in solution.

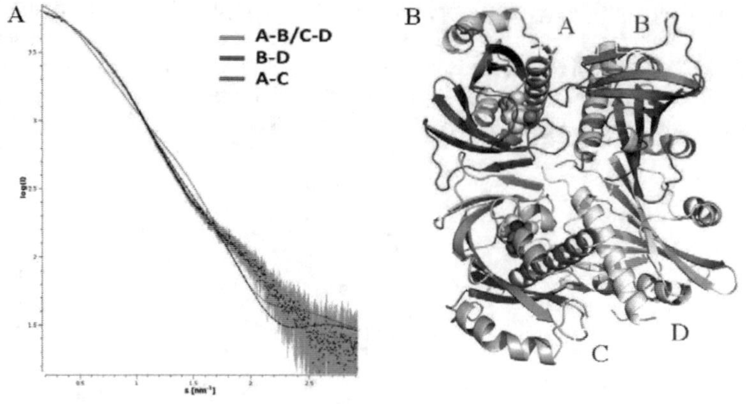

Figure 8. (A) Experimental scattering data (black dots) with error bars calculated from Poisson counting statistics, with fits from ensemble B-D (blue) (χ=1.15), A-C (red) (χ=4.02) A-B/C-D (green) (χ=0.72). (B) tetrameric configuration from the crystal structure with monomers A,B,C and D labelled (physiologically relevant dimer A-B shown in colour).

4. Conclusions

The new facility is intended to complement already existing SAXS facilities at the ESRF. With the high level of automation for data collection and processing at the bio-SAXS station experiments are easy to undertake, reliable and efficient. An additional benefit of building this new facility at the ESRF is the collaboration with the already established small angle neutron scattering facility at the ILL to allow users, via a single proposal, to access both facilities for appropriate experiments.

Feedback from users is positive, the number of proposals for experiments at ID14-3 beamline is increasing and the already published articles show that the bio-SAXS facility of ID14-3 at the ESRF is already providing results for world leading research.

5. References

[1] **http://www.esrf.eu/UsersAndScience/UserGuide/Applying/ProposalGuidelines/MXnon-**BAGproposal/macromolecular-crystallography-rolling-applications

[2] Wakatsuki S, Belrhali H, Mitchell E P, Burmeister W P, McSweeney S M, Kahn R, Bourgeois D, Yao M, Tomizaki T and Theveneau P 1998 *J. Synchrotron Rad.* **5** 215

[3] Clerici M, Mourão A, Gutsche I, Gehring N H, Hentze M W, Kulozik A, Kadlec J, Sattler M and Cusack S 2009 *EMBO J.* **28** 2293

[4] Santiago J, Dupeux F, Round A, Antoni R, Park S-Y, Jamin M, Cutler S R, Rodriguez P L and Márquez J A 2009 *Nature* ***462**, 665-668*

Acknowledgments

The authors acknowledge scientists participating in studies performed on ID14-3 beamline leading to the publications in references [3] and [4] used as experiment examples in this paper. Figures 7 and 8 reprinted by permission from Macmillan Publishers Ltd: [The EMBO Journal]. Figure 9 first published by Nature publishing Group, reference [4].

Recent Development of a 36 meter Small-Angle Neutron Scattering BATAN Spectrometer (SMARTer) in Serpong Indonesia

Edy Giri Rachman Putra[1,2], Bharoto[1], Baek Seok Seong[2]

[1]Neutron Scattering Laboratory, National Nuclear Energy Agency of Indonesia, Gedung 40 Kawasan Puspiptek Serpong, Tangerang 15314, Indonesia

[2]Neutron Science Division, HANARO Center, Korea Atomic Energy Research Institute, 1045 Daedeok-daero, Yuseong-gu, Daejon 305-353, Republic of Korea

Corresponding author: giri@batan.go.id

Abstract. The 36 meter small-angle neutron scattering (SANS) spectrometer BATAN (SMARTer) in Serpong, Indonesia has been revitalised for several years. The work on replacing, upgrading and improving the control system and the experimental method were conducted in order to setup the spectrometer back in operation. Two main personal computers, one for handling and controlling the mechanical system and another one for acquiring neutron data were employed at the spectrometer. The standard and established SANS data reduction and analysis programs, such as GRASP and NIST Igor have been implemented to subtract the raw scattered neutron data with the backgrounds and then analyse the corrected data. The scattering data of ferrofluids samples, Fe_3O_4 and $MnZnFe_2O_4$ have been obtained using SANS spectrometers in BATAN Serpong, Indonesia and HANARO-KAERI, Republic of Korea for inter-laboratory comparison and investigation of proposed research interest. The results were comparable from both scattering data analysis.

1. Introduction

Small-angle neutron scattering (SANS) is a powerful technique for characterizing the static and dynamic-structures of materials in the nanometer scale range of 1 – 500 nm. Information on the average size and its distribution, spatial correlation, as well as shape and internal structure of materials can be obtained from SANS scattering intensity profiles. The quantitative analyses on number or volume density from investigated structures in the surrounding medium can be determined from the absolute scale of scattering intensity. Recently, SANS becomes a valuable technique for characterization in many areas of research, i.e. alloys, ceramics, magnetic, polymers, colloids, vesicles, including biological materials such as proteins, viruses, etc.

A 36 meter SANS spectrometer owned by the National Nuclear Energy Agency of Indonesia (BATAN) was constructed in the neutron-guide hall (NGH) of the 30 MW G. A. Siwabessy research reactor in 1992[1]. It is the second largest SANS spectrometer that operates in the Asia-Pacific region nowadays and detail specification of this instrument has been reported elsewhere[2,3]. Nevertheless, the spectrometer was not well utilized until 2004 due to a shortage in staff members, instrument failures and an undefined long-term research program. A five years in-house work plan was proposed in 2005 to replace, change and upgrade gradually the electronics and mechanics system, computer software of the instrument[4].

The original main computer IBM PS2/70 for data acquisition and control system running on AIX (Advanced Interactive eXecutive) was getting out-of-date and entirely halted at the end of 2003 then the computer finally crashed. Due to the unavailability of the syntax codes for running and controlling all the motor drivers as well as for acquiring the neutron counts from 2-dimensional detector, a new main control system is definitely required. Modifying the data acquisition program based on the time-to-digital converter (TDC) system from the original syntax codes provided by the National Laboratory Risø, Denmark had to be worked out first for acquiring the neutron counts from the detector. In the meantime, a new motor controller program and its hardware were designed and developed to handle motor drivers of the instrument which is controlled by a dedicated computer for device control via ISA (Industry Standard Architecture) interface connection. Shortly, when the new control system has been implemented, re-alignment of the collimation system, detector movements, and beam stop adjustment works were also accomplished. Those initial developments on data acquisition and control system were applied in 2006 for running the SANS experiments[4].

Seeing as the number of SANS experiment[2-8] and the research interest using SANS BATAN spectrometer (SMARTer) significantly increased in the last few years, expanding the applicable conditions and performance of instruments as well as experimental methods are demanded. A major development has been made by improving the data acquisition system, employing the established SANS data reduction and analysis, upgrading the control system and designing a focusing lens device. In this paper, we report the latest development and the inter-laboratory result of SMARTer on magnetic fluids or ferrofluilds samples.

2. Instrument development

2.1. Replacement of the SMARTer control system

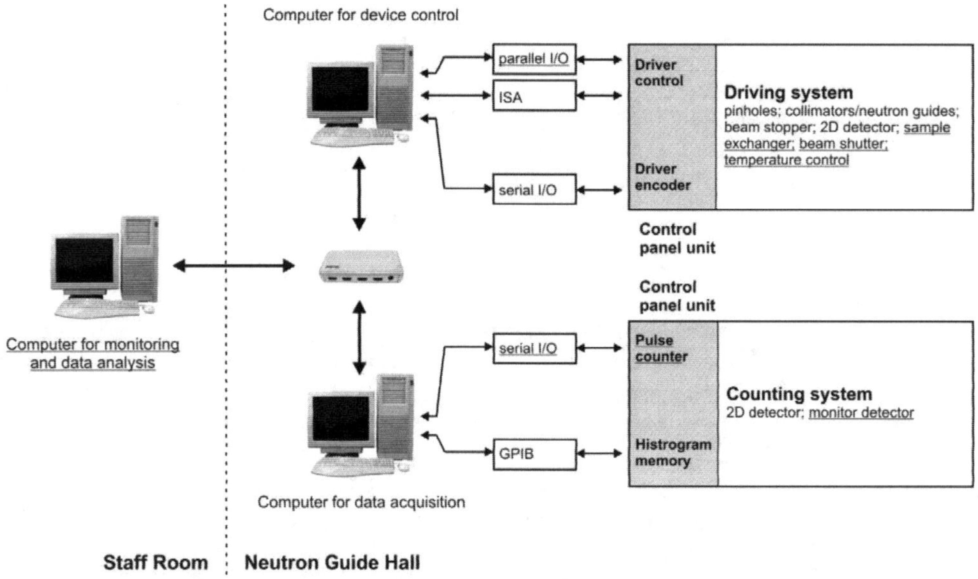

Figure 1. Diagram of the instrument control system of SMARTer

In 2005 the original main computer IBM PS2/70 which run on the AIX operating system was replaced by the IBM compatible computer running on Windows for handling the motor driver system of the instrument[9]. The 17 stepping motors for pinholes, collimators/neutron guides, beam stop and 2-

dimensional main detector movements are controlled by the computer through ISA interface connection. Another dedicated computer running on Linux was also set up for acquiring the neutron counts from the 2-dimensional detector via General Purpose Interface Bus (GPIB) connection. Figure 1 shows the current and in progress development works (underlined) relationship between the devices and the control system of SMARTer. The feedback from encoder of the system is still read by the dedicated computer for device control via the original RS232C interface connection through the CPU system of control panel.

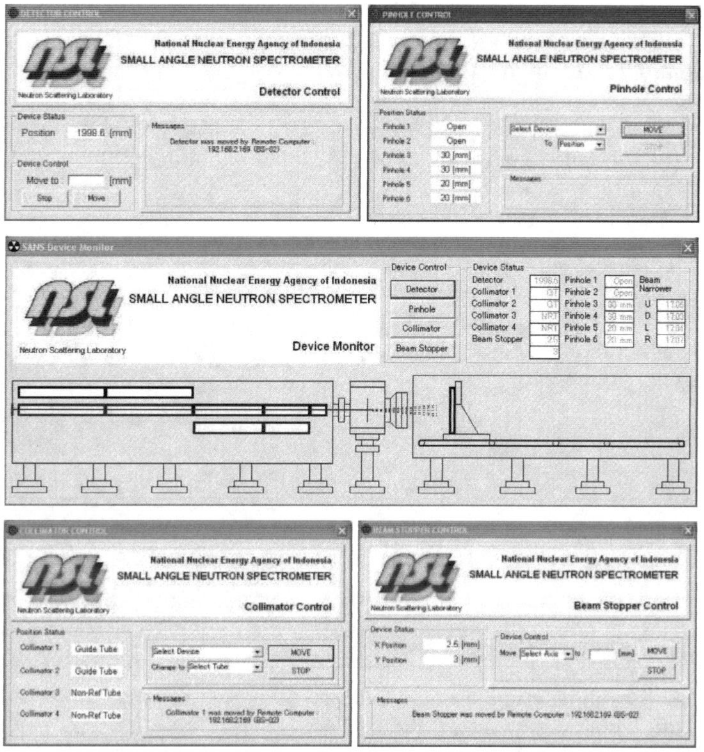

Figure 2. The command menu and current status display on the computer for device control of SMARTer.

A programmable remote control written in Visual Basic was employed in the device control computer for handling the setting configurations of spectrometer. This program is able to display all experimental settings and command menus of spectrometer such as 2-dimensional detector position, pinhole setting, collimator/neutron guide positions, beam stop position and beam narrower setting. The display of the command menu and the current status of the experimental setting configuration are shown in Figure 2.

A microcontroller based on the new control system for SMARTer is now being designed and developed via RS232 (serial) interface connection in order to accommodate and handle more motors or devices. The additional motor devices such as for beam shutter, automatic sample exchanger together with the temperature control for sample environment system are being implemented into the new control system. Recently, a new programmable remote control is also established on the computer for device control and it can also be controlled directly from the computer for data acquisition via a Local Area Network (LAN), Figure 1. The advantage of this new system offers that the SANS experiment based on preset count is able to be run automatically and continuously for ordered samples with a set of sequencers. The experiment can also be monitored from the staff room using a dedicated computer for monitoring and data analysis.

2.2. Real time data acquisition and modifying the raw data format

The previous data acquisition system running on Linux and using a command text mode has been replaced recently by the new one running on Windows with some additional mode applications. The new system which is provided in command window mode is set up for handling the neutron counts from 2-dimensional detector and monitor detector for preset count measurement, respectively via GPIB and serial interface connections, Figure 1. This newest system is also dealing with the computer for device control for automatic and continuous experimental settings via a LAN interface. The additional mode applications of the new system includes 2-dimensional pattern of a real-time data acquisition in linear and log-scales, a command window menu for input parameter of experimental settings, run and experimental numbers for identifying the running experiment and file names, and the format of the output data file, Figure 3.

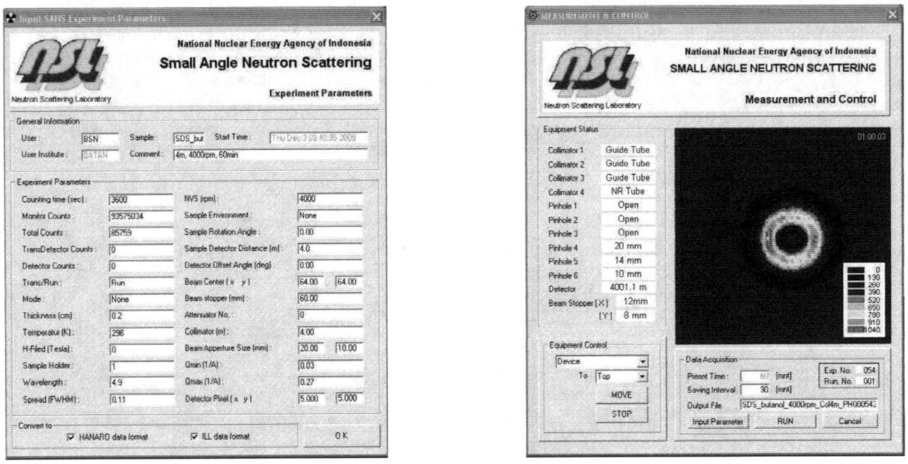

Figure 3. The command menu (left) and the experimental setting display (right) of the new real-time data acquisition system of SMARTer

The standard and established data reduction and analysis package programs have to be applied in order to verify the experimental data of SMARTer for inter-laboratory comparison with other established laboratories. For that reason, the output data file of SMARTer can be saved into HANARO and ILL (the Institut Laue-Langevin) data format. A NIST (the National Institute of Standards and Technology) Igor data reduction[10] modified by T.H. Kim of KAIST (Korean Advance Institute of Science and Technology), Republic of Korea is able to subtract the data of SMARTer in HANARO format, while GRASP (Graphical Reduction and Analysis SANS Program)[11] is able to subtract the data of SMARTer in D22 format of ILL. Other SANS spectrometers such as D11 (ILL), SINQ_SANS I and II (PSI, the Paul Scherrer Institut); V4 (HMI, the Helmholtz-Zentrum Berlin), NG7 & NG3 (NIST), etc. were also supported by GRASP. A NIST Igor data analysis is regularly utilised for analysing the corrected data of SMARTer for solution system such as micellar solution, colloids, polymers and proteins[12].

A set of experiment on ferrofluids samples was examined using SMARTer and HANARO SANS spectrometer. The data was treated using GRASP and then analysed using NIST Igor data analysis for inter-laboratory comparison.

2.3. Focusing collimation

The simulation and analytical calculation for developing a focusing lens collimation of the SANS spectrometer has been carried out. Here, 50 biconcave MgF_2 lenses are required to be implemented for increasing the neutron intensity at the sample position and also for covering a lower momentum transfer q range. The specification of lens is as follows: the curvature radius = 25.0 mm, the center thickness = 1.00 mm, and the outer diameter = 40.0 mm. This calculation is appropriate for the sample to detector distance = 9 m symmetrical optical arrangement or 18 m collimation length with 20 mm ϕ (diameter) sample aperture and a neutron wavelength = 0.57 nm setting configurations. Focusing of the neutron beam by insertion MgF_2 biconcave lens system was achieved and succeeded by S.M. Choi of NIST[13] and now has been implemented at other SANS spectrometers, such as SANS-J of JAEA (Japan Atomic Energy Agency)[14], SANS-U of the University of Tokyo[15], SANS-I of PSI (Paul Scherrer Institut)[16], Quokka of ANSTO (Australian Nuclear Science and Technology Organisation)[17].

3. Experiment for inter-laboratory comparison

The improvement of the experimental method was demonstrated on ferrofluids samples, Fe_3O_4 and $MnZnFe_2O_4$. Those samples were performed using HANARAO SANS spectrometer for inter-laboratory comparison. The detector was set at 4 m from the sample position and shifted laterally for 0.2 m. This setting covers a momentum transfer q range of $0.007 < q$ ($Å^{-1}$) < 0.15 using a neutron wavelength of 6.38Å and sample aperture, ϕ = 12 mm. Detail of the specification of this spectrometer is reported elsewhere[18]. While in order to cover similar q range, the detector of SMARTer was positioned at 4 and 8 m with the neutron wavelength of 4.91 Å and sample aperture, ϕ = 12 mm. The 2-dimensional pattern of ferrofluids samples are shown in Figures 4a, 4c for ferrofluids Fe_3O_4 and Figures 5a, 5c for ferrofluids $MnZnFe_2O_4$. Those scattering data was then subtracted from the backgrounds using GRASP and averaged radially into a 1-dimensional scattering pattern, Figures 4b, 4d and Figures 5b, 5d. All experimental data were then analysed using a NIST Igor data analysis.

A theoretical fractal calculation was fitted on the experimental data of ferrofluids sample of Fe_3O_4 taken using SMARTer as well as HANARO SANS, Figures 4b and d. The calculation is in a good agreement with the experimental data. From the fitting result it can be obtained that a primary particle of magnetite has a radius of c.a. 50 Å which is a nature of the magnetic particle size with a single magnetic domain (superparamagnetic). The mass fractal dimension D related to the structure of the fractal dimension is a positive constant applies only in the regime $\xi >> q^{-1} >> a$, where a is a typical chemical or bond distance related to local structure and ξ is the correlation length or average diameter of a scatter. Here, we obtained that the mass fractal dimension is close to 1 which indicates the magnetite particle of Fe_3O_4 aggregated to form a long-chain or fiber-like structure. A log-normal distribution spherical model calculation of dilute solution was fitted on the experimental data of $MnZnFe_2O_4$ sample. The best fitting values are given as follow; the median radius of a primary particle of magnetite is c.a. 58 Å with the size of distribution σ is c.a. 0.12 – 0.14 which implies that the magnetic particle is entirely monodisperse. The experimental result from SMARTer is definitely as good as the result from HANARO SANS which is analysed by applying the same SANS data reduction and analysis package programs.

4. Conclusion

The SANS BATAN spectrometer, SMARTer has been developed for improving its performance. The latest major developments were as follow (1) Replacement of a new user-friendly data acquisition system with some additional mode applications (2) Upgrading the control system for handling the device motor control using parallel interface connection (3) Setup the SANS data reduction and analysis using established package programs (4) Initiating a focusing SANS development using optical lenses. As described above, the spectrometer is now remarkably more powerful and useful for carrying out many experiments in materials science and biology research.

Acknowledgment

This work was supported by BATAN under the Neutron Beam Utilization of G.A. Siwabessy Reactor for Materials Science Researches (in-house) project 2005 – 2007 and in part by HANARO, KAERI R&D Program of the Ministry of Science and Technology, Republic of Korea 2006 – 2008. The focusing lens development is in part supported by the IAEA through the coordinated research contract 14317/R0, entitled “SANS BATAN: Improvement of the Neutron Intensity by Focusing Optics” 2008

- 2010 and research incentive program - Kementerian Negara Riset dan Teknologi Indonesia 2009. The authors acknowledge to E. Santoso for designing and developing the hardware a new device control system.

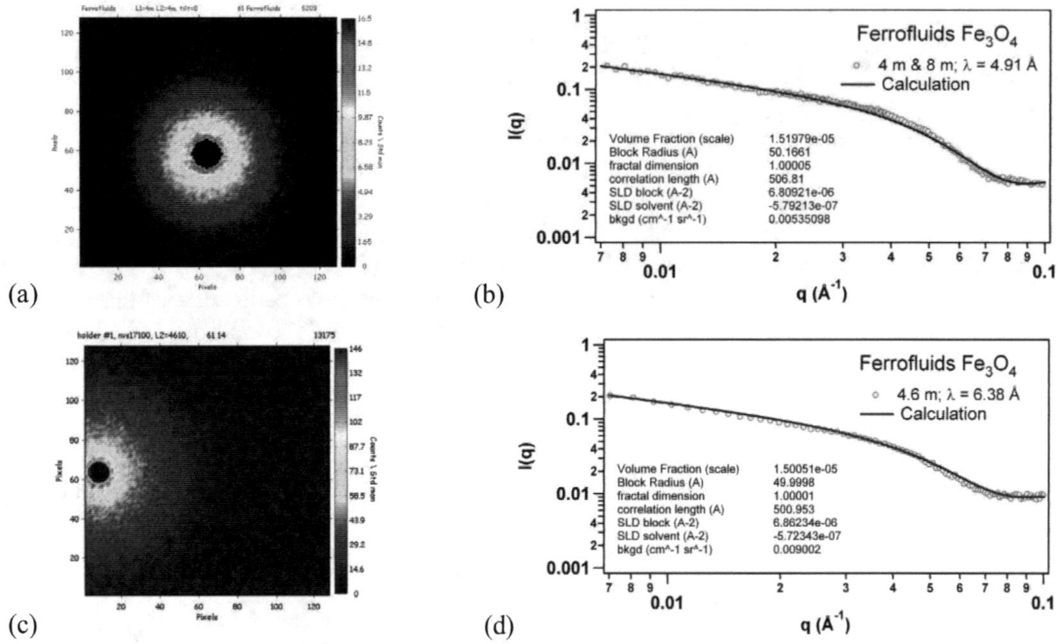

Figure 4. The corrected neutron data on ferrofluids Fe_3O_4 sample in 1- and 2-dimensional profiles were taken using SMARTer: a-b and HANARO SANS: c-d.

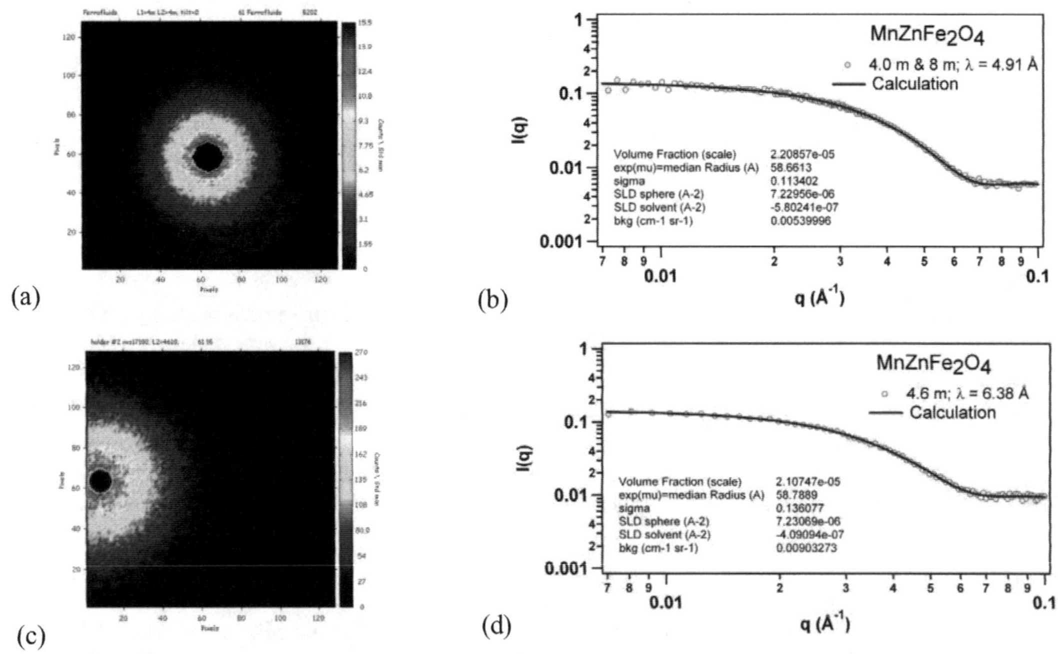

Figure 5. The corrected neutron scattering data on ferrofluids $MnZnFe_2O_4$ sample in 1- and 2-dimensional profiles were taken using SMARTer: a-b and HANARO SANS: c-d.

References

[1] Marsongkohadi, Ridwan, *Neutron News* **7**, 12 - 16 (1996)
[2] E.G.R. Putra, Bharoto, E. Santoso, Y.A. Mulyana, *Neutron News* **18**(1), 23 - 29 (2007)
[3] E.G.R. Putra, A. Ikram, E. Santoso, B. Bharoto, *J. Appl. Cryst.* **40**, s447 - s452 (2007)
[4] E.G.R. Putra, Bharoto, E. Santoso, A. Ikram, *J. Nucl. Instrum. Method Phys. Res.* **A 600**, 198 - 202 (2009)
[5] E.G.R. Putra, A. Ikram, *J. Nucl. Instrum. Method Phys. Res.* **A 600**, 288 - 290 (2009).
[6] E.G.R. Putra, B.S. Seong, A. Ikram, *J. Nucl. Instrum. Method Phys. Res.* **A 600**, 291 - 293 (2009)
[7] E.G.R. Putra, A. Ikram, J. Kohlbrecher, *Pramana J. Phys.* **71**(5), 1045 - 1050 (2008)
[8] A. Patriati, E.G.R. Putra, Proceeding the 2nd International Conference on Mathematics and Natural Sciences (ICMNS) 2008, edited by B. Prijambudi et al., Institut Teknologi Bandung, Indonesia, 806 - 812 (2008)
[9] E. Santoso, E.G.R. Putra, *Atom Indonesia* **34**(2), 79 - 87 (2008)
[10] S. R. Kline, *J. Appl. Cryst.* **39**, 895 - 400 (2006)
[11] C. Dewhurst, "GRASP: Graphical Reduction and Analysis SANS Program for Matlab", http://www.ill.eu/instruments-support/instruments-groups/groups/lss/grasp/, Institut Laue-Langevin 2001 – 2009.
[12] E.G.R. Putra, A . Ikram, *Neutron News* **19**(4), 28 - 33 (2008)
[13] S.M. Choi, J.G. Barker, C.J. Glinka, Y.T. Chen, P.L. Gammel, *J. Appl. Cryst.* **33**, 793 - 796 (2000)
[14] S.Koizumi, H. Iwase, J. Suzuki, T. Oku, R. Motokawa, H. Sasao, H. Tanaka, D. Yamaguchi, H. M. Shimizu, T. Hashimoto, *J. Appl. Cryst.* **40**, s474 - s479 (2007)
[15] S. Okabe, M. Nagao, T. Karino, S. Watanabe, T. Adachi, H. Shimizu, M. Shibayama, *J. Appl. Cryst.* **38**, 1035 - 1037 (2005)
[16] http://kur.web.psi.ch/sans1/sanstechdata.html
[17] E.P. Gilbert, J.C. Schulza, T. J. Noakesa, *Phys.* **B 385-386**(2), 1180 - 1182 (2006)
[18] B.S. Seong, Y.S. Han, C.H. Lee, J.S. Lee, K.P. Hong, K.N. Park, H.J. Kim, *Appl. Phys* **A74**, s201 - s203 (2004)

Journal of Physics: Conference Series **247** (2010) 012011 doi:10.1088/1742-6596/247/1/012011

On the formation of mixed vacancy-copper clusters in neutron-irradiated Fe-Cu alloys

U Birkenheuer[1], A Ulbricht[1], F Bergner[1], A Gokhman[2]

[1] Forschungszentrum Dresden-Rossendorf, P.O. 510119, 01314 Dresden, Germany
[2] South Ukrainian Pedagogical University, 65020 Odessa, Ukraine

E-mail: U.Birkenheuer@fzd.de

Abstract. Setting out from the results found in a set of small-angle neutron scattering (SANS) experiments for neutron-irradiated Fe-Cu model alloys, a rate theory model for the simulation of the irradiation-induced time-evolution of Cu-rich precipitates in these model alloys is presented which follows the idea that the precipitate clusters are *mixed* Cu-vacancy aggregates. This is done by explicitly allowing the defect clusters to absorb vacancies. The resulting Vacancy-Coupled Copper Clustering (V3C) model is calibrated by SANS experiments on two different Fe-Cu model alloys neutron-irradiated at four different doses. Quantitative agreement with the SANS experiments could be achieved by introducing a dependence of the Fe-Cu interface energy on the amount of vacancies in the mixed precipitate clusters. Phenomenologically, this energy can be seen as a function of the weight-percentage of Cu in the iron matrix. An empirical expression for this dependence is suggested. In addition, the new V3C model is used to gain some preliminary insight into the time-evolution of the chemical composition of the mixed Cu-vacancy clusters, confirming qualitatively the experimental findings. The relation of our ansatz to the heterogeneous Cu-precipitation mechanism proposed by others for neutron-irradiated Fe-Cu alloys of low Cu content is discussed.

1. Introduction

Rate theory (RT) modeling plays an important role in the field of neutron-irradiation induced damage of reactor pressure vessel (RPV) steels. In fact, regardless of the inherent limitations RT is one of the few methods which allows – in a multi-scale modeling framework – to close the gap between the processes happening on the atomic level and the properties accessible to experiments such as transmission electron microscopy (TEM) or small-angle neutron scattering (SANS). In particular RT is capable of modeling defect evolutions for periods of up to 40 years and more of reactor operation time. There is a continuing interest in the application of RT models to the Fe-Cu system under neutron irradiation [1-5]. To calibrate such models experimental data on both, reactor pressure vessel steels as well as especially designed Fe-based model alloys, are vital. As part of the experimental program related to the international REVE (reactor for virtual experiments) project a set of well-defined model alloys was fabricated and neutron-irradiated under several irradiation conditions [6]. Within the integrated project PERFECT (prediction of irradiation damage effects on reactor components) of the 6th European Framework program (FP6) these samples were investigated by a series of complementary experimental techniques, including TEM, SANS, PAS (positron annihilation spectroscopy) and TAP (tomographic atom probe). The details of the SANS experiments will be reported in a companion paper [7]. Four different neutron fluences were investigated at one and the

Journal of Physics: Conference Series **247** (2010) 012011 doi:10.1088/1742-6596/247/1/012011

same neutron flux. The material matrix comprises pure iron and a set of binary, ternary and quaternary model alloys with compositions quite close to real reactor pressure vessel (RPV) steels. In this investigation, we will focus on the two binary Fe-Cu model alloys Fe-0.3%Cu and Fe-0.1%Cu.

The aim of the present paper is to suggest modifications to the standard RT models for Cu-rich RPV steels which allow to reproduce the whole set of SANS data, in particular the volume fraction of the defect clusters and the peak radius of the size distribution function (SDF), for both model alloys and each of the four irradiation conditions applied (see table 1). The dependence of the SANS data on the neutron fluence was found to be quite complex, suggesting that a pure Cu precipitation model might not be sufficient to explain the observations. And in fact, as already reported by Radiguet *et al.* [8] it is impossible to parameterize a rate theory model for *pure* Cu precipitates which would be able to describe the experimental results even qualitatively. A heterogeneous nucleation process is thus suggested in Ref [8] with the Cu-containing precipitate clusters being formed in a second stage on top of the nanovoids primarily produced by the applied ion irradiation. The same result, namely the formation of mixed Cu-vacancy agglomerates, can also be obtained by successive vacancy absorption. Therefore, we investigated whether explicitly taking into account the absorption of iron vacancies by the copper-rich precipitates allows to set up a RT model which is able to reproduce the size distributions observed in the SANS experiments.

Table 1. Irradiation conditions for the Fe-0.1%Cu and Fe-0.3%Cu model alloys.

Temperature, T	300 °C
Neutron flux, φ (E > 1 MeV)	0.95×10^{18} n/m^{2}s
Dose rate, G_{dpa} (E > 1 MeV)	1.40×10^{-7} dpa/s
Dose	0.026, 0.051, 0.10, 0.19 dpa

2. The rate theory model

Standard rate theory models for the formation of copper-rich precipitates in irradiated reactor pressure vessel steels as described, for example, in Refs. [1-2,4,9-10] consist of the following three essentials: balance equations for the evolution of each of the mobile point defects, vacancies and self-interstitial atoms (SIAs), a set of master equations for the evolution of the immobile defect clusters up to a given maximum size and a model for the irradiation enhancement of the Cu mobility in the iron matrix. Additionally, the conservation of the total number of Cu atoms is exploited. This model is denoted as CD-P model in Ref. [2]. A more realistic picture of the irradiation induced processes is obtained by taking into account the formation of voids and SIA loops as is done (for example) in the CD-P-VIC model described in Ref. [2]. Yet, the basic problems remain, since the Cu-containing precipitates are still described as if they were *pure* Cu clusters. Thus we will focus on the less complex CD-P model here. The balance equations in this model read

$$\frac{dC_A}{dt} = G_A - k_{\text{vi}} C_{\text{v}} C_{\text{i}} - k_A C_A \ , \quad A = \text{v,i} \ , \tag{1}$$

where C_{v} and C_{i} are the concentrations of the vacancies and SIAs, respectively, measured in point defect per lattice site. Here, G_A are the generation rates of the point defects due to irradiation, $k_{\text{vi}} C_{\text{v}} C_{\text{i}}$ is the recombination rate of the vacancies and self-interstitials, and $k_A C_A$ the loss rates of each type A of point defect at external sinks such as grain boundaries or dislocations in the iron matrix where the decay rates k_A are proportional to the dislocation density ρ. The classical master equations for these clusters are of the general form (e.g. Refs. [1,4,9,11])

$$\frac{dC_n}{dt} = J_{n-\frac{1}{2}} - J_{n+\frac{1}{2}} \quad \text{with} \quad J_{n+\frac{1}{2}} = k_n \cdot \left(C_{\text{Cu}} C_n - K_{n+1}(T)\, C_{n+1} \right) \tag{2}$$

where C_n is the concentration of the n-atomic defect clusters, measured in cluster per lattice site. Here, clusters containing two or more Cu atoms are assumed to be immobile. The velocity $J_{n+\frac{1}{2}}$ for the growth reaction $Cu_n + Cu \leftrightarrow Cu_{n+1}$ consists of an absorption part $k_n C_{Cu} C_n$ and an emission part $\alpha_{n+1} C_{n+1} := k_n K_{n+1}(T) C_{n+1}$ which is set up by the detailed balance principle. The equilibrium constants $K_n(T)$ do neither depend on the mobility of the solute monomers, nor on the concentrations of the involved species. They are solely determined by the thermodynamic o the cluster growth reactions at the given temperature *T*. In particular, none of the details of the more sophisticated CD-P-VIC model has any impact on these constants.

Besides the concentration $C_{Cu} \equiv C_1$ of the Cu atoms in the iron matrix and the number of clusters per lattice site of given size *n* and adjacent sizes $n \pm 1$, the reaction rate $\frac{d}{dt} C_n$ depends on a couple of material parameters $p_1, p_2, \ldots$, the most important ones being the coherent specific Fe-Cu interface energy $\gamma_{\text{Fe-Cu}}$ entering the equilibrium constants $K_n(T)$ and the irradiation enhanced Cu diffusion coefficient D^*_{Cu} entering the reaction rate coefficients k_n. Assuming a vacancy assisted diffusion mechanism for the Cu atoms the irradiation enhancement of the Cu diffusivity can be modeled by [2,12]

$$D^*_{Cu} = D_{Cu} \cdot \frac{C_v(t)}{C_v^{eq}} \quad , \tag{3}$$

where C_v is the actual, irradiation-induced concentration of the vacancies in the iron matrix, while C_v^{eq} and D_{Cu} are the thermal vacancy concentration and the copper diffusion coefficient in the unirradiated material, respectively.

Within the above framework the system of point defects and the copper subsystem are often treated independently [2,4,10] by replacing $C_v(t)$ with the steady-flux solutions C_v^{sf} of the balance equations (1). This is justified by the fact, that the typical time-scale of the point defect system is orders of magnitudes smaller than that of the copper system. Yet, as already mentioned above, there are indications for the copper precipitates in Fe-Cu model alloys of moderate Cu content actually being mixed defect clusters which contain copper and vacancies together. These indications come from both, theoretical studies [13-15] and experimental data, e.g. based on the ratio of magnetic and nuclear contributions to SANS (the so-called A-ratio) [9,16]. Recent results of atom probe experiments on Cu containing Fe-based model alloys [17] support this view. They also found the precipitates not being pure Cu clusters.

The structural optimizations (based on empirical inter-atomic potentials) reported in Refs. [14] and [15] show that the uptake of solute Cu atoms and/or vacancies by mixed Cu-vacancy clusters is always energetically favored (positive *binding* energies throughout), as is the formation of such clusters from pure voids and pure Cu clusters. We repeated this kind of structural optimizations with the more recent inter-atomic Fe-Cu potential developed by Pasianot *et al.* [18] and found the same overall behavior of the energy potential surface of the Cu_nV_m clusters as in the investigations before [14-15].

To allow for such mixed Cu-vacancy clusters, the defect clusters must be able to absorb vacancies. This is explicitly taken into account in our V3C (Vacancy-Coupled Cu-Clustering) model by letting the defect clusters act as additional vacancy sinks. To this end, the simple decay rate k_v in equation (1) is replaced by

$$k_v(t) = k_v^0 + \Delta k_v(t) = z_v D_v \rho \; + \; 4\pi D_v \sum_{n \geq 2} R_n C_n(t) / V_{Fe} \quad . \tag{4}$$

Here, D_v is the diffusion coefficient of the vacancies in the iron matrix, z_v the dislocation sink strength bias for vacancies, R_n the radius of an *n*-atomic defect cluster, and V_{Fe} the atomic volume of bcc iron. The new sink term $S_v = k_v(t) C_v$ consists of a static contribution $S_v^0 = k_v^0 C_v$ due to the given dislocation network in the material, and a dynamic contribution $\Delta S_v = \Delta k_v(t) C_v$ which depends on the actual defect cluster distribution (and describes diffusion limited vacancy absorption by spherical defect clusters). Because the latter term imposes the time modulation of the copper subsystem onto the point defect system, the balance equations cannot be solved independently of the copper subsystem anymore. Instead, the balance equations and the master equations have to be solved simultaneously. In fact, the balance equations can still be solved in an adiabatic fashion with a slowly varying quasi-steady-flux solution $C_v^{\text{q-sf}}(t)$ being used in equation (3). However, regardless of this detail, the new V3C model constitutes a twofold coupling between the point defects and the copper subsystem, a forward coupling which is mediated via the vacancy dependence of the diffusion coefficient D^*_{Cu} entering the master equations (2) and a backward coupling which is caused by the Cu-cluster dependence of the vacancy decay rate $k_v(t)$ entering the balance equations (1). For the self-interstitial atoms only a static contribution $S_i = k_i C_i$ to the sink term is considered, because the (oversized) Cu clusters are assumed not to be able to absorb SIAs in any relevant amount.

Table 2. Material parameters adopted for the rate theoretical simulation. Here k is the Boltzmann constant and T the irradiation temperature (in K)

Parameter	Value	Reference
Interstitial migration energy, $E_{\text{m,i}}$	0.4 eV	[19]
Interstitial pre-exponential factor, $D_{0,\text{i}}$	0.05×10^{-4} m²/s	[19]
Interstitial diffusion coefficient at 300 °C, D_i	1.52×10^{-9} m²/s	$D_\text{i} = D_{0,\text{i}} \exp(-E_{\text{m,i}}/kT)$
Vacancy migration energy, $E_{\text{m,v}}$	1.3 eV	[19]
Vacancy pre-exponential factor, $D_{0,\text{v}}$	0.5×10^{-4} m²/s	[19]
Vacancy diffusion coefficient at 300 °C, D_v	1.85×10^{-16} m²/s	$D_\text{v} = D_{0,\text{v}} \exp(-E_{\text{m,v}}/kT)$
Copper migration energy, $E_{\text{m,Cu}}$	2.29 eV	[2]
Copper pre-exponential factor, $D_{0,\text{Cu}}$	7.2×10^{-6} m²/s	this work
Copper diffusion coefficient at 300 °C, D_{Cu}	5.27×10^{-26} m²/s	$D_{\text{Cu}} = D_{0,\text{Cu}} \exp(-E_{\text{m,Cu}}/kT)$
Vacancy formation energy, $E_{\text{f,v}}$	1.64 eV	a
Non-configurational vacancy entropy, ΔS_v	3 k	a
Thermal vacancy concentration at 300 °C, C_v^{eq}	7.62×10^{-14}	$C_\text{v}^{\text{eq}} = \exp(\Delta S_\text{v}/k - E_{\text{f,v}}/kT)$
Copper demixing temperature, Ω	6255 K	[2]
Non-configurational copper entropy, ΔS_{Cu}	0.866 k	[2]
Solubility limit of Copper at 300 °C, $C_{\text{Cu}}^{\text{eq}}$	4.33×10^{-5}	$C_{\text{Cu}}^{\text{eq}} = \exp(\Delta S_{\text{Cu}}/k - \Omega/T)$
Vacancy-interstitial recombination radius, r_{rec}	5.74 Å	[19]
Recombination rate constant at 300 °C, k_{vi}	9.31×10^{11} / s	$k_{\text{vi}} = 4\pi r_{\text{rec}} (D_\text{v}+D_\text{i})/V_{\text{Fe}}$
Dislocation density, ρ	0.9×10^{14} / m²	[7]
Interstitial sink strength bias, z_i	1.2	[2]
Interstitial decay rate at dislocations at 300 °C, k_i	1.64×10^5 / s	$k_\text{i} = z_\text{i} D_\text{i} \rho$
Vacancy sink strength bias, z_v	1.0	[2]
Vacancy decay rate at dislocations at 300 °C, k_v	$1{,}67 \times 10^{-2}$ / s	$k_\text{v} = z_\text{v} D_\text{v} \rho$
Cascade efficiency factor, η	0.366	[20]
Fraction of cascade point defect clusters, $f_{\text{cl},A}$	0.5	[21] [22]
Point defect production rate, G_A	2.56×10^{-8} / s	$G_A = (1 - f_{\text{cl},A})\eta G_{\text{dpa}}$

[a] For temperatures of 270-300 °C the listed parameters are essentially equivalent to the parameter pair $E_{\text{f,v}} = 1.6$ eV [2] and $\Delta S_\text{v} = 2.2\ k$ [23].

Although the absorption of vacancies by Cu-rich precipitates is fully taken into account, the suggested V3C model contains some idealizations. The precipitate clusters are assumed to be perfect absorbers whose properties do not change when taking up vacancies. In particular, the Cu clusters do neither grow upon vacancies absorption nor do they reemit any vacancies in our model. Both effects could have some impact on the cluster formation, but to include them in a rate theory model requires to describe the time-evolution of the chemical composition of each defect cluster. Work is in progress to setup such an extended rate theory. The aim of the present study is to demonstrate that already with the above idealizations the V3C model is able to reproduce the experimental observations reasonably.

3. Computations

The irradiation conditions of the investigated model alloys Fe-0.1%Cu and Fe-0.3%Cu are summarized in table 1, the material parameters for the RT simulation in table 2. The bcc lattice constant a_{Fe} of Fe is chosen to 2.8665 Å, that of bcc Cu to a_{Cu} = 2.9607 Å [24]. The parameters listed here are essentially the same as those used in our previous studies on Cu precipitation in model alloys of about 0.4% Cu [10,25]. The thermodynamic data for the Cu precipitation is taken from Ref. [2]. However, the pre-exponential factor for Cu diffusion suggested therein (0.63×10^{-4} m²/s) had to be reduced by about one order of magnitude, in order to set the time-scale properly such that the Ostwald ripening stages of the defect clusters start at the experimentally observed doses (see below). The resulting value of 5.27×10^{-26} m²/s for the thermal copper diffusion coefficient D_{Cu} at 300°C is reasonably close to the values obtained from the Arrhenius parameters used in other simulation studies, but still noticeably larger than the extrapolations of the available experimental data down to 300 °C (see table 3 for details). In particular, our Cu diffusion parameter is very close to the one obtained from MD (molecular dynamics) simulations on vacancy assisted Cu diffusion in bcc iron based on the ACKLAND97 embedded atom potential for Fe-Cu [24].

The interface energy $\gamma_{\mathrm{Fe\text{-}Cu}}$ of the defect clusters in the Cu-rich model alloy with 0.3wt% Cu are calculated using the entropy extension of the Cahn-Hilliard expression suggested by Mathon *et al.* [9],

$$\gamma_{\mathrm{Fe\text{-}Cu}}^{0.3\%} = \gamma_{\mathrm{CH}} := 1.08k\left[\tfrac{1}{2}\Omega - T(1+\tfrac{1}{2k}\Delta S_{\mathrm{Cu}})\right]/a_{\mathrm{Cu}}^2 \quad , \tag{5}$$

where Ω and ΔS_{Cu} are the demixing temperature and the non-configurational substitution entropy of Cu in Fe which are also used to calculated the solubility limit $C_{\mathrm{Cu}}^{\mathrm{eq}}$ of Cu (see table 2). With the parameters given in table 2 this results in $\gamma_{\mathrm{Fe\text{-}Cu}}^{0.3\%}$ = 0.39 J/m² at 300 °C. The interface energy of the Cu-poor model alloy had to be fitted (see next section).

Table 3. Thermal diffusivity of Cu in α-Fe extrapolated to 300 °C (for the meaning of the symbols see table 2)

Method	$E_{\mathrm{m,Cu}}$ [eV]	$D_{0,\mathrm{Cu}}$ [m²/s]	$D_{\mathrm{Cu}}^{300°\mathrm{C}}$ [m²/s]	Reference
RT simulation	2.29	0.63×10^{-4}	46.0×10^{-26}	[2]
RT simulation	2.29	0.072×10^{-4}	5.27×10^{-26}	This work
RT simulation	2.53	7.08×10^{-4}	4.01×10^{-26}	[26]
MD simulation	2.31	0.052×10^{-4}	2.54×10^{-26}	[27]
Exp., α-Fe	2.56	3.9×10^{-4}	1.20×10^{-26}	[28]
Exp., α-Fe(ferro)	2.53	0.47×10^{-4}	0.26×10^{-26}	[29]
Exp., α-Fe(para)	2.94	300×10^{-4}	0.04×10^{-26}	[30] [a,b]

[a] No Arrhenius-like behavior was observed for α-Fe(ferro) in that study.
[b] The experimental data points reported in Ref. [31] are very close to the Arrhenius fits for γ-Fe and α-Fe(para) given by Salje *et al.* [30].

4. Results and Discussion

The experimental data used to find a suitable parameter set for the rate theory model are shown in figures. 1 and 2. While the *simulated* cluster size distribution functions of the Cu-rich model alloy Fe-0.3%Cu exhibit distinct maxima at all four neutron fluences with peak radii and volume fractions quite close to the experimental values, no copper precipitation could be observed with the Cahn-Hilliard interface energy for the Cu-poor model alloy Fe-0.1%Cu. Instead, the concentration of the Cu clusters at the critical nucleation size of about 25 Cu atoms turned out to be seven orders of magnitude smaller than for Fe-0.3%Cu (with a critical nucleation size of about 10 atoms), in total agreement with classical nucleation theory, and the simulated system remained in the deterministic growth stage until cluster sizes far larger than the experimentally observed ones were reached. Precisely the same observations were made by Radiguet *et al.* [8] who report Cu cluster densities in the order of 10^{16} / m³ when applying the CD-P-VIC rate theory model to irradiated Fe-0.1%Cu model alloys. They suggest a heterogeneous nucleation mechanism (see next section for a more detailed discussion on that issue).

Because the equilibrium constants $K_n(T)$ determine the critical cluster size and the concentration of the critical clusters for a given Cu content, there is little choice in adopting a RT model based on the master equations (2), as does our V3C model and also the CD-P-VIC model used by Radiguet *et al.* The thermodynamics of the cluster formation must be different. One way to solve the problem is to let the interface energy depend on the composition of the defect clusters which in turn depends on the amount of Cu in the model alloy. Alternatively, one could also modify the solubility limit Cu in the Fe matrix, as a function of the defect cluster composition. Yet, since the equilibrium constants are much more sensitive to the interface energy than to the solubility limit, we decided for the interface energy.

By reducing the Fe-Cu interface energy for Fe-0.1%Cu to 73% of the Cahn-Hilliard value, i.e. to 0.29 J/m^2 at 300 °C, we were able to reproduce the experimentally observed peak radii and volume fractions at all four neutron fluences for the Fe-0.1%Cu system as well (see figures. 1 and 2). All other parameters are the same as for Fe-0.3%Cu. In particular, there is only one further adjusted parameter beside the interface energy reduction for Fe-0.1%Cu, the common pre-factor $D_{0,\mathrm{Cu}}$ for the thermal Cu diffusivity in both alloys. A total of sixteen independent experimental data points could be reproduced this way (see figures. 1 and 2) giving confidence that the chosen parameterization is reasonable. Based on the above observations we suggest, to model the temperature and composition dependence of the Fe-Cu interface energy in a multiplicative way, by following Mathon *et al.* [9] for the temperature dependence but adopting a simple linear dependence on the weight percentage w of Cu in the material:

$$\gamma^{w\%}_{\mathrm{Fe\text{-}Cu}}(T) = \gamma_{\mathrm{CH}}(T) \times (1.35w + 0.595) \quad \text{for} \quad 0.1 \leq w \leq 0.3 \qquad (6)$$

with γ_{CH} as defined in equation (5).

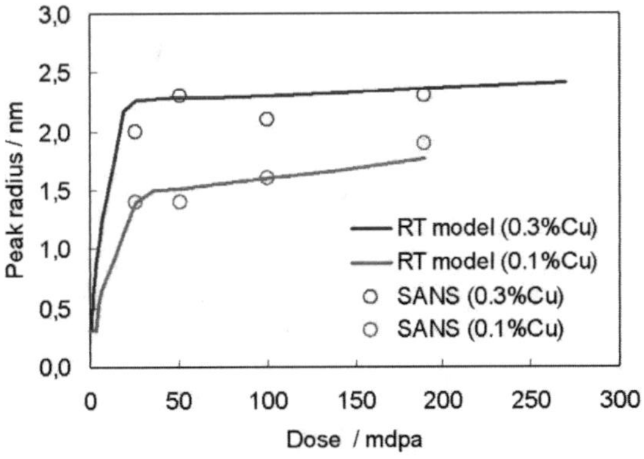

Figure 1 Calculated peak radius of the Cu cluster distribution as a function of dose and comparison with experimental results obtained by SANS for both model alloys.

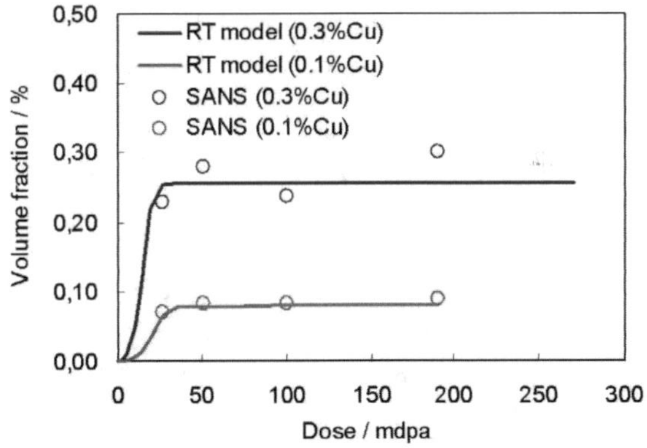

Figure 2 Calculated volume fraction the Cu clusters as a function of dose and comparison with experimental results obtained by SANS for both model alloys.

Taking into account that both model alloys were subject to the same irradiation-induced vacancy production rate it is likely that on average the defect clusters of the Cu-poor model alloy take up more vacancies than the defect clusters in the Cu-rich material (simply because the term $D^*_{\mathrm{Cu}} C_{\mathrm{Cu}}$ is about three times lower in Fe-0.1%Cu than in Fe-0.3%Cu while $D_v C_v$ is roughly the same in both samples). Thus, in view of the quite substantial oversize of copper compared to iron the suggested reduction of the repulsive interface energy with increasing amount of vacancies in the defect clusters seems reasonable. On the other hand, at 300 °C, interface energies in the order of 0.4 J/m^2 are reported for bare Cu clusters [2,9] while for bare voids (vacancy clusters) the interface energies lay in the range of 1-2 J/m^2 [19]. The Cahn-Hilliard expression from equation (5), for example, applied to an iron-void interface (using $E_{f,v}$ = 1.64 eV and ΔS_v = 3 k from table 2) yields a value of 1.47 J/m^2. Thus, a first guess for the interface energy of a mixed cluster would be some value in between these two limiting cases, but a thorough analysis of the underlying thermodynamics actually reveals that in the present case, the really relevant quantity, the effective interface energy for Cu exchange between mixed defect clusters and the iron matrix which is defined via the Gibbs-Thomson relation

$$\frac{\partial}{\partial n}\left[\gamma_{\text{Fe-Cu}}(x)\,A(\text{Cu}_n\text{V}_m)\right] =: \frac{2\mu^{\text{Cu}}_{\text{Fe-Cu}}(x)V_{\text{Fe}}}{R(\text{Cu}_n\text{V}_m)} \quad \text{with} \quad x = \frac{n}{n+m} \tag{7}$$

where $A(\text{Cu}_n\text{V}_m)$ is the surface area of a (spherical) mixed Cu_nV_m cluster, $R(\text{Cu}_n\text{V}_m)$ its radius and x the chemical composition of the cluster (measured as the fraction of Cu atoms here). It is sort of a chemical potential, rather than an energy, and can exhibit a quite different x-dependence than the interface energy $\gamma_{\text{Fe-Cu}}(x)$ itself. In particular, $\mu^{\text{Cu}}_{\text{Fe-Cu}}(x>0)$ can be lower than the value $\mu^{\text{Cu}}_{\text{Fe-Cu}}(x=0)$ for bare Cu clusters even if $\gamma_{\text{Fe-Cu}}(x)$ behaves the opposite.

4.1. On the composition of the defect clusters

The V3C model enables us to monitor the total amount of vacancies and Cu atoms taken up by all defect clusters. Based on this information a rough estimate on the average chemical composition of the precipitates can be given, as done in figure. 3 where the average atom (or mole) fraction x_{Cu} of copper in the defect clusters is shown as a function of the neutron damage. It is computed as the ratio between the total vacancy and solute Cu atom losses

$$L^{\text{v}}_{\text{p}}(t) := \int_0^t \Delta k_{\text{v}}(t')C_{\text{v}}(t')\,dt' \quad \text{and} \quad L^{\text{Cu}}_{\text{p}}(t) := \sum_{n\geq 2} n\,C_n(t) = C_{\text{Cu}}(0) - C_{\text{Cu}}(t) \quad , \tag{8}$$

respectively. Interestingly, a clear maximum in the Cu fraction is found for both model alloys, Fe-0.1%Cu and Fe-0.3%Cu, at about 20-30 mdpa, in nice qualitative agreement with the compositions extracted from the measured A-ratios of the SANS data using the relation [16]

$$A^{\text{SANS}} = \left(\frac{6}{9.5 - 7.7x_{\text{Cu}}}\right)^2 + 1 \quad . \tag{9}$$

(see figure. 3). Growing initially essentially as voids, the defect clusters smoothly transform into Cu clusters with increasing neutron dose until the copper in the iron matrix is subsumed and the Ostwald ripening stage starts where only vacancies can still be taken up in total by the precipitates.

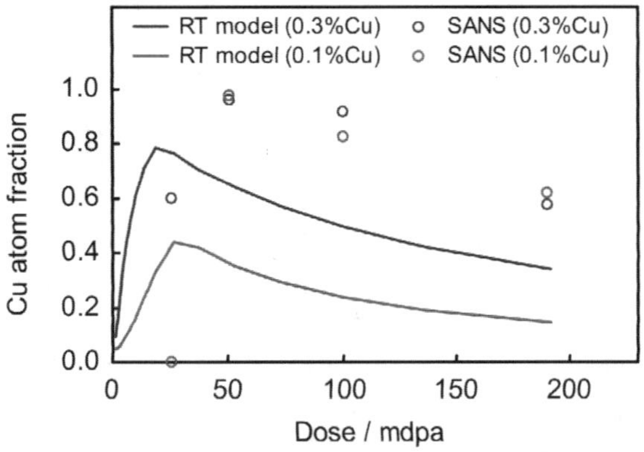

Figure 3 Calculated average atom fraction x_{Cu} of Cu in the mixed Cu-vacancy clusters as a function of dose and comparison with experimental results derived from the *A*-ratio of SANS data for both model alloys. Here, $x_{\text{Cu}} = 0$ means pure voids and $x_{\text{Cu}} = 1$ pure Cu clusters.

It is worth mentioning that the cluster compositions shown in figure. 3 are *predictions* of our V3C model, since the A-ratios were not used for the RT calibration. Understanding why quantitative agreement could not be achieved is subject of further investigations. It might be related to the ability of the mixed Cu-vacancy clusters to also emit vacancies in an substantial amount. However, it is known that such mixed clusters tend to be copper coated voids [14-15], which might partly suppress vacancy emission.

As already mentioned, Radiguet *et al.* [8] suggest a "heterogeneous" nucleation process for Fe-0.1%Cu where Cu-containing precipitate clusters only form on top of already existing nano-voids. In principle, this type of heterogeneous nucleation is included in a rate theory model which allows for the formation of mixed Cu-vacancy clusters. And indeed, we find preferentially void formation at the beginning and increasing Cu absorption later on, which is in line with the heterogeneous processes put forward in Ref. [8]. Yet, since our V3C model was initially designed for small admixtures of vacancies to Cu-rich precipitates we assumed the defect clusters to be perfect vacancy absorbers while keeping the properties of pure Cu clusters. Therefore, the formation of pure voids is not possible in the V3C model. Clearly this limitation has to be removed in future applications.

Also it should be mentioned that when using the mixed Cu-vacancy formation energies from our structural optimization in a rate theory model for small two component clusters, there is no indication for Cu-precipitation (pure or mixed) to start *only* after some nanovoids have already been formed. More detailed rate theory models are required to clarify that issue.

5. Summary

Based on the experience that rate theory models for Cu-precipitation which treat the precipitate clusters as pure Cu clusters are not able to reproduce the experimentally observed behavior of Fe-Cu model alloys under neutron-(or ion) irradiation for Cu contents well below 0.3%, a rate theory model is set up which allows to take explicitly into account the absorption of iron vacancies by Cu-rich precipitate clusters. This way, mixed Cu-vacancy agglomerates can be formed as indicated by SANS and TAP measurements. The idea of switching to multi-component precipitates is not unique. It is also followed in the heterogeneous nucleation mechanism suggested by Radiguet *et al.* [8] for Fe-based model alloys of low Cu content.

In our Vacancy-Coupled Copper Clustering (V3C) model only two parameters, the Cu thermal diffusivity in iron and the reduction of the coherent interface energy in Fe-0.1%Cu compared to Fe-0.3%Cu were used in order to adjust the RT model. Since the obtained value for the thermal Cu diffusion coefficient is the same for both materials and lies well in the range of reported diffusivities, effectively only one parameter was required to reasonably reproduce the entire set of experimental data on the cluster distribution obtained by means of SANS measurements on two neutron-irradiated Fe-Cu model alloys at four different doses.

An analytical expression for the interface energy of Cu-rich precipitates in Fe as a function of temperature and the Cu content of the iron matrix is given. It shows a reduction of the interface energy with increasing amount of vacancy in the defect cluster, a somewhat surprising result, at first glance, since voids exhibit significantly larger interface energies than bare Cu clusters, but which can be rationalized by a more sound thermodynamic analysis of the underlying copper exchange processes.

Some rough, but promising estimates on the evolution of the overall chemical composition of the defect clusters could be given as well, corroborating qualitatively the changes in compositions with increasing neutron damage found in recent SANS experiments. The main trend: high vacancy contents first, increasing Cu absorption later is in line with the heterogeneous nucleation process suggested for the Fe-0.1%Cu model alloy by Radiguet *et al.* [8].

Acknowledgements

This work was partly supported within the European Integrated Project PERFECT under Contract No. F6O-CT-2003-508840 and by the Bundesministerium für Wirtschaft under Contract No. 150 1315. Special thanks to A. Almazouzi for providing his SANS raw data for further analysis. We also thank A. Barbu for the many discussions and the providing of his rate theory code MFPVIC for comparison.

Journal of Physics: Conference Series **247** (2010) 012011 doi:10.1088/1742-6596/247/1/012011

References

[1] Golubov S I, Osetsky Yu N, Serra A and Barashev A V 1995 *J. Nucl. Mater.* **226** 252
[2] Christien F and Barbu A 2004 *J. Nucl. Mater.* **324** 90
[3] Barashev A V, Golubov S I, Bacon D J, Flewitt P E J and Lewis T A 2004 *Acta Mater.* **52** 877
[4] Odette G R and Wirth B D 2005 Radiation Effects in Fission and Fusion Reactors *Handbook of Materials Modeling Part A* ed Yip S (Springer, Dordrecht) p 999
[5] Koptelov E A 2007 *J. Nucl. Mater.* **366** 238
[6] Malerba L, van Walle E, Domain C, Jumel S and van Duysen J-C 2002 State of advancement of the international REVE project: Computational modeling of irradiation-induced hardening in reactor pressure vessel steels and relevant experimental validation programme *Proc. of ICONE 10: 10th International Conference on Nuclear Engineering (Arlington, USA, 14-18 April 2002, ASME) paper 22260.*
[7] Bergner F, Lambrecht M, Ulbricht A and Almazouzi A 2010 *J. Nucl. Mater.* **399** 129
[8] Radiguet B, Barbu A and Pareige P 2007 *J. Nucl. Mater.* **360** 104
[9] Mathon M H, Barbu A, Dunstetter F, Lorenzelli N and de Novion C H 1997 *J. Nucl. Mater.* **245** 224
[10] Gokhman A, Boehmert J and Ulbricht A 2003 *Radiat. Eff. Def. Solids* **158** 783
[11] Schmelzer Jr J, Lembke U and Kranold R 2000 *J. Chem. Phys.* **113** 1268
[12] Odette G R 1983 *Scrip. Metall.* **19** 1183
[13] Osetsky Yu N and Serra A 1996 *Philos. Mag. A* **73** 249
[14] Takahashi A, Soneda N, Ishino S and Yagawa G 2003 *Phys. Rev. B* **67** 024104
[15] Kulikov D, Malerba L and Hou M 2006 *Philos. Mag.* **86** 141
[16] Bergner F, Ulbricht A, Gokhman A and Erak D 2008 *J. Nucl. Mater.* **373** 199
[17] Meslin E, Radiguet B, Pareige P, Barbu A 2010 *J. Nucl. Mater.* **399** 137
[18] Pasianot R C and Malerba L 2007 *J. Nucl. Mater.* **360** 118
[19] Odette G R 1998 Modeling irradiation embrittlement in reactor pressure vessel steels *Neutron Irradiation Effect in Reactor Pressure Vessel Steels and Weldments* ed Davies M (International Atomic Energy Agency, Vienna) IAEA IWG-LMNPP-98/3, p 438
[20] Broeders C H M and Konobeyev A Yu 2004 *J. Nucl. Mater.* **328** 197
[21] Stoller R E 2000 *J. Nucl. Mater.* **276** 22
[22] Calder A F and Bacon D J 1993 *J. Nucl. Mater.* **207** 25
[23] Burton J J 1972 *Phys. Rev. B* **5** 2948
[24] Ackland G J, Bacon D J, Calder A F and Harry T 1997 *Philos. Mag. A* 75 713
[25] Gokhman A, Bergner F, Ulbricht A and Birkenheuer U 2008 *Def. Diffus. Forum* **277** 75
[26] Golubov S I, Serra A, Osetsky Yu N and Barashev A V 2000 *J, Nucl. Mater.* **277** 113
[27] Marian J, Wirth D B, Odette G R and Perlado J M 2004 *Comput. Mater. Sci.* **31** 347
[28] Lazarev V A and Golikov V M 1970 *Physics of Metals and Metallography-USSR* **29** 154
[29] Anand M S and Agarwala R P 1966 *J. Appl. Phys.* **37** 4248
[30] Salje G and Feller-Kniepmeier M 1977 *J. Appl. Phys.* **48** 1833
[31] Rothman S J, Peterson N L, Walter C M and Nowicki L J 1968 *J. Appl. Phys.* **39** 5041

Journal of Physics: Conference Series **247** (2010) 012012 doi:10.1088/1742-6596/247/1/012012

A novel real space scattering theory: efficient characterization of colloidal crystals

C. Cabrillo

E-mail: ccabrilo@foton0.iem.csic.es
Instituto de Estructura de la Materia, CSIC, Serrano 123, E-28006 Madrid, Spain

E. Enciso

Facultad de Ciencias Químicas, Departamento de Química Física I, Universidad Complutense, E-28040, Madrid, Spain

M. J. Capitán

Instituto de Estructura de la Materia, CSIC, Serrano 123, E-28006 Madrid, Spain

A. Cabañas

Facultad de Ciencias Químicas, Departamento de Química Física I, Universidad Complutense, E-28040, Madrid, Spain

M. J. Torralvo

Facultad de Ciencias Químicas, Departamento de Química Inorgánica, Universidad Complutense, E-28040, Madrid, Spain

J. Álvarez

Departamento de Física de la Materia Condensada, Facultad de Ciencias, Universidad Autónoma de Madrid, Fco. Tomás y Valiente 7, E-28049, Madrid, Spain

F. J. Bermejo

Instituto de Estructura de la Materia, Unidad Asociada CSIC, Facultad de Ciencia y Tecnología, Universidad del País Vasco / EHU, P. Box 644, E-48080, Bilbao, Spain

Abstract. Recent advances in self-organized 3D ordered structures of submicron particles as colloidal crystals demand a precise quantitative characterization of the produced nano-structures. Small angle scattering is the technique of choice for such a task but a comprehensive quantitative modeling of the measurements is far from being straightforward. We have developed a theory based in the pair distances distribution which take into the account orientational, positional and staking disorder as well as finite size effects. We show also how the radial scattering length density of the constituent particles, essential for a comprehensive modeling of the experimental data, can be estimated from the position of the form factor local minima. The results reduce to sums of analytical functions over the distribution of pair distances and as such, are suitable for easy (automatic) parallelization.

1. Introduction

Monodisperse latex particles, i.e., microspheres of controlled size and variable composition, can be induced to self-organize into 3D ordered structures (colloidal crystals). This remarkable behaviour is not only outstanding from the fundamental point of view [1, 2, 3, 4] but it is also of a great applicability. The ordered 3D arrays can be used as templates for nano-structured porous materials made of a variety of compounds (inorganic oxides, carbon, metals or polymers [5]) with a broad range of applications: optical switches [6], chemical sensors [7] or photonic band gap materials [8]. For these applications the main effort is addressed to obtain large single crystal possessing long-range order, with a low density of defects and small fluctuations of the particle positions. Many properties of colloidal crystals are correlated with the crystal structure, and according to this, it is important to elucidate the mechanism of crystal formation and to develop methodological tools for structural analysis of these systems (see, for instance, [9] for the use of subcritical adsorption isotherms for identifying structural information of colloidal crystals).

Small angle scattering is probably one of the most used techniques for structural characterization of colloidal crystals [10, 11, 12, 13, 14] due to its microscopic sensibility to the structural details. Given its high informative content and the complexity of the structures studied it is not an easy task a quantitative description of the measurements. The traditional approach to the theoretical developments is to work in the the reciprocal space making recourse of the powerful analytical properties of the imaginary exponentials [15]. Orientational average is left to the latest stage of the calculation in order to preserve analytical advantages of the exponentials. Following this line, elaborated calculations have produced significative advances in the quantitative description of the diffraction patterns [16, 17]. However, the production of higher quality crystals is still hampered by the lack of numerically efficient quantitative assessments of the measurements. Here, we propose an alternative theoretical route, akin to the Debye approach (see pag. 49 in [15]), that begins with the orientational average so that the theoretical development is based on the pair distances distribution. Since the final expressions are expressed as sums of analytical terms they profit of the automatic parallelization capabilities of modern multicore processors. These formulae are complemented with a method to extract from the diffractogram an effective radial scattering length density of the constituent particles, essential for a comprehensive modeling of the experimental data.

2. Theory

Let us start by assuming an arrangement of scattering entities of finite size, each characterized by a scattering length $b_i(\mathbf{Q})$ being $\mathbf{Q}$ the scattering vector for the given radiation. The sample is such that all the orientations are well represented or else they are efficiently explored during the measurement (for instance, by an adequate rotation of the sample). Under such circumstances the singly scattered radiation by such an object is given by,

$$I(Q) = N\langle\overline{|b(\mathbf{Q})|^2}\rangle + \langle\sum_{i\neq j}\overline{b_i(\mathbf{Q})b_j^*(\mathbf{Q})e^{-i\,\mathbf{Q}\cdot(\mathbf{r}_i-\mathbf{r}_j)}}\rangle\,, \tag{1}$$

where the $\mathbf{r}_i$ vectors point to the positions of the "centers" (the centers of mass, for instance) of the N entities illuminated by the radiation, the bar denotes orientational average and the angular brackets denote ensemble average over the rest of needed statistical distributions, typically center positions and size. As a first approximation we will assume that the various distributions are uncorrelated, a reasonable assumption for colloidal crystal of quasi-spherical particles. Then,

$$I(Q) = N\left(\langle\overline{|b(\mathbf{Q})|^2}\rangle - |\langle\overline{b(\mathbf{Q})}\rangle|^2\right) + N|\langle\overline{b(\mathbf{Q})}\rangle|^2 S(Q)\,, \tag{2}$$

where $S(Q)$ is the orientationally averaged structure factor given by

$$S(Q) = 1 + \frac{1}{N}\sum_{i\neq j}^{N} \langle \text{sinc}(Q\, d_{i,j})\rangle \,, \tag{3}$$

being $d_{ij} = |\mathbf{r}_i - \mathbf{r}_j|$. Disordered systems with large size dispersion are not accurately reproduced by Eq. (2) since correlations between size and pair distances develop. Dilution of the system leads to $S(Q) \to 1$ so that the $I(Q) = N\langle\overline{|b(\mathbf{Q})|^2}\rangle = NP(Q)$ being $P(Q)$ the average form factor of the constituent entities. For small enough scattering length variance, $P(Q) \simeq |\langle\overline{b(\mathbf{Q})}\rangle|^2$, the first term in Eq. (2) vanishes, and $I(Q)$ can be approximated by $NP(Q)S(Q)$. In such a case, measuring the scattering of a dilute version of the system allows for the estimation of the structure factor by simple division.

We will now address the various effects shaping the structure factor, i.e., affecting the pair distances distribution, leaving for the next section those related with the scattering length.

2.1. Structure factor

Characteristic of an ideal crystal is a definite deterministic set of pair distances $\{d_{ij}\}$. In a real crystal, on the contrary, the d_{ij} are statistically distributed. Let as assume that the corresponding probability distributions, g_{ij}, are of the Schulz (or Gamma) kind, that is,

$$\gamma(x;\alpha,\bar{x}) = \frac{1}{\Gamma(\alpha)}\left(\frac{\alpha}{\bar{x}}\right)^{\alpha} x^{\alpha-1}\exp\left(-\frac{\alpha x}{\bar{x}}\right), \tag{4}$$

where $\bar{x}$ is the mean value of the random variable x and α is a shape factor such as the variance is given by $\sigma^2 = \bar{x}^2/\alpha$. The inverse of the squared root of shape factor, $\sigma_p = 1/\sqrt{\alpha}$, is in fact the dispersion normalized to the mean. The subindex p refers to polydispersity since, following Pusey [18], we will use it as an index of polydispersity within the context of colloidal particles. It should not be confused with the polydispersity index coefficient (PDI) a more common characterization of polydispersity (see section 3). The advantage of the Schulz distribution is that allows for analytical integration of products of trigonometric functions with powers of the random variable. In particular,

$$\langle \text{sinc}(Q\,d_{ij})\rangle = \int \gamma(d_{ij};\alpha_{ij},\bar{d}_{ij})\,\text{sinc}(Q\,d_{ij})\,\mathrm{d}\,d_{ij} =$$
$$\alpha_{ij}\left(1+\frac{(Q\bar{d}_{ij})^2}{\alpha_{ij}}\right)^{-\frac{\alpha_{ij}-1}{2}} \frac{\sin\left[(\alpha_{ij}-1)\arctan\left(\frac{Q\bar{d}_{ij}}{\alpha_{ij}}\right)\right]}{(\alpha_{ij}-1)Q\bar{d}_{ij}}\,. \tag{5}$$

Choosing different relations between the α_{ij}'s and the $\bar{d}_{ij}$'s we can choose the type of disorder. Thus, disorder of the first type, such us thermal spreading, lead to the same dispersion for all the $\bar{d}_{ij}$ since on average a lattice with long range order is preserved and the positions just spread with a given σ around the sites of the lattice. Therefore $\alpha_{ij} = \bar{d}_{ij}^2/(4\sigma^2)$, that is, α_{ij} is quadratic on the distance. In disorder of the second type, typical of colloidal crystals, no average lattice with long range order is defined but only the mean and dispersion of the first neighbors distances. For a d_{ij} corresponding to positions of a linear chain in the crystal of first neighbors n sites apart, the distribution is that of a random variable sum of n variables following the distribution of the close neighbor distance, d. Being such a distribution a Schulz one with scale parameter α, d_{ij} follows simply $\gamma(d_{ij}; n\alpha, \bar{d}_{ij})$. Since $\bar{d}_{ij} = nd$, we have $\alpha_{ij} = (\alpha/d)\bar{d}_{ij}$, i.e., α_{ij} is linear with the distance. As a first approximation we will assume the same proportionality for every direction in the crystal (or more correctly paracrystal). This will suffice to grasp the structure factor lineshapes characteristic of this type of disorder.

Typically for colloidal crystals the structure will be of the closed packaged kind such as the fcc lattice. Let us suppose that the colloidal particles are quasi-spherical with a mean diameter d and a polydispersity σ_p. As far as the lattice is concerned, we will assume that the lattice parameters are those corresponding to a diameter d_l, not necessarily equal to d. This extra degree of freedom allows up to a certain extend to deal with deformed particles by action of stress forces during self-assembly of the particles. To calculate the $S(Q)$ of a fcc paracrystal we first generate the discrete pair correlation function for the fcc structure, i.e., a lookup table with entries of the form $g_n = \{\bar{d}_{ij}, m_{ij}\}$ where m_{ij} is the multiplicity of $\bar{d}_{ij}$ per particle. $S(Q)$ is then calculated as,

$$S(Q) = 1 + \sum_{n}^{M} g_n(2)\langle \text{sinc}\ [Q\, g_n(1)\ ]\rangle\,, \tag{6}$$

where $g_n(1)$ and $g_n(2)$ denote the first and second elements of the g_n entry, M is the size of the table and the average is calculated using Eq. (5) with $\alpha_{ij} = \bar{d}_{ij}/(d\sigma_p)$. However, generating the lookup table for not very small structures is rather time consuming since its size goes as the square of the number of sites and we need to explore a rather dense set of crystal sizes to match the effective grain size of our polycrystalline sample. The table for the infinite fcc structure is far less expensive to calculate since only the distances to one of the points of the Bravais lattice have to be calculated. The table will be of a finite size and therefore suitable only above a minimum Q_{cut} but very large tables can be generated rather quickly. The m_{ij}'s of a finite crystal decay to zero as the d_{ij}'s approach the size of the crystal. They are also below those of the infinite crystal so that the ratio between both multiplicities also decay to zero. Such a decay depends on the crystallite shape but once normalized to the maximum pair distance it follows a universal curve independent of the size. We have explored master curves for many different crystallite shapes and found that all of then fit well with the mono-parametric family (see fig. 1),

$$f_a(x) = \exp\left(-\frac{a\,x}{1-x}\right)\,, \tag{7}$$

where a is the slope at the origin. Therefore, from the lookup table corresponding to the infinite crystal, we can approximate the structure factor of a crystal of size s by,

$$S(Q) = 1 + \sum_{n}^{M} f_a\left(\frac{g_n(1)}{s}\right) g_n(2)\langle \text{sinc}\ [Q\, g_n(1)\ ]\rangle\,. \tag{8}$$

From excluded volume arguments, we can expect a preference for closed packaged structures in colloidal crystals. However, the fcc is not but one of an infinite family of such structures. All of them are generated by staking successive 2D triangular arrays of the spherical particles (see chapter 7 in ref.[15]). Given one of this triangular arrays denoted by layer A, the next layer, will be laterally displaced to the centers of the triangles in order of minimizing the volume. There are two equivalent ways of doing this. If, as in [19], we define the $\hat{\mathbf{x}}$ axis along one of the lines formed by triangle sides and the $\hat{\mathbf{y}}$ axis as counterclockwise perpendicular within the array plane, a translation in units of the diameter of the spheres of the 2D array by $t_+ = (1/3)\hat{\mathbf{x}} + (\sqrt{3}/6)\hat{\mathbf{y}}$ will lead to one of the possibilities arbitrarily denoted by B while the other is implemented by $t_- = (1/3)\hat{\mathbf{x}} - (\sqrt{3}/6)\hat{\mathbf{y}}$ denoted by C. Alternate application of both kinds of translations leave the array invariant, i.e., $t_+t_- = 1$ while successive application of the same translation is equivalent to the other, i.e., $t^2_{+,-} = t_{-,+}$. Thus, for a third layer there are two nonequivalent possibilities: apply the alternate translation coming back to the original lateral position or apply the same translation going to the alternate lateral position. The series ABCABC$\cdots$ and ACBACB$\cdots$ corresponds to twins fcc structures while the series ABABAB$\cdots$ and ACACAC$\cdots$

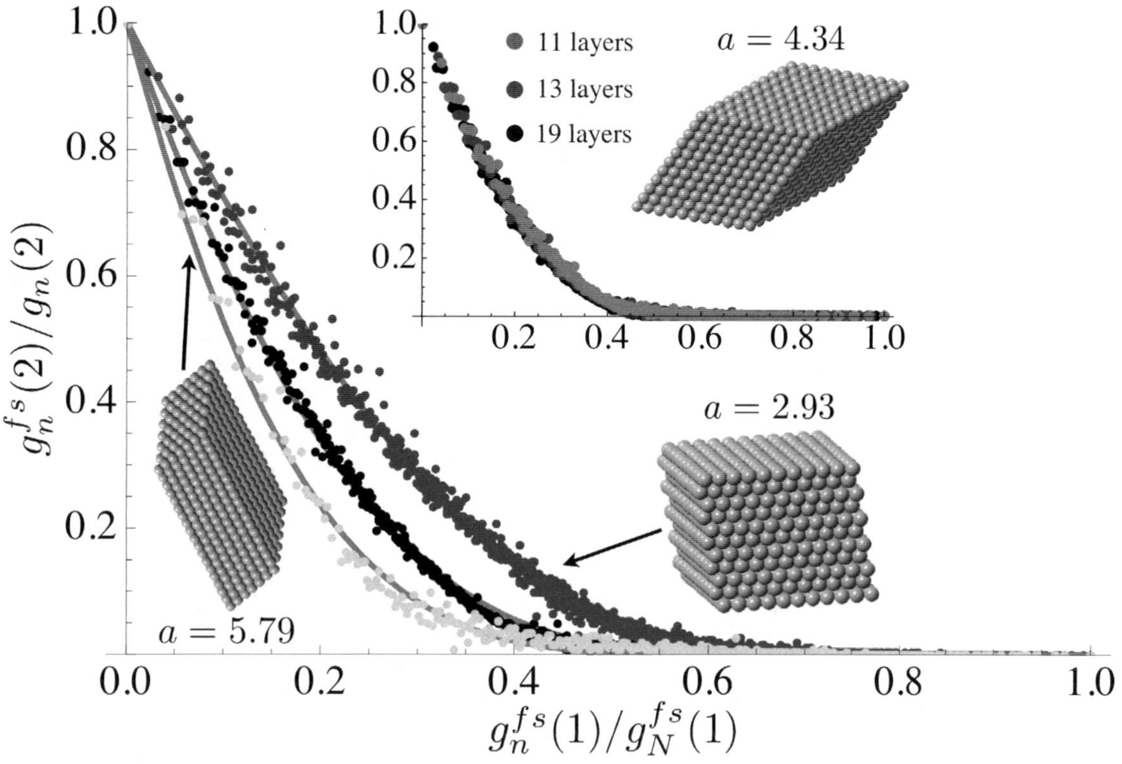

Figure 1. (color online) Ratio between the multiplicities of finite size crystals, $g_n^{fs}(2)$, and those of the corresponding infinite crystal, $g_n(2)$, as function of the pair distance normalized to the longest for three different crystallites. The corresponding fits with f_a are also shown. The data set in black corresponds to a crystallite with the symmetry of a symmetric primitive cell (upper right corner). For this crystallite the scaling with the crystallite size is shown for three different sizes in the inset. The lower left crystallite is of the same type but with and elongated aspect ratio.

yield the same hcp crystal. Any alternate series will represent a closed packed structure. The 2D lattice is generated by the lattice vectors $\mathbf{a} = \hat{\mathbf{x}}$ and $\mathbf{b} = (1/2)\hat{\mathbf{x}}+(\sqrt{3}/2)\hat{\mathbf{y}}$. The third component necessary to span the 3D structure is given by $\mathbf{c} = \sqrt{2/3}\,\hat{\mathbf{z}}$, again in terms of the diameter being $\hat{\mathbf{z}}$ perpendicular to the layers. Now, any given structure is determined giving the set of probabilities P_k that two layers separated by a distance $k\sqrt{2/3}$ being identical. For the closed packed structures P_k can be calculated as a function of k parametrized by the probability β that layers n and $n+2$ have the same lateral position. More explicitly, $P_k(\beta) = 1/3 + q_1\rho_1^{|k|} + q_2\rho_2^{|k|}$ being $\rho_{1,2}$ the roots of the equation $\rho^2+(1-\beta)\rho+1-2\beta = 0$ and q_1 and q_2 determined by the the conditions $P_0(\beta) = 1$ and $P_1(\beta) = 0$ (see page 49 in[15]. See also [20, 21]) Each closed packed structure is, then, determined by a unique parameter, β, with $\beta = 0$ for the fcc lattice, $\beta = 1$ for the hcp and $\beta = 0.5$ for a random stacking sequence. We need, now, to construct the $S(Q)$ from the pair distances of the infinite crystal. From any sphere in the reference layer, $k = 0$, any distance to any other sphere in a no displaced layer adopt the form $\sqrt{(i + j/2)^2 + (3/4)j^2 + (2/3)k^2}$ where i, j, k are the indexes of the lattice vectors $\mathbf{a}, \mathbf{b}, \mathbf{c}$. To a displaced layer we have, irrespectively of being a t_+ or a t_- displacement, $\sqrt{(i + j/2 + 1/2)^2 + (3/4)(j + 1/3)^2 + (2/3)k^2}$. If we generate a lookup table, G_n^0, for the "squared" 2D distances $(i+j/2)^2+(3/4)j^2$ and G_n^1 for the displaced version $(i + j/2 + 1/2)^2 + (3/4)(j + 1/3)^2$, we finally have,

$$\begin{aligned} S(Q) = \sum_k \Bigg\{ & P_k(\beta) \sum_n \Bigg[f_a\left(\sqrt{G_n^0(1) + (2/3)k^2}/s\right) \\ & G_n^0(2) \left\langle \mathrm{sinc}\left[Q\sqrt{G_n^0(1) + (2/3)k^2} \right] \right\rangle \Bigg] + \\ & (1 - P_k(\beta)) \sum_m \Bigg[f_a\left(\sqrt{G_m^1(1) + (2/3)k^2}/s\right) \\ & G_m^1(2) \left\langle \mathrm{sinc}\left[Q\sqrt{G_m^1(1) + (2/3)k^2} \right] \right\rangle \Bigg] \Bigg\} . \end{aligned} \quad (9)$$

The average over the statistical distribution of stacking faults given by the $P_k(\beta)$'s do not allow making recourse of a complete 3D pair distances table. At any rate, Eq. (9) is amenable to automatic parallelization allowing a fast calculation for rather large lookup tables.

2.2. Scattering length terms

In order to compute Eq. (2), we now need to calculate $\langle\overline{|b(\mathbf{Q})|^2}\rangle$ and $\langle\overline{b(\mathbf{Q})}\rangle$. Assuming isotropic particles the orientational average is automatic. For spheres of constant scattering length density ρ and radius R, the corresponding scattering length is $\rho\, b^{sph}(Q, R) = \rho(4\pi R^3/3)\; 3\,(\sin(QR) - QR\cos(QR))/(QR)^3$. Again, using a gamma distribution for the radii both $\langle b(Q)\rangle$ and $\langle |b(Q)|^2\rangle$, although cumbersome, admit analytic expressions [22, 23]. If the particles are isotropic but with a varying scattering length profile, they can be approximated by a finite number, M, of shells of constant ρ so that the scattering length reads, $b(Q) = \sum_{m=1}^{M} c_m\, b^{sph}(Q, mR/M)$, where the coefficients fulfill $c_m = \rho_m - \rho_{m+1}$, being ρ_m the length density of the m shell ($\rho_{M+1} = 0$). This decomposition allow us to calculate averages over radii distributions. The problem now is how to determine the number of shells, R and the c_m's coefficients. In principle we could think of some fitting procedure provided we have measurements of the average form factor from small angle scattering of the particles in dilution. However, the radial profile and radius of the particles in dilution can be substantially different from those of the dried counterparts present in the crystal. Nevertheless, in many cases we can estimate some of the minima of the average form factor since they are particularly conspicuous features in the diffraction patterns. But the minima of the form factor are dominated by the zeros of $b(Q)$ event for rather large dispersion in the radii. Let us suppose we know an estimation of the M first minima which in turn are a good approximation to the first M zeros of $b(Q)$. We can construct a radial profile with M shells given rise to a $b(Q)$ passing by the zeros in the following way. First, notice that the c_m's must fulfill the linear system:

$$\begin{aligned} 0 &= c_1 b^{sph}(Q_1, R/M) + \cdots + c_m b^{sph}(Q_1, R) \\ &\;\;\vdots \\ 0 &= c_1 b^{sph}(Q_M, R/M) + \cdots + c_m b^{sph}(Q_M, R)\,, \end{aligned} \quad (10)$$

being $Q_1, \cdots, Q_M$ the estimated zeros. Therefore, the matrix of coefficients, $A_{i,j} = b^{sph}(Q_i, jR/M)$, of the linear system must have null determinant. Finding the first zero of $det(A)$ give us the minimum R to accomplish the task. Solving, then, Eq. (10) determine the c_m's up to a constant factor. In practice, since the zero of $det(A)$ is determined numerically, it is better to solve the underdetermined system corresponding to only $M - 1$ zeros. Having estimated the zeros of $b(Q)$ from experimental data does not make sense the very high numerical precision necessary to $det(A) = 0$ in order to obtain a non null solution of Eq. (10).

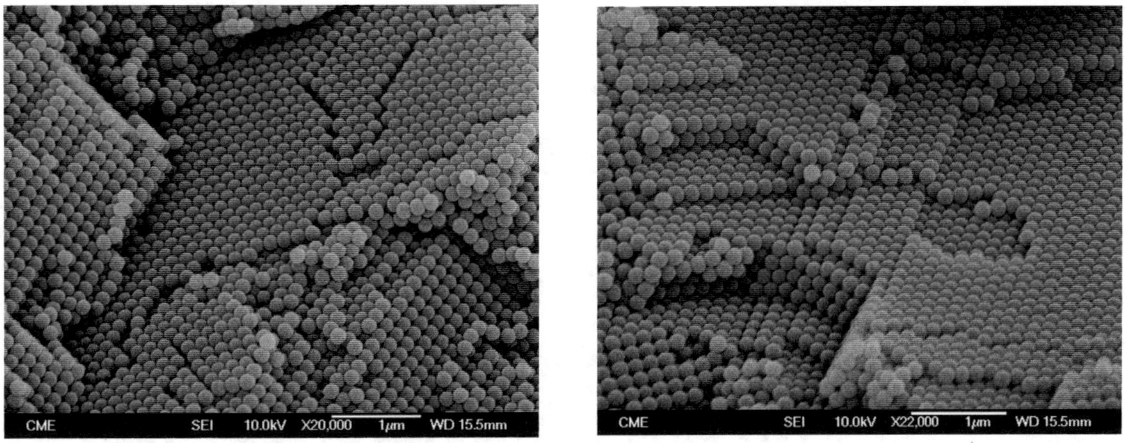

Figure 2. SEM pictures of colloidal crystals prepared in our lab.

3. Sample preparation

Spherical particles of Polystyrene-Methacrylic acid (PS-MAA) latexes were prepared from surfactant-free emulsion by a standard copolymerization process of styrene and methacrylic acid [9, 24]. Small changes of the original recipe, by increasing the hydrophilic monomer to 9.2 wt% of the total monomer, allowed a reduction of particle size to less than 200 nm which facilitated allocation of the corresponding diffraction patterns within the SAXS Q-range window. The synthesized PS-MAA particles were spheres with a smooth surface morphology. A disk centrifuge photosedimentometer (DCP) (Brookhaven BI-DCP) operating in the line start mode yielded a diameter of $d = 195 \pm 7$ nm. The particle size dispersion expressed by the polydispersity index coefficient (PDI), that is, the ratio d_v/d_n where d_v and d_n are the volume and number average diameters, respectively amounts to PDI = 1.007. This value qualifies the sample as mono disperse (PDI < 1.05). On the other hand, the standard deviation normalized to the mean yielded $\sigma_p = 0.036$. When the suspension was dried the particles self-assemble in nice crystals, as it corresponds to monodisperse spheres (fig. 2).

4. Comparison with the experiments

The diffraction measurements were performed in the BW4 line of the HASYLAB at the DESY synchrotron in Hamburg, Germany. The wavelength used were 1.38 Å. The setup was the USAXS configuration corresponding to a distance from the sample to the detector of 8 m with a beam size of 400 μm. The detector was a marCCD165 camera with and active area of 165 mm of diameter. The pixel size was 79.1 μm what yields in this setup a ΔQ resolution of 4.5×10^{-5} Å. With such a high resolution there is no need of including it in the calculations. The sample was introduced as a powder in a standard quartz sample holder for liquids. Fig. 3 shows the diffraction pattern obtained by angular average of the raw 2D digital patterns actually yielded by the machine. It is a particularly useful example since it allows for a comprehensive study of the structure from the first diffraction peak at low Q's, to the complete damping of the structure factor at large Q's. The modeling of the data has been done making recourse of the MINUIT [25] package in its FORTRAN version complemented with manual tweaking and partial fits with Mathematica when necessary.

Centering the modeling just to a few oscillations at large Q's so that the effect of the $S(Q)$ can be neglected, a fit to the data yields a constant background (subtracted from now on in all the shown diffractograms) as well as preliminary values for the mean diameter, $d = 181.1$ nm and polydispersity, $\sigma_p = 0.019$. Adding a simple background representing the tail of the form factor

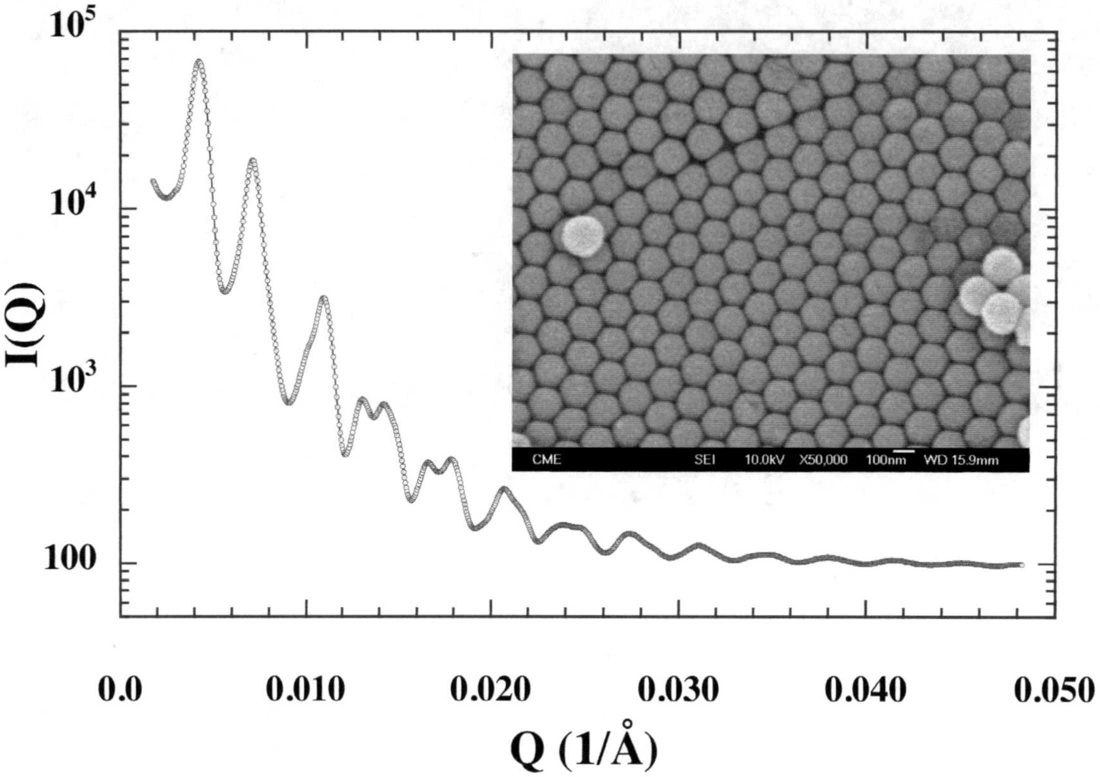

Figure 3. Diffraction pattern of one of our colloidal crystals along with a close view of a SEM image.

of a distribution of very large size aggregates, namely a pure Porod dependence ($\sim 1/Q^4$), we proceed with a first attempt of a complete fit assuming spherical particles with constant scattering length density, ρ. As shown in fig. 4, the model fails to fit the three main peaks, firstly because the background is too high at those Q's and secondly because the minima of the form factor do not match those of the data.

Figure 4 also illustrate how the minima of the data are dominated by the minima of the form factor. A simple graphical estimation of the positions of the thirteen minima was done. The obtained values were then used as an input to Eq. 10 yielding an underdetermined system corresponding to the twelve first zeros of $b(Q)$. The obtained $b(Q)$ showed indeed, a zero at the position of the thirteen minimum. No background can be determined unambiguously from the data. In particular, a pure Porod contribution, although fitting the highs of most of the minima as stated above, fails at low Q's. So the Porod profile, which can be understood as the diffuse scattering coming from large disordered aggregates, must be corrected with some other contributions reflecting the diffuse scattering coming from smaller aggregates and any other kind of disorder not included in the theory. To fix the matter in a simple manner we added to the Porod dependence, a broad distribution of finite diameter spherical particles. In fact, we used the broadest possible distribution of diameters giving a null probability of zero diameter, that is, one with $\alpha = 2$. The amplitudes of both contributions plus the mean diameter of the broad distribution were then chosen to fit of the highs of the experimental minima. A mean diameter of just 19 nm led to a reasonable background as shown in fig. 5.

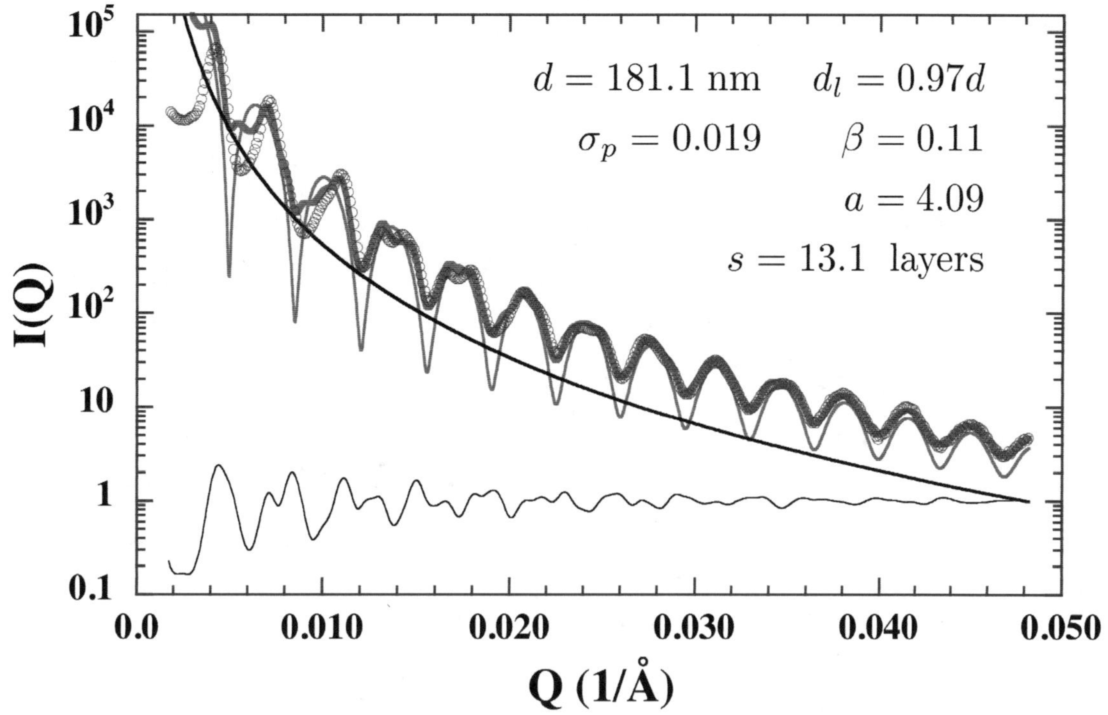

Figure 4. Comparison of the model with constant ρ (thickest red line) to the data (blue circles) for the values shown of the fitted parameters. Also shown are the form factor (thick red line), the background (thick black line) and the $S(Q)$ (thin black line). d is the mean diameter of the particles, σ_p is the diameter dispersion normalized to the mean, d_l is the lattice constant of a triangular layer of the closed packaged structure, β is the probability of two alternate layers being in the same lateral position ($\beta = 0$ implies fcc, $\beta = 1$ implies hcp), a parametrizes the decay of pair distances multiplicities and s is an effective grain size.

As shown in fig. 5, except for the first peak, excellent fits can then be obtained. The high of the first peak, on the other hand, is anomalously low and incompatible with the height of any reasonable form factor. This points to some kind of shadowing of the signal at the lowest angles. Certainly, comparing the model and the signal leads to a multiplicative correction akin of some kind of shadow (see fig. 6). That the experimental setup produced such a shadow in some way we were not aware of seems improbable. Preliminary analysis seems to indicate it is due to specular reflection over crystalline planes. Indeed, below a small enough angle of incidence, the planes shown in fig. 2 will look like a uniform medium. Reflection on such a surface could well produce a similar effect to the mask pictured in the inset of fig. 6 as attested by the aspect of the reflectivity in the Fresnel approximation for this kind of systems [26]. At any rate, further study will be necessary to clarify the matter.

5. Discussion

Table 1 summarizes the final values of the fitted parameters. Some comments regarding their relevance and corresponding consequences seems in order.

Beginning from the size distribution the mean diameter obtained is below that given by DCP (195 nm). This is something to be expected since DCP is giving the hydrodynamic size in

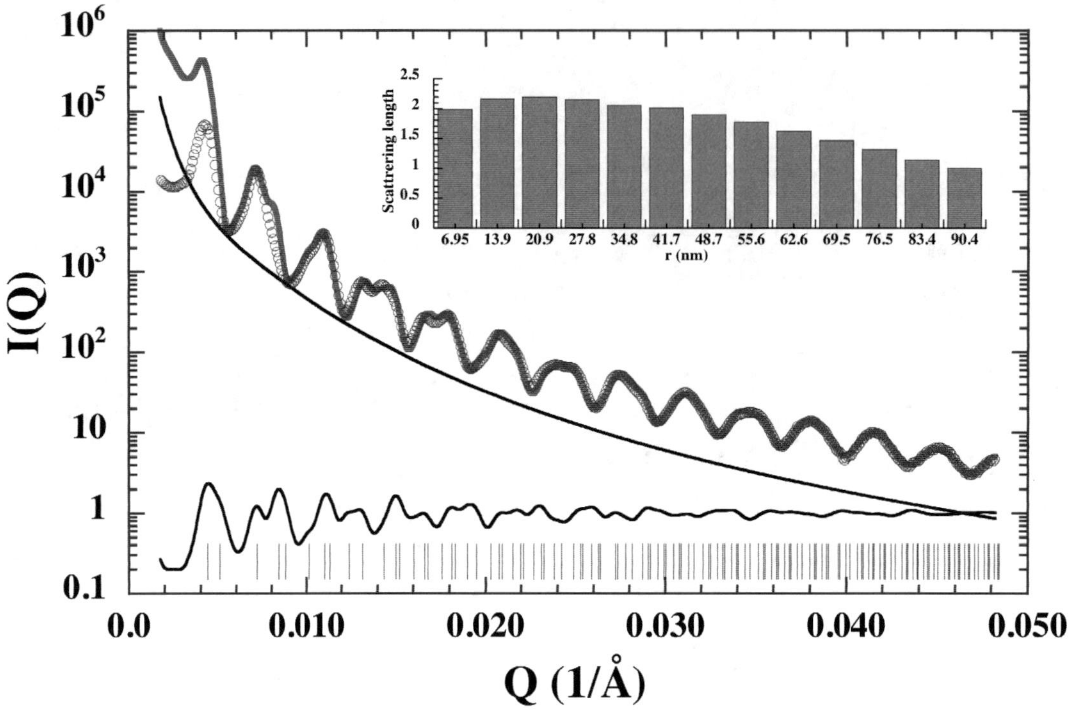

Figure 5. Comparison of the model (thick red line) to the data (blue circles) for a radial distribution of ρ as shown in the inset. Also shown are the background (up thick black line) and the $S(Q)$ (down thick black line). The values of the fitted parameters are the same than in the constant ρ case except for the diameter now slightly reduced: $d = 180.8$ nm. The vertical lines mark the positions of the Bragg peaks of a pure ($\beta = 0$) fcc structure.

suspension known to be larger than the size of the dried particle [27]. The fitted σ_p is also half of DCP value pointing to a segregation out of the crystal of particles with sizes far from the mean. On the other hand, not surprisingly [28, 29, 11], the SEM images show deformed particles towards dodecahedral cells within the crystalline parts (see inset in fig. 3). Estimation of the size of the particles from the SEM picture lead to a major diameter of $d_M = 176 \pm 5$ nm and a minor diameter $d_m = 154 \pm 5$ with a corresponding σ_p of 0.03. Since the fitted diameter corresponds to that of a spherical particle representing an orientational average of the real particles the sensible comparison is with d_M which certainly is compatible with the fitted diameter. The value of σ_p, however, depart substantially from the that of table 1 but still is lower than the one obtained by DCP.

A key point of the theory to successfully model the data is the allowance of a lattice parameters independent of d. Although the value d_l is only smaller than d by a 3 %, the fit is very sensible to it and certainly a $d_l = d$ is ruled out. It is rewarding that this simple isotropic modification with respect to the ideal case of spherical particles is enough to grasp the effect of the particle deformation. The polydispersity, on the other hand, do not need to be altered. The fitted value of β indicates a clear preponderance of the fcc structure giving further support to the rational of Norris et al.[30] in favor of such a lattice for this kind of colloidal crystals.

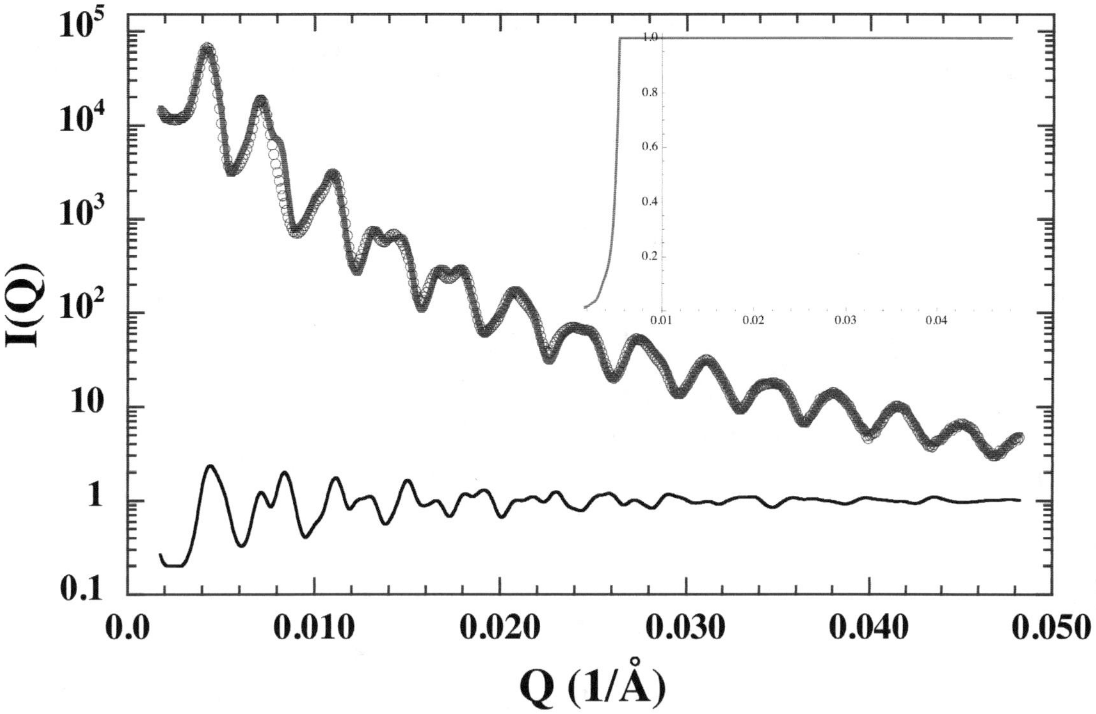

Figure 6. As in the previous figure but with the model masked by the "shadowing" function shown in the inset.

With respect to the effective crystallite size the fit render a grain size of $s = 13.1$ ordered triangular layers before any dislocation breaks the coherence of the crystal. Although substantial, such a coherence length is not enough to separate the first two Bragg peaks (see fig. 5) which would be otherwise split in the infinite crystal. As a matter of fact, we have also checked that the experimental lineshapes cannot be reproduced with disorder of the first kind (too narrow lineshapes at high Q's).

In regarding the shape of the grains, the modeling is fairly sensible to the type of crystallite but only up to a certain degree. From the a value corresponding to the symmetric primitive cell shown in fig. 1, i.e., $a = 4.34$, to above, decently good fittings are obtained choosing the adequate value of s, although the best values are near $a = 4.34$ (the final quoted value of $a = 4.09$ yields only a marginally better fit than $a = 4.34$). As shown in the figure for the case $a = 5.79$, values of a above 4.34 can be obtained from the symmetric primitive cell case simply changing the aspect ratio of the crystallite. The character of the boundary planes of the symmetric primitive cell case, which are all $\{111\}$, are still retained. However, the case $a = 2.93$, corresponding to the best approximation to a pure cubic crystallite, does not fit at all. We see, then, that there is more information in the lineshapes than the mere overall size of the grains given by the s parameter. The densest planes (the $\{111\}$ planes are the densest of a fcc structure) are preferred for the grain boundaries. This makes sense given the feeble interactions among the colloidal particles. Indeed, the claveage planes pictured in the SEM images (fig. 2) in which only planes $\{111\}$ and $\{100\}$ (the next densest) are seen, show this preference.

In regarding the scattering length density, in view of the apparently slight deviation in the minima, it is rather surprising how large is the deviation with respect to the constant case. It makes a strong case towards the necessity of comprehensive modeling of the diffractograms.

Table 1. Final fitted values

Size distribution parameters	Lattice parameters	Effective crystallite parameters
$d = 180.4$ nm	$d_l = 0.97$ d	$a = 4.09$
$\sigma_p = 0.019$	$\beta = 0.11$	$s = 13.1$ layers

Certainly, from the changing rate of polymerization as well as the increasing area during the particle formation we cannot expect a final constant density. At any rate, it should be taken into the account that ρ corresponds to a representative spherical particle representing the average of the real entities and as such it is also affected by the deformation of the real particles.

A part from the deviation of the main peak possibly owing to specular reflection in the cleavage planes of the crystallites, the second main deviation is the shoulder appearing in the second peak not present at all in the data. We do not know the origin of such deviation. We have detected texture in the isolated 2D images actually yield by the diffractometer but it disappears when added up due to the wandering of the X-ray beam. Still, that does not guarantee a negligible influence of texture something that would require a three dimensional scanning of the sample orientation. Perhaps, the detailed structure of the specular refection due to the three dimensional variations of the scattering owing to the crystal structure can lead to extinction of the shoulder. A more accurate account of the particle deformation could contribute also, to correct the deviation.

Finally some words about the computational efficiency. To compute one instance of the I(Q) shown in fig. 4 takes just 0.75" in a laptop with a dual core processor at 2.2 GHz. The code was written in FORTRAN compiled with automatic parallelization. Probably the mayor efficiency enhancement comes from the treatment of the grain size. Based on expression (7) it allows to rely in a table of distances to a given origin which grows as the number of particles and therefore as the cube of the size representing a huge improvement with respect to the pair distances table which grows as the size to the sixth.

6. Conclusion

In conclusion, we have developed a scattering theory capable of a comprehensive quantitive account of the small angle scattering diffractograms of colloidal crystals. We have shown how critically it depends not only in the structure of the crystal but in the internal structure of the scattering particles, at the same time giving a method to extract the radial distribution of scattering length from the diffraction pattern. The implementation proposed is numerically efficient smoothing the way to quantitative characterization and comparison of samples prepared under different conditions.

Acknowledgements

Work supported in part by grant MAT-2007-65711-C04-02 and MAT2007-65711-C04-01 from the Spanish Ministerio de Ciencia e Innovación and by grant GR58/08-921556 under the Universidad Complutense de Madrid/ Banco de Santander financial scheme. We also thank the Microscopy Centre from the Universidad Complutense de Madrid for technical assistance.

References

[1] Pusey P N, van Megan W, Bartlett P, Ackerson B J, Rarity J G and Underwood S M 1989 *Physical Review Letters* **63** 2753
[2] Zhu J, Li M, Rogers R, Meyer W, Ottewill R, Russell W and Chaikin P 1997 *Nature* **387** 883
[3] vanBlaaderen A, Ruel R and Wiltzius P 1997 *Nature* **385** 321
[4] vanBlaaderen A and Wiltzius P 1997 *Advanced Materials* **9** 833

[5] Ozin G A and Arsenault A C 2005 *Nanochemistry: A Chemical Approach to Nanomaterials* 1st ed (Cambridge, U. K.: RSC Publising)
[6] Pau G, Kesavamoorthy R and Asher S A 1997 *Physical Review Letters* **78** 3860
[7] Holtz J H and Asher S A 1997 *Nature* **389** 829
[8] Vlasov Y A, Bo X Z, Sturm J C and Norris D J 2001 *Nature* **414** 289
[9] Carbajo M C, Climent E, Enciso E and Torralvo M J 2005 *Journal of Colloid Interface Science* **284** 639
[10] Sirota E B, Ou-Yang H D, Sinha S K, Chaikin P M, Axe J D and Fujii Y 1989 *Physical Review Letters* **62** 1524
[11] Rieger J, Hädicke E, Ley G and Lindner P 1992 *Physical Review Letters* **68** 2782
[12] Petukhov A V, Aarts D G A L, Dolbnya I P, de Hoog E H A, Kassapidou K, Vroege G J, Bras W and Lekkerkerker H N W 2002 *Physical Review Letters* **88** 208301
[13] Petukhov A V, Dolbnya I P, Aarts D G A L, Vroege G J and Lekkerkerkerl H N W 2003 *Physical Review Letters* **90** 028304
[14] Men Y, Rieger J, von Roth S, Gehrke R and Kong X 2006 *Langmuir* **22** 8285
[15] Guinier A 1994 *X-Ray Diffraction In Crystals, Imperfect Crystals, and Amorphous Bodies* 1st ed (New York: Dover Publications, Inc.)
[16] Matsuoka H, Tanaka H, , Hashimoto T and Ise N 1987 *Physical Review* B **36** 1754
[17] Matsuoka H, Tanaka H, Lizuka N, Hashimoto T and Ise N 1990 *Physical Review* B **41** 3854
[18] Pusey P N 1991 *Liquids Freezing and Glass Transition, second part* (*Les Houches Summer School Session* vol 51) ed Hansen J P, Levesque D and Zinn-Justin J (Elsevier Science) p 763
[19] Ashcroft N W and Mermin N D 1976 *Solid State Physics* 1st ed (New York: Harcourt College Publishers)
[20] Loose W and Ackerson B J 1994 *Journal of Chemical Physics* **101** 7211
[21] Jiang J, Burger C, Li C, Li J, Lin M Y, Colby R H, Rafailovich M H and Sokolov J C 2007 *Macromolecules* **40** 4016
[22] Aragon S R and Pecora R 1976 *Journal of Chemical Physics* **64** 2395
[23] Griffith W L, Triolo R, and Compere A L 1987 *Physical Review* A **35** 2200
[24] Carbajo M C, Gómez A, Torralvo M J and Enciso E 2002 *Journal of Materials Chemistry* **12** 2740
[25] Minuit home web page: URL `http://lcgapp.cern.ch/project/cls/work-packages/mathlibs/minuit/home.html`
[26] Zhou X L and Chen S H 1995 *Physics Reports* **257** 223
[27] Pich A, Lu Y and Adler H 2003 *Colloid Polymer Science* **281** 907
[28] Chevalier Y, Pichot C, Graillat C, Joanicot M, Wong K, j Maquet, Lindner P and Cabane B 1992 *Colloid Polymer Science* **270** 806
[29] Wang Y, Kats A, Juhué D, Winnik M, Shivers R R and Dinsdale C J 1992 *Langmuir* **8** 1435
[30] Norris D J, Arlinghaus E G, Meng L, Heiny R and Scriven L E 2004 *Advanced Materials* **16** 1393

The SNS/HFIR Web Portal System for SANS

Stuart I. Campbell, Stephen D. Miller, Jean-Christophe Bilheux, Michael A. Reuter, Peter F. Peterson, James A. Kohl, James R. Trater, Sudharshan S. Vazhkudai, Vickie E. Lynch

Oak Ridge National Laboratory

Mark L. Green

Tech-X Corporation

campbellsi@ornl.gov

Abstract. The new generation of neutron scattering instruments being built are higher resolution and produce one or more orders of magnitude larger data than the previous generation of instruments. For instance, we have grown out of being able to perform some important tasks with our laptops. The data sizes are too big and the computational time would be too long. These large datasets can be problematic as facility users now begin to struggle with many of the same issues faced by more established computing communities. These issues include data access, management, and movement, data format standards, distributed computing, and collaboration with others. The Neutron Science Portal has been designed, and implemented to provide users with an easy-to-use interface for managing and processing data, while also keeping an eye on meeting modern computer security requirements that are currently being imposed on institutions. Users can browse or search for data which they are allowed to see, run data reduction and analysis applications, perform sample activation calculations and perform McStas simulations. Collaboration is facilitated by providing users a read/writeable common area, shared across all experiment team members. The portal currently has over 370 registered users; almost 7TB of experiment and user data, approximately 1,000,000 files cataloged, and had almost 10,000 unique visits last year. Future directions for enhancing portal robustness include examining how to mirror data and portal services, better facilitation of collaborations via virtual organizations, enhancing disconnected service via "thick client" applications, and better inter-facility connectivity to support cross-cutting research.

1. Background

With the advent of the Spallation Neutron Source (SNS) at Oak Ridge National Laboratory, an opportunity occurred to revisit how to provide software and services via the facility. Under the guidance of Ian Anderson, then Director of the Experimental Facilities Division (XFD), a number of Neutron Science Software Initiative (NeSSI) workshops were held in 2003 and 2004 to solicit feedback from the community on its requirements and wishes for next generation software [1]. The primary outcome was the desire to have a portal system which would help manage and process data for users. For easy access, it was decided that the Neutron Science Portal (portal) would have a web interface via a web browser not requiring users to install software. Portal design work commenced in

2004 and 2005 with an operational portal deployed by April 28, 2006 when first neutrons were detected at SNS. These first data were incorporated into the portal and are still available today.

Since the early days, the portal has grown considerably. It holds data from SNS [5] and High Flux Isotope Reactor (HFIR) [6] as well as the Low Energy Neutron Source (LENS) at the University of Indiana Bloomington [7], some example data from the NPDF instrument at the Lujan Center [8], and the entire IPNS [9] data repository. The portal has approximately 370 registered users, contains almost 4TB of data (much of it is compressed) in almost a million files. On an average day, the portal has about 20 individual visitors, receives about 4500 web site "hits", and performs about 260 compute jobs. Thus for SNS and HFIR users, the portal has established itself as a regular fixture as part of performing experiments and data analysis.

2. What is the portal?

A user accesses the portal via a web browser. In order to do this and to utilize portal computing resources, the user must first obtain an XCAMS computer user account [13], which must be requested and granted. Once approved, users can login to the portal whereupon they will have access to their "workspace", consisting of pointers to experiment data that they are allowed to view. Data are stored by facility, instrument, proposal IDs in order to support providing a multi-facility data repository.

In addition to providing data access, the portal also provides access to applications and computing resources. These same resources are available both to on-site users at SNS and HFIR, and to users external to the facility. The portal provides a suite of applications which the user is able to launch remotely, but the applications appear locally. This capability is made available via NX remote desktop technology [14] accessed via the browser. The NX remote desktop feature enables the portal to present an application to the user as though it runs locally, while also taking advantage of remote computing resources. The application backend can run on a portal server, create and submit jobs to a job queue, and return results to the GUI and the user can access at their sites.

Segmenting computation from the user interface layer enables utilization of a wide array of computing resources. GUI applications run by users at SNS on analysis computers can run locally. Applications launched by remote users can be run on separate portal servers and computation servers at SNS. Other resources include the Oak Ridge Institutional (OIC) cluster [15] and the NSTG cluster [16]. Both of these resources are not managed by SNS staff and thus have specific rules for accessing these resources which could be difficult for most users to deal with. To facilitate using these resources, community accounts are utilized where a job for a portal user is run as the community user on these resources. The Application Manager is the intelligent subsystem within the portal which handles transparent execution of user jobs on a variety of aforementioned target environments. In addition, it handles seamless data staging in and out of these resources, job monitoring, logging, failover and recovery.

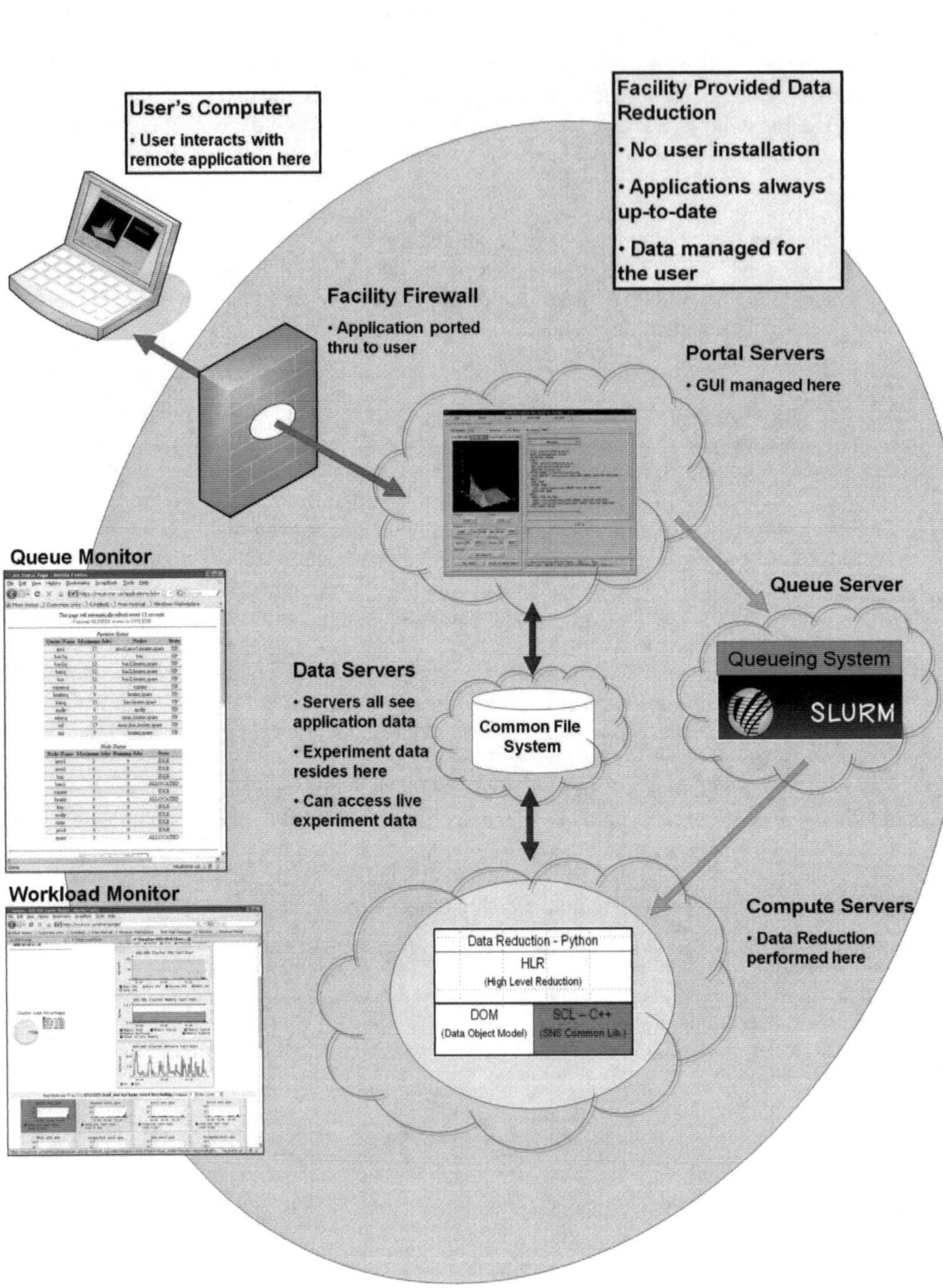

Figure1: Portal Based Data Reduction showing "thin client" remote application access.

Thus the portal includes data browsing and search, but also provides Graphical User Interface (GUI) tools that are connected to powerful computing resources. Applications include experiment data histogramming tools, data reduction tools, data visualization tools, and McStas [4] Monte Carlo simulation. Due to the ability to run applications remotely, the portal also supports hosting of community produced applications such as ISAW [17], DAVE [18], DANSE [19] applications, EXPGUI [20], FullProf [21], GSAS [22]. Database tools such as the Cambridge Structural Database

[23] and the Inorganic Crystal Structure Database [24] are also available. Experiment support tools such as a Sample Activation Calculator are also available which enables users to compute the estimated radioactive dose rate that specified samples will produce for a given instrument.

The portal development effort has produced a significant code base. Currently the composite portal development effort has produced over 650,000 lines of code since the onset of portal development. Data reduction applications account for approximately half of this code base. Software development follows a rigorous process for defining the content of each release. A centralized software build process is utilized to coordinate building correct versions for test and deployment. Testing is done to ensure that features meet requirements, and upon successful completion, the new version is deployed. Software are maintained in a subversion (SVN) [25] repository with a web based issue tracking system called Trac [26] placed atop SVN.

3. Small Angle Scattering Application

Figure 2 shows the Small Angle Scattering Reduction GUI as seen from the portal. It should be noted that the data shown is pre-normalised commissioning data. The vertical lines that can be seen on the plot result from the fact that the detector bank for EQ-SANS at the SNS consists of interleaving detector tubes. The vertical red lines are dead tubes that will be rejected in the data reduction. The bottom right plot displays the counts against neutron time-of-flight of the non-rejected pixels.

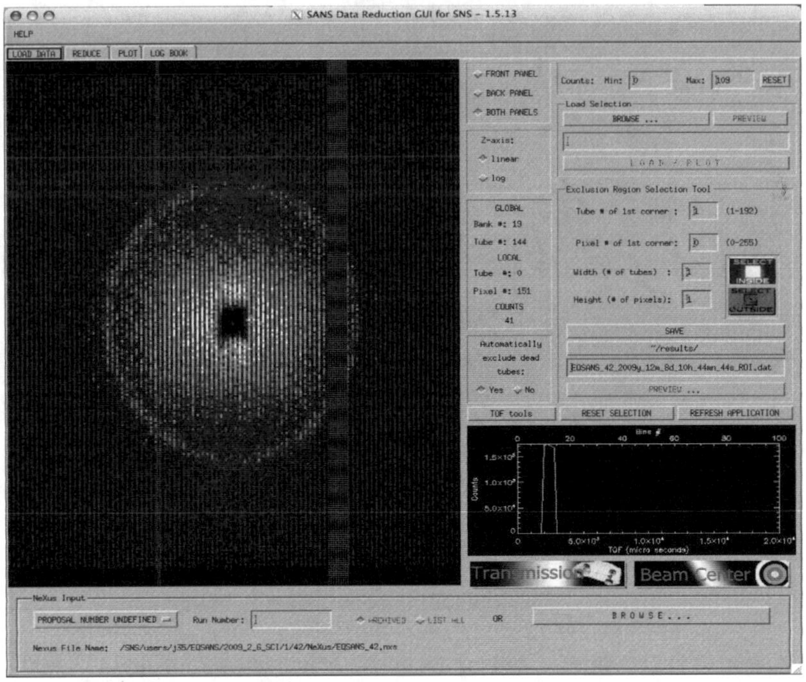

Figure 2: A screenshot of the SANS Data Reduction GUI as seen via the portal. This is the main view of the reduction application.

There has been a great deal of time and effort involved in fitting such a large number of options and functionality into a user interface that is both small and intuitive to use. A tab based design was implemented to try and optimise the use of screen real estate, and arranged / sequenced in a left to right fashion to indicate the order of the operations and user interactions. The experimental data is entered by specifying the run number of the measurement, the system locates the file and passes data onto the compute resources that will be used for the calculation.

The results of the reduction can either be output directly to the users home directory (inside a results directory) or to a shared directory for that measurement. The latter enables users to easily share data and results within the experimental team.

There is a great deal of functionality contained within the application, some examples of which are fitting, beam centre calculations (figure 3), region selection and transmission calculation (figure 4).

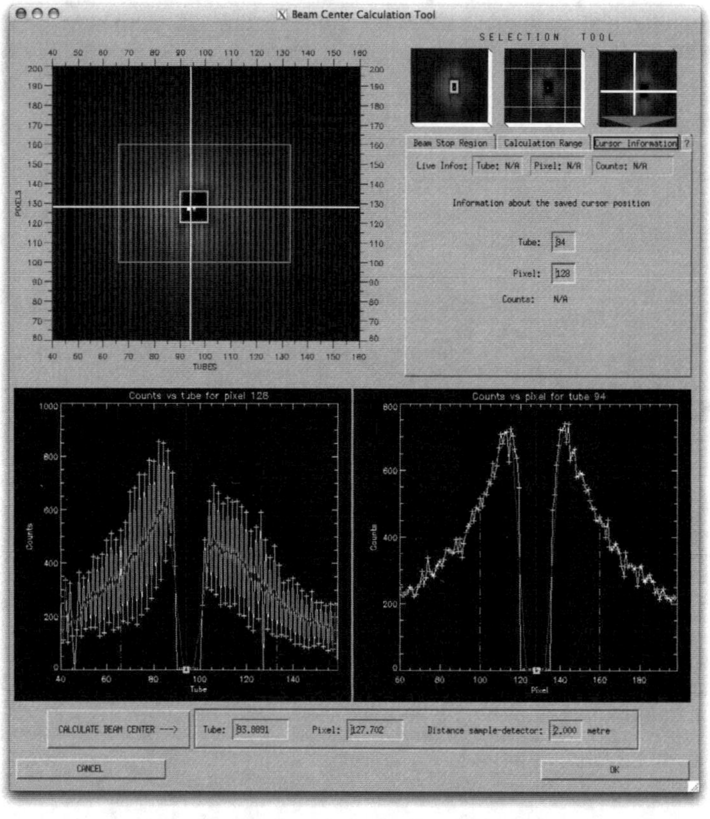

Figure 3: A screenshot of the beam centre calculation tool GUI. This allows the user to determine the beam centre in a user friendly manner.

Every operation that is performed on the GUI is logged to a journal, which can be sent to the developers with a single click from the GUI itself. This helps both the users and developers by easily being able to capture any problems that may arise.

Although at the moment the user runs this application via the portal, it is possible that this GUI forms part of a 'thick client' infrastructure that would allow the user to run the application on their local systems but be able to utilise the computational resources at the facility. Included in this is the intelligence to determine which jobs or calculations should be performed remotely and what is more efficient to execute using local resources.

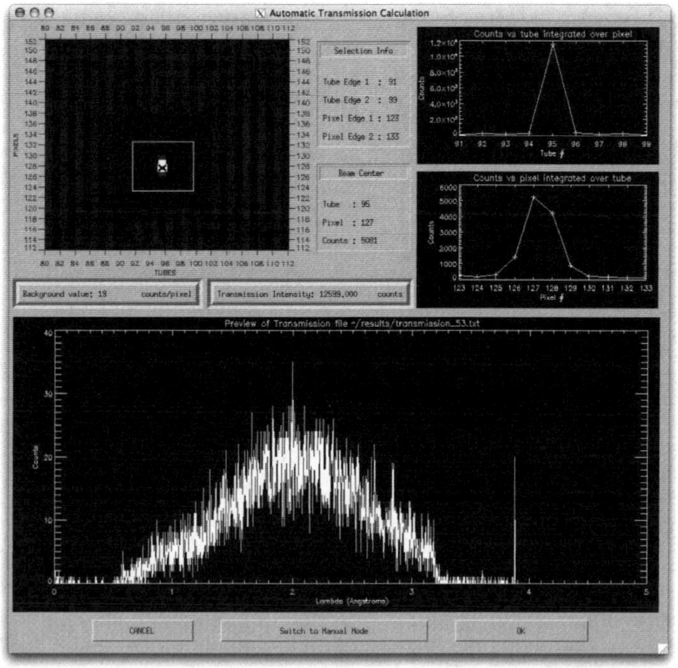

Figure 4: A screenshot of the automatic transmission calculation GUI.

4. Why is the portal usage growing?

A simple answer for why the portal usage is growing could be down to the fact that the data volumes being generated are continually growing and it is becoming difficult for users to move it between systems easily. The facility creates the data and thus has significant influence over how convenient it is for users to acquire, access, and process data. In the past, users collected raw data at their instruments, but this is no longer the primary mode of operation at SNS, which has invested considerable effort into these major capabilities that significantly improve convenience:

1. Experiment metadata are automatically captured and associated with the experiment data. These data are translated into a self-describing file format called NeXus [27], which utilizes the underlying HDF-5 scientific data format [28].
2. The live cataloging process takes experiment data as it is streamed from the instruments, catalogs pertinent metadata, and stores the data for access. The entire process is quick and places data in the centralized repository in a matter of seconds to minutes.
3. There is a high level of integration between user experiment support systems and the data management system. For instance, the data management system integrates with the proposal system (IPTS) [29] in order to capture and associate user and experiment information with experiment data.
4. The facility curates the data on behalf of the user, thus addressing worries users may have for where they store their data, and the integrity of this data.

These functions are all conducted automatically on behalf of the user, in a transparent manner and their data is available for access via the portal using a basic web browser. Thus the complexities of preparing, moving, cataloging, and discovering data are all handled for the user. The portal offers both data browsing and data search for discovering data. A path structure had to be determined in order to facilitate locating data contained within a multi-facility data repository. The generalized path stores data by facility, instrument, proposal ID, collection, and run number. In this way, users can uniquely locate their data across facilities and instruments. The search interface offers users a powerful means for quickly locating data. Search is particularly useful in helping those users who have performed a number of experiments to locate their data. The search interface enables users to locate data by a number of metadata tags such as proposal ID, run number, Principle Investigator (PI) name, sample name, and date. But the user can only locate data which they have permission to access according to the proposals with which they are affiliated. Thus, in addition to the XCAMS authentication system, there is an authorization layer that grants fine-grained access. The facility is capable of providing this data privacy as it has knowledge of the experiment team members via the proposal system. Additional users can be added to the proposal as the PI desires using the proposal check-in utility of the portal.

Once data are located there are a number of things which a user can do such as inspect the data, examine the metadata, or visualize the data. The portal facilitates examining a number of standard file formats such as ASCII text, XML, NeXus, along with a number of imaging formats such as jpg. In particular, one can inspect the NeXus metadata of interest by looking at the attribute groups. The data within the NeXus files can also be visualized. The ISAW visualization tools have been adapted to work via the portal, but needed to be adapted to handle the case where the data could be too large to send over the internet. Thus the visualization tool operates in a client-server mode where only the necessary data are sent over the network link. To enhance data exploration, data can be viewed in tabular form, 1D or 2D plot. The 2D plot can present the data in either linear or log color scales.

The portal also provides users the ability to both upload and download data, and users can freely move data to and from the portal. This begs the question why users do not simply take all of their data home with them to reduce and analyze. They can certainly do so. however, users do not usually take the data with them out of convenience, but out of necessity. One feature of the NeXus files is that they utilize internal compression for the data much like a ".zip" file. The difference being that the metadata within the file are still readable. The HDF-5 utilities perform compression and decompression on behalf of the user. It is common to achieve 90-95% compression with the NeXus files. Thus copying a NeXus file to one's laptop and reading the data can produce an out-of-memory error. For instance, a typical 40MB to 50MB NeXus histogram file expands to approximately 4GB. The largest instrument at the SNS, TOPAZ [30] the data file will expand to almost 15GB. Thus the data are simply too large to process efficiently via typical user laptops, especially when one considers that data reduction typically requires processing a number of files concurrently – perhaps as many as six.

Realizing that the data are large, the facility also provides another convenience – facility produced and maintained data reduction applications are also made available via the portal. Integrated within the portal is a mechanism which enables GUI applications to be displayed locally on one's own computer, however the application is actually running back on facility computers. Note that paging techniques do exist which would make it possible to process such data on a laptop, however there is a question of the feasibility for processing very large data sets this way. Utilizing this web based "thin client" approach has a number of significant advantages:

1. The user always has access to the most recent version of the application without having to perform installation.
2. Data are kept proximal to computing thus the amount of time to move data is minimized.
3. The facility can provide access to substantial computing resources which users may not have available to themselves otherwise.

The portal is useful for providing individual researchers convenient access to computing and data, however it is already providing more than that by facilitating collaboration. In the past, data sharing required the interested parties to figure out where to place the data in a mutually accessible, and safe location. Depending upon the size of the data, finding such a location is not always easy else we would simply email it to our collaborators. Within the portal data hierarchy, each experiment proposal subdirectory has within it a 'shared' subdirectory to facilitate collaboration. These are the only read-write subdirectories within the repository as all others are read-only. In addition, only the experiment team that can access the experiment data can access these shared areas. Thus experiment team members can place data in an area common among themselves in order to share data. The shared areas are the fastest growing regions in the data repository. All of the above features ensure that the portal attracts a steady influx of users.

5. Future Plans

There are several notable growth areas for the portal ranging from data, computing and networking, and new application development. Integration with X-ray sources to make X-ray data available for co-analyses with neutron scattering data has good scientific impact potential. Working with the X-ray facilities is one step towards developing a user facility network (UFNet) which could be used to facilitate cross-cutting scientific research. Other portals exist in the X-ray and neutron scattering communities, and new portals are likely on the horizon. Inter-operating with these portals will be important but will bring significant challenges in federating their respective user authentication systems.

The future also looks to hold a new way of working by enabling users to utilise the ever-increasing portable computing power available to them. To illustrate, the web browser thin client approach does not require the user to install applications and relies upon the server-side for most of the work. But a "thick client" would enable a user to perform work on their local computer while also providing a remote interface to portal infrastructure enabling data movement and remote computation at the discretion of the user. Thus computing via a thick client would shift work away from a remote server onto a user's own computer based upon what makes sense to be done locally. Work has begun in producing a thick client via an Small Business Innovation Research (SBIR) project lead by Mark Green of Tech-X Corporation. Similarly, the ISAW package is also being adapted to leverage remote data and computing resources.

Successful next generation applications will enable scientists to more fully focus on their research and less on the nuances of computing, yet will provide access world class cyberinfrastructure and scientific computing applications. Users will have more freedom of choice and flexibility for how and where they perform data analysis and computing. The future of computing for facility users looks bright.

Acknowledgements

The authors would like to thank J.K. Zhao for the data used in the figures.

References

[1] NeSSI: http://neutrons.ornl.gov/workshops/nessi/NeSSI_agenda_linked.pdf
[2] SQRL: http://sqrl.eecs.utk.edu/
[3] NSTG: http://www.teragrid.org/gateways/projects.php?id=56
[4] McStas: http://www.mcstas.org/
[5] SNS: http://neutrons.ornl.gov/facilities/facilities_sns.shtml
[6] HFIR: http://neutrons.ornl.gov/hfir/hfir_facilities.shtml
[7] LENS: http://www.iucf.indiana.edu/lens/
[8] Lujan: http://lansce.lanl.gov/lujan/
[9] IPNS: http://www.pns.anl.gov/
[10] Tech-X Corporation: http://www.txcorp.com/
[11] UTK SEELAB: http://www.cs.utk.edu/~seelab/
[12] Diamond Synchrotron Facility: http://www.diamond.ac.uk/
[13] ORNL XCAMS: https://xcams.ornl.gov/xcams/
[14] Nomachine: http://www.nomachine.com/

[15] OIC Clusters: https://oic.ornl.gov/
[16] NSTG Cluster: http://www.ornl.teragrid.org/guides/userguide.html
[17] ISAW: http://www.pns.anl.gov/computing/isaw/index.shtml
[18] DAVE: http://www.ncnr.nist.gov/dave/
[19] DANSE: http://danse.us/
[20] EXPGUI: http://www.ncnr.nist.gov/programs/crystallography/software/expgui/expgui.html
[21] FullProf: http://www.ill.eu/sites/fullprof/php/programs.html
[22] GSAS: http://www.ccp14.ac.uk/solution/gsas/
[23] CSD: http://www.ccdc.cam.ac.uk/products/csd/
[24] ICSD: http://icsdweb.fiz-karlsruhe.de/
[25] SVN: http://subversion.tigris.org/
[26] Trac: http://trac.edgewall.org/
[27] NeXus: http://www.nexusformat.org/Main_Page
[28] HDF: http://www.hdfgroup.org/HDF5/
[29] IPTS: http://www.ornl.gov/sci/iums/ipts/
[30] TOPAZ : http://neutrons.ornl.gov/instrument_systems/beamline_12_topaz/index.shtml

The isotropic correlation function of plane figures: the triangle case.

Salvino Ciccariello
Universitá di Padova, Dipartimento di Fisica "G. Galilei",
via Marzolo 8, I-35131 Padova, Italy
E-mail: ciccariello@pd.infn.it

Abstract. The knowledge of the isotropic correlation function of a plane figure is useful to determine the correlation function of the cylinders having the plane figure as right-section and a given height as well as to analyze the out of plane intensity collected in grazing incidence small-angle scattering from a film formed by a particulate collection of these cylinders. The correlation function of plane polygons can always be determined in closed algebraic form. Here we report its analytic expression for the case of a triangle. The expressions take four different forms that depend on the relative order among the sides and the heights of the triangle.

1. Introduction

In small-angle scattering (SAS) theory samples are conceived of being made up of two/three homogeneous phases[1, 2]. In this way the collected scattering intensities are used to determine few structural parameters of the observed samples. In principle the procedure is very simple. Restricting ourselves for definiteness to the case of a two-phase particulate sample and assuming that the particle shape is unique, then the effective scattering density of the sample reads

$$n(\mathbf{r}) = (n_1 - n_2) \sum_{j-1}^{N} \rho({\mathcal{R}^{-1}}_{\hat{\Omega}_j}(\mathbf{r}/D_j) - \mathbf{R}_j), \tag{1}$$

where $(n_1 - n_2)$ is the contrast and the sum is performed over the N particles of the samples. Moreover, D_j, $\mathbf{R}_j$ and $\hat{\Omega}_j$ respectively denote the size, the mass-centre position of the jth particle and the three Euler angles specifying the rotation which the jth particle has undergone to. $\mathcal{R}_{\hat{\Omega}_j}$ denotes the corresponding rotation matrix. Finally, $\rho(\mathbf{r})$ is the function that specifies the particle shape since it is equal to one or zero depending on whether the tip of $\mathbf{r}$ falls inside or outside the particle of unit size ($D = 1$) with its centre (of mass) set at the origin. The observed scattering intensity $I(\mathbf{q})$ [with $q = |\mathbf{q}| = (4\pi/\lambda)\sin(\theta/2)$, λ denoting the ingoing beam particle wave-length and θ the scattering angle] is simply proportional to $|\tilde{n}(\mathbf{q})|^2$, the square modulus of the Fourier transform (FT) of Eq. (1). The fit of the observed $I(\mathbf{q})$ to $|\tilde{n}(\mathbf{q})|^2$, calculated by the procedure sketched below, allows one to fix the values of the parameters present in Eq. (1). In practice, before proceeding to the best fit analysis, it is necessary to exploit any information on the physical structure of the sample. To this aim one first considers the convolution of $n(\mathbf{r})$ and then uses the property that its FT yields the scattering intensity. The convolution involves the double sum $\sum_{j,l=1}^{N} T_j * T_l$ where T_j denotes the jth addend of Eq. (1). In handling the sum, it is convenient to separate isotropic from anisotropic samples. In the first case, the sum of the terms involving particles with different orientations is assumed to average to zero provided we substitute the remaining convolutions with their isotropic components. These, in turn, are obtained by suitably scaling the expression

$$\gamma_3(r) \equiv \frac{1}{4\pi V}\int d\hat{\omega}\int \rho(\mathbf{r}_1)\rho(\mathbf{r}_1 + r\hat{\omega})dv_1, \tag{2}$$

which defines the isotropic (auto)correlation function (CF) of the particle of unit size and volume V. The FT of $\gamma_3(r)$ is the particle (isotropic) form-factor. The scattering intensity is

$$I(q) \propto \sum_j D_j^6\tilde{\gamma}_3(D_jq) + \sum_{j\neq l} D_j^3 D_l^3\tilde{\gamma}_3(D_jq)\overline{\tilde{\gamma}_3(D_lq)}e^{-i\mathbf{q}\cdot(\mathbf{R}_j-\mathbf{R}_l)}, \tag{3}$$

where the tilde and the over-bar respectively denote the FT and the complex conjugate. Further approximations as the decoupling [1] and the local monodisperse one [3] allow us to simplify somewhat more expression (3). In any case, the final expression requires the knowledge of $\tilde{\gamma}_3(D_jq)$ which is algebraically known only for the spherical shape. It can numerically be evaluated by a one-dimensional integral for the parallelepiped[4, 5], the circular cylinder[6], the rotational ellipsoid[7] and the tetrahedron[8] because the CFs of these shapes are known in terms of transcendental functions. In all the other cases it requires a 3-5 dimensional integral, a condition that makes any numerical analysis with other particle shapes much more awkward. Recently it has been shown that the CF of a cylinder of arbitrary right section is related to the CF of the plane section by a generalized Abel transform[9, 10]. Thus, if one algebraically knows the correlation function of a plane figure, the CF of the associated cylinder is obtained by a quadrature and its FT by a further integration. Generalizing the fact that the CFs of the regular triangle, square[12], pentagon[13] and hexagon[14] are algebraically known[1], Ciccariello[11] has recently shown that the CF of any plane polygon can be determined in closed algebraic form. By this result it is now practically possible to consider cylindrical shapes more involved than the circular ones, see *e.g.* [16].
Surprisingly the algebraic expression of the CF of the simplest polygonal shape, namely that of a generic triangle, as yet is not known. In this paper we fill up the gap. This will be done in the next section. Before concluding this section we wish to emphasize that the mentioned property of the CF of a plane polygon can also be used in the case of some anisotropic samples. In fact, in the case of fiber samples (see, *e.g.*, Ref.[17]), one uses the anisotropic CF of a circular cylinder. By the aforesaid result it becomes now possible to consider the correlation function of a polygonal cylinder angularly averaged over the only directions perpendicular to the cylinder axis. This result can also be applied to the out of plane diffuse component of the intensity collected in grazing incidence small-angle scattering (GISAS) experiments[18, 19, 20] when the specular region is avoided so as to make the kinematical approximation satisfactory. See, in particular, the discussion reported in sect. III.A of [19].

Before passing to the illustration of the results relevant to the triangle, we recall three basic relations worked out in Ref.[11] to which one should defer for details. They generalize those of the three dimensional(3D) case and are respectively related to the values at the origin of the first, second and third derivative of the 2D CF. They read

$$\gamma'(0) = -L/\pi S, \tag{4}$$

$$\gamma''(0^+) = \frac{1}{2\pi S}\sum_{i=1}^{M}[(\pi - \psi_i)\cot\psi_i + 1], \tag{5}$$

[1] Very recently it has been determined the algebraic expression of the CF of a regular polygon whatever the side number[15].

Journal of Physics: Conference Series **247** (2010) 012014 doi:10.1088/1742-6596/247/1/012014

and

$$\gamma'''(0^+) = \frac{1}{4\pi S}\int_L \frac{1}{R^2{}_{curv}(\tau)}d\tau. \tag{6}$$

Here L and S are the perimeter length and the surface area, the ψ_is are the angles at the possible M vertices of the boundary of the figure and $R_{curv}(\tau)$ is the curvature radius at the point of the boundary with curvilinear coordinate τ.

2. The triangle case

According to Ref.[11] the distances that determine the intervals where $\gamma(r)$ takes an appropriate expression are the lengths of the sides and the heights of the considered triangle. Hence we must first establish the order existing among these distances. To this aim, given a triangle, we convey to name a the shortest side, c the longest one and b the third one so as to always have $a < b < c$. Vertices opposite to sides a, b and c are named A, B and C, and the angles at vertices A, B and C as α, β and γ (the context will avoid any confusion with the CF's symbol γ). The height from A to a is h_a and similarly for h_b and h_c (see figure 1). Owing to condition $a < b < c$ the following inequalities

$$a < b < c, \quad h_c < h_b < h_a \quad \text{and} \quad \alpha < \beta < \gamma \tag{7}$$

always hold true. To order the six lengths: a, b, c, h_a, h_b and h_c we need to know if $a < h_a$ or $a > h_a$ and if the triangle is acute or obtuse. Hence the four distinct shapes shown in Fig. 1.

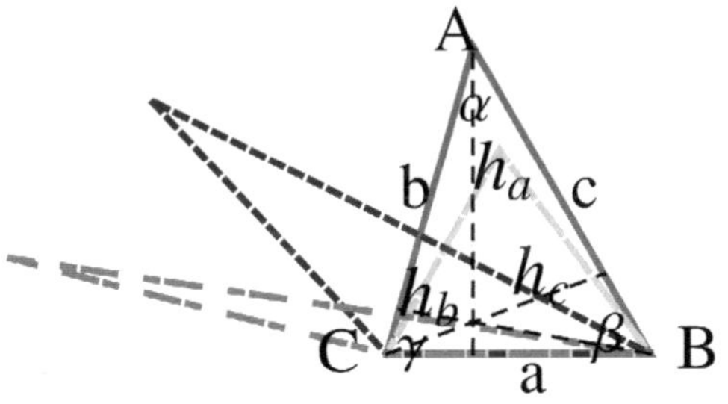

Figure 1. The triangle's four distinct cases $\mathcal{A}$, $\mathcal{B}$, $\mathcal{C}$ and $\mathcal{D}$. Red continuous sides → shape $\mathcal{A}$, green dashed sides → shape $\mathcal{B}$, blue short-dashed sides → shape $\mathcal{C}$ and magenta dot-dashed sides → shape $\mathcal{D}$. The side lengths of the shown $\mathcal{A}$, $\mathcal{B}$, $\mathcal{C}$ and $\mathcal{D}$ triangles are $(1, 1.5, 1.611)$, $(1, 1.06, 1.127)$, $(1, 1.5, 2.239)$ and $(1, 1.5, 2.470)$, respectively.

Their names respectively are:

$$\begin{array}{lllll}
\mathcal{A} & \rightarrow & h_c < h_b < a < h_a < b < c, & \gamma < \pi/2 \\
\mathcal{B} & \rightarrow & h_c < h_b < h_a < a < b < c, & \gamma < \pi/2 \\
\mathcal{C} & \rightarrow & h_c < h_b < a < h_a < b < c, & \gamma > \pi/2 \\
\mathcal{D} & \rightarrow & h_c < h_b < h_a < a < b < c, & \gamma > \pi/2.
\end{array} \tag{8}$$

These inequalities show that we have six r-intervals. In case $\mathcal{A}$ they are: $[0,\, h_c]$, $[h_c,\, h_b]$, $[h_b,\, a]$, $[a,\, h_a]$, $[h_a,\, b]$ and $[b,\, c]$. They will be referred to as $I\mathcal{A}$, $II\mathcal{A}$, $III\mathcal{A}$, $IV\mathcal{A}$, $V\mathcal{A}$ and $VI\mathcal{A}$, respectively. The same convention is adopted in cases $\mathcal{B}$, $\mathcal{C}$ and $\mathcal{D}$. Since intervals $I\mathcal{A}$, $I\mathcal{B}$, $I\mathcal{C}$ and $I\mathcal{D}$ coincide we shall simply speak of interval I. The same happens for intervals II and VI. In each of the subcases listed in Eq.(8) we can apply the procedure expounded in ref.[11]

and determine the expression of the second derivative of the (isotropic) CF of the considered triangle. The procedure is straightforward though somewhat cumbersome. After putting

$$\mathcal{A}(\alpha,\beta,\gamma) \equiv 3 + (\pi-\alpha)\cot\alpha + (\pi-\beta)\cot\beta + (\pi-\gamma)\cot\gamma, \tag{9}$$

$$\mathcal{B}(\alpha,\beta,\gamma) \equiv (\pi-2\alpha)\cot\alpha + (\pi-2\beta)\cot\beta + 2(1+\gamma\cot\gamma), \tag{10}$$

$$\mathcal{C}(\alpha,\beta,\gamma) \equiv 4 + 2(\pi-2\alpha)\cot\alpha + \pi\cot\beta + \pi\cot\gamma, \tag{11}$$

$$c(\alpha,\beta) \equiv \cot\alpha\, +\, \cot\beta \tag{12}$$

and

$$\Omega_2(x) \equiv -2x\sqrt{1-x^2} - 2\arccos(x), \tag{13}$$

the final expressions of $\gamma''(r)$ are

$$\begin{aligned}
\gamma''_{I}(r) &= \mathcal{A}(\alpha,\beta,\gamma)/(2\pi S), & 0<r<h_c; \quad (14)\\
\gamma''_{II}(r) &= [c(\alpha,\beta)\Omega_2(h_c/r) + \mathcal{A}(\alpha,\beta,\gamma)]/(2\pi S), & h_c<r<h_b; \quad (15)\\
\gamma''_{IIIA}(r) &= [c(\alpha,\beta)\Omega_2(h_c/r) + c(\alpha,\gamma)\Omega_2(h_b/r) + \mathcal{A}(\alpha,\beta,\gamma)]/(2\pi S), & h_b<r<a; \quad (16)\\
\gamma''_{IVA}(r) &= [c(\alpha,\beta)\Omega_2(h_c/r) + c(\alpha,\gamma)\Omega_2(h_b/r) + \mathcal{C}(\alpha,\beta,\gamma)]/(4\pi S), & a<r<h_a; \quad (17)\\
\gamma''_{VA}(r) &= [c(\alpha,\beta)\Omega_2(h_c/r) + 2c(\beta,\gamma)\Omega_2(h_a/r) + c(\alpha,\gamma)\Omega_2(h_b/r) & (18)\\
&\qquad +\mathcal{C}(\alpha,\beta,\gamma)]/(4\pi S), & h_a<r<b;\\
\gamma''_{IIIB}(r) &= \gamma''_{IIIA}(r), & h_b<r<h_a; \quad (19)\\
\gamma''_{IVB}(r) &= [c(\alpha,\beta)\Omega_2(h_c/r) + c(\beta,\gamma)\Omega_2(h_a/r) + c(\alpha,\gamma)\Omega_2(h_b/r) & (20)\\
&\qquad +\mathcal{A}(\alpha,\beta,\gamma)]/2\pi S, & h_a<r<a;\\
\gamma''_{VB}(r) &= \gamma''_{VA}(r), & a<r<b; \quad (21)\\
\gamma''_{IIIC}(r) &= \gamma''_{II}(r), & h_b<r<a; \quad (22)\\
\gamma''_{IVC}(r) &= \gamma''_{IVA}(r), & a<r<h_a; \quad (23)\\
\gamma''_{VC}(r) &= \gamma''_{IVC}(r), & h_a<r<b; \quad (24)\\
\gamma''_{IIID}(r) &= \gamma''_{IIIC}(r), & h_b<r<h_a; \quad (25)\\
\gamma''_{IVD}(r) &= \gamma''_{II}(r), & h_a<r<a; \quad (26)\\
\gamma''_{VD}(r) &= \gamma''_{IVA}(r), & a<r<b; \quad (27)\\
\gamma''_{VI}(r) &= [c(\alpha,\gamma)\Omega_2(h_b/r) + c(\beta,\gamma)\Omega_2(h_a/r) + \mathcal{B}(\alpha,\beta,\gamma)]/(4\pi S), & b<r<c. \quad (28)
\end{aligned}$$

From these expressions the calculation of $\gamma'(r)$ is straightforward after noting that the primitive of $\Omega_2(h/x)$ is function $\Omega_1(h,x)$ reported below

$$\Omega_1(h,x) \equiv \frac{2h\sqrt{x^2-h^2}}{x} - 2x\arccos(h/x). \tag{29}$$

In fact, integrating Eq. (28) over the interval $[r,\, c]$ with $b<r<c$ one finds that

$$\begin{aligned}
\gamma'_{VI}(r) &= [\mathcal{B}(\alpha,\beta,\gamma)(r-c) + c(\alpha,\gamma)(\Omega_1(h_b,r) - \Omega_1(h_b,c)) + \\
&\quad c(\beta,\gamma)(\Omega_1(h_a,r) - \Omega_1(h_a,c))]/4\pi S, \quad b<r<c,
\end{aligned}$$

where we used the property that $\gamma'(r)$ is continuous at $r=c$ and therefore equal to zero there. It is now convenient to put

$$\hat{\Omega}_1(h_a,r,b) \equiv \Omega_1(h_a,r) - \Omega_1(h_a,b). \tag{30}$$

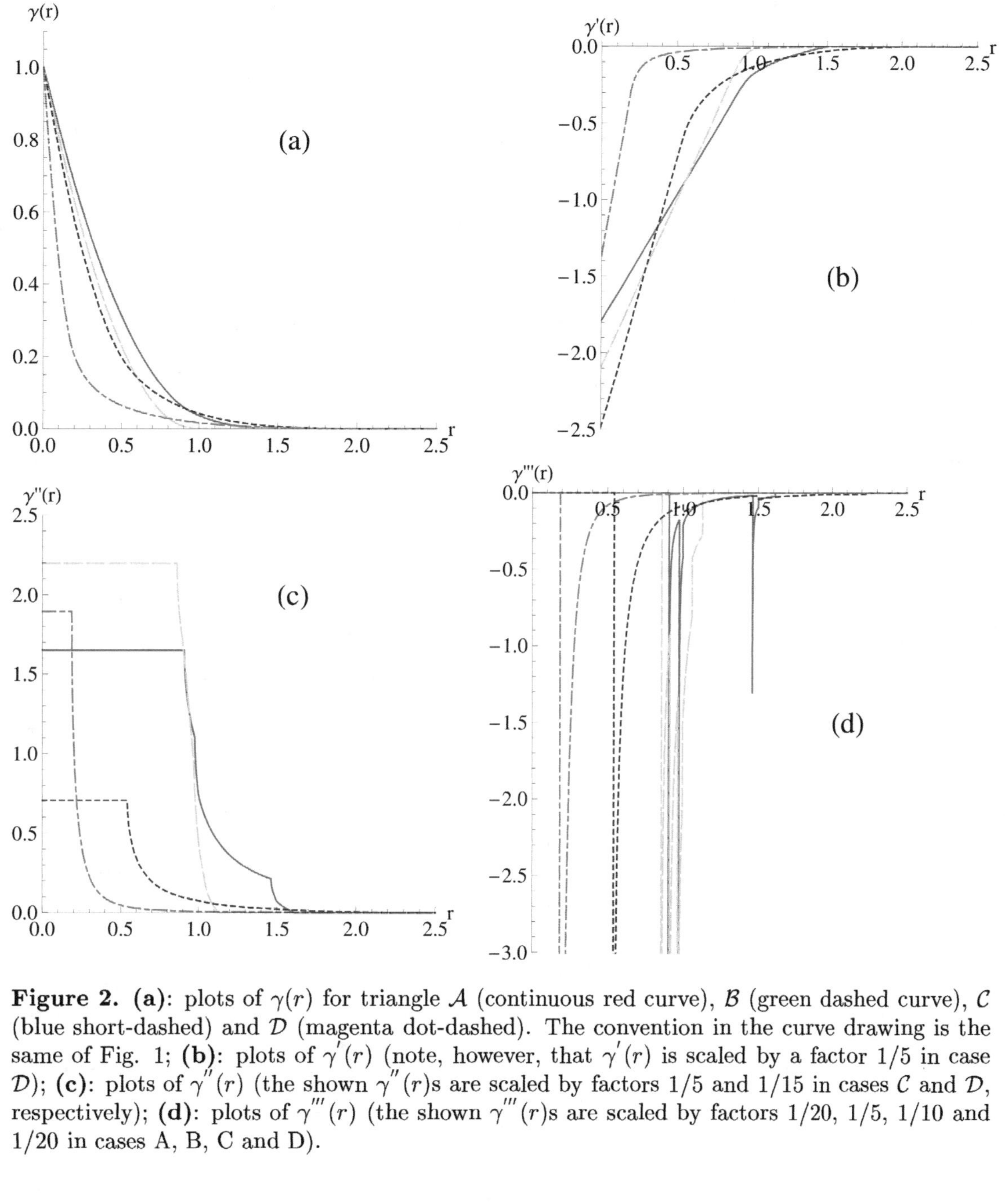

Figure 2. (a): plots of $\gamma(r)$ for triangle $\mathcal{A}$ (continuous red curve), $\mathcal{B}$ (green dashed curve), $\mathcal{C}$ (blue short-dashed) and $\mathcal{D}$ (magenta dot-dashed). The convention in the curve drawing is the same of Fig. 1; **(b)**: plots of $\gamma'(r)$ (note, however, that $\gamma'(r)$ is scaled by a factor 1/5 in case $\mathcal{D}$); **(c)**: plots of $\gamma''(r)$ (the shown $\gamma''(r)$s are scaled by factors 1/5 and 1/15 in cases $\mathcal{C}$ and $\mathcal{D}$, respectively); **(d)**: plots of $\gamma'''(r)$ (the shown $\gamma'''(r)$s are scaled by factors 1/20, 1/5, 1/10 and 1/20 in cases A, B, C and D).

The previous derivative expression then reads

$$\gamma'_{VI}(r) \;=\; \frac{\mathcal{B}(\alpha,\beta,\gamma)(r-c)+c(\alpha,\gamma)\hat{\Omega}_1(h_b,r,c)+c(\beta,\gamma)\hat{\Omega}_1(h_a,r,c)}{4\pi S}, \quad b<r<c, \tag{31}$$

In the same way, integrating Eq. (18) over the interval $[r,\, b]$ with $h_a < r < b$ and imposing the continuity at $r = b$, one finds that

$$\gamma'_{V\mathcal{A}}(r) \;=\; \frac{1}{4\pi S}[2c(\beta,\gamma)\hat{\Omega}_1(h_a,r,b)+c(\alpha,\gamma)\hat{\Omega}_1(h_b,r,b)+c(\alpha,\beta)\hat{\Omega}_1(h_c,r,b) \tag{32}$$

$$+\mathcal{C}(\alpha,\beta,\gamma)(r-b)]+\gamma'_{VI}(b), \quad h_a<r<b,$$

and then, step by step, that

$$\gamma'_{IVA}(r) = \frac{1}{4\pi S}[c(\alpha,\gamma)\hat{\Omega}_1(h_b,r,h_a)+c(\alpha,\beta)\hat{\Omega}_1(h_c,r,h_a)+\mathcal{C}(\alpha,\beta,\gamma)(r-h_a)] \quad (33)$$
$$+\gamma'_{VA}(h_a), \quad a<r<h_a;$$

$$\gamma'_{IIIA}(r) = \frac{1}{2\pi S}[c(\alpha,\gamma)\hat{\Omega}_1(h_b,r,a)+c(\alpha,\beta)\hat{\Omega}_1(h_c,r,a)+\mathcal{A}(\alpha,\beta,\gamma)(r-a)] \quad (34)$$
$$+\gamma'_{IVA}(a), \quad h_b<r<a;$$

$$\gamma'_{IIA}(r) = \frac{c(\alpha,\beta)\hat{\Omega}_1(h_c,r,b)+\mathcal{A}(\alpha,\beta,\gamma)(r-h_b)}{2\pi S}+\gamma'_{IIIA}(h_b), \quad h_c<r<h_b; \quad (35)$$

$$\gamma'_{IA}(r) = \frac{\mathcal{A}(\alpha,\beta,\gamma)(r-h_c)}{2\pi S}+\gamma'_{IIA}(h_c), \quad 0<r<h_c. \quad (36)$$

The expressions of $\gamma'(r)$ are similarly obtained in the other three cases and are reported in the appendix. In the same way, from the resulting expressions of the first derivative, using the fact that a primitive of $(\Omega_1(h,x)$ is

$$\Omega(h,x)\equiv -\pi x^2/2+3h\sqrt{x^2-h^2}+(x^2+2h^2)\arcsin(h/x), \quad (37)$$

starting from the outermost r-interval and imposing the continuity at the outermost end points one obtains the algebraic expression of the CF of a generic triangle. For instance, the CF expression in the outermost interval is

$$\gamma_{VI}(r) = \Big[\mathcal{B}(\alpha,\beta,\gamma)(r-c)^2/2+c(\alpha,\gamma)(\Omega(h_b,r)-\Omega(h_b,c)-\Omega_1(h_b,c)(r-c))+$$
$$c(\beta,\gamma)(\Omega(h_a,r)-\Omega(h_a,c)-\Omega_1(h_a,c)(r-c))\Big]/4\pi S. \quad (38)$$

This expression, after putting

$$\hat{\Omega}(h_a,r,b)\equiv\Omega(h_a,r)-\Omega(h_a,b)-\Omega_1(h_a,b)(r-b), \quad (39)$$

simplifies into

$$\gamma_{VI}(r) = \frac{\mathcal{B}(\alpha,\beta,\gamma)(r-c)^2/2+c(\alpha,\gamma)\hat{\Omega}(h_b,r,c)+c(\beta,\gamma)\hat{\Omega}(h_a,r,c)}{4\pi S}, \quad b<r<c. \quad (40)$$

In the remaining interval, for case $\mathcal{A}$ one finds

$$\gamma_{VA}(r) = \frac{1}{4\pi S}\Big[\mathcal{C}(\alpha,\beta,\gamma)(r-b)^2/2+c(\alpha,\beta)\hat{\Omega}(h_c,r,b)+c(\alpha,\gamma)\hat{\Omega}(h_b,r,b)+ \quad (41)$$
$$+2c(\beta,\gamma)\hat{\Omega}(h_a,r,b)\Big]+\gamma'_{VI}(b)(r-b)+\gamma_{VI}(b), \quad h_a<r<b,$$

$$\gamma_{IVA}(r) = \frac{1}{4\pi S}\Big[\mathcal{C}(\alpha,\beta,\gamma)(r-h_a)^2/2+c(\alpha,\gamma)\hat{\Omega}(h_b,r,h_a)+c(\alpha,\beta)\hat{\Omega}(h_c,r,h_a)\Big] \quad (42)$$
$$+\gamma'_{VA}(h_a)(r-h_a)+\gamma_{VA}(h_a), \quad a<r<h_a,$$

$$\gamma_{IIIA}(r) = \frac{1}{4\pi S}\Big[\mathcal{A}(\alpha,\beta,\gamma)(r-a)^2/2+c(\alpha,\gamma)\hat{\Omega}(h_b,r,a)+c(\alpha,\beta)\hat{\Omega}(h_c,r,a)\Big] \quad (43)$$
$$+\gamma'_{IVA}(a)(r-a)+\gamma_{IVA}(a), \quad h_b<r<a,$$

$$\gamma_{IIA}(r) = \frac{1}{2\pi S}\Big[\mathcal{C}(\alpha,\beta,\gamma)(r-h_b)^2/2+c(\alpha,\beta)\hat{\Omega}(h_c,r,h_b)\Big] \quad (44)$$
$$+\gamma'_{IIIA}(h_b)(r-h_b)+\gamma_{IIIA}(h_b), \quad h_c<r<h_b,$$

$$\gamma_{IA}(r) = \frac{\mathcal{A}(\alpha,\beta,\gamma)(r-h_c)^2}{4\pi S}+\gamma'_{IIA}(h_c)(r-h_c)+\gamma_{IIA}(h_c), \quad 0<r<h_c. \quad (45)$$

The expressions relevant to the other cases are given in the appendix. Figures 2a-d show the plots of $\gamma(r)$ and its first three derivatives for the four triangle cases. The CFs look each other rather similar because the different supports cannot be appreciated on the figure scale. The first derivatives show greater difference since their behaviours about $r = 0$ signal that the specific length (i.e. the ratio perimeter/surface, see Eq.(4)) increases as one passes from case $\mathcal{A}$ to $\mathcal{D}$ (note that the magenta curve is scaled). The differences become more evident for the second derivatives and dramatic for the third one. Here the cuspids signal algebraic singularities $\propto (r-h)^{-1/2}$ and they involve the only heights which meet the opposite side within the triangle. This explains why Fig. 3d shows three spikes in cases $\mathcal{A}$ and $\mathcal{B}$ and only one spike in cases $\mathcal{C}$ and $\mathcal{D}$.

3. Conclusions

We have briefly described the cases where the algebraic knowledge of the CF of a plane figure can usefully be applied to analyze observed scattering intensities. In particular we have reported here the explicit expression of the CF of a general triangle. Besides their intrinsic interest for the shown singularities, which are important from the point of view of the figure reconstruction, these expressions allow us to evaluate the CF of the associated cylinder by a simple quadrature.

Appendix A. Expressions for cases $\mathcal{B}$, $\mathcal{C}$ and $\mathcal{D}$

For completeness we give the expressions of the CF and its 1st derivative in remaining cases $\mathcal{B}$, $\mathcal{C}$ and $\mathcal{D}$. For the first derivative one finds

$$\gamma'_{VI\mathcal{B}}(r) = \gamma'_{VI}(r), \qquad b < r < c; \tag{A.1}$$

$$\gamma'_{V\mathcal{B}}(r) = \frac{1}{4\pi S}\Big[2c(\beta,\gamma)\hat{\Omega}_1(h_a,r,b) + c(\alpha,\gamma)\hat{\Omega}_1(h_b,r,b) + c(\alpha,\beta)\hat{\Omega}_1(h_c,r,b) + (r-b)\mathcal{C}(\alpha,\beta,\gamma)\Big] + \gamma'_{VI}(b), \qquad a < r < b; \tag{A.2}$$

$$\gamma'_{IV\mathcal{B}}(r) = \frac{1}{2\pi S}\Big[c(\alpha,\beta)\hat{\Omega}_1(h_c,r,a) + c(\beta,\gamma)\hat{\Omega}_1(h_a,r,a) + c(\alpha,\gamma)\hat{\Omega}_1(h_b,r,a) + \mathcal{A}(\alpha,\beta,\gamma)(r-a)\Big] + \gamma'_{V\mathcal{B}}(a), \qquad h_a < r < a; \tag{A.3}$$

$$\gamma'_{III\mathcal{B}}(r) = \frac{1}{2\pi S}\Big[c(\alpha,\beta)\hat{\Omega}_1(h_c,r,h_a) + c(\alpha,\gamma)\hat{\Omega}_1(h_b,r,h_a) + \mathcal{A}(\alpha,\beta,\gamma)(r-h_a)\Big] + \gamma'_{IV\mathcal{B}}(h_a), \qquad h_b < r < h_a; \tag{A.4}$$

$$\gamma'_{II\mathcal{B}}(r) = \frac{c(\alpha,\beta)\left(\Omega_1(h_c,r) - \Omega_1(h_c,h_b)\right) + \mathcal{A}(\alpha,\beta,\gamma)(r-h_b)}{2\pi S} + \gamma'_{II\mathcal{B}}(h_b), \qquad quad \quad h_c < r < h_b; \tag{A.5}$$

$$\gamma'_{I\mathcal{B}}(r) = \frac{\mathcal{A}(\alpha,\beta,\gamma)(r-h_c)}{2\pi S} + \gamma'_{II\mathcal{B}}(h_c), \qquad 0 < r < h_c; \tag{A.6}$$

in case $\mathcal{B}$,

$$\gamma'_{VI\mathcal{C}}(r) = \gamma'_{VI}(r), \qquad b < r < c; \tag{A.7}$$

$$\gamma'_{V\mathcal{C}}(r) = \frac{1}{4\pi S}[c(\alpha,\gamma)\hat{\Omega}_1(h_b,r,b) + c(\alpha,\beta)\hat{\Omega}_1(h_c,r,b) + \mathcal{C}(\alpha,\beta,\gamma)(r-b)] + \gamma'_{VI\mathcal{C}}(b), \qquad h_a < r < b; \tag{A.8}$$

$$\gamma'_{IV\mathcal{C}}(r) = \frac{1}{4\pi S}[c(\alpha,\gamma)\hat{\Omega}_1(h_b,r,h_a) + c(\alpha,\beta)\hat{\Omega}_1(h_c,r,h_a) + \mathcal{C}(\alpha,\beta,\gamma)(r-h_a)] + \gamma'_{V\mathcal{C}}(h_a), \qquad a < r < h_a; \tag{A.9}$$

$$\gamma'_{IIIC}(r) = \frac{c(\alpha,\beta)\hat{\Omega}_1(h_c,r,a)+\mathcal{A}(\alpha,\beta,\gamma)(r-a)}{2\pi S}+\gamma'_{IVC}(a), \quad h_b<r<a; \tag{A.10}$$

$$\gamma'_{IIC}(r) = \frac{c(\alpha,\beta)\hat{\Omega}_1(h_c,r,h_b)+\mathcal{A}(\alpha,\beta,\gamma)(r-h_b)}{2\pi S}+\gamma'_{IIIC}(h_b), \quad h_c<r<h_b; \tag{A.11}$$

$$\gamma'_{IC}(r) = \frac{\mathcal{A}(\alpha,\beta,\gamma)(r-h_c)}{2\pi S}+\gamma'_{IIC}(h_c), \quad 0<r<h_c; \tag{A.12}$$

in case $\mathcal{C}$, and

$$\gamma'_{VI\mathcal{D}}(r) = \gamma'_{VI}(r), \quad b<r<c; \tag{A.13}$$

$$\gamma'_{V\mathcal{D}}(r) = \frac{1}{4\pi S}[c(\alpha,\gamma)\hat{\Omega}_1(h_b,r,b)+c(\alpha,\beta)\hat{\Omega}_1(h_c,r,b)+\mathcal{C}(\alpha,\beta,\gamma)(r-b)] \tag{A.14}$$
$$+\gamma'_{VI\mathcal{D}}(a), \quad a<r<b;$$

$$\gamma'_{IV\mathcal{D}}(r) = \frac{\mathcal{A}(\alpha,\beta,\gamma)(r-a)+c(\alpha,\beta)\hat{\Omega}_1(h_c,r,a)}{2\pi S}+\gamma'_{V\mathcal{D}}(a), \quad h_a<r<a; \tag{A.15}$$

$$\gamma'_{III\mathcal{D}}(r) = \frac{\mathcal{A}(\alpha,\beta,\gamma)(r-h_a)+c(\alpha,\beta)+\hat{\Omega}_1(h_c,r,h_b)}{2\pi S}+\gamma'_{IV\mathcal{D}}(h_a), \quad h_b<r<h_a; \tag{A.16}$$

$$\gamma'_{II\mathcal{D}}(r) = \frac{c(\alpha,\beta)\hat{\Omega}_1(h_c,r,h_b)+\mathcal{A}(\alpha,\beta,\gamma)(r-h_b)}{2\pi S}+\gamma'_{III\mathcal{D}}(h_b), \quad h_c<r<h_b; \tag{A.17}$$

$$\gamma'_{I\mathcal{D}}(r) = \frac{\mathcal{A}(\alpha,\beta,\gamma)(r-h_c)}{2\pi S}+\gamma'_{II\mathcal{D}}(h_c), \quad 0<r<h_c \tag{A.18}$$

in case $\mathcal{D}$.
For the CF one finds

$$\gamma_{VI\mathcal{B}}(r) = \gamma_{VI}(r), \quad b<r<c \tag{A.19}$$

$$\gamma_{V\mathcal{B}}(r) = \frac{1}{4\pi S}\Big[2c(\beta,\gamma)\hat{\Omega}(h_a,r,b)+c(\alpha,\gamma)\hat{\Omega}(h_b,r,b)+c(\alpha,\beta)\hat{\Omega}(h_c,r,b) \tag{A.20}$$
$$+\mathcal{C}(\alpha,\beta,\gamma)\frac{(r-b)^2}{2}\Big]+\gamma'_{VI}(b)(r-b)+\gamma_{VI}(b), \quad a<r<b;$$

$$\gamma_{IV\mathcal{B}}(r) = \frac{1}{2\pi S}\Big[c(\beta,\gamma)\hat{\Omega}(h_a,r,a)+c(\alpha,\gamma)\hat{\Omega}(h_b,r,a)+c(\alpha,\beta)\hat{\Omega}(h_c,r,a) \tag{A.21}$$
$$+\mathcal{A}(\alpha,\beta,\gamma)\frac{(r-a)^2}{2}\Big]+\gamma'_{V\mathcal{B}}(a)(r-a)+\gamma_{V\mathcal{B}}(a), \quad h_a<r<a;$$

$$\gamma_{III\mathcal{B}}(r) = \frac{1}{2\pi S}\Big[c(\alpha,\beta)\hat{\Omega}(h_c,r,h_a)+c(\alpha,\gamma)\hat{\Omega}(h_b,r,h_a)+\mathcal{A}(\alpha,\beta,\gamma)\frac{(r-h_a)^2}{2}\Big] \tag{A.22}$$
$$+\gamma'_{IV\mathcal{B}}(h_a)(r-h_a)+\gamma_{IV\mathcal{B}}(h_a), \quad h_b<r<h_a;$$

$$\gamma_{II\mathcal{B}}(r) = \frac{1}{2\pi S}\Big[c(\alpha,\beta)\hat{\Omega}(h_c,r,h_b)+\mathcal{A}(\alpha,\beta,\gamma)\frac{(r-h_b)^2}{2}\Big] \tag{A.23}$$
$$+\gamma'_{III\mathcal{B}}(h_b)(r-h_b)+\gamma_{III\mathcal{B}}(h_b), \quad h_c<r<h_b;$$

$$\gamma_{I\mathcal{B}}(r) = \frac{\mathcal{A}(\alpha,\beta,\gamma)(r-h_c)^2}{4\pi S}+\gamma'_{II\mathcal{B}}(h_c)(r-h_c)+\gamma_{II\mathcal{B}}(h_c), \quad 0<r<h_c \tag{A.24}$$

in case $\mathcal{B}$,

$$\gamma_{VI\mathcal{C}}(r) = \gamma_{VI}(r), \quad b<r<c; \tag{A.25}$$

$$\gamma_{V\mathcal{C}}(r) = \frac{1}{4\pi S}\Big[c(\alpha,\beta)\hat{\Omega}(h_c,r,b)+c(\alpha,\gamma)\hat{\Omega}(h_b,r,b)+\mathcal{C}(\alpha,\beta,\gamma)(r-b)^2/2\Big] \tag{A.26}$$

$$
\begin{aligned}
& +\gamma'_{VI}(b)(r-b)+\gamma_{VI}(b), && h_a<r<b; \\
\gamma_{IVC}(r) &= \frac{1}{4\pi S}\Big[c(\alpha,\beta)\hat{\Omega}(h_c,r,h_a)+c(\alpha,\gamma)\hat{\Omega}(h_b,r,h_a)+\mathcal{C}(\alpha,\beta,\gamma)(r-h_a)^2/2\Big] && \text{(A.27)} \\
& +\gamma'_{VC}(h_a)(r-h_a)+\gamma_{VC}(h_a), && a<r<h_a; \\
\gamma_{IIIC}(r) &= \frac{1}{2\pi S}\Big[c(\alpha,\beta)\hat{\Omega}(h_c,r,a)+\mathcal{A}(\alpha,\beta,\gamma)\frac{(r-a)^2}{2}\Big] && \text{(A.28)} \\
& +\gamma'_{IVC}(a)(r-a)+\gamma_{IVC}(a), && h_b<r<a; \\
\gamma_{IIC}(r) &= \frac{1}{2\pi S}\Big[c(\alpha,\beta)\hat{\Omega}(h_c,r,h_b)+\mathcal{A}(\alpha,\beta,\gamma)\frac{(r-h_b)^2}{2}\Big] && \text{(A.29)} \\
& +\gamma'_{IIIC}(h_b)(r-h_b)+\gamma_{IIIC}(h_b), && h_c<r<h_b; \\
\gamma_{IC}(r) &= \frac{\mathcal{A}(\alpha,\beta,\gamma)(r-h_c)^2}{4\pi S}+\gamma'_{IIC}(h_c)(r-h_c)+\gamma_{IIC}(h_c), \quad 0<r<h_c && \text{(A.30)}
\end{aligned}
$$

in case $\mathcal{C}$, and

$$
\begin{aligned}
\gamma_{VI\mathcal{D}}(r) &= \gamma_{VI}(r), && b<r<c; \quad \text{(A.31)} \\
\gamma_{V\mathcal{D}}(r) &= \frac{1}{4\pi S}\Big[c(\alpha,\beta)\hat{\Omega}(h_c,r,b)+c(\alpha,\gamma)\hat{\Omega}(h_b,r,b)+\mathcal{C}(\alpha,\beta,\gamma)\frac{(r-b)^2}{2}\Big] && \text{(A.32)} \\
& +\gamma'_{VI}(b)(r-b)+\gamma_{VI}(b), && a<r<b; \\
\gamma_{IV\mathcal{D}}(r) &= \frac{1}{2\pi S}\Big[c(\alpha,\beta)\hat{\Omega}(h_c,r,a)+\mathcal{A}(\alpha,\beta,\gamma)\frac{(r-a)^2}{2}\Big] && \text{(A.33)} \\
& +\gamma'_{V\mathcal{D}}(a)(r-a)+\gamma_{V\mathcal{D}}(a), && h_a<r<a; \\
\gamma_{III\mathcal{D}}(r) &= \frac{1}{2\pi S}\Big[c(\alpha,\beta)\hat{\Omega}(h_c,r,h_a)+\mathcal{A}(\alpha,\beta,\gamma)\frac{(r-h_a)^2}{2}\Big] && \text{(A.34)} \\
& +\gamma'_{IV\mathcal{D}}(h_a)(r-h_a)+\gamma_{IV\mathcal{D}}(h_a), && h_b<r<h_a; \\
\gamma_{II\mathcal{D}}(r) &= \frac{1}{2\pi S}\Big[c(\alpha,\beta)\hat{\Omega}(h_c,r,h_b)+\mathcal{A}(\alpha,\beta,\gamma)\frac{(r-h_b)^2}{2}\Big] && \text{(A.35)} \\
& +\gamma'_{III\mathcal{D}}(h_b)(r-h_b)+\gamma_{III\mathcal{D}}(h_b), && h_c<r<h_b; \\
\gamma_{I\mathcal{D}}(r) &= \frac{\mathcal{A}(\alpha,\beta,\gamma)(r-h_c)^2}{4\pi S}+\gamma'_{II\mathcal{D}}(h_c)(r-h_c)+\gamma_{II\mathcal{D}}(h_c), \quad 0<r<h_c && \text{(A.36)}
\end{aligned}
$$

in case $\mathcal{D}$.

Acknowledgment
I am grateful to Dr. Wilfried Gille since his critical reading of the manuscript led to a considerable improvement in the formulae's presentation.

References

[1] Guinier A and Fournét G 1955 *Small-Angle Scattering of X-rays* (New York: John Wiley)
[2] Debye P, Anderson HR and Brumberger H 1957 *J Appl Phys* **20** 518
[3] Pedersen JS, Vyskocil P, Schönfeld B and Kostorz G 1997 *J Appl Cryst* **30** 975
[4] Goodisman J 1980 *J Appl Cryst* **13** 132
[5] Gille W 1987 *Exp Tech Phys* **35** 93
[6] Ciccariello S 1991 *J Appl Cryst* **24** 509
[7] Burger C and Ruland W 2001 *Acta Cryst A* **57** 482
[8] Ciccariello S 2005 *J Appl Cryst* A**38** 97
[9] Gille W 2002 *Preprint Dept of Physics, Halle University*, unpublished
[10] Ciccariello S 2003 *Acta Cryst* A**59** 506

[11] Ciccariello S 2009 *J Math Phys* **50** 103527
[12] Sulanke R 1961 *Math Nach* **23** 51
[13] Harutyunyan HS 2007 *Scientific Letters of the Yerevan State University* **1** 17
[14] Aharonyan NG and Ohanyan VK 2005 *J of Contemporary Mathematical Analysis (Armenian Academy of Sciences)* **40** 43
[15] Harutyunyan HS and Ohanyan VK 2009 *Adv Appl Prob* **41** 358
[16] Gille W, Aharonyan NG and Ohanyan VK 2009 *J Appl Cryst* **42** 326
[17] Burger C 2010 *Polymer Reviews* in the press
[18] Sinha SK, Sirota EB, Garoff S and Stanley HB 1988 *Phys Rev B* **38** 2297
[19] Rauscher M, Salditt T and Spohn H 1995 *Phys Rev B* **52** 16855
[20] Lazzari R 2002 *J Appl Cryst* **35** 406

Synaptic Vesicles Studied by SAXS: Derivation and Validation of a Model Form Factor

S Castorph[1], L Arleth[2], M Sztucki[3], U Vainio[4], S K Ghosh[1], M Holt[5], R Jahn[5] and T Salditt[1]

[1] Institut für Röntgenphysik, Georg-August-Universität Göttingen, Göttingen, D
[2] Biophysics, Faculty of Life Sciences, University of Copenhagen, Frederiksberg, DK
[3] European Synchrotron Radiation Facility, Grenoble, F
[4] Hamburger Synchrotronstrahlungslabor at Deutsches Elektronen-Synchrotron, Hamburg, D
[5] Max Planck Institut für Biophysikalische Chemie, Department of Neurobiology, Göttingen, D

E-mail: `scastor@gwdg.de, tsaldit@gwdg.de`

Abstract. We discuss different spherically symmetric and anisotropic form factor models and test them against high resolution synchrotron based small-angle x-ray scattering (SAXS) data from synaptic vesicles (SVs), isolated from rat brain. Anisotropy of the model form factors is found to be a key ingredient for the description of the native synaptic vesicle structure. We describe changes in structural parameters due to protease digestion of SVs, and present SAXS data of SVs recorded under different pH conditions.

1. Introduction

The release of neurotransmitters from neurons, in response to stimulation, forms the basis of biochemical communication in the nervous system. Neurotransmitters are stored in small membranous organelles, synaptic vesicles (SVs), within the presynaptic terminal. These vesicles undergo an elaborate cycle of fusion with the plasma membrane (releasing neurotransmitter), followed by retrieval and reformation and transport back to the plasma membrane for further rounds of fusion [1].

In recent years there has been enormous progress in our knowledge of the molecular composition and structure of SVs [2]. Recently, we have determined the average radial density profile $\rho(r)$, as well as the polydispersity function $p(R)$, on an absolute scale from small angle x-ray scattering (SAXS), with no free prefactors [3]. The density profile of the vesicle with an inner and outer protein layer has confirmed a previous modeling study of SVs based on biochemical studies of protein inventory and stochiometry as well as the crystal structures of the constituent proteins [2]. In addition, the best model fits to the SAXS curves required the presence of laterally anisotropic structures on the vesicle surface, indicative of large protein clusters.

In this work we present the full derivation of the form factors used and give a more detailed account of the comparison between different models as well as a more comprehensive overview of different SV data sets. In particular, we present different spherically symmetric form factors, which are then tested and falsified. In particular both radially symmetric and asymmetric density profiles consisting of three or five coupled Gaussians fail to describe the structure of

Journal of Physics: Conference Series **247** (2010) 012015 doi:10.1088/1742-6596/247/1/012015

SVs. Contrarily, an anisotropic form factor representing a symmetric bilayer profile, consisting of three Gaussians, with Gaussian chains attached to the outside and inside describes the SV structure well. A form factor where the attached Gaussian chains are replaced by hard spheres shows qualitatively the same result. Further, we investigate the structural changes associated with protease digestion of protein residues on the surface of SVs, and compare scattering patterns from SVs recorded under different pH conditions.

2. Materials and Methods

2.1. Purification of Synaptic Vesicles

Synaptic vesicles from rat brain were purified following a purification protocol by Jahn et al., [2]. Samples were kept on water/ice mixtures until investigation by electron microscopy (EM). Samples for the SAXS measurements were snap frozen for transportation to the synchrotron. All samples retained the capacity to acidify, which is an important indication that the SV membranes are functionally intact [4]. The resulting SV stock solutions had a protein concentration in the range of $5.47 - 6.45$ μg/μl.

For the protease digested SVs, one sample was split in two following the usual purification. To one sample 0.1 μg Trypsin per μg protein was added and incubated at 37 °C for 60 minutes. The other sample was incubated the same time on ice as a control. Both samples were centrifuged at 3×10^5 g (average) for two hours. The pellets were then re-suspended in aqueous buffer of 100 mM KCl, 25 mM HEPES and 1 mM DTT at pH 7.40. Both the centrifugation and re-suspension followed similar steps from the usual purification protocol.

2.2. Preparation of DOPS Vesicles

Small unilamellar lipid vesicles were prepared by dissolving 1,2-Dioleoyl-sn-Glycero-3-Phosphatidilserine (DOPS) supplied by Avanti Polar Lipids (Alabaster, AL) at a concentration $c = 20$ mg/ml in ultrapure water (Milli-Q), and sonication to clarity (Sonoplus, Germany).

2.3. Small-Angle X-Ray Scattering

SAXS experiments were performed at the high brilliance undulator beamline ID-2 of the European Synchrotron Radiation Facility (ESRF) in Grenoble, France, and beamline B1 at the Doris III storage ring of HASYLAB at Deutsches Elektronen-Synchrotron, Hamburg, Germany. Both beamlines were operated under the beamline's standard conditions at 12.4 keV [5] and 9.0 keV [6] photon energy, respectively. The SV samples were at total protein concentrations of 2.74 μg/μl in aqueous buffer of 100 mM KCl, 25 mM HEPES and 1 mM DTT at pH 7.40. The samples were kept in a glass flow through capillary, or wax sealed glass capillary, with a diameter of 1.5 mm and a wall thickness of 0.01 mm.

At beamline ID-2, the diffraction patterns were recorded with a FReLoN CCD detector positioned 0.85 and 5 m behind the sample in an evacuated detector tube. Data were collected over a q-range from 0.016 to 5.5 nm^{-1}. Typical exposure times were 0.1 seconds. The 2D isotropic (powder average) diffraction pattern was corrected for the CCD dark current, offset of the analogue to digital converter, spacial distortion, detector sensitivity (flat-field), and was calibrated to the absolute scale (water reference) employing the procedure described in [5]. Data recorded at the two different detector distances were combined to give a corrected scattering curve $I(q)$ covering more than two orders of magnitude in q. Radiation damage was ruled out by comparison of scattering patterns recorded with different exposure times from 0.01 to 10 seconds. For the standard accumulation time, the absorbed dose during exposure was about 6.5×10^3 Gy. Dilution series revealed no measurable inter-particle correlations or aggregation for samples of a total protein concentration between 6.45 μg/μl and 1.29 μg/μl.

At beamline B1, the diffraction patterns were recorded with a multiwire proportional chamber gas detector (Gabriel), or a Pilatus 300k silicon single photon counting pixel array detector positioned 935 and 3635 mm behind the sample. Data were collected over a q-range from 0.08 to 2.9 nm^{-1}. Typical exposure times were 30 minutes. The 2D isotropic (powder average) diffraction pattern was corrected for the dark current, fluctuations in primary intensity, spacial distortion, detector sensitivity / flat-field (only for Gabriel detector), and was calibrated to the absolute scale (glassy carbon reference) employing a similar procedure as described in [5]. Data recorded at the two different detector distances were combined to give a corrected scattering curve $I(q)$. Detector resolution effects could be ignored at both the ID-2, ESRF, and the B1, HASYLAB, measurements.

2.4. *Cryo Electron Microscopy*

To determine the relative size polydispersity $p(R)$ of the SV population independently from SAXS, cryo electron microscopy (cryo-EM) measurements were performed on vitrified SV solutions. In total, 559 SV were analyzed and R was determined from the diameter of the SV, measured from bilayer surface to bilayer surface. Independently, $p(R)$ was also confirmed by dynamic light scattering (data not shown).

2.5. *Scattering Form Factor Models*

The scattering cross section for a dilute, polydisperse system of particles of radius R with the number size distribution $p(R)$, the volumes $V(R)$, and the scattering form factor $P(q,R)$ is given by [7]

$$\frac{d\sigma(q)}{d\Omega} = \Delta\rho^2 \int_0^\infty p(R)V(R)^2 P(q,R)\, dR\,. \qquad (1)$$

$V(R)$ is the dry volume of the particle defined as the total volume V_{tot} minus the volume of the solvent core V_{core}. In the SV models, a bimodal size distribution function $p(R)$ was employed as shown in Fig.2 (B). The branch of $p(R)$ around approximately $R = 20$ nm has been determined by cryo-EM (smoothed) [3]. The diameter of the SV was measured from bilayer surface to bilayer surface. The branch of $p(R)$ around approximately $R = 200$ nm was calculated as a freely varied Gaussian contribution with mean μ and standard deviation σ to account for the trace number of larger membranous particles in the sample. In the DOPS vesicle models, $p(R)$ was a freely varied Gaussian only. $\Delta\rho = M/V$ denotes the difference between the scattering length density of the solvent and the average scattering length density of the decorated bilayer. M will be used below as the total excess scattering length of a particle.

The form factor models derived in this work are built from a central bilayer profile [8, 9, 10] with added contributions on the inside (lumen) and outside of the SV model, accounting for the numerous proteins associated with the SV membrane. The spherically symmetric electron density profile of the bilayer is modeled by three concentric Gaussians [11], representing the head groups of the two lipid leaflets and the hydrophobic core. Note that protein residues associated with the head groups and trans-membrane protein segments are included in this contribution. The larger proteins, or protein clusters, of the inner and outer protein shells, which can be clearly seen in cryo-EM images [3] are (1) not explicitly integrated into the model (also an asymmetric bilayer profiles has been tested), or modeled by coronas of (2) concentric Gaussians, (3) hard spheres, (4) Gaussian chains [12] attached to the inner and outer sides of the bilayer, respectively [13, 14, 15, 16, 3]. The contribution of the hard spheres, or Gaussian chains, coronas explicitly introduces an in-plane structure to the models, breaking the spherical symmetry. The individual hard spheres or Gaussian chains are assumed to be perfectly uncorrelated, forming an ideal gas on the sphere. The extension of the additional concentric Gaussians, the hard spheres and the Gaussian chains in the radial direction might be interpreted as the thickness of the protein layers, whereas the lateral extension (parallel to the membrane tangent plane) of the hard spheres or Gaussian chains may reflect the in-plane size of individual proteins, protein clusters and/or distinct lipid micro domains in a coarse grained sense. For the calculation of polydisperse populations, the local structure of the concentric bilayer profile and, depending on the model, the additional terms describing the protein coronas were kept constant for all population members. Least-squares fitting was performed using the "lsqnonlin" routine of MATLAB Optimization Toolbox (Version 7.5.0.342 (R2007b), The MathWorks Inc.), dedicated to solve non-linear least-squares problems. The quality of a model fit $f(x)$ with p free model parameters to N data points of experimentally estimated photon counts y_i with empirical variances σ_i^2 was assessed by the value of the *reduced* χ^2 given by

$$reduced\ \chi^2 = \frac{\sum_{i=1}^{N} \frac{(y_i - f(x_i))^2}{\sigma_i^2}}{N - p - 1} . \tag{2}$$

The variances σ_i^2 were calculated by propagating the counting errors from the photon counts of the individual detector pixels through the data reduction and correction process. Cross-correlations between pixels were neglected.

In addition to the parameters of the different models given in Tabs. 1, 2, and 3 three parameters reflecting the freely varied Gaussian component of $p(R)$, accounting for larger membranous particles in the SV samples, and a small constant background, were subject to optimization during the fitting procedure of the form factor models to $I(q)$. The part of $p(R)$ representing the size distribution of the SVs as obtained by cryo-EM [2, 3] was kept constant for all fits. For both branches of $p(R)$, the same form factor was used. In particular the bilayer profile parameters, the parameters of the additional concentric Gaussians, or the density of hard spheres or Gaussian chains per surface area ($N_c^{in}/(4\pi(R-D-R_g^{in})^2)$ and $N_c^{out}/(4\pi(R+R_g^{out})^2)$) were kept constant. The effective number of free parameters of the different models are 7 (3 Gaussians, symmetric profile), 9 (3 Gaussians, asymmetric profile), 11 (3 Gaussians forming symmetric profile, with 2 additional concentric Gaussians), and 12 (3 Gaussians forming symmetric profile, with hard spheres or Gaussian chains coronas attached). However, it should be emphasized that part of the parameters were in practice only free to vary within relatively narrow intervals due to the structural constraints to these parameters imposed by the model. Details on the different form factors are given in the Appendix.

3. Results

Table 1. Parameter values of the decorated bilayer as obtained from the optimized isotropic SV SAXS model fits shown in Fig. 1. The models consist of concentric Gaussian distributed scattering densities. Parameter values of the corresponding optimized bimodal polydispersity distributions of the SVs are given in the text.

Model	$\rho^*_{c\ in}$ $t_{c\ in}\sqrt{2\pi}$	ρ^*_{in} $t_{in}\sqrt{2\pi}$	ρ^*_{tail} $t_{tail}\sqrt{2\pi}$	ρ^*_{out} $t_{out}\sqrt{2\pi}$	$\rho^*_{c\ out}$ $t_{c\ out}\sqrt{2\pi}$
`3 Gaussians` `(asymmetric)`	– –	2.24 a.u. 1.14 nm	−1 a.u. 0.97 nm	0.05 a.u. 8.88 nm	– –
`3 Gaussians` `(symmetric)`	– –	2.29 a.u. 1.64 nm	−1 a.u. 2.38 nm	2.29 a.u. 1.64 nm	– –
`5 Gaussians`	0.06 a.u. 2.10 nm	1.43 a.u. 1.61 nm	−1 a.u. 2.11 nm	1.43 a.u. 1.61 nm	0.40 a.u. 1.52 nm

* Normalized to $\rho_{tail} = -1$.

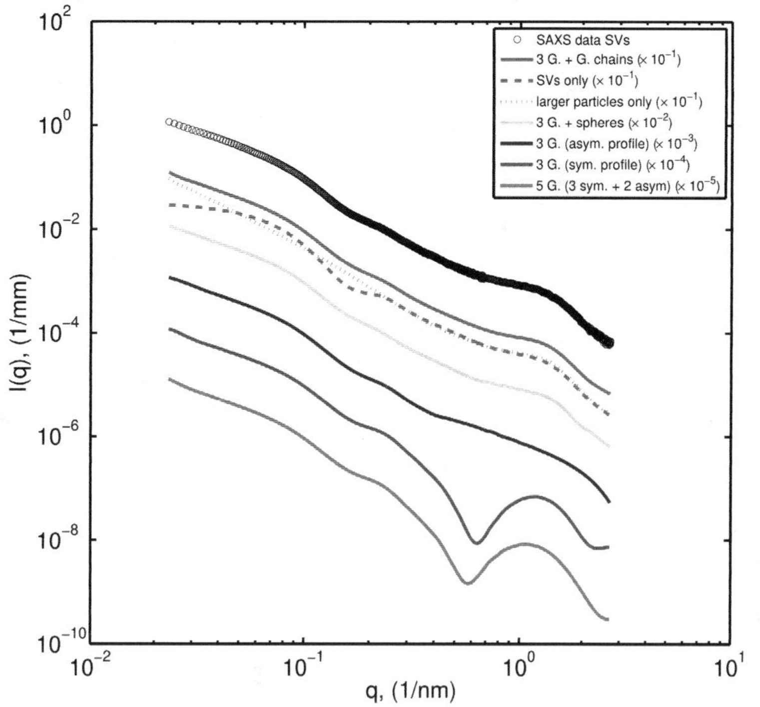

Figure 1. SV SAXS data (○ , black). Fit of anisotropic model (three Gaussians forming symmetric bilayer profile, with Gaussian chains attached) to the SV data (——, red), and contributions to the model calculation originating from the SVs (- - - -, red) and the larger membranous particles (······, red). Corresponding electron density profile and size distribution are given in Fig. 2. Fit of anisotropic model (three Gaussians forming symmetric bilayer profile, with hard spheres attached) to the SV data (——, cyan). Fit of isotropic model (three Gaussians forming asymmetric bilayer profile) to the SV data (——, blue). Fit of isotropic model (three Gaussians forming symmetric bilayer profile) to the SV data (——, dark green). Fit of isotropic model (three Gaussians forming symmetric bilayer profile, plus one additional Gaussian attached to each sides of the profile) to the SV data (——, magenta). Curves partly shifted downwards for clarity, as detailed in the key.

Figure 1 shows data $I(q)$ vs q from SVs in aqueous buffer (○ , black), and least-square fits of isotropic and anisotropic form factor models to the data. The isotropic form factors consist of concentric Gaussians forming symmetric or asymmetric scattering density profiles. Three Gaussians forming a symmetric profile (solid dark green line , *reduced* $\chi^2 = 602$), five Gaussians of which three are forming a symmetric core profile (solid magenta line, *reduced* $\chi^2 = 578$), three Gaussians forming an asymmetric profile (solid blue line, *reduced* $\chi^2 = 34.1$). A bimodal size polydispersity distribution function $p(R)$ was used, with one branch corresponding to cryo-EM data on the size distribution of the SVs, and a second branch corresponding to larger membranous particles, modeled by a freely varied Gaussian distribution.

In addition to the parameters given in Tab. 1 the following parameters have been obtained from the fits (model with three Gaussians (symmetric profile), model with five Gaussians,

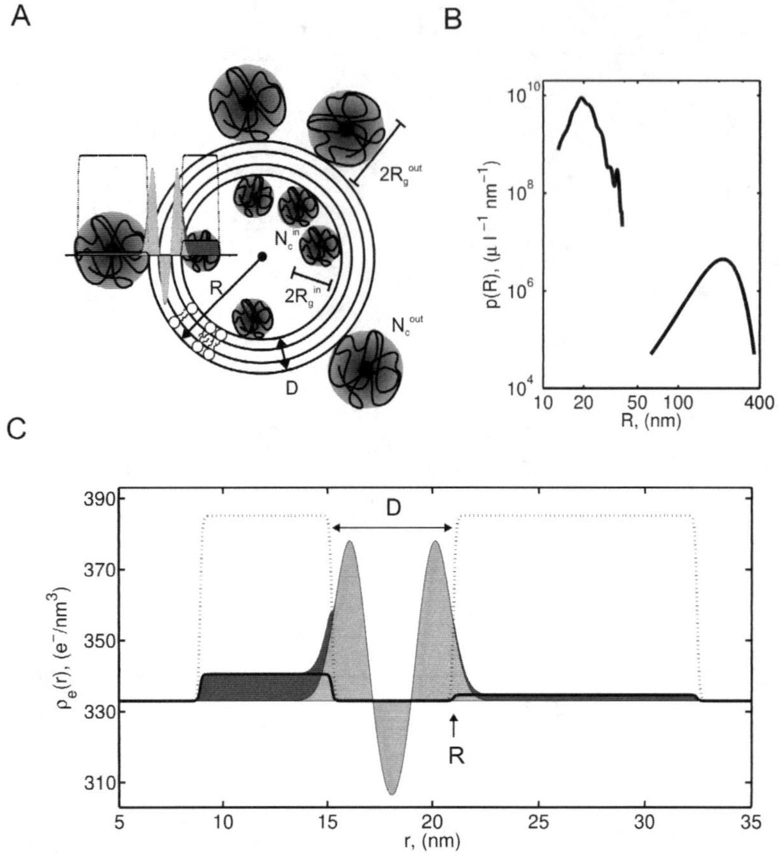

Figure 2. *(A)* Sketch of a real space model corresponding to an optimized scattering form factor consistent with the measured SAXS data from SVs shown in Fig. 1. For details on the model parameters see Table 2, model Gaussian chains coronas, and [3]. *(B)* Bimodal size distribution function $p(R)$ employed in the form factor calculation. Binning size 1 nm. *(C)* Electron density profile $\rho_e(r)$ across a SV membrane corresponding to least-square fit given in Fig. 1 (——red). Typical maximum local contribution of Gaussian chains (······, black), and spherically averaged contribution of Gaussian chains (——, black) are shown. Light blue area: lipid bilayer. Darker blue area: spherically averaged contribution of protein layers. Both $\rho_e(r)$ and $p(R)$ have been calibrated to absolute scales. Adapted from [3].

model with three Gaussians (asymmetric profile)): mean radius larger particles ($\mu = 160.7$ nm, $\mu = 180.1$ nm, $\mu = 127.0$ nm), width distribution larger particles ($\sigma = 45.2$ nm, $\sigma = 50.2$ nm, $\sigma = 51.7$ nm), number larger particles (1.55 %, 1.61 %, 2.47 %) and a small constant background (7×10^{-5} mm^{-1}, 3×10^{-5} mm^{-1}, -3×10^{-5} mm^{-1}). The excess scattering densities ρ_i (relative to aqueous buffer) have been scaled to arbitrary units such that $\rho_{tail} = -1$. Parameters are explicitly defined in the Appendix.

Only the form factor consisting of three Gaussians forming an asymmetric profile is capable to describe the SV SAXS data, at least qualitatively. However, the parameter values of the profile do not correspond to a physically meaningful model. The maximum excess scattering density of the inner leaflet is about 45 times higher than for the outer leaflet. The model parameters would suggest the structure of a lipid monolayer with a thickness of about 3 nm. However, EM

Journal of Physics: Conference Series **247** (2010) 012015 doi:10.1088/1742-6596/247/1/012015

Table 2. Parameter values of the decorated bilayer as obtained from the optimized anisotropic SV SAXS model fits shown in Fig. 1. The models consist of three concentric Gaussians (symmetric bilayer profile), with coronas attached to both sides. Parameter values of the corresponding optimized bimodal polydispersity distributions of the SVs are given in the text.

Model	$\rho^*_{in,out}$ ρ^*_{tail}	$t_{in,out}\sqrt{2\pi}$ $t_{tail}\sqrt{2\pi}$	ρ^*_c	R^{in}_g R^{out}_g	$N^{in}_c/(4\pi(R-D-R^{in}_g)^2)$ $N^{out}_c/(4\pi(R+R^{out}_g)^2)$
`Hard spheres coronas`#	0.18 a.u. −1 a.u.	2.9 nm 0.6 nm	0.28 a.u.	2.4 nm 5.6 nm	13.75×10^{-3} nm^{-2} 0.42×10^{-3} nm^{-2}
`Gaussian chains coronas`	1.63 a.u. −1 a.u.	1.8 nm 2.1 nm	1.81 a.u.	3.2 nm 5.7 nm	7.09×10^{-3} nm^{-2} 0.47×10^{-3} nm^{-2}

* Normalized to $\rho_{tail} = -1$.
# $R^i_g = R^i$, $i = in,\ out$.

shows a more or less roughly symmetrical bilayer structure of about 5 nm thickness. Further, the indirect Fourier transformation method [17] applied to the SAXS data, followed by the numerical deconvolution of the pair distance distribution function [18, 19, 20] supports the view that a spherical symmetric model is not suited to describe the data (data not shown). Elliptically deformed models consisting of a bilayer profile, similar to the isotropic models above, have been also falsified (data not shown).

The anisotropic models consist of a symmetric core profile (three concentric Gaussians) with either attached hard spheres (solid cyan line, *reduced* $\chi^2 = 4.18$), or with attached Gaussian chains (solid red line, *reduced* $\chi^2 = 2.84$) on both sides of the bilayer profile. The dashed and dotted red lines depict the contributions of the two branches of the bimodal size distribution function $p(R)$ of the model consisting of three Gaussians with Gaussian chains coronas. Again a bimodal size polydispersity distribution function $p(R)$ was used, as defined above.

In addition to the parameters given in Tab. 2 the following parameters have been obtained from the fits (model hard spheres coronas, model Gaussian chains coronas): mean radius larger particles (μ = 232.5 nm, μ = 210.1 nm), width distribution larger particles (σ = 54.5 nm, σ = 50.2 nm), number density of larger particles (0.68 %, 0.86 %) and a small constant background. Details on the form factors are given in the Appendix.

Figure 2 (*A*) shows a sketch of a real space model corresponding to the least-squares fit of the form factor consisting of three Gaussians (symmetric profile) with Gaussian chain coronas shown in Fig. 1 (red lines). Figure 2 (*B*) shows the corresponding size distribution function $p(R)$, and (*C*) the electron density profile $\rho_e(r)$ deduced from the optimized model calculation.

Fig. 3 shows data $I(q)$ vs q from a control sample consisting of DOPS vesicles, 20 mg/ml in water (○ , blue), and how they compare to SV data (○ , black). While the data looks qualitatively similar towards both lower and higher q values for the measured interval, the scattered intensities differ up to approximately two orders of magnitude at intermediate q-values. A least-square fit of an isotropic model (three Gaussians, symmetric bilayer profile, and mono-modal Gaussian distributed polydispersity of the particles) to the DOPS data (- - - -, cyan) describes the data well (*reduced* $\chi^2 = 1.07$), although the slight lift-off of the minima towards larger q values of the measured interval, which is usually due to asymmetry of the bilayer profile, is naturally not reproduced by the model calculation employed here [21]. The fit yields for the most frequent radius μ = 48.5 nm, and for the width of the polydispersity distribution σ = 10.2 nm. The same type of isotropic model (but with a bimodal polydispersity

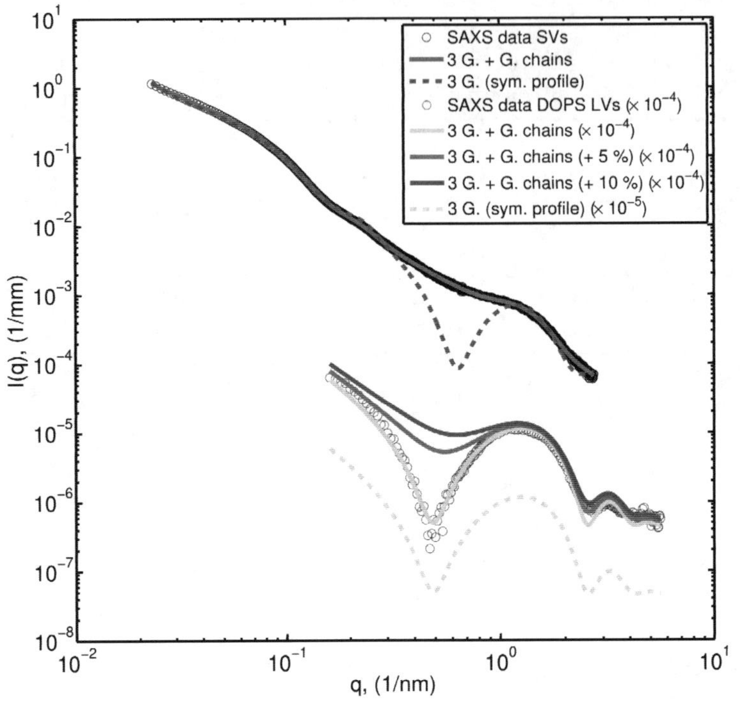

Figure 3. DOPS vesicle data (○ , blue). Fit of isotropic model (three Gaussians, symmetric bilayer profile) to the DOPS data (- - - -, cyan) and anisotropic model with Gaussian chains (——, light green). Anisotropic model with Gaussian chain contributions form SV model added (5 %, ——, brown) and (10 %, ——purple). Curves partly shifted downwards for clarity, as detailed in the key. SVs (○ , black), data identical to data shown in Fig. 1. Fit of isotropic model to the SV data (- - - -, dark green) and anisotropic model with Gaussian chains (——, red).

function) applied to the SV data (- - - -, dark green) fails qualitatively to describe the data, especially in a region around $q \simeq 0.5\ \mathrm{nm}^{-1}$ where data and best-fit deviate by more than one order of magnitude. The fit reveals for the mean of the Gaussian component of the bimodal polydispersity distribution $\mu = 160.7$ nm, for the width $\sigma = 45.2$ nm, and for the number of large particles is 1.54 %.

Least-squares fits of an anisotropic model, similar to the one above but with additional Gaussian chains attached to both sides of the bilayer, to the DOPS data (——, light green) yields almost identical results as the isotropic model (*reduced* $\chi^2 = 1.07$). The size distribution of the liposomes is again assumed to be Gaussian; the fit yields for the polydispersity distribution the most frequent radius $\mu = 48.5$ nm, and a width of $\sigma = 10.2$ nm. In addition, a small constant background was fitted. Interestingly, the number of Gaussian chains in the anisotropic model is zero for the best fit to the DOPS data (within the numerical precision). The qualitative effects of the attached Gaussian chains of the anisotropic model fit to the DOPS data is illustrated by adding 5 % (——, brown) and 10 % (——, purple) of the number of attached Gaussian chains as compared to the corresponding optimized SV model. The parameter values for R_g^{in} and R_g^{out} of the Gaussian chains are taken from the optimized SV model, while the parameter value for

Table 3. Parameter values of the decorated bilayer as obtained from the optimized isotropic and anisotropic SAXS model fits shown in Fig. 3. Parameter values of the corresponding optimized polydispersity distributions of the SVs and DOPS LVs are given in the text.

Sample	Model	$\rho^*_{in,out}$ ρ^*_{tail}	$t_{in,out}\sqrt{2\pi}$ $t_{tail}\sqrt{2\pi}$	ρ^*_c	R_g^{in} R_g^{out}	$N_c^{in}/(4\pi(R-D-R_g^{in})^2)$ $N_c^{out}/(4\pi(R+R_g^{out})^2)$
SVs	isotropic[#]	2.29 a.u. −1 a.u.	1.64 nm 2.38 nm	–	– –	– –
SVs	anisotropic[$]	1.63 a.u. −1 a.u.	1.8 nm 2.1 nm	1.81 a.u.	3.2 nm 5.7 nm	7.09×10^{-3} nm^{-2} 0.47×10^{-3} nm^{-2}
DOPS LVs	isotropic	2.24 a.u. −1 a.u.	0.89 nm 2.81 nm	–	– –	– –
DOPS LVs	anisotropic	2.24 a.u. −1 a.u.	0.89 nm 2.81 nm	1.30 a.u.	3.9 nm 4.0 nm	$< 1 \times 10^{-13}$ nm^{-2} $< 1 \times 10^{-13}$ nm^{-2}
DOPS LVs	anisotropic (+ 5 % G. chains)	2.24 a.u. −1 a.u.	0.89 nm 2.81 nm	2.49 a.u.	3.2 nm 5.7 nm	3.55×10^{-3} nm^{-3} 0.24×10^{-3} nm^{-3}
DOPS LVs	anisotropic (+ 10 % G. chains)	2.24 a.u. −1 a.u.	0.89 nm 2.81 nm	2.49 a.u.	3.2 nm 5.7 nm	7.09×10^{-3} nm^{-3} 0.47×10^{-3} nm^{-3}

[*] Normalized to $\rho_{tail} = -1$
[#] Identical to model fit "3 Gaussians (symmetric)" in Table 1.
[$] Identical to model fit "Gaussian chains coronas" in Table 2.

Table 4. Parameter values as obtained from the optimized anisotropic SAXS model fits shown in Fig. 4

Model fit parameter	Native SVs	SVs protease	Unit
ρ_{in}, ρ_{out}	1.58	1.50	a.u.
ρ_{tail}	−1	−1	a.u.
$t_{in}\sqrt{2\pi}$, $t_{out}\sqrt{2\pi}$	1.9	1.7	nm
$t_{tail}\sqrt{2\pi}$	2.1	2.4	nm
R_g^{in}	2.8	2.6	nm
R_g^{out}	5.7	5.9	nm
$N_c^{in}/(4\pi(R-D-R_g^{in})^2)$	15.04×10^{-3}	8.73×10^{-3}	nm^{-2}
$N_c^{out}/(4\pi(R+R_g^{out})^2)$	0.27×10^{-3}	0.15×10^{-3}	nm^{-2}
ρ_c	1.84	1.73	a.u.
μ	210.5	211.0	nm
σ	49.7	51.1	nm
Number larger particles	0.86	0.68	%
Constant background	3.67×10^{-4}	15.5×10^{-4}	mm^{-1}

ρ_c is calculated relative to $\rho_{in} = \rho_{out}$ to have the same value as for the optimized SV model. A least-squares fit of the anisotropic model to the SV data (——, red, same fit as in Fig. 1) is in excellent agreement with the data, while employing physiological and biochemical meaningful parameter values [3, 2].

Fig. 4 shows the SAXS intensity function $I(q)$ vs. q for a SV sample (black circles) and SV treated with a Trypsin protease (blue circles) and corresponding least-squares fits (*reduced* $\chi^2 = 0.66$, and *reduced* $\chi^2 = 0.51$, respectively) to the form factor model (solid red

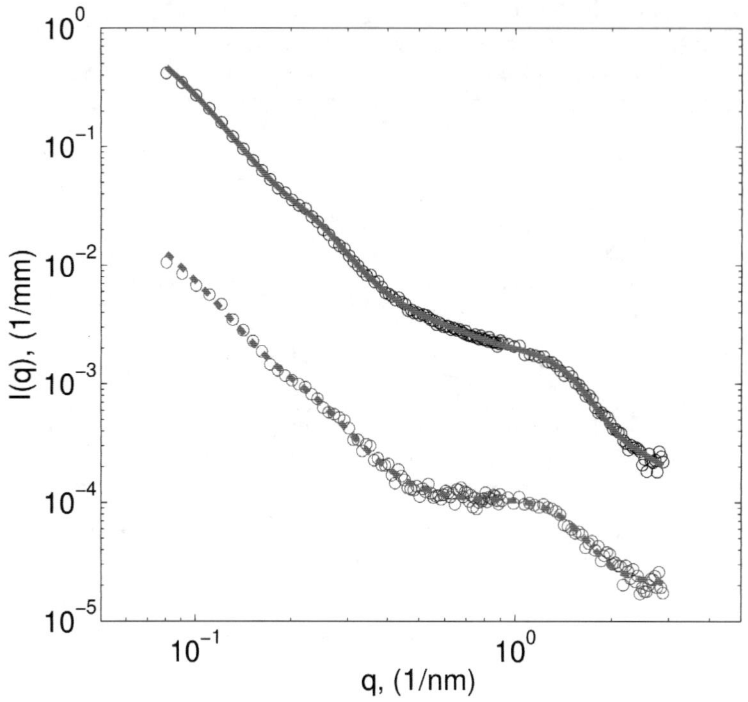

Figure 4. SVs with additional centrifugation in purification protocol (○ black), and SVs treated with protease to remove protein residues facing outwards, also with an additional centrifugation in purification protocol (○ blue, shifted by factor 0.1 for clarity). Fit of anisotropic model with Gaussian chains attached to the SV data (——red, and - - - -red, shifted by factor 0.1 for clarity).

line, and dashed red line, respectively) for bilayer vesicles with Gaussian chains attached to the inside and outside, as detailed in the Appendix. Again a bimodal size polydispersity distribution function $p(R)$ was used, with the left branch corresponding to the cryo-EM data of SVs centered around 21 nm, and the right branch corresponding to the larger membranous particles, modeled by a Gaussian distribution centered around $\mu \approx 210$ nm with a standard deviation of $\sigma \approx 50$ nm. It should be emphasized that due to the resolution of the SAXS technique, our data are relatively insensitive to the exact size distribution of these larger particles. The errors appear to slightly overestimate the actual variances, as the model fit does not over fit the data, even though we obtain $\chi^2 < 1$ for both fits. The parameters of the fits are given in Tab. 4. Most parameter values do not change much when comparing the results from the fit to the SVs (with additional centrifugation step in the purification protocol) data and the protease treated SV sample, also with additional centrifugation step. But, importantly, the number density of Gaussian chains facing the lumen and the outside are both reduced by about a factor of two as an effect of the protease treatment of the sample.

4. Discussion

In combination with complimentary techniques such as cryo-EM and biochemical analysis, we have shown that SAXS is capable of elucidating structures of significant complexity such as SVs.

State of the art beamlines have been used to record the SAXS data, which were found to be highly reproducible [3]. As we show in Fig. 1, several different isotropic models have been discussed and have been falsified. After assessment of numerous samples, we are convinced that the presented effective model representing proteins in a coarse-grained sense by Gaussian chains is well suited to describe the SVs quantitatively at the given resolution.

However, possible ambiguities and model dependencies of the results are important issues, which need to be addressed explicitly. We have therefore compared fits with two independent anisotropic models that share the main features of anisotropic protein clusters, but use different detailed implementations. In particular, the clustered proteins are represented by hard spheres in the first, and by Gaussian chains in the second. Importantly, these two models show qualitatively similar results, and thus support the main conclusion that an anisotropic model is needed to accurately describe the SV data. Of course, to some extent the derived model parameters of the observed clustering effect are different. These differences mark the degree to which the results depend on the specific formulation of the model, see Tab. 2. The main conclusions and existence of protein clusters are, however, maintained in a model independent way.

Models where the proteins are represented by spherically symmetric Gaussian electron densities as depicted in Fig. 3 illustrate the effects that the Gaussian chains parameters, or hard spheres parameters, have on the scattering curve. In an intermediate q-range, the difference in scattering length of the SV data, when compared to unilamellar DOPS vesicles is found to be in excess of two orders of magnitude.

Importantly, least squares model fits to the DOPS data yield almost identical results for a spherically symmetric model with a symmetric density profile of three coupled Gaussians (Fig. 3, dashed cyan line), and a model with Gaussian chains attached to the profile (Fig. 3, solid light green line). Interestingly, the number of Gaussian chains in the later model is zero for the best fit (within the numerical precision). Models with significant amounts of attached Gaussian chains with parameter values similar to the ones from the corresponding optimized SV model can be falsified against the DOPS data Fig. 3, solid brown line, and solid purple line).

The mathematical model utilized to fit the experimental data contains 12 parameters to account for the structural complexity of the SV sample. Given the high complexity of the system, the high number of data points (up to 763) and the wide q-range of the data, a model form factor fit with 12 parameters is perfectly justified. We have verified that the model components are both necessary and sufficient to accurately describe the data, see also the reduced, and alternative model fits in Fig. 1 and Fig. 3. None of the contributions are marginal, neither within the model structure of one particle, nor considering the two branches of the polydispersity distribution $p(R)$.

In addition, the values for several parameters can be constrained fairly well due to information from cryo-EM and additional biochemical knowledge. The fact that SVs are known to have a unilamellar bilayer structure with incorporated and attached proteins introduces several constrains to the model parameters. Further, the size distribution of the larger particles can be estimated, at this point only qualitatively, by cryo-EM. The investigated q-range is not very sensitive to the precise size of particles in the size range of the larger particles. Due to the heterogeneous nature and broad size distribution of these particles, pronounced features at lower q regions cannot be expected. Particular features outside the measurement interval do not appear in the model calculations, either at small or at high q, i.e. the measurement range (after stitching different detector distances) was well adapted. In particular, the heterogeneous nature and broad size distribution of the larger particles suppressed any pronounced features at lower q-ranges.

Let us now consider the relative contributions to the measured SV SAXS curves, of photons scattered from particles belonging to the two branches of the polydispersity function $p(R)$, namely SVs and larger membranous particles, which can be considered as a contamination. Importantly, the relative scattering contributions of particles, and parts of particles, of different sizes depend strongly on the considered value of q.

The number density of the actual SVs is about 100 times higher as the number density of the larger particles. The factor of about 10 in size difference of the particles gives a factor of about 100 in favor of the larger structures considering the relative surface areas of the particles. This is because the part of the reciprocal space studied here contains mainly information about length scales in the order of the thickness of the protein covered bilayer. This leads to approximately equal numbers of photons scattered into the measured q-range from particles belonging to either branch of $p(R)$. Comparing this result to the relative number of photons scattered into the entire reciprocal space (q-integrated) from the two branches of $p(R)$ this means that about 99 % of the scattering intensity due to the larger particles is expected to be scattered into lower q-regions, as compared to the q-interval measured here. The results of these considerations correspond well with the contributions from the two branches of $p(R)$ in the model fit calculation shown in Fig. 1, (red lines).

SVs which are treated with Trypsin from the outside show distinct scattering curves when compared to data from native SVs (open black, and blue circles, in Fig. 4). Least-squares fits to a form factor model (symmetric bilayer profile with attached Gaussian chains) result in a reduction of scattering mass attributed to the Gaussian chains, both inside and outside. This may be due to the fact that the attached Gaussian chains facing the lumen and the outside of the SV model account to some extent in a similar manner for structures on the SV sticking both towards the inside and the outside. Alternatively, the protease treatment may also cause disintegration of the protein clusters on the SVs, thus affecting the organization of proteins facing outwards and inwards. Further, although it seems highly unlikely, it can not be ruled out completely at this point that the protease might have penetrated into the SVs, and subsequently removed protein residues from both sides of the SV membrane. Electron micrographs of the protease treated SVs show no evidence of penetration of the protease into the SVs (data not shown). However, EM is not capable of strictly excluding such a scenario. In other experiments, we have used protease treatment of SVs many times (using harsher conditions than those shown here) followed by checking the intravesicular fragments of synaptotagmin, synaptophysin, and synaptogyrin. For all these proteins, fragments of the expected molecular mass were detectable (that are degraded when the membranes are solubilized), showing that there is efficient protection of the vesicle interior. Importantly, breaking of spherical symmetry still remains an essential ingredient in order to obtain a good fit.

The results from the fit of the SVs with an additional centrifugation step in the purification protocol as compared to the results obtained from SV samples following the usual purification protocol, see Tab. 3 and [3] show different number densities of the Gaussian chains attached to both sides of the bilayer profile. The number density of Gaussian chains facing the inside of the SVs $N_c^{in}/(4\pi(R-D-R_g^{in})^2)$ is increased by about a factor of two, while the number density of Gaussian chains facing the outside $N_c^{out}/(4\pi(R+R_g^{out})^2)$ is decreased by about a factor of two. At the moment the significance and possible reasons of this discrepancy remains unclear, since the usual SV purification protocol already contains several centrifugation steps similar to the additional one needed to remove the Trypsin and the protein fragments from the protease treated SVs.

5. Summary and perspective

In summary, we have discussed different spherically symmetric and anisotropic form factor models and have tested them against high resolution SAXS data from SVs, isolated from rat brain. Anisotropy of the model form factors is found to be a key ingredient for the description of the native SV structure. After protease digestion of the surface of the SVs (unspecific protein removal), a significant reduction of the anisotropic terms is observed. This compares well with the observation, that the anisotropic terms in the form factor vanish in least-squares fits of

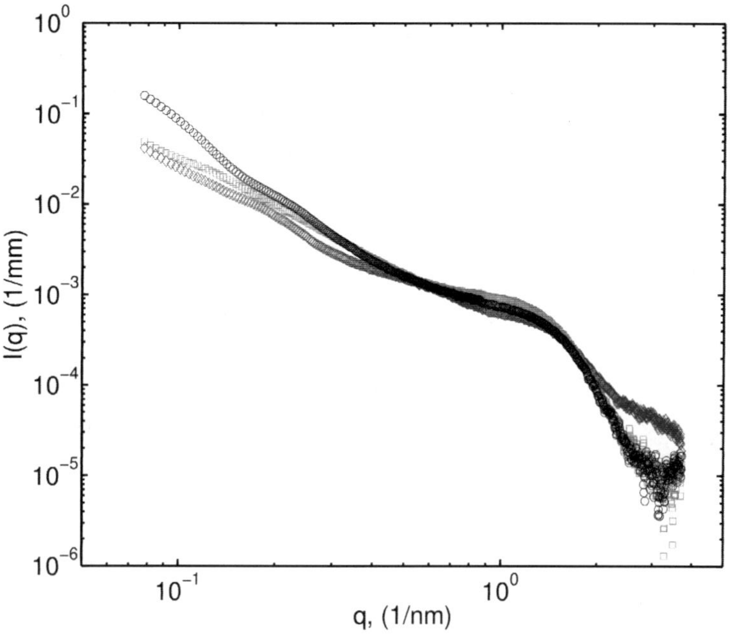

Figure 5. SVs at pH 7.40 (○ black), SVs at low (2.1) pH (◇blue), and SVs at high (12.9) pH (□red).

SAXS data from small unilamellar vesicles composed of DOPS. These vesicles are spherically symmetric on the experimentally resolved length scales. Besides the excellent agreement with the SAXS data, the suggested SV form factor model is consistent with published electron microscopy, biochemical and physiological data. Furthermore, we have presented SAXS data of SVs recorded under different pH conditions, and have described changes in structural parameters due to protease digestion of SVs.

In terms of further structural analysis of SVs, changes in structure as a function of physiological conditions (pH, ion concentration, osmotic pressure etc.) are obvious candidates for investigation.

As a first example, Fig. 5 shows data from SV samples in HB100 buffer at pH 2.1, pH 7.4 and pH 12.9. The SAXS curves show distinct features depending on the pH conditions. The overall slope of the scattering curves is decreased at both high and low pH conditions as compared to neutral pH. This might be due to changes in overall shape and/or size of the SVs. Also aggregation (docking) of SVs might lead to this effects. Resolved changes in the higher q-regions of the recorded data indicate distinct changes in the local structure of the protein decorated bilayer of the SVs as a function of pH conditions. The underlying detailed structural changes are to be further investigated and remain unclear at present. However, the SVs seem to retain well defined structures even under extreme pH conditions. More subtle changes of the chemical environment of the SV samples like different calcium concentrations in the range of 0 to 100 μM, or magnesium concentrations around 200 mM did not show resolvable differences in the corresponding SAXS curves (data not shown).

A particular advantage of the method is the lack of invasive sample preparation steps, such as the fixation and staining needed for electron microscopy, which can introduce artifacts into the analysis. Owing to the small sample quantities needed for high brilliance synchrotron SAXS (on the order of 10^{-4} to 10^{-3} μl), a large number of samples can be collected and compared in a batch. Comparable sample series would consume too much time and material using other techniques (in-house SAXS, transmission electron microscopy etc.). Our present SAXS data has employed scattering intensities up to $q = 2.7$ nm^{-1}. By increasing the final vesicle concentration, through modifications in the purification protocol, this range could be extended. For example, with a theoretical concentration increase of 100-fold, the exploitable q-range should be extended by a factor of three, reaching a resolution of about 0.8 nm. Structural studies of vesicles isolated from wild-type animals could also be augmented by using so-called "knock-in" and "knock-out" animals, to genetically manipulate essential components of the system (Holt and Jahn, manuscript in preparation). Furthermore, SAXS could be employed to determine time-resolved structural changes resulting from membrane fusion, in combination with microfluidic devices.

With the advent of x-ray lasers delivering ultra short and extremely intensive x-ray pulses, time resolved scattering experiments at individual organelles and fusion intermediates might eventually become possible. Such experiments would undoubtedly shed light on the biophysical principles of membrane merger.

Acknowledgments

The authors thank Maria Druminski for help with the SV preparations and Dietmar Riedel for EM measurements. SC thanks Otto Glatter for enlightening discussions regarding inversion of scattering data. This work was financially supported by the "Excellence Cluster Initiative 171/Deutsche Forschungsgemeinschaft (DFG) Research Center 103 – Center for Molecular Physiology of the Brain", as well as DFG SFB 803 and the Max Planck Society. European Synchrotron Radiation Facility and Hamburger Synchrotronstrahlungslabor at Deutsches Elektronen-Synchrotron are acknowledged for beamtime and excellent working conditions.

Appendix A. Scattering Form Factor Models

Appendix A.1. Isotropic Form Factor Models

Equation A.2 (below) gives the final form factor of the isotropic models used in the fitting procedures. The models are built from a central symmetric or asymmetric bilayer profile, with added spherically symmetric protein shells on the inside (lumen) and outside of the SV. The bilayer electron density profile is modeled by three concentric Gaussians [11], representing the headgroups of the two lipid leaflets and the hydrophobic core. Note amino acid residues associated with the headgroups and trans-membrane protein segments are included in this contribution. The inner and outer protein shells are modeled by concentric Gaussians attached to the inner and outer sides of the bilayer, respectively. The excess scattering length density of the bilayer profile is given by

$$\rho(r) = \sum_i \rho_i \exp\left(-\frac{(r-R_i)^2}{2t_i^2}\right) , \qquad \text{(A.1)}$$

with the peak position R_i, amplitude ρ_i and width t_i with $i \in$ *in*, *out*, *tail*, *inner*, *outer*, for each of the three Gaussians representing the headgroups of the two leaflets and the tail region, and the inner and outer protein shell, respectively. The (characteristic) radius R is defined as $R = R_{out} + t_{out}\sqrt{2\pi}/2$ mimicking a SV with the outer lipid bilayer surface at $r \simeq R$. To reduce the number of model parameters, we choose $R_{tail} = R - (t_{out} + t_{tail}/2)\sqrt{2\pi}$, $R_{in} = R - (t_{out} + t_{tail} + t_{in}/2)\sqrt{2\pi}$. Thus, the total thickness D of the bilayer is characterized by $D = \sqrt{2\pi}(t_{in} + t_{tail} + t_{out})$. Further, we choose $R_{inner} = R - (t_{out} + t_{tail} + t_{in} + t_{inner}/2)\sqrt{2\pi}$ and $R_{outer} = R + (t_{outer}/2)\sqrt{2\pi}$ so that the overall thickness of the synaptic membrane is characterized by the total thickness D of the bilayer structure and the thickness of the inner and outer protein layers. The total excess scattering length with respect to the aqueous buffer is

β_b. Depending on the specific model, we choose (1) $t_{in} = t_{out}$ (symmetric bilayer), (2) $t_{in} \neq t_{out}$ (asymmetric bilayer), or (3) $\rho_{inner} = \rho_{outer} = 0$ (no protein shells). Note that the Gaussians representing the bilayer profile and the protein shells interpenetrate to some extent due to their tails.

The form factor corresponding to the above model can be calculated in kinematic scattering theory, yielding the following result

$$P(q,R) = \frac{1}{M^2}\beta_b^2 F_b^2(q,R) \; . \tag{A.2}$$

The symbols and functions are given below: $M = \beta_b$ denotes the excess scattering length. The normalized amplitude of the self-correlation term of the bilayer profile and the isotropic protein shells is given by

$$F_b(q,R) = \frac{1}{\beta_b}\sum_i \beta_{b\ i} F_{b\ i}(q,R_i) \; , \tag{A.3}$$

with $i \in$ *in*, *out*, *tail*, *inner*, *outer*, and

$$F_{b\ i}(q,R_i) = \frac{1}{\beta_{b\ i}} 4\sqrt{2}\pi^{3/2} t_i \rho_i \exp\left(-\frac{t_i^2 q^2}{2}\right)\frac{1}{q}[t_i^2 q\cos(qR_i) + R_i \sin(qR_i)] \; , \tag{A.4}$$

where $\beta_{b\ i} = \rho_i \frac{4\pi}{3}((R_i + t_i\sqrt{2\pi}/2)^3 - (R_i - t_i\sqrt{2\pi}/2)^3)$ denotes the excess scattering mass of one peak of the profile [11].

Appendix A.2. Anisotropic Form Factor Models

Equation A.5 (below) gives the final form factor for the anisotropic model with Gaussian chains coronas used in the fitting procedure. The model is built from a central bilayer profile with added protein coronas on the inside (lumen) and outside of the SV model.

The bilayer electron density profile is modeled by three concentric Gaussians [11], representing the headgroups of the two lipid leaflets and the hydrophobic core. Note amino acid residues associated with the headgroups and trans-membrane protein segments are included in this contribution. The inner and outer protein shells are modeled by Gaussian chains [12] attached to the inner and outer sides of the bilayer, respectively [13, 14, 15, 16]. The excess scattering length density of the bilayer profile is given by Eq. A.1 with $i \in in, out, tail$, for each of the three Gaussians representing the headgroups of the two leaflets and the tail region, respectively. The (characteristic) radius R, R_{tail}, R_{in}, D and β_b are defined as for the the isotropic form factor models. To reduce the number of model parameters further, we choose $t_{in} = t_{out}$ (symmetric bilayer). There are N_c^{in} and N_c^{out} Gaussian chains distributed randomly and without correlations forming the inner and outer protein shell, respectively. The individual Gaussian chains are characterized by their root-mean-square radius of gyration, R_g^{in} and R_g^{out}, and their common average excess scattering length density ρ_c. The distance between the inner headgroup maximum of the bilayer profile and the center of mass of the Gaussian chains facing the lumen is taken to be $t_{in}\sqrt{2\pi}/2 + R_g^{in}$, and the distance between the outer headgroup maximum and the center of mass of the Gaussian chains facing outwards is taken to be $t_{out}\sqrt{2\pi}/2 + R_g^{out}$. This limits the penetration of the Gaussian chains into the bilayer, although there is some remaining overlap, mostly due to the extending tails of the bilayer profile [13]. The form factor corresponding to the above model can be calculated in kinematic scattering theory, yielding the following result

$$\begin{aligned} P(q,R) = \frac{1}{M^2} \times [\quad & \beta_b^2 F_b^2(q,R) \\ + & \sum_{i=in,out} N_c^i \beta_c^{i\ 2} P_c^i(q) \\ + & \sum_{i=in,out} 2N_c^{i\ 2} \beta_b \beta_c^i S_{b\ c}^i(q,R) \\ + & \sum_{i=in,out} N_c^i(N_c^i - 1)\beta_c^{i\ 2} S_c^i(q,R) \\ + & S_c^{in\ out}(q,R) \prod_{i=in,out} N_c^i \beta_c^i \ ] \ . \end{aligned} \quad (A.5)$$

The symbols and functions are given below: $M = \beta_b + N_c^{in}\beta_c^{in} + N_c^{out}\beta_c^{out}$ denotes the excess scattering length, with $\beta_c^i = \rho_c R_g^{i\ 3} 4\pi/3$ and $i = in, out$ representing the total excess scattering length of a single chain on the outside and on the inside of the bilayer profile, respectively. The normalized amplitude of the self-correlation term of the bilayer profile is given by Eq. A.3 with Eq. A.4 where $\beta_{b\ i}$ is defined as for the isotropic form factor models.

The self-correlation terms of the Gaussian chains are given by Debye functions

$$P_c^i(q) = \frac{2[\exp(-x^i) - 1 + x^i]}{x^{i\ 2}} \ , \quad (A.6)$$

with $x^i = q^2 R_g^{i\ 2}$ and $i = in, out$ for the inner and outer chains, respectively. The interference cross-terms $S_{b\ c}^{in}(q,R)$ and $S_{b\ c}^{out}(q,R)$ between the bilayer and the Gaussian chains on the inside and outside, are given by

$$S_{b\ c}^i(q,R) = F_b(q,R)\psi^i(x^i) \frac{\sin(q[R_{tail} \mp (D/2 + R_g^i)])}{q[R_{tail} \mp (D/2 + R_g^i)]} \ , \quad (A.7)$$

with $i = in, out$ and $\psi^i(x^i) = [1 - \exp(-x^i)]/x^i$ the effective form factor amplitude of the Gaussian chains [22]. The product of the scattering form factor amplitudes of the chains belonging to one of the chain layers with the scattering form factor amplitude of a infinite thin shell is equivalent to a convolution of the corresponding scattering density distributions in real space [14] and accounts for the fact that the Gaussian shells are located on a spherical shell.

$$S_c^i(q,R) = \left(\psi^i(x^i) \frac{\sin(q[R_{tail} \mp (D/2 + R_g^i)])}{q[R_{tail} \mp (D/2 + R_g^i)]} \right)^2 \ , \quad (A.8)$$

with $i = in, out$. The interference term between the chains of the inner and outer shells is taken into account by

$$S_c^{in\ out}(q,R) = \prod_{i=in,out} \psi^i(x^i) \frac{\sin(q[R_{tail} \mp (D/2 + R_g^i)])}{q[R_{tail} \mp (D/2 + R_g^i)]} \ . \quad (A.9)$$

The anisotropic form factor, where spherical particles were placed on the inside and outside of a symmetric bilayer profile instead of the Gaussian chains, is similar to that given in Eq. A.5, except that P_c^i in Eq. A.6 and ψ^i in Eq. A.7 are replaced by the scattering form factor of spheres and the scattering form factor amplitude of spheres, respectively. The spheres are of radii R^{in} and R^{out}, the number of free parameters is the same as for the model with the attached Gaussian chains.

References

[1] Südhof T C 2004 *Annu. Rev. Neurosci.* **27** 509 – 547
[2] Takamori S, Holt M, Stenius K, Lemke E A, Grønborg M, Riedel D, Urlaub H, Schenck S, Brügger B, Ringler P, Müller S A, Rammner B, Gräter F, Hub J S, De Groot B L, Mieskes G, Moriyama Y, Klingauf J, Grubmüller H, Heuser J, Wieland F and Jahn R 2006 *Cell* **127** 831 – 846
[3] Castorph S, Riedel D, Arleth L, Sztucki M, Jahn R, Holt M and Salditt T 2010 *Biophys. J.* **98** 1200–1208
[4] Hell J W, Maycox R P, Stadler H and Jahn R 1988 *EMBO J.*, **7** 3023 – 3029
[5] Narayanan T, Diat O and Boesecke P 2001 *Nucl. Instrum. Methods Phys. Res., Sect. A* **467** 1005 – 1009
[6] Haubold H G, Gruenhagen K, Wagener M, Jungbluth H, Heer H a A, Rongen H, Brandenberg G, Moeller R, Matzerath J, Hiller P and Halling H 1989 *Rev. Sci. Instrum.* **60** 1943 – 1946
[7] Guinier A and Fournet G 1955 *Small Angle Scattering of X-rays* (New York: Wiley Interscience)
[8] Bouwstra J A, Gorris G S, Bras W and Talsma H 1993 *Chemistry and Physics of Lipids* **64** 83 – 98
[9] Pabst G, Rappolt M, Amenitsch H and Laggner P 2000 *Phys. Rev. E* **62** 4000 – 4009
[10] Pabst G, Koschuch R, Pozo-Navas B, Rappolt M, Lohner K and Laggner P 2003 *J. Appl. Cryst.* **36** 1378 – 1388
[11] Gradzielski M, Langevin D, Magid L and Strey R 1995 *J. Phys. Chem.* **99** 13232 – 13238
[12] Debye P 1947 *J. Phys. Colloid Chem.* **51** 18 – 32
[13] Pedersen J S and Gerstenberg M C 1996 *Marcromolecules* **29** 1363 – 1365
[14] Pedersen J S 2000 *J. Appl. Cryst.* **33** 637 – 640
[15] Pedersen J S 2001 *J. Chem. Phys.* **114** 2839 – 2846
[16] Arleth L and Vermehren C *J. Appl. Cryst.* Submitted
[17] Glatter O 1977 *J. Appl. Cryst.* **10** 415–421
[18] Glatter O 1981 *J. Appl. Cryst.* **14** 101–108
[19] Glatter O and Hainisch B 1984 *J. Appl. Cryst.* **17** 435–441
[20] Mittelbach R and Glatter O 1998 *J. Appl. Cryst.* **31** 600–608
[21] Kucerka N, Gallová J, Uhríková D, Balgavý P, Bulacu M, Marrink S and Katsaras J 2009 *Biophys. J.* **97** 1926–1932
[22] Hammouda B 1992 *Journal of Polymer Science B* **30** 1387 – 1390

Spin-Echo Small Angle Neutron Scattering analysis of liposomes and bacteria

Léon F van Heijkamp[1], Ana-Maria Sevcenco[1,2], Diane Abou[1], Remko van Luik[1], Gerard C Krijger[1], Peter-Leon Hagedoorn[2], Ignatz M de Schepper[1], Bert Wolterbeek[1], Gerben A Koning[3] and Wim G Bouwman[1]

[1] Faculty of Applied Sciences, Department of Radiation, Radionuclides & Reactors, Delft University of Technology, Mekelweg 15, 2629 JB Delft, The Netherlands
[2] Faculty of Applied Sciences, Department of Biotechnology, Delft University of Technology, Julianalaan 67, 2628 BC Delft, The Netherlands
[3] Laboratory Experimental Surgical Oncology, Section Surgical Oncology, Department of Surgery, Erasmus Medical Center, POBox 1738, 3000 DR Rotterdam, The Netherlands

E-mail: `l.f.vanheijkamp@tudelft.nl`, `w.g.bouwman@tudelft.nl`

Abstract. Two types of liposomes, commonly used in drug delivery studies, and *E. coli* bacteria, all prepared in H_2O, were resuspended in D_2O and measured with Small Angle Spin-Echo Neutron Scattering (SESANS). Modeling was performed using correlation functions for solid spheres and hollow spheres. The signal strength and curve shape were more indicative of hollow particles, indicating that the H_2O-D_2O exchange occurred too fast to be observed with the available time resolution. Fitting the particle diameter and membrane thickness of the hollow sphere model to the data, gave results which were in good agreement with Dynamic Light Scattering (DLS) data and literature, showing as a proof-of-principle that SESANS is able to investigate such systems.

SESANS may become a good alternative to conventional tritium studies or a tool with which to study intracellular vesicle transport phenomena, with possible *in vivo* applications. Calculations show that a substantial change in numbers of a mixed system of small and large biological particles should be observable. A possible application is the destruction by external means of great numbers of liposomes in the presence of tumor cells for triggered drug release in cancer treatment. Since SESANS is both non-invasive and non-destructive and can handle relatively thick samples, it could be a useful addition to more conventional techniques.

1. Introduction

The delivery of drugs or contrast agents to tumors and their uptake by tumor cells is of great interest to the medical community to aid the diagnosis and treatment of cancer [1, 2, 3, 4, 5]. In order to develop improved methods to identify and destroy malignant tumor tissue, research has focused on the transport of substances across the bi-lipid layer using encapsulation by liposomes [1, 6, 7, 8, 9, 10]. These lipid vesicles have a similar composition to cellular mebranes and are therefore also used as a model in transport studies. The cellular transport of water is an important quantity for medical diagnostic tools using cellular contrast agents. Much work has been done studying two key processes: the internalization of small cells and particles by larger cells [11, 12, 13, 14] and the rate at which water can diffuse in and out of cells [15, 16, 17].

For the development of cancer treatment it is desirable to be able to observe both liposomal uptake by tumor cells and cellular water exchange as directly as possible in living patients, using non-invasive methods. Many observations can be done down to the cellular level in a laboratory or inside living patients, using methods as Dynamic Light Scattering (DLS) and Magnetic Resonance Imaging (MRI). Water transport is often studied using tritiated water (T_2O, HTO) as a tracer. Changing concentrations of this isotope produce local changes in levels of radioactivity. This technique is used to observe metabolic processes in laboratory animals [11], but not generally applied *in vivo* to larger organisms because of its radioactivity, which increases the chance of cancer when applied internally. For water transport studies with liposomes, the tritiated liposomes have to be separated from the medium. Since it is a weak β-particle emitter, experiments require either large quantities or the very sensitive liquid scintillation detectors. All this makes it unsuitable for *in vivo* application, whereas SESANS does not have these limitations. Furthermore, the exchange rate of this super-heavy water with normal, light water (H_2O) provides little information on cellular structure.

Small Angle Neutron Scattering has been employed in the past to study cellular components, such as ribosomes [12], DNA (including helical structure) [13], protein [14] and lipid structures [18, 19] and protein hydration [20]. Using Spin-Echo Small Angle Neutron Scattering (SESANS) microstructures of 20 nm to 20 µm can be observed, corresponding to sub-cellular and cellular sizes. The sample thickness depends on the neutron transmission of the investigated substance. The sensitivity of neutron scattering experiments to the hydrogen-deuterium contrast is often exploited in watery samples. Typical sample thicknesses in SESANS are one centimeter when containing mostly light water, and up to several centimeters when using mostly heavy water, or deuterium oxide (D_2O, HDO). Neutron scattering is non-invasive and a SESANS measurement exposes samples only to small doses of radiation, because it requires limited neutron intensities with moderate wavelengths. For stable biological samples, this technique can be easily applied *in vitro* when containing either light or heavy water or both. SESANS can in principle also be applied to organisms *in vivo* without deuterating, if there is enough contrast and no more than a few centimeters of water in the beam.

Water transport across cell membranes can be studied *in vitro* with SESANS by observing changes in contrast, due to local changes in light and heavy water concentration. Although deuterium is a non-radioactive isotope, large amounts of this heavy water can still be harmful to organisms. Some normal reactions in cells, especially cell division, are disrupted by the stronger hydrogen bonds formed. Toxic levels of deuteration range from around 30% to 50% for rodents and large mammals respectively [21, 22]. Since it is not possible to replace large amounts of water with D_2O without any harmful effects, an *in vivo* SESANS study focussing on water exchange in human cells will be difficult.

SESANS experiments were performed on liposomes and bacteria, as prototypes for small and large cells. The aim was to observe both as whole cells and investigate water transport through the membranes. All samples were prepared or grown in H_2O-based buffers or growth media and finally resuspended in D_2O. The liposomal polydispersity was determined using DLS. Fitting the measurement data showed the average cell diameters, cell-wall thicknesses and the inner contrast with respect to the medium, which were in good agreement with expected values from other techniques and literature. To our knowledge, this was the first time observing entire cells using neutrons.The rate of hydrogen-deuterium exchange could only be observed if the timescale was within the temporal resolution of the technique, which was not the case for any of the measured liposomes. A neutron source with a higher flux and larger wavelength should significantly increase the temporal resolution of the SESANS setup. Anticipating future developments, time-dependent measurements may become possible so that water exchange may be observed as well.

Model calculations predicted that large-scale removal of liposomes from a sample, through destruction by external means or biological process, should be observable with SESANS. A large drop in the liposome population of an investigated system would result in a change of curve shape and more noticeably, a change in the polarization saturation level, due to a change in the scattering power of the whole sample. SESANS measurements performed so far on small amounts of tumor cells exhibited too little scattering for a significant signal. Measurements on combined systems of liposomes with much higher concentrations of tumor cells are to be performed in the future.

The aim of this paper is to show as a proof of principle that SESANS has the potential to contribute to studies on biological bulk-phenomena, without disturbing the investigated systems. It may therefore become a useful addition to the broad range of techniques already available, with possible applications in the field of cancer research.

2. Biological background

2.1. Liposomes

A liposome is a spherical vesicle that consists of a bilayered phospholipid membrane, entrapping an aqueous phase. In this study the lipid vesicles only contain H_2O or D_2O or both. The membrane is made of natural substances: lipids and cholesterol that confer biocompatibility, non-toxicity, and biodegradable properties. The phospholipids are amphiphilic, consisting of a hydrophilic phosphate head group and two hydrophobic tails, which causes them to self-organize into lipid bilayers in an aqueous environment. Over the last decade, liposomes have largely been used in the medical field as a tool for drug delivery for therapeutic purposes and, more recently, for molecular imaging in disease diagnoses.

Liposomes are classified as Large Multilamellar Vesicles (LMV) or as Large and Small Unilamellar Vesicles (LUV and SUV). The SUV is the main interest of the present study, because it is the most used type of liposome in the medical field. The optimum liposome diameter for an efficient extravasation out of the blood into a tumor is known to be around 100 nm. This is an important property for passive drug-targeting of the diseased area, especially for tumor-targeting of multiple types of cancers.Stealth® liposomes are grafted with PolyEthylene Glycol (PEG)-chains and offer additional advantages, such as avoiding self-aggregation of the particles and enhancing the in vivo half-life by reducing the interaction with the blood plasma proteins. For this study, DSPC and DPPC SUV PEGylated liposomes have been synthesized with a diameter D of about 100 nm and a membrane thickness T of about 5 nm [23] and characterized using DLS and SESANS. DSPC and DPPC liposomes posses different membrane rigidity and therefore different water exchange properties.

2.2. Escherichia coli

The microorganism *Escherichia coli* (*E. coli*) is a well known species of bacterium, which was discovered by German pediatrician and bacteriologist Theodor Escherich in 1885. *E. coli* is placed on the bacteria branch in the phylogenetic tree of life, as a gram negative bacterium. *E. coli* cells can grow easily under both aerobic and anaerobic conditions and by fermentation. Optimal growth occurs at 37°C, but in some cases *E. coli* can also multiply at temperatures below 25°C and up to 49°C. The cells may have flagella and are about 2 µm long and 0.8 µm in diameter [24], with a volume of 0.6-0.7 μm^3. Strains that possess flagella can swim and are motile. There are many different types of *E. coli* based on the antigen type: the somatic (O) antigen, the capsular (K) antigens and the flagellar (H) antigen. There are over 170 O antigens, over 100 K antigens and over 50 H antigens.

Bacteria do not posses a cell nucleus or any other large structures, but have their components scattered throughout the cytoplasm. Since none of these are observable with the techniques used in this study, *E. coli* will also be considered as empty vesicles. The strain K12 was selected, since it is safe to handle and easy to grow, with cell division occurring every 20 minutes under

Table 1. Coherent neutron scattering lengths, component volumes and scattering length densities for DPPC, its head and tails, and cholesterol [25]. By comparison, DSPC consists of $C_7H_{16}NO_4P$ (head group) and $C_{37}H_{72}O_4$ (tails), giving $C_{44}H_{88}NO_8P$.

Molecule	Formula	b [fm]	V_m [Å^3]	ρ_s [nm^{-2}]
DPPC	$C_{40}H_{80}NO_8P$	27.63	1,232	$2.2 \cdot 10^{-5}$
head	$C_{10}H_{18}NO_8P$	60.1	326	$18.4 \cdot 10^{-5}$
tails	$C_{30}H_{62}$	32.4	891	$-3.6 \cdot 10^{-5}$
Cholesterol	$C_{27}H_{46}O$	13.25	629	$2.1 \cdot 10^{-5}$

Table 2. Absolute scattering length density differences of phases under consideration.

Phases	$\vert\Delta\rho_s\vert$ [nm^{-2}]
H_2O – D_2O	$6.89 \cdot 10^{-4}$
lipid – H_2O	$0.76 \cdot 10^{-4}$
lipid – D_2O	$6.13 \cdot 10^{-4}$

absolutely optimal conditions. It is also easily-manipulated, making it one of the most-studied prokaryotic model organisms, and an important species in biotechnology.

2.3. Contrast of lipid bilayers

Coherent neutron scattering occurs in materials due to fluctuations in scattering length density (SLD), defined as the total bound coherent scattering length b divided by the occupied volume V_m. In this system, the largest SLD-difference arises between either H_2O inside and D_2O outside the liposomes and bacteria, or between their lipid bilayers and the D_2O medium. For the lipid bilayers the average SLD is taken, so that the SLD-profile over the membrane is represented by a constant. Using the chemical formulas and the mass or number densities of the compounds, the overall SLD of lipid bilayers can be calculated according to

$$\rho_s = \frac{\sum_i n_i b_i}{\sum_i n_i V_{m,i}}. \tag{1}$$

Estimating the mass densities of the organic compounds to be between 0.8 and 1.0 kg/m^3 and using tabulated scattering lengths [26], yields values around ρ_s=$2 \cdot 10^{-5}$ nm^{-2} for both liposomes, ranging from $1.5 \cdot 10^{-5}$ to $1.9 \cdot 10^{-5}$ nm^{-2} for DSPC and from $1.8 \cdot 10^{-5}$ to $2.3 \cdot 10^{-5}$ nm^{-2} for DPPC.This result is irrespective of the $\frac{1}{3}$ molar cholesterol, since it has a similar SLD. The few molar % contributions of the PEG chains are neglected. Literature values [25] of molecular volumes for DPPC and cholesterol support these assertions, giving ρ_s=$2.2 \cdot 10^{-5}$ nm^{-2} and ρ_s=$2.1 \cdot 10^{-5}$ nm^{-2} using eq. 1, as summarized in table 1. Lacking the necessary data to calculate the SLD for the membrane of *E. coli*, the lipsome values have been used under the assumption that both membranes have an SLD of the same order of magnitude. H_2O and D_2O have similar mass densities of ρ=1.0 and 1.1 kg/m^3 respectively, but massively different scattering length densities of ρ_s=$-5.6 \cdot 10^{-5}$ and $+6.33 \cdot 10^{-4}$ nm^{-2}. The resulting SLD-differences between the lipid membrane, H_2O and D_2O are listed in table 2.

3. Experimental

3.1. Sample preparation

3.1.1. Liposomes. *1,2-distearoyl-sn-glycero-3-phosphorylcholine* (DSPC, $C_{44}H_{88}NO_8P$), *1,2-dipalmitoyl-sn-glycero-3-phosphocholine* (DPPC, $C_{40}H_{80}NO_8P$) and *1,2-distearoyl-sn-glycero-3-phosphoethanolamine-N-[methoxy-(polyethylene glycol)-2000]* (DSPE-PEG2000-OMe) were supplied by Lipoid AG, Cham, Switzerland. The used cholesterol ($C_{27}H_{46}O$) was purchased from Avanti polar Inc., Alabaster, AL, USA.

DSPC and DPPC liposome batches were prepared independently. For the first batch a total compound amount of 200 μmol was used with DSPC:cholesterol:DSPE-PEG2000-OMe ratios of 1.75:1:0.25 dissolved 1:1 in chloroform ($CHCl_3$) and methanol (MeOH). The second batch consisted of 100 μmol with DPPC:cholesterol:DSPE-PEG2000-OMe ratios of 1.85:1:0.15, which were dissolved 1:1 in dichloromethane (CH_2Cl_2) and MeOH. Both mixtures were vacuum-dried to obtain thin lipidic films, which were hydrated and resuspended in MilliQ. These mixtures were subsequently extruded through polycarbonate membrane filters with a pore size of 200 nm (ten times) and 100 nm (ten times), using Lipofast extruder (Avestin, Canada). The temperature during the extrusion was set around 50°C. The extruded solution was ultra-centrifuged for 60 minutes at 4°C at 45 krpm (0.30 Mg) and at 40 krpm (0.23 Mg) respectively for both batches, using an ultracentrifuge Beckman L7, rotor 50 Ti. The pellets were separated from the supernatant and injected into cuvettes with 5 ml D_2O.

3.1.2. E. coli. Wild-type *E. coli* K12 strains were grown at 37°C overnight in 200 ml Luria Bertani medium in 500 ml Erlenmeyer flasks under shaking conditions until a standard optical density was reached (O.D.=2 using a wavelength of 600 nm). The cells were spun down using a Jouan CR 4.11 centrifuge at 3.5 krpm at 4°C for 60 minutes. Wet pellets were re-suspended in D_2O after washing with 10 ml D_2O to eliminate the excess of water in between the cells. Centrifugation was repeated and the obtained new pellets were re-suspended in 2 ml D_2O each after vortexing with a home-made bead beater at 2.5 krpm at 4°C for 30 minutes. Four homogeneous pellets totaling 0.4 g were injected into a cuvet with 4 ml D_2O.

3.2. The SESANS Technique

3.2.1. Principles. Spin-Echo Small Angle Neutron Scattering (SESANS) is a non-destructive technique that probes density correlations in real space directly from 20 nm up to 20 μm. SESANS uses spin-echo to encode small scattering angles of neutrons when traversing a sample and it measures the effective loss of polarization of a neutron beam as a function of spin-echo length. This method is used to study the structure of suspensions of mesoscopic particles and its range of accessible length scales makes it suitable for a variety of materials, including large colloids [27] and granular matter [28]. The used SESANS setup consists of a monochromatic, polarized neutron beam traveling in the horizontal x-direction through two magnetic field regions, in between which a sample is placed. Experiments are performed by varying the magnetic field strength to probe correlations along the vertical spin-echo length z. The technique is not sensitive to the y-direction and can be described classically, as the net Larmor precession of neutrons of a polarized neutron beam [29], or quantum-mechanically, as the vertical splitting and recombination of two neutron eigenstates, due to their different interaction with a magnetic field [30, 31]. The polarization P of the beam, normalized with the empty beam polarization P_0, depends on the scattered fraction of the beam $\Sigma\,\ell$ and on the normalized, dimensionless function $G(z)$:

$$\frac{P}{P_0}(z) = \mathrm{e}^{\Sigma\,\ell\,[G(z)-1]} \tag{2}$$

where ℓ is the neutron path length through the sample, i.e., the sample thickness, Σ is the average number of times a neutron scatters per unit length and $G(z)$ is a measure of the microstructure.

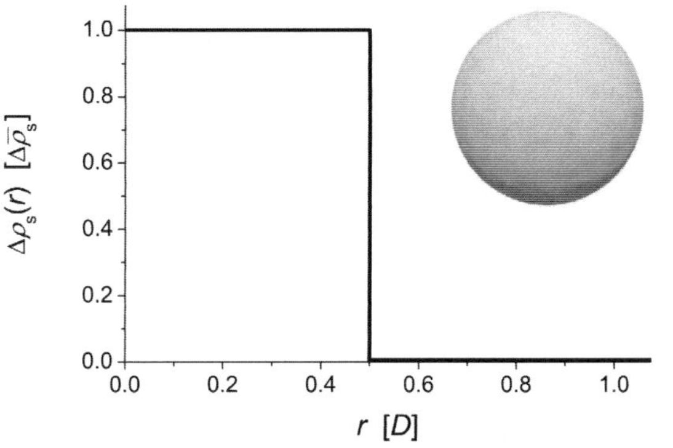

Figure 1. A solid sphere of diameter D has a homogeneous SLD-distribution $\rho_s(r)$ inside the sphere, where the average difference $\overline{\Delta\rho_s}$ is with respect to the SLD of the medium outside the sphere.

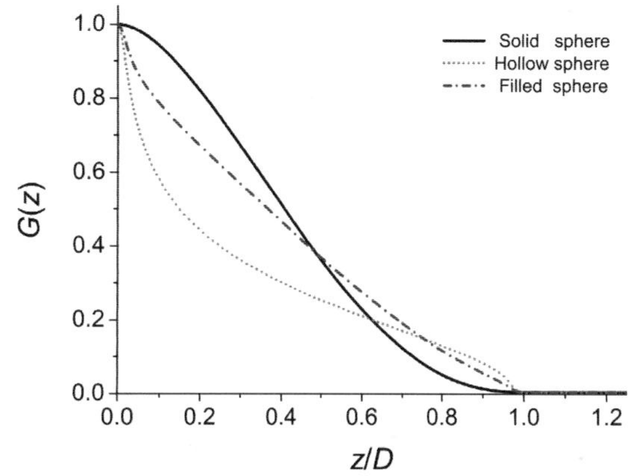

Figure 2. Examples of the projected correlation functions for solid, hollow and filled spheres, using $T = \frac{1}{10}D$ for the shell thickness and $s = 0.1$ for the core contrast. See the text for further details.

It is the projection along the neutron beam of the density auto-correlation function, which correlates SLD-fluctuations throughout the sample and is denoted $\gamma(r)$ for isotropic systems, such as here. Long-range correlations disappear beyond distances z_{sat}, so that $G(z \geq z_{\text{sat}}) = 0$, causing the polarization signal to saturate. The amount of scattering Σ depends on beam and sample properties:

$$\Sigma = \lambda^2 \left\langle \Delta\rho_s^2 \right\rangle \xi \overset{\text{binary}}{=} \lambda^2 \phi_V (1-\phi_V) \overline{\Delta\rho_s}^2 \xi \tag{3}$$

with neutron wavelength λ, SLD-variance $\langle \Delta\rho_s^2 \rangle$ and correlation length ξ. For a two-phase system the mean-squared SLD-difference can be expressed as the squared average difference $\overline{\Delta\rho_s}^2$ (contrast) between the two phases, multiplied with the volume fraction ϕ_V of the suspended phase and the medium. The correlation length ξ is defined as the integral over $\gamma(r)$ along x at $z=0$ and can be interpreted as the average length that neutrons travel through one phase before encountering another phase, making ξ a measure of typical length scale.

3.2.2. Correlation Functions. For dilute systems, the auto-correlation function $\gamma(r)$ gives the shared volume fraction of an object with an overlapping copy of itself. Analytical expressions for $\gamma(r)$ and its projection $G(z)$ have been derived for several dilute, monodisperse [32] and polydisperse [33] model systems such as solid and multi-layered spheres, using geometry and calculus.

A solid sphere of diameter D has a constant density distribution $\rho_s(r)$ inside volume $V_{\text{sphere}} = \frac{1}{6}\pi D^3$. The density correlation function is equal to the shared volume fraction of two identical spheres when separated by a distance $0 \leq r \leq D$ from center to center. This lens shaped volume consists of two equal sphere caps, so that $\gamma_{\text{SS}}(r) = 1 - 1\frac{1}{2}\frac{r}{D} + \frac{1}{2}\left(\frac{r}{D}\right)^3$ is readily derived. The analytical expression for the projected correlation function is

$$G_{\text{SS}}(z) = \Re\left\{ \sqrt{1-\left(\frac{z}{D}\right)^2} \left[1+\tfrac{1}{2}\left(\frac{z}{D}\right)^2\right] + 2\left(\frac{z}{D}\right)^2 \left[1-\tfrac{1}{4}\left(\frac{z}{D}\right)^2\right] \ln\frac{\frac{z}{D}}{1+\sqrt{1-\left(\frac{z}{D}\right)^2}} \right\}, \tag{4}$$

where $\Re$ denotes the real part. The correlation length of a solid sphere is

$$\xi_{\text{SS}} = \tfrac{3}{4}D. \tag{5}$$

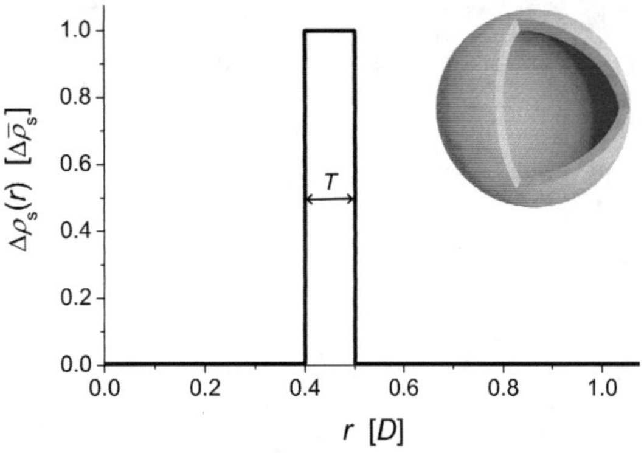

Figure 3. A hollow sphere of diameter D and shell thickness T has a constant SLD-difference $\overline{\Delta\rho_\mathrm{s}}$ distributed over a spherical shell, with respect to the medium on the inside and the outside.

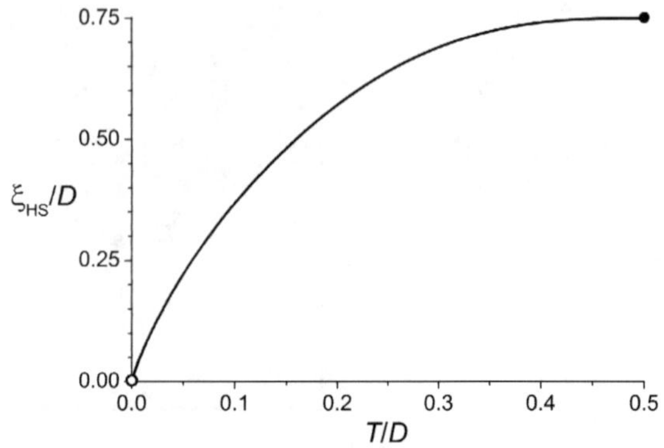

Figure 4. The hollow sphere correlation length ξ_HS of diameter D as a function of shell thickness T, between the limits of an infinitely thin shell ($T\downarrow 0$) and a solid sphere ($T=\frac{1}{2}D$).

A hollow sphere of outer diameter D and shell thickness T has its density homogeneously distributed in a spherical shell of volume $V_\mathrm{shell} = \pi D^2T - 2\pi DT^2 + \frac{4}{3}\pi T^3$ and is obtained by subtracting the density distribution of a sphere from that of a larger copy. The elaborate correlation function has been derived analytically by auto-convolution of the density distribution of the hollow sphere [32]. The projected correlation function is

$$G_\mathrm{HS}(z) = \frac{\frac{2}{3}}{f\left(\frac{D-2T}{D}\right)} F\left(\frac{2z}{D}, \frac{D-2T}{D}\right). \tag{6}$$

The normalization of $G_\mathrm{HS}(z)$ is performed by the smooth dimensionless function

$$f(\sigma) = (1-\sigma)\left(1-\sigma^3\right) + \tfrac{1}{2}\left(1-\sigma^2\right)^2 \ln\frac{1+\sigma}{1-\sigma}. \tag{7}$$

with argument $0 \le \sigma < 1$, which is the ratio between inner and outer diameter $\frac{D-2T}{D}$. When considering a fixed diameter D, then for an infinitely thin shell $T\downarrow 0$ gives $\lim f(\sigma\uparrow 1) = 0$, and for $T=\frac{1}{2}D$ the shell fills the sphere, so that $f(\sigma=0) = 1$. From this function the correlation length has been derived, which is given by

$$\xi_\mathrm{HS} = \left(\frac{2T}{D}\left[1 + \tfrac{2}{3}\frac{D^2 - 2DT + T^2}{D^2 - 2DT + \frac{4}{3}T^2}\ln\frac{D-T}{T}\right]\right)\tfrac{3}{4}D. \tag{8}$$

For the limiting case of a shell of half the diameter ξ_HS becomes equal to ξ_SS in nearly constant fashion, and for a hollow sphere with an infinitely thin membrane ξ_HS can be shown to approach zero nearly linearly. Figure 4 shows the dependency of the correlation length of a hollow sphere on the relative shell thickness. $F(\zeta,\sigma)$ expresses the correlations between different parts of the geometry of a hollow sphere in a dimensionless function

$$\begin{aligned} F(\zeta,\sigma) = {} & H_0(2,\zeta) - H_0(1{+}\sigma,\zeta) + H_0(1{-}\sigma,\zeta) + \sigma^4 H_0\left(2,\frac{\zeta}{\sigma}\right) \\ & - \sigma^4 H_0\left(\frac{1{+}\sigma}{\sigma},\frac{\zeta}{\sigma}\right) + \sigma^4 H_0\left(\frac{1{-}\sigma}{\sigma},\frac{\zeta}{\sigma}\right) \\ & - 2\sigma^3 h_1(1{-}\sigma,\zeta) + \tfrac{3}{8}\left(1{-}\sigma^2\right)^2\left[h_0\left(1{+}\sigma,\zeta\right) - h_0(1{-}\sigma,\zeta)\right]. \end{aligned} \tag{9}$$

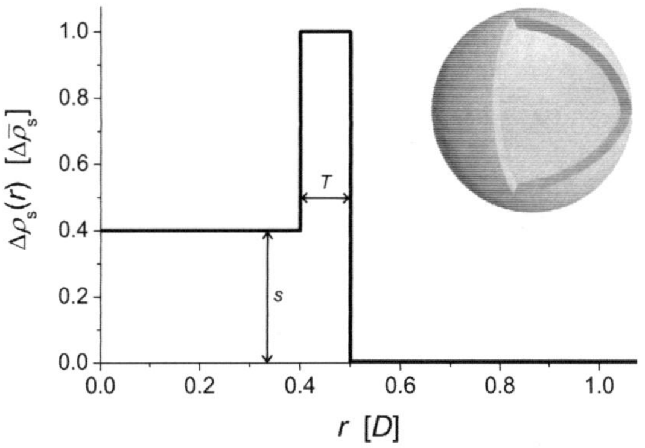

Figure 5. A filled sphere has separate SLD-differences $\Delta\rho_s(r)$ for its core and its shell, characterized by the parameter s. It is equal to a hollow sphere for $s = 0$ and to a solid sphere for $s=1$.

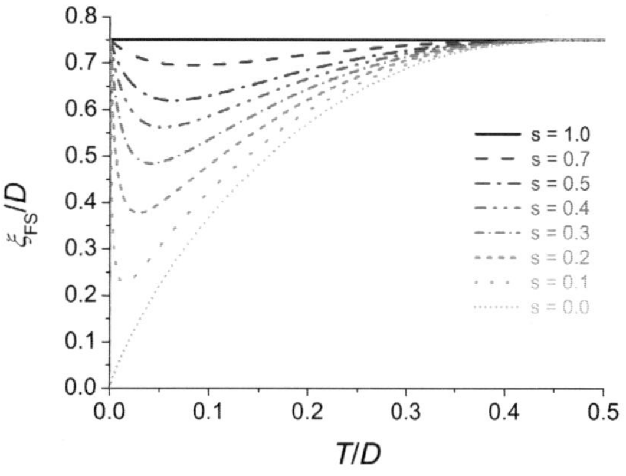

Figure 6. The correlation length ξ_{FS} of a filled sphere of diameter D as a function of shell thickness T, for $s = 0$ (hollow sphere), $s = 1$ (solid sphere) and six intermediate values. At $T = 0$ a filled sphere only has a core, which is also a solid sphere, except for an empty core ($s=0$).

which is constructed from

$$h_1(\alpha,\zeta) = 2\Re\left\{\sqrt{\alpha^2-\zeta^2}\right\} \tag{10}$$

$$h_0(\alpha,\zeta) = \Re\left\{\ln\frac{2\alpha+h_1(\alpha,\zeta)}{2\alpha-h_1(\alpha,\zeta)}\right\} \tag{11}$$

$$H_0(\alpha,\zeta) = -\tfrac{3}{8}\zeta^2\left(1-\tfrac{1}{16}\zeta^2\right)h_0(\alpha,\zeta) + \left(1-\tfrac{3}{8}\alpha+\tfrac{1}{64}\alpha^3+\tfrac{3}{128}\alpha\zeta^2\right)h_1(\alpha,\zeta). \tag{12}$$

An intermediate model is that of the filled sphere, which has a solid spherical core with a shell around it. A filled sphere has the density distribution of a hollow sphere filled with a third phase. The corresponding correlation function is constructed from the hollow sphere model by addition of a dimensionless parameter s, defined as the difference in SLD between core and medium normalized with the $\Delta\rho_s$ between shell and medium:

$$s = \frac{\rho_s^{\text{core}} - \rho_s^{\text{medium}}}{\rho_s^{\text{shell}} - \rho_s^{\text{medium}}} = \frac{\Delta\rho_s^{\text{core}}}{\Delta\rho_s^{\text{shell}}}. \tag{13}$$

The projected correlation function for a filled sphere then becomes

$$G_{FS}(z) = \frac{F\left(\frac{2z}{D},\frac{D-2T}{D}\right) + \frac{s}{1-s}H_0\left(2,\frac{2z}{D}\right) - s\left(\frac{D-2T}{D}\right)^4 H_0\left(2,\frac{2z}{D-2T}\right)}{\frac{3}{2}\left[f\left(\frac{D-2T}{D}\right)+\frac{s}{1-s}-s\left(\frac{D-2T}{D}\right)^4\right]}. \tag{14}$$

The behaviour of this correlation function is very close to that of hollow spheres and solid spheres, when s is close to 0 and 1 respectively. How quickly it deviates with s from these two cases, depends on the value of T. The difference between a filled sphere and a hollow sphere is negligable for $0 \leq s \lesssim \frac{2T}{D}$. The correlation length of a filled sphere is

$$\xi_{FS} = \frac{(1-s)\,f\left(\frac{D-2T}{D}\right) + s - s(1-s)\left(\frac{D-2T}{D}\right)^4}{1-(1-s^2)\left(\frac{D-2T}{D}\right)^3}\,\tfrac{3}{4}D. \tag{15}$$

An example of the projected correlation functions for dilute, monodisperse solid, hollow and filled spheres is shown in figure 2, using a shell thickness $T = \frac{1}{10}D$ for the hollow and filled spheres and a value of $s = \frac{1}{10}$ for the filled spheres, giving their core 10% of the $\Delta\rho_s$ between shell and medium.

3.3. SESANS Experiments

3.3.1. Experiments & Sample stability. SESANS measurements were performed with a neutron wavelength λ=2.09 Å on the DSPC and DPPC batches of liposomes and on *E. coli*. All samples were measured in quartz cuvettes of thickness ℓ=1.00 cm, with volume fractions of respectively ϕ_V=0.02 and ϕ_V=0.09 for the liposome and *E. coli* samples. The DSPC liposome concentration was found by phosphate determination according to Rouser; the *E. coli* concentration was calculated from the pellet mass, assuming weight densities equal to H_2O and assuming no losses during the final suspension.

Repeat measurements and DLS showed the liposome samples to remain stable for several days. An onset of sedimentation was observed for the *E. coli* after one day, most likely because of cell death as the D_2O environment contained no nutrients. Measurements with lower *E. coli* concentrations provided similar data with regards to particle dimensions.

3.3.2. Modeling. Two cases were considered for curve fitting: no H_2O-D_2O exchange versus full exchange. All fits were performed using non-linear least-squares regression with 95% confidence bounds. The first case could be represented by droplets of H_2O in a D_2O environment, since the effect of the cell membrane on the neutron signal would then be negligible. This model consisted of the projected correlation function $G(z)$ for dilute, monodisperse solid spheres and curve fitting was performed for the amount of scattering and particle size, i.e., the saturation level P_{sat} and diameter D. In the second case, the H_2O would be extremely diluted throughout the D_2Oinside and outside the cells, making the cell membrane contribute significantly to the signal. This system was modeled as hollow spheres filled with D_2O using the $G(z)$ for dilute, monodisperse spherical shells. Fitting was performed for the saturation level P_{sat}, the outer diameter D and membrane thickness T.

3.3.3. Assumptions. All samples were treated as isotropic systems and modeled with analytical expressions for monodisperse spherical particles. Since the concentrations used were dilute to semi-dilute, excluded volume effects were insignificant. Samples without sedimentation were isotropic suspensions, since all particles had rotational freedom. Particle size distributions were not considered to have a substantial effect and not included in the modeling, based on the very small degree of liposomal polydispersity as observed with DLS and the fact that *E. coli* multiply through cell-division, causing a uniform and highly monodisperse population under standard conditions.

In order to avoid over-parameterization certain assumptions were made. The liposomes were assumed to be spherical since there was no cause for deformation. As a first approximation, the aspect ratio of the *E. coli* cells was disregarded. However, *E. coli* are more rod-like than spherical with an aspect ratio of 2.5. The correlation function of a rod differs from that of a sphere, although arguably less for a hollow body than for a solid one.

4. Results

4.1. Characterization of liposomes with DLS

To confirm the consistency of the two liposome formulations, dynamic light scattering was used to establish their size distributions in order to characterize them by their average diameter and polydispersity. Due to faster degradation, there was not enough *E. coli* material available

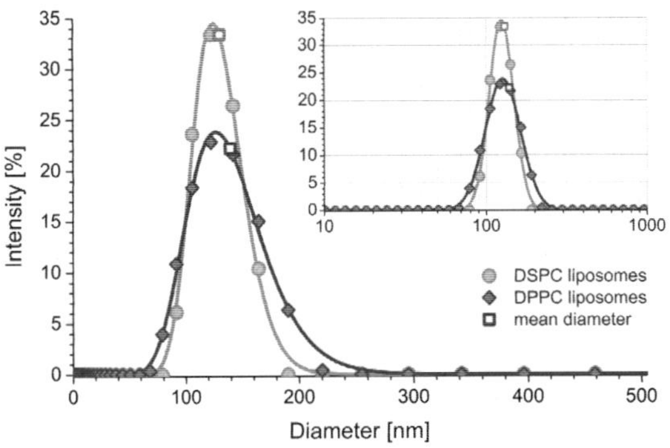

Figure 7. Distribution of DSPC and DPPC liposome diameters acquired by DLS. The inset shows approximately normal size distributions on a log-scale. Curves were fit using a lognormal distribution of arbitrary amplitude. Mean liposomal diameters are shown for comparison.

Table 3. Fit results of the size distribution acquired by DLS of DSPC and DPPC liposomes, with a lognormal distribution of sizes. Fit parameters μ and σ were converted to a mean diameter and polydispersity index.

Parameter	Meaning	DSPC liposomes	DPPC liposomes
μ	fit value	4.854 ± 0.004	4.905 ± 0.006
σ	fit value	0.174 ± 0.004	0.251 ± 0.005
D_{mode} [nm]	peak	124.4 ± 0.5	126.7 ± 0.8
D_{median} [nm]	middle	128.3 ± 0.5	135.0 ± 0.8
D_{mean} [nm]	average	130.2 ± 0.5	139.3 ± 0.9
PDI [-]	width	0.094 ± 0.002	0.138 ± 0.003

to use DLS to confirm their monodispersity. Both liposome size distributions had modes (peak-positions) around 125 nm with an approximately normal distribution of diameters on a logarithmic scale, as shown in the inset of figure 7.

The average diameter and distribution width were acquired by fitting a lognormal distribution of diameters to the DLS data, using arbitrary amplitude and obtaining parameters μ and σ, see Table 3. The distribution width was expressed by the polydispersity index (PDI), where a large PDI would cause the mean and median (the diameter halving the population) to shift away from the mode. The median and mean diameters of the DSPC liposomes were close to the mode and with a PDI just below 0.1, the DSPC liposomes were effectively monodisperse particles with a diameter of 130 nm. The DPPC liposomes had a slightly wider distribution with a PDI of 0.14, resulting in an average size of 140 nm.

4.2. SESANS Results

4.2.1. Liposomes. Less than 4% scattering was observed for the liposome suspensions in D_2O, requiring more than half a day of measurement time to reach an acceptable signal quality such as shown in figures 8 and 9. For both the DSPC and the DPPC batch, the polarization saturated at spin-echo lengths z_{sat} above 150 nm and at a saturation level P_{sat}=0.965, where both quantities contained length scale information. Since the liposomes were approximately monodisperse, z_{sat} corresponded to the diameter of the liposomes, which also determined the saturation level through the correlation length (see eq. 3 and eq. 2 for $G(z)$=0). The amount of scattering provided a constant for the product of contrast and correlation length $\overline{\Delta\rho_{\text{s}}}^2\xi$, where ξ_{SS} and ξ_{HS} were calculated according to eq. 5 and eq. 8 for the solid and hollow sphere models.

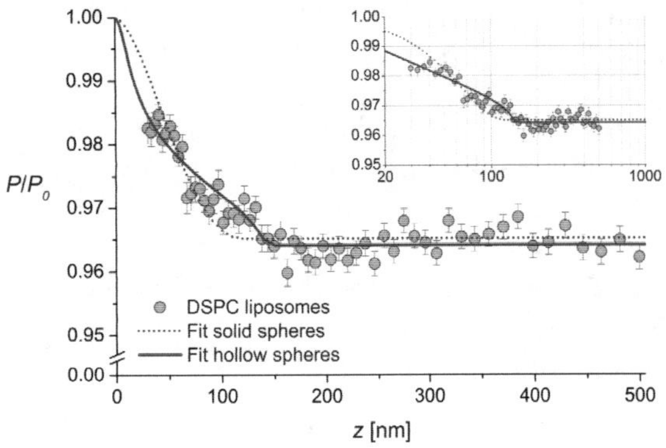

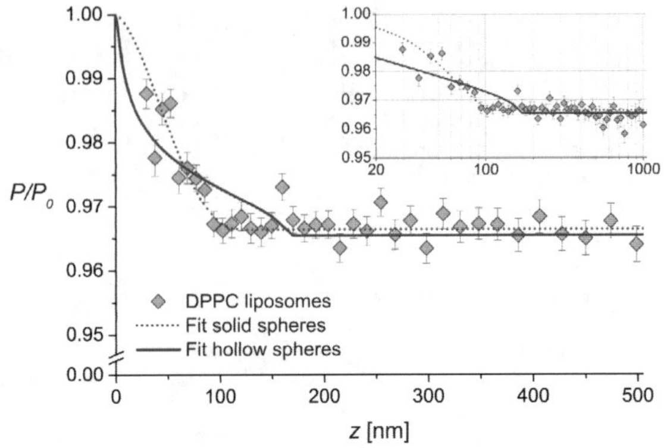

Figure 8. SESANS data of DSPC liposomes. **Figure 9.** SESANS data of DPPC liposomes. Both figures include log-scale insets. The curves are fits for the solid and hollow sphere models.

The difference in SLD between particle and medium was retrieved as

$$|\Delta\rho_{\rm s}| = \sqrt{\frac{-\ln P_{\rm sat}}{\ell\,\lambda^2\,\phi_{\rm V}\,(1-\phi_{\rm V})\,\xi}} \tag{16}$$

For solid spherical particles of $\approx$150 nm the correlation length was $\xi_{\rm SS}$=110 nm, yielding $|\Delta\rho_{\rm s}|$=1.9·10^{-4} nm^{-2}, which was a factor 3.6 smaller than the H_2O-D_2O SLD-difference. For hollow spheres of outer diameter $\approx$150 nm and shell thickness $\approx$5 nm, the correlation length $\xi_{\rm HS} \approx 25$ nm, which gave $|\Delta\rho_{\rm s}|$=4.1·10^{-4} nm^{-2}, a factor 1.5 smaller than calculated between membrane and D_2O. The scattered intensity therefore supported the notion that liposomes were better described as hollow than as solid spheres.

4.2.2. E. coli. About four times more concentrated than the liposome samples, the *E. coli* suspension in D_2O exhibited an equally larger scattering power. In spite of only six hours measurement time, an adequate SESANS signal quality could therefore be reached. The polarization saturated around 88% at spin-echo lengths above 700 to 800 nm, as shown in figure 10. Considering the *E. coli* as solid spherical particles of $\approx$750 nm gave a correlation length $\xi_{\rm SS}$ $\approx$560 nm, yielding $|\Delta\rho_{\rm s}|$=0.80·10^{-4} nm^{-2}, which was a factor 8.6 smaller than the H_2O-D_2O SLD-difference. For hollow spheres of outer diameter $\approx$750 nm and shell thickness $\approx$5 nm, the correlation length $\xi_{\rm HS}$ $\approx$35 nm, which gave $|\Delta\rho_{\rm s}|$=3.3·10^{-4} nm^{-2}, a factor 1.8 smaller than calculated between membrane and D_2O. The amount of scattering therefore indicated that modeling *E. coli* as hollow spheres was more appropriate than as solid spheres.

4.2.3. Curve fitting. Describing the SESANS liposome data with the solid sphere model gave reasonable fits, yielding an effective size of 120 nm for both batches with a relative error of 8%, see table 4. Fitting the hollow sphere model to the data produced an apparently better fit for DSPC and worse for DPPC liposomes, with diameters of 150 and 170 nm and shell thicknesses around 10 nm, as shown in table 4. The relative errors were 7% and 15% for the diameters and 40% and 90% for the shell thicknesses, showing that the data were not very sensitive for T. This was supported by the small correlations of T with the other parameters, as seen in table 5, which also shows that cross correlations between parameters were acceptably low for both models. The largest interdependency was observed between the saturation level $P_{\rm sat}$ and the particle diameter D, which were connected through the correlation length.

Modeling the *E. coli* data as monodisperse, solid spheres produced a diameter of 720 nm with a relative error of 15%, but did not capture the initial decay of the measured data, see figure 10.

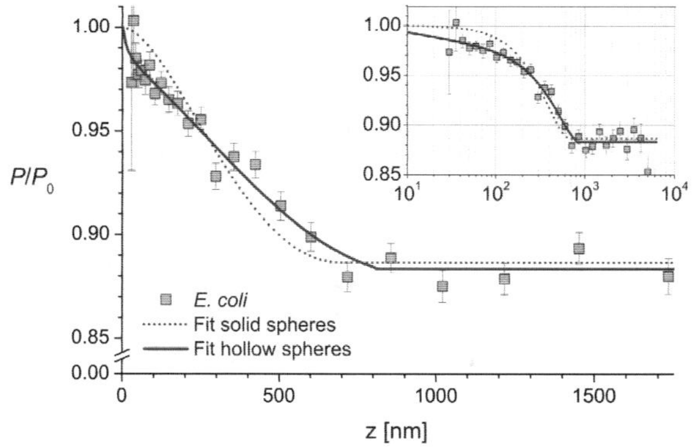

Figure 10. SESANS measurement of *E. coli* and fit curves for solid and hollow spheres. The inset shows the same data on a log-scale.

Table 4. Fit results of SESANS measurements of liposomes and *E. coli*, modeled as solid, hollow and filled spheres. Parameters are the polarization saturation level (scattering power) and particle dimensions. Filled sphere *E. coli* values were obtained in the hollow sphere limit, with s approaching zero. Goodness-of-fit indications are given by χ^2_k, where k equals the number of measured points minus the number of fitted parameters.

Parameter	DSPC liposomes Solid sphere	DSPC liposomes Hollow sphere	DPPC liposomes Solid sphere	DPPC liposomes Hollow sphere	*E. coli* Solid sphere	*E. coli* Filled sphere
P_{sat} [-]	0.965±0.001	0.964±0.001	0.966±0.001	0.965±0.001	0.866±0.009	0.883±0.005
D [nm]	122±9	152±11	121±11	169±25	718±106	818±98
T [nm]	–	13±5	–	10±11	–	8±44
s [-]	–	–	–	–	–	0.06±0.26
g.o.f.	$\chi^2_{58} = 2.27$	$\chi^2_{57} = 1.37$	$\chi^2_{48} = 1.38$	$\chi^2_{47} = 2.08$	$\chi^2_{28} = 3.22$	$\chi^2_{26} = 0.97$

The hollow sphere model seemed to produce better fits, with a sharper initial decay. Minimizing χ^2_{27} to 1.51 produced a saturation level P_{sat}=0.885±0.007, an outer diameter D=815±133 nm, but an unphysical shell thickness of T=120±30 nm. These hollow sphere values are not shown in table 4. Using the filled sphere model helped to avoid this pitfall, producing the same saturation level and outer diameter, as well as a realistic shell thickness. The extra degree of freedom in this model was parameter s, the ratio between the $\Delta\rho_s$ values of the core and the shell, both with respect to the medium. A jump in the value for T to the unphysical range appeared when $s<0.0735$, which was established by fitting with s fixed increasingly closer to zero.

As shown in table 4, fitting with filled spheres without constraints produced a physical but imprecise value of T=7.7±43.5 nm. The value s=0.064 was small enough to effectively represent a hollow sphere, as the core-medium contrast had a negligible contribution to the overall 3-phase contrast, corresponding to a mere 0.36% of the shell-medium contrast. Since the hollow and filled sphere models gave identical curves at these parameter values, these *E. coli* dimensions were taken as an appropriate fit. Relative errors of the diameter and shell thickness were 12% and 565%, with small correlations between T and the rest of the model. The very large uncertainty for T meant, that even though it could be established to be small compared to D, it could not be quantified accurately from this data. The cross correlations between all parameters were similar to those of the liposome fits, see table 5.

Table 5. Correlations between parameters of solid and hollow sphere models fitted to SESANS data of DSPC and DPPC liposomes and *E. coli* bacteria. P_{sat} is the polarisation saturation level, D is the particle diameter and T the shell thickness for hollow spheres.

	DSPC liposomes					DPPC liposomes					*E. coli*				
	Solid sphere		Hollow sphere			Solid sphere		Hollow sphere			Solid sphere		Hollow sphere		
	P_{sat}	D	P_{sat}	D	T	P_{sat}	D	P_{sat}	D	T	P_{sat}	D	P_{sat}	D	T
P_{sat}	1	0.4	1	0.5	0.2	1	0.4	1	0.4	0.3	1	0.5	1	0.6	0.2
D	0.4	1	0.5	1	0.0	0.4	1	0.4	1	0.2	0.5	1	0.6	1	0.1
T	–	–	0.2	0.0	1	–	–	0.3	0.2	1	–	–	0.2	0.1	1

5. Discussion

5.1. Measurement results

The DSPC and DPPC liposomes gave close to identical fit results using the solid sphere model and similar, but less accurate fit results for the hollow sphere model. The hollow sphere model exhibited a larger sensitivity to the shorter spin-echo lengths within the resolvable range, than the solid sphere model. Therefore the differences between the hollow sphere results of DSPC and DPPC were attributed to the number of sampled spin-echo points below 150 nm. The statistical quality of the measurements did not allow for any extension of the modeling to include concentration effects, anisotropy, aspect ratios or polydispersity. The amounts of scattering from the DSPC and DPPC liposomes and from the *E. coli* cells suggested that these samples with D_2O medium were better described by hollow spheres than by solid spheres. The SESANS observations were in better agreement with the contrast between lipid membranes and D_2O than with the contrast between H_2O and D_2O. The shells with mostly hydrogen-atoms were therefore more likely enclosing D_2O-cores than H_2O-cores.

Results of curve fitting of the liposome data were ambiguous, finding a better hollow sphere fit for DSPC and a better solid sphere fit for DPPC, despite the strong similarities between the raw data of both samples. Comparison with DLS results showed best agreement between solid spheres, with diameters of 120 nm, and the modes of the size distribution at 125 nm. The hollow sphere models produced 15% and 20% larger particle sizes of 150 and 170 nm for DSPC and DPPC than the mean diameters of 130 and 140 nm as observed by DLS. However, since the DLS characterisations were performed after the SESANS measurements and liposome samples are only semi-stable, it is possible that the size distributions had shifted to somewhat smaller averages. The shell thicknesses of the hollow spheres were determined to be between 0 and 20 nm by fitting, where the data did not allow for a more accurate assessment. The findings were nevertheless in reasonable agreement with the expected value of 5 nm, which is commonly reported for the thickness of a lipid bilayer membrane without PEGylation, and with the expected value of about 10 nm for a membrane with PEG-brushes grafted to each side.

E. coli cells are known to have a cell length of 2 µm, a diameter of 0.8 µm and a cell wall thickness of about 5 nm. The filled sphere model agreed best with these expected values close to the limit of a hollow sphere, i.e., near zero contrast between inside and outside, finding a diameter of 820±100 nm and a thickness of 8±40 nm. The large uncertainties found in the diameter and thickness were due to the aspect ratio of the rod-like bacteria and because the cell wall is sandwiched between a thin plasma membrane and a thick outer membrane capsule. Since *E. coli* multiply through cell-division, a high degree of monodispersity was expected, but this could not be confirmed with DLS due to lacking supplies of the fast degrading bacteria and their shape, which complicated analysis.

Journal of Physics: Conference Series **247** (2010) 012016 doi:10.1088/1742-6596/247/1/012016

5.2. Outlook

When a change in SESANS curve shape and/or polarization saturation level is observed in experiments, a change in structure, concentration or contrast can in principle be deduced. It is therefore worthwhile to investigate if SESANS could be applied to study such a change in a biological system.

Liposomes can be destroyed by external factors, such as ultrasound, or taken up and broken down through endocytosis by much larger cells, such as tumor and white blood cells, or macrophages. Typical diameters of tumor cells are $5 \leq D \leq 7$ µm in rats and $10 \leq D \leq 20$ µm in humans, whereas macrophages are about 13 µm in rats and about 21 µm in humans. SESANS measurements have already been performed on tumor cells from the rat pancreatic CA20948 tumor cell line with an approximate diameter of 6 µm, but only low concentrations were available, causing too little scattering for a significant signal to be discerned. However, we can still show through a simple calculation that a change in liposome population *in situ* may constitute a noticeable signal change.

Consider a dilute system of small liposomal vesicles and of much larger cells, so it can be considered a bimodal distribution of spherical particles, say types A and B. Take their size distributions as approximately monodisperse and let A and B have similar contrasts with respect to the medium. The shape and level of a SESANS curve then depend only on whether these particles are solid or hollow in nature and on the two scattering contributions. If these contributions are not comparable, then the curve will merely show the signal of the particles which scatter the most: the presence or absence of the other particles cannot be observed. If the amounts of scattering are comparable, then the measured signal is sensitive to the removal of smaller particles from the system. This can affect both the shape of the curve, through the disappearance of a kink at the diameter of the smaller particle, and the saturation level, as the total amount of scattering is altered.

The scattering contributions for both types A and B depend on the two correlation lengths and volume fractions, and will be comparable for similar products $\phi_V\,(1-\phi_V)\cdot\xi$. In the dilute regime cross-correlations can be neglected, which gives a simple description of the signal as

$$\frac{P}{P_0}(z) = \exp\left\{\ell\,\lambda^2\,\overline{\Delta\rho_s}^2\left[\phi_{V,A}\,\xi_A\,(G_A(z)-1) + \phi_{V,B}\,\xi_B\,(G_B(z)-1)\right]\right\} \qquad (17)$$

For solid spheres the correlation length scales with the diameter and is therefore much larger for the large particles. The correlation length of large hollow spheres is however much smaller than of small hollow spheres with the same shell thickness.

An example calculation is shown in figure 11 for a system of 50 mg liposomes, 25 mg rat tumor cells and 5.5 grams of D_2O, corresponding to volume fractions of 1.0% and 0.5% for the liposomes and tumor cells. This gave a total sample volume of 5.0 ml. A standard sample thickness of ℓ=1.00 cm was used for the calculation and a wavelength λ=6 Å, was taken to increase signal strength. Liposomes and tumor cells were modeled as hollow spheres of 130 nm and 6 µm with a lipid bilayer membrane thickness of 5 nm, giving correlation lengths of 24 nm and 43 nm respectively. For the $\overline{\Delta\rho_s}$ between membrane and D_2O the value $6.13\cdot10^{-4}$ nm^{-2} from table 2 was used.

The situation afterwards was calculated by removal of 90% of the liposomes, corresponding to exactly $1.8\cdot10^5$ liposomes per tumor cell. The tumor cell and D_2O fractions (mole, mass and volume), as well as the total sample volume were hardly affected with less than 1% difference and were therefore neglected. In the unlikely case of massive endocytosis by tumor cells, the tumor cell volume would only change significantly when internalizing more than ~10,000 liposomes. When modelling with solid spheres about 100 times less liposomal and tumor material was required for good signal strength, but the differences before and after were much less pronounced.

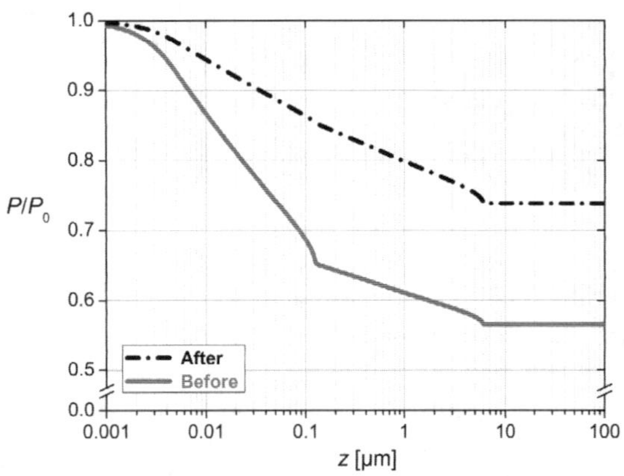

Figure 11. Predictive calculation of the SESANS signal of a mixed liposome and a rat tumor cell system, both modeled as hollow spheres. Signals are shown before and after decimating the liposome population, and consist of scattering contributions of both particles as well as the correlations up to spin-echo lengths z of 0.13 µm and 6.0 µm, which are the diameters used for the liposomes and the tumor cells. Further details are given in the text.

6. Conclusions

Liposomal vesicles and *E. coli* bacteria have been observed in their entirety by SESANS. Samples have been prepared in H_2O and resuspended in D_2O; a state of relative neutron invisibility for the membrane and other cellular components, leaving solid spherical droplets of H_2O in D_2O. However, the liposomes and *E. coli* data from SESANS are better described with the projected correlation function for a hollow sphere, as they exhibit little scattering and a sharp initial decay of the polarization curves. It is therefore evident that the H_2O/D_2O-exchange has occurred before or during the initial stages of the SESANS measurements, i.e., within at most half an hour. To observe H_2O/D_2O-exchange measurement times need to be reduced drastically, requiring higher neutron fluxes, preferably in combination with a larger wavelength than available for this study. Fulfilling these prerequisites should make SESANS a viable alternative to *in vitro* tritium studies, since tritiated water has more undesirable properties than deuterated water.

Fit results using the hollow sphere model are in reasonable agreement with DLS for liposomes and in good agreement with literature on *E. coli*, but better statistics are required for accurate quantitative fitting, especially in the low spin-echo range. Although no new insights have been obtained on liposomes and *E. coli* themselves, the results constitute a proof-of-principle that SESANS is able to non-invasively observe entire microscopic biological entities with typical sizes between 50 nm and 20 µm in bulk. Under the right conditions it may be possible to observe changes in a system composed of two different biological species, as has been shown for a mix of small particles (e.g., liposomes) and large cells (e.g., tumor cells). Considerable change in signal requires great numbers of small particles to disappear per large cell. Here SESANS may be an additional tool for drug delivery studies, where liposomes are destroyed by external means to release drugs at the tumor site.

Acknowledgments

The authors thank Chris Duif (NPM^2/R^3, TU Delft, The Netherlands) for assisting with the SESANS measurements, Ulla Woroniecka (RIH/R^3, TU Delft, The Netherlands) for synthesizing liposomes and Krishna Kowlgi (DCT, TU Delft, The Netherlands) for assisting with DLS measurements.

This work is part of the research programme of the *Stichting voor Fundamenteel Onderzoek der Materie (FOM)*, which is financially supported by the *Nederlandse Organisatie voor Wetenschappelijk Onderzoek (NWO)*.

References

[1] Koning G A and Krijger G C 2007 *Anti-Cancer Agents in Medicinal Chemistry* **7** 425–440
[2] Park J 2002 *Breast Cancer Res* **4** 95–99
[3] Kong G, Anyarambhatla G, Petros W, Braun R, Colvin O, Needham D and Dewhirst M 2000 *Cancer Research* **60** 6950–6957
[4] Vasey P, Kaye S, Morrison R, Twelves C, Wilson P, Duncan R, Thomson A, Murray L, Hilditch T, Murray T, Burtles S, Fraier D, Frigerio E and Cassidy J 1999 *Clinical Cancer Research* **5** 83–94
[5] Copland M, Baird M, Rades T, McKenzie J, Becker B, Reck F, Tyler P and NMDavies 2003 *Vaccine* **21** 883–890
[6] Myhr G 2009 *Medical Hypotheses* **69** 1325–1333
[7] Wang X, Yang L, Chen Z and Shin D M 2008 *CA: A Cancer Journal for Clinicians 58* **58** 97–110
[8] Dreher M R and Chilkoti A 2007 *JNCI Journal of the National Cancer Institute* **99** 983–985
[9] Ponce A M, Viglianti B L, Yu D, Yarmolenko P S, Michelich C R, Woo J, Bally M B and Dewhirst M W 2007 *JNCI Journal of the National Cancer Institute* **99** 53–63
[10] Dromi S, Frenkel V, Luk A, Traughber B, Angstadt M, Bur M, Poff J, Xie J, Libutti S K, Li K C P and Wood B J 2007 *Clinical Cancer Research* **13** 2722–2727
[11] McCabe B J and Previs S F 2004 *Metabolic Engineering* **6** 25–35
[12] Briganti G, R Giordano AND P L and Pedone F 1997 *Physica B: Condensed Matter* **234-236** 225–227
[13] Hammermann M, Brun N, Klenin K, May R, Toth K and Langowski J 2000 *Journal of Applied Crystallography* **33** 526–529
[14] Thiyagarajan P, Henderson S and Joachimiak A 1996 *Structure* **4** 79–88
[15] Paganelli C and Solomon A 1957 *The Journal of General Physiology* **41** 259–277
[16] Mazur P 1963 *The Journal of General Physiology* **47** 347–369
[17] Herbst M and Goldstein J 1989 *AJP - Cell Physiology* **256** C1097–C1104
[18] Bu Z, Wang L and Kendall D A 2003 *Journal of Molecular Biology* **332** 23–30
[19] Labischinski H, Goodell E, Goodell A and Hochberg M 1991 *The Journal of Bacteriology* **173** 751–756
[20] Svergun D, Richard S, Koch M H J, Sayers Z, Kuprin S and Zaccai G 1998 *Proceedings of the National Academy of Sciences of the United States of America* **95** 2267–2272
[21] Czajka D M, Finkel A J, Fischer C S and Katz J J 1961 *AJP - Legacy* **201** 357–362
[22] Schloerb P R, Friis-Hansen B J, Edelman I S, Solomon A and Moore F D 1950 *The Journal of Clinical Investigation* **29** 1296–1310
[23] Nagle J F and Tristram-Nagle S 2000 *Biochimica et Biophysica Acta* **1469** 159–195
[24] Takeuchi S, DiLuzio W R, Weibel D B and Whitesides G M 2005 *Nano Letters* **5**(9) 1819–1823
[25] Pencer J, Jackson A, Kuerka N, Nieh M P and Katsaras J 2008 *Eur Biophys J* **37** 665–671
[26] Sears V F 1992 *Neutron News* **3** 26–37
[27] Krouglov T, Bouwman W G, Plomp J, Rekveldt M T, Vroege G J, Petukhov A V and Thies-Weesie D M 2003 *Journal of Applied Crystallography* **36** 1417–1423
[28] Andersson R, Bouwman W, Luding S and de Schepper I 2008 *Physical Review E* **77** 051303
[29] Rekveldt M T, Plomp J, Bouwman W G, Kraan W H, Grigoriev S and Blaauw M 2005 *Review of Scientific Instruments* **76** 033901
[30] Gähler R, Golub R, Habicht K, Keller T and Felber J 1996 *Physica B* **229** 1–17
[31] Bouwman W G, Plomp J, de Haan V O, Kraan W H, van Well A A, Habicht K, Keller T and Rekveldt M T 2008 *Nuclear Instruments and Methods in Physics Research A* **586** 9–14
[32] Andersson R A, van Heijkamp L F, de Schepper I M and Bouwman W G 2008 *Journal of Applied Crystallography* **41** 868–885
[33] van Heijkamp L F, de Schepper I M, Strobl M, Tromp R H, Heringa J R and Bouwman W G 2010 *J, Phys. Chem. A* **114** 2412–2426

Small angle X-ray scattering and transmission electron microscopy study of the *Lactobacillus brevis* S-layer protein

Pentti Jääskeläinen[1], Peter Engelhardt[2], Ulla Hynönen[3], Mika Torkkeli, Airi Palva[3] and Ritva Serimaa

Department of Physics, POB 64, 00014 University of Helsinki, Finland

[1] Department of Biomedical Engineering and Computational Science, P.O.Box 2200, FI-02015 Aalto University School of Science and Technology, Finland

[2] Haartman Institute, Department of Pathology, P.O. Box 21, FIN-00014 University of Helsinki, Finland

[3] Department of Basic Veterinary Sciences, Division of Microbiology, FIN-00014 University of Helsinki, Finland

E-mail ritva.serimaa@helsinki.fi

Abstract. The structure of self-assembly domain containing recombinant truncation mutants of *Lactobacillus brevis* surface layer protein SlpA in aqueous solution was studied using small-angle X-ray scattering and transmission electron microscopy. The proteins were found out to interact with each other forming stable globular oligomers of about 10 monomers. The maximum diameter of the oligomers varied between 75 Å and 435 Å.

1. Introduction

Surface (S-) layer is the outermost layer of cell envelope that is present in most bacteria and in all Archaea. It consists of identical proteins that are attached to each other and to the underlying cell wall, forming a two-dimensional crystalline surface. In Gram-positive bacteria, S-layer attaches to the underlying peptidoglycan layer, in Gram-negative bacteria, S-layer attaches to the outer membrane in the cell wall, and in Archaea, S-layer attaches to the underlying plasma membrane, forming the only cell wall component in addition to the membrane. The bacterial surface (S) layer consists of proteins consisting usually of two structural domains. One domain is involved in the attachment of the S-layer subunit to the underlying cell wall and the other is involved in the S-layer self assembly into two-dimensional crystal structure [1, 2]. The structures and functions of most S-layer proteins and the mechanisms leading to the S-layer protein self assembly are poorly understood.

Two-dimensional crystal structures of S-layers on solid supports or at the air-water interface have been studied by electron microscopy [3, 4, 5, 6], scanning tunneling microscopy [10] and atomic force microscopy [8, 9]. Lattice constants have been obtained but the resolution has not been high enough to determine the detailed shape and size of the proteins. X-ray crystallography has been utilized to solve the protein structure in few cases in which three-dimensional crystals of S-layer proteins have been obtained.

Small-angle X-ray scattering, SAXS, is a powerful method for determining the low resolution shape of proteins or to study aggregation of proteins in solution. Recently, Pavkov et al. [10] determined the high resolution structure of the N-terminal domain of SbsC of *Geobacillus steareothermophilus* in a three-dimensional crystal using X-ray crystallography. They used also SAXS to determine the rod-like shape of the C-terminal region of SbsC. The organization of S-layer proteins of Gram-positive *Clostrium difficile* was also studied using crystallography and SAXS [11]. The *C. difficile* S-layer consists of low- and high-molecular-weight S-layer proteins. A truncated mutant of the low molecular weight S-layer protein was crystallized and the three-dimensional structure was determined with the resolution of 2.4 Å. The shapes of the low-molecular-weight S-layer protein, both full length and the crystallized truncation mutant, and a non-covalent complex of low- and high-molecular-weight S-layer protein were determined using SAXS. Both the protein and the complex were elongated in shape.

The S-layer protein SlpA of *Lactobacillus brevis* ATCC 8287, a Gram-positive bacterium with potentially probiotic properties, consists of 435 amino acids and has a molecular weight of 46 kDa. Precipitation in aqueous solution and cell wall binding of recombinant SlpA (rSlpA) truncation mutants was studied by Åvall-Jääskeläinen et al [12]. The truncation mutants comprising amino acids 190-423 formed a precipitate, while truncation mutants consisting of amino acids 1-145, 1-189 or 1-290 did not form a precipitate in the solution. The truncation mutants containing N-terminal amino acids 1–145 were found to attach to the cell wall while the truncation mutants without N-terminal amino acids did not bind to the cell wall. The N-terminal part of SlpA was shown to bind a polysaccharide receptor within the cell wall. Self-assembly products of rSlpA truncation mutants were studied using transmission electron microscopy (TEM). The truncation mutants of rSlpA consisting of amino acids 179–435 ($rSlpA_{179–435}$) formed crystalline self-assembly products similar to that formed by full length SlpA while $rSlpA_{190–435}$, $rSlpA_{167–435}$ and $rSlpA_{149–435}$ did not form regular lattice structures. This lead to the conclusion that amino acids 179–435 are responsible for *L. brevis* S-layer self-assembly [12].

The purpose of this work was to study the behaviour of the self-assembly domain containing truncation mutants of *L. brevis* surface layer proteins $rSlpA_{179–435}$ and $rSlpA_{167–435}$ in aqueous solution to obtain information on the protein-protein interactions that would be important for formation of the S-layer. These structural domains have appeared to be more soluble than the full length SlpA and the smaller of them, $rSlpA_{179–435}$, formed a crystalline precipitate with about the same structure as the full length SlpA.[12] The same 4-5 mg/ml protein solution was studied using both SAXS and transmission electron microscopy (TEM), although the solution was diluted into a concentration of 1 mg/ml for TEM.

2. Experimental

2.1. Preparation of samples for SAXS

Sample 1. 1.3 mg of rSlpA (residues 167–435) in water was lyophilized o/n and dissolved in 10 mM Tris-HCl pH 7.5 to a final concentration of 4 mg/ml. The solution was centrifuged (16000 g, 20 minutes, +4 °C) and the resulting pellet was discarded.

Sample 2. 1.3 mg of rSlpA (residues 179–435) in water was lyophilized o/n and dissolved in 10 mM Tris-HCl pH 7.5 to a final concentration of 5 mg/ml. The solution was centrifuged (16000 g, 20 minutes, +4 °C) and the resulting pellet was discarded.

Sample 3. 1.3 mg of rSlpA (residues 179–435) in water was lyophilized o/n, dissolved in 5 M guanidium hydrochloride (denaturation) and dialyzed against 10 mM Tris-HCl pH 7.5 o/n at +4 °C (renaturation). After dialysis the volume was adjusted with 10 mM Tris-HCl pH 7.5 to a final concentration of 5 mg/ml. The solution was centrifuged (16000 g, 20 minutes, +4 °C) and the resulting pellet was discarded.

2.2. Electron microscopy

The sample 1 was placed on EM grid (Quantifoil 400 mesh holey carbon) and vitrified using Vitrobot (FEI company). The excess water was removed by blotting the grid twice for 1 seconds and the sample was by plunged into liquid ethane/propane mixture. The sample was then imaged in 18 K temperature using JEOL JEM-3200FSC Liquid Helium Cryo-TEM with 300 kV accelaration voltage and field emission gun. Images were taken with a GATAN Ultrascan 4000 CCD camera with a pixel size of 14 μm.

The samples 2 and 3 were placed on EM grids (Quantifoil 400 mesh holey carbon) and vitrified using Vitrobot (FEI company). The excess water was removed by blotting the grid once for 1.5 seconds and the sample was by plunged into liquid ethane/propane mixture and then imaged in liquid nitrogen temperature using FEI Tecnai 12 electron microscope with 120 kV acceleration voltage and LaB_6 filament. The images were taken using GATAN Ultrascan 1000 CCD camera with a pixel size of 14 μm.

2.3. SAXS experiments

The sample 1 was measured at the beamline X33 at EMBL/DESY, Hamburg. The wavelength was 1.5 Å. The range of the magnitude of the scattering vector q was from 0.0088 to 0.6 Å^{-1}. Here q was defined as $q = (4\pi \sin\theta)/\lambda$, where 2θ is the scattering angle and λ is the wavelength. The intensity was recorded with a photon counting Pilatus 1M pixel detector (67 x 420 mm^2). The measurement time was 8 times 30 s. A solution of bovine serum albumin (BSA) (molecular mass 66 kDa) was measured for calibration of intensity at zero angle. The preliminary data treatment (masking, integration, correction for detector response) was done using the software at the beamline.

The SAXS measurements for samples 2 and 3 were performed using a conventional X-ray tube with CuK_α radiation (1.54 Å) monochromatized by using a collimating Montel multilayer (Incoatec). The point focus beam geometry was used. The intensity curves were measured with an HI-STAR area detector (Bruker AXS) at a sample-to-detector distance of 50 cm. The scattering vector scale was calibrated on the basis of the known diffraction peaks of silver behenate [13]. The range of q was 0.02–0.42 Å^{-1}. The samples and their buffer backgrounds were measured for 2 hours each.

3. Results

3.1. Electron microscopy results

Electron micrographs of sample 1 (Figure 1) show that the $rSlpA_{167–435}$ proteins form globular particles in the solution. Similar particles were also seen in electron micrographs of samples 2 and 3. The seemingly longest diameters of 264 particles for sample 1, 400 particles for sample 2, and 160 particles for sample 3 were measured by hand. The mean particle sizes were 200 Å for sample1, 267 Å for sample 2, and 228 Å for sample 3. Histograms of the longest particle diameters for samples 1-3 are given in Figure 2. Most of the particles were isolated in the solution, but few large agglomerates were also seen. However, the number of such agglomerates was small and their effect on SAXS intensity curves was assumed negligible. The edges of the particles were darker than the central part in the electron micrographs, indicating that the central part of the particles is generally more sparse than the edges. Some of the largest particles could be separate particles overlapping in the image and it was difficult to see particles smaller than 75-150 Å, depending on the image. Thus, it is possible that there are particles smaller than those seen in the distributions present in the samples.

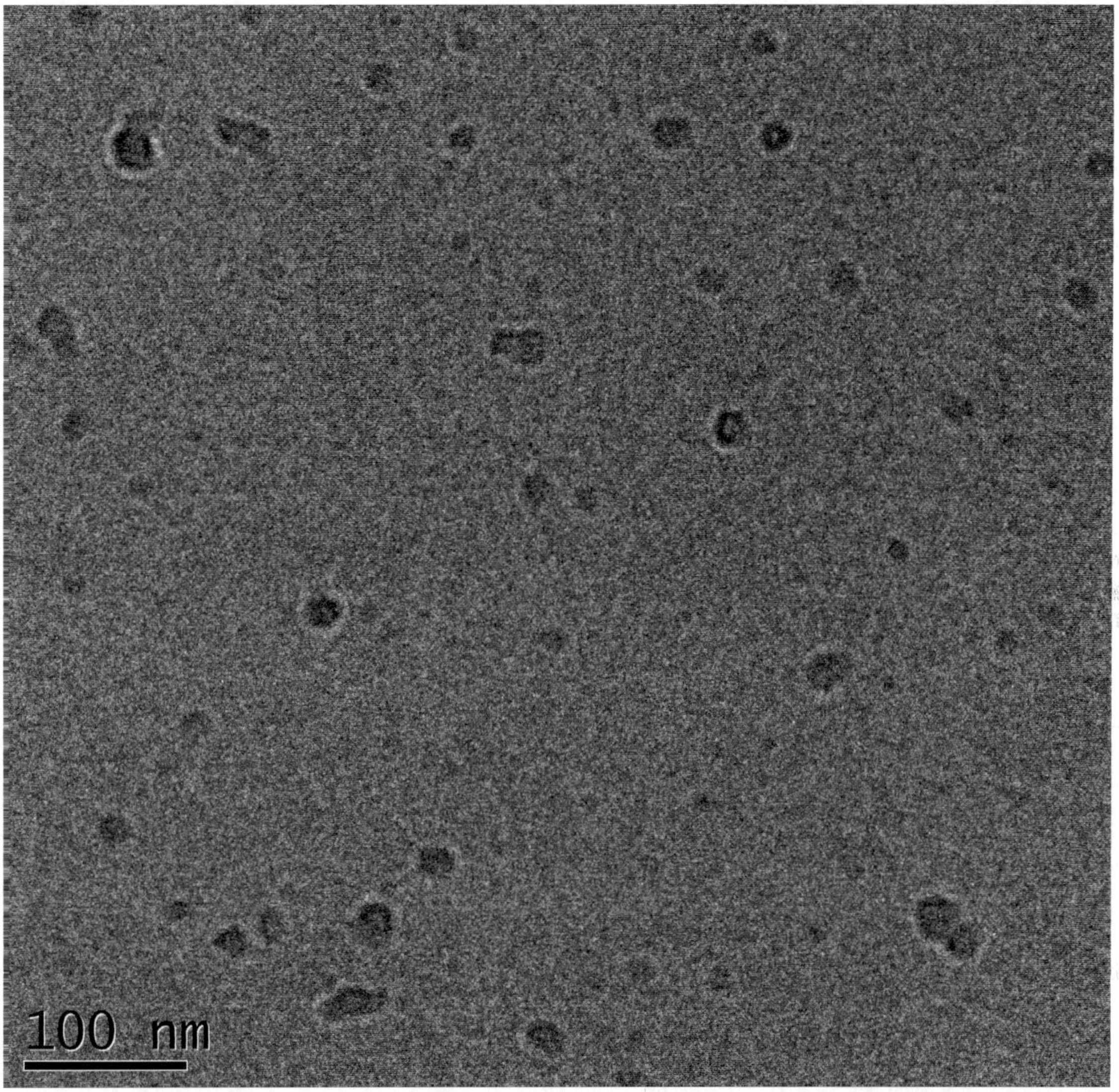

Figure 1. Electron micrograph of rSlpA$_{167–435}$ particles in solution (sample 1).

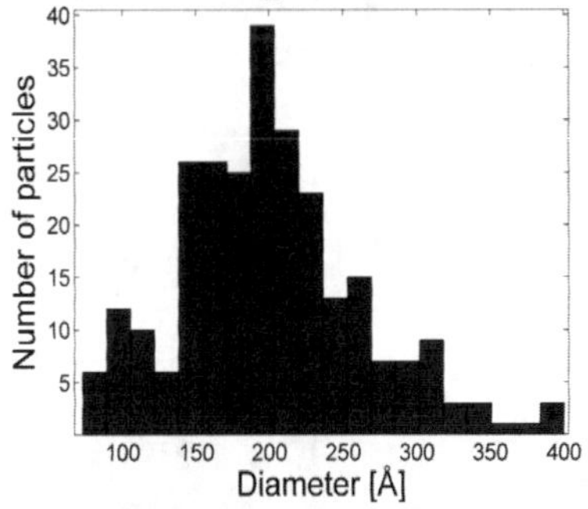

Figure 2a. Distribution of the longest distances measured from the rSlpA$_{167–435}$ particles in sample 1.

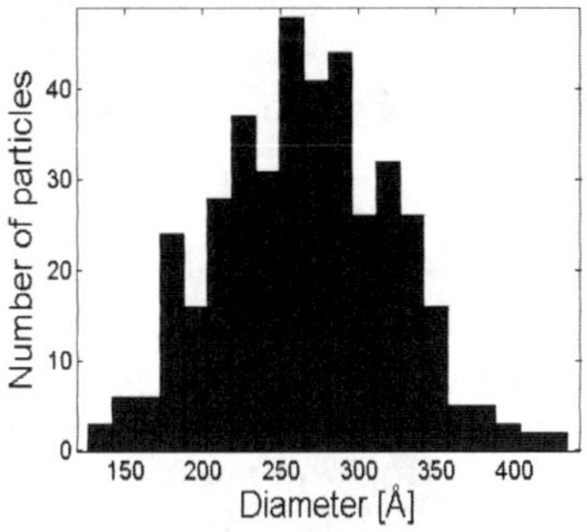

Figure 2b. Distribution of the longest distances measured from the rSlpA$_{179–435}$ particles in sample 2.

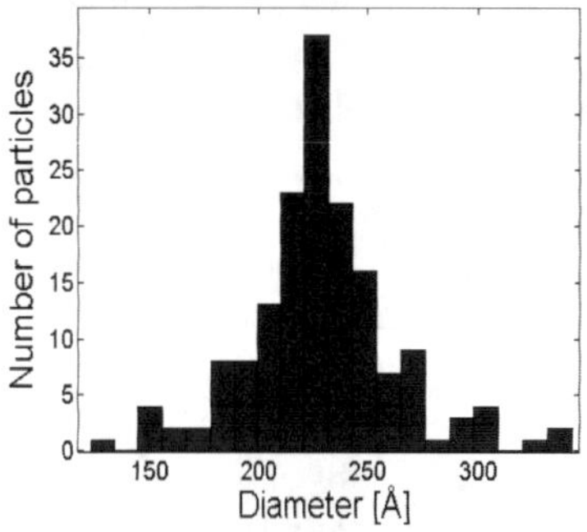

Figure 2c. Distribution of the longest distances measured from the rSlpA$_{179–435}$ particles in sample 3.

3.2. SAXS results

The SAXS intensity of sample 1 (Figure 3) obeyed Guinier law upto 0.022 $Å^{-1}$ and Porod law in the interval 0.22 $Å^{-1} < q < 0.27$ $Å^{-1}$. The distance distribution function was computed from the SAXS intensity using the program GNOM [14, 15]. For sample 1 an excellent solution was obtained using the maximum particle dimension of 300 Å. The obtained R_g value was 88 ± 2 Å. The volume of the particles was determined to be 350 nm^3. The molecular weight of the rSlpA monomer is 28.61 kDa, which corresponds to the volume of 35.3 nm^3, assuming that the protein density is 0.81 Da/$Å^3$. This indicates that the average $rSlpA_{167-435}$ particle in solution contains 10 monomers.

The intensities of samples 2 and 3 were almost equal (Figure 3), differing slightly from the intensity of sample 1 at low q values. Distance distributions were also computed and for sample 2 an excellent solution was obtained using the maximum particle dimension of 250 Å. The obtained R_g value was 80 ± 2 Å. This result corresponds well to the mean particle diameter obtained by EM. However, since EM shows that there are also larger particles, the longest diameter was varied on the basis of the EM results. For instance, another distance distribution function with the longest distance of 345 Å was also found out to fit well to the experimental SAXS data (good solution). The corresponding R_g value was 117 ± 5 Å, which was at the upper limit of what fit into the Guinier region observed in the intensity. For the denatured/renatured sample 3, an excellent solution was obtained with a maximum particle dimension of 220 Å, with R_g value of 75 ± 1 Å.

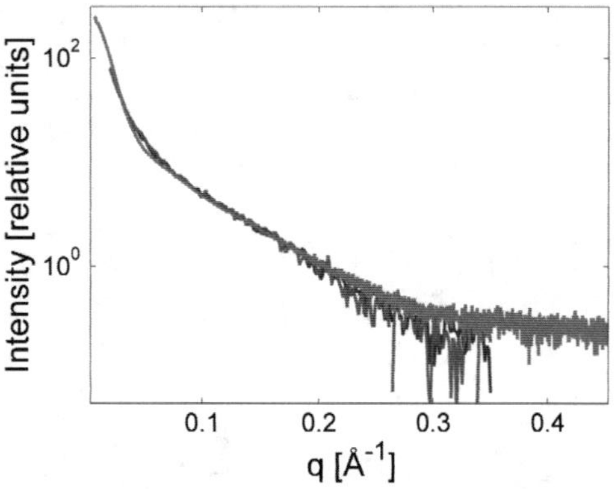

Figure 3. SAXS intensities of samples 1 (red), 2 (blue), and 3 (green).

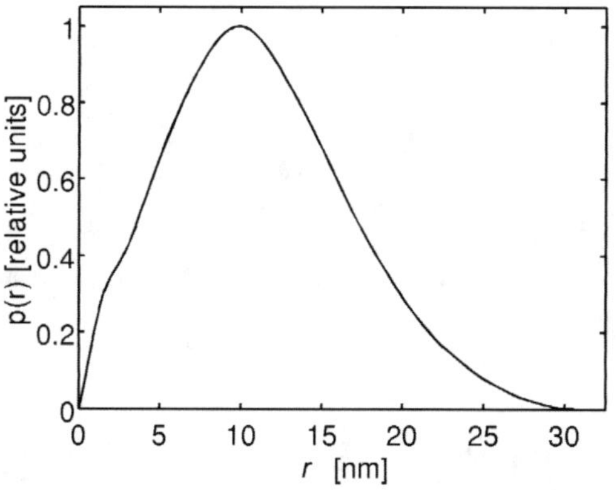

Figure 4. Distance distribution function calculated from the intensity of sample 1.

Attempts were made to model the SAXS intensity of samples 1-3 with spherical particles or spherical shells of varying sizes and to solve the volume distribution using the program GNOM [14, 15]. No good fitting results were obtained for any of the samples.

4. Discussion

In previous studies it has been observed that the full length SlpA and the self-assembly domain truncation mutant $rSlpA_{179-435}$ form aggregates readily in aqueous solution. However, the precipitation of the full length SlpA in solution is much stronger than that of the truncation mutants (U. Hynönen, personal communication) and the cell wall binding domain remains soluble without precipitation. The N-terminal cell wall binding domain is hydrophilic and carries a net positive charge and the self-assembly domain carries a negative net charge [12]. That is why it is of interest to study these domains separately in solution. In this study the behaviour of the self assembly domain truncation mutants $rSlpA_{167-435}$ and $rSlpA_{179-435}$ were studied in aqueous solution.

The proteins were observed to form aggregates, the largest diameters of which varied between 73 Å and 400 Å, with a mean particle size of 200 Å for sample 1 on the basis of EM results. There was more variation in the particle size of sample 2 than in the denatured/renatured sample 3 and sample 1.

SAXS intensities of all the samples resembled each other closely and no essential differences in the aggregation of the samples 1-3 were observed by SAXS. We were not yet successful in preventing the aggregate formation and are thus not able to determine the detailed shapes of the monomers.

The outer regions of both $rSlpA_{167–435}$ and $rSlpA_{179–435}$ aggregates were darker than the central part in the electron micrographs, indicating that the central part may be less dense than the edges. This kind of attachment may arise from the natural property of the SlpA subunits to attach to each other both in aqueous solution and on the cell wall surface.

It is of interest that the aggregates of SlpA self-assembly domains are able to stay soluble regardless of their function to form a large, crystalline layer in the bacterial cell wall. The full length SlpA forms weakly crystalline precipitates in solution [16]. The self-assembly into crystalline structures is much more effective in the presence of cell wall [12, 16]. It may thus be energetically more favorable for rSlpA particles to form oligomeric aggregates rather than crystalline arrays in solution enviroment.

Unlike *L. brevis* SlpA, S-layer protein domain responsible for S-layer self-assembly, $rSbsC_{447-1099}$, of *Geobacillus steareothermophilus* [7] and low molecular weight S-layer proteins and a complex of low and high molecular weight S-layer proteins *Clostridium difficile* [8] appear to be monomeric in aqueous solution. Both *G. stearothermophilus* and *C. difficile* S-layers require the presence of calsium to form S-layer while in case of *L. brevis*, calsium is not required for the S-layer formation. Thus, the formation of *L. brevis* $rSlpA_{179–435}$ oligomers could be due to hydrophobic interactions between the protein monomers in aqueous solution.

S-layers are capable of forming crystalline coatings on the bacterial cell surface and at water-air interface [1]. We are also studying using SAXS the structure of S-layer of *Lactobacillus brevis* directly on the intact cells and as reassembled arrays on purified *L. brevis* cell wall fragments. All these samples have showed periodic order with a spacing of about 80 Å. [16, 17]. However, differences in the degree of order in S-layers on the intact cells and reassembled on cell walls were observed [16, 17]. These results indicate that the key point in the understanding and controlling the self-assembly of S-layer proteins are the intermolecular interactions both between the S-layer protein monomers and between S-layer proteins and the cell wall.

5. Conclusions

In a previous study [12] about the same oblique lattice was observed in the precipitates of the length rSlpA and the recombinant protein $SlpA_{179–435}$, but the precipitate of the slightly larger N-terminally truncated protein, $rSlpA_{167–435}$, did not show any regular lattice structure. According to SAXS and EM results, both $rSlpA_{167–435}$ and $rSlpA_{179–435}$ formed stable oligomers in aqueous solution, presumably via hydrophobic interaction. The oligomers of both truncation mutants showed similar variation of size and shape. SAXS curves were nearly equal, which indicates no signifigant change in protein conformation between the truncation mutants.

Acknowledgements
The electron micrographs were taken in National Nanomicroscopy Center of Aalto University's School of Science and Technology in Espoo, Finland. Portions of this research were carried out at the light source facility X33 at EMBL, HASYLAB/DESY. DESY is a member of Helmholtz Association (HGF). This work is supported by Academy of Finland project 133186. PJ thanks Centre of Excellence in Computational Complex Systems Research for partial support.

References

[1]Sleytr U B, Egelseer E M, Ilk N, Pum D and Schuster B 2007 S-layers as a basic building block in a molecular construction kit *FEBS Journal* **274** 323–34
[2]Engelhardt H 2007 Are S-layers exoskeletons? The basic function of protein surface layers revisited *J. Struct. Biol.* **160** 115–24
[3]Norville J E, Kelly D F, Knight Jr. T F, Belcher A M and Walz T 2007 7 Å projection map of the

S-layer protein sbpA obtained with trehalose-embedded monolayer crystals *J. Struct. Biol.* **160** 313–23
[4]Smit E, Oling F, Demel R, Martinez B, and Pouwels P H 2001 The S-layer protein of Lactobacillus acidophilus ATCC 4356: Identification and characterization of domains responsible of S-protein assembly and cell wall binding *J. Mol. Biol.* **305** 245–57
[5]Lembcke G, Baumeister W, Beckmann E and Zemlin F 1993 Cryoelectron microscopy of the surface protein of Sulfolobus shibatae *Ultramicroscopy* **49** 397–406
[6]Bingle W H, Engelhardt H, Page W J and Baumeister W 1987 Three-dimensional structure of the regular tetragonal surface layer of Azotobacter vinelandii *J. Bacteriol.* **169** 5008–15
[7]Beveridge T J 1990 High-Resolution Topography of the S-Layer Sheath of the Archaebacterium Methanospirillum hungatei Provided by Scanning Tunneling Microscopy. *J. Bacteriol.* **172** 6589–95
[8]Moreno-Flores S, Kasry A, Butt H J, Vavilala C, Schmittel M, Pum D, Sleytr U B and Toca-Herrera J L 2008 From native to non-native two-dimensional protein lattices through underlying hydrophilic/hydrophobic nanoprotrusions *Angew. Chem. Int. Ed.* **47** 4707–10
[9]Györvary E S, Stein O, Pum D and Sleytr UB 2003 Self-assembly and recrystallization of bacterial S layer proteins at silicon supports imaged in real time by atomic force microscopy *J. Microsc.* **212** 300–6
[10]Pavkov T, Egelseer E M, Tesarz M, Svergun D I and Sleytr U B 2008 The structure and binding behavior of the bacterial cell surface layer protein SbsC *Structure* **16** 1226–37
[11]Fagan R P, Albesa-Jove D, Qazi O, Svergun D I, Brown K A and Fairweather N F 2009 Structural insights into the molecular organization of the S-layer from Clostridium difficile *Molecular Microbiology* **71** 1308–22
[12]Åvall-Jääskeläinen S, Hynönen U, Ilk N, Pum D, Sleytr U B and Palva A 2008 Identification and characterization of domains responsible for self-assembly and cell wall binding of the surface layer protein of Lactobacillus brevis ATCC 8287 *BMC Microbiol.* **8** 165–79
[13]Huang T C, Toraya H, Blanton T N and Wu Y 1993. X-ray Powder Diffraction Analysis of Silver Behenate, a Possible Low-Angle Diffraction Standard *J. Appl. Cryst.* **26** 180–4
[14]Svergun D I 1992 Determination of the regularization parameter in indirect-transform methods using perceptual criteria *J. Appl. Cryst.* **25** 495–503
[15]Svergun D I 1991 Mathematical methods in small-angle scattering data analysis *J.Appl.Cryst.* **24** 485–92
[16]Jääskeläinen P, Engelhardt P, Torkkeli M, Hynönen U, Vilén H, Heikkonen J, Palva A and Serimaa R Manuscript in preparation
[17]Vilen, H, Hynönen U, Badelt-Lichtblau H, Ilk N, Jääskeläinen P, Torkkeli M and Palva A 2009 Surface Location of Individual Residues of SlpA Provides Insight into the Lactobacillus brevis S-Layer *J. Bacteriol.* **191** 3339–49

Journal of Physics: Conference Series **247** (2010) 012018 doi:10.1088/1742-6596/247/1/012018

Morphology transition of raft-model membrane induced by osmotic pressure: Formation of double-layered vesicle similar to an endo- and/or exocytosis

Teruaki Onai, and Mitsuhiro Hirai*.

Department of Physics, Gunma University, Maebashi 371-8510, Japan

mhirai@fs.aramaki.gunma-u.ac.jp

Abstract. The effect of osmotic pressure on the structure of large uni-lamellar vesicle (LUV) of the lipid mixtures of monosialoganglioside (G_{M1})-cholesterol-dioleoyl-phosphatidylcholine (DOPC) was studies by using wide-angle X-ray scattering (WAXS) method. The molar ratios of the mixtures were 0.1/0.1/1, 0/0.1/1, and 0/0/1. The ternary lipid mixture is a model of lipid rafts. The value of osmotic pressure was varied from 0 to 4.16×10^5 N/m^2 by adding the polyvinylpyrrolidone (PVP) in the range from 0 to 25 % w/v. In the case of the mixtures without G_{M1}, the rise of the osmotic pressure just enhances the multi-lamellar stacking with deceasing the inter-lamellar spacing. On the other hand, the mixture containing G_{M1} shows the structural transition from a uni-lamellar vesicle to a double-layered vesicle (a liposome including a smaller one inside) by the rise of osmotic pressure. In this morphology transition the total surface area of the double-layered vesicle is mostly as same as that of the LUV at the initial state. The polar head region of G_{M1} is bulky and highly hydrophilic due to the oligosaccharide chain containing a sialic acid residue. Then, the present results suggest that the existence of G_{M1} in the outer-leaflet of the LUV is essentially important for such a double-layered vesicle formation. Alternatively, a phenomenon similar to an endo- and/or exocytosis in cells can be caused simply by a variation of osmotic pressure.

1. Introduction

In this decade, the formation of lipid microdomain in mammalian plasma membrane, so-called a lipid raft [1], has been attracting intensive interests since lipid rafts are assumed to have functions as platforms of membrane-associated events such as signal transduction, cell adhesion, lipid/protein sorting and so on [1-3]. Although the steady-state existence, size, and shape of lipid rafts (liquid-ordered microdomains) in plasma membranes still remain the subject of debate, lipid rafts can be considered essentially as a dynamic assembly of a variety of lipids and proteins. As knowledge about lipid raft functions in cellular signaling has been accumulated, the present agreement has been reached on the fact that raft domains coalesce upon cross-linking to form signaling and sorting platforms [4]. A common feature of lipid rafts is their peculiar lipid composition, being rich in glycosphingolipids (GSLs), sphingomyelin and cholesterol. Gangliosides, major components of GSLs, are acidic lipids composed of a ceramide linked to an oligosaccharide chain containing one or more sialic acid residues, which are abundant in central nervous systems. Functions of lipid rafts are assumed to relate closely to the peculiar features of GSL molecules both in ceramide and oligosaccharide portions that can form complex hydrogen bonding networks (hydrogen bond donor and acceptor) [5].

Journal of Physics: Conference Series **247** (2010) 012018 doi:10.1088/1742-6596/247/1/012018

By using small-angle X-ray scattering (SAXS), small-angle neutron scattering (SANS) and neutron spin-echo (NSE), we studied functional properties of gangliosides and aggregates of gangliosides with other lipids including raft-model membranes (ganglioside/cholesterol/phospholipid) under various conditions. In the case of ganglioside micelles the hydration of the polar head region sensitively changes depending on temperature with accompanying the conformational change of sugar chains [6-10] and the change of dissociation degree of sialic acids [11, 12]. Ganglioside-cholesterol binary mixtures show a maximum miscibility of cholesterol molecules against gangliosides, the cholesterol-dependent micelle-to-vesicle transitions [13], and the Ca^{2+}-induced vesicle-to-lamellar transitions accompanying the formation of an interdigitated structure between the sugar heads of gangliosides in the opposing bilayers [14]. In the case of small uni-lamellar vesicle (SUV) consisting of ganglioside-cholesterol-phospholipid ternary mixtures gangliosides and ganglioside-cholesterol rich regions show an asymmetric distribution at the outer leaflets of the vesicle bilayers [15, 16]. On the dynamics of the lipid aggregates observed by the NSE experiments, the dehydration and bending of the ganglioside sugar heads suppress the undulation of the micellar structure of gangliosides [17]. In the case of the SUVs of ganglioside-cholesterol-phospholipid ternary mixtures (as a model of intact neuronal cell membrane including lipid rafts), the bending modulus takes the smallest value at the lipid composition of [ganglioside]/[cholesterol]/[glycerophospholipid] $\approx$ 0.1/0.1/1 that is a similar composition as in intact neuronal cell membrane including rafts [18]. The coexistence of ganglioside and cholesterol enhances the permeability of water across the bilayer by K^+ ions [19]. The above results suggest that lipid rafts (especially ganglioside-cholesterol rich microdomains) can afford specific properties to membrane structure and dynamics that would affect functions of membrane proteins through the interaction between lipid and protein (hydrophobic coupling between a protein and a surrounding lipid bilayer) [20].

On the other hand, traffics in cells through a local deformation of membranes are known to take place in a highly crowed molecular environment where the total macromolecular concentration in a cytoplasm, for example, is approximately 350 mg/ml [21]. Structural stability and morphology transition of model biomembranes have been studied under a crowed molecular environment by using the so-called "osmotic stress" method. This method changes the osmotic pressure by adding a high molecular weight neutral polymer (osmolyte) into lipid aggregates to discuss about the hydration force between lipid bilayers [22-27]. As one of high molecular weight osmolytes, the use of PVP is well established [24] as well as dextran [23]. PVP is also used as a nontoxic additive in various industrial products such as medicines, cosmetics, etc. In the present study, we employed the osmotic stress method using PVP to investigate the effect of the change of osmotic pressure on the structure of large uni-lamellar vesicle (LUV) composed of monosialoganglioside (G_{M1}) / cholesterol / dioleoylphosphatidylcholine (DOPC). We found the transition from the uni-lamellar vesicle to the double-layered lamellar vesicle depending on osmotic pressure. The present results indicate that under a crowed molecular environment such as a cytoplasm lipid-rafts are able to respond dynamically to a local perturbation of the osmotic pressure, suggesting a new insight on the functions of lipid-rafts from a physical point of view.

2. Material and methods

2.1. Materials

According to the method described elsewhere [28-31], G_{M1} was extracted from bovine brain and purified by using a liquid column chromatography. A single band of G_{M1} on the thin layer chromatography (TLC) plate (Kiesel gel 60; E. Merck, Darmstadt) was confirmed by a fluorometric method reported previously [31]. The purity determined by the TLC-densitometry of G_{M1} used for the experiments was >95%. Cholesterol, DOPC, and polyvinylpyrrolidone (PVP, Mt. 40,000), purchased from SIGMA Chemical Co., were used without further purification. All other chemicals used were of analytical grade or better.

2.2. *Preparation of LUV solutions*

G_{M1}-cholesterol-DOPC LUVs were prepared as follows. G_{M1}, DOPC and cholesterol were separately dissolved in the chloroform/methanol mixture solvent (1/1 (v/v)). These solutions were mixed with the required molar ratios. After mixing, to remove the organic solvent, the lipid mixture solutions were dried up under a nitrogen stream and annealed *in vacuo* for overnight at 45 °C. The dried mixtures were dissolved in 10 mM HEPES (*N*-(2-hydroxymethyl) piperazine-*N'*-(2-ethane-sulfonic acid)) buffer (pH 7.0) to become 5 % w/v of total lipid concentration, and were vortexed for several minutes. Large uni-lamellar vesicle (LUV) solutions were prepared by an extrusion method using the LiposoFast Basic extruder system (Avestin, Canada) with a polycarbonate filter (pore diameter of 50 nm from Nucleopore, Plesanton, USA). The above suspensions were subjected to ~40-times passes through the filter. The LUV suspensions with an odd number of passes were used for the X-ray scattering measurements to avoid a kind of contamination such as large multi-lamellar vesicles. The PVP solutions with different contents (10, 20, 30, 40, 50 % w/v) were also prepared by using the same HEPES buffer. The above G_{M1}-cholesterol-DOPC (or DPPC) LUV solutions and the PVP solutions were mixed by 1:1 in v/v. Finally, we obtained the LUV solutions served for the X-ray measurements. The values of the osmotic pressure at 5 %, 10 %, 15 %, 20 %, and 25 % PVP (w/v) are 1.27 x10^4, 4.3 x10^4, 1.09 x10^5, 2.28 x10^8, 4.16 x10^5 (N/m^2), respectively.

2.3. *X-ray scattering measurements and analysis*

Wide-angle X-ray scattering (WAXS) measurements were performed by using the spectrometer installed at BL-40B2 beam port of the 8 GeV synchrotron radiation source (SPring-8) at the Japan Synchrotron Radiation Research Institute (JASRI), Harima, Japan. Scattered X-ray was recorded by an imaging plate system of R-AXIS IV from RIGAKU Co. The X-ray wavelengths and the sample-to-detector distances were 0.75 Å for 51 cm camera length and 1.0 Å for 4089 cm camera length, respectively. Sample cells composed of a pair of thin-quartz windows with 1 mm path length were used. While the measurements, the sample solutions were moved oscillatory and slowly to avoid radiation damage on the sample due to the small beam size (~0.1 mm^2). Under the present measurement conditions, some radiation damages were negligible as reported previously [32. 33]. Small-angle scattering (SAXS) measurements were also performed by using the synchrotron radiation small-angle X-ray scattering spectrometer installed at BL10C beam port of the 2.5 GeV synchrotron radiation source (PF) at the High Energy Accelerator Research Organization (KEK), Tsukuba, Japan. The X-ray wavelength used was 1.49 Å and the sample-to-detector distances were 80 cm and 190 cm. Scattered X-ray was recorded by the one-dimensional position-sensitive proportional-counter (PSPC) from RIGAKU Co. The exposure time for each measurement was 4 minutes at PF and 30 seconds at SPring-8. In the above SAXS and WAXS measurements, the temperatures of the samples were controlled at 25 °C.

The background correction of WAXS data was described in detail in elsewhere [33]. The distance distribution function $p(r)$ was obtained by Fourier transform of the observed scattering intensity $I(q)$ as

$$p(r) = \frac{1}{2\pi^2} \int_0^\infty rqI(q)\sin(rq)dq \qquad (1)$$

where q=(4π/λ)sin(θ/2), θ and λ are the scattering angle and the X-ray wavelength. In the present model-fitting analysis, we used the scattering function $I(q,R)$ of a multi-shelled spherical particle with a size-distribution as shown previously [16].

$$I(q) = \int_{R_{\min}}^{\infty} I_s(q,R)D(R)dR \qquad (2)$$

Journal of Physics: Conference Series **247** (2010) 012018 doi:10.1088/1742-6596/247/1/012018

where R_{min} is a lower limit of particle radius, namely, in the present case the bilayer thickness of the LUV; $D(R)$ is the number distribution function of the particle radius R; $I_s(q, R)$ is the spherical averaged scattering function of an ellipsoidal particle with the radius R composed of n shells (i-th shell with average excess scattering density $\overline{\rho}_i$ (so-called contrast), radius R_i and V_i). $I_s(q, R)$ and $D(R)$ are given as follows.

$$I_S(q,R) = \int_0^1 \left[3\left\{ \overline{\rho}_1 V_1 j_1(qR_1)/(qR_1) + \sum_{i=2}^{n} (\overline{\rho}_i - \overline{\rho}_{i-1}) V_i j_1(qR_i)/(qR_i) \right\} \right]^2 dx \tag{3}$$

where j_1 is the spherical Bessel function of the first rank. R_i is defined by

$$R_i = r_i\left(1 + x^2\left(v_i^2 - 1\right)\right)^{1/2} \tag{4}$$

where r_i and v_i are the semi-axis and its ratio of i th ellipsoidal shell, respectively. For a spherical-shelled particle ($v_i = 1$, $R_i = r_i$), Eq. 3 is simplified as

$$I_S(q,R) = 9\left\{ \overline{\rho}_1 V_1 j_1(qR_1)/(qR_1) + \sum_{i=2}^{n} (\overline{\rho}_i - \overline{\rho}_{i-1}) V_i j_1(qR_i)/(qR_i) \right\}^2 \tag{5}$$

As a number distribution function of particle radii $D(R)$, we adopted the following Gaussian distribution function which is used in many cases of SUV systems [34].

$$D(R) = \frac{1}{\sqrt{2\pi}\sigma} \exp\left\{ -\frac{(R-\overline{R})^2}{2\sigma^2} \right\} \tag{6}$$

In Eq. 6, $\overline{R}$ and σ are the average radius and the standard deviation, respectively.

3. Results and discussion

3.1. WAXS curves depending on PVP concentration

Fig. 1 shows the osmotic pressure dependence of the wide-angle X-ray scattering (WAXS) curves of LUVs at 2.5 % w/v DOPC, pH 7.0, 25 °C, where A, [G_{M1}]/[cholesterol]/[DOPC] = 0.1/0.1/1; B, [cholesterol]/[DOPC] = 0.1/1; C, only DOPC. The osmotic pressure was raised from 0 to 4.16×10^5 N/m^2 by adding the PVP in the concentration range from 0 to 25 % w/v. In Fig. 1A, the change of the

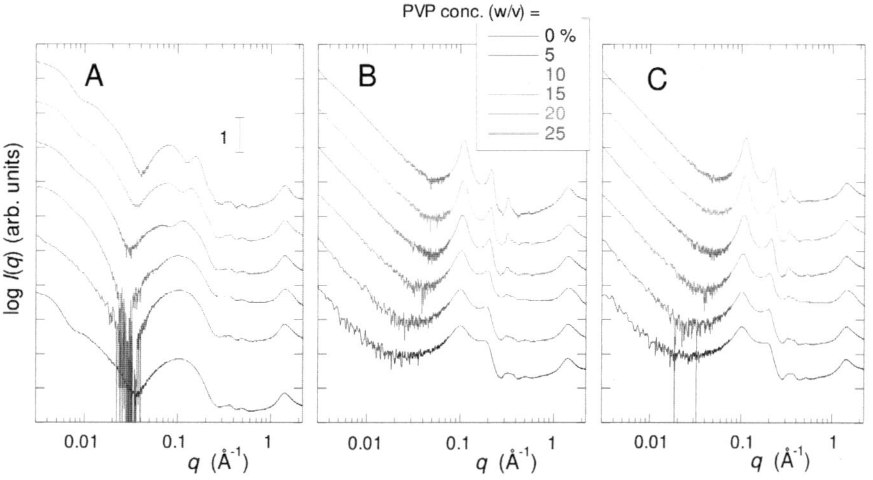

Fig. 1. Osmotic pressure dependence of WAXS curves of LUVs at 2.5 % w/v DOPC, pH 7.0, 25 °C, where A, [G_{M1}]/[cholesterol]/[DOPC] = 0.1/0.1/1; B, [cholesterol]/[DOPC] = 0.1/1; C, DOPC only. The data obtained under two different camera settings (0.75 Å for 51 cm camera length and 1.0 Å for 4089 cm camera length) were merged at q = ~0.07 Å^{-1}. The osmotic pressure was changed by PVP concentration from 0 to 25 % w/v.

WAXS curve below q = ~0.02 Å^{-1} mostly reflects the structural change in the shape of the LUV. The saturating profile of the scattering intensity below q = ~0.005 Å^{-1} except PVP = 5 % w/v indicates that the particle takes a globular shape. The shoulder at q = ~0.01 Å^{-1} is ascribed to the spherical shape of the particle. The change of the WAXS curve in the q range from 0.04 Å^{-1} to 0.2 Å^{-1} depends on the bilayer structure (thickness) of the membrane. At the initial state (PVP = 0 % w/v), there exists a single broad hump with the maximum at q = ~0.11 Å^{-1}. With raising the PVP concentration, the single hump becomes to be separated into two humps. Figs. 2A and 2B show the Guinier plot (ln $I(q)$ vs. q^2) and the radius of gyration, R_g, depending on the osmotic pressure, respectively. The R_g value decreases after having increased once at PVP = 5 % w/v. The above results show that the application of the osmotic pressure results in the structural transition from the LUV to another type LUV that is a double-layered lamellar vesicle (a liposome including a smaller one inside). The broad peak at q = ~1.4 Å^{-1} evidently shows that the hydrocarbon-chain packing of lipid molecules is at a liquid crystalline phase (fluid phase).

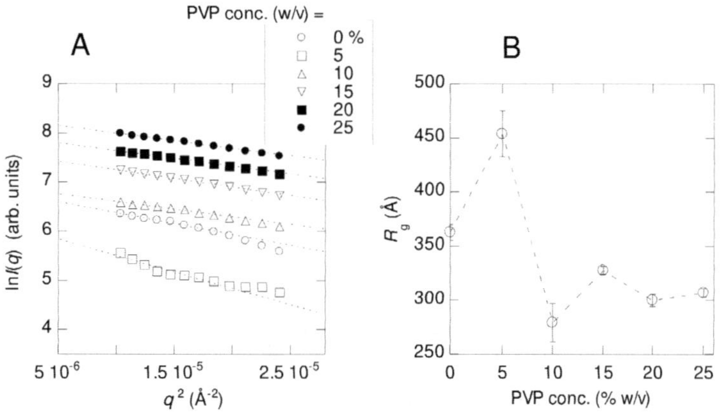

Fig. 2. Osmotic pressure dependence of Guinier plot (ln $I(q)$ vs. q^2) and the radius of gyration, R_g, of [G_{M1}]/[cholesterol]/[DOPC] = 0.1/0.1/1 LUV as shown Fig. 1A.

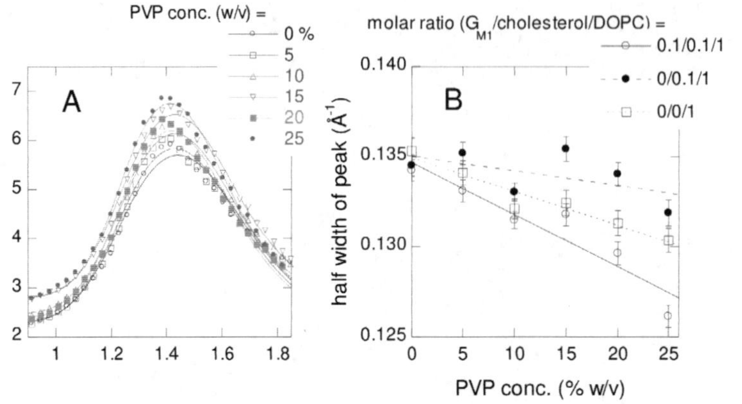

Fig. 3. Change of the high-*q* peak and its half width corresponding to the hydrocarbon chain packing depending on osmotic pressure. A, peak profiles of [G_{M1}]/[cholesterol]/[DOPC] = 0.1/0.1/1 LUV fitted using the log-normal function; B, half widths of the half maximum of the peaks [G_{M1}]/[cholesterol]/[DOPC] = 0.1/0.1/1, 0/0.1/1, and 0/0/1 depending on the PVP concentration.

Fig. 3A shows the peak profiles of the [G_{M1}]/[cholesterol]/[DOPC] = 0.1/0.1/1 LUV in the high-*q* region of Fig. 1A. The solid lines in Fig. 3A are the fitting curves obtained by using the so-called log-normal function. The half width of the peak is plotted in Fig. 3B. As shown in Fig. 3B, the half-width indicates the decreasing tendency depending on the increase of the osmotic pressure, suggesting that the chain packing becomes to be tighter. In the cases of [cholesterol]/[DOPC] = 0.1/1 and 0/1, although the hydrocarbon-chain packing of lipid molecules is at a liquid crystalline phase in the whole PVP concentration range, the phenomena observed in the *q* range below 0.2 Å^{-1} is quit different from that on Fig. 1A. The rise of the osmotic pressure induces the multi-lamellar stacking. As shown in Fig. 4A, the lamellar spacing mostly holds at PVP = 5 % w/v and begins to decrease gradually from for 62.8 Å to 57.3 Å for [cholesterol]/[DOPC] = 0.1/1 and from 61.9 Å to 56.5 Å for [cholesterol]/[DOPC] = 0/1. With the rise of the osmotic pressure, the full width of the half maximum (FWHM) of the first lamellar peak at q = ~0.1 Å^{-1} decreases (Fig. 4B), and the third lamellar peak appears above PVP = 15 % w/v. Thus, it indicates that the lamellar ordering becomes higher with the rise of the osmotic pressure.

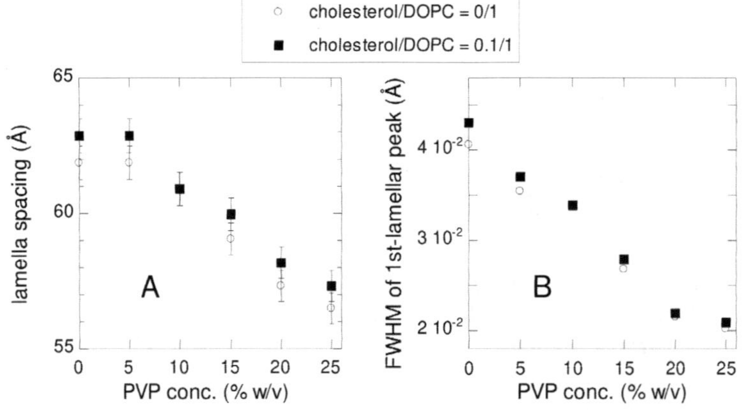

Fig. 4. Lamellar spacing and width of the first lamellar peak of [cholesterol]/[DOPC] = 0.1/1 and 0/1 mixtures obtained from Fig. 1B and 1C.

3.2. Distance distribution function depending on PVP concentration

Fig. 5A shows the PVP concentration dependence of the distance distribution function $p(r)$ of [G_{M1}]/[cholesterol]/[DOPC] = 0.1/0.1/1 LUV obtained by using Eq. 1. The $p(r)$ function of LUV is characterized by the following viewpoints. In the short distance region (0-100 Å) in the insert of Fig. 5A, the profile of the $p(r)$ function at PVP = 0 % w/v has a small peak at ~16 Å followed by a dent at ~40 Å. This is ascribed to the presence of the negative average excess scattering density (so-called negative contrast) region of the bilayer against the average scattering density of the solvent namely, the negative contrast of the hydrophobic tail region within the bilayer. With raising the osmotic pressure, the dent becomes gradually shallow and changes to show a positive value followed by a shoulder at ~50 Å. In the long distance region above 150Å, the $p(r)$ functions at PVP = 0, 15, 20, 25 % w/v have broad rounded maxima at ~450-550 Å and show mostly bell-shapes, which are typical characteristics of $p(r)$ functions for globular particles. With the rise of the osmotic pressure, the shoulder or hump appears at ~100 Å, suggesting a generation of double-layer of the bilayer membranes. This understanding is supported by the following modeling analyses. Fig. 5B shows the maximum diameter D_{max} of the solute particle estimated from the intercept of the $p(r)$ functions at $p(r)$ = 0 for r > 100 Å. The D_{max} value decreases after having increased once at PVP = 5 % w/v, which agrees well with the change of the R_g in Fig. 2B. The $p(r)$ profiles at PVP = 5 and 10 % w/v deviate from a symmetrical bell-shape, suggesting the deformation of the LUV structure from a spherical shape due to the osmotic pressure (below 4.3 x10^4 N/m^2). In other words, the LUV structure once deforms from a sphere and recovers a spherical shape as a double-layer vesicle, as suggested by the following modeling analyses.

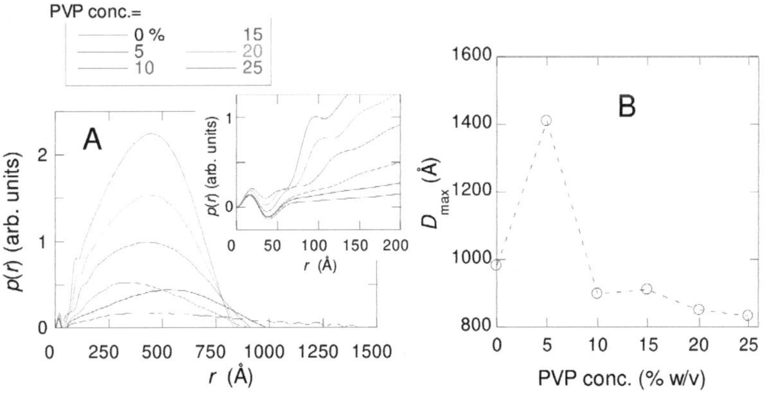

Fig. 5. Distance distribution function $p(r)$ and maximum diameter D_{max} of [G_{M1}]/[cholesterol]/[DOPC] = 0.1/0.1/1 LUV depending on osmotic pressure. A, $p(r)$; B, D_{max}. The insert in A expands the short distance region of $p(r)$. D_{max} values are obtained from the intercepts of the $p(r)$ functions at $p(r)$ = 0 for r > 100 Å since the intercepts below r = 100 Å are attributable not to the particle sizes but to the internal structure of the bilayer as explained in Fig. 7.

3.3. Modeling analyses

3.3.1. Basis of structure modeling of lipid aggregates and characteristics in scattering functions

When analyzing structures of particles in solutions, at first variations in shape and size of particles have to be considered. Fig. 6 shows the scattering functions of particles with different shapes, where we assumed three kinds of representative structure that lipid molecular aggregates could take, namely, spheroid, disk and rod. In Fig. 6, the volumes of the particles are constant, and the axial ratios of the spheroid model are set to be 1 for sphere, 2 for prolate ellipsoid, and 0.5 for oblate ellipsoid, respectively. The radius of the sphere is given as 368.6 Å that is obtained from the shell-model fitting shown below. It is well known that the scattering functions below ~0.02 $Å^{-1}$ (small-angle scattering regions) reflect the shape of the particles. It is clear that the experimental scattering curves below ~0.02 $Å^{-1}$ in Fig. 1A cannot be described by a disk or a rod. The significant deviation of the model scattering functions from the experimental ones in the *q* range from 0.04 $Å^{-1}$ to 0.2 $Å^{-1}$ is attributable to the scattering density fluctuation within a particle. In the present case, each lipid molecule (G_{M1}, DOPC) has a polar head group and two hydrophobic hydrocarbon tails that bind to the backbone of sphingosine (G_{M1}) or glycerol (DOPC) by ester bonds. The electron densities of head, backbone, and tail portions are different, that is, the average excess scattering density from the average scattering density of water (so-called contrast) is positive in the head and backbone portions, and negative in the tail portion. Then, as shown previously [15, 18, 19], the observed scattering curves of the aggregates of the present ternary lipid mixtures are reasonably described by an ensemble of spheres composed of multi-shells given by Eqs. 5 and 6.

Fig. 7A and 7B show the model scattering functions of spherical multi-shelled vesicles with different size distributions (different polydispersities) and those distance distribution functions, respectively. The polydispersity defined by $\sigma / \overline{R} \times 100$ in Eq. 6 is varied from 1 % to 30 %. The insert in Fig. 7A shows the size distribution functions given by Eq. 6. In Fig. 7, the structural parameters of the bilayer are as same as those obtained from the shell-model fitting in Fig.8, namely, a lipid bilayer forming a vesicle is given to consist of six shells with different scattering densities and widths; 1st, 2nd, 4th, and 5th shells correspond to the head and backbone portions of the outer and inner leaflet within the bilayer; 3rd shell, the tail portion. The most inner shell, the sixth shell, is a core of water pool that radius is variable depending on the vesicle size. With increasing the polydispersity, the ripple profile in the scattering function becomes to be smeared to show a shoulder at ~0.01 $Å^{-1}$ and a broad rounded peak at ~0.04-0.2 $Å^{-1}$. It is clear that the rounded peak results from the internal structure of the bilayer. Such smearing occurs in the long distance region of *p*(*r*) function, whereas, the short distance region is not affected by the polydispersity. Alternatively, it can be understood that the short distance region *p*(*r*) function reflects well the internal structure of the bilayer even under the presence of polydispersity.

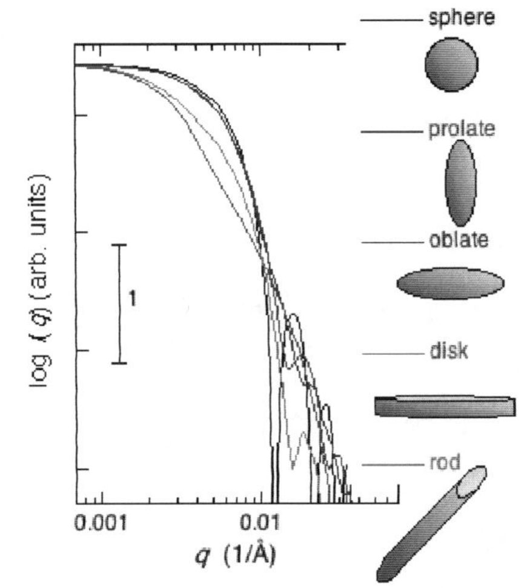

Fig. 6. Model scattering functions depending on particle shapes such as spheroids (sphere, prolate ellipsoid, oblate ellipsoid), disk and rod. The volumes of the particles are set to be constant. Sphere,

radius 368.6 Å ; prolate ellipsoid, short axis 292.5 Å, axial ratio 2.0; oblate ellipsoid, short axis 464.4 Å, axial ratio 0.5; disk, radius 577.8 Å, height 200 Å; rod, radius 200 Å, height 1669.3 Å.

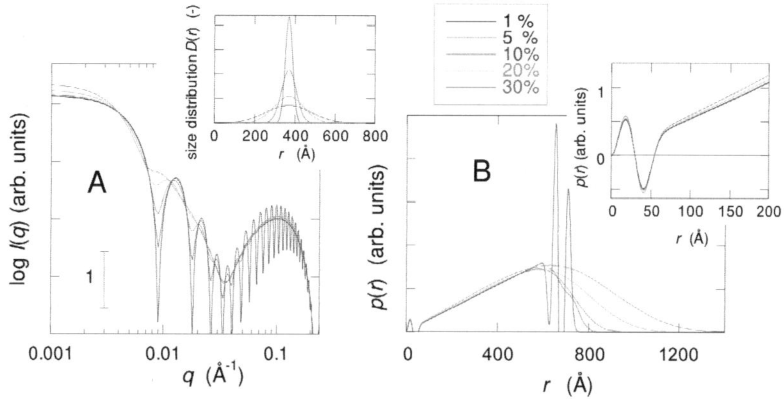

Fig. 7. Model scattering functions of multi-shelled spherical particles and those distance distribution functions depending on size distribution functions. A, scattering functions *I*(*q*); B, distance distribution functions *p*(*r*). The insert in A shows the size distribution functions *D*(*r*) used for the calculations. The insert in B expands the short distance region of *p*(*r*). The spherical particle used is a model of lipid liposome consisting of six shells as same as in Fig.8.

3.3.2. Structure determined by multi-shell model fitting

The above simulations shown in Figs. 6 and 7 suggest that the scattering curves below ~0.02 $Å^{-1}$ and in the range of ~0.04 $Å^{-1}$ - 0.2 $Å^{-1}$ are dominated by the vesicle shape and size and the intra-membrane structure, respectively. The experimental scattering curves can be reproduced by the scattering functions of the multi-shelled vesicle model structure with a size-distribution function. Fig. 8 shows the theoretical scattering functions and size-distribution functions of the multi-shell models that were optimized to describe the experimental data of [G_{M1}]/[cholesterol]/[DPPC] = 0.1/0.1/1 LUV by the iteration fitting using Eqs. 2-6. In Fig. 8, A and B correspond to the LUVs at the PVP concentrations of 0 % w/v and of 25 % w/v, respectively. In Fig. 8A, the experimental data of the LUV at PVP = 0 % w/v was fitted by using a single-bilayer model consisting of six shells with different scattering densities and widths as explained in the above paragraph. The agreement between the theoretical scattering function and the experimental data is good enough. The reasonability of the present single-bilayer modeling is also supported by the previous results of the similar lipid mixture [15, 16]. On the other hand, the experimental data of the LUV at PVP = 25 % w/v was not able to be reproduced by a single-bilayer model in spite of the evidence that the LUV takes a spherical shape as shown in Fig. 5A. Only a double-bilayer model can qualitatively reproduce the experimental data in Fig. 8B. The agreement of the theoretical scattering function of the LUV at PVP = 25 % w/v with the experimental one is not so good. However, as shown in Fig. 9A, the characteristics of the obtained model scattering functions can qualitatively reproduce those of the experimental p(r) functions in Fig. 5A, especially in the short distance region from 0 to ~100 Å. Namely, at PVP = 0 % w/v, the model p(r) function has the small peak at ~16 Å, the negative dent at ~40 Å followed by the broad rounded sub-maximum. The model p(r) function at PVP = 25 % w/v has the small peak at ~18 Å, the positive dent at ~40 Å, the second small peak at ~50 Å that is seen as a shoulder in the insert of Fig. 5A, and the broad rounded maximum at ~450 Å. Such qualitative agreements would support the present modeling analysis. Fig. 9B summaries the contrast profiles of the optimized LUV models of [G_{M1}]/[cholesterol]/[DPPC] = 0.1/0.1/1 at PVP = 0 & 25 % w/v. At PVP = 0 % w/v, the internal bilayer structure indicates the asymmetric distribution of lipid components due to the preferential location of G_{M1} molecules at the outer-leaflet of the membrane, which agrees with the previous results [15, 16]. The most outside region is resulted from the oligosaccharide portions of G_{M1} molecules. On the other hand, at PVP = 25 % w/v, there exists two bilayers across the thin solvent layer. The outer bilayer shows the similar asymmetry as at PVP = 0 % w/v. The width of the inner bilayer is smaller than that of the outer one.

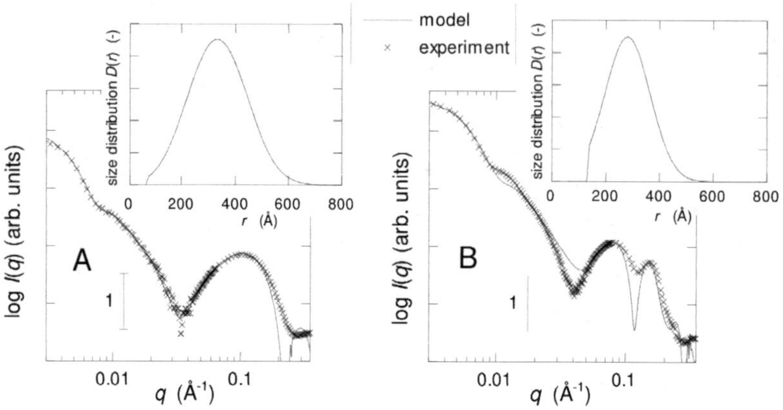

Fig. 8. Model scattering functions and size-distribution functions of $[G_{M1}]$/[cholesterol]/[DOPC] = 0.1/0.1/1 LUV obtained by the shell-model fitting. A, at PVP = 0 % w/v; B, at PVP = 25 % w/v. The inserts in A and B show the size-distribution function *D*(*r*). The experimental data shown in Fig. 1A are also plotted by the marks.

Thus, it can be assumed that G_{M1} molecules locate at the outer-leaflet of the outer-bilayer in the transition process to the double-layered vesicle. It should be mentioned that the total surface area of the hydrophobic tail regions of the double-layered vesicle is mostly as same as that of the LUV at PVP = 0. In addition, we have already confirmed that the above transition proceeds reversibly against the change of the osmotic pressure (data not shown).

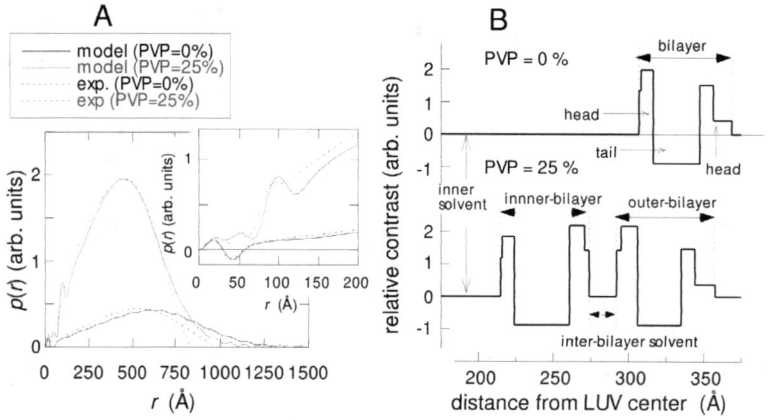

Fig. 9. Distance distribution functions obtained from the model scattering functions in Fig. 8A and the contrast profiles of $[G_{M1}]$/[cholesterol]/[DOPC] = 0.1/0.1/1 LUV (at PVP = 0 % w/v and 25 % w/v) obtained by the shell-model fitting. The experimental functions in Fig. 5A are plotted by the broken lines. A, distance distribution functions; B, contrast profiles. The insert in A expands the short distance region of *p*(*r*).

4. Conclusion

The present results indicate that a small quantity of ganglioside molecules can afford a LUV of lipid mixtures an ability of the reversible morphology transition from a liposome to a double-layered liposome depending on the osmotic pressure, which is reasonably supported by multi-shell model simulation and fitting analyses. In the previous study [35], we found that the small uni-lamellar vesicles (SUVs) of [G_{M1}]/[cholesterol]/[DOPC] = 0.1/0/1, 0.1/0.1/1 and 0.2/0.2/1 show only a minor deformation from a spherical shape by the rise of the osmotic pressure under the same experimental conditions. Combined with the previous results, it is suggested that the morphology transition from a liposome to a double-layered liposome depends on the vesicle size, namely, there would exist a critical radius or curvature of the liposome. In the present measurements, the critical value of the osmotic pressure is in the range from 1.27×10^4 to 4.3×10^4 (N/m^2) (0.125-0.424 atm), which seems to be small enough for some kind of perturbation occurred in living cells. On the other hand, not only for an endo- and/or exocytosis on plasma membranes but also for intracellular membrane-traffic systems, creation and annihilation of liposome always happen in cells. Alternatively, the present finding suggests that membranes containing specific lipids such as glycosphingolipids are able to create or annihilate of liposome spontaneously and locally due to a small-perturbation in intra-cellular environments. Further results and analyses of the osmotic pressure dependence of the morphology transition to a double-layered liposome on lipid molar ratio, LUV size, and others will be reported in elsewhere.

Acknowledgements

This work was performed under the approval of the PF program advisory committee (Proposal No. 2008G095) and of the JASRI program advisory committee (Proposal No. 2008A1035 & 2009B1730).

References

[1] K. Simons, E. Ikonen, Nature 387 (1997) 569.
[2] K. Simons, E. Ikonen, Science 290 (2000) 1721.
[3] S. Hakomori, Trend. Glycosci. Glycotech. 13 (2001) 219.
[4] R.G. W. Anderson, K. Jacobson, Science. 296 (2002) 1821.
[5] I. Pascher, Biochem. Biophys. Acta 455 (1976) 433.
[6] M. Hirai, T. Takizawa, S. Yabuki, Y. Nakata, K. Hayashi, Biophys. J. 70 (1996) 1761.
[7] M. Hirai, T. Takizawa, S. Yabuki, T. Hirai, K. Hayashi, J. Phys. Chem. 100 (1996) 11675.
[8] M. Hirai, S. Arai, T. Takizawa, S. Yabuki, Y. Nakata, Thermochim. Acta. 308 (1998) 93.
[9] M. Hirai, T. Takizawa, Biophys. J. 74 (1998) 3010.
[10] T. Hayakawa, M. Hirai, Eur. Biophys. J. 31 (2002) 62.
[11] M. Hirai and T. Takizawa, S. Yabuki, K. Hayashi, J. Chem. Soc. Faraday Trans. 92 (1996) 4533.
[12] M. Hirai, T. Takizawa, S. Yabuki, K. Hayashi, J. Phys. Chem. B. 103 (1999) 1013.
[13] T. Hayakawa, M. Hirai, Mol. Cryst. and Liq. Cryst. 367 (2001) 631.
[14] T. Hayakawa, M. Hirai, J. Appl. Cryst. 36 (2003) 489.
[15] M. Hirai, H. Iwase, T. Hayakawa, M. Koizumi, H. Takahashi, Biophys. J. 85 (2003) 1600.
[16] M. Hirai, H. Hirai, M. Koizumi, K. Kasahara, K. Yuyama, N. Suzuki, Physica B 385&386 (2006) 868.
[17] M. Hirai, H. Iwase, T. Hayakawa, J. Phys. Soc. Jpn. 70 (2001) 420.
[18] M. Hirai, M. Koizumi, H. Hirai, T. Hayakawa, K. Yuyama, N. Suzuki, K. Kasahara, J. Phys.: Condens. Matter, 17 (2005) s2965.
[19] M. Hirai, T. Onai, M. Koizumi, H. Hirai, K. Kasahara, K. Yuyama, N. Suzuki, K. Inoue, J. Appl. Cryst. 40 (2007) s159.
[20] J.A. Lundbaek, P. Birn, A.J. Hansen, R. Sogaard, C. Nielsen, J. Girshman, M.J. Bruno, S.E. Tape, J. Egebjerg, D.V. Greathouse, G.L. Mattice, R.E. Koeppe, and O.S. Andersen, J. Gen.

Physiol., 123 (2004) 599.
[21] D.S., Goodsell, Trend Biochem. Sci., 16 (1991) 203.
[22] D.M. LeNeven, R.P. Rand, Biophys. J. 18 (1977) 209.
[23] L.J. Lis, M. McAlister, N. Fuller, R.P. Rand, Biophys. J. 37 (1982) 657.
[24] T.J. McIntosh, S.A. Simon, Biochemistry 25 (1986) 4058.
[25] R.P. Rand, V.A. Parsegian, Biochim. Biophys. Acta 988 (1989) 351.
[26] T.J. McIntosh, S.A. Simon, D. Needham, C.H. Huang, Biochemistry 31 (1992) 2020.
[27] S.T. Nagle, H.I. Petrache, J.F. Nagle, Biophys. J. 75 (1998) 917.
[28] L. Svennerholm, P. Fredman, Biochim. Biophys. Acta 617 (1980) 97.
[29] Y. Hirabayashi, T. Nakao, M. Matsumoto, J. Chromatogr. 445 (1988) 377.
[30] T. Momoi, S. Ando, Y. Nagai, Biochim. Biophys. Acta 441 (1976) 488.
[31] T. Hayakawa, M. Hirai.Anal. Chem. 75 (2003) 6728.
[32] M. Hirai, M. Koizumi, T. Hayakawa, H. Takahashi, S. Abe, H. Hirai, K. Miura, and K. Inoue, Biochemistry 43 (2004) 9036.
[33] M. Hirai, H. Iwase, T. Hayakawa, K. Miura, K. Inoue, J. Synchrotron Rad. 9 (2002), 202.
[34] P. Balgavy, M. Dubnickova, N. Kucerka, M. A. Kiselev, S. P. Yaradaikin, and D. Uhrikova, *Biochim. Biophys. Acta.* 1512 (2001) 40.
[35] T. Onai, M. Hirai, J. Physics: Conference Series 83 (2007) 012016.

Journal of Physics: Conference Series **247** (2010) 012019 doi:10.1088/1742-6596/247/1/012019

New lamellar phase with pores in the chain-melting regime of an anionic phospholipid dispersion

Francesco Spinozzi[1], **Paolo Mariani**[1], **Lydia Paccamiccio**[1] **and Lia Q. Amaral**[2]

[1] Facoltà di Scienze, Dipartimento SAIFET, Sezione di Scienze Fisiche, Università Politecnica delle Marche and CNISM, Ancona, Italy
[2] Instituto de Física, Universidade de São Paulo, São Paulo, Brasil

E-mail: f.Spinozzi@univpm.it

Abstract. The anionic phospholipid DMPG (dimyristoyl phosphatidylglycerol) may exhibit in water, instead of a unique melting transition of the hydrocarbon chains, a "melting regime" for pH values above 5, where the phosphate groups are deprotonated, and for low ionic strength, where charge screening is weak. The chain-melting process of DMPG starts at T_m^{on} (onset of the melting regime at $\sim 20°C$), but the complete fluid phase exists only above T_m^{off} (offset of the melting regime at $\sim 30°C$). In a recent paper we developed a SAXS model for a bilayer with pores to explain SAXS results obtained for concentrations up to 70 mM DMPG (F. Spinozzi, L. Paccamiccio, P. Mariani, and L. Q. Amaral, *Langmuir*, in print, 2010). A new lamellar phase with pores, starting 3°C above T_m^{on} and existing up to 4°C above T_m^{off}, was also identified at the higher investigated DMPG concentrations (up to 300 mM DMPG). In this paper we focus in more detail the SAXS curves obtained in the concentration interval 70-300 mM DMPG. The slope of the scattering profile in the very small q range, as well as the anomalous increase in the intensity of the bilayer band centered around 0.12 Å^{-1} after T_m^{off}, have been in particular analyzed. By using a model of water-penetrated bilayers, the volume fractions of DMPG and water molecules inside the bilayer was derived as a function of temperature.

1. Introduction

Phospholipid molecules in aqueous dispersions may aggregate in the form of closed vesicles for small phospholipid concentrations (typically up to 100 mM or about 10% in weight). Vesicles can be unilamellar (a single bilayer) or multilamellar (several bilayers in an onion structure). Neutral phospholipids generally form multilamellar vesicles while charged phospholipids tend to form unilamellar vesicles due to electrostatic repulsion. Unilamellar vesicles of various sizes can be produced depending on the particular phospholipid and on the specific sample treatment used (agitation, sonication, extrusion, etc.) [1].

Phospholipids with two hydrocarbon chains and detergents with one hydrocarbon chain form respectively vesicular and micellar aggregates above a critical micellar concentration and above the so-called Krafft temperature, which defines the transition from ordered to disordered chains in the aggregates [2]. In biological membranes the order-disorder transition of the chains, called "main transition", has received particular attention after the discovery that it is similar to the transition occurring in detergents [3].

In this paper we focus a particular system with an abnormal melting transition of the chains: dimyristoyl phosphatidylglycerol (DMPG), which may exhibit a "melting regime", instead of a

Journal of Physics: Conference Series **247** (2010) 012019 doi:10.1088/1742-6596/247/1/012019

unique melting transition. This abnormal behavior is observed for pH values above 5, where the phosphate groups are deprotonated, and for low ionic strength, where charge screening is weak. The chain-melting process of DMPG starts at T_m^{on} (onset of the melting regime at $\sim$ 20°C), but the complete fluid phase exists only above T_m^{off} (offset of the melting regime at $\sim$ 30°C). A review on the properties of DMPG can be found in [4].

DMPG dispersions (50 mM DMPG in 10 mM Hepes with 2 mM NaCl) were previously investigated as a function of temperature by small angle X-ray scattering (SAXS) down to $q \sim 0.03$ Å^{-1} (q being the scattering vector modulus given by $q = 4\pi \sin\theta/\lambda$, where 2θ is the scattering angle and λ is the X-ray wavelength) [5]. At all the investigated temperatures, SAXS profiles show a broad band centered at 0.12 Å^{-1} (dependent on the bilayer electron density profile and called "bilayer band") and a q^{-2} behavior at lower q values, typical of unilamellar vesicles. However, an anomalous decrease in electron density contrast in the bilayer band characterized the melting region. Additional X-ray results extended the q-range, from $q \sim 0.007$ Å^{-1} to the wide angle region (WAXS), and DMPG concentration, from 10 to 80 mM DMPG [6]. WAXS analysis showed that the Bragg peak typical of ordered chains disappears above T_m^{on}, while the lower q range revealed only in the intermediate phase the presence of a band (called IP band), which indicates the existence of a mesoscopic structure with a repeating of around 370 Å, interpreted as in-plane correlated cavities/pores on the vesicle surface. This proposal was corroborated by observation of DMPG giant unilamellar vesicles with phase contrast optical microscopy [6]. The effect of addition of salt to 50 mM DMPG dispersions was investigated in more two papers; the disappearance of the IP band in SAXS [7] was correlated with disappearance of transparency in the melting regime, as detected by light absorbance, DSC and fluorescence microscopy [8].

Studies varying DMPG concentration indicated the existence of another type of structure above 70 mM, with possible destruction of the vesicles [6]. This made us start both to develop a SAXS model for a bilayer with pores and to study by SAXS higher DMPG concentrations, in the interval 70-300 mM. A model has been indeed constructed, able to explain SAXS curves in the melting regime up to 70 mM DMPG, while a new lamellar phase with pores was found at the higher concentrations, starting 3°C above T_m^{on} and existing up to 4°C above T_m^{off}. Part of this study, focused in the SAXS pore model and on SAXS curves obtained with 70 mM and 150 mM DMPG, as well as WAXS, DSC and optical microscopy results with 150 mM DMPG, has been recently published in Langmuir [9].

In this paper we focus in more detail the SAXS curves obtained in the whole concentration interval (70-300 mM DMPG), and analyze changes in the slope of the curves observed in the very small q range, as well as the abnormal behavior of the intensity of the bilayer band above T_m^{off}. The formation of pores in natural as well as model phospholipid bilayers is a key biophysical process, which may be associated with many phenomena, such as the anomalously large permeability of water, the formation of local defects that act as nucleation processes in membrane fusion and the permeability of ions through the membranes.

It should be noticed that the DMPG peculiar thermal and structural behavior at the transition from gel to fluid lipids opens several questions on the biological relevance. Indeed, about 10-40% of all naturally occurring lipids are negatively charged and in prokaryotes the most abundant anionic phospholipid is phosphatidylglycerol [10]. Such an intermediate regime, at least locally and characterized by perforated membrane microdomains, could also occur in natural lipid systems and cells could make use of a similar concept to activate structural changes and functionality in their membranes. Moreover, triggering similar membrane changes by controlling temperature, pH and salt concentration could potentially be used in targeted drug delivery [11].

2. Experimental

The sodium salt of DMPG (Avanti Polar Lipids, Birmingham, USA) and bidistilled water were used to prepare samples with DMPG concentrations in the interval 70-300 mM, as previously described [9]. The same buffer (10 mM Hepes pH 7.4 with 2 mM NaCl) was used throughout, but pH remained always above the apparent pK of DMPG, ensuring that DMPG can be assumed to be fully deprotonated and effects are due to changes in DMPG concentration.

Scattering curves were obtained at the SAXS beamline of LNLS Synchrotron (Campinas, Brazil) with $\lambda = 1.608$ Å^{-1} and $0.008 < q < 0.25$ Å^{-1}, using a linear position sensitive detector. A thermal bath was used for temperature variation from 12°C up to 55°C. Samples were conditioned in a sample holder with flat mylar walls and 1 mm thickness. Data, with acquisition time 10-20 minutes, were normalized for monitor integral counts (to compensate for oscillations in the beam intensity), sample attenuation and corrected for the SAXS detector response (measured with a radioactive source). The measured scattering due to the buffer, which has no structure in the SAXS region was subtracted from all scattering curves.

3. SAXS results at various concentrations

Visual observation of the DMPG dispersions under temperature variation showed that the melting regime is transparent only up to 70 mM DMPG, but for higher concentrations the sample is milky in all phases. SAXS data were obtained in different concentrations (70, 100, 150, 200 and 300 mM DMPG) varying the temperature. The whole ensemble of measured curves is shown in Figure 1 (note that in the previous paper [9] only representative results for 70 mM, 300 mM and 150 mM were displayed).

The broad band at around 0.12 Å^{-1}, observed at all concentrations and temperatures, is related to the inner bilayer structure [5, 6], and therefore called "bilayer band". The position, $q_{\rm max}$, and intensity, $I_{\rm max}$, of the bilayer band are shown in Figure 2 as a function of temperature, for all measured concentrations. As previously described, the complex thermal dependence of $q_{\rm max}$ and $I_{\rm max}$ can be explained in term of variations of the bilayer electron density profile [5], attributed to the appearance/disappearance of holes/pores in the melting transition [6], but a more careful analysis of these results will be presented later on.

Besides the bilayer band at ~ 0.12 Å^{-1}, at smaller q (around 0.01 Å^{-1}) correlation peaks, which change with both temperature and concentration, are observed. With 70 mM DMPG, the intermediate region starts at $T_{\rm m}^{\rm on} \sim 20$°C, where a broad band of the same type as the IP band observed at smaller concentrations occurs [6]. A model for the porous bilayers, in the form of bicelles with in-plane correlated pores (see Fig. 7), has been developed [9] and could fit the whole SAXS curves observed at 70 mM DMPG in the temperatures where the IP band exists. This confirms that the IP broad band is associated to the 2D correlation among pores, as proposed qualitatively in [6]. Still with 70 mM, the sharper peak detected above 23°C is attributed to a new lamellar phase, called $L_{\rm p}$ (lamellar phase with pores) in the previous paper [9], which changes its characteristics above $T_{\rm m}^{\rm off} \sim 30$°C.

SAXS curves observed at DMPG concentration higher than 70 mM evidence that the sharp peak of the new $L_{\rm p}$ phase becomes well defined at 23°C. Moreover, other peaks appear also in the gel and fluid phases, indicating the emergence of a correlation among bilayers in the direction perpendicular to their plane. The SAXS curves always change at $T_{\rm m}^{\rm on} \sim 20$°C, but the curves for $20 - 21$°C do not present the defined peak of the $L_{\rm p}$ phase, which exists above 23°C.

The evolution of the position of this sharp peak with concentration was analyzed in [9] in terms of the lipid surface fraction parameter $\alpha = c_V(d/t)$, where c_V is the lipid volume fraction in the sample, t the lipid bilayer thickness and $d = 2\pi/q_{\rm peak}$ the repeat distance from a correlation peak centered at $q_{\rm peak}$ [6]. In all cases, at 25°C the parameter α remains < 1, indicating the presence of bilayers with pores and/or with finite size. Calculations of α for other temperatures showed that the value $\alpha = 1$ expected for a normal lamellar bulk phase L_α is reached only for

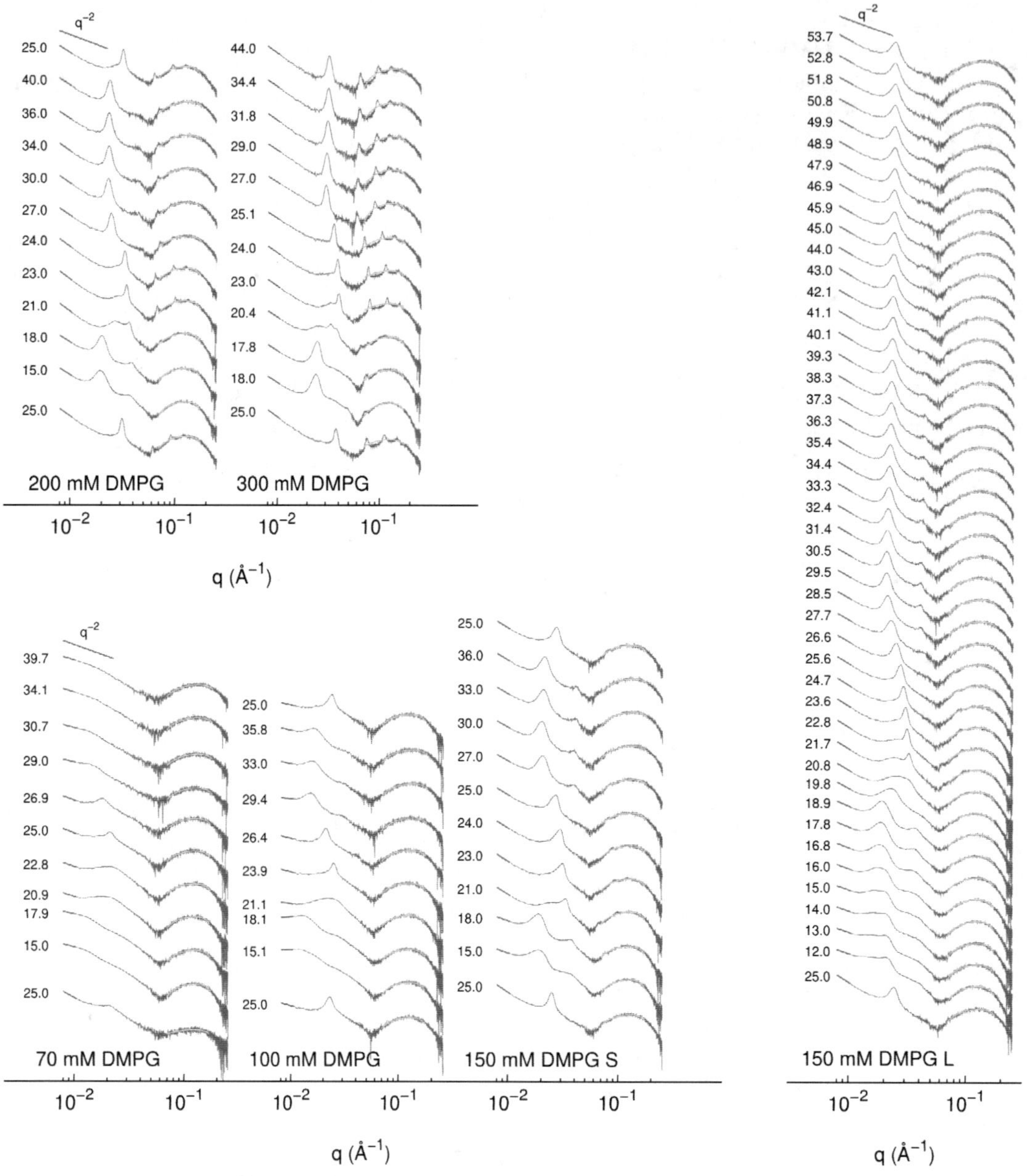

Figure 1. Log-log plot of SAXS curves for DMPG samples at different concentrations (shown at the bottom of each column) and temperatures (reported on the left of each curve in °C). For 150 mM, a short (150 mM DMPG S) and a long (150 mM DMPG L) series of measurements with two independent samples are shown. Red lines refer to the determination of the initial slope up to $q = 0.012$ Å^{-1}. Green curves are the fits with a parabola used to obtain the position, q_{max}, and intensity, I_{max}, of the bilayer band.

300 mM DMPG in the gel phase ($T = 17.8$°C).

In the previous study [9] an extensive analysis was made on the different results obtained for 150 mM DMPG, in order to characterize the new L_{p} phase. Several properties of the new

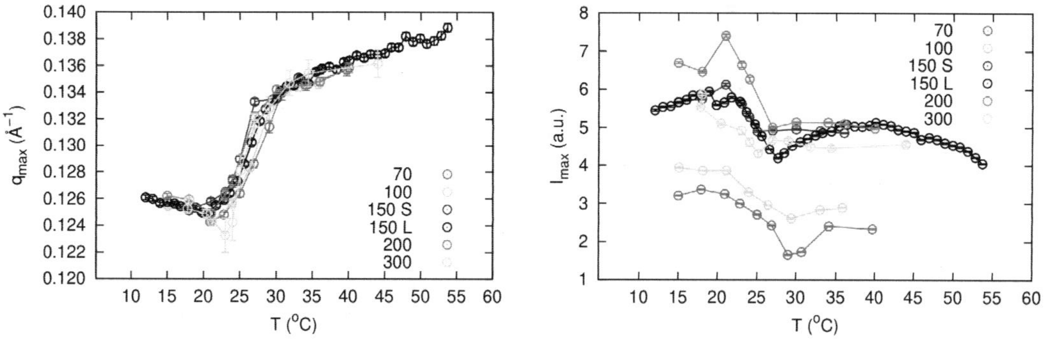

Figure 2. Temperature dependence of position, $q_{\rm max}$ (left panel), and intensity, $I_{\rm max}$ (right panel), of the SAXS bilayer band for DMPG samples at different concentrations (reported in mM).

lamellar phase $L_{\rm p}$ were indeed studied by SAXS, WAXS, DSC and optical microscopy. In particular, the temperature-induced phase sequence Ripple Gel - IP - $L_{\rm p}$ - fluid was defined. Moreover, it was suggested that all the transitions to and out of the intermediate melting region should correlate with strong fluctuations of the bilayer, together with opening and closing of pores of yet unknown dynamics [9].

The variation of the position of the correlation peaks with temperature for all measured concentrations is shown in Figure 3. At temperatures where both the broad and sharp peaks are present, only the position of the sharper one was considered in Figure 3 (note that for concentrations larger than 70 mM there are peaks both below $T_{\rm m}^{\rm on}$ and above $T_{\rm m}^{\rm off}$, while they are not well defined at 70 mM).

The repeat distance d decreases with concentration, from 300 Å at 70 mM to 160 Å at 300 mM within the melting regime, from 450 Å to 250 Å before $T_{\rm m}^{\rm on}$ and from 400 Å to 200 Å above $T_{\rm m}^{\rm off}$. The change at $T_{\rm m}^{\rm on}$ is sharp for 70, 100 and 150 mM, but not for 300 mM DMPG. Moreover, the melting interval becomes smaller with concentration, although not converging yet to $T_{\rm m} = 23°$C, the unique melting temperature observed for electrostatically screened DMPG dispersions with 500 mM salt [7], the same temperature where the new phase appears. Note that a decrease of the melting interval with temperature is also suggested by the changes in position and intensity of the bilayer band ($q_{\rm max}$ and $I_{\rm max}$), as reported in Figure 2, where $T_{\rm m}^{\rm off}$ approaches $T_{\rm m}^{\rm on}$ when the DMPG concentration increases.

The intensity of the bilayer band ($I_{\rm max}$) merits to be further analyzed also as a function of DMPG molar concentration, C. The observed increase in intensity can be in fact partially explained by the trivial dependence with the number of scattering unities. This is shown in the Figure 4, where the relative intensity $I_{\rm max}(C)/I_{\rm max}(C = 70$ mM) is shown as a function of C for several temperatures, together with the linear behavior expected for a change only in the number of scattering unities (with a constant form factor and interference function). At all temperatures, the intensity drops for the higher concentrations (200 mM, and specially 300 mM), indicating an increase of interference among scattering unities, consistent with the observed higher diffraction orders. For 70, 100 and 150 mM samples, a linear behavior similar to what expected is detected at $T = 28°$C, near $T_{\rm m}^{\rm off}$, while for lower temperature the intensity is lower than expected, and for higher temperature it is higher than expected. This fact is indicative of changes in the bilayer structure while passing from 70 mM to 100 mM and 150 mM, specifically in the melting regime. This point will be further discussed.

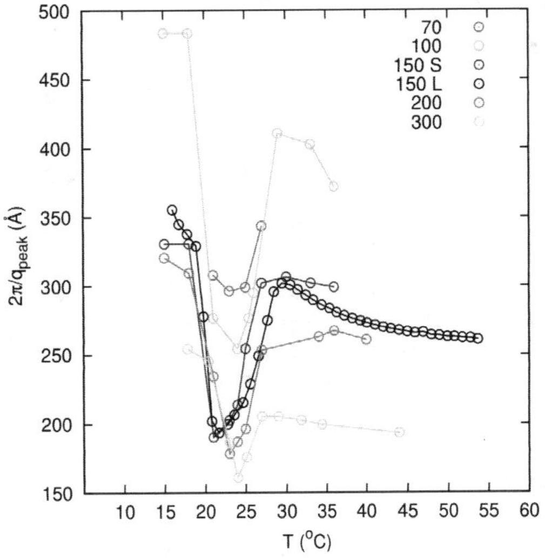

Figure 3. Peak positions ($d = 2\pi/q_{\mathrm{peak}}$) as a function of T for all measured DMPG concentrations.

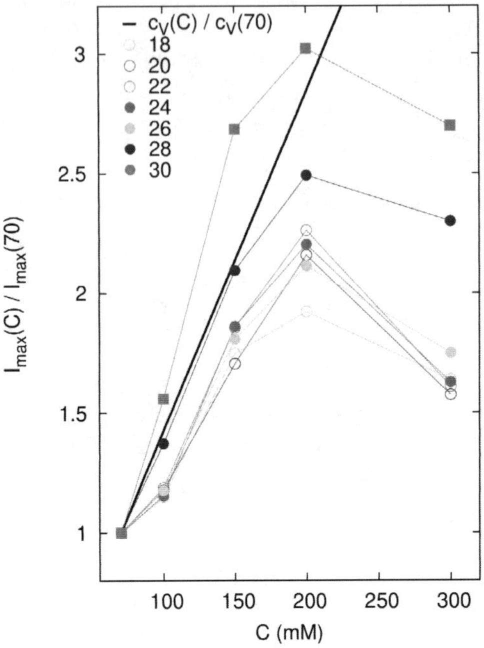

Figure 4. Relative intensity of the bilayer band with respect to the value for 70 mM DMPG as a function of DMPG concentration C for several temperatures. Black line refers to the behavior expected for a change only in the number of scattering unities.

4. Slope of the log-log SAXS curves at small q

All the measured SAXS curves have been analyzed in order to define the slope of the $\log[I(q)]$ versus $\log(q)$ at very low q (up to 0.012 Å^{-1}). Fitting lines are shown in Figure 1 while the changes in the measured slope with temperature and concentration are displayed in Figure 5.

It can be seen that there is a large variation of the slope with concentration. Values close to -2, the fingerprint of large isolated bilayers, occur both for 70 mM DMPG in the whole temperature range and for 150 mM DMPG, but only in the low temperature region. Slopes from -4 to -3 characterize samples with the higher DMPG concentrations as well as the 150 mM DMPG sample at temperature above T_m^{on}, indicating possible existence of correlated bilayers building an apparently more symmetrical object.

In order to better understand these results, extensive simulations have been performed with a structural model based on the presence of lamellar bicelles of radius R, as used also in the model developed in [9] and elsewhere [12].

For a single bicelle with rims, simulations varying R are shown in Figure 6, where also the line with slope -2 is reported for comparison. Very clearly, the slope -2 (which is characteristic for an "infinite bilayer") exists only for q values above a q^* value defined by the intersection of the line with q^{-2} behavior and the initial region with constant intensity. The insert of Figure 6 shows q^* as a function of R values from all the simulations. A clear straight line, also shown in

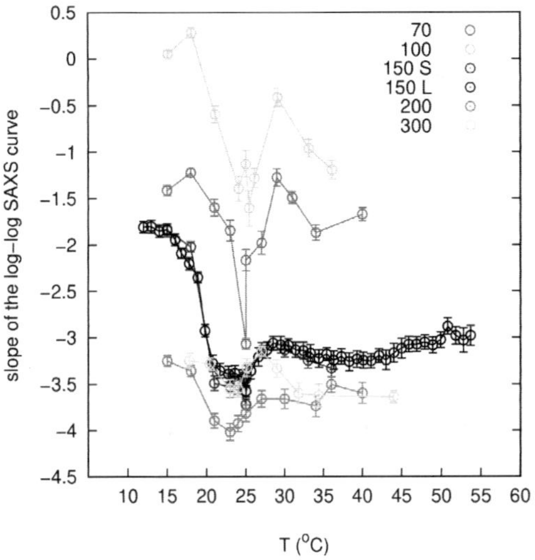

Figure 5. Slopes of the log-log SAXS curves obtained for the lowest q region (from 0.008 to 0.012 Å^{-1}).

the insert of Figure 6, gives the relationship between q^* and R, according to:

$$q^* = \frac{k}{R^\beta} \tag{1}$$

where $k = 1.9 \pm 0.1$ and $\beta = 1.03 \pm 0.01$.

This result gives a practical way to obtain the R value when the change in slope can be experimentally detected, and also a minimum value for R when the slope -2 is experimentally observed in a defined q range. The present experimental SAXS curves arrive only to $q =$ 0.008 Å^{-1}: the -2 slope observed for example at 70 mM DMPG means that in this case the bicelle radius R should be larger than 250 Å. However, it could be much larger than 250 Å, as only measurements at much smaller q values could give the radius of a large bicelle.

In order to understand the larger (absolute) slopes observed in the lamellar L_{p} phase, it should be reported that the fit to the L_{p} peaks with Lorentzian interference functions gave a width compatible with lamellar correlation of around $N \sim 20$ bilayers [9]. The high absolute values for slopes then indicate that SAXS is sensing a global object whose shape is cylindrical or spheroidal, due to the presence of largely correlated bilayers. The indication is that the slope of the SAXS curve is sensible to the total size of correlation, which may go well over 1000 Å. On the other hand, the analysis of the experimental SAXS curves show that the very small slopes detected in few conditions are determined by the influence of nearby correlation peaks distorting the initial slope.

Therefore, the measured slopes support the conclusions that the IP phase corresponds to large uncorrelated bilayers, with the IP band due to two-dimensional correlation between pores in the bilayer surface, while the L_{p} phase corresponds to lamellar correlation between bilayers. At 70 mM there is a match at 25°C between the IP band and the peak of the L_{p} phase, as evidenced also by the values of the α parameter related to the IP band and the L_{p} peak, displayed in [9].

5. Analysis of the bilayer band

The variations of I_{max} with temperature for 150 mM DMPG have been analyzed in [9] in the context of the pore model. The formation of the L_{p} phase is proposed to come from a balance between bilayer inflation, as visually observed in giant vesicles by fluorescence microscopy [8]

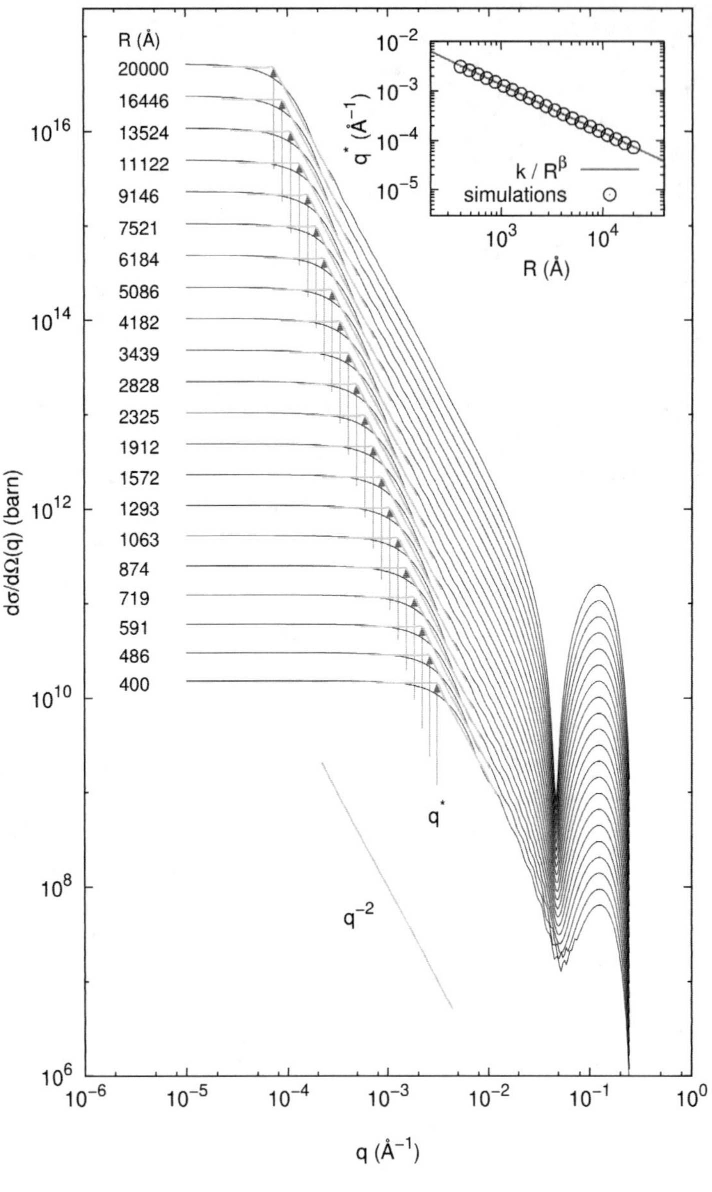

Figure 6. Simulated X-ray differential scattering cross sections for a single bicelle with rims, varying the inner radius R reported on the left of each curve. Red arrows refer to the intersection, occurring at q^*, between the line with q^{-2} behavior and the initial region with constant intensity, both shown as green lines. The insert shows the linear relationship between $\log(q^*)$ and $\log(R)$.

and available interlamellar space. The defined change of $I_{\rm max}$ from the IP value at 20°C to the $L_{\rm p}$ value at 23°C, as observed in Figure 2, is not directly associated with the parameters of the DMPG electron density function, and is probably connected to a change from the larger pores in IP to smaller pores in $L_{\rm p}$.

A clear correlation also exists between the temperature variation of $I_{\rm max}$ and of the intensity of the sharp lamellar peak of the new phase $L_{\rm p}$. Since peak intensities are defined by the intensity of the form factor of the bilayer with pores in the q region where there is the peak, a series of simulations were made in [9], showing the strong dependence on the electron density contrasts, particularly between the flat and pore regions of the bilayer. Such simulations indicate that the transition from the $L_{\rm p}$ to the fluid state occurs via transfer of lipids from the pore borders to the flat region [9]. The complete study of 150 mM DMPG indicates clearly that pores open at $T_{\rm m}^{\rm on}$ and seal at $T_{\rm m}^{\rm off}$. On the other hand, the recovery of $I_{\rm max}$ occurs only about 4°C above $T_{\rm m}^{\rm off}$, after the transition as detected by DSC has been completed [9].

The minimum of $I_{\max}$ at $T_{\rm m}^{\rm off}$, followed by a maximum value only 4°C above $T_{\rm m}^{\rm off}$, was qualitatively ascribed to water remaining inside the bilayer [9]. Fits to $I_{\max}$ with the pore model are however difficult, both because the number of model parameters and the fact that there is not a simple analytical expression for $I_{\max}$. Simulations showed that $I_{\max}$ depends directly on the size and number of pores, and particularly on the fraction of the bicelle surface occupied by pores, but depend also on the electron density distribution in the round parts of pores [9].

In this paper, the specific point of the recovery of $I_{\max}$ after $T_{\rm m}^{\rm off}$ is focused, using a simple model of a planar infinite bilayer without pores. The idea is to analyze the volume fractions of DMPG and water molecules inside the bilayer, in order to define whether the recovery in $I_{\max}$ is indeed due to water still remaining inside the bilayer above $T_{\rm m}^{\rm off}$, and being expelled as $I_{\max}$ increases.

Tests were also previously made with a simple bicelle model without pores described by three levels of electron density contrasts. The region for $q > 0.05$ Å^{-1} [9] was considered, which includes the minimum of the SAXS curve around 0.06 Å^{-1} which is influenced by the bicelle radius R. Results showed that a bicelle radius $R = 953$ Å was obtained and the parameters of the bilayer profile remained practically the same as those of an infinite bilayer [9]. It can be then concluded that in the region where $I_{\max}$ is measured, for $q > 0.07$ Å^{-1}, it is possible to use, for $T > T_{\rm m}^{\rm off}$ $(= 28 \pm 1°{\rm C})$, a simple model of bilayer to deal with the amount of water that could remain among lipid molecules after pore sealing. A cartoon representation of this simple model, compared with the pore model of Ref. [9] is reported in Fig. 8

The scattering intensity of a infinitely wide plane bilayer is given by the basic equation [13]

$$I(q) \quad = \quad \kappa r_{\rm e}^2 \frac{c_V}{t}\frac{2\pi}{q^2} A_f^2(q) + B, \tag{2}$$

where c_V is the volume fraction of the hydrated lipid molecules and t is the bilayer thickness. $r_{\rm e}$ is the classical radius of the electron, whereas κ and B are a scaling factor and a background, both due to the instrumental calibration. $A_f(q)$ is the Fourier transform of the electron density profile $\rho(z)$ along the direction z perpendicular to the bilayer plane. According to [9], $\rho(z)$ has been modeled by assuming four domains (indexed with $i = 0, 1, 2, 3$) of constant electron density, corresponding to solvent, head group, alkyl chains and terminal domains, respectively, with smooth transitions between two adjacent levels. The hydrated lipid volume fraction is calculated as a function of the nominal lipid molar fraction $x_{\rm L}$ of the sample,

$$c_V \quad = \quad \frac{\nu x_{\rm L}}{(1 - x_{\rm L})\nu_{\rm W} + x_{\rm L}(\nu - N_{\rm H}\nu_{\rm W})}, \tag{3}$$

where ν is the volume of one hydrated lipid molecule, $\nu_{\rm W}$ is the volume of a water molecule in the bulk solvent (with a well-known dependence on temperature [14]) and $N_{\rm H}$ is the total number of hydration water molecules per lipid molecule dispersed in the bilayer. The main point of this simple model is an explicit determination of ν in terms of $N_{\rm H}$, which is also used to calculate the domain electron densities ρ_i. Details are hereafter reported.

The volume of one hydrated lipid molecule is written as a sum of polar head, alkyl chains and terminal group volumes, $\nu = \nu_1 + \nu_2 + \nu_3$. Regarding the location of the $N_{\rm H}$ water molecules associated to one lipid molecule, we make the hypothesis that $N_{\rm H_{pol}}$ are in the polar head domain and $N_{\rm H_{par}}$ are in the paraffinic moiety, with the condition $N_{\rm H} = N_{\rm H_{pol}} + N_{\rm H_{par}}$. The paraffinic moiety of one lipid molecule is then constituted by $N_{\rm CH_2}$ methylenes, $N_{\rm CH_3}$ methyls and $N_{\rm H_{par}}$ water molecules. Each of these three groups (indexed by $k = {\rm CH_2}, {\rm CH_3}, {\rm H_{par}}$) will be, on average, located in both paraffinic domains 2 and 3, according to three corresponding fractions, x_k, which are defined as the ratio between the number of k-groups in the domain 2

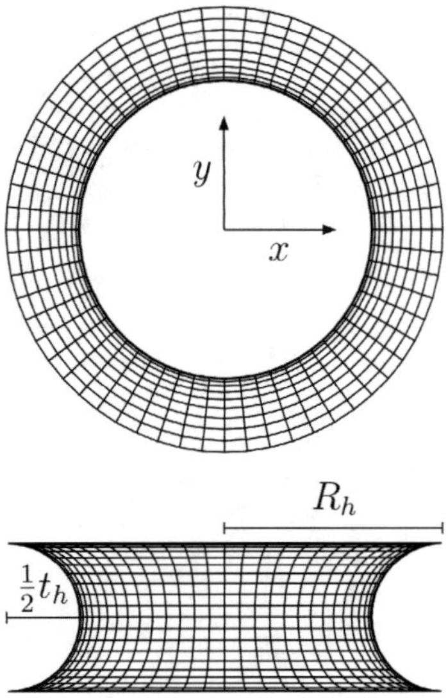

Figure 7. Geometry of the pore as used in the pore model developed in ref. [9].

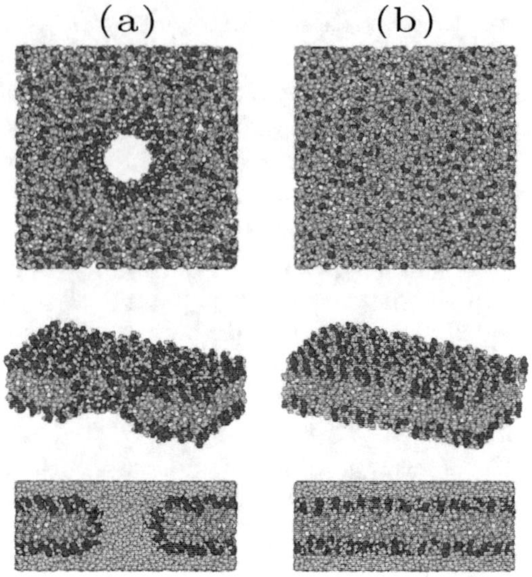

Figure 8. Molecular representation of the bilayer (a) in the pore model from ref. [9] and (b) in the simple model presented in this paper. DMPG has red polar heads, green CH_2 groups and yellow CH_3 groups, while water is cyan. The bilayer is seen from its perpendicular axis in the upper part of the figure, in perspective in the middle part and its profile is seen in the bottom part of the figure, where bulk water molecules have been added for clarity. The bilayer profile in (a) has one pore, as sketched also in the TOC of [9].

and the total number of k-groups in the paraffinic moiety. In this framework, the three domain volumes per lipid molecule can be written as

$$\nu_1 = \nu_{\mathrm{pol}} + N_{\mathrm{H_{pol}}} \nu_{\mathrm{W}} / d_{\mathrm{pol,W}} \tag{4}$$

$$\nu_2 = N_{\mathrm{CH_2}} \nu_{\mathrm{CH_2}} x_{\mathrm{CH_2}} + N_{\mathrm{CH_3}} \nu_{\mathrm{CH_3}} x_{\mathrm{CH_3}} + N_{\mathrm{H_{par}}} \nu_{\mathrm{W}} x_{\mathrm{H_{par}}} / d_{\mathrm{par,W}} \tag{5}$$

$$\nu_3 = N_{\mathrm{CH_2}} \nu_{\mathrm{CH_2}} (1 - x_{\mathrm{CH_2}}) + N_{\mathrm{CH_3}} \nu_{\mathrm{CH_3}} (1 - x_{\mathrm{CH_3}}) + N_{\mathrm{H_{par}}} \nu_{\mathrm{W}} (1 - x_{\mathrm{H_{par}}}) / d_{\mathrm{pol,W}} \tag{6}$$

In these equations, ν_{pol}, $\nu_{\mathrm{CH_2}}$ and $\nu_{\mathrm{CH_3}}$ represent the group volumes, whose values as a function of temperature and with the fraction of α conformation in the chains (x_α) are well known from literature data [15]. Moreover, $d_{\mathrm{pol,W}}$ and $d_{\mathrm{par,W}}$ are the relative mass density of hydration water molecules in the polar and paraffinic moieties.

By the knowledge of the volumes ν_i of the domains, it is straightforward to calculate the domain thicknesses R_i as a function of the bilayer thickness t:

$$R_i = \frac{\nu_i}{\nu} \frac{t}{2}. \tag{7}$$

The area per lipid molecule is easily calculated by $A = 2\nu/t$. The domains' electron densities are $\rho_i = \mathrm{e}_i/\nu_i$ where the domain number of electrons e_i are calculated on the basis of the domain

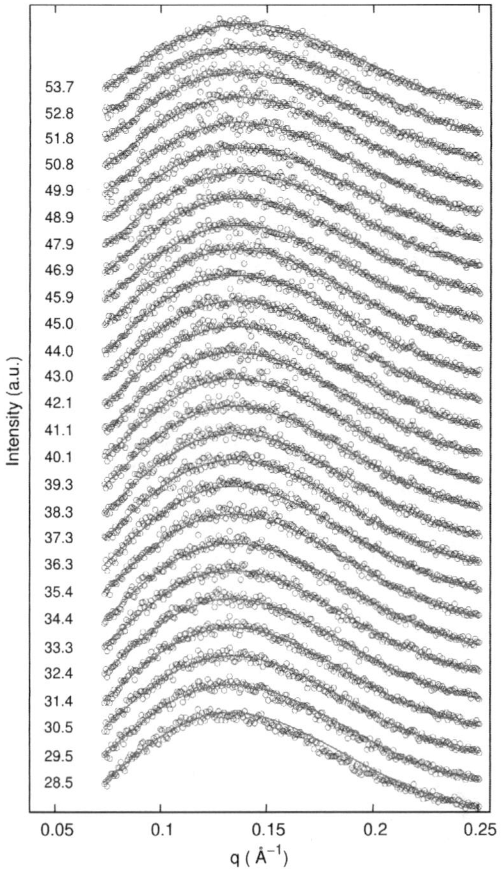

Figure 9. Global fits of the SAXS curves for 150 mM DMPG, long series, at $T >$ 28°C using the model of a water-penetrated infinite bilayer. Temperature values are reported on the left of each curve in °C. Curves are stacked by a factor 1.2.

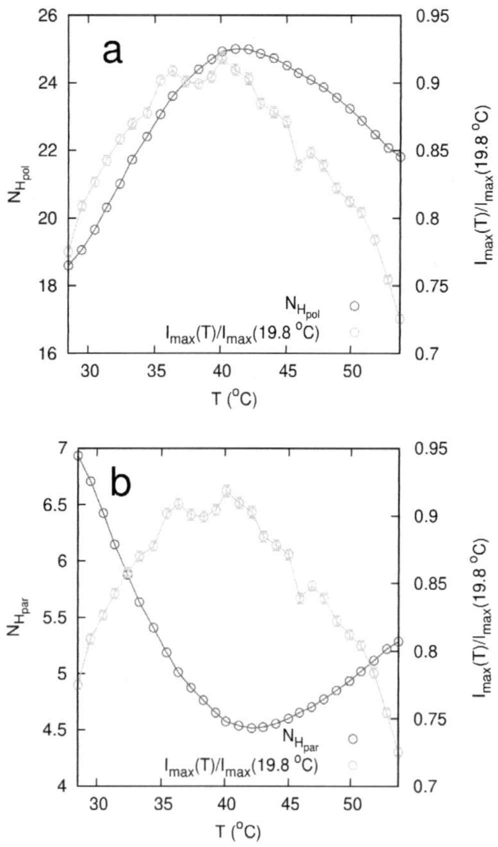

Figure 10. Temperature dependence of the number of water molecules associated to each DMPG molecule in the polar, $N_{\mathrm{H_{pol}}}$ (a), and paraffinic, $N_{\mathrm{H_{par}}}$ (b), moieties obtained by the analysis shown in Figure 9. In both graphs, the relative intensity of the bilayer band, $I_{\mathrm{max}}(T)/I_{\mathrm{max}}(T_{\mathrm{m}}^{\mathrm{on}} = 19.8°\mathrm{C})$ is also reported in green for comparison.

composition (see Eqs. 4-6),

$$\begin{aligned} \mathrm{e}_1 &= \mathrm{e}_{\mathrm{pol}} + N_{\mathrm{H_{pol}}}\mathrm{e}_{\mathrm{W}} + x_{\mathrm{Na^+}}\mathrm{e}_{\mathrm{Na^+}} && (8) \\ \mathrm{e}_2 &= N_{\mathrm{CH_2}}\mathrm{e}_{\mathrm{CH_2}}x_{\mathrm{CH_2}} + N_{\mathrm{CH_3}}\mathrm{e}_{\mathrm{CH_3}}x_{\mathrm{CH_3}} + N_{\mathrm{H_{par}}}\mathrm{e}_{\mathrm{W}}x_{\mathrm{H_{par}}} && (9) \\ \mathrm{e}_3 &= N_{\mathrm{CH_2}}\mathrm{e}_{\mathrm{CH_2}}(1-x_{\mathrm{CH_2}}) + N_{\mathrm{CH_3}}\mathrm{e}_{\mathrm{CH_3}}(1-x_{\mathrm{CH_3}}) + N_{\mathrm{H_{par}}}\mathrm{e}_{\mathrm{W}}(1-x_{\mathrm{H_{par}}}) && (10) \end{aligned}$$

where $\mathrm{e_{pol}}$, $\mathrm{e_{CH_2}}$, $\mathrm{e_{CH_3}}$, $\mathrm{e_{Na^+}}$ and $\mathrm{e_W}$ are the number of the electrons of $\mathrm{DMPG^-}$, CH_2, CH_3, Na^+ and water groups, respectively. Here $x_{\mathrm{Na^+}}$ is the fraction of Na^+ counterion dissolved in the domain 1. Notice that the possible electrostriction effect of Na^+ is indirectly taken into account by variations of the relative mass density of water in the domain 1, $d_{\mathrm{pol,W}}$.

In summary, the free parameters of the model, which will be dependent on temperature and then specific of each SAXS curve, are the bilayer thickness, t, the number of water molecules in the polar and in the paraffinic moieties, $N_{\mathrm{H_{pol}}}$ and $N_{\mathrm{H_{par}}}$, and the $x_{\mathrm{CH_2}}$, $x_{\mathrm{CH_3}}$ and $x_{\mathrm{H_{par}}}$ parameters which define the fraction of CH_2, CH_3 and water groups located in the paraffinic domain 2. Note that for $T > 28°\mathrm{C}$, the fraction of lipids in α conformation can be reasonably fixed to $x_\alpha = 1$. Moreover, a few tests have shown that the sodium fraction $x_{\mathrm{Na^+}}$ does not significantly change with temperature. Hence, a unique average value $x_{\mathrm{Na^+}}$ can be determined by fitting all the curves to the model. Other common parameters, that can be considered independent on temperature, are the relative mass densities of the water in the polar and in the paraffinic moieties $d_{\mathrm{pol,W}}$ and $d_{\mathrm{par,W}}$. The variances of the error function that smoothes the transitions between two adjacent levels of electron density [9] have been fixed to 1 Å. Finally, a unique instrumental calibration constant, κ (Eq. 2) should be found by the analysis. Notice that all the other parameters of the models, such as thicknesses and domain electron densities are calculated from the fitting parameters.

The dataset of $N_C = 27$ curves of the 150 mM DMPG sample ($x_{\mathrm{L}} = 0.00298$), long series, at increasing temperature from 28.5 to 53.7°C, in the range $q > 0.07$ Å^{-1}, has been fitted with the method reported above. In order to avoid dramatic variations or oscillations with temperature of the six single curve fitting parameters, (t, $N_{\mathrm{H_{pol}}}$, $N_{\mathrm{H_{par}}}$, $x_{\mathrm{CH_2}}$, $x_{\mathrm{CH_3}}$ and $x_{\mathrm{H_{par}}}$, here indicated as X_j), a regularization method has been exploited [16]. The merit functional to be minimised is constituted by a standard reduced χ^2 of the dataset and a regularization term,

$$\chi^2 = \frac{1}{N_C}\sum_{c=1}^{N_C}\frac{1}{N_q}\sum_{i=1}^{N_q}\left\{\frac{I_{c,\mathrm{exp}}(q_i) - I_c(q_i)}{\sigma_c(q_i)}\right\}^2 + \alpha_{\mathrm{reg}}\sum_{c=1}^{N_C-1}\sum_{j=1}^{6}\left(1 - \frac{X_{c+1,j}}{X_{c,j}}\right)^2 \tag{11}$$

where N_q is the number of q-points of each curve, $\sigma_c(q)$ is the experimental uncertainty of the c-curve and α_{reg} is a regularization parameter whose value is chosen in order to get a relative weight of the second term of 10-20% with respect to the first term.

Table 1. Parameters obtained by the analysis of the bilayer band. 27 SAXS curves at 150 mM DMPG, long series, for $T > 28.5°\mathrm{C}$ have been considered.

κ (a.u. cm^{-1})	$x_{\mathrm{Na^+}}$	$d_{1,\mathrm{W}}$	$d_{\mathrm{W},\alpha}$
25.5 ± 0.3	0.25 ± 0.07	1.076 ± 0.002	1.08 ± 0.02
$x_{\mathrm{CH_2}}$	$x_{\mathrm{CH_3}}$	$x_{\mathrm{H_{par}}}$	t (Å)
0.71 ± 0.03	0.46 ± 0.03	0.60 ± 0.08	45 ± 1

Figure 9 shows the fits obtained with the SAXS curves of 150 mM DMPG for $T > 28°\mathrm{C}$ using such a model. The quality of the fit is very high. The common parameters are reported in Table 1, together with the average value of the single curve fitting parameters which resulted practically independent on temperature. The more interesting results of this analysis are the number of water molecules associated to each DMPG molecule in the polar, $N_{\mathrm{H_{pol}}}$, and paraffinic, $N_{\mathrm{H_{par}}}$, moieties. Their temperature dependence, shown in Figure 10, suggests that at $T_{\mathrm{m}}^{\mathrm{off}}$ water is still inside the bilayer, probably reminiscent of the pores present in the phases occurring at lower temperatures. On heating up to around 40°C, water molecules are expelled from the paraffinic region, so that they may contribute to the observed increased hydration of the polar heads. At the higher temperatures, water molecule are also released from the polar region, probably due to the increased thermal disorder.

6. Conclusions

In solution SAXS data, obtained at different DMPG concentrations (from 70 to 300 mM) and as a function of temperature across the main transition, have been analyzed to derive information on the structural modifications accompanying the changes from vesicles to bulk lamellar structures induced by concentration and/or temperature.

At one side, simulated and measured slopes of the SAXS curves at very low q support the conclusions that the IP phase corresponds to large uncorrelated bilayers, characterized by two-dimensional correlated water pores within the DMPG lamellae, while the L_{p} phase is characterized by largely correlated bilayers. On the other side, the model fitting of the bilayer band centered at 0.12 Å^{-1} after $T_{\mathrm{m}}^{\mathrm{off}}$ showed that at this temperature water is still inside the DMPG bilayer, probably due to a defective pore closing. On heating, water molecules are released first from the paraffinic region and then from the polar heads, probably due the increased thermal disorder. It is well known that in concentrated solutions, amphiphiles may form bulk lamellar structures, and other complex liquid crystalline structures, largely studied in the case of detergents and soaps [17]. The change from vesicles to bulk lamellar structures is expected to occur with increase in phospholipid concentration, but this transformation is in general not studied in detail mostly because the region of smaller concentration is a physico-chemical problem, while the higher concentration region is a condensed matter problem. We are persuaded that this paper will give some contribution to the understanding of the transformations occurring on this charged phospholipid system, which may have biological significance.

Acknowledgements

Financial support from FAPESP and USP made this work possible. Research undertaken at the Brazilian Synchrotron Light Source, proposal D11A-SAS 2266/03 (under responsibility of LQA, who thanks both financial and experimental support). Present financial support by INCT project on Complex Fluids (CNPq/MCT/FAPESP).

References

[1] Lasic D D 1988 *Biochem. J.* **256** 1–11
[2] Jacobs R E, Hudson B and Andersen H C 1975 *PNAS* **72** 3993–3997
[3] Cevc G 1987 *Biochemistry* **26** 6305–6310
[4] Lamy-Freund M T and Riske K A 2003 *Chem. Phys. Lipids* **122** 19–32
[5] Riske K A, Amaral L Q and Lamy-Freund M T 2001 *Biochim. Biophys. Acta* **1511** 297–308
[6] Riske K A, Amaral L Q, Döbereiner H G and Lamy M T 2004 *iophys. J.* **86** 3722–3733
[7] Fernandez R, Riske K A, Amaral L Q, Itri R and Lamy M 2008 *BBA-Biomembranes* **1778** 907916
[8] Riske K A, Amaral L Q and Lamy M T 2009 *Langmuir* **25** 1008310091
[9] Spinozzi F, Paccamiccio L, Mariani P and Amaral L Q 2010 *Langmuir* **26** 6484–6493
[10] Cronan J E 2003 *Annu. Rev. Microbiol.* **57** 203–224
[11] Atwood J L and Steed J 2004 *Encyclopedia of supramolecular chemistry* vol 1 (J. L. Atwood & J. W. Steed)
[12] Nieh M P, Glinka C J, Krueger S, Prosser R S and Katsara J 2002 *Biophys. J.* **82** 2487–2498
[13] Glatter O and Kratky O 1982 *Small Angle X-ray Scattering* (Academic Press) ISBN 0-12-286280-5
[14] Robert C W 1974-75 *Handbook of Chemistry, Physics, 55th ed.* (Cleveland, Ohio: CRC Press)
[15] Luzzati V and Spegt P A 1967 *Nature* **215** 701–704
[16] Glatter O 2002 *Neutron, X-rays and Light. Scattering Methods Applied to Soft Condensed Matter* ed Lindner P and Zemb T (North-Holland) pp 103–124
[17] Luzzati V 1968 *X-ray diffraction studies of lipid-water systems. In Biological Membranes, p. 71-123* (Academic Press, New York: D. Chapman)

SANS investigation of assembly state of proteasome activator 28 and the 20S proteasome

Masaaki Sugiyama[1], Eiji Kurimoto[2,3], Hiroki Sahashi[2], Eri Sakata[2], Yukio Morimoto[1], Keiji Itoh[1], Kazuhiro Mori[1], Toshiharu Fukunaga[1], Yasufumi Minami[4] and Koichi Kato[2,5]

[1]Research Reactor Institute, Kyoto University, Osaka 590-0494, Japan
[2]Graduate School of Pharmaceutical Sciences, Nagoya City University, Nagoya 467-8603, Japan
[3]Graduate School of Pharmacy, Meijo University, Nagoya 468-8503, Japan
[4]Department of Biotechnology, Maebashi Institute of Technology, Gunma 371-0816,Japan
[5]Okazaki Institute for Integrative Bioscience and Institute for Molecular Science, National Institutes of Natural Sciences, Okazaki 444-8787, Japan

E-mail: sugiyama@rri.kyoto-u.ac.jp

Abstract. The state of proteasome activator 28 (PA28) and the formation of the proteasomal complex in an aqueous solution are investigated with small-angle neutron scattering (SANS). The most appropriate state of PA28, which well reproduces the observed SANS profile, is the dissociation equilibrium between dimer and monomer with dissociation degree of 0.5. In addition, it is revealed that the packing of PA28 in the dimer is same as that in a crystal. It is also revealed that the proteasomal complex in which two PA28s connects to both basal planes of the 20S proteasome is spontaneously formed in the mixture solution of PA28 and the 20S proteasome.

1. Introduction

Small-angle neutron scattering (SANS) enable us to analyze a nano scale particle structure in an aqueous solution[1]. In addition, since neutron is a very gentle probe for samples, we have little or no worry about protein damage caused by the neutron irradiation. Therefore, SANS is one of the most suitable methods to investigate the formation and disruption of assembly state of proteins, which are deeply related with expression and loss of their functions, in an aqueous solution.

We have been focusing on the proteasomal systems consisting of several proteins, of which combinations depend upon their functions. The main component in the system is the 20S proteasome (PRS) with the molecular mass of ~700 kDa known as a protein degradation machinery in cells[2, 3]. PRS has a hollow cylindrical shape constructed with four rings (α-β-β-α rings), each of which consists of seven subunits[4, 5]. The function of this huge protease complex is regulated through the attachment of other protein complexes termed proteasome activators (PAs): PAs identify proteins which are degraded by PRS. For example, proteasome activator 28 (PA28) with the molecular mass of ~200 kDa has a truncated cone shape[6] consisting of highly homologous α- and β-subunits[7]. PA28 connects to both basal planes of PRS[8] and this protease complex (PRS + two PA28; PRS2PA28) contributes to the processing of antigenic proteins into peptides for presentation via the MHC class I pathway[9, 10, 11, 12]. Therefore, PA28, PRS and PRS2PA28 are very important in our immune system.

For a better understanding of the assembly mechanisms in this proteasomal system, it is necessary to reveal the state of PA28, PRS and PRS2PA28 in an aqueous solution. Therefore, as the first step in our research, using a SANS method, we characterized the state of PA28 and then examined the formation of PRS2PA28 in a solution of PA28 mixtured with PRS.

2. SANS experiment and simulation method

2.1. Sample preparation

The cDNAs of mouse PA28α and PA28β were kindly gifted by Dr. Keiji Tanaka (Tokyo Metropolitan Institute of Medical Science). The PA28 α-subunit lacking the loop segment (residue Val69 to Lys97) was generated using standard PCR and genetic-engineering techniques. A mutated α-subunit and a wild-type β-subunit were subcloned into the expression vectors pET21d and pET23a, respectively. Each subunit was separately expressed using *Escherichia coli* (BL-21) and purified by ammonium sulfate precipitation, hydroxyapatite chromatography, and ion-exchange chromatography. The hetero-oligomer of PA28 was constructed by mixing the two subunits and purified by gel filtration chromatography. Hereafter, this hetero-oligomeric PA28 will be simply referred to as PA28. PA28 was concentrated to 5.0 mg/mL in a buffer solution composed of the following: 50 mM Tris-HCl (pH 7.5), 1 mM dithiothreitol, and 99.9% D_2O.

The strain of *S. cerevisiae* for expression of FLAG-tagged PRS was also kindly gifted by Dr. Keiji Tanaka. Cells were grown to an absorbance of 600 nm at 2.0 in YPD medium at 27°C. The grown cells were harvested and suspended in 50 mM Tris-HCl buffer (pH 7.5) containing 100 mM NaCl and 10% glycerol, then lysed by mixing with glass beads. The extract was centrifuged at 26,740 g for 30 min at 4°C, then the supernatant was subjected to the affinity chromatography using anti-FLAG M2-agarose beads (Sigma). Anti-FLAG M2-agarose beads, to which the 26S proteasome had been bound, were incubated with 50mM Tris-HCl buffer (pH 7.5) containing 500mM NaCl, 10% glycerol and 10% Triton X-100 for 30 min at 25°Cto dissociate the 19S components from the 26S proteasome. After washing the column, PRS was eluted with 3×FLAG peptide and the concentration was tuned to be 3.8 mg/mL in same buffer.

In order to examine the formation of PRS2PA28 in a aqueous solution, both prepared samples were mixed with the molar ratio of [PA28]:[PRS]=2.6:1. The concentration of the mixture sample was also tuned to be 4.1 mg/mL.

2.2. SANS experiment

SANS experiments of three samples were performed with a SANS-U spectrometer of the Institute for Solid State Physics, University of Tokyo, installed at JRR-3 of the Japan Atomic Energy Agency, Tokai, Japan[13]. The SANS data were measured at two sample-detector distances (SDLs) of 8 and 2 m. The total q-range covered was 0.008-0.2 Å^{-1} using 7.0 Å neutrons. The measurement times for the low (SDL=8 m) and middle (SDL=2 m) q-ranges were 3 and 1 hr, respectively, and temperature was kept constant at 25 °C. The observed SANS intensities were corrected for background, cell, buffer scattering, and transmission factor.

2.3. SANS simulation

The SANS profiles are simulated as follows[14, 15]. Firstly, a appropriate protein structure, monomer, dimer and/or their modulated ones, is fabricated using with atomic coordinates downloaded from Protein Data Bank: the examined structure models are described later. Because the samples were in the aqueous solution with heavy water and SANS cannot ignore hydrogen, we should make two improvements on the downloaded atomic coordinates. One is to add protons to the atomic coordinates at the most probable position, and the other is to replace all exchangeable protons with virtual hydrogen atoms and remove the dissociatable protons/deuterons under the experimental condition (pH 7.5 and 25°C). The virtual hydrogen atom has an average scattering length depending on the ratio of light water and heavy water in the solution. Both procedures are simultaneously carried out using the computer program CNS[16] with a neutron-dedicated parameter file[17, 18].

Next, to reduce the calculation time, the amino acid residues are replaced with simple beads which have the volumes of the corresponding amino-acid residues[19] and the scattering contrast. With this simplified protein structure, SANS intensity $I(q)$ is calculated as follows,

$$I(q) = \sum_i |f_i(q)|^2 + 2\sum_{i>j} f_i(q)f_j(q)\frac{\sin(q\cdot|\boldsymbol{r}_i - \boldsymbol{r}_j|)}{q\cdot|\boldsymbol{r}_i - \boldsymbol{r}_j|} + I_{inc}, \quad (1)$$

$$f_i(q) = 3v_i\triangle\rho_i\frac{\sin(qR_i) - (qR_i)\cos(qR_i)}{(qR_i)^3}, \quad (2)$$

$$I_{inc} = \frac{1}{4\pi}\sum_k \sigma_k^{inc} + N_\text{x}\left(< b_\text{x}^2 > - < b_\text{x} >^2\right), \quad (3)$$

$$< b_\text{x}^2 > = c_{\text{H}_2\text{O}}b_\text{H}^2 + c_{\text{D}_2\text{O}}b_\text{D}^2, \quad (4)$$

$$< b_\text{x} >^2 = (c_{\text{H}_2\text{O}}b_\text{H} + c_{\text{D}_2\text{O}}b_\text{D})^2, \quad (5)$$

where $f_i(q)$, $\boldsymbol{r}_i$, v_i, $\triangle\rho_i$ and $R_i(=3v_i^{1/3}/(4\pi))$ are the form factor, the center of mass, the volume, the scattering contrast and average radius of the i-th amino acid, respectively, σ_k^{inc} is the incoherent scattering cross section of the k-th atom, N_x is the total number of replaceable hydrogen atoms, $c_{\text{H}_2\text{O}}$ and $c_{\text{D}_2\text{O}}$ are the ratios of light water and heavy water in the solvent, and b_H and b_D are the scattering lengths of proton and deuteron, respectively.

3. Results and analysis

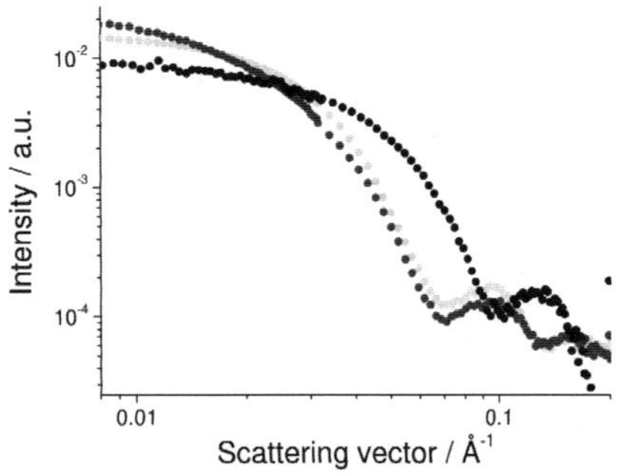

Figure 1. SANS profiles of PA28 (black circles), PRS (cyan circles) and their mixture (blue circles), respectively.

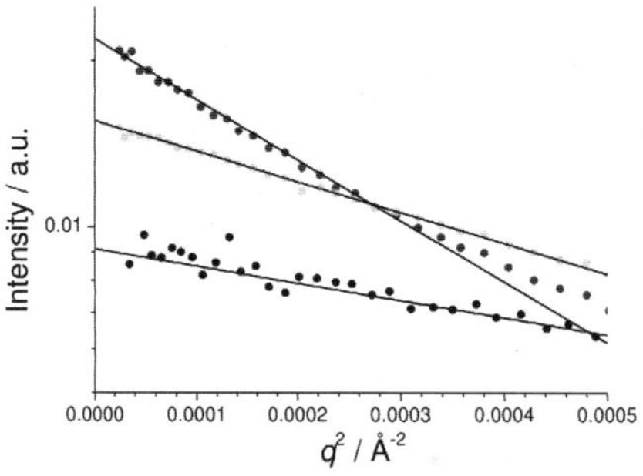

Figure 2. Guinier plots of SANS profiles of PA28 (black circles), PRS (cyan circles) and their mixture (blue circles), respectively. The straight lines show the results of the least square fittings with Guinier formula.

Figure 1 show SANS profiles of PA28, PRS and their mixture, respectively, and Figure 2 does their Guinier plots. The straight lines in Fig. 2 were the results of the least square fitting of Guinier formula; $I(q) = I(0)\exp(-(Rg^2/3)q^2)$, where Rg is a gyration radius. Rgs are found to be 44.2±0.9 Å, 61.9±0.7 Å and 87.2±0.8 Å of PA28, PRS and their mixture, with Guinier

Table 1. Concentrations and the results of Guinier analysis of PA28, PRS and their mixture.

	PA28	PRS	Mixture
Concentration [mg/mL]	5.0	3.8	4.1
Rg [Å]	44.2±0.9	61.9±0.7	87.2±0.8
$I(0)/I_{PRS}(0)$	0.58	1	1.42

formula, respectively. The ratios of the SANS intensities of PA28 and the mixture at q=0 Å^{-1} to that of PRS, which should be the monodisperse system, are 0.58 and 1.42, respectively.

3.1. PA28 in an aqueous solution

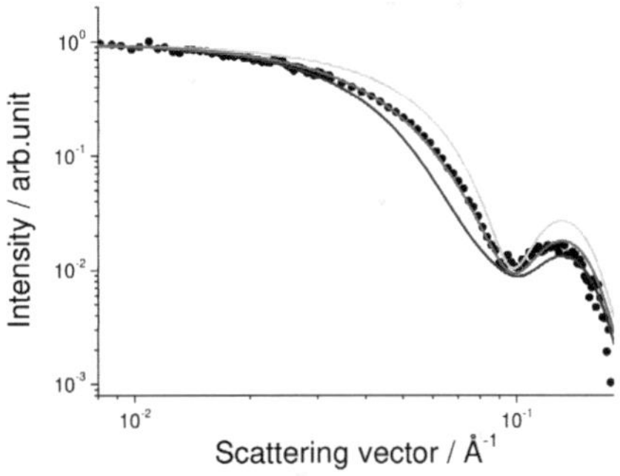

Figure 3. Simulated and experimental SANS profiles of PA28. Green, blue and red lines indicate the simulated SANS profiles of s-PA28, d-PA28 and state of the dissociation equilibrium between d-PA28 and s-PA28 at 0.5 (see text), respectively. Closed circles denote the experimental SANS profile.

We start from the examination of the state of PA28 in an aqueous solution by simulating the SANS profiles and the gyration radii of several models: the atomic coordinates of PA28[6] was downloaded from PDB. As shown in Fig. 3, the simulated SANS profile of a single PA28 (s-PA28) cannot reproduce the experimental one (see green line and closed circles). In addition, Rg of s-PA28 was calculated to be 33.7 Å with the simulation. This is quite smaller than the experimental value (44.2±0.9 Å). Therefore, it was considered that PA28 makes aggregation in an aqueous solution.

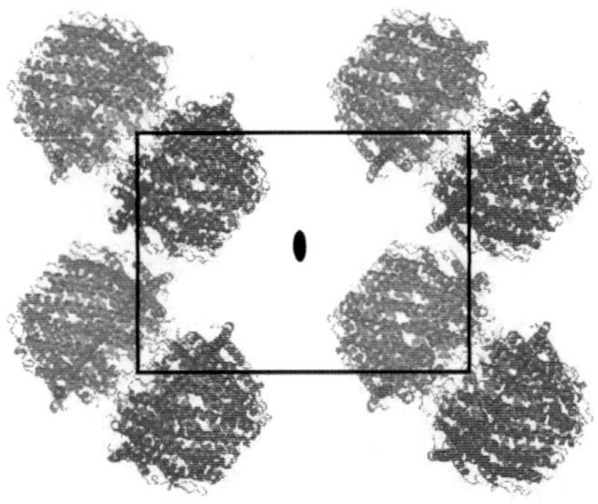

Figure 4. The projection of crystal structure of PA28 to *ac*-plane[6]. The crystal structure belongs to the space group $C2$ with the lattice parameters, a=162.600, b=134.300, c=116.200.

The crystal structure of PA28[6] gave us a good starting point for considering the assembly state in an aqueous solution. As shown in Fig.4, two PA28s closely locate in the crystal, facing the planes each of which connects to PRS. Therefore, we assumed that PA28 could make a

Journal of Physics: Conference Series **247** (2010) 012020 doi:10.1088/1742-6596/247/1/012020

similar dimer in an aqueous solution; a dimer of PA28 will be abbreviated to d-PA28. The SANS profile of d-PA28 is also shown in Fig. 3 (see blue line) and its Rg was calculated to be 47.4 Å. The SANS profile and the gyration radius of d-PA28 became closer to the experimental ones than those of s-PA28. However, there were still discrepancies both in the experimental and simulated ones; The simulated SANS profile of d-PA28 showed the deviation from the experimental one between 0.03 Å^{-1} and 0.08 Å^{-1}. In addition, Rg of d-PA28 was slightly larger than the experimental one.

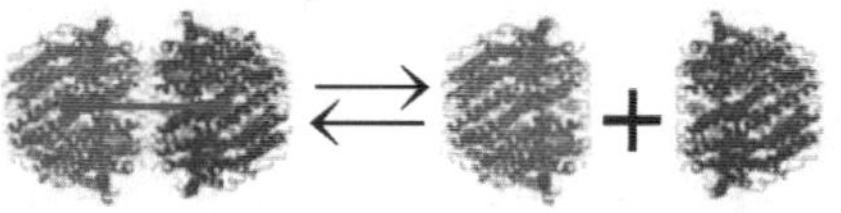

Figure 5. Dissociation equilibrium state of PA28 in an aqueous solution: d1-PA28 $\rightleftharpoons$ s-PA28 + s-PA28. Red circles and a red arrow indicate the centers of mass of s-PA28s and the distance between them l, respectively.

Here, we noticed that the experimental SANS profile and Rg are almost the average between those of d-PA28 and s-PA28. Therefore, it is rational that the state of PA28 in an aqueous solution is in the dissociation equilibrium between d-PA28 and s-PA28 as shown in Fig. 5. In other words, d-PA28 and s-PA28 could coexist in an aqueous solution. Furthermore, to make two s-PA28s get closer is another possibility to fit Rg of d-PA28 to the experimental one: the out side chains around the connecting plane are random coils and can move freely in aqueous solution. Based on above mentioned consideration, we made the further simulation with two parameters: one was the dissociation degree, $x(=[\text{s-PA28}]/(2[\text{d-PA28}]+[\text{s-PA28}]))$, and the other was the distance of the centers of mass between two PA28s in d-PA28, l (see a red arrow in Fig. 5). l is varied from where two PA28s slightly touch (68 Å) as in the crystal to where two PA28s are embedded by the edge of columns of α-helix each other (59 Å).

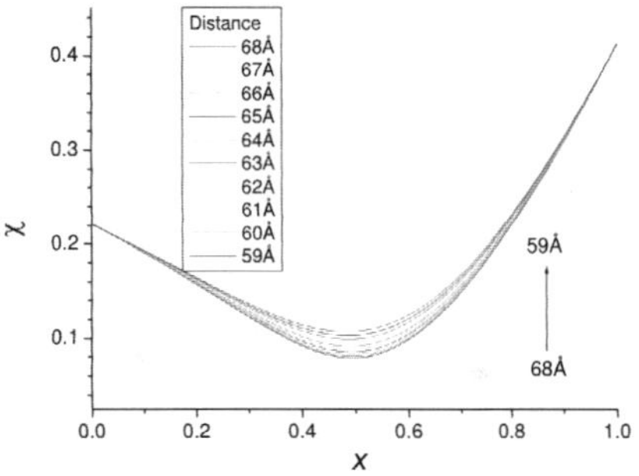

Figure 6. Relation between the error sum of the squares χ and the dissociation degree x in the distance l of 68-59Å.

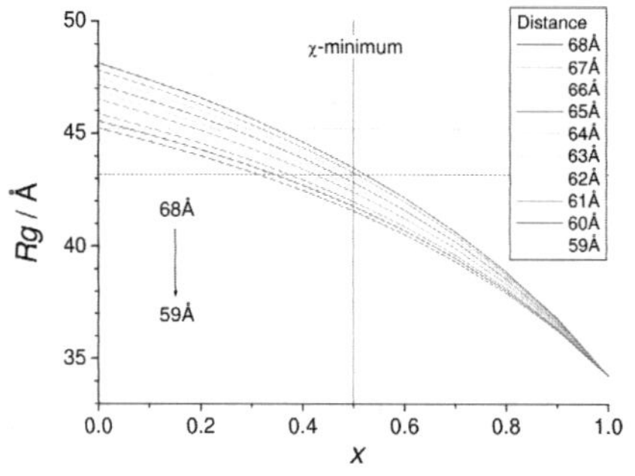

Figure 7. Relation between the gyration radius Rg and the dissociation degree x in the distance l of 68-59Å. The zone between two broken lines indicates the gyration radii satisfying the experimental one (44.2±0.9 Å).

Figure 6 shows the relation between the error sum of the squares χ and x in l of 68-59Å: $\chi^2 = \Sigma_i[W_i\{(I_i^{exp} - I_i^{sim})/I_i^{exp}\}^2]$, where I_i^{exp} and I_i^{sim} are the experimental and simulated SANS intensities at the i-th data point, respectively, and W_i is the weight function, which is the normalized inverse of experimental error at the i-th data point. As shown in Fig. 6, χ becomes minimum at x=0.5, [d-PA28]:[s-PA28]=1:2, for all distances.

Figure 7 shows the relation between Rg and x in l of 68-59Å. The zone between two broken lines indicates the gyration radii satisfying the experimental one (44.2±0.9 Å). Therefore, considering that the error sum of squares becomes minimum at x=0.5 shown in a black straight line, the most appropriate distance between two s-PA28s is 67±1 Å, which is almost same as the distance in the crystal. The simulated SANS profile of the resulted state (x=0.5 and l=67 Å) as PA28 in an aqueous solution well reproduces the experimental one as shown in Fig. 3 (see red line).

The obtained dissociation ratio ([d-PA28]:[s-PA28]=1:2) is also supported with $I_{obs}(0)$ obtained with Guinier analysis (Table 1). The obtained $I_{obs}(0)$ is described with the number of particles in a solution, N, and the intensity of one particle at q=0 Å^{-1}, $I_p(0)$, as follows.

$$I_{obs}(0) = NI_p(0) = N\left|\int_{V_p,\boldsymbol{q}=0}(\rho_p(\boldsymbol{r}) - \rho_s)\exp(\imath\boldsymbol{q}\cdot\boldsymbol{r})d\boldsymbol{r}\right|^2, \tag{6}$$

where V_p, $\rho_p(\boldsymbol{r})$ and ρ_s are the particle volume, the scattering length density of particle and that of solvent, respectively. In the samples with same optical paths, N is proportional to d/M_w, where d and M_w are the density in weight percent and the molecular weight of the particles, respectively. In addition, $I_p(0)$ is expressed as follows, $I_p(0) = |(\rho_p - \rho_s)V_p|^2$: ρ_p is the average of the scattering length density of particle. Here, since ρ_p is almost same (~0.3 fmÅ^{-3}) and V_p is proportional to M_w in proteins, $I_p(0)$ is proportional to M_w^2 in the same solvent. As a result, the obtained $I_{obs}(0)$ is also described in the solution with same solvent and optical path as follows.

$$I_{obs}(0) \propto dM_w. \tag{7}$$

Considering that PRS should be monodisperse in the solution, $I_{PA28}(0)/I_{PRS}(0)$ are calculated to be 0.38, 0.75 and 0.56 for the monodisperse solutions of s-PA28 and d-PA28 and the solution with [d-PA28]:[s-PA28]=1:2, respectively, where ds are indicated in Table 1 and Mws of s-PA28 and PRS are 4×10^5 and 7×10^5, respectively. Therefore, it means that the solution of PA28 is not a monodisperse system of s-PA8 nor d-PA28 but in the dissociation equilibrium between d-PA28 and s-PA28. In addition, $I_{PA28}(0)/I_{PRS}(0)$ calculated with the dissociation ratio ([d-PA28]:[s-PA28]=1:2) also shows a good agreement with the observed one (0.58).

In conclusion, the state of PA28 in an aqueous solution is the dissociation equilibrium between d-PA28 and s-PA28. In addition, the structure of d-PA28 could be almost same as that in the crystal.

3.2. Mixture of PA28 and the 20S proteasome

Based on the above shown result, it is considered that the following two steps could be necessary to fabricate the 20S proteasome-PA28 complex (PRS2PA28).

$$\text{d-PA28} \rightleftharpoons \text{s-PA28} + \text{s-PA28}, \tag{8}$$

$$\text{s-PA28} + \text{s-PA28} + \text{PRS} \rightleftharpoons \text{PRS2PA28}. \tag{9}$$

Here, it is important to clarify whether the second step proceeds spontaneously or with help of any chaperon when PA28 and PRS coexist in the solution. We began to examine it with the SANS results. The simulated Rgs of PRS and PRS2PA28 were calculated to be 60.6 Å and 85.6 Å, respectively. These values show good agreement with the observed Rgs of PRS and the *mixture* of PA28 and PRS as indicated in Table 1, respectively. This result means that PRS2PA28 was fabricated in the *mixture* of PA28 and PRS.

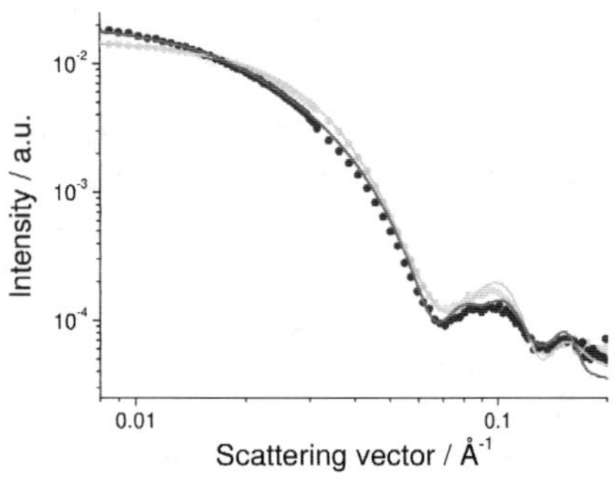

Figure 8. Simulated and experimental SANS profiles. Cyan and blue circles show PRS and the *mixture* of PA28 and PRS, respectively. Green and red curves denote the simulated SANS profiles of PRS and PRS2PA28, respectively.

Figure 8 shows the simulated SANS profiles of PRS and PRS2PA28, and the experimental ones of PRS and the *mixture* of PA28 and PRS, respectively. The simulated SANS profile of PRS2PA28 reproduced the experimental one of the mixture of PA28 and PRS. This result strongly supports that PRS2PA28 was spontaneously fabricated in the *mixture* of PA28 and PRS.

4. Conclusion and further remarks

The state of PA28 in an aqueous solution is the dissociation equilibrium between d-PA28 and s-PA28. This is a suitable solution to store surplus PA28s for association to PRSs when they are introduced. In addition, it is revealed that PA28s spontaneously connect to PRS and make PRS2PA28.

However, the reaction rate of the association of PA28s to PRS has not been clarified yet. In other words, the association rate of equation (9) is unclear. As shown in Fig. 8, there is little deviation between the simulated SANS of PRS2PA28 and the experimental one of the mixture of PA28 and PRS in the q range of 0.03-0.05 Å^{-1}. This suggested that a small amount of PRS could remain: all PRS could not be introduced to PRS2PA28. This is also supported with analysis of $I_{obs}(0)$. As indicated in Table 1, $I_{mixture}(0)/I_{PRS}(0)$ is 1.42 even though $I_{PRS2PA28}(0)/I_{PRS}(0)$ should be 1.70 with equation (7). This means that not all PRS and PA28 become PRS2PA28. Therefore, it is important to clarify the association rate of equation (9) and the change of dissociation degree of equation (8) when PRS is introduced into a solution of PA28. For the clarification, we have to estimate four partial SANS intensities, SANS intensities of s-PA28, d-PA28, PRS and PRS2PA28, in one SANS profile. However, it is difficult without any other information. Therefore, in order to solve this difficulty, SANS experiments with the deuterated PA28 and employing a contrast variation method are now in progress.

Acknowledgments

The cDNAs of mouse PA28α, PA28β, and the strain of *S. cerevisiae* for expression of FLAG-tagged PRS were kindly provided by Dr. Keiji Tanaka of Tokyo Metropolitan Institute of Medical Science. The authors would like to thank Professor Toshiyuki Chatake of Kyoto University for his kind help with the operation of the CNS program. This work is partially supported by Creative Scientific Research (No. 16GS0417), Grant-in-Aid for Scientific Research, Ministry of Education, Culture, Sports, Science and Technology, as well as by Grant-in-Aid for Scientific Research of the Ministry of Education, Science and Culture, Japan (Nos. 18076003, 19540427, 20059029, 20107004, and 21540417). Y. Morimoto and K.K. are supported by Targeted Proteins Research Program from the Ministry of Education, Culture, Sports, Science and Technology of Japan. E.S. is a recipient of Japan Society for the Promotion of Science Research Fellowships for Young Scientists. The SANS experiments using SANS-U of the Institute for Solid State Physics, the University of Tokyo were performed under Proposal Nos. 6574 and 7846.

References

[1] Feigin L A and Svergun D I 1987 *Structure Analysis by Small-Angle X-ray and Neutron Scattering* (Plenum Press: New York, 1987), chapeter 3
[2] Arrigo A P, Tanaka K, Goldberg A L and Welch W J 1988 textitNature **331** 192
[3] Baumeister W, Walz J, Zühl F and Seemüller E 1998 *Cell* **92** 367
[4] Groll M, Ditzel L, Löwe J, Stock D, Bochtler M, Bartunik H D and Huber R 1997 *Nature* **386** 463.
[5] Unno M, Mizushima T, Morimoto Y, Tomisugi Y, Tanaka K, Yasuoka N and Tsukihara T 2002 *Structure* **10** 609
[6] Knowlton J R, Johnston S C, Whitby F G, Realini C, Zhang Z, Rechsteiner M and Hill C P 1997 *Nature* **390** 639
[7] Zhang Z, Krutchinsky A, Endicott S, Realini C, Rechsteiner M and Standing K G 1999 *Biochemistry* **38** 5651.
[8] Förster A, Masters E I, Robinson H and Hill C P 2005 *Mol. Cell* **18** 589
[9] Groettrup M, Soza A, Eggers M, Kuehn L, Dick T P, Schild H, Rammensee L G, Koszinowski U H and Kloetzel P M 1996 *Nature* **381** 166
[10] Murata S, Udono H, Tanahashi N, Hamada N, Watanabe K, Adachi K, Yamano T, Yui K, Kobayashi N, Kasahara M, Tanaka K and Chiba T 2001 *EMBO J.* **20** 5887
[11] Sijts A, Sun Y, Janek K, Kral S, Paschen A, Schadendorf D and Kloetzel P M 2002 *Mol. Immunol.* **39** 165
[12] Kloetzel P M 2004 *Biochim. Biophys. Acta* **1695** 217
[13] Okabe S, Nagao M, Karino T, Watanabe S, Adachi T, Shimizu H and Shibayama M 2005 *J. Appl. Cryst.* **38** 1035
[14] Sugiyama M, Hamada K, Kato K, Kurimoto E, Okamoto K, Morimoto Y, Ikeda S, Naito S, Furusaka M, Itoh K, Mori K and Fukunaga T 2009 *Nucl. Instrum. Methods Phys. Res.* A **600** 272
[15] Sugiyama M, Kurimoto E, Morimoto Y, Sahashi H, Sakata E, Hamada K, Itoh K, Mori K, Fukunaga T and Kato K 2009 *J. Phys. Soc. Jpn.* **78** 124802.
[16] Brünger A T, Adamas P D, Clore G M, deLano W L, Gros P, Grosse-Kunstleve R W, Jiang J S, Kuszewski J, Nilges M, Pannu N S, Read R J, Rice L M, Simonson T and Warren G L 1998 *Acta Cryst.* **D54** 905
[17] Ostermann A, Tanaka I, Engler N, Niimura N and Parak F G 2002 *Biophys. Chem.* **95** 183
[18] Chatake T, Kurihara K, Tanaka I, Tsyba I, Bau R, Jenney F E, Adams M W W and Niimura N 2004 *Acta Cryst.* D**60** 1364
[19] Jacrot B 1976 *Rep. Prog. Phys.* **39** 911

XIV International Conference on Small-Angle Scattering (SAS09) IOP Publishing
Journal of Physics: Conference Series **247** (2010) 012021 doi:10.1088/1742-6596/247/1/012021

Small-angle and wide-angle X-ray scattering study on the bilayer structure of synthetic and bovine heart cardiolipins

Hiroshi Takahashi[1,3,4], Tomohiro Hayakawa[2,3], Kazuki Ito[4], Masaki Takata[4], Toshihide Kobayashi[3]

[1] Biophysics Laboratory, Department of Chemistry and Chemical Biology, Gunma University, Maebashi, Gunma, 371-8510, Japan
[2] Life Science Laboratory, Advanced Materials Laboratories, Sony Corporation, Yushima, Bunkyo-ku, Tokyo, 113-8510, Japan
[3] Lipid Biology Laboratory, RIKEN, Wako, Saitama 351-0198, Japan
[4] Structural Materials Science Laboratory, RIKEN SPring-8 Center, Sayo, Hyogo 679-5148, Japan

E-mail: htakahas@chem-bio.gunma-u.ac.jp

Abstract. Cardiolipin (CL) is a membrane phospholipid containing four fatty acid chains. CL plays an important role in energy transformation in mitochondria. The disorder of CL biosynthesis is involved in a genetic disease, Barth syndrome. Alteration of fatty acid composition of CLs has been found in Barth syndrome patients, i.e., the decrease of unsaturated fatty acid chains. In this study, we investigated how the degree of saturation alters the structure of CL bilayers by using X-ray scattering. Bovine heart CL and two synthetic CLs were compared. Fatty acid compositions of these three CLs have different saturation. Small-angle X-ray scattering data showed that the decrease of the number of double bonds in the unsaturated fatty acid chains causes to thicken the CL bilayers. In addition, wide-angle X-ray scattering data suggested that the decrease reduces the degree of disorder of the hydrophobic region in a liquid crystalline phase. These results may be related to the dysfunction of mitochondria in Barth syndrome.

1. Introduction

Cardiolipin (CL) is a unique membrane lipid [1]. It is a phospholipid dimer that contains four fatty acids. Two phosphatidyl groups are esterified to a single glycerol. CL was first isolated from heart [2] and is almost exclusively located in the inner membrane of mitochondria, where it represents the 25-30% of total mass [3]. The function of mitochondria is creation of adenosine triphosphate (ATP) used as a common chemical energy molecule for living things. It is well-known that CL is required for the ATP production [1,4]. The ATP synthesis is associated with several protein complexes embedded in the inner mitochondrial membrane, so-called mitochondrial respiratory chain complexes (I - IV). Biochemical and biophysical studies have revealed that CL is needed to stabilize the native structure of the complexes III and IV [5,6].

There are several diseases related to the dysfunction of mitochondria. One of such example is Barth syndrome [7]. The syndrome patients suffer from delayed growth, weakness of heart muscle, and disorder of immune system. Barth syndrome is an X-linked disease associated with mutations in the *tafazzin* gene that codes an enzyme involved in CL synthesis [8]. In fact, recent biochemical studies have revealed that Barth syndrome induces the change of fatty acid composition of CL [9,10]. Normal heart CL contains mainly 18:2 unsaturated fatty acid (linoleic acid) chains. By contrast, in CLs of Barth syndrome patients, various fatty acid chains are found and the amount of saturated fatty acid chains increases particularly [9,10]. Thus, the disorder of CL synthesis seems to be a cause of the mitochondrial dysfunction observed in Barth syndrome. The molecular mechanism of the disease, however, remains unknown yet.

Theoretical and experimental studies have indicated that the physical properties of phospholipid bilayers affect the membrane protein functions [11,12]. In transmembrane protein-lipid interaction, one important factor is a matching of hydrophobic part lengths between the proteins and lipid bilayers. If the hydrophobic part lengths differ each other under non-stress conditions, this mismatch does not induce the hydrophobic parts of either lipid or protein to expose water but induces conformational change of either protein or lipid or both to compensate the mismatch. The conformational change of protein affects its activity. In other words, the thickness of lipid bilayer parts of biomembranes affects transmembrane protein's activity [12,13]. This concept has been supported by several studies using a model system where an ion pump protein is incorporated into a model membrane made from a single phospholipid species. Such studies have revealed that the activity of ion pump depends on the number of carbon atoms in the fatty acid chains of phospholipid [14,15].

The final goal of our study is to clarify the molecular mechanism how the disorder of CL synthesis causes the mitochondrial dysfunction in Barth syndrome. As the first step, in this study, we investigated how the degree of saturation of the hydrocarbon chain alters the structure of CL bilayers by using X-ray scattering. We used three different CLs, i.e., natural bovine heart CL (Heart CL), synthetic tetraoleoyl-CL (TOCL) and synthetic tetramyristoyl-CL (TMCL). The principal fatty acid component of the heart CL is linoleic acid having two *cis* double bonds [9,10]. TOCL is a mono-unsaturated CL and TMCL is a saturated CL. The latter two CLs were chosen as CLs found in Barth syndrome patient contain saturated and monounsaturated fatty acids.

2. Experimental

2.1. *Materials*

All lipid powder samples used in this study, natural cardiolipin extracted from bovine heart tissue (Heart CL), synthetic 1',3'-bis[1,2-dioleoyl-*sn*-glycero-3-phospho]-*sn*-glycerol (tetraoleoyl-cardiolipin, TOCL) and 1',3'-bis[1,2-dimyristoyl-*sn*-glycero-3-phospho]-*sn*-glycerol (tetramyristoyl-cardiolipin, TMCL) were purchased from Avanti Polar Lipids, Inc. (Alabaster, AL). All lipids were used without further purification.

2.2. *X-ray scattering measurements*

Lipid film was formed from a chloroform solution of lipids under a stream of nitrogen gas, dried in high vacuum overnight, and then hydrated and vortexed with buffers. The buffer solution (pH 7.4) was 20 mM 4-(2-hydroxyethyl)-1-piperizineethanesulfonic acid (HEPES) buffer with 100 mM NaCl and 10 mM ethylenediaminetetraacetic acid (EDTA). Because it has been reported that the phase behaviour of CL is sensitive to the presence of Ca^{2+} ions [16], EDTA was added to the buffer to remove free Ca^{2+} ions. The lipid concentration of the samples was 50 mM. This corresponds to about 1,000 water molecules per a lipid. To the best of our knowledge, there is no paper reporting the numbers of bound water or interlamellar water molecules for CLs. Judging from the literature data of other phospholipids, for example, about 30 waters per an egg phosphatidylcholine [17], it can be assumed that 50 mM CL solution contains excess water. After mixing about 5 minutes by vortex mixer at room temperature, the sample solutions were visually clear and homogenous.

X-ray scattering measurements were carried out at RIKEN Structural Biology Beamline I (BL45XU) at SPring-8. The optical system of the beamline has been described elsewhere

[18]. X-ray scattering data were collected with a RAXIS-IV imaging-plate area detector (Rigaku). The sample-to-detector distance was about 550 mm. The digital two-dimensional intensity data were transformed into one-dimensional intensity data as a function of $q = 4\pi\sin\theta/\lambda$, where 2θ is the scattering angle and λ is the wavelength of X-ray (0.090 nm), by circular averaging using a FIT2D software developed by Dr. A. Hammersley (http://www.esrf.fr/computing/scientific/FIT2D/). The scattering angle (2θ) was calibrated with silver behenate powder crystals [19]. Samples were measured in a sample cell with a path length of 1.5 mm and a pair of thin quartz windows (thickness of 20 μm each).

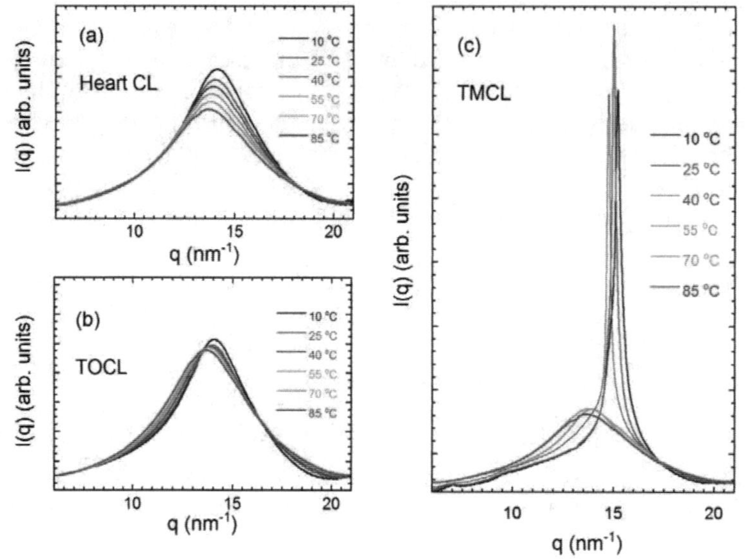

Figure 1. Temperature dependence of wide-angle X-ray scattering profiles for (a) Heart CL, (b) TOCL, and (c) TMCL.

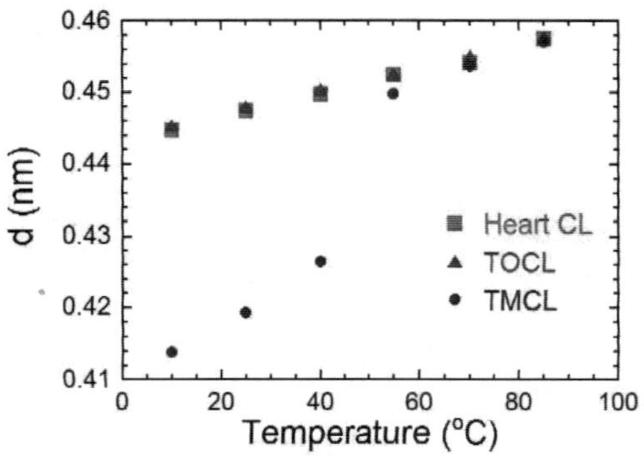

Figure 2. Wide-angle spacings (d) plotted versus temperature. (■) Heart CL, (▲) TOCL, and (•) TMCL.

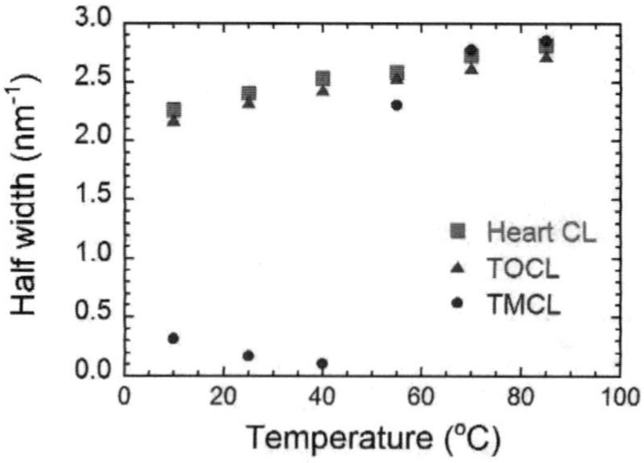

Figure 3. Half width at half maximum of the wide-angle bands plotted versus temperature. (■) Heart CL, (▲) TOCL, and (•) TMCL.

Sample solutions were transferred to the sample cell by syringe through the hole with the diameter of 1 mm of the cell. It is known that dehydration affects the phase behaviour of CLs [20]. To prevent evaporation of water through the hole during measurements (the samples were heated up to 85 °C), the hole was tightly covered by using Parafilm®. The scattering data were not used when any volume changes of the sample solution or any bubbles were detected by visual inspection after measurements. The sample cell was attached to a sample folder. The cell was in good thermal contact with the folder whose temperature was controlled to ±0.01 °C with a high-precision thermoelectric device. The exposure time was 60 sec.

3. Results

3.1. *Wide-angle X-ray scattering*

Figure 1 shows X-ray scattering profiles of the wide-angle region for three different CLs, Heart CL, TOCL, and TMCL. The scattering from the hydrocarbon chain packing appears in this region. Figure 2 presents the *d*-spacings calculated from the maximum positions. Figure 3 displays the values of half width at half-maximum determined by fitting Lorentz shape function with liner background.

Except for TMCL at low temperatures (below 40 °C, Fig. 1(c)), only broad band centred at about $q = 14\ \mathrm{nm}^{-1}$ ($d = 0.45$ nm) was observed, indicating that the hydrocarbon chains are melted, i.e., they are in a liquid crystalline phase. The sharp peaks observed for TMCL showed that TMCL is in a gel phase at low temperatures. By using differential scanning calorimetry, the transition temperature was measured for TMCL dispersed into the same HEPES buffer used in the present X-ray study (data not shown). The temperature was 41.5 °C, agreeing with the disappearance of the sharp diffraction peak above 40 °C.

Both *d*-spacings and the values of half width increased with rising of temperature for all three CLs in the liquid crystalline phase. This indicates that the increasing temperature induces conformational disorder of the hydrocarbon chains. At the same temperature, the *d*-spacings of TMCL in the liquid crystalline phase are almost the same as those of other CLs, however, the half width of Heart CL was slightly wider than that of TOCL. The latter result suggests that the increase of number of *cis* double bonds decreases the structural regularity in the melted chains.

3.2. *Small-angle X-ray scattering*

Small-angle X-ray scattering profiles obtained at different temperatures are shown in Fig. 4. In the figure, $q^2I(q)$ is plotted as a function of q, where $I(q)$ is one-dimensional intensity data transformed from two-dimensional data recorded on imaging plate by circular averaging, i.e. it corresponds to the data obtained by using a one-dimensional detector, and q^2 corresponds to the Lorentz-polarization (LP) factor for unoriented samples. Therefore, $q^2I(q)$ is proportional to the square of the Fourier transform of one-dimensional electron density profile of the sample, i.e., form factor [21,22].

In the electron density profiles of phospholipid bilayers, the polar head group regions have the highest values. Based upon model calculations, Engelman [23] have shown that, if the highest peak-to-peak distance is kept constant, the modulation of model electron density affects mainly the shape of the first band of the form factor, however, hardly affects the peak position of second band. Hence, the position of second maximum in the $q^2I(q)$ curve is a good measure of the average bilayer thickness. Figure 5 represents the bilayer thickness values estimated by this method. For the liquid crystalline phase, the decreases of bilayer thickness (D_B) were not linear but exponential. Table 1 shows the parameters estimated by fitting the bilayer thickness with an equation of $D_B = D_0\exp(E/k_BT)$, where D_0 and E are parameters, k_B is the Boltzmann constant, and T is absolute temperature. Although the carbon number of the chains is the same 18, the bilayer thickness of Heart CL is thinner than that of TOCL. This indicates that the existence of *cis* double bond in the hydrocarbon chains greatly reduces the bilayer thickness.

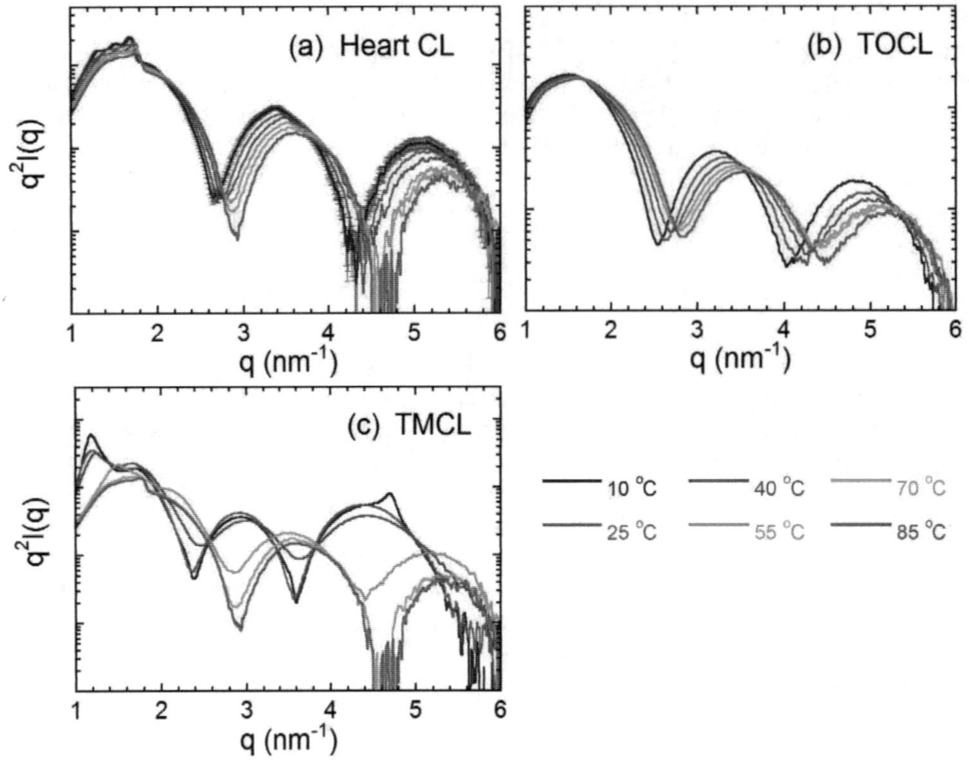

Figure 4. Temperature dependence of corrected small-angle X-ray scattering intensity curves ($q^2I(q)$) for (a) Heart CL, (b) TOCL, and (c) TMCL. Error bars (the square root of counts) are shown for the curve of Heart CL at 10°C. Because the other curves have almost the same error values, these error bars are omitted for clarity.

For TMCL at 10 °C, two diffraction peaks were observed at q = 1.176 nm^{-1} and 4.703 nm^{-1}. These are the first and fourth lamellar diffraction peaks, respectively, corresponding to the lamellar spacing of 5.38 nm. This value is fairly close to that reported for TMCL dispersed into sodium phosphate buffer at 2 °C (5.42 nm) [24]. Above 25 °C, the fourth peak disappeared and the lamellar spacing shifted toward to shorter direction with increasing temperature (D = 5.34 nm at 25 °C, D = 5.13 nm at 40 °C).

For Heart CL, some broad diffraction peaks with asymmetric shape appeared in the range of q = 1.2 to 1.8 nm^{-1}. Although the peak positions were not reproducible perfectly, the broad peaks in this region appeared in all our experiments (see Appendix). At present, the assignment of these diffraction peaks remains unresolved.

4. Discussion

From biological viewpoints, the most important result of this study is that the bilayer thickness of Heart CL is thinner than that of TOCL. In other words, the reduction of the number of double bonds induces to thicken the cardiolipin membranes. It has been pointed out that bilayer thickness has a large effect on the activities of many membrane proteins [12-15]. i.e., there is an optimal thickness for protein's activity. In the CLs extracted from normal heart muscle, the main components are tetralinoleoyl-CL containing four linoleic acid (18:2), and trilinoleoyl-oleoyl-CL containing three linoleic acid (18:2) and one oleic acid (18:1) [9]. On the other hand, in the CLs extracted from the heart muscle of Barth syndrome patients, dominant components are CLs containing palmitic acid (16:0), stearic acid (18:0), oleic acid (18:1), and so on. [9,10]. Based upon these data and the result of this study, it is likely that the thickness of inner mitochondrial membrane of Barth syndrome patients is thicker than that of normal. This would be the origin for dysfunction of membrane protein(s) in mitochondria in Barth syndrome.

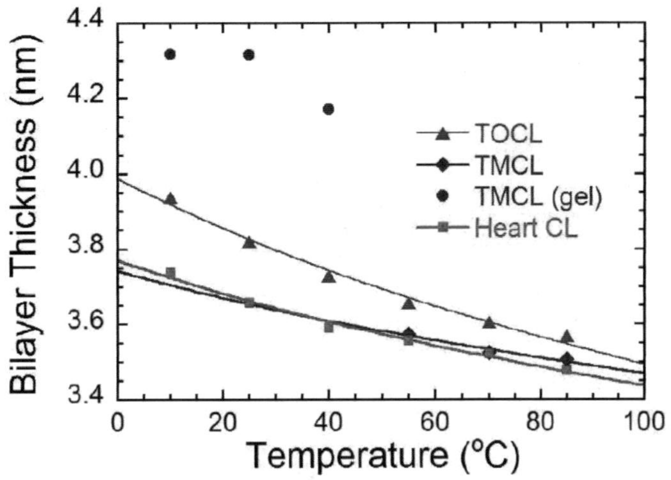

Figure 5. Estimated values of bilayer thickness from maximum positions of the second band of $q^2I(q)$ curves plotted versus temperature. (■) Heart CL, (▲) TOCL, and (●) TMCL.

Table 1. Fitting parameters of the bilayer thickness (see text)

	Heart CL	TOCL	TMCL
D_0 (nm)	2.67	2.43	2.82
E (J)	6.84×10^{-22}	9.77×10^{-22}	5.60×10^{-22}

In this study, the bilayer thickness was estimated according to a simple method proposed by Engelmen [23]. Here we discuss the reliability of this method. Recently, Lewis *et al.* [24] reported the phase behaviour and structure of TMCL. In their study, the electron density profiles of TMCL bilayers were reconstructed from X-ray diffraction data of the multilamellar vesicles by using Caillé [25] and paracystalline [26] theories for disorder of the lamellar stacking. From the reconstructed electron density profiles, the phosphate-phosphate distances across the bilayer (d_{pp}) were estimated to be 4.35 nm at 35 °C and 3.78 nm at 55 °C, respectively [24]. Our bilayer thickness values (4.17 nm at 40 °C, 3.57 nm at 55 °C) are slightly thinner than those of Lewis *et al.*, but the difference before and after the melting transition is in good agreement with each other (about 0.6 nm). Hence, the bilayer thickness estimation method used here is reliable for the relative comparison. However, careful treatments should be needed for the estimation of absolute values.

At room temperature, the difference of estimated bilayer thickness between TOCL and Heart CL was about 0.2 nm. The similar value has been reported for the difference between dioleoyl-phosphatidylcholine and dilinoleoyl-phosphatidylcholine [27]. Lewis and Engelman [28] found a linear increase of bilayer thickness with the number of carbon of acyl chain for saturated and monounsaturated phosphatidylcholines. The degree of increase was 0.175 nm per carbon atom. If the same increase is assumed for CLs, the bilayer thickness of CL having four 16-carbon saturated fatty acid (palmitic acid) chains that appears in Barth syndrome patients, is about 0.3 nm thicker than Heart CL based on the results of TMCL. As shown by the width of scattering band in wide-angle region, the two *cis* double bonds in one hydrocarbon chain introduce more randomness in the hydrophobic core in the bilayers by comparison with only one double bond. The randomness leads expansion of the occupied area per lipid [29]. Thereby, the reduction of bilayer thickness takes place, although the same numbers of carbon atoms are contained in the chains.

Lewis *et al.* [24] studied the effects of different buffer solution on TMCL's polymorphic phase behaviour. They have reported that the melting transition temperatures for TMCL dispersed into Tris and phosphate buffers are 38.9 °C and 41.2 °C, respectively. They have also shown the existence of a subgel phase with an orthorhombic subcell that transforms into the gel phase at about 25 °C on heating. In this study, HEPES buffer was used and the melting transition temperature was 41.5 °C. A broad peak was detected at around 28 °C in our differential scanning calorimetric measurements (data not shown). This peak may correspond to a subgel-to-gel phase transition. However, an orthorhombic subcell was not clearly observed in the present X-ray study. The peak shape of TMCL at 10 °C was slightly asymmetric, i.e., a small broad shoulder peak appeared at the small-angle side. In addition, the half width of the TMCL's sharp diffraction peak at low temperatures decreased with increasing temperature (Fig.3). These results might be related to the formation of the subgel phase. However the subgel phase is not a main issue of the present study and we leave it to future work.

Another problem that requires further detailed future study is the assignment of diffraction peaks appeared in the small-angle region of heart CL. The three peaks were asymmetric and have some shoulders. These peaks should be resolved into more than three by high resolution measurements.

Our study predicts that the mitochondrial membrane from Barth syndrome patients is thicker than that of normal subject. Future biophysical and cell biological analysis of CL in both model and biomembranes will clarify the molecular significance of CL in Barth syndrome.

Acknowledgements
This work was supported by Barth Syndrome Grant 2007-2008 from the Barth Syndrome Foundation and Lipid Dynamics Project of RIKEN. The synchrotron radiation experiments were performed at BL45XU in SPring-8 with the approval of RIKEN (Proposal No. 20080023 and 20090067).

[1] Haines T H 2009 *Biochim. Biophys. Acta* **1788** 1997
[2] Pangborn M C 1947 *J. Biol. Chem.* **168** 351
[3] Hostetler K Y and van Den Bosch H 1972 *Biochim. Bipohys. Acta* **260** 380
[4] Robinson N C, Zborowski J and Talbert L H 1990 *Biochemistry* **29** 8962
[5] Zhang M, Mileykovskaya, E and Dowhan W 2002 J. *Biol. Chem.* **277** 43553
[6] Pfeiffer K, Gohil V, Stuart R A, Hunte C, Brandt U, Greeberg M L and Schägger H 2003 *J. Biol. Chem.* **278** 52873
[7] Barth P G, Scholte H R, Berden J A, van der Klei-van Moorsel JM, Luyt-Howen I E M, van't Veer-Korthof E T, van der Harten J J and Sobotka-Plojhar M A 1983 *J. Neurol. Sci.* **62** 327
[8] Schlame M and Ren M 2006 *FEBS Lett.* **580** 5450
[9] Schlame M, Towbin J A, Heerdt P M, Jehle R, DiMauro S, Blanck T J J 2002 *Ann. Neurol.* **51** 634
[10] Schlame M, Ren M, Xu Y, Greenberg M L and Haller I 2005 *Chem. Phys. Lipids* **138** 38
[11] Mouritsen O G 2005 *Life as a Matter of Fat: The Emerging Science of Lipidomics* (Berlin: Springer)
[12] Heimburg T 2007 *Thermal Biophysics of Membranes* (Weinheim: Wiley-VCH)
[13] Mouritsen O G and Bloom M 1984 *Biocphys. J.* **46** 141

[14] Lee A G 1998 *Biochim. Biophys. Acta* **1376** 381
[15] Cornelius F 2001 *Biochemistry* **40** 8842
[16] Rand R P, Sengupta S 1972 *Biochim. Biophys. Acta* **255** 484
[17] Nagle J F, Tristram-Nagle S, 2000 *Biochim. Biophys. Acta* **1469** 159
[18] Fujisawa T, Inoue K, Oka T, Iwamoto H, Uraga T, Kumasaka T, Inoko Y, Yagi N, Yamamoto M and Ueki T 2000 *J. Appl. Cryst.* **33** 797
[19] Huang T C, Toraya H, Blanton T N and Wu Y 1993 *J. Appl. Cryst.* **26** 180
[20] Alessandrini A, Muscatello U 2009 *J.Phys. Chem. B* **113** 3437
[21] Glatter O and Kratky O 1982 *Small Angle X-ray Scattering* (New York: Academic Press)
[22] Takahashi H and Ito K 2007 *J. Phys.Conference Ser.* 83 012022
[23] Engelman D M 1971 *J. Mol. Biol.* **58** 153
[24] Lewis R N A H, Zweytick D, Pabst G, Lohner K and McElhaney R N 2007 *Biophys. J.* **92** 3166
[25] Zhang RT, Suter R M and Nagle J F 1994 *Phys. Rev. E.* **50** 5047
[26] Guinier A 1963 *X-ray diffraction in crystals, imperfect crystals, and amorphous bodies* (San Francisco : W. H. Freeman)
[27] Rawick W, Olbrich K C, McIntosh T, Needham D and Evans E 2000 *Biophys. J.* **79** 328
[28] Lewis B A and Engelman D M 1983 *J. Mol. Biol.* **166** 211
[29] Hub J S, Salditt T, Rheinstädter M C and de Groot B L 2007 *Biophys. J.* **93** 3156

Appendix

Figure 6 represents small-angle X-ray scattering intensity curves of heart CL at 25 °C. The two different curves were obtained from two different samples prepared under the same condition. The arrows indicate the corresponding position of the beamstop edge. Except for the peak at q = 0.8 nm^{-1} and broad peak(s) in q = 1.2 to 1.8 nm^{-1}, the two curves are almost identical to each other. Judging from the position of the beamstop, the peaks (s) in q = 1.2 to 1.8 nm^{-1} would not be due to the scattering from the beamstop. At present, the origin of these peaks cannot be assigned; at least, the existence of these peaks does not affect the estimation of the bilayer thicknesses.

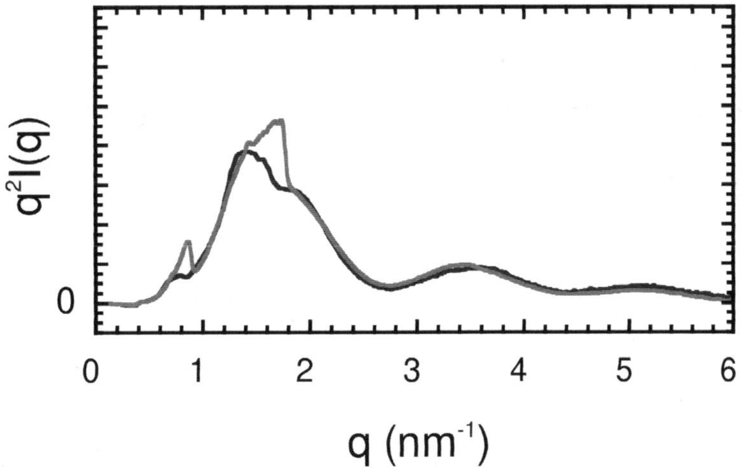

Figure 6. Reproducibility of small-angle X-ray scattering intensity curve of heart CL. These curves were recorded at 25 °C. The arrow indicates the position of the beamstop edge. Differing from Fig.4, the vertical axis is linear.

Spinodal Decomposition in Ni-Nb-Y Metallic Glasses Analyzed by Quantitative Anomalous Small-Angle X-ray Scattering

G.Goerigk[1] and N.Mattern[2]

[1]Institut für Festkörperforschung, Forschungszentrum Jülich, Postfach 1913, D-52425 Jülich, Federal Republic of Germany, [2]Leibniz-Institute IFW Dresden, Institute for Complex Materials, P.O. Box 270116, D-01171 Dresden, Federal Republic of Germany

G.Goerigk@fz-juelich.de

Abstract. Phase separated Ni-Nb-Y metallic glasses were prepared by means of rapid quenching from the melt. Quantitative Anomalous Small-Angle X-ray Scattering experiments were performed at the K-absorption edges of Nickel, Niobium and Yttrium. The paper outlines, that the separation of the pure-resonant scattering contribution is obtained by employing the Gaussian algorithm on a vector equation. Moreover the quantitative analysis of the Resonant Invariant at the K-absorption edges of Niobium and Yttrium provided the chemical concentrations of Yttrium and Niobium in the random density fluctuations, which are attributed to spinodal decomposition. The results are compared to a partially crystallized sample annealed at 773 K over 30 minutes.

1. Introduction

In the last 3 decades the Synchrotron Radiation (SR) provided the small-angle X-ray scattering technique with major improvements. Among other improvements SR provides a photon flux, which is many orders higher compared to classical X-ray sources giving access to systems with only weak SAXS contributions like highly diluted chemical solutions or solid state systems like glasses or amorphous alloys, which are to a high degree homogeneous. Additionally the continuous energy spectrum of SR allows energy tunability in the vicinity of the K- and L_{III}–absorption edges of most of the elements. This technique - known as Anomalous Small-Angle X-ray Scattering (*ASAXS*) – is based on the anomalous variations of the atomic scattering factors near the absorption edges and allows the element-specific structural and quantitative characterization of the sample under investigation. After the exploratory studies of the 1980s [1-3] *ASAXS* became in the last decade a precise quantitative technique (*q-ASAXS*), which combines the structural analysis with the quantitative analysis of chemical concentrations (i.e. fluctuations) of the different atomic species of a multi-component system by addressing the elements via the X-ray absorption edges [4-10].

The ternary Ni-Nb-Y system exhibits an extended miscibility gap in the liquid [11]. By means of rapid quenching technique the decomposed melt can be frozen in to a phase separated metallic glasses [12]. The critical temperature of liquid-liquid decomposition depends on the composition. For Ni content < 60 at% a hierarchical heterogeneous microstructure is obtained with size distribution from

10 nm up to 500 nm which can be analyzed by transmission electron microscopy (TEM). Such microstructures represent a coarsened state of spinodal decomposition, growth of the melts, and secondary decomposition within the liquids. For such phase separeated Ni-Nb-Y glasses SAXS curves with fractal q-dependence have been observed [13]. For alloys with Ni > 60 at% early stages of phase separation can be obtained due to reduced critical temperature having fluctuation in nm dimensions with almost no contrast in TEM images. By means of *ASAXS* quantitative parameters of the fluctuation can be determined.

In this presentation, the application of the Gauss algorithm to the vector equation established by *ASAXS* measurements at three X-ray energies is demonstrated with the aim to separate the pure-resonant scattering contribution. Moreover the question shall be addressed, how to determine the amount of Yttrium respective Niobium atoms localized in the periodic fluctuations of Ni-Nb-Y metallic glasses undergoing spinodal decomposition with respect to the total amount of Y- respective Nb-atoms in the alloy, by quantitative analysis of what was introduced as the Resonant Invariant in a former publication [7].

2. ASAXS measurements

The remarkable possibilities of the *ASAXS* technique are based on the energy dependence of the atomic scattering factors giving selective access to the specific SAXS contributions of nanophases, which are built up by different chemical constituents in composites like for instance metallic glasses. In general the atomic scattering factors are complex quantities and energy dependent:

$$f_Z(E) = f_{0,Z} + f'_Z(E) + if''_Z(E) \tag{1}$$

where Z represents the atomic number. In this presentation we will deal with the atomic scattering factors of Ni, Y and Nb: $f_Z =: f_{Ni}(E), f_Y(E), f_{Nb}(E)$. When performing a SAXS experiment at energies in the vicinity of the absorption edge of one of the elements, three different scattering contributions can be distinguished as a consequence of Eq.(1)[1]:

$$I(q,E) = \Delta f_A^2 |A(q)|^2 + 2\Delta f_A \operatorname{Re} A(q)[\Delta f_{0,Z} + f'_Z(E)]\operatorname{Re} B(q) + [(\Delta f_{0,Z} + f'_Z(E))^2 + f''_Z(E)^2]|B(q)|^2 \tag{2}$$

$A(q)$ represents the scattering amplitude of the non-resonant scattering atoms and $B(q)$ represents the scattering amplitude of the resonant scattering atoms in the energy range of the absorption edge. The quantity $\Delta f_{0,Z} = f_{0,Z} - \rho_m V_Z$ is the electron density difference between the atom with the volume V_Z and the surrounding matrix of the alloy with the electron density ρ_m. q is the magnitude of the scattering vector $[= (4\pi/\lambda)\sin\Theta]$, where 2Θ is the scattering angle and λ the X-ray wavelength. For instance in the case of the Ni-Nb-Y alloys $[\Delta f_{0,Y} + f'_Y(E)]\operatorname{Re} B(q)$ represents the scattering amplitude (real part) of the inhomogeneously distributed Y-atoms in the alloy, when performing SAXS measurements in the energy range of the K-absorption edge of Yttrium, while $\Delta f_A \operatorname{Re} A(q)$ represents the scattering amplitude (real part) of the non-resonant scattering atoms i.e. Ni and Nb, with the contrast, Δf_A, of the non-resonant scattering atoms with respect to the entire matrix.

When measuring the scattering curves at three energies in the vicinity of the absorption edge of the atoms with atomic number Z the following vector equation can be established:

$$M_{ij}(E_i) \otimes A_j(q) = I_i(q, E_i) \tag{3a}$$

$$\begin{pmatrix} \Delta f_A^2 & 2\Delta f_A[\Delta f_{0,Z} + f'_Z(E_1)] & [(\Delta f_{0,Z} + f'_Z(E_1))^2 + f''_Z(E_1)^2] \\ \Delta f_A^2 & 2\Delta f_A[\Delta f_{0,Z} + f'_Z(E_2)] & [(\Delta f_{0,Z} + f'_Z(E_2))^2 + f''_Z(E_2)^2] \\ \Delta f_A^2 & 2\Delta f_A[\Delta f_{0,Z} + f'_Z(E_3)] & [(\Delta f_{0,Z} + f'_Z(E_3))^2 + f''_Z(E_3)^2] \end{pmatrix} \begin{pmatrix} |A(q)|^2 \\ \mathrm{Re}(A(q)B(q)) \\ |B(q)|^2 \end{pmatrix} = \begin{pmatrix} I(q,E_1) \\ I(q,E_2) \\ I(q,E_3) \end{pmatrix} \quad (3b)$$

The vector on the right side represents the scattering curves measured at the three different energies E_1, E_2, E_3 while the vector on the left side represents the squared, respectively multiplied, scattering amplitudes. The matrix, $M_{ij}(E_i)$, contains the atomic scattering factors, which are energy dependent in case of the resonant scattering atoms. Combining the matrix with the vector on the right side of the equal sign yields the so-called extended matrix:

$$\begin{array}{ccc|c} \Delta f_A^2 & 2\Delta f_A[\Delta f_{0,Z} + f'_Z(E_1)] & [(\Delta f_{0,Z} + f'_Z(E_1))^2 + f''_Z(E_1)^2] & I(q,E_1) \\ \Delta f_A^2 & 2\Delta f_A[\Delta f_{0,Z} + f'_Z(E_2)] & [(\Delta f_{0,Z} + f'_Z(E_2))^2 + f''_Z(E_2)^2] & I(q,E_2) \\ \Delta f_A^2 & 2\Delta f_A[\Delta f_{0,Z} + f'_Z(E_3)] & [(\Delta f_{0,Z} + f'_Z(E_3))^2 + f''_Z(E_3)^2] & I(q,E_3) \end{array} \quad (4)$$

When applying the Gaussian algorithm (elimination procedure) to the extended matrix the so-called triangle matrix is obtained,

$$\begin{array}{ccc|c} a_{11} & a_{12} & a_{13} & b_1 \\ 0 & a_{22} & a_{23} & b_2 \\ 0 & 0 & a_{33} & b_3 \end{array} \quad (5)$$

where the $a_{i,j}, b_j$ are written in the matrix of Eq.(6):

$$\begin{array}{ccc|c} G & F(E_1) & H(E_1) & I(q,E_1) \\ 0 & F(E_2)-F(E_1) & H(E_2)-H(E_1) & I(q,E_2)-I(q,E_1) \\ 0 & 0 & H(E_3)-H(E_1)-\dfrac{F(E_3)-F(E_1)}{F(E_2)-F(E_1)}(H(E_2)-H(E_1)) & I(q,E_3)-I(q,E_1)-\dfrac{F(E_3)-F(E_1)}{F(E_2)-F(E_1)}(I(q,E_2)-I(q,E_1)) \end{array} \quad (6)$$

with

$$\begin{aligned} G &= \Delta f_A^2 \\ F(E_i) &= 2\Delta f_A(\Delta f_{0,Z} + f'_Z(E_i)) \\ H(E_i) &= (\Delta f_{0,Z} + f'(E_i))^2 + f''(E_i)^2 \end{aligned} \quad (7)$$

When introducing the formula of Eq.(7) into the matrix of Eqs.(5-6) the squared Fourier transform (FT) of the form factor, $|B(q)|^2$, of the spatial distribution of the atoms with atomic number Z can be calculated:

$$|B(q)|^2 = \left[\frac{I(q,E_1)-I(q,E_2)}{f'_Z(E_1)-f'_Z(E_2)} - \frac{I(q,E_1)-I(q,E_3)}{f'_Z(E_1)-f'_Z(E_3)}\right]\cdot\frac{1}{F(E_1,E_2,E_3)}$$

$$F(E_1,E_2,E_3) = f'_Z(E_2)-f'_Z(E_3) + \frac{f''^2_Z(E_1)-f''^2_Z(E_2)}{f'_Z(E_1)-f'_Z(E_2)} - \frac{f''^2_Z(E_1)-f''^2_Z(E_3)}{f'_Z(E_1)-f'_Z(E_3)} \quad (8)$$

For this elimination procedure the matrix element M_{11} of the starting matrix - the so-called Pivotelement – must not be zero. After the first step the new 2x2 submatrix, $M_{i,j}$ with the new Pivotelement $M'_{22} = F(E_2) - F(E_1)$ is obtained. Again the new Pivotelement must be not zero or in other words the difference $M'_{22} = f'(E_2) - f'(E_1)$ obtained from Eq.(7) must be large enough to gain significant accuracy from the employed algorithm.

Due to Eq. (8) *ASAXS* provides a technique to access directly the scattering of the Z-atoms i.e. Ni or Y or Nb depending at which X-ray absorption edge the *ASAXS* measurements are performed. The structural information of the distribution of the Z-atoms (Ni or Y or Nb) in the alloy can be obtained from the analysis of $|B(q)|^2$, which we will call form factor for reason of convenience. More generally spoken, Eq. (8) provides a method, which gives access to the pure-resonant scattering contribution of the selected constituent of an alloy by measuring the small-angle scattering at only three suitable energies [5-10].

In addition to the structural information, which can be obtained from $|B(q)|^2$, important quantitative information related to the amount of inhomogeneously distributed Z-atoms can be deduced from the integral Q_Z:

$$Q_Z(E) = |\Delta f_Z(E)|^2 \int_Q |B(q)|^2 d^3q \tag{9}$$

In analogy to the so-called invariant [14], we will call $Q_Z(E)$ the Resonant Invariant of the inhomogeneously distributed resonant scattering Z-atoms. The invariant, $Q_Z(E)$, as defined in Eq. (9), is related to the number density of inhomogeneously distributed Z-atoms, $\overline{v_Z}$, as was outlined in detail in a previous publication [7,10]:

$$Q_Z(E) = |\Delta f_Z(E)|^2 \int_Q |B(q)|^2 d^3q = (2\pi)^3 r_0^2 |\Delta f_Z(E)|^2 \overline{v_Z}\left(\frac{1}{V_Z} - \overline{v_Z}\right) \tag{10a}$$

$$\frac{dQ_Z(E)}{dq} = |\Delta f_Z(E)|^2 4\pi |B(q)|^2 q^2 \tag{10b}$$

where $V_Z = 4\pi/3R_Z^3$ is the volume of the Z-atoms with atom radius R_Z and r_0 is the classical electron radius. Eq. (10b) represents the first derivative of the resonant invariant with respect to q. Finally from the quadratic Eq. (10a) two values for the number density of the Z-atoms can be calculated:

$$\overline{v_Z} - \overline{v_Z}^2 V_Z = \frac{V_Z \cdot Q_Z(E)}{(2\pi)^3 \cdot r_0^2 |\Delta f_Z(E)|^2} = \frac{V_Z}{(2\pi)^3 \cdot r_0^2} \int_Q |B(q)|^2 d^3q \tag{11a}$$

$$\overline{v_Z} = \frac{1}{2V_Z} \pm \sqrt{\frac{1}{4V_Z^2} - \frac{1}{(2\pi)^3 r_0^2} \int_Q |B(q)|^2 d^3q} \tag{11b}$$

In this presentation only one of the two solutions is significant i.e. the negative sign. It should be mentioned that the number density calculated from Eq.(11b) includes the Z-atoms, which are homogeneously distributed in the alloy. These Z-atoms provide an isotropic scattering contribution at small q-values similar to the isotropic scattering of the alloys matrix atoms but with a different amount. As will be discussed in more detail below, the amount of the isotropic scattering contribution originating from these remaining Z-atoms in the matrix can be neglected.

3. Experimental

Pre-alloyed ingots were prepared by arc-melting elemental Ni, Nb and Y with purities of 99.9% or higher in a Ti-gettered argon atmosphere. To ensure homogeneity, the samples were remelted several times. From these pre-alloys, thin ribbons (3 mm in width and 30 μm in thickness) with nominal compositions $Ni_{68}Nb_{16}Y_{16}$ were prepared by single-roller melt spinning under argon atmosphere. The casting temperature was 1923 K. The chemical compositions were determined by the titration technique. The resulting values were $Ni_{67.9}Nb_{17.3}Y_{14.8}$ for the as-prepared ribbons. These composition exhibit only slight deviations from the nominal value of up to ~1 at.%. For convenience, the nominal compositions are used in the following sections. Additionally, a partly crystallized sample, c-$Ni_{68}Nb_{16}Y_{16}$, was produced by heating part of the ribbons in a differential scanning calorimeter (DSC 7, Perkin Elmer) just above the temperature of the first exothermic crystallization event at 773 K over 30 min.

Anomalous small angle X-ray scattering (*ASAXS*) was measured with the JUSIFA beam line [15,16] at the DORIS storage ring at HASYLAB/DESY Hamburg. *ASAXS* sequences were performed for both samples $Ni_{68}Nb_{16}Y_{16}$ and c-$Ni_{68}Nb_{16}Y_{16}$ with three X-ray energies in the vicinity of the K-absorption edges of Nickel at 8333 eV and with four X-ray energies at the K-absorption edges of Yttrium at 17038.4 eV and Niobium at 18985.6 eV. The tables in Fig.1a-c provide the related anomalous dispersion corrections of Nickel, Yttrium and Niobium at the energies used in the experiment based on the calculations of Cromer and Liberman [17,18]. These were used to permit the separation of the Ni-, Y- and Nb-related pure-resonant scattering described in detail by Eqs. (1)-(8). The scattering intensity is calibrated into macroscopic scattering cross sections in units of cross section per unit volume $[cm^2/cm^3]=[cm^{-1}]$. Transmission measurements were performed with a precision of better 10^{-3} ($\Delta t/t=2.8\times10^{-4}$) using a special (windowless) photodiode (Hamamatsu S2387-1010N).

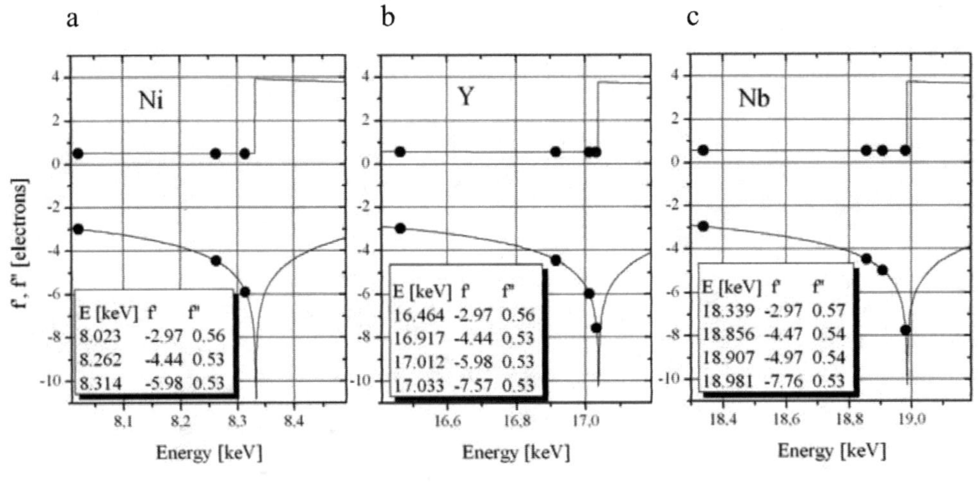

Figure 1: Anomalous dispersion corrections of Nickel, Niobium and Yttrium obtained by Cromer-Liberman calculations [17,18]. The dots represent the X-ray energies, where the *ASAXS* measurements were performed. The minimum distance of 5 eV to the K-absorption edges was kept, to omit as far as possible the Resonant Raman Scattering, which is for all components of the alloys significant, due to the high chemical concentrations.

Background measurements took 15 min followed by measurements of a calibration standard (glassy carbon, 5 min) and subsequent measurements of the sample frames (15 min). This measurement cycle was repeated for the four different energies. The measurements were performed at two sample-detector distances (935 and 3635 mm) covering a q-range between 0.005 and 0.6 $Å^{-1}$. After the completion of the *ASAXS* measurements at four energies, the complete cycle of four energies was repeated four times for accumulation of intensity. In total for the samples and the background measurements a beam time of 1h and for the related reference measurements a beam time of 20 min was accumulated at each energy and distance. This strategy was used in numerous publications emerging from the JUSIFA beam line in the last decade (see examples [5-8]) and compromises between the goal of accumulating sufficient scattering intensity (reducing the statistical error bars of the scattering curves) and reducing the influence of the significant beam instabilities of DORIS (2nd generation) as much as possible.

4. Results and Discussion

The SAXS curves, $d\sigma/d\Omega(q)$, of the two samples - $Ni_{68}Nb_{16}Y_{16}$, and c-$Ni_{68}Nb_{16}Y_{16}$ - measured at the energy E=8029 eV, 304 eV below the Ni-K edge are shown in Figure 2. The increasing intensity below $q < 0.2$ nm^{-1} probably originates from surface scattering also present in the Ni-Nb-Y data. Additionally, the presence of a maximum in the curves gives clear evidence for the existence of correlated fluctuations in electron density for the Ni-Nb-Y alloys. Variation of the energy at the Y- and Nb- K-edges changes the cross section of the SAXS curves in opposite directions indicating that the inhomogeneities are related to clusters enriched in Y or in Nb in agreement with atom probe tomography results [19]. The occurrence of the maximum in the SAXS curves is due to the high density of electron density fluctuations with a dominant correlation length. Using the relationship correlation length ζ and peak maximum in reciprocal distances $\zeta = 2\pi/q_{max}$ one obtains ζ=12.2 nm.

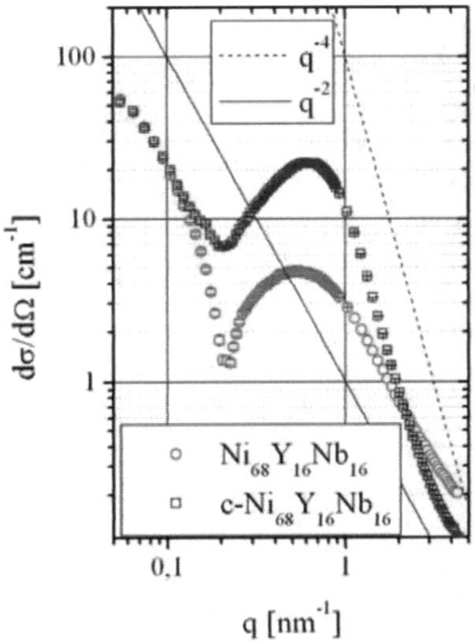

Figure 2: Total scattering curves of the alloy $Ni_{68}Nb_{16}Y_{16}$ with different heat-treatment: as-quenched (red circles) and heat-treated above the first crystallization point at 773 K but below the second crystallization point (blue squares). Above the first crystallization point the asymptotic behaviour of the scattering curves changes from q^{-2} to q^{-4} indicating smooth interfaces probably between the small crystallites and a surrounding amorphous phase.

The influence of partial crystallization on the SAXS curves is shown by the comparison of the scattering curves of the alloy $Ni_{68}Nb_{16}Y_{16}$ far below the K-absorption edge at about 8 keV. ($Ni_{68}Nb_{16}Y_{16}$ as quenched, c-$Ni_{68}Nb_{16}Y_{16}$: heat treated at 773 K i.e. beyond the 1st crystallization event). The heat treated sample consists of a nanocystalline Ni_2Y phase and a remaining amorphous Ni-Nb phase. The mean crystallite size is about 5 nm as can be seen from the TEM image in Figure 3b. The lattice parameter of the Ni_2Y phase indicates that to some extend Nb is still solved. No inhomogeneous structures could be resolved from the TEM image of the as-cast sample (Figure 3a). In the SAXS curves a dramatic change takes place for the sample, which was heat treated beyond the 1st crystallization temperature. Though the correlation maximum remains unchanged on the q-axis, it is shifted on the y-axis by nearly an order of magnitude probably due to a higher electron density contrast caused by a larger concentration gradient. Additionally the asymptotic behaviour changes from q^{-2} to q^{-4} indicating the formation of small crystallites with smooth interfaces to the surrounding phase. Obviously the fluctuation length of the as-quenched state determines the size of the nanocrystals. The phase separation of the glass is also the reason for different crystallization steps of the two phases (the amorphous Ni-Nb phase crytallizes into $NbNi_3$ at higher a temperature of about 900K). The *ASAXS* sequences of the alloys at three selected X-ray energies in the vicinity of the K-absorption edge of Ni show systematic but very small declination of the cross section in the range between 0.2 and 2 nm^{-1} indicating small differences of Ni-amount in the concentration fluctuations, which constitute the correlation maxima of the alloy. A closer analysis of the resonant invariant obtained from the *ASAXS* sequences near the Ni-K-edge could not provide reliable numbers for a Ni-gradient (concentration), indicating that the difference of the Ni-concentration in the fluctuations is too small i.e. below the resolution of this experiment. Anyway, because the Nickel component serves as the matrix, the important quantitative parameters of decomposition are related to the concentration fluctuations of Yttrium and Niobium and thus can be only obtained from quantitative *ASAXS* (*q-ASAXS*) measurements at the K-absorption edges of Yttrium and Niobium.

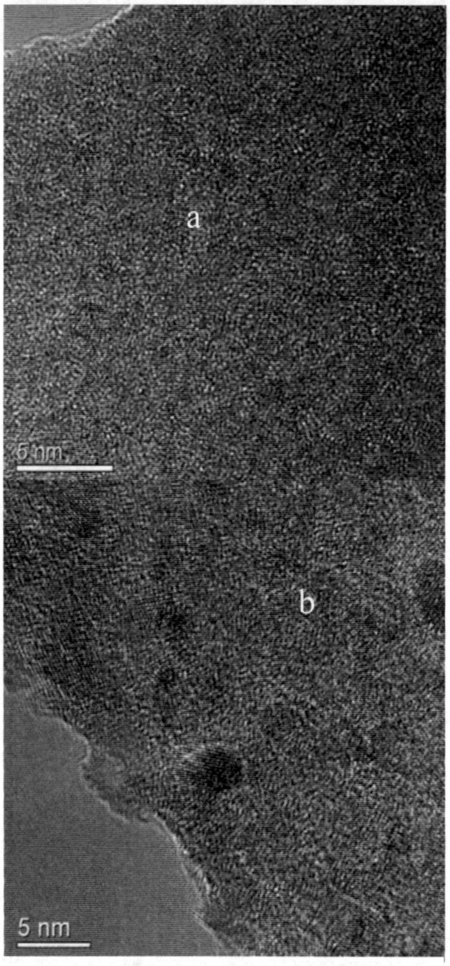

Figure 3: TEM pictures of the alloy $Ni_{68}Nb_{16}Y_{16}$ as-quenched (a) and partially crystallized (b).

Detailed quantitative information was obtained from the measurements at the K-absorption edges of Yttrium and Niobium. Figure 4a-b summarizes the results obtained from the Yttrium K-edge (17038 eV) for the two samples $Ni_{68}Nb_{16}Y_{16}$ and c-$Ni_{68}Nb_{16}Y_{16}$. The black squares in Figure 4 represent the total scattering of the as-quenched sample (a) and the heat-treated sample (b). The correlation maximum appears for both samples at about 0.6 nm^{-1} as known already from the *ASAXS* measurements at the Ni-K-edge. The total scattering curves are superimposed by a strong isotropic scattering contribution, which originates mainly from the fluorescence of the Ni-component (68at%). By calculating the separated scattering curves, $I(q,E_1)-I(q,E_3)$, this isotropic fluorescence contribution is removed revealing a q^{-4} behaviour in the asymptotic part of the separated scattering curve (blue triangles) of the heat-treated (crystallized) sample, which is already known from the measurements at the Ni-K-edge. As can be seen from the error bars of the separated scattering curve of the as-quenched sample (blue triangles) the asymptotic behaviour above 1 nm^{-1} is not clear i.e. beyond the resolution and needs a closer consideration by analyzing the pure-resonant scattering contribution with larger q-bins (see below). The solid lines in Figure 4a and 4b represent the q^{-2} and q^{-4} behaviour respectively.

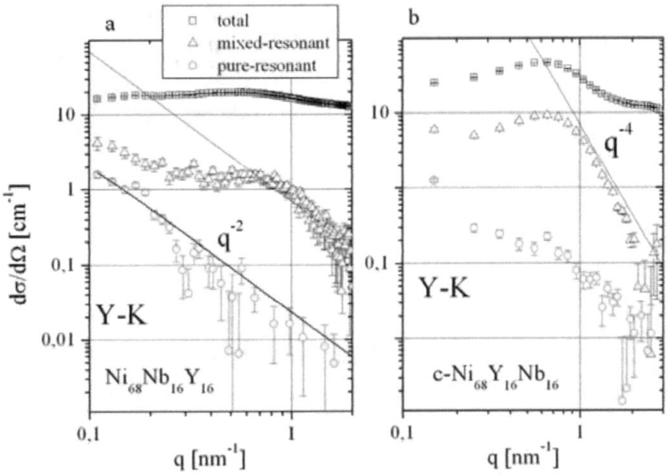

Figure 4: Total scattering curves (squares), separated scattering curves (triangles) and the form factors of the pure-resonant scattering contribution (circles) of the Y-related concentration fluctuations. Figures 6a-b represent the different heat-treatment: as-quenched (a) and partially crystallized (b). Above the first crystallization point the asymptotic behaviour of the scattering curves changes from q^{-2} to q^{-4} indicating smooth interfaces probably between the small crystallites and a surrounding amorphous phase. The solid line passing through the red circles of the pure-resonant scattering in Figure 6a represents a model function (see text).

From Eq. 8, the pure-resonant scattering (or more precisely the form factor of the pure-resonant scattering) was calculated for both samples. Because the error bars of the pure-resonant scattering become rather large at higher q-values, especially for the as-quenched sample, the number of q-bins was reduced in comparison to the mixed-resonant and the total scattering curves, with the aim to obtain smaller error bars at q-values larger 0.5 nm^{-1}. The procedure was as follows: the cross-section of the pure-resonant form factor was averaged over neighbouring q-bins. The result gives the cross-section at the centre of the averaged q-interval. The error of the averaged cross-section is the averaged squared error calculated from the errors of the single q-bins. Though the error bars are still large at higher q-values, the pure-resonant scattering for both samples was clearly resolved. Especially at higher q-values of about 1 nm^{-1} the pure-resonant scattering contribution of the crystallized sample is nearly an order of magnitude higher compared to the as-quenched sample, indicating a larger amount of Yttrium atoms localized in the crystalline phase. Ni_2Y has more Y than the average alloy, even when some Nb is solved

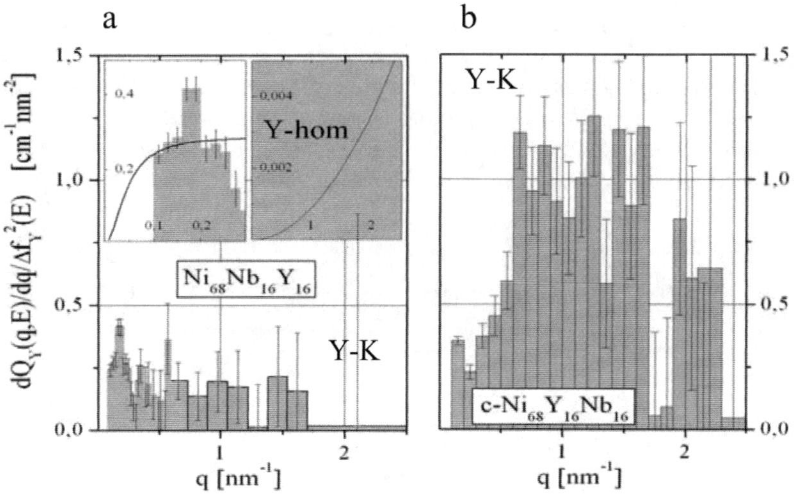

Figure 5: The first derivative of the Resonant Invariant (Eq.10b) of $Ni_{68}Nb_{16}Y_{16}$ as-quenched (a) and partially crystallized (b) obtained from *ASAXS* sequences at the Y-K-edge (17038 eV). The integrals (grey area) represent the Resonant Invariant (Eq.10a) from which the Y-concentrations can be calculated. The right inset represents the upper limit of 16at% homogeneously distributed Y, contributing to the Resonant Invariant. The left inset explains the contribution, which is missing due to the q-resolution. For reason of convenience the division by 4π is not written at the vertical axis.

The latter was analyzed more precisely by employing the Resonant Invariant of Eq. 10. Figure 5 shows the first derivative of the Resonant Invariant (Eq. 10b) of the as-quenched sample (a) and the crystallized sample (b). The grey areas represent the integrals, Q_Z (i.e. the Resonant Invariant), which are introduced into Eq. 11 for the calculation of the amount of Yttrium atoms, $\overline{\nu_Y}$, which enter the concentration fluctuations of the as-quenched sample and the crystallites of the heat-treated sample respectively. From Eq. 11 an amount of $2.13(28)\cdot 10^{21} cm^{-3}$ Yttrium (excess) atoms, which enter the crystallites, was calculated corresponding to about 19% of the Y-atoms in the alloy.

A completely different result was obtained from the analysis of the Resonant Invariant of the as-quenched sample. As can be seen already from a simple comparison of the integrals of both samples in Figure 5, the amount of Yttrium atoms, which enter the concentration fluctuations must be much more smaller. From Eq. 11 a concentration of $0.39(20)\cdot 10^{21} cm^{-3}$ Yttrium atoms was deduced, which corresponds to an amount of only 4% Yttrium atoms.

Though this result is significant, it is near the resolution limits of the experiment as can be seen from the error bars in Figure 5a. For a more quantitative discussion of the result, the limited integration area has to be extended onto the whole regime $0 \le q < \infty$. Because no comprehensive description of the scattering curves of a ternary alloy undergoing spinodal decomposition is at hand, a model function with a q^{-2}-behaviour (Ornstein-Zernike originally for binary systems, Eq. 12) was employed. The model function serves only as a guide for the eye with the aim to estimate the amount of the 'missing' scattering contributions at higher and lower q-values i.e. beyond the resolution of the experiment.

$$i(q) = \frac{const}{c + q^2} \tag{12}$$

At q=0 the model function reaches a finite value defined by thermodynamic parameters of the alloy (Eq.12) and the first derivative of the Resonant Invariant approaches 0 with $q \to 0$. The amount of missing contribution to the Resonant Invariant can be neglected as can be seen from the left inset in Figure 5a, which shows a magnified part of the Resonant Invariant in the low q-range. The missing part below the resolution at q-values smaller q_{min} is the area between the grey histogram and the solid line of the model function.

The right inset in Figure 5a represents the contribution of homogeneously distributed Y-atoms with a concentration of 16at% to the Resonant Invariant. Although this corresponds to the maximum possible amount – a considerable portion of the Y-atoms are localized in the fluctuations – the contribution is many orders of magnitude smaller than the contribution detected from the Y-atoms in the concentration fluctuations and thus can be also neglected.

Considering the scattering contributions beyond the q-resolution at higher q-values the result must be interpreted as a lower limit as is demonstrated by of the model function (Figure 5a). At higher q-values a cut-off of the scattering function is expected, because otherwise the integral of Eq. 11 would become infinite and the invariant would not exist. As can be seen from Figure 5a the cut-off of the pure-resonant form factor cannot be resolved due to the large error bars and if the cut-off is located at q-values beyond 2 nm^{-1}, the concentration of Yttrium atoms must be higher. In the case of the crystallized sample the scattering curve at larger q-values is deep in the Porod regime and no significant contribution can appear in the integral at higher q-values.

Table I: Structural and quantitative parameters of the two alloys (one of them with different heat-treatment). Column ζ is the correlation length. $\bar{\nu}(Y), \bar{\nu}(Nb)$ give the amount of Yttrium respective Niobium atoms localized in the concentration fluctuations calculated from the Resonant Invariant at the related K-absorption edges. The two columns c_Y and c_{Nb} contain the amount in percent of the complete amount of Yttrium or Niobium atoms respectively. The latter concentrations were obtained from the averaged atomic volume, which was calculated from the average atomic weight (68.999 g/mol) and the average density (8.147 g/cm^3) of the alloy. Additionally the corresponding volume fractions, Φ_Y and Φ_{Nb}, have been calculated (see text).

Sample	ζ	$\bar{\nu}(Y)$	c_Y	Φ_Y	$\bar{\nu}(Nb)$	c_{Nb}	Φ_{Nb}
	[nm]	10^{21}[cm^{-3}]	[%]	[%]	10^{21}[cm^{-3}]	[%]	[%]
$Ni_{68}Nb_{16}Y_{16}$	12.2	0.39(2)	4	0.6	1.3(3)	12	1.9
c-$Ni_{68}Nb_{16}Y_{16}$	12.2	2.1(3)	19	3.0	1.8(3)	16	2.5

The same arguments hold for the interpretation of the results obtained from the *ASAXS* measurements at the Nb-K-edge. Again the asymptotic behaviour of the pure-resonant scattering turns from q^{-2} to q^{-4} , when applying a heat-treatment beyond the 1st crystallization temperature. The amount of Niobium atoms in the fluctuations of the as-quenched sample calculated from the Resonant Invariant turns out to be higher (with respect to Yttrium) corresponding to about 12% of the complete

Niobium amount in the alloy (Table 1). Surprisingly the amount of Niobium atoms is only slightly higher for the partially crystallized sample i.e. 16% of the complete Nb-amount, indicating that the spatial distribution of Niobium in the alloy does not change much, when the partial crystallization (Ni_2Y-phase) takes place. This may be attributed to the fact, that the Niobium component crystallizes beyond the 2nd crystallization temperature (beyond 900 K), which we have not yet reached.

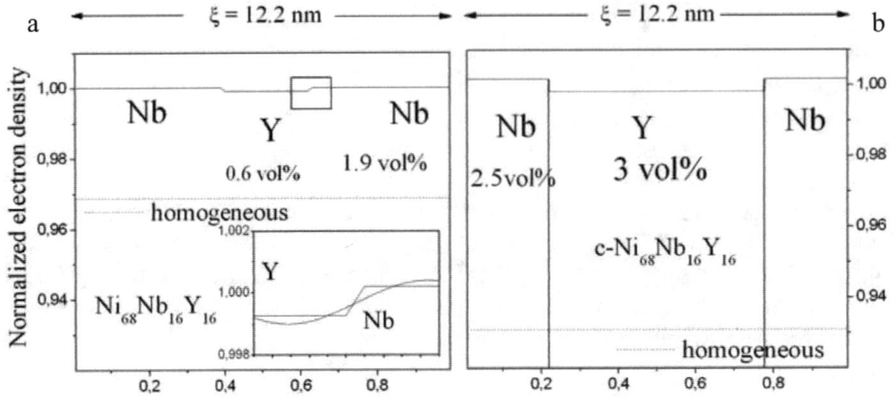

Figure 6: Normalized electron density of the as-quenched sample (a) and the partially crystallized alloy (b).The step functions represent the volume fractions of the two phases, Y-rich and Nb-rich respectively. The dashed lines show the normalized electron density of homogeneously distributed atoms in the two alloys.

From the concentrations of Y and Nb deduced from the Resonant Invariant in the as-quenched and the partially crystallized state respectively the related volume fractions, Φ_Y and Φ_{Nb}, can be estimated using the averaged atom volumes of the alloy for V_Y and V_{Nb}. In a further step the normalized electron densities have been calculated. The results are summarized in Figure 6. In Figure 6a the normalized electron density of the as-quenched sample is depicted within one correlation period. The dashed line represents the average electron density resulting from homogeneously distributed atoms in the alloy, which cover nearly 97% of the entire volume. The step function shows the different volume fractions of inhomogeneously distributed Yttrium, 0.6%, and Niobium, 1.9%, respectively. The integral of the step function over one correlation period amounts to 1. The inset shows a magnification of the step function with the oscillating function giving a more realistic sketch of a continuously changing concentration over the correlation length.

Again the dashed line in Figure 6b represents the average electron density resulting from homogeneously distributed atoms in the partially crystallized alloy. Because the amount of inhomogeneously distributed Y and Nb increased to 3% and 2.5% respectively (black step function) the value of the volume fraction covered by homogeneously distributed material is reduced to about 93%. The area between the two vertical lines, the step function and the dashed line depicts the volume fraction of the additional amount of Yttrium (excess) atoms in the crystallized phase with respect to the surrounding matrix, while the intersection of the vertical lines with the x-axis represent the volume fraction of the spatial extension of the crystallites, i.e. 55%.

In both Figures the contrast between the Y-rich and the Nb-rich phases is negative explaining the declination of the scattering cross section, when approaching the K-absorption edge of Niobium, while the scattering cross section becomes larger, when approaching the K-absorption edge of Yttrium.

5. Conclusions

In the last years Anomalous Small-Angle X-ray Scattering became a precise quantitative method (*q-ASAXS*) for element specific structural analysis on the mesoscopic length scale. Due to the precision of the techniques small-angle scattering contributions of $Ni_{68}Nb_{16}Y_{16}$ alloys in the resolution regime $10^{-3} < \Delta I/I < 10^{-2}$ were reliably separated from a superimposed, orders of magnitude larger inelastic background caused by fluorescence and Resonant Raman Scattering. The separated pure-resonant scattering contributions were obtained by employing the Gauss algorithm to a vector equation, which represents the ASAXS measurements at different X-ray energies.

The pure-resonant scattering contributions of inhomogeneously distributed Nickel, Niobium and Yttrium atoms in $Ni_{68}Nb_{16}Y_{16}$ alloys with different heat-treatment have been separated from the total scattering SAXS curves. The separated scattering curves obtained from the as-quenched sample, show the typical behaviour of concentration fluctuations (Ornstein-Zernike) with a correlation maximum corresponding to a correlation length of 12.2 nm. From the Resonant Invariant an amount of 4% and 12% of the Yttrium atoms and Niobium atoms respectively was found to be localized in the fluctuations corresponding to volume fractions of 0.6 and 1.9% respectively. This compares to about 19% (3% volume fraction) of enriched Yttrium atoms localized in the crystalline phase of the sample, which was heat-treated beyond the 1st crystallization temperature. Surprisingly the amount of enriched Niobium atoms does not change much due to the latter heat-treatment (from 12% to 16%) indicating that the crystallization temperature for the Niobium phases is not yet reached. A model calculation suggests a large volume fraction of 55% covered by small crystallites beyond the 1st crystallization temperature surrounded by an amorphous shell enriched with Nb with a volume fraction of 45%.

The example demonstrates the capability to determine quantitatively the extent of spinodal decomposition in metallic glasses by Anomalous Small-Angle X-ray Scattering.

References

[1] Stuhrmann H B 1985 *Adv. Polym. Sci.* **67**, 123
[2] Simon J P, Lyon O and de Fontaine D J 1985 *Appl. Cryst.* **18**, 230
[3] Stuhrmann H B, Goerigk G and Munk B 1991 In: Ebashi S, Koch M, Rubenstein E, editors. *Handbook on Synchrotron Radiation* **Vol. 4**. Amsterdam: Elsevier; 555.
[4] Goerigk G and Williamson D L 2001 *J. Non-Cryst. Solids* **281**, 181
[5] Goerigk G, Schweins R, Huber K and Ballauff M 2004 *Europhys. Lett.* **66(3)**, 331
[6] Goerigk G and Williamson D L 2006 *J. Appl. Phys.* **99**, 084309
[7] Goerigk G, Huber K and Schweins R 2007 *J. Chem. Phys.* **127**, 154908
[8] Bota A, Varga Z and Goerigk G 2007 *J. Phys. Chem. B* **111**, 1911
[9] Bota A, Varga Z and Goerigk G 2008 *J. Phys. Chem. C* **112**, 4427
[10] Goerigk G and Mattern N 2009 *Acta Mater.* **57**, 3652
[11] Mattern N, Zinkevich M, Löser W, Behr G and Acker J 2008 *J. Phase Equil. Diff.* **29**, 141
[12] Mattern N, Kühn U, Gebert A, Gemming T, Zinkevich M, Wendrock H and Schultz L 2005 *Scripta Materialia* **53**, 271
[13] Mattern N, Gemming T, Goerigk G and Eckert J 2007 *Scripta Mater.* **57**, 29
[14] Glatter O and Kratky O. 1982 *Small-Angle X-ray Scattering*, Academic Press, London
[15] Haubold H G, Gruenhagen K, Wagener M, Jungbluth H, Heer H, Pfeil A, Rongen H, Brandenburg G, Moeller R, Matzerath J, Hiller P and Halling H 1989 *Rev. Sci. Instrum.* **60**, 1943
[16] Goerigk G 2006 *'Electronic and Computer Upgrade at ASAXS Beamline JUSIFA', HASYLAB Annual Report* p. 77
[17] Cromer D T and Liberman D 1970 *J.Chem.Phys.* **53**, 1891
[18] Cromer D T and Liberman D 1981 *Acta Cryst.* **A37**, 267
[19] Mattern N, Goerigk G, Vainio U, Miller M K, Gemming T and Eckert J 2009 *Acta Mat.* **57(3)**, 903

Journal of Physics: Conference Series **247** (2010) 012023 doi:10.1088/1742-6596/247/1/012023

Pore structure characterization and in-situ diffusion test in nanoporous membrane using SANS

P Strunz[1,2], D Mukherji[3], J Šaroun[1,2], U Keiderling[4], J Rösler[3]

[1] Nuclear Physics Institute, CZ-25068 Řež near Prague, Czech Republic
[2] Research Center Řež, CZ-25068 Řež near Prague, Czech Republic
[3] IfW, TU Braunschweig, D-38106 Braunschweig, Langer Kamp 8, Germany
[4] Helmholtz Centre Berlin for Materials and Energy GmbH, Glienickerstr. 100, D-14109 Berlin, Germany

E-mail: strunz@ujf.cas.cz

Abstract. Using a selective phase dissolution technique, nano-porous membrane can be produced from simple two-phase metallic alloys. It contains through-thickness elongated channel-like pores of only a few hundred nanometer width and has a number of prospective applications. Knowledge of microstructural parameters is essential for membrane optimization. Non-destructive characterisation of the pore microstructure was carried out by small-angle neutron scattering technique. The combined results from pinhole and double-crystal facilities enabled to determine microstructural parameters of the nanoporous membrane (pore-to-pore distance, raft thickness, pore volume fraction, specific interface). The contrast variation using D_2O and H_2O helped to conclude on scattering length density of both γ' pore walls as well as the original γ-phase matrix. The kinetics experiment showed that the pores are filled instantly by liquid. The subsequent emptying of pores by evaporation was observed.

1. Introduction

A novel process was developed at TU Braunschweig in the past for the production of nano-porous membrane from simple two-phase metallic alloys [1]. The highlight of the membrane fabrication [2, 3] is a selective phase dissolution technique, which isolates one phase by dissolving the other phase. The fabrication essentially exploits, in the first step, the self-assembly of nano-sized precipitates to form a network structure in the nano-scale. Common Ni-base superalloys with Ni_3Al (γ') precipitates have been used for this purpose. Subjecting Ni-superalloys to thermo-mechanical treatment induces self-assembly of the γ' precipitates (which is commonly known as γ' rafting). Rafting actually produces a complex interconnected network of both the matrix and the precipitate phases. In the second step of fabrication, one of the interconnected phases is extracted from the bulk by selective electrochemical phase dissolution, thus leaving the remaining undissolved phase in a form of a scaffold. Thereby, a nano-structured porous membrane is produced.

Membrane thickness can be varied from 100 μm to 1 mm. It contains through-thickness elongated channel-like pores of only a few hundred nanometer width. An example can be seen in Fig. 1. This unique porous material with fine open porosity and a high degree of regularity is suitable for many applications in filtering, including separation of gas from liquid. Being metallic, the membranes can be easily heated using electric current and can be sterilized in-situ. Prospective applications can be:

Journal of Physics: Conference Series **247** (2010) 012023 doi:10.1088/1742-6596/247/1/012023

catalytic substrate, miniature heat exchangers and separation processes like gas permeable membrane or tap water filter in hospitals to filter bacteria.

The development of the porous membrane focuses on the optimization of both steps of fabrication (the rafting as well as the selective phase dissolution). Knowledge of microstructural parameters is essential for this purpose. There is a need for a detailed microstructural characterization of the starting alloy, in particular understanding the complex self-assembly / rafting as well as the phase dissolution process, with the help of in-situ measurements. Non destructive characterisation of the pore structure and the pore walls in the final membane is also indispensable. Small angle neutron scattering (SANS) is an invaluable tool for this purpose.

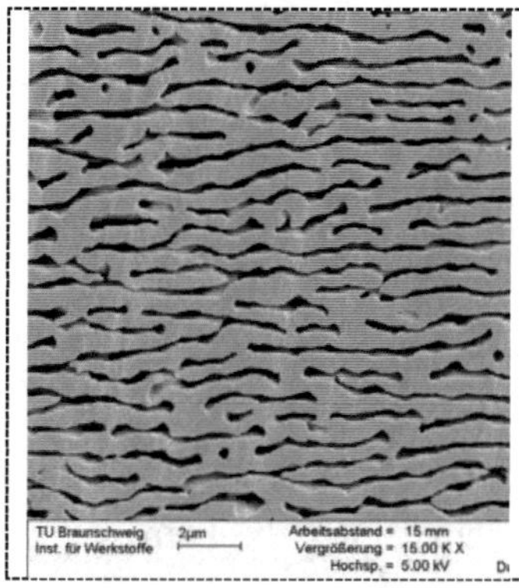

Figure 1. An example of porous membrane produced from Ni superalloy.

In the past, the etching efficiency and the penetration depth (which limits the ultimate thickness of the porous membrane) during the electrochemical dissolution step was successfully investigated by double-crystal SANS [4]. In the present study, we intended to determine microstructural parameters of a Ni_3Al (γ') porous membrane (where the matrix phase from the superalloy was leached out). Contrast variation experiments with complementary double crystal (DC) and pin-hole SANS facilities were done on membranes containing through and through pores.

Additionally, the kinetics of the H_2O and D_2O diffusion through the membrane was studied. D_2O lowers the scattering contrast between pores and γ' walls as the liquid fill into the pores. In contrast, H_2O increases the scattering contrast. The diffusion rate of the liquid through the pores could in principle be determined through a time–resolved experiment to study the extent of pore filling. The diffusion of liquids and gasses is an important question for the prospective applications of the porous metallic membrane. Knowledge of the diffusion can also help to optimize the fabrication of the membrane, because the electrolyte has to penetrate the porous channel and reach the etching front in order to continue the electro-dissolution process.

2. Experimental

The sample for the SANS experiment was a porous membrane fabricated from the single-crystal Ni-base superalloy CMSX-4. The average scattering length density (SLD) of the solid alloy was calculated as $\rho_{alloy} = 67.27\times10^9$ cm^{-2}. This is the average of three values (67.164×10^9 cm^{-2} [5], 67.434×10^9 cm^{-2} [6], 67.220×10^9 cm^{-2} [7]), where the SLD values were computed from the nominal composition of CMSX4 found in the respective literature.

The CMSX4 alloy was subjected to the following heat treatment: 1573K/2.5h + 1583K/6h, gas-fan quenched + 1413K/6h + 1123K/24h, which led to large volume fraction (over 50%) of cubic Ni_3Al-type γ'-precipitates. Afterwards, the alloy was exposed to a uniaxial tensile creep load (at 1273K and 170MPa) along the [001] direction. This exposure produced the rafted microstructure [2]. Subsequently, the matrix phase is removed by electrochemical selective phase dissolution [2, 3]. This procedure resulted in a 0.71 mm thick porous membrane based on the Ni_3Al phase.

The small angle neutron scattering measurements were carried out at the V4 instrument of the BENSC facility at the Helmholtz-Zentrum Berlin für Materialien und Energie (formerly, HMI),

Berlin, Germany [8] and at the neutron double-bent-crystal (DBC) SANS diffractometer MAUD at NPI Řež [9].

At the DBC SANS facility MAUD, the following parameters were used: λ=2.1 Å, Q range 0.003-0.021 nm^{-1} (scattering vector magnitude $Q=|\mathbf{Q}|=4\pi\sin\theta/\lambda$, 2θ is the full scattering angle). As the samples are single crystals containing oriented precipitates and pores, anisotropic scattering was expected. The rafts and pores (as well as the [100] direction) were aligned to be vertical in case of DBC SANS measurements (Fig. 2a). This orientation ensures the visibility of interparticle interference peaks on 1D position sensitive detector (PSD).

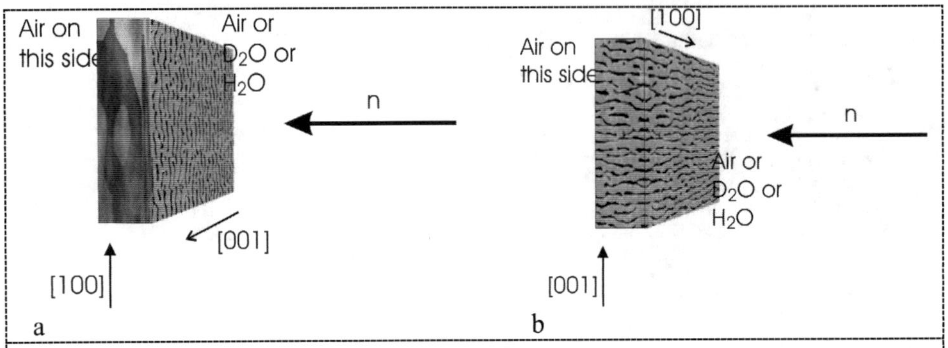

Figure 2. Sample orientation for (a) DBC and (b) pinhole SANS experiment.

The geometry of the V4 pinhole SANS facility was set up with the aim to cover lower Q-values, i.e. the larger sizes during the intended kinetics experiment: the data were collected with a sample-to-detector distance (SDD) equal either to 4 m or 15.8 m and a wavelength λ = 7.5Å. The rafts and pores of the membrane were aligned to be horizontal for pinhole SANS experiment (Fig. 2b). Therefore, the streaks of the highest intensity are aligned in vertical direction on the 2D PSD.

In order to enable contrast variation method, a special cell was constructed. It contains reservoirs on both sides of the membrane, separated by the membrane. A fluid (D_2O or H_2O) could be filled in the reservoir in such a way, that it comes in contact with the porous membrane on one side only. The reservoir on the other side was used to collect the fluid flowing through the membrane (however, this reservoir was never filled to the level of the neutron beam). In this way, the fluid was allowed to fill the pores. This cell was also used in the subsequent kinetics experiment testing the fluid diffusion through the pores under ambient pressure.

3. Results

The measurement at the double-bent-crystal SANS facility MAUD (NPI Řež) shows Bragg-like scattering on the aligned pore walls (ordered rafts) which causes an interparticle interference maximum in the scattering data. The scattering curves obtained for empty pores and for the pores filled by D_2O are displayed in Fig. 3.

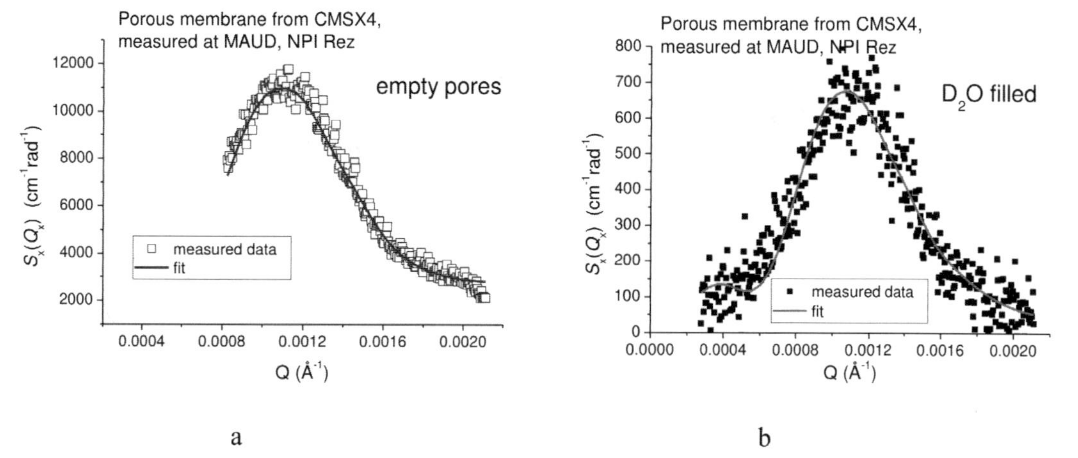

Figure 3. Double-Bent-Crystal SANS data, facility MAUD (NPI Řež). $S_x(Q_x)$ is the cross-section $d\Sigma/d\Omega(Q_x,Q_y)$ integrated over the vertical angular component. (a) empty pores, (b) pores filled by D_2O.

Pinhole 2D SANS scattering curves with data for empty pores, for D_2O-filled and for H_2O filled pores are shown in Fig. 4.

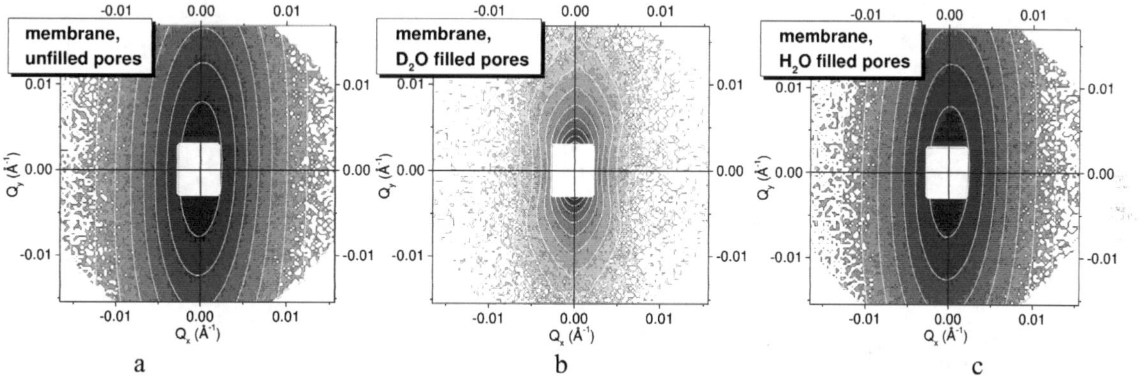

Figure 4. Pinhole SANS data, V4, BENSC, HZ Berlin. The grey scale map shows measured 2D data and the white equi-intensity lines depict the fitted curves. Data for empty pores (left), for D2O-filled pores (middle) and for H_2O filled pores (right) of the membrane. 2D cross-section $d\Sigma/d\Omega(Q_x,Q_y)$ is shown.

SANS data were processed using the NOC program [10]. The procedure is based on the numerical simulation of a scattering profile using a 3D microstructural model of the particle system. The modeled scattering curve thus also covers the interparticle-interference effect. A model of agglomerated γ' -precipitates (rafts) forming the pore walls, was employed. The model allows a distribution in pore width as well as the pore wall thickness. Fig. 5 displays the models corresponding to the optimum fits to the DBC and pin-hole SANS data (see also the lines in Figs. 3 and 4).

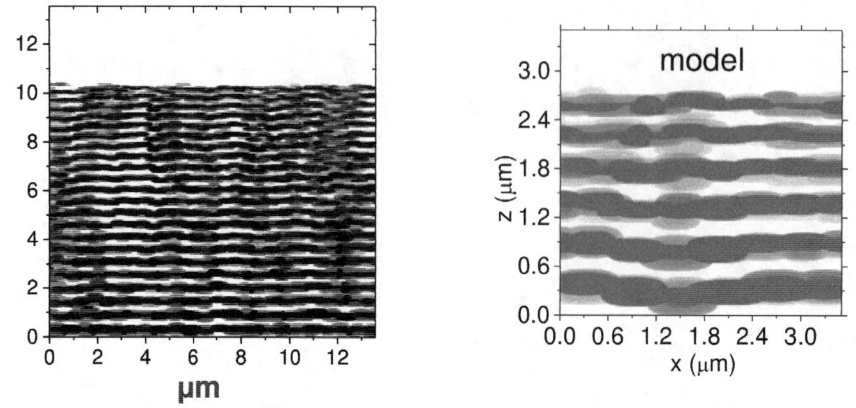

Figure 5. Slice through the 3D model corresponding to the optimum fit to (a) Double-Bent-Crystal SANS data and (b) pinhole SANS data. Black and gray: γ' rafts; white: pores.

4. Microstructural parameters

From the DBC SANS data it was possible to determine the average distance between the longitudinal pores. It was found to be 480 nm. By fixing this value, the pinhole SANS data could be evaluated in more detail. The average thickness of the pore wall was found to be 280 nm and its volume fraction as 64% (this value was obtained from the 3D model of the microstructure and is referred to as the geometrical volume fraction throughout the text). Consequently, the volume fraction of the pores should be around 36%. The determined specific interface between pore walls and the pores is 49000 cm^2/cm^3.

From the measurement with empty pores and by using the geometrical volume fraction obtained above, the SLD of pore wall material (Ni_3Al based composition) was determined as 73.046×10^9 cm^{-2}. When performing a contrast variation experiment using H_2O filling, a very similar value 73.022×10^9 cm^{-2} is calculated for the SLD magnitude of the pore wall. On the other hand, the wall SLD value of 74.251×10^9 cm^{-2} is obtained for D_2O-filled pores, which differs form the two above mentioned values. The discrepancy could be in principle caused by incomplete leaching of the matrix material, such that a small amount of matrix (Ni-solid solution of Ni-Cr-Mo-W-Re etc.) is left on top of the Ni_3Al-based pore walls. It would mean that the total scattering is caused by a combination of scattering from the regions containing pores and regions with remaining "rafted precipitates + matrix" microstructure. Numerical simulation showed that the scattering from sample with empty pores and that from the pores filled with D_2O or with H_2O could be explained by this hypothesis by assuming that 2.5% of the matrix phase remained un-leached on the pore walls. In this case, the fit yeilds a SLD value for the pore walls (γ') as 73.97×10^9 cm^{-2} and the SLD value for the remaining γ matrix as 55.74×10^9 cm^{-2}. Scattering contrast of γ' precipitates in γ-phase matrix in CMSX4 alloy would then be equal to 18.23×10^9 cm^{-2}.

However, the previously mentioned discrepancy in the SLD values can also be corrected if it is assumed that the pores are not fully filled. When supposing that 0.46% of all pores are not filled in both cases (i.e. for H_2O or D_2O filling of the membrane), then the SLD of pore walls is equal to 73.046×10^9 cm^{-2} in all three cases (i.e. H_2O filled, D_2O filled and empty pores). The reason for non-perfect filling could be that not all the pores are open towards the surface which is in contact with the liquid in the reservoir. However, in such a case, they have to be open towards the rear surface. It is important to note here that as the pores are created by a surface leaching process, all pores are open and must be connected to at least one surface. The membranes used in the present experiment were prepared by electro-polishing from both surfaces, so it is a possibility that a small solid layer can be left at the centre of the membrane separating the otherwise interconnected pores (i.e. for some pores a full penetration was not obtained during the leaching process). However, the numbers of such disconnected pores are relatively small. Except these ≈0.5% of disconnected pores (i.e. pores connected to the rear surface and not connected to the front surface and vice a versa), all other pores (i.e. more than 99%) are connected through and through the membrane. It confirms a high permeability of the membrane. In this case, the value 73.05×10^9 cm^{-2} for the SLD of rafts and their geometrical volume fraction can be used to back-calculate the SLD of the original γ matrix (i.e. solid solution of Ni-Cr-Mo-W-Re etc. that existed at the place of the pores before the selective phase dissolution step). Its magnitude is found to be equal to 57.33×10^9 cm^{-2}. The scattering contrast of γ' precipitates in γ-phase matrix of CMSX4 would then be 15.7×10^9 cm^{-2}.

It is presently not possible to say which amongst the two phenomena is the true explanation of the observed effect. The combination of both causes is also quite possible. Therefore, only a range for the SLD values of the pore walls ($74.0\text{-}73.0\times10^9$ cm^{-2}) and the γ-phase matrix ($55.7\text{-}57.3\times10^9$ cm^{-2}) can be given. The scattering contrast of γ' precipitates in γ-phase matrix thus lies in the range from 15.7 to 18.3×10^9 cm^{-2}. The scattering contrast value relates to the CMSX4 alloy in the given heat treatment condition only. As the best estimate for the contrast, the value $(17.0\pm1.3)\times10^9$ cm^{-2} can be used. It is well known that determination of the scattering contrast between precipitates and matrix in superalloys is not a trivial task. Particularly, because the matrix and the precipitate compositions in multicomponent commercial Ni-superalloys like CMSX4 are not easy to determine. Usually, it can be done only approximatelly which is insufficient for further evaluation on absolute scale. The present method of selective phase dissolution, contrast variation and combination of DBC with pinhole data shows possibility to determine the scattering contrast in superalloys relatively accuratelly.

5. Kinetics experiment

It was found that the pores are occupied very quickly, almost immediately, after the fluid (D_2O or H_2O) was filled into the cell reservoir. This time was much shorter than the time between the reservoir filling and the measurement start, i.e. a time span of less than 20s. A similar test using the more viscous silicon oil also produced the same result. This indicates that a strong capillary force exists in the narrow and high aspect-ratio (depth: width) pores in the membrane. It was therefore interesting to study also the rate of removal of liquid from the pores. By removing the fluid from the reservoir cell and exposing thus both membrane surfaces to air, the evaporation of liquid from the pores started. In-situ integral intensity monitoring was carried out to monitor the evaporation process. The sample-to detector distance 4m and wavelength 7.5 Å was the selected geometry for this measurement. This is the asymptotic region of the scattering curve and thus not influenced by multiple scattering which is possibly affecting the low Q-magnitudes. The integral SANS intensity evolution after D_2O removal from the reservoir can be seen in Fig. 6. Due to the enormous scattering from the freed pores, the scattering intensity increase can be clearly observed as the evaporation proceeds. For the initial phase of evaporation, the integral intensity can be assumed to increase linearly with time. The slope of the increase - when compared with the known integral intensity of the empty membrane - gives information on the actual amount of pores emptied due to evaporation. It could be estimated that around 1.0 μm total depth of pores (i.e. about 0.5 μm depth per surface) is emptied each minute. Moreover, a sign of saturation is visible after 20 minutes of evaporation. It means that the evaporation rate starts to slow down in the depth of 10 μm due to the increasingly difficult transport of the vapour through the high aspect ratio pores.

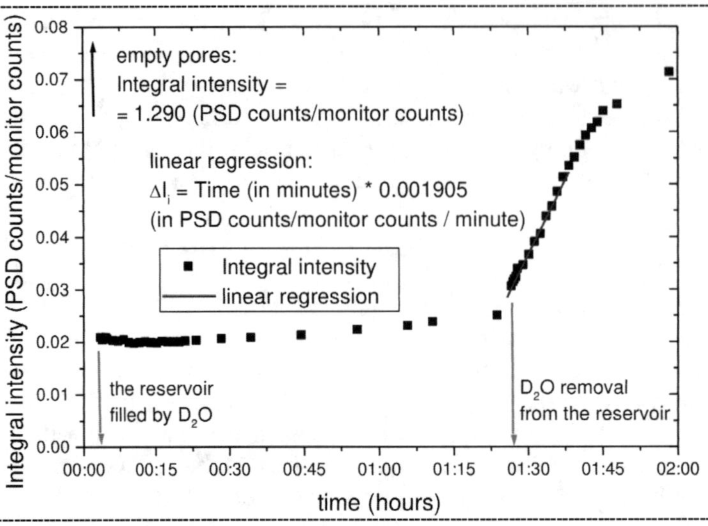

Figure 6. Evolution of the integral SANS intensity before and after D_2O removal from the reservoir.

6. Conclusions

Although only a rough estimation of the diffusion is obtained, the measured data was useful for microstructural characterization of the membrane / pores and for the determination of the scattering length density of the porous membrane walls.

The combined SANS results from pinhole and double-bent-crystal facility enabled us to determine microstructural parameters of the nanoporous membrane (pore-to-pore distance, raft thickness, pore volume fraction, specific interface). The contrast variation using D_2O or H_2O helped to conclude on SLD value of both γ' rafts as well as the original γ-phase matrix. The advantage of complementarities of both types of facilities has to be stressed.

The kinetics experiment showed that the pores are filled instantly (less than 20s) by D_2O, H_2O or silicon oil. This is attributed to strong capillary effects. The subsequent emptying of pores by evaporation, which is a much slower process, was observed from the gradual change (increase) of the integral intensity with time. The observation provides qualitative information on the diffusion process through the high aspect ratio pores and indicates that diffusion slows down when evaporation occurs at larger depths (> 10 μm). This is likely to have consequences on the electrolytic phase dissolution step in the membrane fabrication.

Acknowledgments

This research project was supported by EC under the FP6 (Key Action: Strengthening the European Research Infrastructures), contract No. RII3-CT-2003-505925 (NMI3). Some team-members (P. Strunz, J. Šaroun) gratefully acknowledge support by MSM2672244501 and IRP AV0Z10480505 projects.

References

[1] Rösler J and Mukherji D 2003 *Adv. Eng. Mater.* **5** 916-918
[2] Rösler J, Näth O, Jäger S, Schmitz F and Mukherji D 2005 *Acta Materialia* **53** 1397-1406

[3] Mukherji D, Pigozzi G, Schmitz F, Näth O, Rösler J and Kostorz G 2005 *Nanotechnology* **16**, 2176-2187
[4] Strunz P, Mukherji D, Näth O, Gilles R, Rösler J 2006 *Physica B* **385–386** 626–629
[5] Lukáš P, Čadek J, Kunz L, Svoboda M, Klusák J 2005 *Kovové Materiály* **43** 5–19
[6] Wilson B 2002 *Journal of Undergraduate Research (University of Florida)* **3** (7), http://www.clas.ufl.edu/jur/200205/papers/paper_wilson.html
[7] Glatzel U, Mueller A 1994 *Scripta Metall. Mater.* **31** 285-290
[8] Keiderling U, Wiedenmann A 1995 *Physica B* **213-214** 895
[9] Strunz P, Šaroun J, Mikula P, Lukáš P and Eichhorn F 1997 *J. Appl. Cryst.* **30** 844-848
[10] Strunz P, Gilles R, Mukherji D, Wiedenmann A 2003 *J. Appl. Cryst.* **36** 854-859

Journal of Physics: Conference Series **247** (2010) 012024 doi:10.1088/1742-6596/247/1/012024

A Microbeam Small-Angle X-ray Scattering Study on Enamel Crystallites in Subsurface Lesion

N Yagi[1,2,4], N Ohta[1], T Matsuo[1], T Tanaka[3], Y Terada[3], H Kamasaka[3], and T Kometani[3]
[1]Japan Synchrotron Radiation Research Institute, 1-1-1 Kouto, Sayo, Hyogo 679-5198, Japan.
[2]RIKEN SPring-8 Center, 1-1-1, Kouto, Sayo, Hyogo 679-5148, Japan.
[3]Ezaki Glico Co. Ltd., 4-6-5 Utajima, Nishiyodogawa-ku, Osaka 555-8502, Japan.

E-mail: yagi@spring8.or.jp

Abstract. The early caries lesion in bovine tooth enamel was studied by two different X-ray diffraction systems at the SPring-8 third generation synchrotron radiation facility. Both allowed us simultaneous measurement of the small and large angle regions. The beam size was 6μm at BL40XU and 50μm at BL45XU. The small-angle scattering from voids in the hydroxyapatite crystallites and the wide-angle diffraction from the hydroxyapatite crystals were observed simultaneously. At BL40XU an X-ray image intensifier was used for the small-angle and a CMOS flatpanel detector for the large-angle region. At BL45XU, a large-area CCD detector was used to cover both regions. A linear microbeam scan at BL40XU showed a detailed distribution of voids and crystals and made it possible to examine the structural details in the lesion. The two-dimensional scan at BL45XU showed distribution of voids and crystals in a wider region in the enamel. The simultaneous small- and wide-angle measurement with a microbeam is a powerful tool to elucidate the mechanisms of demineralization and remineralization in the early caries lesion.

1. Introduction

Early caries lesion in tooth enamel is caused by acids produced by bacteria in dental plaque and dentally observed as a opaque white spot. In a cross section along the direction perpendicular to the enamel surface, a caries lesion appears as an enamel decay with a relatively intact surface layer and a mineral loss in the subsurface region [1]. When not treated, the decay spreads into the dentin and forms a cavity. Over the past three decades, many studies have been made on demineralization of the early subsurface lesion under different conditions. Most of these studies used transversal microradiography (TMR) [2] to measure the mineral content. The TMR analysis measures the mineral density from microradiography of thin sections of enamel and is regarded as the standard method to evaluate the degree of demineralization [3,4]. However, since evidence on the crystal structure of hydroxyapatite (HAp), which comprises tooth enamel, cannot be obtained by this method, the apparent change in mineral content could be partially due to changes in other substances such as precipitates of calcium phosphate. The enamel HAp is arranged in rod-like units called crystallites. The morphology of HAp crystallites was studied by high-resolution transmission electron microscopy and decrease in

[4] To whom any correspondence should be addressed.

the HAp crystal size by demineralization was reported [5]. Such morphological changes in the submicron order are expected to change small-angle X-ray scattering (SAXS).

In the present study, we used SAXS with a microbeam to investigate the HAp crystallites in the subsurface lesion [6]. Wide-angle X-ray diffraction (WAXD) from HAp was recorded simultaneously. A 6μm microbeam was small enough to study the 100μm thick subsurface lesion and enabled us to compare the crystallite structure with the TMR results at the spatial resolution required for the analysis of the lesion. In another experiment, use of a 50μm beam enabled us to map two-dimensionally the HAp crystallites and the voids in them. In addition, we suggest the possibility of using laboratory equipments for similar studies.

2. Materials and Methods

2.1. Preparation of artificial subsurface lesion

Bovine incisors were obtained from a slaughterhouse (Osaka, Japan). Enamel blocks (approx. 7 mm × 7 mm × 3 mm depth) were cut out (Figure 1a) and embedded in resin (GC CORP, Japan). The blocks were polished with a wet polishing paper (3M wrapping paper 800, 1000 and 2000 grid, USA) to expose a sound and flat enamel surface. Then, sound and demineralized zones were made on the same enamel surface as follows. First, half of each block surface was covered with a nail varnish as described [7]. Then, subsurface lesions were formed by demineralization in a two-layer system of 8% methyl cellulose gel (Fluka, USA) and 0.1 M lactate buffer (pH 4.6) at 37°C for 14 days [4]. After lesions were made in the demineralized zone by the acid treatment, the nail varnish was washed off with acetone to reveal the intact, sound zone.

After the demineralization treatment, 150μm–thick enamel slices were cut out from the middle of each block perpendicularly to the surface with a water-cooled diamond-coated wire saw (Well Diamond Wire Saws Co., Ltd, Germany) (Figures 1a, 1b).

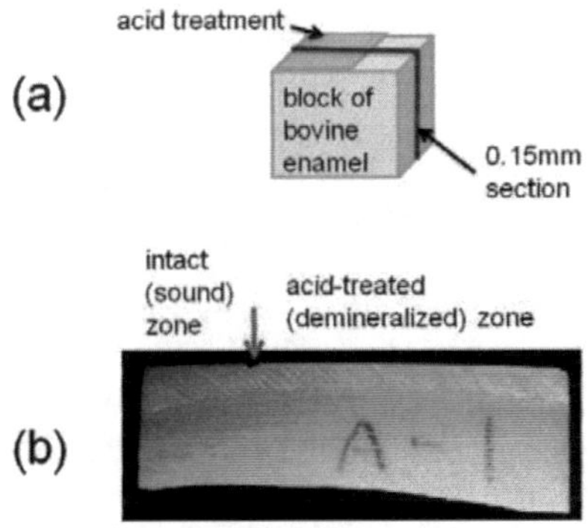

Figure 1. Sample preparation. (a) One surface of a block of bovine enamel was polished and partially treated with lactate. Then a section with a thickness of 0.15 mm was cut out. (b) Photograph of a sample. The white region at the top is enamel, while the darker region at the bottom is dentin. The oblique lines in the enamel are due to polishing. There is a nick between the region treated with lactate (demineralized zone) and that not treated (sound zone).

2.2. Simultaneous WAXD and SAXS measurement at BL40XU

The experiments were carried out at the BL40XU beamline (Figure 2) [8] of the SPring-8 third-generation synchrotron radiation facility with an X-ray energy of 15.0 keV (bandwidth about 3%). An X-ray microbeam was obtained by placing a 5μm collimating pinhole on the focused beam [9]. The beam expanded to about 6 μm at the sample because of Fresnel diffraction. A guard pinhole with a diameter of 200μm was placed in front of the sample. The distance between the two pinholes was about 80 mm (with a small ion chamber between them). To remove the scatter from the edge of the collimating pinhole, it is necessary to make this distance as large as possible. On the other hand, the beam becomes larger with distance from the collimating pinhole because of diffraction. Thus, the distance of 80 mm was chosen as a compromise in the current setup. The X-ray flux was about 3×10^{11} photons/sec. The X-ray detector for SAXS consisted of an X-ray image intensifier (V5445P,

Hamamatsu Photonics, Hamamatsu, Japan) [10] coupled with a tandem lens to a cooled CCD camera (ORAC-II-ER, Hamamatsu Photonics). The WAXD pattern was recorded with a CMOS flatpanel detector (C9728DK, Hamamatsu Photonics) [11]. An ionization chamber (S-1329, Oken, Tokyo, Japan) was also inserted between the flatpanel detector and the vacuum tube to measure the transmission of the X-rays through the sample. The distance from the sample to the SAXS detector was about 2m and to the WAXD detector about 100mm. The reciprocal spacing was calibrated with a powder diffraction pattern of LaB_6 for WAXD and the meridional reflections from dried chicken tendon collagen (1st order at 63 nm) for SAXS. The SAXS and WAXD patterns were recorded simultaneously. The reciprocal spacing is expressed as $q=4\pi\sin\theta/\lambda$, where 2θ is the scattering angle and λ is the wavelength of the X-ray.
An enamel section was mounted vertically so that its enamel edge was more or less horizontal (Figure 1b). The X-ray beam passed perpendicularly through the slice. The sample was moved upwards with a 5μm step so that the X-ray beam scanned across the enamel from the surface towards dentin. At each step, SAXS and WAXD patterns were recorded. The region to the depth of 150 μm was scanned. The sound and demineralized zones of each sample were studied.

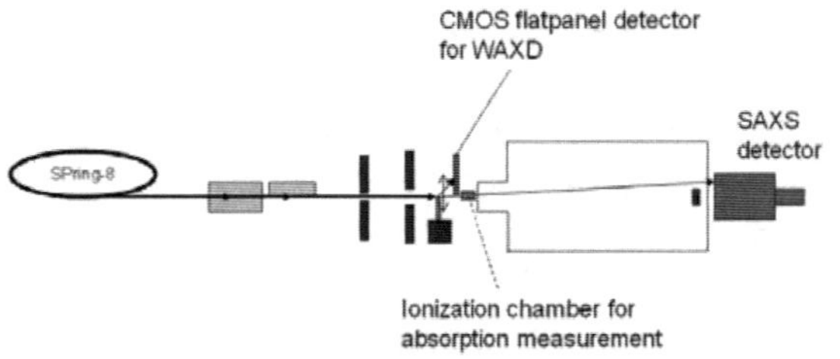

Figure 2. The experimental setup at BL40XU. The X-ray beam from SPring-8 was focused with two mirrors and collimated by a 5μm pinhole. A 200μm pinhole was used to remove scatter from the edge of the 5μm pinhole. The distance between the 5μm pinhole and the sample was about 100 mm. A small-angle scattering, wide-angle diffraction and transmittance were simultaneously measured.

2.3. Simultaneous WAXRD and SAXS measurement at BL45XU
BL45XU has tandem helical undulators as an X-ray source and uses a diamond crystal as a beam splitter [12]. In the present experiment, the protein crystallography hutch with three sets of diamond double-crystal monochromators was used. One set of the monochromators was used to obtain a 13.8 keV X-ray. A bent cylindrical mirror was used to focus the X-ray beam. The beam size without a pinhole was approximately 250 μm × 250μm. A 0.1mm thick tantalum pinhole with a diameter of 50 μm was inserted at 335 mm upstream of the specimen (Figure 3a). The sample-to-the-detector distance was 450 mm. At 15mm upstream to the specimen was a guard pinhole with a diameter of 0.8 mm. With a beam stop with a diameter of 2 mm, the scatters from the edge of the pinhole fell within the beam stop. Thus, the insertion of the 50μm pinhole did not affect the background in the small-angle region significantly. The flux through the 50μm pinhole was about 3×10^9 cps. The X-ray detector was Jupiter210 (RIGAKU, Tokyo, Japan), which comprises of four CCDs with tapered fibers. The detection area was 200 mm × 200 mm. The pixel size was 100 μm × 100 μm with 2×2 binning. In order to cover a wide q-range, the detector was moved so that the beam stop is located near the bottom of the detection area (Figure 3b). A fiber diffraction pattern from a dried chicken tendon collagen (Figure 3c) shows the 3rd-order meridional reflection at about 1/21 nm^{-1} and the meridional reflection at 1/0.28 nm^{-1} (Figure 3b) showing that this setup covers a q-range of about two orders of magnitude. Since the SAXS intensity exceeds the WAXD intensity by nearly three orders of magnitude, the software for the measurement control was designed to allow multiple exposures with different exposure time. Thus, images with different exposure times were recorded from the same spot in the sample to cover the large intensity range.

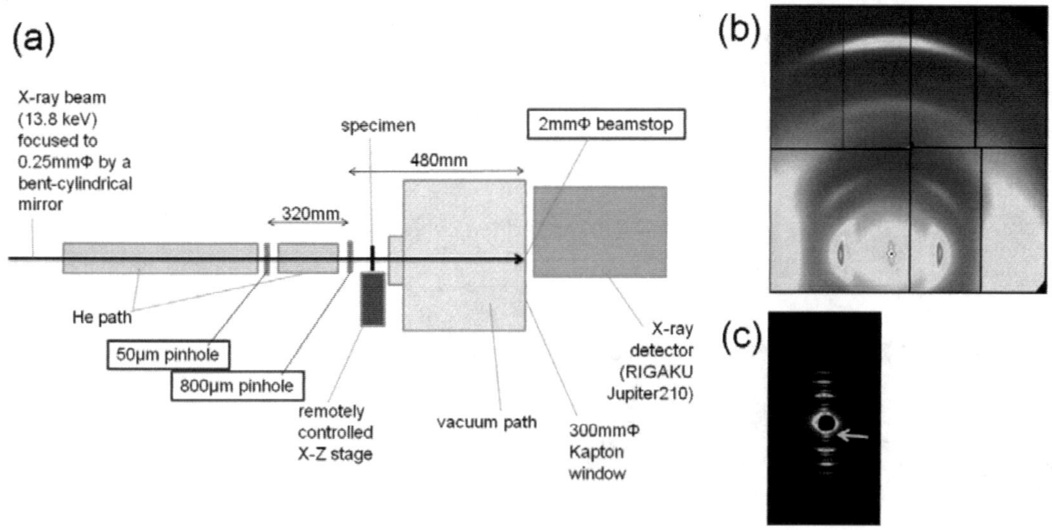

Figure 3. The SAXS/WAXD recording at the BL45XU protein crystallography hutch. (a) Arrangement in the experimental hutch. A 50µm pinhole and a 800µm pinhole were placed at 335 mm and 15 mm upstream of the specimen, respectively. Helium paths were employed to avoid attenuation of the beam by air. The area detector (Jupiter210, RIGAKU, Tokyo, Japan) was placed off-center to cover a larger q-range. (b) A wide-angle diffraction pattern from collagen (dried chicken tendon). The strong arc at the top is the 0.28nm meridional reflection. The black lines are unusable regions between segments of the CCDs. (c) A small-angle diffraction pattern from collagen. The peak just outside of the beam stop (pointed by an arrow) is the third order meridional reflection from the 63nm periodicity.

3. Results

3.1 Small-angle X-ray scattering (SAXS)

The SAXS patterns from the sound enamel showed a weak equatorial scatter, that is, at right angles to the c-axis of the HAp crystallites, which is the long axis of the needle-shaped crystallites (Figure 4a). These patterns often showed superposition of two sets of diffraction diagrams with different orientations, which were given by two regions with different crystallites orientations [13]. The scatter is likely to be due to the density difference between the needle-shaped HAp crystallites and the matrix between them that is composed of lighter non-mineral substances including water and carbonate.

In the demineralized zone, the equatorial scatter was greatly enhanced in the region 30-60 µm from the surface (Figure 4d). Since it has been known that decalcification by acid treatment creates longitudinal voids in the matrix regions by reducing the size of the HAp crystallites [14], the SAXS enhancement is interpreted to be due to formation of voids in the enamel. This scatter from voids is similar to the SAXS from carbon fiber [15] and cellulose [16]. The SAXS from a powdered enamel sample was interpreted in the same manner [17]. Although markedly enhanced, the general shape of the equatorial streak was unchanged, showing that the general texture of the apatite was unchanged by demineralization. The equatorial scatter is rather featureless and hard to interpret. If the voids can be assumed to be cylindrical with different diameters, detailed analysis can be made. However, since the voids seen in electron micrographs are all different in size and shape, it is hard to use standard software for the analysis. Qualitatively, the scattering profile from the demineralized zone is more steep in the small-angle region ($q < 0.2$ nm^{-1}) than that from the sound zone, suggesting that there are a population of larger voids in the demineralized zone (Figure 5a). If the voids are assumed to be infinitely long cylinders parallel to the long axis of the crystallites, the intensity along the equator is proportional to $(J_1(qR)/(qR))^2$ (where R is the radius of the cylinder), which asymptotically decreases with q^{-3} in the large qR region (Figure 5b). After the Lorenz factor, the observed intensity is expected to decrease with q^{-4}. The slopes of the curves (Figure 5a) are much less steeper than expected. This may imply that there is contribution from region with $qR < 3$ in Figure 5b, suggesting the presence of rather small voids with diameters of tens of nm. However, in electron micrographs [14], the voids are very irregular in the cross-sections and too close to each other to diffract independently. These factors must be affecting the scattering intensity and hamper more precise interpretation.

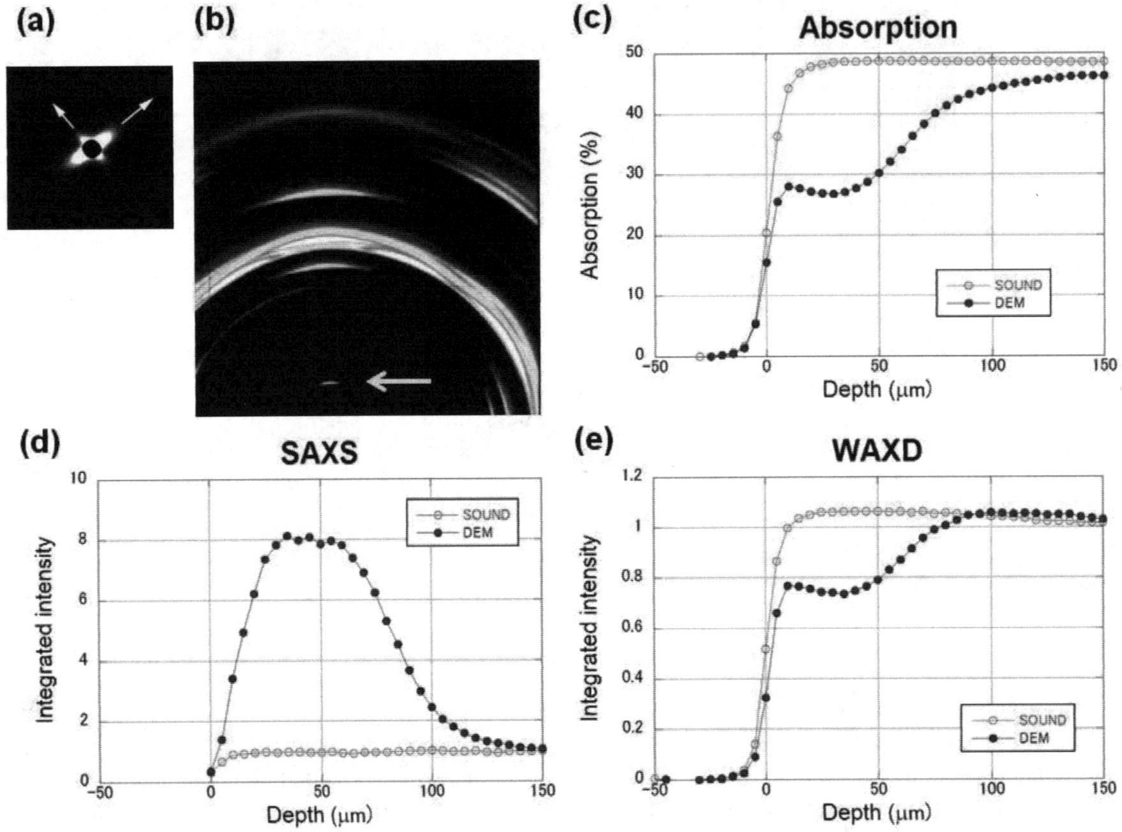

Figure 4. Diffraction patterns and intensity profiles from bovine tooth enamel. (a) SAXS pattern recorded by an X-ray image intensifier. The two arrows indicate the directions of the equators in the two crystalline domains. (b) WAXD pattern recorded by a CMOS flat panel detector. The arrow indicates the (100) equatorial reflection. (c) Absorption measured by an ion chamber along the direction perpendicular to the enamel surface. (d) Integrated intensity in the small-angle region. (e) Integrated intensity of the (100) equatorial reflection. In (c)-(d), SOUND indicates the data obtained from the intact region and DEM that from the acid-treated (demineralized) region.

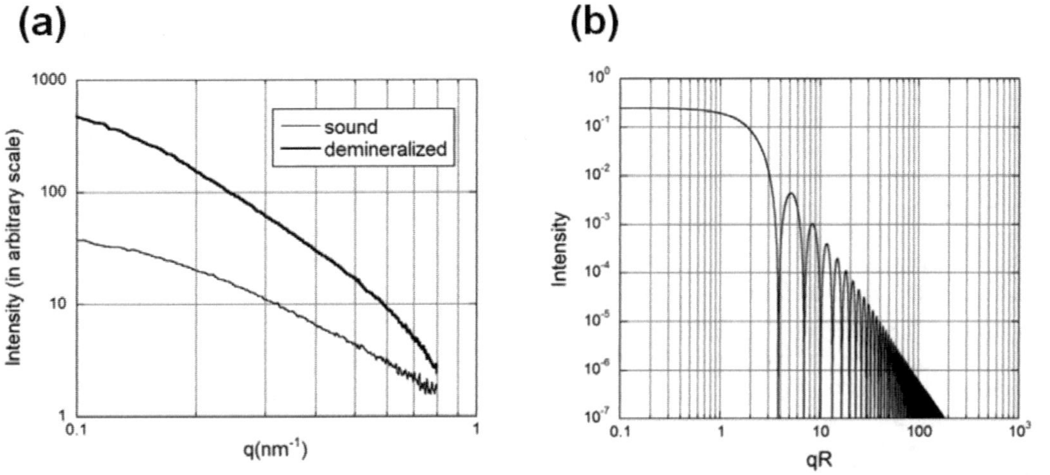

Figure 5. (a) Intensity profiles of the small-angle scattering from bovine enamel. The thin and thick lines are from the sound and the demineralized zones in enamel, respectively. The profiles were obtained by integrating along the arc. (b) Calculated scattering intensity profile from a solid cylinder. R represents the radius of the cylinder.

3.2 Wide angle X-ray diffraction (WAXD)

The WAXD pattern from a slice of bovine enamel shows diffraction from the HAp crystallites (Figure 3a). The HAp microcrystals (crystallites) in the enamel are 0.05-0.2 μm in diameter and up to a few μm in length [18,19]. The long axis corresponds to the c-axis of the unit cell of the HAp crystal. The crystallites are arranged with their c-axes nearly parallel to each other but their rotation around the c-axis is random. Thus, the WAXD pattern from the enamel resembles that of fibrous materials and the diffraction spots can be indexed on a hexagonal three-dimensional lattice [20]. The diffraction spots are arced because the crystallites are not pointing to the same direction throughout the path of the X-ray beam in the sample. However, the widths across the arcs in the radial direction are similar to the X-ray energy spread (about 3%), suggesting that the perfection in each crystallite is high. In most samples, the c-axis of the crystallite (that is, the meridian) was inclined by 30-60 degrees from the surface of the enamel that was more or less parallel to the surface of the tooth. Although difficult to observe in the partial pattern in Figure 3b, as was the case with the SAXS, two sets of diffraction diagrams with different orientations were often superimposed, with the c-axes tilted by about 90 degrees from each other ("two fiber axes" patterns) [13]. In reciprocal space, the rings of the equatorial reflections always intercept the Ewald sphere unless the c-axis becomes nearly parallel to the X-ray beam. On the other hand, the meridional reflections are spots in reciprocal space and easily deviate from the Ewald sphere when the c-axis is tilted. Thus, the intensity of the equatorial reflections is much less affected by the tilt of the c-axis and gives a more reliable measure of the amount of HAp crystals. In the present study, the two-dimensionally integrated intensity of the (100) equatorial reflection (at a Bragg spacing of 0.815 nm) after background subtraction was measured as an index of the amount of the HAp crystallites.

When the integrated intensity of the (100) peak was plotted against the distance of the X-ray beam from the surface, it was almost constant in the sound zone but lower within about 100 μm below the surface in the demineralized zone (Figure 4e). This result is similar to that by TMR. The radial widths (full-width at the half-maximum) of the (100) and (002) peaks were 0.0006 and 0.0021 nm^{-1}, respectively, in the control enamel. Within the resolution of the present study (the width is determined mostly by the energy bandwidth as explained above), these values did not change significantly along the depth or by demineralization (data not shown).

3.3 SAXS/WAXD measurements

In some experiments the SAXS and the WAXD were measured simultaneously (Figure 2). In this case, the WAXD was measured with a CMOS flatpanel detector [11] which could record only less than a half of the diffraction pattern (Figure 4b). However, since the direction of the equator was usually confined to the azimuth of about 45 degrees from the surface, by adjusting the orientation of the detector, it was possible to monitor the intensity of one (100) diffraction spot (Figure 4e). A simultaneous measurement of X-ray transmission through the sample with an ionization chamber (Figure 4c) is equivalent to TMR. The SAXS data were obtained at about 3m downstream with an X-image intensifier and a CCD camera (Figure 4d). This simultaneous measurement ensures that the transmission, SAXS and WAXD data are obtained at the identical spot of the same sample, facilitating comparison of these data.

As exactly the same part of the sample was interrogated by the three techniques, two interesting features can be pointed out. One is the surface layer with higher density that is clearly seen with TMR [1]. This is less distinct in the absorption (Figure 4c) because of the lack of spatial resolution that is apparent in the gradual rise of absorption at the edge. Although the FWHM of the microbeam is about 6 μm, there is a considerably long tail due to Fresnel fringes from the pinhole. Also, the enamel sample may not be set up with its surface edge exactly parallel to the X-ray beam. Compared with the absorption, the surface layer has even lower intensity in the WAXD data (Figure 4e). This may show that the surface layer contains a lower amount of the HAp crystallites per density, but a study at higher spatial resolution will be required to confirm this.

The other interesting observation is that, compared with the absorption and the SAXS intensity, the WAXD intensity in the demineralized region seems to return to the sound level in the shallower region: the WAXD intensity recovers at around 80 μm, while the changes in the absorption and the SAXS intensity persist to more than 100 μm. It can be also noticed that, up to 80 μm from the surface, the decrease of absorption in the demineralized zone is larger than the decrease of the WAXD intensity. These are discussed in detail below.

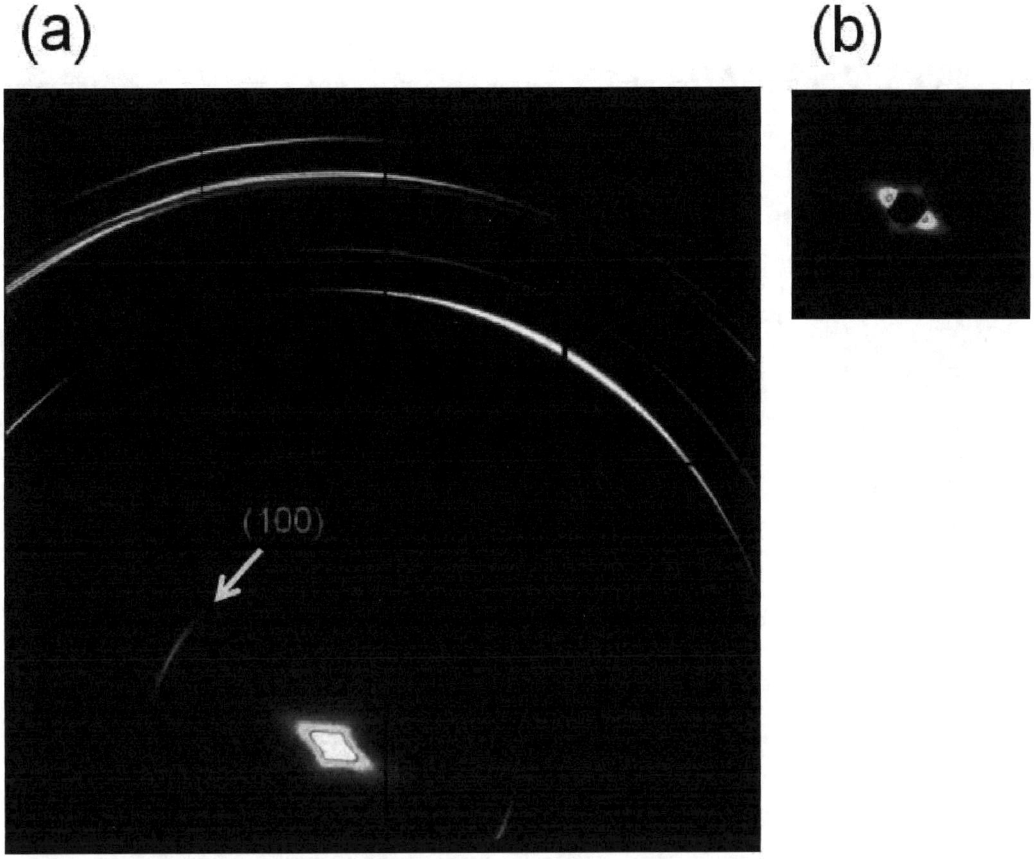

Figure 6. An X-ray diffraction pattern from bovine sound enamel recorded by a large CCD detector (Jupiter210) at BL45XU. (a) The entire diffraction pattern. (b) An enlarged image of the small-angle region.

3.4 2D mapping of the SAXS and WAXD

At BL45XU, by scanning a sample with the 50-μm pinhole, two dimensional maps of the SAXS and WAXD intensity were created. The SAXS map (Figure 7b) shows the integrated intensity in q=0.35-3.2 nm^{-1} summed in all directions. In the demineralized region, the surface of the enamel showed much stronger SAXS than in the sound zone. The width of the high intensity region is about 200 μm, wider than that measured with a 6μm beam, because of the 50μm beam size in this experiment. The WAXD map (Figure 7a) was created from the integrated intensity of the (100) reflection after background stripping.

Although not striking, compared with the SAXS map it is apparent that the surface edge in the demineralized zone has receded slightly. This is due to the lower intensity in this region (Figure 4e). In other areas, there is a gradual intensity difference which may represent changes in the thickness of the slice.

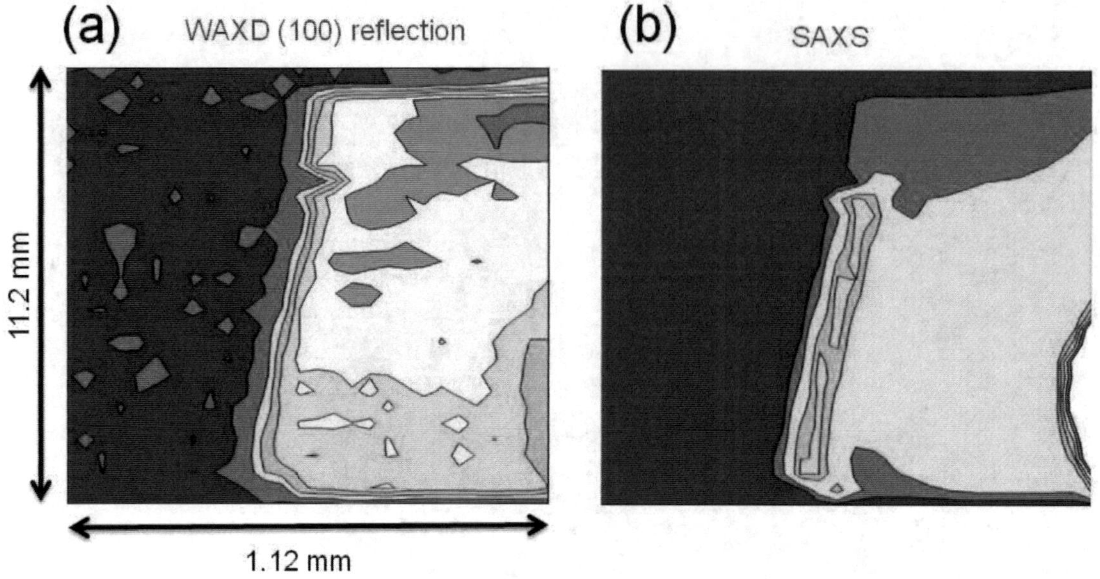

Figure 7. Results from the 2D scan with the 50μm beam (horizontally 0.02mm×56points, vertically 0.2mm×56points) . The region below the nick is the demineralized zone. Note the difference in the scales in horizontal and vertical directions.

4. Discussion

4.1 SAXS/WAXD measurements at the two beamlines

Two different types of the SAXS/WARD microbeam measurement were performed on the tooth enamel with an sound and demineralized regions. The measurement at BL40XU employed a 6μm beam and used different detectors for the small and wide angle regions. On the other hand, at BL45XU the beam was 50 μm and a single detector was used to record both small and wide angles. The former method can resolve the detailed profile of the subsurface lesion which extends only about 100μm below the surface. The latter can be used to study the global differences of SAXS and WAXD in a large area. One advantage of the microbeam at BL40XU is its high flux, above 10^{12} cps with the 6um beam which was made possible by the low energy resolution (about 2%). On the other hand, BL45XU is monochromatic (energy resolution about than 0.01%) and hence the diffraction peaks are sharper. Although this is not very important in the present study in which only integrated intensity was measured, it can be vital when one is trying to measure the width of a peak precisely.

The simultaneous SAXS/WAXD measurement is a useful technique when relating structures at different scales. In the present study on the subsurface lesion, the demineralized WAXD profile was different from the sound profile only within 100 μm from the surface, while it extended deeper in the SAXS profile. This comparison is only possible when the two profiles are obtained from exactly the same part of the sample. These results suggest that the materials in the enamel other than the HAp contribute significantly to the absorption and the SAXS. The matrix between the crystallites is filled with non-crystalline materials such as carbohydrates, lipids and proteins. Loss of these materials decreases absorption, and the voids due to the loss may give rise to small-angle scattering without affecting the diffraction from the HAp.

It is common to use two detectors to record the small and wide angle regions of the scattering [21]. However, the part of the scattering that can be recorded by the wide angle detector is usually limited to either linear or a part of the two-dimensional scattering. Thus, it is ideal to use a large detector that can cover a large q-range. This has been realized at BL40B2 at SPring-8 where an image plate detector with an area of 30cm × 30cm (RIGAKU RAXIS-VII) is used to cover a q-range of 0.35 to 18 nm^{-1} [22]. The detector used at BL45XU has a smaller area (20cm × 20 cm) and thus the beam stop had to be located near an edge to cover a wide q-range. However, since this detector has a faster readout than the image plate detector, it was possible to do a 2D scan. An ideal detector is a CCD-based detector with a larger area such as ADSC Quantum-315 or MAR MX-325 which has an area larger than 30cm × 30cm. A large detector area will enable us to use a longer camera length and improve the small-angle resolution. If a 2D scan is not required, use of a larger but slower detector can be considered: for example, RIAGKU RAXIS-V has a detection area of 40cm × 40cm [23]. Two technical points should be reminded in a SAXS/WAXD experiment with a single detector: (1) the size of the exit window of the vacuum tube becomes larger for a larger detector. In the cameras described here, 130μm thick Kapton film is used as a window material. A special care should be taken to make a window as large as 40cm × 40cm. (2) A larger detector size means a large data size. The amount of data created by a 2D scan could become challenging.

4.3 Future prospects
Subsurface lesion is quite common and it may be possible to reverse the process before a cavity is formed. In fact, it has been suggested that saliva may have such a function: Ca and phosphate in saliva may be able to remineralize the lesion [24]. Chewing gums which are claimed to facilitate this process are in the market. It is anticipated that the remineralization rebuilds the HAp crystals that are lost by acids. However, the remineralization may be due to accumulation of precipitates of calcium phosphate. The present X-ray method confirms that the mineralization definitely involves increase of the HAp crystals. Thus, it will be a powerful tool to examine caries lesion in future. In this context, it should be pointed out that the study can be made to some extent with a laboratory X-ray apparatus. In order to evaluate the degree of demineralization and remineralization, a linear scan with a beam size of 50μm may be sufficient as evident in Figure 7. An X-ray tube with a linear focus smaller than 50μm is available with a reasonable flux. It should be noted that for a linear scan the beam does not have to be point-collimated. A linearly collimated beam may distort the 2D diffraction pattern. However, it may be used for the purpose of evaluating SAXS because measurement of total scattering intensity suffices. Also, as the (100) reflection is an isolated peak, a short linear beam does not affect the reliability of the WAXD result. Although the synchrotron microbeam experiment is ideal to obtain the most reliable data, a laboratory-based method can make it a more general tool in dentistry.

Acknowledgements
We thank Dr. Takaaki Hikima and Dr. Hiroyasu Masunaga for help at BL45XU.

References
[1] Arends J and Christoffersen J 1986 *J. Dent. Res.* **65** 2
[2] Damen J, Exterkate R, and ten Cate J 1977 *Adv. Dent. Res.* **11** 415
[3] Featherstone J and Zero D 1992 *J. Dent. Res.* **71** 804
[4] ten Cate J, Dundon K, Vernon P, Damato F, Huntington E, Exterkate R, Wefel J S, Jordan T, Stephen K W, and Roberts A J 1996 *Caries Res.* **30** 400
[5] Featherstone J D 1979 *J. Ultrastr. Res.* **67** 117
[6] Yagi N, Ohta N, Matsuo T, Tanaka T, Terada Y, Kamasaka H, To-o K, Kometani T, and Kuriki T 2009 *J. Synchrotron Rad.* **16** 398
[7] Meyer-Lueckel H, Tschoppe P, and Kielbassa A 2006 *J. Oral Rehab.* **33** 760
[8] Inoue K, Oka T, Suzuki T, Yagi N, Takeshita K, Goto S, and Ishikawa T 2001 *Nucl. Instrum. Meth.* **A467-468** 674
[9] Ohta N, Oka T, Inoue K, Yagi N, Kato S, and Hatta I 2005 *J. Appl. Cryst.* **38** 274
[10] Amemiya Y, Ito K, Yagi N, Asano Y, Wakabayashi K, Ueki T, and Endo T 1995 *Rev. Sci. Instrum.* **66** 2290
[11] Yagi N and Inoue K 2007 *J. Appl. Cryst.* **40** s439

Journal of Physics: Conference Series **247** (2010) 012024 doi:10.1088/1742-6596/247/1/012024

[12] Kumasaka T, Yamamoto M, Yamashita E, Moriyama H, and Ueki T 2002 *Structure* **10** 1205
[13] Hirota F 1986 *J. Dent. Res.* **65** 978
[14] Simmelink J and Abrigo S 1989 *Adv. Dent. Res.* **3** 241
[15] Gupta A, Harrison I, and Lahijani J 1994 *J. Appl. Cryst.* **27** 627
[16] Crawshaw J and Cameron R 2000 *Polymer* **41** 4691
[17] Gutierrez P, Piña C, Lara V, and Bosch P 2005 *Arch. Oral Biol.* **50** 843
[18] Arends J and Jongebloed W L 1978 *J. Biol. Buccale* **6** 161
[19] Kerebel B, Daculsi G, and Kerebel L M 1979 *J. Dent. Res.* **58 (Spec Issue B)** 844
[20] Trautz O, Klein E, Fessenden E, and Addelston H 1953 *J. Dent. Res.* **32** 420
[21] Bras W, Dolbnya I P, Detollenaere D, van Tol R, Malfois M, Greaves G N, Ryan A J, and Heeley E 2003 *J. Appl. Cryst.* **36** 791
[22] Hatta I, Ohta N, Inoue K, and Yagi N 2006 *Biochimica Biophysica Acta* **1758** 1830
[23] Yamamoto M, Kumasaka T, Yamazaki H, Sasaki K, Yokozawa Y, and Ishikawa T 2001 *Nuclear Instruments and Methods A* **467-468** 1160
[24] Dowd F J 1999 *Dent. Clinics of North America* **43** 579

Journal of Physics: Conference Series **247** (2010) 012025 doi:10.1088/1742-6596/247/1/012025

Effect of water on self-assembled tubules in β-sitosterol + γ-oryzanol-based organogels

Ruud den Adel, Patricia C.M. Heussen and Arjen Bot

Unilever Research and Development Vlaardingen, Olivier van Noortlaan 120, NL-3133 AT Vlaardingen, The Netherlands

ruud-den.adel@unilever.com

Abstract. Mixtures of β-sitosterol and γ-oryzanol form a network in triglyceride oil that may serve as an alternative to the network of small crystallites of triglycerides occurring in regular oil structuring. The present x-ray diffraction study investigates the relation between the crystal forms of the individual compounds and the mixture in oil, water and emulsion. β-Sitosterol and γ-oryzanol form normal crystals in oil, in water, or in emulsions. The crystals are sensitive to the presence of water. The mixture of β-sitosterol + γ-oryzanol forms crystals in water and emulsions that can be traced back to the crystals of the pure compounds. Only in oil, a completely different structure emerges in the mixture of β-sitosterol + γ-oryzanol, which bears no relation to the structures that are formed by both individual compounds, and which can be identified as a self-assembled tubule (diameter 7.2±0.1 nm, wall thickness 0.8±0.2 nm).

1. Introduction

Over the years, it has been realized that complex molecules may form solid phases that are ordered in a fundamentally different way on a supra-molecular scale than according to simple crystal lattices (for which the simple, face-centered or body-centered cubic structures may serve as the archetypical example). Among these solid phases are those that do not form a space-filling structure. Some systems are known to form helical or twisted ribbons, resulting in the formation of fibrillar structures instead of forming a structure that expands in three dimensions [1]. This type of molecular aggregation is sometimes referred to as self-assembly. The exact 'design rules' for obtaining a particular type of supra-molecular ordering are not yet established, but the chirality of the molecule does seem to play a role, as most examples of self-assembly involve chiral molecules [1].

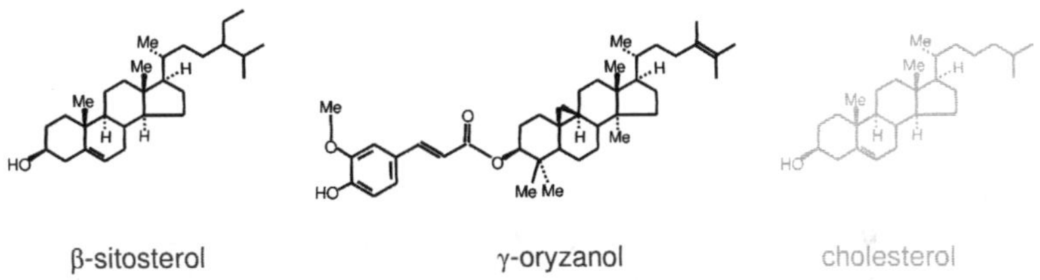

Figure 1: Chemical structure of the main components in β-sitosterol and γ-oryzanol. For comparison, the structure of cholesterol is also drawn.

Cholesterol (see Figure 1) is an example of a molecule that forms a fibrillar structure under specific conditions: the molecules aggregate to form a helical ribbon of which the wall is slowly filled up to form a solid tubule [2]. Many cholesterol derivatised compounds seem to show fibrillar aggregation as well: cholesterol substituted with anthracene [3-5], azobenzene [6,7], carbamate [8], stilbene [9-10], squaraine [9] and amino acid [11] groups have the ability to gel selected organic phases by themselves. Most of these successful organogelling agents contain an aromatic ring in the appended group. Next to this, a range of dicholesteryl compounds have been identified with various linking groups that show the capability to gel organic phases [12].

Mixtures of sterol molecules have been investigated less extensively. A system that received some attention is the mixture of γ-oryzanol with β-sitosterol in triglyceride oils (see Figure 1). The individual compounds cannot gel the oil [13]. The mixture can form tubules in oil with a diameter of only 7.2±0.1 nm and a wall thickness of 0.8±0.2 nm [14-16]. The triglyceride oil can be found both inside and outside the tubules. The exact molecular structure of the wall is not known, but circumstantial evidence allows some speculation: It has been shown that these molecules can stack, but that they do not order exactly parallel due to the presence of a hydrogen bridge [17]. As a consequence, a curved supermolecular structure can form, which may be characterized best as a tubule on longer length scales (see Figure 2). The wall of the tubule is assumed to consist of stacked sterol ring systems, and the hydrogen bond is located at the exterior wall of the tubule. As a consequence, the ferulic acid moieties in the oryzanol molecules would stick out of the tubule [15]. The tubules may stick together, most likely at locations that are relatively depleted in oryzanol as the ferulic acid moieties at the exterior wall seem to reduce aggregation of the tubules [15,17]. If sufficient aggregation of tubules occurs, the system would indeed form the firm, slightly hazy gel that has been observed experimentally [13]. The mostly transparent, slightly hazy appearance of these gels is explained by the small diameter of the tubules that form the gel.

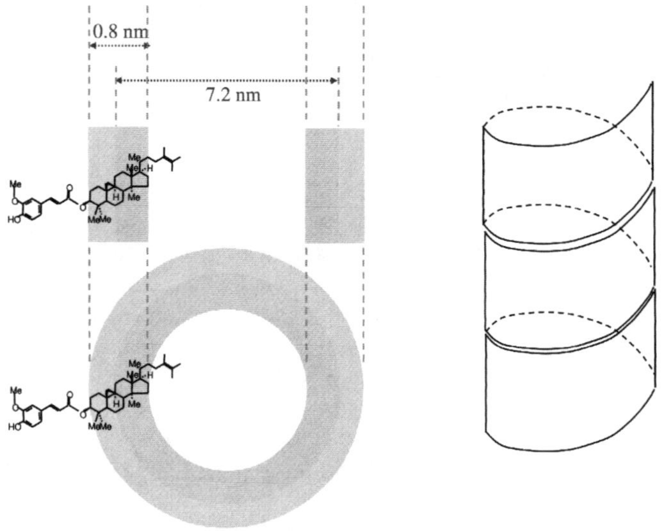

Figure 2: Proposed structure of the sterol+oryzanol tubules, with a diameter of 6.7-8.0 nm and a wall thickness of 0.8-1.2 nm. The dimensions of the tubule are derived from SAXS measurements, the arrangement of molecules in the wall is just hypothesis for now. On the right hand side, a schematic representation of a helical ribbon is shown.

Some alternatives to β-sitosterol have been identified, such as the chemically very similar compounds ergosterol, cholestanol, stigmasterol and cholesterol [15]. The identification of such alternatives, however, did not allow pin-pointing obvious molecular characteristics required to form the tubules. Until now, only two have been found. First, the presence of the hydroxyl moiety in the sterol (i.e. replacing β-sitosterol by 5α-cholestane does not result in the formation of a tubule anymore

[13]). Secondly, the number of double bonds in the sterol ring system seems to affect the diameter of the tubules: more double bonds correlate with more narrow tubules (see Figure 3). Interestingly, the chemical details of the alkyl chains do not seem to affect the formation of the tubules very much. A likely explanation is that these flexible alkyl chains probably blend in with the surrounding liquid triglyceride phase, and do not contribute much to the formation of the tubules in the first place.

No food-grade alternatives to γ-oryzanol have been identified yet. Replacing the ferulic acid moiety in γ-oryzanol by the fatty acids from sunflower oil prevents tubule formation [15]. A limited number of alternative esters are known to gel organic phases outside the realm of food-related ingredients (and food grade oils) [3-11].

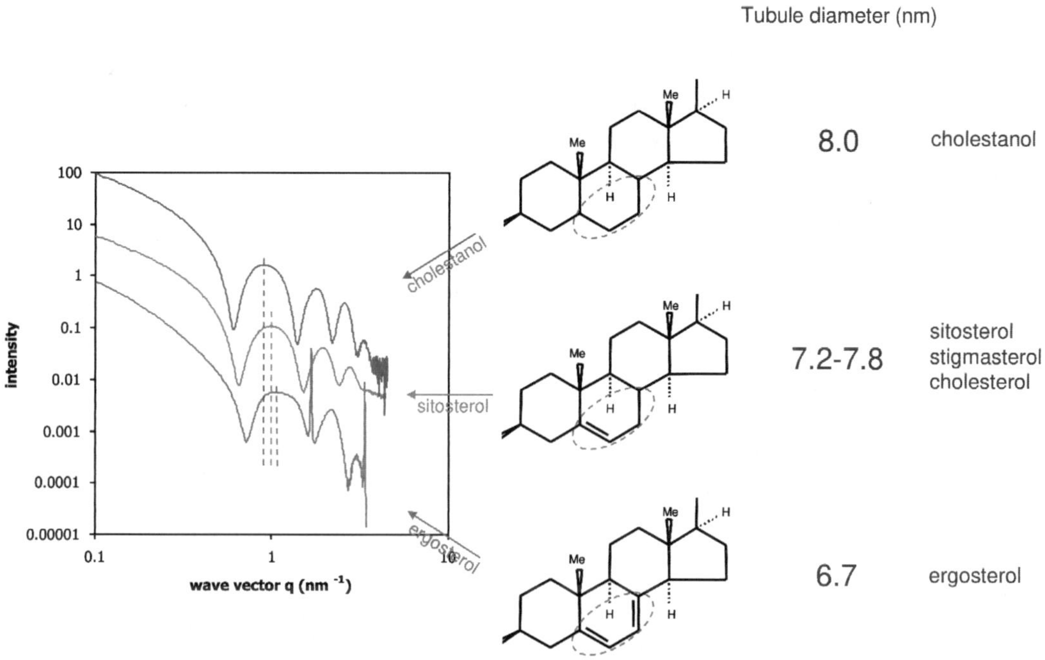

Figure 3: Effect of sterol ring structure on tubules diameter. The sharp peaks in the ergosterol + oryzanol system are due to separate crystallisation of ergosterol crystals.

The end vision for these organogelling systems would their application in the structuring of food emulsions [17-19]. Significant hurdles will have to be overcome to achieve this. Amongst these is the ability to combine the structuring functionality of these systems with the presence of water. Some initial scoping in this field has already been published [20]. The present paper will lay the ground work for establishing a deeper understanding of the effect of water on the tubules, the building blocks of these organogelling systems.

2. Experimental

2.1. Materials

In the present experiments γ-oryzanol (Tsuno Rice Fine Chemicals) and tall oil sterol (Unilever, 78.1% β-sitosterol, 10.3% campesterol, rest minor but structurally very similar sterols [15]) were used in combination with refined sunflower oil (Cargill).

Samples were prepared by mixing the ingredients until all powder had dissolved in the sunflower oil (at ~100°C). Emulsions were prepared by mixing in the appropriate amount of water (no additional emulsifiers used). The samples were cooled and stored for 1 week at 5°C.

2.2. Measurement techniques

Small-angle and wide-angle x-ray scattering (SAXS, WAXS) experiments were performed at the high-brilliance ID2 beamline of the European Synchrotron Radiation Facility (ESRF) in Grenoble, France. Details of the experimental set-up are given elsewhere [14,21]. SAXS data were collected in the range $0.069<q/nm^{-1}<4.3$ and WAXS data in the range $3.6<q/nm^{-1}<33.1$, where $q = 4\pi \cdot \sin\theta / \lambda$ is the wave vector (and θ the scattering angle and λ the wavelength of the incoming x-ray radiation). Scattering data from separately obtained water and sunflower oil reference samples were subtracted from the emulsion and dispersion data to obtain curves representing the sitosterol and/or oryzanol.

A number of complementary x-ray diffraction (XRD) measurements were performed at Unilever R&D Vlaardingen using a Bruker D8-Discover in a θ/θ configuration. Data was collected in the range $0.64<q/nm^{-1}<7.7$ [15]. Data analysis was described elsewhere [15,22].

Fourier Transform Infrared (FTIR) experiments were performed on a Bio-Rad FTS-6000 spectrometer (V 4.0) which was equipped with a deuterated triglycine sulphate (DTGS) detector. The accessory used was a Micro-ATR equipped with a Si-crystal. For both samples and background, 64 interferograms with a resolution of 4 cm^{-1} were co-added and Fourier transformed. The spectral information was obtained over a frequency window from 4000 to 400 cm^{-1}.

Differential Scanning Calorimetric (DSC) experiments were performed on a Perkin-Elmer Pyris 1. The samples (15-30 mg) were weighed into stainless steel sample pans and subjected to a heating-cooling-heating cycle between 20 and 120°C at a scanning rate of 10°C/min.

3. Results and discussion

3.1. Kinetics of sitosterol+oryzanol tubule formation

A previous study demonstrated that the tubule structure vanishes at the melting temperature of the gel, which showed that no mesophase structure persists above the melting temperature [15]. A similar experiment investigating the kinetics of tubule formation showed that most of the tubules (>85%) were formed within the time resolution of that particular experiment (~10 min at 5°C). This is in line with macroscopic observations that the gels develop very quickly, once gelation is initiated [13].

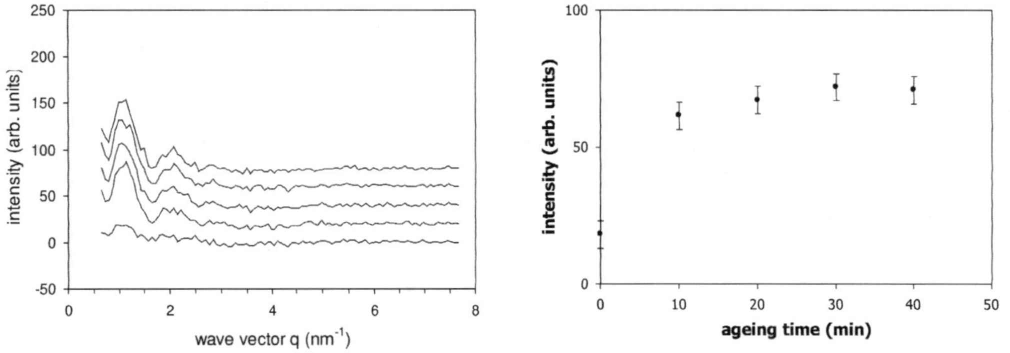

Figure 4: (Left panel) XRD curves for organogels composed of 3.2% sitosterol + 4.8% oryzanol in sunflower oil (after correction for contribution from sunflower oil) as a function of storage time at 5°C after cooling from 100°C at a rate of 10°C/min. Vertical shifts were applied to the curves for clarity; (Right panel) Time dependence of the first maximum in the XRD curves near $q=1$ nm^{-1}.

3.2. Molecular structure of powder, lipid and aqueous slurries, and emulsions

3.2.1. *Sitosterol.* A number of SAXS curves were obtained for β-sitosterol in powder form, in lipid and aqueous slurries, and in an emulsion with 30% water (see Figure 5a and Table 1). It is not expected that the present preparation procedure produces neat single crystals, nor is it our aim to derive information on units cells from the data. The slurries and the emulsions were stored for 1 week before measurement. The data were compared to literature data [23-25] which distinguished three types of sitosterol crystals: anhydrous crystals (most pronounced and characteristic peaks at $d = 2\pi/q_i$ = 1.76, 0.880 and 0.523 nm), hemi-hydrates (0.5 mol water/mol sitosterol, most pronounced and characteristic peaks at d = 1.88 and 0.481 nm) and monohydrates (1 mol water/mol sitosterol, most pronounced and characteristic peaks d = 1.76 and 0.503 nm). Next to that, all crystal forms typically exhibit Bragg reflections for 1.2 and 0.59 nm. Note that the data in the present experiment accesses a wider q-range than the literature data ($0.4<d<2$ nm).

The close resemblance of the sitosterol chemical structure to that of cholesterol (see Figure 1) allows an interpretation of the main features of the diffractograms, despite the fact that a precise crystallographic analysis of sitosterol has not been done. The sitosterol crystals are expected to form a bilayer structure of about 3.8 nm thickness (cf. Reference [26]). The position of the characteristic sterol ring system in these structures is defined quite accurately, whereas the alkyl chains tend to have more conformational freedom. The relative orientation of the molecules in the various crystal forms seems to be determined by the need to maximise the hydrogen bonding between the molecules, explaining some of the differences between anhydrous and hydrated forms [26].

The presence of crystallographic reflections for the powder at 0.89 and 0.48 nm in Figure 5a, suggests the presence of anhydrous and hemi-hydrate sitosterol in the powder, respectively. On the other hand, the absence of a peak at 0.50 nm suggests that mono-hydrated sitosterol is not present in the powder [23].

Infrared spectroscopy (IR) can be used to investigate hydrogen bonding in these systems (see Figure 6) [27]. Generally, intermolecular hydrogen bonds should appear as medium-intensity bands in the 3550-3450 cm^{-1} wave number range, and intramolecular hydrogen bonds as medium intensity bands in the 3570-3540 cm^{-1} wave number range [28]. Focussing on the 3800-3000 cm^{-1} wave number range, the β-sitosterol powder only shows a broad band in around ~3350 cm^{-1}, and a very weak band at the wave numbers for intermolecular hydrogen bonding, ~3420 cm^{-1} (Figure 7a). The spectra for sitosterol and oryzanol are in agreement with those of Qian and Xu [29] and Bucci et al [30], respectively.

In oil slurries, the pattern of crystallographic reflections suggests a mixture of anhydrous and hemi-hydrated sitosterol too (Figure 5a). However, some features become much more pronounced at both relatively large distances (d=5.19 nm) and small distances (d=0.46, 0.43, 0.22 and 0.21 nm), the intermediate range being relatively unaffected. It should be noted that most of the peaks did occur already in the powder spectrum, but at insufficient intensities to report them in Table 1. The infrared spectrum of a sitosterol slurry in oil shows a very small peak at 3441 cm^{-1}, indicating some very weak intermolecular hydrogen bonding (Figure 7a).

In aqueous slurries, the absence of peaks at d = 1.88 and 0.48 nm suggests that the sample does not contain a hemi-hydrate, and the absence of peaks at d = 0.88 and 0.52 nm suggests (not surprisingly) that the sample is not anhydrous (Figure 5a). Although Christiansen et al [23] do not list a peak at 0.75 nm as characteristic for their mono-hydrate form, Argay et al [24] do list the 0.75 nm peak in their mono-hydrate crystal. The combination of data suggests therefore the presence of sitosterol monohydrate, possibly as a mixture of the forms observed by Christansen et al [23] and Argay et al [24]. The infrared spectrum is similar to that of the powder (Figure 7a).

The most pronounced peaks for sitosterol dispersed in an o/w emulsion with 30% water coincide with those for the aqueous slurries, except that some of the characteristic mono-hydrate peaks tend to be weaker than in the aqueous slurry (Figure 5a). This suggests that the alternative hydrate forms more easily in emulsions, and one can speculate that the triglycerides play a role in this. Generally, however, peaks in the emulsion are more pronounced than in the oil slurry, which can be explained by the reduced solubility of sitosterol in oil in the presence of water from the emulsion [31].

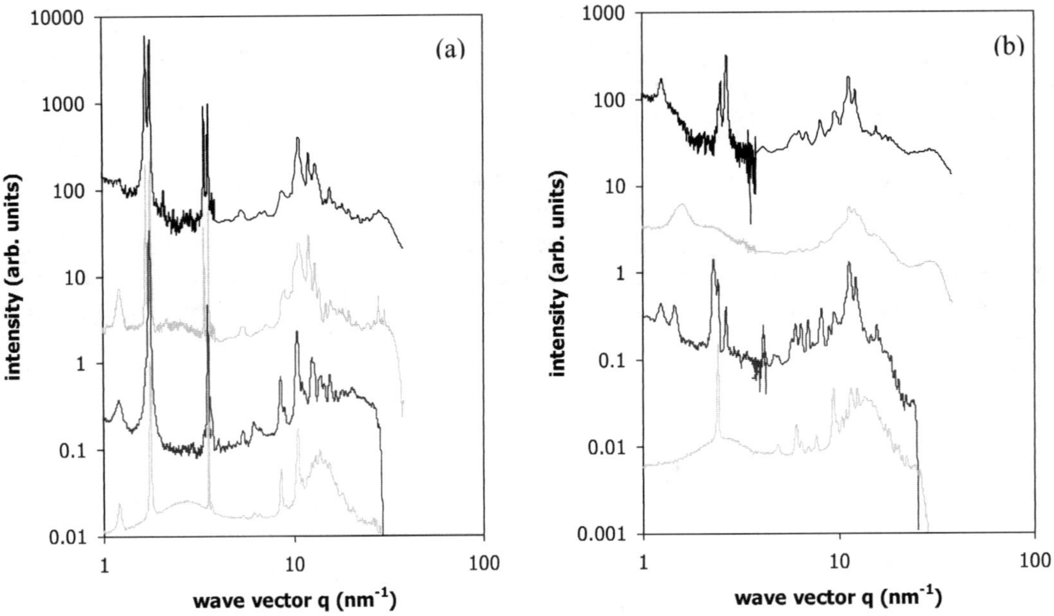

Figure 5: SAXS data for (a) sitosterol and (b) oryzanol, dissolved/dispersed in a number of different phases. From top to bottom: (—) powder, (—) 16% in oil, (—) 16% in water, and (—) 16% in an o/w emulsion containing 30% water after 1 week storage (20°C for oil slurry, 10°C for water slurry and emulsion).

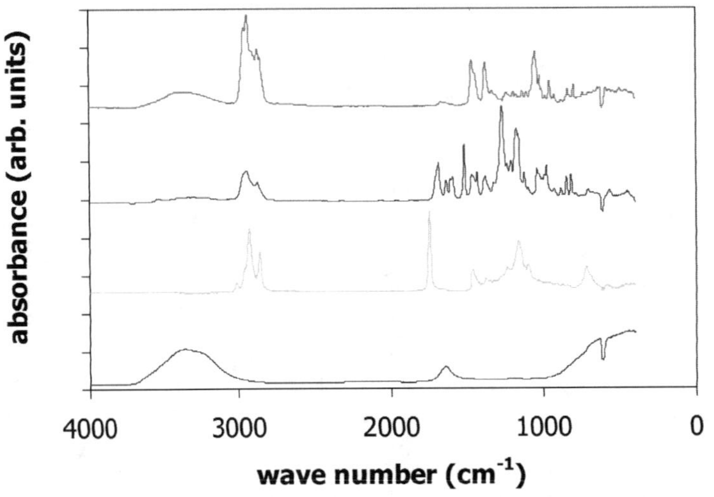

Figure 6: IR spectra of the raw materials. From top to bottom: (—) sitosterol powder, (—) oryzanol powder, (—) sunflower oil and (—) water.

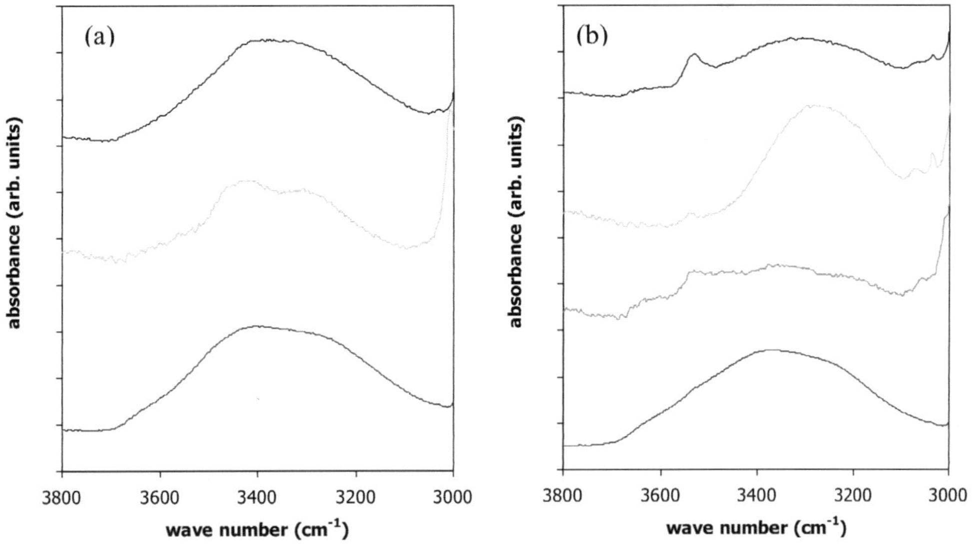

Figure 7: IR spectra for (a) sitosterol and (b) oryzanol, dissolved/dispersed in a number of different phases. From top to bottom: (—) powder, (, —) in oil, (—) in water. Two types of crystals were observed in the case of oryzanol: ( ) transparent, (—) whitish.

3.2.2. *Oryzanol.* Cholesteryl esters form complex crystal structures, sometimes dominated by the intermolecular interactions between the sterol ring systems (as in so-called 'type-II monolayers') or by the balance between the interaction of the esterified acid moiety with the sterol ring system on the one hand and the interaction of the sterol ring systems on the other hand (as in 'type-I monolayers') [26]. Systems that are dominated by interactions between the esterified acid moieties tend to organise in bilayers, but this seems to happens mainly in systems where the esterified moiety consist of long fatty acid chains [26].

The SAXS data that was obtained for γ-oryzanol powder, in lipid and aqueous slurries, and in an emulsion with 30% water is shown in Figure 5b (see also Table 1). Bragg peaks for γ-oryzanol powder confirm that the main Bragg reflection is moved to much smaller d values around 2.5 nm, suggesting that no bilayer structure is formed. The absence of reference data makes it difficult to interpret the data in terms of hydration state, although it seems likely that the two peaks at d=2.50 and 2.34 nm should be assigned to different hydration states of the oryzanol crystals. It should be noted however, that the oryzanol powder cannot be wetted very well in water, more or less in contrast to sitosterol powder. Infrared spectroscopy on γ-oryzanol powder shows a weak band indicating intramolecular hydrogen bonding at 3532 cm^{-1} (Figure 7b).

Oil slurries do not show very pronounced Bragg reflections, signalling the high solubility of oryzanol in the triglyceride oil. The reflections in the spectrum of the aqueous oryzanol slurries match the powder spectrum best, because the oryzanol does not dissolve very well in water. Like in the sitosterol case, we can see the main reflection moving to somewhat bigger d values – it is tempting therefore to identify these peaks at d = 2.60 en 2.74 nm as features of an oryzanol hydrate (Figure 5b). The infrared spectra suggest the presence of two different types of crystals in the dispersion. Transparent crystals contain traces of water, whereas the whitish crystal appear to be anhydrous. Both show very weak intramolecular hydrogen bonding peak as the powder (Figure 7b).

The diffractogram for emulsions is simpler than for the aqueous solutions, because a large fraction of the oryzanol dissolves in the oil (Figure 5b). The infrared spectrum for the aqueous dispersion is dominated by water, and does not show any pronounced intermolecular of intramolecular features (Figure 7b).

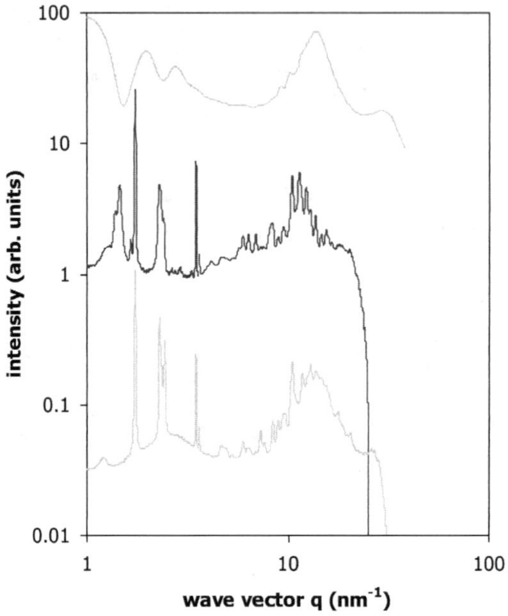

Figure 8: SAXS data for a 40:60 mixture of sitosterol:oryzanol dissolved/dispersed in a number of phases. From top to bottom: ( ) 16% in oil, ( ) 16% in water, and ( ) 16% in an o/w emulsion containing 30% water after 1 week storage (20°C for oil slurry, 10°C for water slurry and emulsion). Note that one additional decade of the q-range is depicted here compared to Figure 5.

3.2.3. *Sitosterol+oryzanol mixtures.* The experiments reported until now were intended to have the data available to determine whether oryzanol or sitosterol show any special compound formation in triglyceride oil, or in water, or in their mixture. Furthermore, the effect of water on such mixtures was considered to be of interest. The data is shown in Figure 8 (see also Table 1).

The data on oryzanol + sitosterol in triglyceride oil shows the broad interference pattern of self-assembled tubules that has been presented before [14,15]. It is of interest to note that the scattering pattern does not show any of the sharp reflections that are characteristic for the diffractograms of the pure compounds. Instead, the data shows very smooth, broad features, except for a few very weak wiggles in the WAXS region of the spectrum (d = 0.68 and 0.62 nm). Thus, it can be concluded that the molecular order in the tubules is very different from that in the pure components. The absence of additional peaks in WAXS is a strong argument to support the claim that no bilayer is formed in the wall of the tubules. In addition, the infrared spectrum of sitosterol+oryzanol in oil shows a small peak at 3441 cm^{-1}, indicating intermolecular hydrogen bonding (Figure 9). It can be shown that this is not a feature of any of the single components in the mixture by subtracting the contributions of sitosterol, oryzanol and sunflower oil from the spectrum, because the resulting corrected spectrum is essential identical in this frequency window to the uncorrected spectrum. This peak was already observed in the spectrum for the sitosterol oil slurry, but at *much* lower intensity. It is interesting to note that the intermolecular peak can be found at a different position than the intramolecular peak for oryzanol powder and in the aqueous sitosterol + oryzanol slurry. The data on the aqueous slurry contains a mixture of sitosterol and oryzanol crystals, as can be concluded from inspection of Table 1.

The scattering data on the emulsion reveals reflections that can be traced back to sitosterol crystals in most cases, and to oryzanol crystals in a smaller number of other cases. This can be explained by the high solubility of oryzanol in the oil phase of the emulsion, leading to greater contribution of sitosterol crystals to the diffractogram. More importantly, however, we see no evidence for the presence of tubule formation under the present conditions. This indicates that the tubule structure is extremely sensitive to the presence of water.

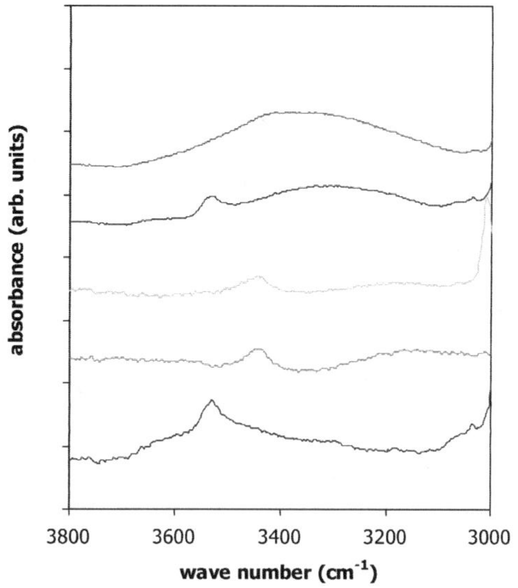

Figure 9: From top to bottom, IR spectra of the (—) sitosterol, (—) oryzanol, ( ) sitosterol+oryzanol in oil, (—) sitosterol+oryzanol in oil (individual components subtracted), and (—) sitosterol+oryzanol in water (water subtracted). The mixture in oil shows a small peak at 3441 cm^{-1}, which is ascribed to intermolecular hydrogen bonding.

Table 1: Most pronounced Bragg spacings d = 2π/q_i in SAXS and WAXS diffractograms for sitosterol, oryzanol and a 40:60 mixture of sitosterol+oryzanol (total sterol(esters)s at 16% by weight). A more or less equal number of weaker reflections are present in the diffractograms. Numbers in bold refer to the strongest reflections.

d (nm)	sitosterol	oryzanol	sitosterol+oryzanol
powder	**3.76**, **3.57**, 3.04, **1.88**, **1.78**, 1.19, 0.89, 0.73, **0.59**, **0.52**, **0.48**, **0.40**, 0.32, 0.24	4.95, 2.50, 2.34, 1.53, 0.99, 0.90, 0.77, **0.56**, **0.52**, 0.40	not applicable
dispersed in oil	5.19, **3.76**, **3.55**, **1.88**, **1.78**, 1.17, 0.89, 0.71, **0.59**, **0.52**, **0.48**, 0.46, 0.43, 0.40, 0.32, 0.22, 0.21	3.90, 0.76, 0.56, 0.52	**6.68**, **3.19**, **2.28**, 0.68, 0.62
dispersed in water	5.19, **3.61**, **1.81**, 1.18, 1.04, **0.75**, **0.60**, **0.51**, 0.45, **0.41**, 0.38, 0.35, 0.31, 0.24	5.07, 4.33, **2.74**, **2.60**, 2.36, 1.50, 1.34, 1.04, 0.97, 0.90, 0.77, 0.66, **0.56**, **0.51**, 0.46, 0.40	5.15, 4.30, **3.59**, 2.72, 2.60, 1.80, 1.05, 0.98, 0.90, 0.77, 0.66, **0.60**, **0.56**, **0.51**, 0.49, 0.46, 0.43, 0.41
dispersed in emulsion	5.19, **3.59**, **1.79**, **0.75**, **0.60**, 0.51, 0.49, 0.46, 0.41, 0.35, 0.28, 0.24	**2.61**, 1.28, 1.03, 0.82, 0.67, 0.54, 0.50	5.19, **3.59**, 2.72, 2.59, 1.80, **0.60**

3.2.4. *Behaviour at higher temperatures.* The fact that the mixture of sitosterol and oryzanol in the presence of water does not behave differently from the single components at low temperatures (below ~70°C) is also clear from the behaviour at a differential calorimetric scan (DSC) in the emulsion. The data is shown in Figure 10.

For sitosterol, there are clear features at ~45°C, ~80°C and ~100°C. The monohydrate is the most stable from below 45°C, the hemi-hydrate between 45 and 80°C, and the anhydrous form between 80 and 110°C [23] (but note that the small differences between the x-ray diffractograms after chilled storage for one week reported in the previous sections and the ones reported by Christiansen et al [23]). Sitosterol is completely dissolved above ~110°C. There is qualitative agreement with the phase behaviour of hydrated cholesterol as a function of temperature [32]. Oryzanol shows three features: peaks at ~70°C, ~100°C and ~115°C. Although reference data is lacking, it seems likely that the transitions are also related to (de)hydration phenomena as oryzanol shares many of its structural features with sitosterol (see Figure 1). The emulsion with the sitosterol + oryzanol mixtures shows the same behaviour as the single components, i.e. a mono-hydrate to hemi-hydrate transition for the sitosterol at ~45°C and a transition at ~70°C in oryzanol.

Surprisingly, however, the mixture does behave differently at higher temperatures (above ~70°C). No further transitions are observed for the sitosterol+oryzanol mixture dispersed in emulsion, whereas the individual components do show a number of transitions in what is assumed to be their anhydrous state.

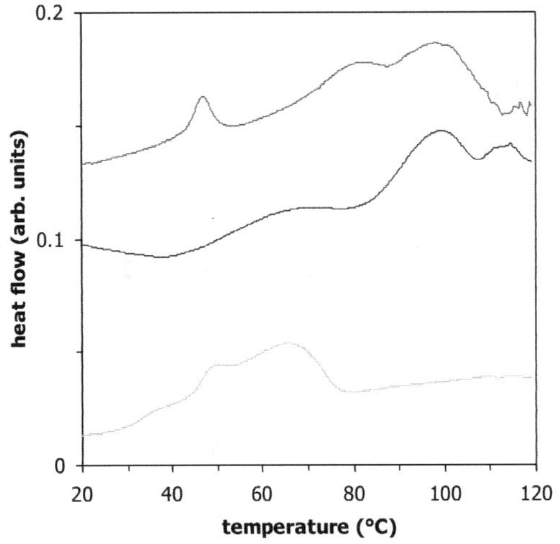

Figure 10: DSC data for (—) sitosterol, (—) oryzanol and ( ) mixed sitosterol+oryzanol in an emulsion containing 30% water after 1 week storage at 10°C. The scanning rate was 10°C/min.

4. Conclusions

The mixture of γ-oryzanol with β-sitosterol forms a network of tubules in oil that may serve as an alternative to the network of small crystallites of triglycerides occurring in regular oil structuring [17-19,33-37]. The tubule diameter is found to be 7.2±0.1 nm, the wall thickness 0.8±0.2 nm [15]. The results are consistent with a previously proposed helical ribbon structure. The present experiments demonstrate that the tubules form rapidly upon sufficiently deep cooling.

The x-ray diffraction experiments have shown that for the present experimental conditions 16% sitosterol and oryzanol form regular crystals in oil, in water, and in emulsions. The crystals are sensitive to water. The mixture of sitosterol + oryzanol (16% in total, ratio 40:60) forms crystals in water and in emulsions that can be traced back to the crystals of the pure compounds. Only in oil, a completely different structure is formed, which bears no relation to the structures that are formed by both individual compounds, and which can be identified as a self-assembled tubular structure involving both molecules. Obviously, the fact that different molecular structures are formed in the presence of water does not exclude the possibility that such systems do develop firmness as well (cf Reference [20]).

5. Acknowledgement

The authors would like to thank T. Narayanan, M. Sztucki and S. Callow (ESRF) for support during experiments on the ID2 beam line in Grenoble. Christos Regkos is acknowledged for additional experimental support.

References

[1] M. Žinic, F. Vogtle and F. Fages, In: Fages F, editor. Low molecular mass gelators, Topics in Current Chemistry, 256. Berlin, Springer, p. 39-76 (2005).
[2] F.M. Konikoff, D.S. Chung, J.M. Donovan, D.M. Small and M.C. Carey, J Clin Invest **90**, 1155-1160 (1992).
[3] Y.C. Lin, B. Kachar and R.G. Weiss, J Am Chem Soc **111**, 5542-5551 (1989).
[4] P. Terech, I. Furman, R.G. Weiss, H. Bouas-Laurent, J.P. Desvergne and R. Ramasseul, Faraday Disc **101**, 345-358 (1995).
[5] P. Terech, E. Ostuni and R.G. Weiss, J Phys Chem **100**, 3759-3766 (1996).
[6] K. Murata, M. Aoki, T. Suzuki, T. Harada, H. Kawabata, T. Komori, F. Ohseto, K. Ueda, and S. Shinkai, J Am Chem Soc **116**, 6664-6667 (1994).
[7] K. Sakurai, Y. Ono, H.J. Jung, S. Okamoto, S. Sakurai and S. Shinkai, J Chem Soc - Perkin Trans II **2001**, 108-112 (2001).
[8] L. Lu, T.M. Cocker, R.E. Bachman and R.G. Weiss, Langmuir **16**, 20-34 (2000).
[9] C. Geiger, M. Stanescu, L.H. Chen and D.G. Whitten, Langmuir **15**, 2241-2245 (1999).
[10] R. Wang, C. Geiger, L. Chen, B. Swanson and D.G. Whitten, J Am Chem Soc **122**, 2399-2400 (2000).
[11] Y.G. Li, K.Q. Liu, J. Liu, J.X. Peng, X.L. Feng and Y. Fang, Langmuir **22**, 7016-7020 (2006).
[12] J. Peng, K. Lin, J. Liu, Q. Zhang, X. Feng and Y. Fang, Langmuir **24**, 2992-3000 (2008).
[13] A. Bot and W.G.M. Agterof, J Am Oil Chem Soc **83**, 513-521 (2006).
[14] A. Bot, R. den Adel and E.C. Roijers, J Am Oil Chem Soc **85**, 1127-1134 (2008).
[15] A. Bot, R. den Adel, E.C. Roijers and C. Regkos, Food Biophysics **4**, 266-272 (2009).
[16] N. Duffy, J.C.G. Blonk, C.M. Beindorff, M. Cazade, A. Bot and G.S.M.J.E. Duchateau, J Am Oil Chem Soc **86**, 733-741 (2009).
[17] M. Pernetti, K.F. van Malssen, E. Flöter and A. Bot, Curr Opin Colloid Interface Sci **12**, 221-231 (2007).
[18] M.A. Rogers, Food Res Int **42**, 747-753 (2009).
[19] M.A. Rogers, A.J. Wright and A.G. Marangoni, Soft Matter **5**, 1594-1596 (2009).
[20] A. Bot, Y.S.J. Veldhuizen, R. den Adel and E.C. Roijers, Food Hydrocolloids **23**, 1184-1189 (2009).
[21] T. Narayanan, O. Diat and P. Bösecke, Nucl Instrum Meth Phys Res, Section A **467**, 1005-1009 (2001).
[22] J.M. Deutch, Macromolecules **14**, 1826-1827 (1981).
[23] L.I. Christiansen, J.T. Rantanen, A.K. von Bonsdorff, M.A. Karjalainen and J.K. Yliruusi, Eur J Pharm Sci **15**, 261-269 (2002).
[24] G. Argay, A. Kálmán, S. Vladimirov, D. Zivanov-Stakic and B. Ribár, Z Kristallographie **211**, 725-727 (1996).
[25] B.M. Craven, In: Hanahan D.J. and Small D.M., editors. Handbook of lipid research, volume 4, The physical chemistry of lipids – from alkanes to phospholipds, p. 149-182 (1986).

[26] H. Kawachi, R. Tanaka, M. Hirano, K. Igarashi and H. Ooshima, J Chem Eng Jpn **39**, 869-875 (2006).
[27] M.A. Rogers, T. Pedersen and L. Quaroni, Crystal Growth Design **9**, 3621-3625 (2009).
[28] L. Bellamy, The infrared spectra of complex molecules, Chapman and Hall, London (1975).
[29] J.H. Qian and W.L. Xu, Spec Spectral Anal **23**, 692-694 (2003).
[30] R. Bucci, A.D. Magri, A.L. Magri and F. Marini, Anal Bioanal Chem **375**, 1254-1259 (2003).
[31] R.J. Jandacek, M.R. Webb and F.H. Mattson, J Lipid Res **18**, 203-210 (1977).
[32] C.R. Loomis, G.G. Shipley and D.M. Small, J Lipid Res **20**, 525-535 (1979).
[33] D.W. de Bruijne and A. Bot, In: Rosenthal A.J., editor. Food texture: measurement and perception. Aspen, Gaithersburg, MD, USA, p. 185-227 (1999).
[34] F.G. Gandolfo, A. Bot and E. Flöter, J Am Oil Chem Soc **81**, 1-6 (2004).
[35] H.M. Schaink, K.F. van Malssen, S. Morgado Alves, D. Kalnin and E. van der Linden, Food Res Int **40**, 1185-1193 (2007).
[36] J.F. Toro-Vazquez, M. Alonzo-Macias, E. Dibildox-Alvarado and M.A. Charó-Alonso, Food Biophysics **4**, 199-212 (2009).
[37] L.S.K. Dassanayake, D.R. Kodali, S. Ueno and K. Sato, J Am Oil Chem Soc **86**, 1163-1173 (2009).

Structure and dynamics of glassy charged colloids studied with coherent small angle X-ray scattering

Birgit Fischer[1], Joachim Wagner[2], Christian Gutt[1], Fabian Westermeier[1], Gerhard Grübel[1]

[1] Deutsches Elektronen-Synchrotron, Notkestr. 85, Hamburg, Germany
[2] University of Rostock, Physical Chemistry, Albert-Einstein-Str. 3a, Rostock, Germany

E-mail: birgit.fischer@desy.de

Coherent small angle X-ray scattering methods have been used to investigate the structure and dynamics of charged colloidal systems with different volume concentration. The static structure factor indicates that the degree of ordering depends on both the electrostatic interactions and the volume fraction. All systems showed a compressed exponential form of the dynamic structure factor. The *Q*-dependence of the relaxation time τ hints towards ballistic motion of the colloidal particles in the glassy state. For the sample with the highest volume fraction we observe an influence of the static structure factor on the relaxation rate of the particles indicating a caging effect also for the ballistic type of motion.

1. Introduction

The understanding of the dramatic slowing down and non-exponential relaxations in glasses and dense colloidal suspensions is one of the most challenging problems in condensed matter physics. Recently, the concept of "heterogeneous dynamics" has attracted much attention as it is considered a key ingredient for understanding the drastic slowing down at the glass transition without significant changes in the structure.
A fingerprint of glass-like or jammed systems is that these systems are out-of-equilibrium, so that their properties change continuously with time, while the system slowly evolves towards the equilibrium. Typically, the dynamics become progressively slower as the sample ages and it spends more and more time in the metastable states it visits. The spatial extent of the motions within these systems is extremely reduced and often the dynamic structure factor shows a two-step decay. The commonly accepted physical picture is that the motion of a particle in a disordered, glass-like material is constrained by the "cage" constituted by its neighbors [1]. The dynamics of these systems can be observed with the dynamic structure factor $f(Q,\tau)$. On short times, attributed to the initial fast decay of $f(Q,\tau)$ also referred to as β-decay, the particles diffuse nearly freely within the cage. On longer time scales, attributed to the final decay of $f(Q,\tau)$ or α-decay, the dynamics are frozen, until the particles escape from the cage. This long time motion is subdiffusive and the decay of the correlation function $f(Q,\tau)$ shows a variety of forms.

Indeed both theoretical work and experiments have shown stretched exponential [2], power-law [3] or logarithmic decays [4] as the final relaxation of the dynamic structure factor.
Very recently for glassy systems dynamics faster than exponential have been observed [1, 6]. The dynamic structure factor $f(Q, \tau) \sim \exp[-(\Gamma\tau)^{\gamma}]$ decays as a „compressed" exponential with an exponent larger than one [1, 5, 7]. In addition for a wide variety of soft materials the characteristic time τ for slow relaxation quite often displays an intriguing linear $\Gamma \sim Q$ scaling. This scaling suggests that the final decay of $f(Q, \tau)$ is due to a „ballistic" motion of the particles. Therefore, all models based on cage-escape processes have been ruled out and Cipelletti and co-workers proposed instead that the dynamics may be due to randomly distributed internal stress sources acting on the sample [5]. However, most of the systems investigated up to now have been colloidal suspensions of low or medium concentration. In contrast to hard-sphere colloidal suspensions the phase diagram of charged colloidal systems depends not only on the volume fraction but also on the strength of the direct particle interactions. This allows us to distinguish if the caging-effect depends upon the structure or the volume concentration. Here we investigated the structure and dynamics of three charged colloidal systems of different volume concentration with coherent small angle X-ray scattering methods.

2. Experimental part

2.1. Synthesis

2.1.1. *Materials*
1H,1H,5H-Octafluoropentylmethacrylate (M1), Methacryloxymethyltrimethylsilan (M2) and 2,2,3,3,4,4,5,5,-Octafluorohexanediol-1,6-dimethacrylate (M3) were purchased from ABCR. $(NH_4)_2Fe(SO_4)_2$ and $NaSO_3$ were purchased from Fluka and $K_2S_2O_8$ was purchased from Merck. Tetraethoxysilane (TEOS), ethanol and ammonia (25% by weight) were purchased from Sigma Aldrich.

2.1.2. *Silica particles*
The silica particles were prepared according to a Stöber polycondenstation [8]. A solution of 8 ml TEOS in 500 ml ethanol was prepared. Under stirring at room temperature, the polycondensation was started by the addition of 15 ml aqueous ammonia. After stirring for 24 h, the resulting suspension was filtrated and dialyzed against distilled water for one week. Hereby the suspending medium is exchanged from ethanol to water and impurities of small molecular weight are removed. Afterwards the colloidal suspension was concentrated by vacuum evaporation of the solvent and soft centrifugation of the particles.

2.1.3. *Perfluorinated polyacrylate particles [9]*
The reaction vessel was a 2.5-L three-neck flask equipped with a reflux condenser, a nitrogen inlet tube, and a magnetic stirrer which was set to a stirring speed of 700 rpm during the whole synthesis process. 1.5 l of water was added to the flask and saturated with nitrogen to remove the oxygen in the water. The reaction vessel was heated to 65 °C and a monomer mixture consisting of 0.04 mol of M1, 0.04 mol of M2 and 100 µl of M3 was added. As a redox-system a solution of 400 mg of $NaSO_3$ and 5 mg of $(NH_4)_2Fe(SO_4)_2$ in 10 ml of deionized water was added. The reaction was initialized by adding a solution of 800 mg $K_2S_2O_8$ in 25 ml of deionized water. After stirring for 24 h, the resulting suspension was filtrated and dialyzed against distilled water for one week.

2.1.4. *Core-shell particles [10]*
The preparation of the inorganic/organic hybrid core-shell particles is a two step procedure consisting of a Stöber polycondensation for the fabrication of the silica cores as described in 2.1.2 and covering them by a perfluorinated polymer shell in a subsequent step in a similar manner as described in the section 2.1.3. In this step, the preparation of the shell starts from a dilute suspension of dialyzed silica

spheres in distilled water. Therefore, approximately one quarter of the silica spheres obtained from the first step are stirred in 1500 ml of distilled water and heated to 65 °C under nitrogen atmosphere to remove oxygen. To this suspension a $Fe^{2+}/Fe^{3+}/HSO^{3-}$ redox catalyst prepared from 5 mg $(NH_4)_2Fe(SO_4)_2$ and 300 mg $NaHSO_3$ is added. The polymer shell is produced by the radical induced polymerization of M1. By means of a two channel peristaltic pump, during 24 hours 20.0 ml of the monomer M1 and 400 mg of $K_2S_2O_8$ dissolved in 20 ml distilled water are added via Teflon capillaries immersed in the strongly stirred suspension. Hereby the concentration of radicals and monomer is kept low to suppress the formation of new nuclei forming pure polymer colloids and to promote the growth of long polymer chains entangling the silica spheres. At the end of the reaction, the suspension is filtrated and again dialyzed for one week against distilled water. For all three systems, stray ions are removed by means of a mixed bed ion exchanger leading to opalescent colloidal liquids indicating short range order of the colloidal particles at interdistances comparable to visible light.

Charged colloids can self-organize into liquid-like, crystalline or glassy structures depending upon the concentration of the colloidal suspensions. The colloidal samples are concentrated by gentle centrifugation within the quartz capillaries. For the further investigation of the dynamics we have chosen three different charged colloidal systems. Sample **A** is a silica colloid (an inorganic charged colloid), sample **B** is a perfluorinated polyacrylate (an organic charged colloid) and sample **C** is an inorganic/organic core-shell system consisting of a silica core and a perfluorinated polyacrylate shell. They not only differ in their composition, but also in their volume fraction and size (table 1). In addition, we added to sample **A** 10 μM KCl and to sample **B** 50 μM KCl. Sample **C** is without any added salt.

Table 1. Material, radius *R*, polydispersity $\frac{\Delta R}{R}$, peak position Q_{max} and height $S(Q_{max})$ of the static structure factor, volume fraction ϕ and electrolyte concentration C_{KCl} of sample **A**, **B** and **C**.

Sample	Material	R [nm]	$\frac{\Delta R}{R}$	Q_{max} [nm^{-1}]	$S(Q_{max})$	ϕ	C_{KCl} [μM]
A	Silica	101.8	0.063	0.038	2.2	0.51	10
B	Perfluorinated polyacrylate	107.0	0.054	0.032	4.2	0.47	50
C	Silica (core) perfluorinated polyacrylate (shell)	48.6	0.032	0.049	2.8	0.17	0

2.2. X-ray scattering experiments

X-ray photon correlation spectroscopy (XPCS) experiments were performed using partially coherent X-rays at the Troika beamline (ID10A and ID10C) of the European Synchrotron Radiation Facility (ESRF) in Grenoble, France. A Si (111) monochromator was used to select 8 keV X-rays, with a relative bandwidth of 10^{-4}. A Si mirror downstream of the monochromator suppressed higher-order harmonics of the undulator spectrum. A transversely coherent beam was defined by slit blades with highly polished cylindrical edges. The slit size was 20 microns in both dimensions. Under these conditions, the partially coherent X-ray flux was 10^9 photons/s. The scattered photons were detected by a direct illumination charged coupled device (CCD) camera from Princeton (20×20 μm^2 pixel size used at ID10C) and from Andor (13×13 μm^2 pixel size used at ID10A) which were placed 3.30 and 2.14 m behind the sample, respectively. For each sample a series of 1000 CCD images with typical exposure times of 1 s were measured. The electronic background was determined by dark exposures of the CCD camera and subtracted from the data.

3. Scattering theory

3.1. Static behaviour measured with small angle X-ray scattering (SAXS)

Scattering experiments give access to the form factor and static structure factor of colloidal suspensions by measuring the scattered intensity *I(Q)*:

$$I(Q) \propto P(Q)S(Q)\,. \tag{1}$$

The form factor *P(Q)* is due to the shape of the particles and can be determined from a dilute sample if the interactions of the particles are screened and – as a consequence of the random order under this condition - the static structure factor *S(Q)* equals unity.
For spherical particles the form factor is given by:

$$P(QR) = 9\left(\frac{\sin(QR) - \cos(QR)}{(QR)^3}\right)^2. \tag{2}$$

The static structure factor *S(Q)* describes the interparticle structure of the colloidal dispersion and can be obtained by dividing the scattered intensity *I(Q)* by *P(Q)*.
Since colloidal systems have a certain polydispersity $\frac{\Delta R}{R}$, for the form factor the Schulz-Flory distribution

$$\mathrm{f(R, R_0, Z)} = \frac{1}{\Gamma(Z+1)}\left(\frac{Z+1}{R_0}\right)^{Z+1} R^Z \exp\left(-\frac{Z+1}{R_0}R\right) \tag{3}$$

was taken into account, where R_0 denotes the number averaged size of the particles. The polydispersity $\frac{\Delta R}{R} = \left(\frac{\langle R^2\rangle}{\langle R\rangle^2} - 1\right)^{1/2}$ is related to the parameter *Z* of the Schulz-Flory distribution by $\frac{\Delta R}{R} = (Z+1)^{-1/2}$.

For core-shell particles with an outer shell radius *R* and a ratio of the radii $\beta = R_C/R$ the form factor is more complex than the standard expression for spheres and it can be calculated via:

$$P(Q,R_0,\beta_0,Z_R,Z_\beta,\rho_1,\rho_2) = \int_0^\infty \int_0^1 f_1(R,R_0,Z_R) f_2(\beta,\beta_0,Z_\beta)\alpha(R,\beta,\rho_1,\rho_2)$$
$$\left(\frac{3}{(QR)^3\left[\beta^3\rho_1 + (1-\beta^3)\rho_2\right]}\right)^2 \times (\rho_1\left[\sin(\beta QR) - \beta QR\cos(\beta QR)\right])$$
$$-\rho_2\left[\sin(QR) - QR\cos(QR) - \sin(\beta QR) + \beta QR\cos(\beta QR)\right]^2 d\beta dR\,, \tag{4}$$

where $\rho_1 = \rho_C - \rho_0$ and $\rho_2 = \rho_S - \rho_0$ are the differences in the electron densities between the core ρ_C and the medium ρ_0 and between the shell ρ_S and the medium, respectively. The form factor is weighted by the product of a Schulz-Flory-distribution f_1, which describes the possibility to find a particle with the outer shell radius R and arbitrary core-to-shell ratio and a second, mapped Schulz-Flory-distribution f_2, which describes the probability to find a given core-to-shell ratio β. Finally, $\alpha(P, \beta, \rho_1, \rho_2)$ is the volume weighted contrast of the core-shell particles [11].

3.2. Dynamics via XPCS

A coherent X-ray beam scattered from a disordered system produces a random diffraction or 'speckle' pattern which is related to the exact spatial arrangement of the particles in the sample. If the spatial arrangement of the disorder changes with time the corresponding speckle pattern will also change and the intensity fluctuations of the speckles provide a measure of the underlying dynamics [12].
From the intensity fluctuations the normalized intensity autocorrelation function can be calculated:

$$g_2(Q,\tau) = \frac{< I(Q,t)I(Q,t+\tau) >_t}{< I(Q,t) >_t^{\ 2}} = 1 + A\,|\,f(Q,\tau)\,|^2, \tag{5}$$

where the angular brackets denote the time average and A is the speckle contrast. The intermediate scattering function $f(Q,\tau)$ is related to the structure factor via

$$f(Q,\tau) = \frac{S(Q,\tau)}{S(Q,0)}. \tag{6}$$

In the case of heterogeneous dynamics the decay of the intermediate scattering function $f(Q,\tau)$

$$f(Q,\tau) = \exp\left(-(\Gamma\tau)^\gamma\right) \tag{7}$$

can phenomenologically be described by a compressed exponential form with an exponent γ larger than 1. Usually γ is in the range between 1.5 and 2 [1]. Typically for heterogeneous dynamics this phenomenon is accompanied by a linear dependence of the relaxation rate Γ on Q. Thus the diffusion process is a ballistic type of motion which is a common feature for jammed and hyper-diffusive systems [1].
Equation (5) represents a time average and is therefore not sensitive to temporal heterogeneities. In the case of temporal heterogeneities the dynamics depend on the time interval investigated. In this case the two-time intensity correlation function allows studying non homogenous dynamics and also aging phenomena. It is calculated according to [13]

$$c_2(\tau_1,\tau_2) = \frac{\left\langle I_P(\tau_1)I_P(\tau_2)\right\rangle_{pix}}{\left\langle I_P(\tau_1)\right\rangle\left\langle I_P(\tau_2)\right\rangle_{pix}} - 1 \tag{8}$$

Here, the averages are performed over a region of interest of the CCD detector, all having the same $|Q|$-value. Important variables for the two-time correlation function are the average time or age of the sample given by $(\tau_1+\tau_2)/2$ measured along the diagonal of c_2 with $\tau_1 = \tau_2$ and the time difference $\Delta\tau = \tau_1 - \tau_2$ which is measured perpendicular to the diagonal direction. In an equilibrium system, the two-time correlation functions depend only on the time difference $\Delta\tau$, and hence the two-time correlation contour lines are parallel. In such a situation the usual one-time correlation function is retrieved by averaging the two-time correlation function over different ages.

4. Results and Discussion

4.1. Structure of the colloidal systems

The size and polydispersity of the colloidal particles were obtained from the form factor of the particles. The form factor was determined from a dilute suspension whose electrostatic interactions were screened by 10^{-3} mol·l^{-1} potassium chloride so that the static structure factors equal unity. The overall radius for sample **A** and **B** were determined to 101.8 nm ($\Delta R/R$= 0.063) and 107.0 nm ($\Delta R/R$ = 0.054) with a form factor for spherical particles (equation (2)) under assumption of a Schulz-Flory size distribution (equation (3)), respectively. In the case of the inorganic/organic core-shell particles (sample **C**) the form factor for core-shell systems (equation (4)) was used and the overall radius of the particles was determined as 48.6 nm ($\Delta R/R$ = 0.032). For sample **C** smaller particles can be used compared to sample **B** since the high electron density provided by the silica core yield similar scattering intensities. With the smaller particles lower volume concentrations with strong ordering tendencies were achieved.

The static structure factors were obtained by dividing the scattered intensity of the concentrated suspensions by the diluted suspensions whose electrostatic interactions were screened by 10^{-3} mol·l^{-1} potassium chloride. Finally, the data were rescaled by the ratio of number densities of the diluted to the concentrated sample.

The position of the first peak in $S(Q)$ (figure 1) indicates the different volume fractions of the samples. A least squares fit to the static structure factors applying the rescaled mean sphere approximation (RMSA) yields a good approximation of the volume fraction of the colloidal systems [14]. $S(Q)$ for Sample **A** peaks at the largest value of QR and we estimate via RMSA a volume fraction of ϕ = 0.51. Despite the high volume fraction the peak height of $S(Q)$ is only 2.2, indicating a rather low degree of interparticle correlation. In contrast, Sample **B** is more structured showing a high $S(Q)$ peak value of 4.2. The sample is still in a disordered state as no Bragg peaks are visible in the 2D-diffraction pattern. From the first peak of the structure factor we estimate via the RMSA a volume fraction of ϕ = 0.47. Sample **C** exhibits the lowest QR value and accordingly the lowest volume fraction of ϕ = 0.17. Despite the low volume fraction the structure peak is clearly visible with a peak value of 2.85. Thus according to the Hansen-Verlet criteria ($S(Q)$>2.85) [15], sample **B** and **C** can be assumed to be in the glassy phase. Similar concentrations for glassy materials were also found for charged colloidal system from other groups [16-18]. Despite the low peak value of sample **A** its dynamics also show glass like features.

The interparticle correlation depends not only on the volume fraction, as in the case for hard spheres, where above ϕ = 0.56 the glass transition is reached [19-21], but additionally depends on the long-range electrostatic interactions in the sample. The higher the electrostatic interactions are, the more ordered are these systems. Moreover, the electrostatic interactions depend on the surface charges of the particles. In case of silica particles the charges on the surface stem from hydroxy-groups and for the perfluorinated polyacrylate systems from HSO^{3-}-groups resulting from the initiator during the synthesis process. The latter one provides a significantly higher surface charge than the first one. In both cases the interactions can be screened by adding electrolyte like potassium chloride. If the surface charges are not screened, these high surface charges from HSO^{3-}-groups can result in a highly ordered structure even at comparatively low volume fractions as shown in case of sample **C**. With increasing ionic strength of the medium the electrostatic interactions are screened, resulting in a more and more hard-sphere like potential. For such systems the critical volume fraction for the glass transition is much higher than for a system with long range interactions. The system of weakly charged silica spheres shows even at volume fractions close to a dense packing only a maximum of the static structure peak $S(Q)\approx 2.2$. A summary of the sample properties are given in table 1.

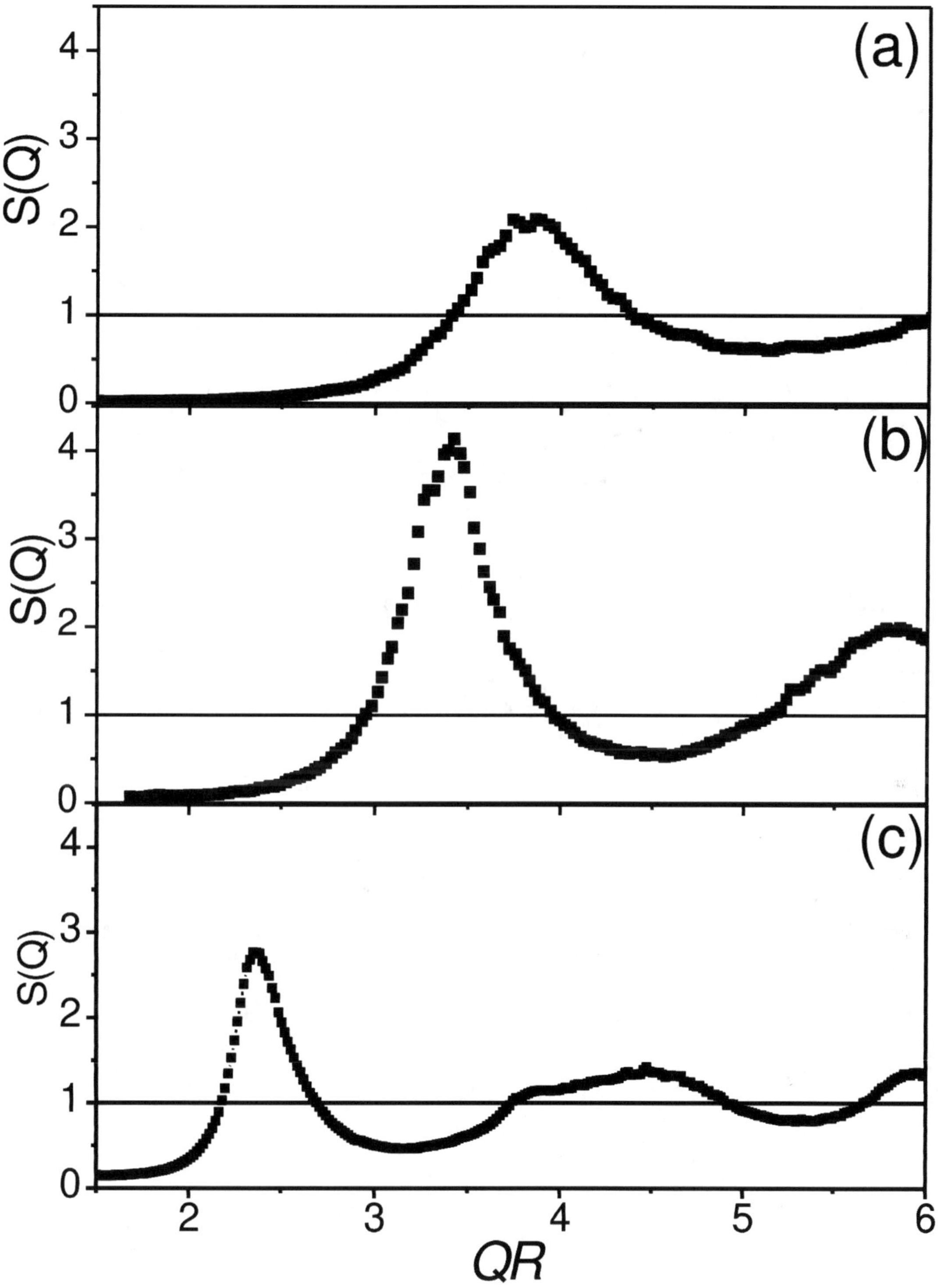

Figure 1. Static structure factors *S(QR)* of the samples **A** , **B** and **C** which differ in composition, size, volume fraction and added electrolyte. The shift to lower *QR* values from top to bottom is due to increasing interparticle distances resulting from decreasing volume fractions of the samples.

4.2. Dynamics

Figure 2 displays the two-time correlation functions of the different samples obtained by applying equation (8) to the series of CCD images at a fixed Q-position. All correlation functions display features typical for out-of equilibrium systems. In the first case for sample **A**, the dynamics progressively slow down with time (figure 2(a)), but at some points in time the motion accelerates again which indicates the heterogeneity of this sample. Contrary to sample **A**, sample **B** does not display heterogeneity (figure 2 (b)), however an aging is visible by a slowing down of the dynamics over the whole time period. This phenomenon of aging of the sample or slowing down of the motion has also been seen by Cipelletti [5]. Sample **C** shows very heterogeneous dynamics with strong fluctuations (figure 2 (c)). In the two-time correlation function of sample **C** obviously a two step process is present and no aging is visible in the spectra.

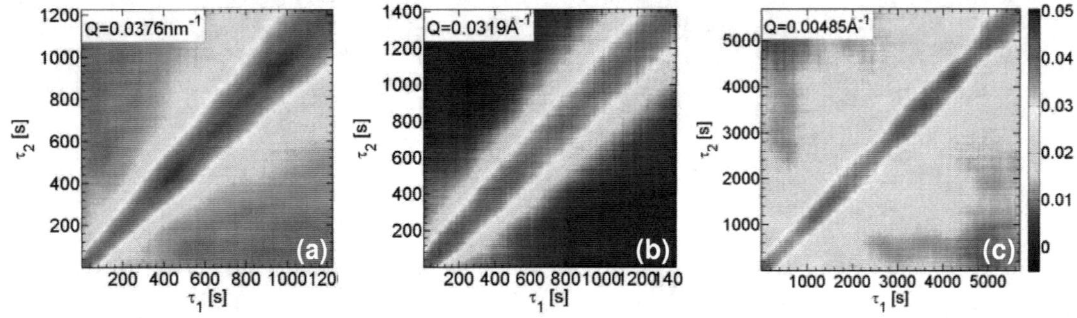

Figure 2. Two-time correlation functions for the three different samples **A**, **B** and **C** show the time evolution of the dynamics in these systems. The two-time correlation functions were calculated for the Q-value at the peak position of the static structure factors.

From the g_2-function and the speckle contrast the dynamic structure factor $f(Q,\tau)$ is calculated. For all three samples the correlation functions $f(Q,\tau)$ are fitted best with a compressed exponential, equation (7), with an exponent γ larger than 1.75. In addition, the compression exponent γ increases with the degree of heterogeneity of the sample as indicated by the fluctuations of the two-time correlation function (figure 2). Sample **B** exhibits the lowest compression exponent with 1.75 ± 0.17 (figure 3 (b)). Sample **A** can be described with a slightly higher compression exponent of about 1.85 ± 0.13 (figure 3(a)). For sample **C**, the two-time correlations shows very heterogeneous dynamics and the compression exponent γ reaches values up to 2.19 ± 0.12 (figure 3(c)). For sample **C** a second exponential decay is clearly visible. For the other two samples at least an indication of a second decay appears, but not as pronounced as for sample **C**. The relaxation rate Γ shows a linear dependence on Q which is a typical indication for ballistic nature. Only for sample **A** the relaxation rate Γ shows a minimum at the peak position of the static structure factor (figure 4(a)), indicating that the dynamics here are still influenced by caging effects. This implies that above a certain volume fraction the nearest-neighbours interactions are locally dominating the dynamics, even if the system shows ballistic dynamics.

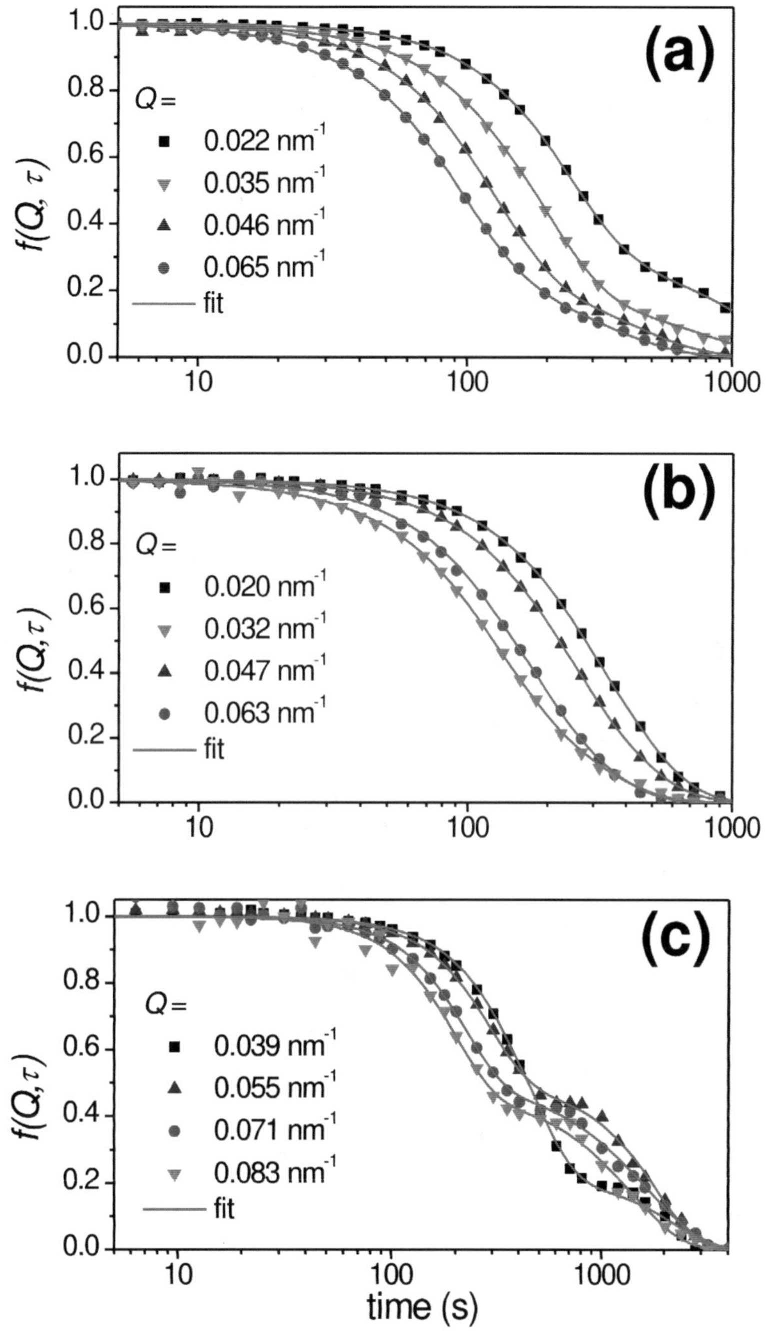

Figure 3. The dynamic structure factor of sample **A**, **B** and **C** for different Q-values. They all show a compressed exponential behaviour for the first decay which can be described by a γ-value of about 1.85 ± 0.13 for sample **A**, 1.75 ± 0.17 for sample **B** and 2.19 ± 0.12 for sample **C**. The γ-value increases when the sample shows a more pronounced heterogeneous behaviour. For sample **C** a second decay is clearly visible in the dynamic structure factor.

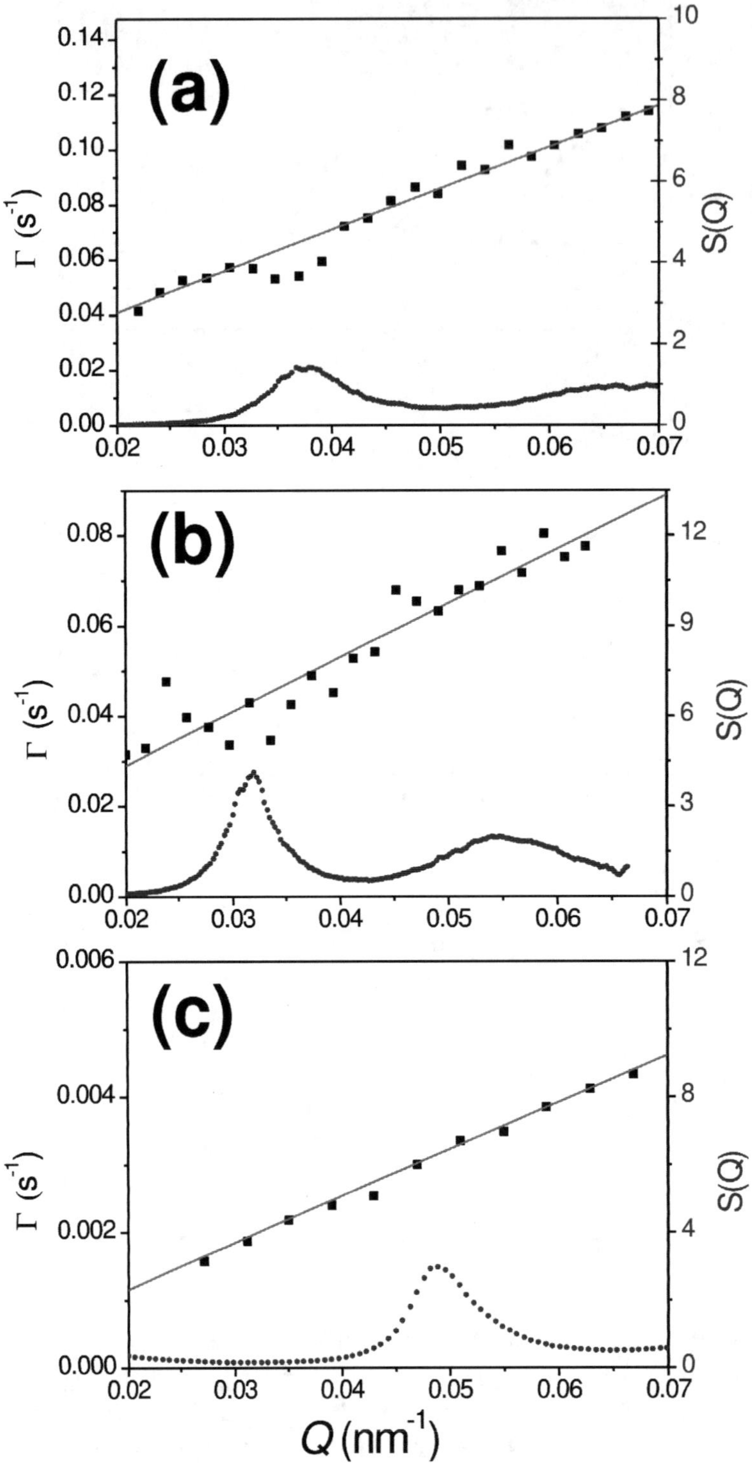

Figure 4. The static structure factor *S(Q)* and the relaxation rate *Γ* of Sample **A**, **B** and **C** vs Q.

5. Conclusion

In this study we compare the dynamics of three charged colloidal systems. All samples showed a high ordering tendency as evidenced from their static structure factor. The two-time correlation functions revealed both heterogeneity of the dynamics and aging of the samples. For all samples we observe a peculiar compressed exponential shape of the dynamic structure factor with a compression factor γ larger than 1.75. The inverse relaxation time Γ shows a linear dependence on Q which can be attributed to a ballistic motion of the colloidal particles. For the highest volume fraction we also observed the caging effect in the ballistic motion, i.e. Γ shows a minimum at the peak position of the static structure factor.

6. Acknowledgement

The authors would like to thank the ESRF and in particular the ID10 staff members, in particular A. Robert and Y. Chushkin for their help during the experimental work.

7. References

[1] Cipelletti L and Ramos L 2005 *J. Phys. Cond. Matter* **17** R253
[2] Götze W 1999 J. Phys. Cond. Matter **11** A1; Götze W 1990 in *Liquids, Freezing and the Glass Transition*, ed. Hansen J – P, Levesque D and Zinn-Justin J Les Houches Series, North Holland, Amsterdam, part 1
[3] Mallamace F, Gambadauro P, Micali N, Tartaglia P, Liao C and Chen S H 2000 *Phys. Rev. Lett.* **84** 5431
[4] Dawson K, Foffi G, Götze W, Sciortino F, Sperl M, Tartaglia P, Voigtmann T and Zaccarelli E 2000 *Phys Rev E* **63** 11401
[5] Cipelletti L, Ramos L, Manley S, Pitard E, Weitz D A, Pashkovski E E and Johansson M 2003 *Faraday Discuss.* **123** 237
[6] Cipelletti L, Manley S, Ball R C and Weitz D A 2000 *Phys. Rev. Let.* **84** 2275
[7] Ramos L and Cipelletti L 2001 *Phys. Rev. Let.* **87** 245503
[8] Stöber W, Fink A and Bohn E 1968 *J. Coll.* Int. Sci. **26** 62
[9] Härtl W and Zhang-Heider X 1997 *J. Col. Int. Sci.* **185** 298
[10] Fischer B, Autenrieth T and Wagner J 2010 *Langmuir* accepted
[11] Wagner J 2004 *J. Appl. Cryst.* **37** 750
[12] Grübel G and Zontone F 2004 *J. Alloys and Compounds* **362** 3
[13] Sutton M, Laaziri K, Livet F and Bley F 2003 *Opt. Expr.* **11** 2268
[14] Hayter J B, Penfold J 1981 *Molecular Physics* **42** 108
[15] Hansen J P and Verlet L 1969 *Phys. Rev.* **184** 151
[16] Beck C, Härtl W and Hempelmann R 1999 *J. Chem. Phys.* **111** 8209
[17] Härtl, W, Versmold H and Zhang-Heider X 1995 *J. Chem. Phys.* **102** 6613
[18] Sirota E B, Ou-Yang, H D, Sinha S K, Chaikin P M, Axe J D and Fujji Y 1989 *Phys. Rev. Lett.* **62** 1524.
[19] Di Cola E, Moussaid A, Szucki M, Narayanan T and Zaccarelli E 2009 *J. Chem. Phys.* **131** 144903
[20] Pusey P N and van Megen W 1986 *Nature* (London) **320** 340
[21] Pusey P N and van Megen W 1987 *Phys. Rev. Lett.* **59** 2083

Traceable size determination of PMMA nanoparticles based on Small Angle X-ray Scattering (SAXS)

G Gleber[1], L Cibik[1], S Haas[2], A Hoell[2], P Müller[1] and M Krumrey[1]

[1]Physikalisch-Technische Bundesanstalt (PTB), Abbestrasse 2-12, 10587 Berlin, Germany
[2]Helmholtz-Zentrum-Berlin für Materialien und Energie (HZB), Albert-Einstein-Strasse 15, 12489 Berlin, Germany

E-mail: gudrun.gleber@ptb.de

Abstract. The size and size distribution of PMMA nanoparticles has been investigated with SAXS (small angle X-ray scattering) using monochromatized synchrotron radiation. The uncertainty has contributions from the wavelength or photon energy of the radiation, the scattering angle and the fit procedure for the obtained scattering curves. The wavelength can be traced back to the lattice constant of silicon, and the scattering angle is traceable via geometric measurements of the detector pixel size and the distance between the sample and the detector. SAXS measurements and data evaluations have been performed at different distances and photon energies for two PMMA nanoparticle suspensions with low polydispersity and nominal diameters of 108 nm and 192 nm, respectively, as well as for a mixture of both. The relative variation of the diameters obtained for different experimental conditions was below ± 0.3 %. The determined number-weighted mean diameters of (109.0 ± 0.7) nm and (188.0 ± 1.3) nm, respectively, are close to the nominal values.

1. Introduction

Nanoparticles are of a growing importance for many areas of modern life, such as medicine (e.g. drug delivery [1]), transport technology (e.g. fuel additives [2]), industrial production (e.g. catalysis [3]) or cosmetics (e.g. sun blocks [4]). Their properties but also their potential risks are strongly size dependent, thus, size determination is a very critical issue [5].

During the last few years, several nanoparticle reference materials have been developed [6], and many different methods have become available to measure nanoparticle size and size distribution [7]. However, the results are not always consistent, even if small uncertainties are indicated for the individual methods. An example is the reference material 8011 of the National Institute of Standards & Technology (NIST) [8] which consists of colloidal gold spheres of a nominal diameter of 10 nm. The average diameter indicated in the report of investigation varies between (8.5 ± 0.1) nm measured by Atomic Force Microscopy (AFM) and (13.5 ± 0.1) nm obtained by Dynamic Light Scattering (DLS).

The aim of the international Joint Research Project "Traceable characterization of nanoparticles" [9] in the framework of the European Metrology Research Programme (EMRP) is to provide new traceable standards and procedures to determine the size, shape and size distribution of nanoparticles with high accuracy. Here traceable refers to the SI units. Several national metrology institutes are participating in this project: the National Physical Laboratory (NPL, UK), the Mittatekniikan keskus (MIKES, Finland), the Bundesamt für Metrologie (METAS, Switzerland), the Istituto Nazionale di Ricerca Metrologica (INRIM, Italy), the Centro Espanol de Metrologia (CEM, Spain), the Institutul National de Metrologie (INM, Romania), the Czech Metrology Institute (CMI), and the Physikalisch-Technische Bundesanstalt (PTB, Germany).

Several different methods will be compared, such as Dynamic Light Scattering (DLS), Small Angle X-ray Scattering (SAXS), Scanning Force Microscopy (SFM) and Scanning Electron Microscopy (SEM). SEM will also be applied in transmission geometry [10]. The scattering methods DLS and SAXS are well suited for measuring the size and size distribution of a particle ensemble, because they deliver information about the huge number of particles from a single measurement. However, DLS is an indirect method relying on the Brownian motion of particles in a suspension of known temperature and viscosity. SAXS in contrast is based on straight-forward X-ray scattering [11]. For almost monodispersed particles, the scattering pattern shows concentric rings with a diameter directly related to the mean diameter of the particles. The achievable uncertainty should be rather low, which was not the case for the SAXS results of small gold nanoparticles (nominal diameter 10 nm) in the NIST report of investigation [8]. Recently, the diameters of these gold nanoparticles have been determined by SAXS with significantly lower uncertainties [12].

2. SAXS measurements

The SAXS measurements presented here for traceable size determination were performed on poly-methyl-methacrylate (PMMA) particles obtained from the company microparticles GmbH, Berlin, Germany. The nominal diameter of the investigated particles in aqueous suspension was 108 nm and 192 nm, respectively. The suspension (nominal concentration 10 % w/v) was injected into glass capillaries with a 1 mm diameter and a wall thickness of 10 µm. In addition, a mixture (approximately equal particle masses) of both particle sizes was prepared and other capillaries were filled with distilled water for background subtraction. After filling up to a few cm, the capillaries were hermetically sealed by welding.

The measurements were conducted with synchrotron radiation at the four-crystal monochromator beamline in the laboratory of PTB at the electron storage ring BESSY II [13]. At this beamline, any photon energy in the range from 1.75 keV to 10 keV is available by using either InSb (111) or Si (111) crystals in the monochromator. Different motor-controlled sets of slits in this 37 m long UHV beamline were used to produce a well-defined monochromatic photon beam with an approximately 0.5 mm diameter and a spectral resolving power of about 10^4.

Up to 15 capillaries can be loaded via a load-lock in the UHV reflectometer allowing for the rotation and the translation of the samples in all degrees of freedom [14]. The last guard slit is already in the reflectometer, only a few cm in front of the sample. Directly behind the reflectometer, the SAXS setup of the HZB is installed, mainly consisting of an adjustable and tiltable, 3 m long support structure, an MAR CCD-based detector of 165 mm diameter and long edge-welded bellows with a 250 mm inner diameter to vary the distance between he sample and the detector continuously without breaking the vacuum [15]. The direct beam is blocked by a beamstop with an area of 1 cm^2. It is located directly in front of the detector and can be translated horizontally and vertically by ± 6 mm. The SAXS setup and the reflectometer are placed on motorized platforms for alignment with respect to the incoming beam. Both instruments are shown in figure 1. The suspension-filled capillaries in the reflectometer sample holder are enlarged in the inset. The entire beamline, including the monochromator, mirrors, slit systems, the reflectometer and the SAXS setup, is completely computer-controlled. The incident photon flux and the sample transmittance are measured by means of calibrated semiconductor photodiodes.

For the traceable size determination of nanoparticles, the position of the maxima and minima on the momentum transfer axis has to be known precisely, while a calibration of the scattered intensity is not required. For the momentum transfer, the scattering angle and the wavelength have to be calibrated. For the wavelength, back-reflection of radiation from silicon crystals can be used [13], relating the wavelength (or photon energy) scale of the monochromator to the precisely known lattice constant of silicon [16]. The resulting relative uncertainty of the wavelength is about 10^{-4}. For the scattering angle, the distance between the detector and the sample as well as the pixel size of the detector has to be known. Both values can be determined by using features of the HZB SAXS setup. The pixel size is obtained by translations of the detector perpendicular to the beam direction. Pictures of the direct but strongly attenuated beam are taken at different displacements which can be measured precisely using a Heidenhain encoder. A pixel size of (78.94 ± 0.04) µm was obtained, close to the manufacturer's specification. The distance can not be measured directly, but strongly-structured scattering patterns can be recorded at different distances and the change of the distance can be measured with the built-in 3 m long Heidenhain encoder with an uncertainty of 20 µm. By triangulating to the source point, the sample-detector distance can be determined with an uncertainty of below 1 cm. The total relative uncertainty for the distance and the pixel size is about 0.5 %. Further improvements have been initiated to reduce these uncertainties.

The measurements were performed at two different distances (2.03 m and 4.34 m) and at photon energies of 6 keV, 8 keV and 10 keV, respectively. Long exposure times of up to 1 h were used to obtain high-quality data sets.

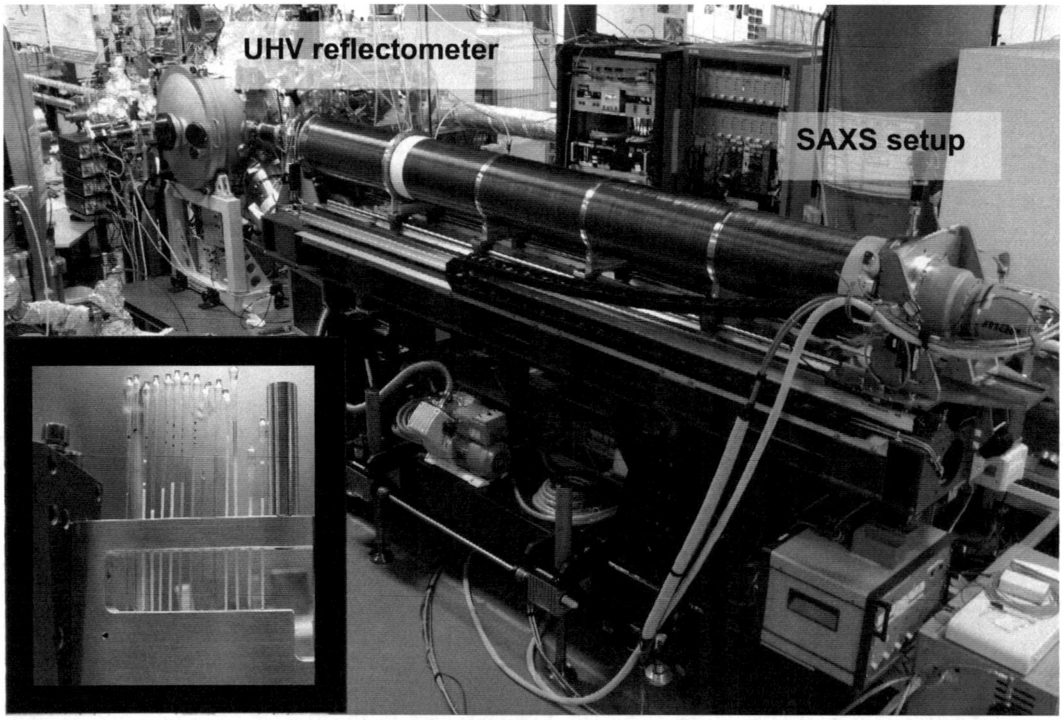

Figure 1. UHV reflectometer for sample positioning and SAXS setup including long edge-welded bellows and a CCD-based large-area detector. The inset shows sealed glass capillaries in the sample holder of the UHV reflectometer.

3. Results

Scattering patterns recorded for 8 keV radiation with the CCD-based detector at a distance of 2.05 m from the samples are presented in figure 2. The concentric rings are easily seen. While the rings are equally spaced for samples containing only one particle size, the obtained structure for the mixture is more complex. Similar pictures were obtained for all investigated photon energies and both distances. From each picture, a dark image was first subtracted to remove the readout noise of the camera. To subtract a background image of a capillary filled with distilled water, the images were normalized with the measurement time, the incident beam intensity, and the transmittance of the sample. The shadow of the beamstop was masked, traces of cosmic rays removed, the center was determined and the flat image was projected to a spherical surface. Then the data were circularly integrated to obtain the scattered intensity as a function of the momentum transfer. The scattered intensity curves were fitted with the program SASfit [17] as the sum of background and particle contributions. For a Gaussian distribution of spherical particles with constant electron density, their mean diameter, the width of the distribution and the product of the particle number and the density contrast are the only free parameters. In the evaluated momentum transfer range, contributions from particle interactions can be neglected for the investigated particles sizes and concentrations. The measured and fitted scattering intensities are also shown in figure 2.

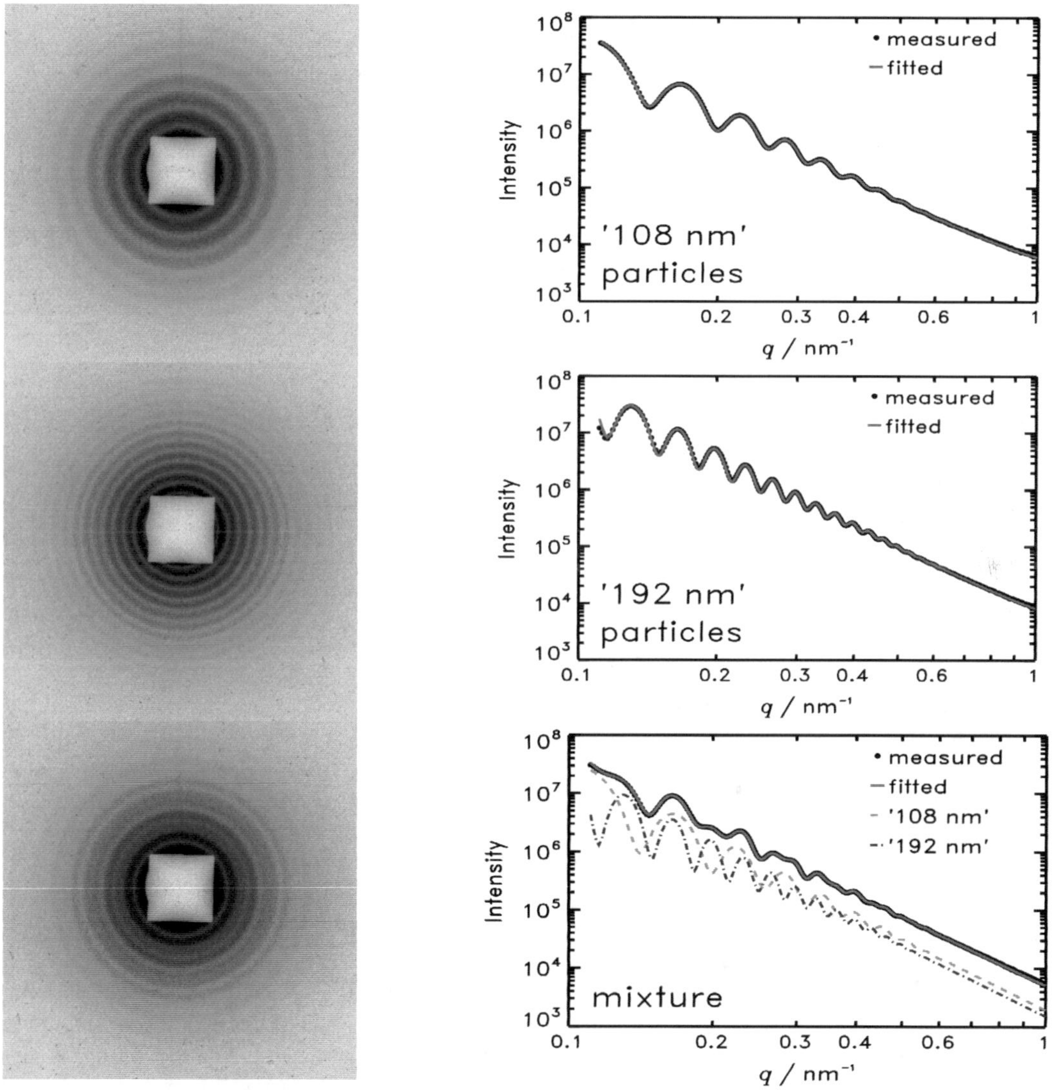

Figure 2. Left: Scattering patterns recorded at 8 keV with the CCD-based detector at a distance of 2.04 m from the sample position, for samples containing particles with different nominal diameters: top: '108 nm', center: '192 nm', bottom: mixture of both diameters.
Right: Corresponding circularly integrated scattered intensity and fit results from SASfit assuming a Gaussian distribution.

All results obtained for the number-weighted mean diameter as well as for the width of the Gaussian distribution curve are summarized in table 1. As the distributions are relatively narrow, there is little difference between Gaussian distribution and LogNorm or Schulz-Zimm distribution as also shown recently for gold nanoparticles [12]. The obtained values for the mean diameter exhibit very little variation below ± 0.3 % for different distances and even different photon energies. The results for the mixture and the individual measurements of both particle types are also in agreement within about 0.1 %.

Table 1. Number-weighted mean diameter and FWHM of the Gaussian distribution curve obtained at different photon energies and different distances between sample and detector for PMMA particles of two sizes as well as for a mixture of both.

Nominal diameter (nm)	Photon energy (keV)	Distance (m)	'108 nm' particles		'192 nm' particles	
			Mean diameter (nm)	FWHM (nm)	Mean diameter (nm)	FWHM (nm)
108	6	2.04	108.9	12.1	-	-
108	6	4.33	108.9	13.4	-	-
108	8	2.05	109.2	11.2	-	-
108	8	4.33	108.9	12.3	-	-
108	10	2.03	109.1	11.1	-	-
108	10	4.33	108.9	12.0	-	-
Mean values			**108.97**	**12.03**		
192	6	2.04	-	-	188.1	11.1
192	6	4.34	-	-	188.0	10.7
192	8	2.04	-	-	188.1	11.5
192	8	4.34	-	-	187.8	11.1
192	10	2.04	-	-	188.0	11.2
192	10	4.34	-	-	187.8	10.7
Mean values					**187.96**	**11.05**
108 & 192	6	2.04	109.2	11.6	188.2	11.2
108 & 192	6	4.34	109.0	10.4	188.2	10.4
108 & 192	8	2.04	109.2	10.9	188.2	11.4
108 & 192	8	4.34	108.7	12.2	187.8	11.2
108 & 192	10	2.04	109.0	11.9	188.2	11.0
108 & 192	10	4.34	108.8	12.0	187.9	11.2
Mean values			**108.98**	**11.52**	**188.09**	**11.06**

The indicated mean diameters are valid for the number-weighted distribution that allows better comparison to microscopic methods. For other weightings, higher values are obtained. However, due to the low polydispersity, the influence of the weighting is small: for the particles with a nominal diameter of 108 nm, the number-weighted mean diameter is 109.0 nm, volume weighting results in a diameter of 109.7 nm, and intensity weighting leads to 110.3 nm, thus, the total variation is about ± 0.6 %. For the larger particles with a nominal diameter of 192 nm, the effect is even smaller with a variation of only ± 0.2 %. The total relative uncertainty for the size determination for the investigated PMMA particles of currently 0.7 % is dominated by the uncertainties of the sample detector distance and contributions from the fitting procedure.

4. Conclusion

The mean diameter of PMMA particles in liquid suspension has been determined using SAXS with a small relative uncertainty of about 0.7 %. To further reduce the uncertainty, the measurement capabilities for the geometrical quantities will be improved. Measurements on nanoparticles in the diameter range between 10 nm and 100 nm have been started to extend the accessible diameter range. In the framework of the JRP project "Nanoparticles", the results will be compared to data obtained by DLS and microscopic methods like SEM and AFM.

References

[1] Olivier J 2005 Drug transport to brain with targeted nanoparticles *NeuroRx.* **2** 118–19
[2] Jung H, Kittelson D B and Zachariah M R 2005 The influence of a cerium additive on ultrafine diesel particle emissions and kinetics of oxidation *Combustion and Flame* **142** 276–88
[3] Jøgensen B, Kristensen S B, Kunov-Kruse A J, Fehrmann R, Christensen C H and Riisager A 2009 Gas-phase oxidation of aqueous ethanol by nanoparticle vanadia/anatase catalysts *Topics in Catalysis* **52** 253–7
[4] Wissing S A and Müller R H 2003 Cosmetic applications for solid lipid nanoparticles (SLN) *Int. J. Pharm.***254** 65–8
[5] Seaton A, Tran L, Aitken R and Donaldson K 2009 Nanoparticles, human health hazard and regulation *Journal of The Royal Society Interface* **7** 119–29
[6] www.nano-refmat.bam.de
[7] Borchert H, Shevehenko EV, Robert A, Mekis I, Kornowski A, Grubel G and Weller H 2005 Determination of nanocrystal sizes: A comparison of TEM, SAXS, and XRD studies of highly monodisperse CoPt3 particles *Langmuir* **21** 1931-6
[8] National Institute of Standards and Technology (NIST) 2007 Report of investigation, Reference Material 8011, Gaithersburg, USA
[9] Euramet, T3.J1.1, http://www.euramet.org/index.php?id=1011
[10] Buhr E, Senftleben N, Klein T, Bergmann D, Gnieser D, Frase C G and Bosse H 2009 Characterization of nanoparticles by scanning electron microscopy in transmission mode *Measurement Science and Technology* **20** 084025
[11] Glatter O and Kratky O 1982 *Small-angle X-ray Scattering* (London: Academic Press)
[12] Bienert R, Emmerling F and Thünemann A F 2009 The size distribution of 'gold standard' nanoparticles *Anal. Bioanal. Chem.* **395** 1651–60
[13] Krumrey M and Ulm G 2001 High-accuracy detector calibration at the ptb four-crystal monochromator beamline *Nucl. Instr. and Meth.* **467–468** 1175–8
[14] Fuchs D, Krumrey M, Müller P, Scholze F and Ulm G 1995 High precision soft X-ray reflectometer *Rev. Sci. Instr.* **66**, 2248-50
[15] Hoell A, Zizak I, Bieder H and Mokrani L 2007 German patent 10 2006 029 449
[16] Massa E, Mana G and Kuetgens U 2009 Comparison of the INRIM and PTB lattice-spacing standards *Metrologia* **46** 249–53
[17] Kohlbrecher J 2009 SASfit. PSI, Laboratory for Neutron Scattering ETHZ & PSI CH-5232 Villigen PSI, Switzerland, 0.92.2 edition http://kur.web.psi.ch/sansl/SANSS0ft/sasfit.html

Journal of Physics: Conference Series **247** (2010) 012028 doi:10.1088/1742-6596/247/1/012028

Fractal Structures on Fe_3O_4 Ferrofluid: A Small-Angle Neutron Scattering Study

Edy Giri Rachman Putra[1,2], Baek Seok Seong[2], Eunjoo Shin[2], Abarrul Ikram[1], Sistin Ari Ani[3], Darminto[3]

[1]Neutron Scattering Laboratory, National Nuclear Energy Agency of Indonesia, Gedung 40 Kawasan Puspiptek Serpong, Tangerang 15314, Indonesia

[2]Neutron Science Division, HANARO Center, Korea Atomic Energy Research Institute, 1045 Daedeok-daero, Yuseong-gu, Daejon 305-353, Republic of Korea

[3]Department of Physics, Faculty on Mathematics and Natural Sciences, Sepuluh Nopember Institute of Technology Kampus ITS Sukolilo, Surabaya 60111, Indonesia

Corresponding author: giri@batan.go.id

Abstract. A small-angle neutron scattering (SANS) which is a powerful technique to reveal the large scale structures was applied to investigate the fractal structures of water-based Fe_3O_4 ferrofluid, magnetic fluid. The natural magnetite Fe_3O_4 from iron sand of several rivers in East Java Province of Indonesia was extracted and purified using magnetic separator. Four different ferrofluid concentrations, i.e. 0.5, 1.0, 2.0 and 3.0 Molar (M) were synthesized through a co-precipitation method and then dispersed in tetramethyl ammonium hydroxide (TMAH) as surfactant. The fractal aggregates in ferrofluid samples were observed from their SANS scattering distributions confirming the correlations to their concentrations. The mass fractal dimension changed from about 3 to 2 as ferrofluid concentration increased showing a deviation slope at intermediate scattering vector q range. The size of primary magnetic particle as a building block was determined by fitting the scattering profiles with a log-normal sphere model calculation. The mean average size of those magnetic particles is about 60 – 100 Å in diameter with a particle size distribution $\sigma = 0.5$.

1. Introduction

Magnetic fluid (ferrofluid) which is a liquid dispersion of magnetic nanoparticles has a potential number of applications in different fields as magnetism, optics, rheology, biophysics, and medicine or biomedical[1]. A magnetic particle has a characteristic size of about 100 Å in diameter with a single magnetic domain (superparamagnetism) coated or stabilized by surfactant molecules to prevent them from aggregation. However, due to the molecular forces such as van der Waals attraction which is similar to a classical mechanism of colloidal coagulation, magnetic particles also have possibility to be aggregated. The present of aggregates and their internal structure greatly influence the properties of ferrofluid[2]. Fractal aggregates from magnetic nanoparticles have been indicated theoretically[3] and experimentally[4] on a specific physicochemical condition of a solution.

Small-angle scattering (SAS), including neutron (SANS) and X-ray (SAXS) is a powerful technique for investigating the inhomogenities in materials on the scale range of 10 – 5000 Å. This technique directly provides valuable structure information of the particles, aggregation phenomena, fractal structure and dimension as well. While by neutron, polarized or unpolarized neutrons, the magnetic structures can be possibly examined[5].

Fractal concept was found and then validated to be important in many structure studies of disordered materials since many processes as well as the structures can be described and understood in terms of non-integral dimensions. The theoretical and experimental small-angle scattering techniques on fractal structures have been developed and well established[6,7]. A characteristic feature of the small-angle scattering from fractal scatterers and many other disorder solids is often obeys a "power-law" scattering in the magnitude of the scattering vector q given as

$$\frac{d\Sigma}{d\Omega}(q) = I(q) \propto q^{-D} \qquad (1)$$

where D related to the structure of the scatterer or fractal dimension is a positive constant. Eq. (1) applies only in the regime of $\xi >> q^{-1} >> a$, where a is a typical chemical or bond distance related to local structure and ξ is the correlation length or average diameter of a scatterer.

In order to find out more detail on aggregation phenomena and fractal structures in magnetic nanoparticles, a series of magnetic Fe_3O_4 coated and stabilized by tetramethyl ammonium hydroxide (TMAH) was prepared as a function of concentration and then characterized by means of SANS. This work is also to distinguish the performance of the 36 meter SANS BATAN spectrometer (SMARTer) for investigating the aggregate structures of magnetic nanoparticle materials for the first time.

2. Experimental method

2.1. Materials and synthesis

The natural magnetite Fe_3O_4 was extracted directly from iron-sands taken from several rivers in East Java, Indonesia, using a magnetic separator. The synthesis was started from a co-precipitation of 20 g of the natural magnetite at ~ 70 °C in 38 ml of 12 Molar (M) chloride acid solution and then stirred rigorously using magnetic stirrer for 20 minutes. The solution was then added by 24 ml of 6.5 M NH_4OH solution and stirred for about 1 hour. After continuous stirring, a dark-brown homogeneous solution obtained. This solution was filtered and then washed several time using distilled water until a clear filtrate solution attained. Finally, for each 5 g of the magnetic particles Fe_3O_4 residue in the filter paper was then dispersed and stabilized in water containing tetra-methyl ammonium hydroxide (TMAH) of surfactant. The final concentrations of ferrofluid were 0.5, 1, 2 and 3 M with respectively the magnetite Fe_3O_4 %-volume concentrations were 0.28, 0.55, 1.11 and 1.67.

2.2. Small-angle neutron scattering (SANS)

The small-angle neutron scattering measurements using the 36 meter SANS BATAN spectrometer were carried out at the neutron scattering laboratory (NSL) in Serpong, Indonesia[8]. All samples were exposed using a neutron wavelength λ of 3.90 Å with the sample-to-detector position of 1.5, 4 and 13 meters. These experimental settings cover the scattering vector of $0.004 < q < 0.3$ Å^{-1} with $q = (4\pi/\lambda) \sin\theta$, where θ is the angle between the incident and scattered beams.

Scattering intensities of each solution sample were subtracted using a standard SANS data reduction program, GRASP[9] by their solvent backgrounds (pure H_2O) and the dark electronic current. The ambient temperature was maintained during the measurements. The corrected scattering intensity data was normalized and then fitted with a combined log-normal spherical and clustered object model calculation using a data analysis program, SASfit[10].

2.3. Data analysis

The scattering intensity in SANS is calculated by

$$I(q) = P(q)\, S(q) + Bkg \qquad (2)$$

where $P(q)$ is the scattering from randomly distributed spherical particles having a radius r_0 and its distribution, volume fraction ϕ and scattering length density difference (contrast) $\Delta\rho$:

$$P(q) = \phi V_p^2 \Delta\rho^2 F(qr_0)^2 f(r) \tag{3}$$

The volume of particle $V_p = (4/3)\pi\, r_0^3$ and spherical form factor is given by

$$F(qr_0) = \frac{3[\sin(qr_0) - qr_0\cos(qr_0)]}{(qr_0)^3} \tag{4}$$

For clustered object model, the spherical particle as a building block aggregate formed a fractal-like cluster with a correlation length ξ corresponding to their overall size and self-similarity dimension D, known as a fractal dimension. Thus, for the interference from building blocks of fractal-like clusters, the structure factor $S(q)$ can be calculated by[7]

$$S(q) = 1 + \frac{\sin[(D-1)\tan^{-1}(q\xi)D\Gamma(D-1)}{(qr_0)^D\left[1 + \dfrac{1}{(q^2\xi^2)^{(D-1)/2}}\right]} \tag{5}$$

where $\Gamma(x)$ is a gamma function. This expression reduces to $S(q) \sim q^{-D}$ when $\xi >> q^{-1} >> a$ as given in Eq. 1, while the unity term in $S(q)$ becomes dominant at large q. Thus the scattered intensity $I(q)$ is dominated only by the form factor $P(q)$, Eq. 2. Hence, by applying a log-normal sphere model calculation which is a good approximation for the aggregates and clustered object models, the term of $f(r)$ in Eq. (3) is given as:

$$f(r) = \frac{N}{c_{LN}}\frac{1}{r^p}\exp\left(\frac{1}{2\sigma^2}(\ln r - \ln r_{med})^2\right) \tag{6}$$

where

$$c_{LN} = \sqrt{2\pi}\, r_{med}^{1-p}\exp\left((1-p)^2\frac{\sigma^2}{2}\right) \tag{7}$$

The initial or primary magnetic nanoparticles size of about 100 Å in diameter can be determined by applying Eq. 6 into Eq. 3 for a spherical object at high q region.

3. Result and discussion

It is necessary to subtract the measured intensity or raw data with respect to the scattering contribution of the backgrounds, i.e. solvent and aperture. By subtracting the measured data with the solvent, i.e. water which contains also the scattering intensity of the cuvette and the aperture scattering. At first, the log-normal sphere and aggregate fractal model calculations have been applied separately to fit the corrected scattering data of ferrofluid, Figure 1. A log-normal sphere model calculation is in agreement with the experimental data at high to intermediate q region as $S(q) \approx 1$, Figure 1a. While at intermediate to low q region, an aggregate fractal model calculation is fit with the data as $I(q) \approx S(q)$ and the primary particle was assumed in a spherical shape and monodisperse, Figure 1b. These preliminary fittings have been worked out in order to obtain the fitting parameters at two regimes. A combined two model calculation was then numerously calculated to find the best fitting on the corrected experimental data, Figure 2.

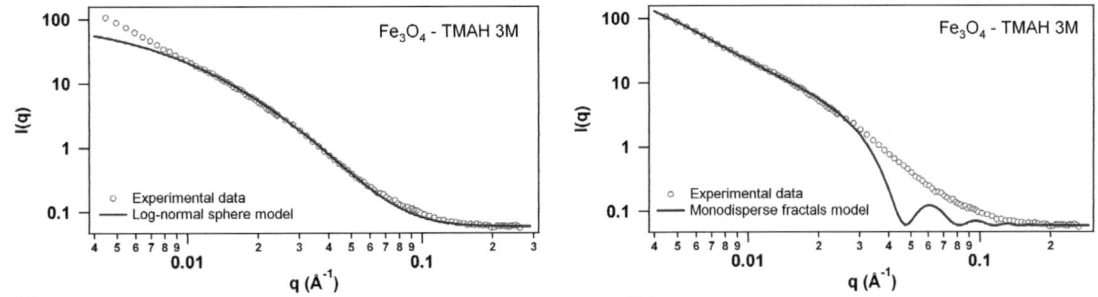

(a) (b)

Figure 1. SANS profiles of ferrofluid sample fitted with two model calculations (a) log-normal sphere at high to intermediate q region and (b) aggregate fractals at intermediate to low q region. The circle symbol is the experimental data and the solid line is the theoretical model calculation.

Figure 2 shows the double logarithmic between scattering intensity $I(q)$ versus q from ferrofluid Fe_3O_4 sample as a function of concentration fitted with the combined two theoretical model calculation. The experimental data showed as the concentration of ferrofluid increased, two regimes, Porod (high to intermediate q range) and power-law scattering (intermediate to low q range) become emerging. It can be understood as the magnetic concentration increased the scattering profile and intensity are also affected by the structure factor. Scattering profile at the lowest concentration of 0.5 M is purely indicating that non-correlated particle $I(q) \approx P(q)$ term is dominant in this sample. A large aggregate cluster formed and is dominantly composed of surfactant molecules and small fraction of the magnetic particles which shows a weak scattering intensity at intermediate to low q range. Here, an individual small magnetic particle is not able to be determined.

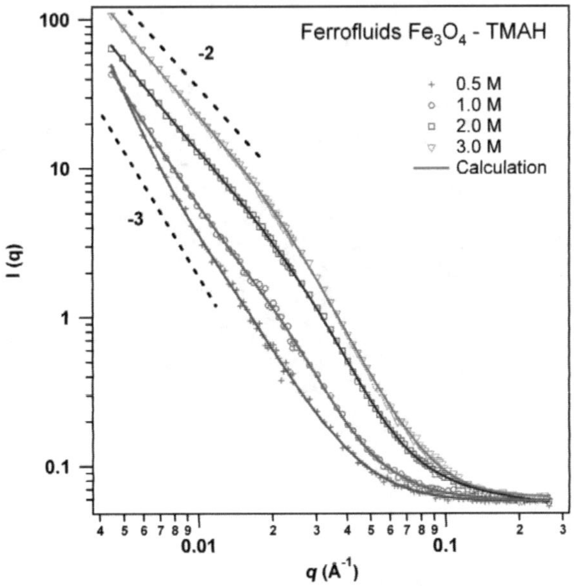

Figure 2. SANS profiles of Fe_3O_4 ferrofluid samples in TMAH fitted with a combined two model calculation, log-normal sphere and aggregate fractals. The symbols are experimental data and the solid line is theoretical model calculation.

As magnetic concentration is increased the primary magnetic particles of the building blocks which can be distinguished from their aggregates in the two regimes, Porod and power-law (mass fractal region) are certainly appearing. The crossover of these two regimes which defined the estimated size of the building block of the large aggregate cluster is fairly fixed at $q = 0.025$ Å^{-1} or about ~ 500 Å in diameter. Meanwhile, from the fitting results, a log-normal sphere model calculation, the size of the magnetic Fe_3O_4 particles in TMAH at a given particle size distribution σ of 0.5 are 95.6, 62.6 and 64.4 Å in diameter for the concentration of 1, 2 and 3 M, respectively. Those values verified that the magnetic Fe_3O_4 particle as an initial magnetic nanoparticle entirely has a particle size of about 100 Å in diameter which is a characteristic size of nanoparticle. From the size of aggregate and primary particle data can be calculated the aggregation number for the particle in cluster from the relation $N_{agg} = (\xi/r_0)^D$,[11] from which we estimate around 30 – 70 small primary magnetic particles aggregated naturally as a building block of the large aggregate cluster due to the condensation reaction[12]. This magnetic particle aggregation is confirmed by enhancing the scattering profile intensity.

On the other hand, the size of the primary magnetic particle is definitely affected by its concentration in the solution. The inter-particle correlation interference becomes dominating as magnetic concentration is increased which indicates the ordering, short and long-range orders of the magnetic particles in the system contributed to the scattering profiles. The repulsion between magnetic particles can be reduced by dilution and then has a possibility to interpenetrate or form a short range order[12]. Therefore, the particles can be appeared as a single large particle or aggregate which composed of several small coated magnetic particles. As the concentration increased, it can be assumed that the possibility of magnetic particles collision is high. Those magnetic particles tend to be aggregated and then coated by the surfactant. It was found that as the magnetic concentration is low, a short range order is exist and showing the high the scattering intensity[12,13].

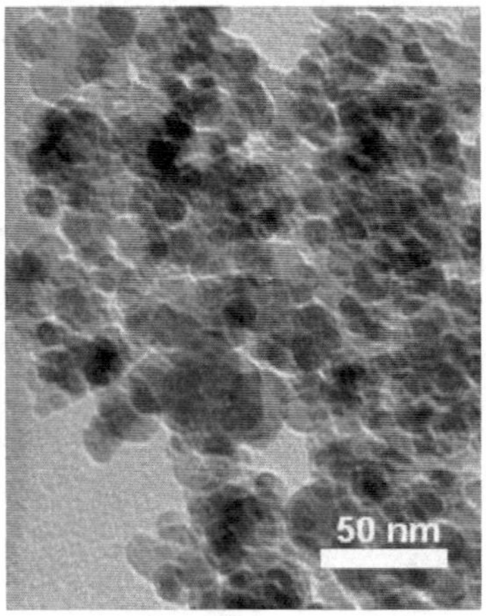

Figure 3. Transmission Electron Microscopy (TEM) image of dried magnetic Fe_3O_4 nanoparticles aggregation in TMAH.

Due to the molecular force of each magnetic particle as a building block, those particles possibly aggregate to form a large aggregate cluster. Then, particle binding may be affected by van der Waals attraction between magnetic cores. The aggregation of magnetic Fe_3O_4 nanoparticles in TMAH which appeared from their SANS scattering profiles at intermediate to low q region is also confirmed from a transmission electron microscopy (TEM) investigation, Fig. 3. However, the initial magnetic nanoparticle as the building blocks of fractal-like clusters and their fractal dimension is not probable to be verified by a microscopy technique. Here from SANS data, the slope at the magnitude of the scattering vector $q < 0.03$ Å^{-1} showing a power law scattering which corresponds to the scattering distribution from fractal aggregates. The aggregate fractal dimension D appeared between 3 and 2, Figure 2 showing that a small particle of building block is likely growing in 2-dimensional space direction to form a disc-like macrostructure.

The fractal dimension is strongly dependent on the mechanism of aggregates formation. The aggregates fractal dimension which determined in this work indicated that a reaction-limited mechanism occurred in magnetic nanoparticles system. In this mechanism, more compact structures formed as a slower restructuring process of the clusters is occurred. This result is dissimilar with the carbon black aggregates in solution. A diffusion-limited cluster aggregation mechanism may also occur in the carbon black aggregates where a particle by a slow Brownian process sticks to the first particle it encounters in a cluster and the cluster growth normally[14]. Then, the extended aggregates of carbon black formed with the fractal dimension less than 2.

4. Conclusion

It was confirmed that a large aggregate cluster has already formed at the lowest magnetic concentration and the mass fractal structure dimension was evidently affected by the magnetic concentrations. The building block was formed by about 30 – 70 small magnetic particles with the size of 60 – 100 Å in diameter and particle size distribution $\sigma = 0.5$. The SANS BATAN spectrometer together with the experimental method including data reduction and analysis have demonstrated its performance in a low q region for revealing the fractal structure aggregates of Fe_3O_4 ferrofluid.

Acknowledgment

This work was supported in part by HANARO, Korea Atomic Energy Research Institute R&D Program of the Ministry of Science and Technology, Republic of Korea 2006 – 2007 and the program of Research Incentive in Nanoscience and Nanotechnology, sponsored by the State Ministry of Research and Technology (KNRT), Republic of Indonesia, 2006 (Darminto as the Principle Investigator).

References

[1] B. Berkovsky, V. Basthovoy, Magnetic Fluids and Applications Handbook, Begell House Inc., New York, 1996.

[2] A. Zubarev, Magnetically Controllable Fluid and their Application, Springer, Berlin, Germany, 2002, 143 – 162.

[3] E.A. Elfimova, *J. Magn. Magn. Mater.* **289** (2005) 219 – 221;
E.A. Elfimova, *J. Magn. Magn. Mater.* **300** (2007) e203 – e205.

[4] M. Balasoiu, M.V. Avdeev, V.L. Aksenov, D. Hasegan, V.M. Garamus, A. Schreyer, D. Bica, L. Vekas, *J. Magn. Magn. Mater.* **300** (2006) e225;
M. Balasoiu, M.V. Avdeev, V.L. Aksenov, *Cryst. Reports* **52** (2007) 505 – 511;
R.C. Woodward, J. Heeris, T.G. St. Pierre, M. Saunders, E.P. Gilbert, M. Rutnakornpituk, Q. Zang, J.S. Riffle, *J. Appl. Cryst.* **40** (2007) s495 – s500;
K. Butter, A. Hoell, A. Wiedenmann, A.V. Petukhov, G.J. Vroege, *J. Appl. Cryst.* **37** (2004) 847 – 856.

[5] W. Wagner, J. Kohlbrecher, Small-angle neutron scattering, in Modern Techniques for Characterization Magnetic Materials, edited by Y. Zhu, Kluwer Academic Publishers, New York, 2005.

[6] J.E. Martin, A.J. Hurd, *J. Appl. Cryst.* **20** (1987) 61 – 78.

[7] J. Teixeira, *J. Appl. Cryst.* **21** (1988) 781 – 785.

[8] E.G.R. Putra, A. Ikram, E. Santoso, Bharoto, *J. Appl. Cryst.* **40** (2007) s447 – s452;
E.G.R. Putra, Neutron News 18 (2007) 23 – 29;
E.G.R . Putra, Bharoto, E. Santoso, A. Ikram, *J. Nucl. Instrum. Method* **A 600** (2009), 198 – 203.

[9] C. Dewhurst, "GRASP: Graphical Reduction and Analysis SANS Program for Matlab",

http://www.ill.eu/fileadmin/ users_files/Other_Sites/lss-grasp/grasp_main.html, Institut Laue Langevin 2001 – 2007.

[10] J. Kohlbrecher, SASfit ver. 0.82, November 2007, PSI.

[11] X.H. Guo, N.M. Zhao, S.H. Chen, J. Teixeira, *J. Biopolymers* **335** (1990) 335.

[12] M.V. Avdeev a,V.L. Aksenov, M. Balasoiu,V.M. Garamus, A. Schreyer, Gy. Török, L. Rosta, D. Bica, L. Vékás, *J. Coll. Interface Sci.* **295** (2006) 100 – 107.

[13] G.Y. Torok, V.T. Lebedev, D. Bica, L. Vekas, M.V. Avdeev, *Romanian Rep. Phys.* **58** (2006) 313 – 318.

[14] J. Lahaye, F. Ehrburger-Dolle, *Carbon* **32** (1994) 1319 – 1324.
P. Bezot, C. Hesse-Bezot, *Carbon* **36** (1998) 467 – 469.

Journal of Physics: Conference Series **247** (2010) 012029 doi:10.1088/1742-6596/247/1/012029

Study of Inverse Ni-based Photonic Crystal using the Microradian X-ray Diffraction

A V Vasilieva[1], N A Grigoryeva[2], A A Mistonov[2], N A Sapoletova[3], K S Napolskii[3], A A Eliseev[3], A V Lukashin[3], Yu D Tretyakov[3], A V Petukhov[4], D Byelov[4], D Chernyshov[5], A I Okorokov[1], W G Bouwman[6] and S V Grigoriev[1]

1 Petersburg Nuclear Physics Institute, Gatchina, 188350, St. Petersburg, Russia

2 Department of Physics, St. Petersburg State University, 198504, St. Petersburg, Russia

3 Department of Materials Science, Moscow State University, 119899, Moscow, Russia

4 Debye Institute, Utrecht University, 3584 CH Utrecht, the Netherlands

5 SNBL European Synchrotron Radiation Facility (ESRF), 38043 Grenoble, France

6 Delft Technical University, 2629 JB Delft, the Netherlands

E-mail: vasilieva@lns.pnpi.spb.ru

Abstract. Inverse photonic nickel-based crystal films formed by electrocrystallization of metal inside the voids of polymer artificial opal have been studied using the microradian X-ray diffraction. Analysis of the diffraction images agrees with an face-centred cubic (FCC) structure with the lattice constant a_0 = 650 ± 10 nm and indicates two types of stacking sequences coexisting in the crystal (twins of ABCABC… and ACBACB… ordering motifs), the ratio between them being 4:5 The transverse structural correlation length L_{tran} is 2.4 ± 0.1 μm, which corresponds to a sample thickness of 6 layers. The in-plane structural correlation length L_{long} is 3.4 ± 0.2 μm, and the structure mosaic is of order of 10°.

1. Introduction

Artificial opals consisting of submicron monodisperse microspheres packed in a face-centered cubic structure and materials on their basis are good candidates for the creation of high quality Photonic Crystals (PhC). They have recently attracted great attention due to their unusual optical properties and promising applications in optical devices. Inverse opals can be synthesized by filling the voids of opal templates and subsequent removing the initial microspheres. Three-dimensionally ordered porous materials are thus obtained. One of the most promising methods of filling the voids is electrodeposition of metal, which enables almost 100% filling. The conducting PhC are of great interest from the viewpoint of multi-functionality and interplay between the optical, magnetic, and electronic transport properties.

The present study is aimed to investigate the structure of the inverse metallic PhCs deposited onto a conductive substrate. The important parameters of this technology are the substrate material, the surface roughness and the potential applied to the substrate. It has been established that PhCs synthesized by the vertical deposition method are often ordered in a face-centred cubic (FCC) structure [1]. However, they are prone to stacking disorder, which may result in the formation of a Random Hexagonal Close Packed (RHCP) structure with almost equal chances of finding FCC and hexagonal close-packed (HCP) stacking sequences [2]. Since photonic properties are based on the light interference phenomena, they may be strongly affected by structural disorder. It is difficult, however, to characterize and manipulate the degree of order. One of important properties of any crystal is the size and shape of structurally coherent blocks; technologically wise these properties describe an effective size of structurally ordered optical device. In this paper we characterize Ni-filled inverse crystal in terms of average structure and also give an estimate of its structural coherence.

2. Synthesis of Samples

The synthesis of samples was performed by a templating technique. At first the colloidal crystal films were formed from monodisperse polystyrene microspheres of ~ 450 ± 10 nm on the mica substrates with a thin thermally evaporated Au layer. Colloidal crystals were deposited at 60 °C from water suspension using the electric-field-assisted vertical deposition technique [2, 3]. During the crystal growth, the electric field of ~ 0.5 V/sm was applied perpendicular to the substrates; as it has been shown in [2], this option allows formation of high-quality photonic crystals. At the second step the voids between the spheres were filled with nickel by electrochemical deposition. To control the filling of voids with metal, the electrodeposition was carried out in the potentiostatic mode at a voltage of -0.9 V relative to the Ag/AgCl reference electrode from the electrolyte having the following composition: 0.1M NiCl, 0.6M $NiSO_4$, 0.1M H_3BO_4, 4M C_2H_5OH. To obtain the inverse structure, the polystyrene spheres were dissolved in toluene during 3 hours.

The preliminary characterization of samples was carried out using the scanning electron microscopy. Figure 1 shows the typical images of the obtained inverse PhC. Panels (a) and (b) represent top view and cross section, respectively.

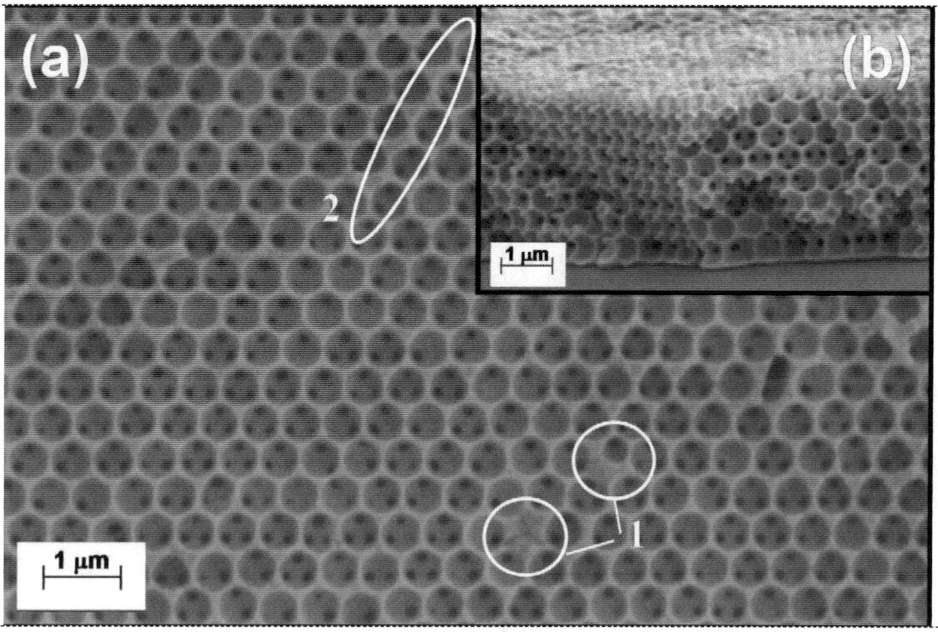

Figure 1. SEM images of the inverse Ni-based PhC: top view (a) and cross section (b).

One can see that the crystal surface bears point (1) and linear (2) defects, which are related to microsphere packing, as well as the cracks related to conditions of the template synthesis [4]. In addition, we observe a surface containing hexagonally ordered spherical voids. It is known that the direction perpendicular to the substrate is always the [111] axis of the FCC structure [5, 6], and the vertical axis (along which the meniscus is moving) corresponds always to the [20-2] crystallographic direction. Thus, one can get the idea of the crystal orientation as early as at the stage of synthesis.

3. Experiment

Microradian X-ray diffraction experiments were performed at the BM-26B Dutch-Belgian beam line (DUBBLE) of the European Synchrotron Radiation Facility (ESRF) in Grenoble, France [7-9]. The monochromatic X-ray beam with the energy E = 15 keV (wavelength λ = 0.826Å, size 0.5 × 0.5 mm^2 and divergence 10 mrad) was used. Diffraction of the synchrotron radiation was registered by a two-dimensional detector (Photonic Science CCD-camera, 4000 × 2700 pixels of 22 μm^2) positioned at a distance of ~8 m downstream of the sample along the Y axis. To improve the resolution, the beam was focused by a set of compound beryllium lenses installed in front of the sample. This setup allows achieving an angular resolution in the order of a few microradians, which is sufficient for collecting detailed information on the structure of large period photonic crystals [10, 11]. The inverse PhC films were mounted on a goniometric head, which allows careful orientation of the sample. The 3D crystal structure is investigated by collecting the X-ray diffraction patterns for different rotation angles of the sample around its [2-20] axis in the range from – 60 to + 60 [12].

4. Results and discussion

The small angle diffraction of synchrotron radiation is a very efficient and distinctive way to determine the structure and to characterize the ordering degree of the inverse PhC. In the small angle diffraction geometry the image on the detector is a section of the Fourier image of the crystal structure by the plane perpendicular to the X-ray beam. Thus, by rotating the crystal and simultaneously recording the diffraction images, one can reconstruct the 3D Fourier image [13], like in a single crystal X-ray diffraction experiment, but on the mesoscopic scale.

Experiments with microradian synchrotron diffraction were performed on the Ni inverse PhC deposited onto the conductive substrate; some of experimental images are shown in Figure 2.

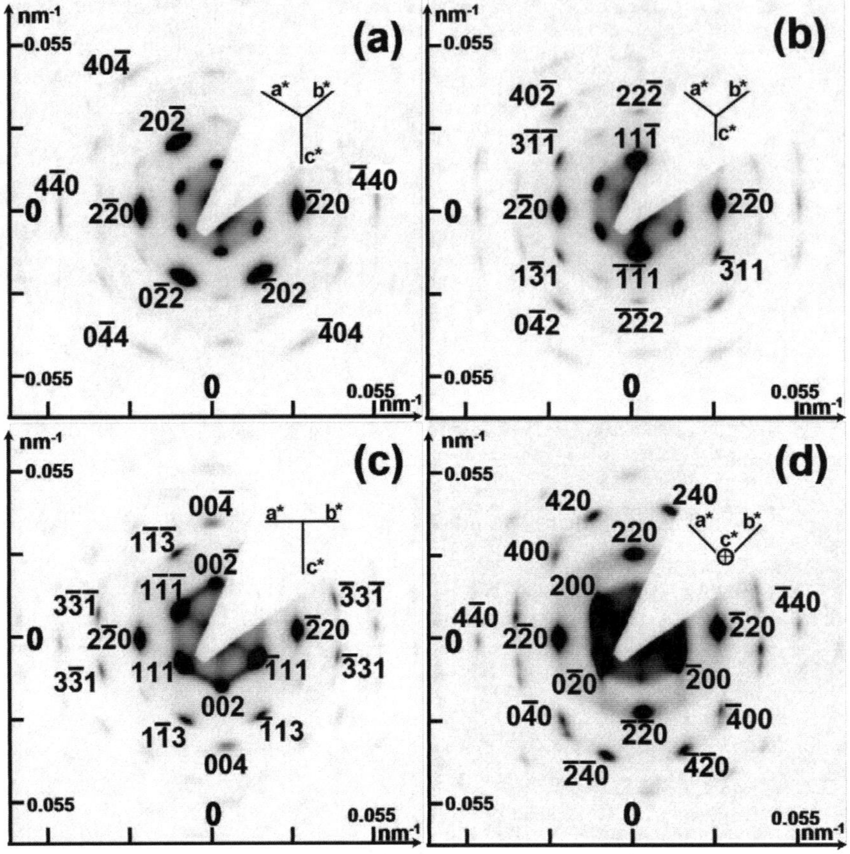

Figure 2. Diffraction patterns of the Ni inverse PhC at various rotation angles: ω = 0° (a), 19° (b), 35° (c), 55° (d) around the [2-20] axis.

The diffraction patterns were recorded at various rotation angles: ω = 0°, 19°, 35°, 55°, where the [2-20] was the rotation axis and zero angle corresponds to the geometry of the inverse PhC surface perpendicular to the beam. These obtained pictures can be identified assuming the face-centered cubic (FCC) structure and accounting for thickness (the SEM shows that the sample consists of 8 layers only) and hexagonal fault effects. Note that peaks of the inner hexagonal rings at ω = 0 (Figure 2a) are much weaker than reflexes of the second ring, which are the (2-20) reflexes of the FCC structure. Accordingly, panels (b, c, d) show the reflection patterns for the reciprocal lattice of FCC structure cut by the (112) (b), (110) (c), and (100) (d) planes. Bragg reflexes appearing in the diffraction pictures can be identified as those belonging to the FCC structure, which is further corroborated by position of the diffraction maximum on the rocking curves (Figure 3)

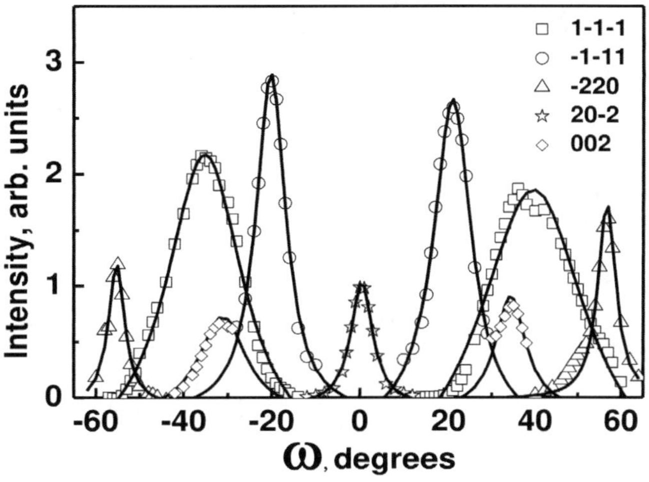

Figure 3. Rocking curve I (ω) of Ni inverse PhC: symbols and lines represent experimental data and Lorentzian approximations of the Bragg peaks, respectively.

Presence of two almost symmetrically positioned diffraction maxima in the rocking curves for the most of reflexes indicates that the crystal is composed from blocks with different sequences of packing layers, ABCABC- and ACBACB-type stacking [1, 14], the volume ratio those twin components can be roughly estimated from the ratio of integral intensities of Bragg reflexions and it is close to 4:5 for the considered case.

Figure 4 shows Q-dependences of individual reflexes, positions of the maxima also correspond to the FCC structure, for example $Q_{(2-20)}^{exp} = 0.02746 \pm 10^{-5}$ nm^{-1} and $Q_{(2-20)}^{cal} = 0.027911$ nm^{-1}, $Q_{(11-1)}^{exp} = 0.01645 \pm 10^{-5}$ nm^{-1} and $Q_{(2-20)}^{cal} = 0.017092$ nm^{-1}, i.e. the inaccuracy is 3 %. High resolution allows us to resolve positions of the (-1-11) and (002) reflections (Fig.4), although they are located close to each other. Note that the experimentally observed ratio of intensities well coincides with the theory, according to which the (002) reflexes appear to be orders of magnitude weaker what the (111) or (220) reflexes. The lattice constant of Ni inverse PhC is 650 ± 10 nm.

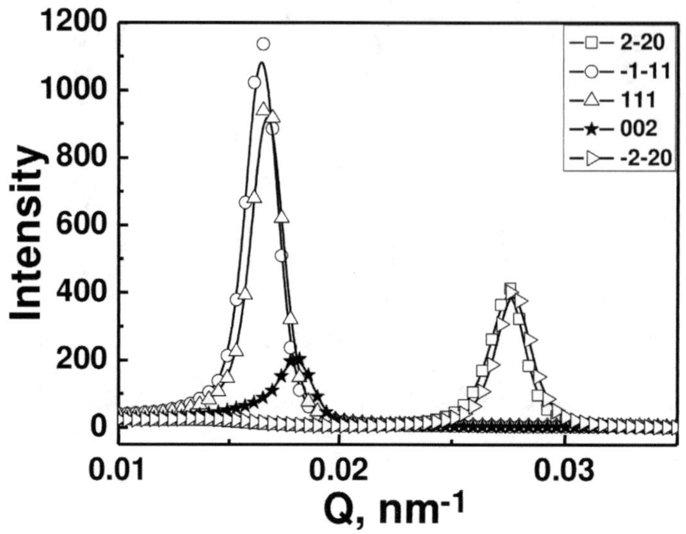

Figure 4. Q-dependence of individual reflexes, which were moved in reflecting plane.

The spatial extent of the periodic order could be derived from the full width at half maximum ΔQ of the diffraction peaks, which can be determined either from a single diffraction pattern (for lateral periodicity) or from the rocking curves (for the periodicity along the beam). The transverse structural correlation length (perpendicular to the scattering vector Q) L_{tran} has been calculated using the equation: $L_{tran} = 2\pi/(Q_{(2\text{-}20)} \sin\Omega)$, when $Q_{(2\text{-}20)}$ is the position of reflex (2-20) in nm^{-1}, Ω is the FWHM of reflection. This transverse structural correlation length L_{coh} defined from the rocking curves is 2.4 ± 0.1 μm, which corresponds to a sample thickness of order of 6 layers, which is two layers smaller than that obtained from the SEM image in Fig.1. One can not pretend to give the exact number of layers obtained from the coherent length but the qualitative agreement has been clearly achieved. The structural correlation length (longitudinal along the Q) obtained from the value of FWHM according to the formula $L_{long} = 2\pi/\Delta Q$ is 3.4 ± 0.2 μm. The mosaicity, i.e. mis-orientation of ABC- ACB- domains within the sample, is of order of $\delta\phi = 10°$. Thus, the degree of disorder has been described by the three parameters: transverse correlation length L_{long}, structural correlation length L_{tran}, and mosaic spread $\delta\phi$ of the structure.

Conclusions

The data presented here show that the electrochemical method of synthesis, indeed, allows one to duplicate the structure of artificial opals and to obtain the inverse photonic crystal. The crystals have predominantly FCC structure with the lattice constant being 650 ± 10 nm and of average size of coherent block about few microns. Among many methods to characterize photonic crystals, the microradian X-ray diffraction of synchrotron radiation provides the most detailed information on average structure, disorder phenomena, and structural coherence. The first two characteristics are normally expressed in terms of certain type of dense-packing structure and stacking defects and diffraction experiment becomes necessary metrological components of photonic crystallography. The last portion of diffraction data, namely, is shape and size of Bragg nodes, seems to be of a lesser use for now. We show that even very simple analysis presented here provides the unique information on the size of structurally coherent crystal blocks. It would be very interesting for a future study to monitor the dependence of structural coherence on materials and technological parameters of fabrication of inverse photonic crystals.

Acknowledgment

The Russian authors thank for partial support the Russian Foundation of Basic Research (Grant No 10-02-00634) and the Federal Special Program of Russian Federation (Projects Nos. 02.740.11.5186). This work is also supported by the Presidium of the Russian Academy of Sciences under the Program for Basic Research and the Programs of the High School of Russian Federation 2.1.1/4661 and "Michail Lomonosov II". We thank The Netherlands Organisation for Scientific Research (NWO) for the access provided to a synchrotron X-ray beamline and K. Kvashnina and D. Detollenaere, station staff members of the Dutch–Belgian Beam Line (ESRF-DUBBLE), for assistance in the synchrotron X-ray experiment.

References

[1] Hilhorst J, Abramova V V, Sinitskii A, Sapoletova N A , Napolskii K S, Eliseev A A, Byelov D V, Grigoryeva N A , Vasilieva A V, Bouwman W G, Kvashnina K, Snigirev A, Grigoriev S V , and Petukhov A V 2009 *J Langmuir* **25** 10408

[2] Napolskii K S , Sapoletova N A , Gorozhankin D F, Eliseev A A, Petukhov A V, Byelov D V, Mistonov A A, Grigoryeva N A, Bouwman W G, Kvashnina K, Snigirev A, Chernyshov D Yu , Vasilieva A V, Grigoriev S V 2009 *J Langmuir* in print DOI: 10.1021/la902793b

[3] Kuai Su-Lan, Hu X-F, Hache A, Truong Vo-Van 2004 *J of Crystal Growth* **267** 317

[4] Plekhanov A I, Kalinin D V, Serdobintseva V V 2006 *J Rossiiskie nanotekhnologii* **1** 245

[5] Baryshev A V, Kosobukin V A, Samusev K B, Usvyat D E, and Limonov M F 2006 *J Phys. Rev. B* **73**, 205118

[6] Gasperino D, Meng L, Norris D J, Derby J J 2008 *J of Crystal Growth* **310** 131–139

[7] Borsboom M, Bras W, Cerjak I, Detollenaere D, van Loon D G, Goedtkindt P, Konijnenburg M, Lassing P, Levine Y K, Munneke B, Oversluizen M, van Tol R and Vlieg E 1998 *J Synch Rad* **5** 518

[8] Bras W, Dolbnya I P, Detollenaere D, van Tol R, Malfois M, Greaves G N, Ryan A J and Heeley E 2003 *J Appl Cryst* **36** 791

[9] Nikitenko S, Beale A M, van der Eerden A M J, Jacques S D M, Leynaud O, O'Brien M G, Detollenaere D, Kaptein R, Weckhuysen B M and Bras W 2008 *J Synch Rad* **15** 632

[10] Grigoriev S V, Napolskii K S, Grigoryeva N A, Vasilieva A V, Mistonov A A, Chernyshov D Yu, Petukhov A V, Byelov D et al 2009*J Phys Rev B* **79** 045123

[11] Thijssen J H J, Petukhov A V,'t Hart D C, Imhof A, van der Werf C H M, Schropp R E I, and van Blaaderen A 2006 *J Adv Mater* **18** 1662

[12] Grigoriev S V, Napolskii K S, Grigoryeva N A, Vasilieva A V, Mistonov A A, Chernyshov D, Petukhov A V, Belov D V et al, 2009 *J Phys Rev B* **79** 045123

[13] Eliseev A A, Gorozhankin D F, Napolskii K S, Petukhov A V, Sapoletova N A, Vasilieva A V, Grigoryeva N A, Mistonov A A, Byelov D V, Bouwman W G, Kvashnina K O, Chernyshov D Yu, Bosak A A and Grigoriev S V 2009 *JETP Lett* **90** 272

[14] Abramova V V, Sinitskii A S, Grigor'eva N A, Grigor'ev S V, Belov D V,. Petukhov A V, Mistonov A A, Vasil'eva A V and Tret'yakov Yu D 2009 *JETP* **109** 29

Small-angle x-ray scattering study on the structure of microcrystalline and nanofibrillated cellulose

Kirsi Leppänen[1], Kari Pirkkalainen[1], Paavo Penttilä[1], Jenni Sievänen[2], Nina Kotelnikova[3] and Ritva Serimaa[1]

[1] Division of Materials Physics, Department of Physics, P.O.B. 64, FI-00014 University of Helsinki, Finland
[2] VTT, Espoo, Finland
[3] Institute of Macromolecular Compounds, Russian Academy of Sciences, Bolshoy pr. 31, St. Petersburg 199004, Russia

E-mail: kirsi.leppanen@helsinki.fi

Abstract. The effects of different solvents on the structure of microcrystalline and nanofibrillated cellulose (MCC, NFC) were studied using small-angle x-ray scattering (SAXS). MCC was immersed in water, ethanol, and acetone, and NFC was immersed only in water and ethanol, but studied also in the form of foam-like water-NFC-gel in wet, air-dried and re-wet states. The solvent affected the average chord length, which reveals the typical length scale of the structure of the sample: 2.4 $\pm$ 0.1 nm was obtained for MCC-water, 2.5 $\pm$ 0.1 nm for re-wet NFC-gel, 1.6 $\pm$ 0.1 nm for MCC-ethanol, 1.2 $\pm$ 0.1 nm for NFC-ethanol, and 1.3 $\pm$ 0.1 nm for MCC-acetone. The specific surface of cellulose increased strongly when MCC and NFC were immersed in the solvents compared to dry cellulose. The specific surface of cellulose was determined to be larger for NFC-water than MCC-water, and slightly larger for dry NFC powder than for dry MCC, which can be explained by the fact that the width of cellulose crystallites perpendicular to the cellulose chain direction was slightly larger in MCC than in NFC on the basis of wide-angle x-ray scattering results.

1. Introduction

The way to produce microcrystalline cellulose (MCC) from native cellulosic material or pulp by acid hydrolysis has been known already for over 50 years [1]. MCC has become an important material for pharmaceutical industry as a tablet excipient and a binder and filler in dosage forms. It is used as a fat replacer and stabilizer in food industry, and a composite material in wood and plastic industry including for example furniture and sport and leisure items [1, 2, 3]. The production methods, properties, and structure of microcrystalline cellulose have been studied widely. In MCC, the diameter of the cellulose fibres is in the scale of few micrometers. These fibres consist of elementary cellulose microfibrils, whose crystalline parts have a width of about 5 nm and a length of about 20 $-$ 30 nm. [4]

Besides MCC, e.g. cellulose whiskers, nanocrystalline and nanofibrillated cellulose (NFC) can be produced from native cellulose using various preparing methods. In whiskers, nanocrystalline and -fibrillated cellulose the diameter of the cellulose fibres can be in the scale of few nanometers [1, 5]. Because of the nanoscale, the mechanical and optical properties are different, and that is why there exists a large interest in nanocrystalline and -fibrillated cellulose as a raw material.

NFC is more attractive compared to synthetic nanomaterials because of its better availability, biodegradability, and lower cost.

For example, it has been found out that in aqueous colloïdal suspensions cellulose whiskers form chiral nematic phases with interesting optical properties. These properties could be used in security and identification papers and iridescent pigments in inks and coatings. [1] It has also been found out that cellulose whiskers can be aligned by a magnetic field and thus they could be used for magnetic alignment of proteins in biomolecular NMR [1]. One more possible application area is in electrolytes. Azizi Samir et al. have studied lithium-conducting nanocomposite polymer electrolytes made from cellulose whiskers and lithium imide salt [6].

We have previously compared the structure of cellulose in native plant material with that of pulp and MCC using x-ray scattering [4]. In that study, MCC was made by mild acid hydrolysis from cotton linter, cellulose from flax fibres and sulphite and kraft cooked wood pulp. Wide-angle x-ray scattering (WAXS) results showed that the width of the cellulose crystallites increased both due to the pulping process and due to the hydrolysis. Furthermore, native spruce wood, flax fibres and cotton linter and the pulp samples made of flax, spruce, pine, and birch wood were immersed in water in order to determine the cross-sectional size and shape and packing of the cellulose crystallites using SAXS. The close range order was the best in the native wood sample, whereas the packing in the other samples was irregular. [4]

Many studies on the effects of various solvents on wood, pulp, and microcrystalline cellulose have been made using chemical methods [7, 8, 9, 10, 11]. The aim of this work was to study the effects of the solvents on the structure of MCC and NFC using small- and wide-angle x-ray scattering. These methods are useful tools for determining the nanostructure of samples. MCC was made of cotton linter by mild acid hydrolysis and NFC was commercial nanocellulose. For SAXS studies dry MCC was immersed in water, ethanol, and acetone, and dry NFC was immersed in water and ethanol. Also foam-like water-NFC-gel in wet, air-dried, and re-wet states was measured by SAXS. The crystalline structures of the dried water-NFC-gel and dry NFC powder were determined by WAXS, so that these structures could be compared with that of MCC.

2. Materials and methods

2.1. Microcrystalline cellulose samples

MCC was prepared from cotton linter with a mild acid hydrolysis in 2.5 M solution of HCl at 105°C for 1 h [12]. Cotton linter was hydrolyzed to the so called level-off degree of polymerization [13]. After hydrolysis MCC was white colored, contained moisture less than 1 mass% and had a powder-like macroscopic morphology. The DP of MCC was 165 [4].

For the SAXS measurements MCC powder was immersed in distilled water, ethanol, and acetone about three weeks before measurements and kept in sealed plastic tubes at room temperature until the measurements were made. MCC-solvent systems were in equilibrium at the time of the measurements because the equilibrium swelling is reached shortly for cellulose-solvent systems [14].

2.2. Nanofibrillated cellulose samples

NFC-gel sample was commercial nanocellulose which was supplied by J. Rettenmaier & Söhne GmbH, Rosenberg, Germany. The product name is Arbocel MF40-10, and it is an aqueous 10 mass% suspension of nanocellulose fibres. The water-NFC-gel had a foam-like macroscopic morphology. NFC-gel was first measured as the original wet suspension, after which the sample was dried in air for 3 hours and measured again. When the gel dried, it formed hard and dry macroscopic aggregates. These aggregates were immersed in distilled water and this mixture was let to swell for one week before a new measurement. So the NFC-gel sample was measured in three states: wet, air-dried and re-wet states.

NFC powder was also produced by J. Rettenmaier & Söhne GmbH. The product name of the powder-like NFC is MF40-100. NFC-water and NFC-ethanol mixtures were prepared by immersing NFC powder in distilled water and ethanol about one week before the SAXS measurements and kept in sealed plastic tubes at room temperature until the measurements were done, so that these samples also reached the equilibrium swelling before the measurements. The carboxyl group content of NFC was 85 mmol/kg [15].

2.3. SAXS

SAXS measurements were conducted at the Department of Physics at the University of Helsinki. A sealed x-ray tube with copper anode and point focus was used. The beam was monochromated with a collimating Montel multilayer to obtain Cu-K_α radiation (wavelength $\lambda = 1.5418$ Å). Bruker HI-Star area detector was used, and the distance between the sample and the detector was 52.5 cm. The q-range (the length of the scattering vector: $q = 4\pi \sin\theta/\lambda$, where θ is the half of the scattering angle) was calibrated with silver behenate and it was $0.03 - 0.4$ Å^{-1}.

The MCC- and NFC-mixtures and NFC-gel samples were measured in aluminum rings (diameter 4 mm, thickness 1.6 mm) between two Mylar-foil layers. The measurement time was 60 min. Transmission values for the samples were determined using primary beam values through a semi-transparent copper beamstop. All the 2D SAXS patterns were isotropic, so the data were radially averaged. The absolute intensity scale was determined using the fact, that the maximum of the absolute intensity curve of Lupolen is 5.5 cm^{-1} around 0.03 Å^{-1}. So the maximum intensity of the scattering curve of Lupolen, which was normalized by time, transmission and optical thickness of the sample and corrected for air-scattering, was set to that value [16]. The intensities of the MCC- and NFC-water and re-wet NFC-gel samples were corrected for water background, but all the other intensities were corrected only for background of an empty aluminum ring with Mylar-foils.

The MCC-water, MCC-ethanol, MCC-acetone, NFC-water, NFC-ethanol, and re-wet NFC-gel samples were weighed before the SAXS measurements in wet state, and after the measurements the samples were dried in air for over 24 hours, and the weights were measured again. Thus the weight fractions of cellulose and matrix (water, ethanol, acetone) could be determined. From these values and known densities of water (1 g/cm^3), ethanol (0.789 g/cm^3), acetone (0.79 g/cm^3), and cellulose (1.6 g/cm^3), the volume fractions of the phases were calculated. The prepared samples in the Al-rings were weighed, not the mixtures, because the sample preparation could have affected the weight fraction of the solvent phase due to the fact that there were excess amounts of solvents in the mixtures and due to the fast evaporation of the solvent. Thus the weighed amounts were very small and there are large relative errors in the weights, which leads to large relative errors also in the determined volume fraction values. In the case of the NFC-gel, the weight fractions of NFC and water were given by the manufacturer, and it was assumed that the gel was homogeneous enough, so that the known fractions were valid in the measurement.

2.4. WAXS

The WAXS measurements were carried out for the dry NFC powder and dried NFC-gel samples using perpendicular transmission geometry and Rigaku rotating anode x-ray tube at the University of Helsinki. A two-dimensional MAR345 image-plate was used as a detector. The beam was monochromated with a Si(*111*) crystal and a totally reflecting mirror and focused to the detector. The wavelength of the radiation was 1.541 Å, and the q-range was $0.6 - 3.8$ Å^{-1}. Intensities were corrected for absorption, geometry of the detector and the background arising from air-scattering.

The crystallinity and the width of cellulose crystallites from the cellulose reflections *1-10*, *110* and *200* were determined from the WAXS patterns of the NFC samples and the analysis was made exactly in the same way as in [4].

2.5. Analysis of SAXS results

To set the experimental intensities onto the absolute scale, the measured intensities were divided by the measurement time t, the optical thickness d and the transmission T of the sample. The optical thickness d of the sample was calculated using linear absorption coefficient μ:

$$d = -\frac{\ln(T)}{\mu}. \tag{1}$$

All the samples were assumed to consist of two phases: in the case of the dry MCC and NFC samples, the two phases were cellulose and air, in MCC/NFC-water cellulose and water, and in MCC/NFC-ethanol and MCC-acetone cellulose and ethanol and cellulose and acetone, respectively. Here cellulose is denoted as phase 1 and air/water/ethanol/acetone as phase 2. The volume fraction of phase 1 is denoted as ψ_1, and that of phase 2 as ψ_2.

It was checked, if the intensities would follow a power law

$$I(q) \propto q^{-\alpha}, \tag{2}$$

at some q-range in order to obtain information on the nanoscale structure of cellulose in the samples. For example, the value $\alpha = 2$ would arise from lamellar structures, and the values $\alpha < 3$ would indicate that the structures may be described as mass fractals (e.g. crumpled paper). The values $3 < \alpha < 4$ would arise from surface fractals with a dense and homogenous inner structure. The value 4 (Porod law) indicates that the phases have smooth and well-defined interfaces. It is essential that the intensities follow the same power law at a wide q-region before such interpretations can be made. [17, 16]

For all the samples a q-range was found where the intensity obeyed the Porod law,

$$I(q) = Aq^{-4} + B, \tag{3}$$

where A is called Porod constant and B is a constant for background deviation [18]. Since the intensities could be determined in absolute scale, the specific surface of phase 1 could be computed as

$$S = \frac{KA*}{2\pi(\Delta\rho)^2\rho d_1}, \tag{4}$$

where $A*$ is the Porod constant determined from the experimental intensity curve, which has been divided by transmission and corrected for background, K is a coefficient, that scales the intensities onto the absolute values, $\Delta\rho$ is the scattering length contrast between the phases, ρ is the density of phase 1, and d_1 is the optical thickness of phase 1 [16].

For the scattering length contrast ($\Delta\rho$) were used values of $1.44 \cdot 10^{11}$ 1/cm^2 (between cellulose and air), $5.0 \cdot 10^{10}$ 1/cm^2 (between cellulose and water), $6.8 \cdot 10^{10}$ 1/cm^2 (between cellulose and ethanol), and $7.0 \cdot 10^{10}$ 1/cm^2 (between cellulose and acetone). The difficulty in the case of the wet samples was to determine the optical thickness of phase 1 (cellulose) reliably.

The intensities of most of the wet samples followed the Guinier law,

$$I(q) = I_0 \exp\left(\frac{-R_g^2 q^2}{3}\right), \tag{5}$$

at small q values. Here R_g is the radius of gyration of the particles [19]. For these samples the invariant Q could be determined as

$$Q = \int_0^\infty I(q) q^2 dq. \tag{6}$$

For computing Q, the intensities were extrapolated to $q = 0$ with the Guinier law (Eq. 5) and to high q with the Porod law (Eq. 3). For these samples, the specific surface of phase 1 was also determined using the invariant Q in equation

$$S = \frac{\pi \psi_2 A}{Q\rho}, \tag{7}$$

where A is the Porod constant, ψ_2 is the volume fraction of phase 2, and ρ is the density of phase 1 [20].

For these same wet samples, also the average chord length of the sample could be determined by using Q and Porod constant A. The relation between these quantities is [19]

$$l_c = \frac{4Q}{\pi A}. \tag{8}$$

The average chord length can also be calculated using the determined specific surface S, but then the volume fractions of the two components must be known:

$$l_c = \frac{4\psi_1\psi_2}{S\rho}. \tag{9}$$

The average chord length of phase 1 is $l_1 = l_c/\psi_2$ and that of phase 2 is $l_2 = l_c/\psi_1$ [19].

The average chord length reflects the typical length scale of the phase, which is dominant in scattering contrast. If the typical length scale of phase 1 is l_1 and the typical length scale of phase 2 is l_2, the average chord length l_c is determined as $1/l_c = 1/l_1 + 1/l_2$. According to Jungnikl et al., the typical length scale for cellulose samples in dry state is the dimension of the collapsed interspace between the cellulose microfibrils, and in wet state l_c reflects the dimensions of the fibrils themselves [21].

For the wet samples cellulose microfibrils were assumed as long rigid rods [22], and the Guinier law for rod-like particles,

$$I(q) = \frac{1}{q} I_{c0} \exp\left(\frac{-R_{gc}^2 q^2}{2}\right), \tag{10}$$

was fitted to the intensities at small q in order to determine the radius of gyration of the cross-section R_{gc} [19].

3. Results

3.1. SAXS

The Porod law (Eq. 3) was fitted to the intensities at about $q = 0.23 - 0.35$ Å^{-1}, and the background constant gained from this fitting was subtracted from all the intensities. The corrected intensity curves of MCC-water, MCC-ethanol, MCC-acetone, and dry MCC samples are presented in Fig. 1, and the corresponding results for the wet, dried, and re-wet NFC-gel, dry NFC powder, NFC-water, and NFC-ethanol are presented in Fig. 2. The intensity curve of MCC-water obeyed the Porod law already from $q = 0.09$ Å^{-1}, and that of NFC-water from $q = 0.12$ Å^{-1}. All the other samples obeyed the Porod law at larger q-values (around the mentioned q-region of $0.23 - 0.35$ Å^{-1}).

The intensity curves of wet NFC-gel and NFC-water correspond to each other very well throughout the studied q-range, whereas the intensity curves of dry NFC powder and dried NFC-gel correspond to each other only at large q-values. The intensity curve of re-wet NFC-gel is different compared to all the other curves. The power law behaviours of all the samples are given in Table 1. At least two different power law regions were determined for all the samples, including the Porod law.

Journal of Physics: Conference Series **247** (2010) 012030 doi:10.1088/1742-6596/247/1/012030

Table 1. The power law behaviour of the MCC and NFC samples. At least two different power laws were found for each sample, so two rows are given for the same sample. Porod law region was found for all the samples, and in some cases the given second power law is directly the Porod law. In any case the Porod law holds always for all the samples in $q > 0.25$ Å^{-1}. FD=fractal dimension (for surface fractal: surface fractal dimension), SF=surface fractal, MF=mass fractal.

Sample	α	q [Å^{-1}]	FD	SF/MF
1. MCC-water	1.2	0.03 – 0.05	1.2	MF
2. MCC-water	3.9	0.09 – 0.35	2.1	Porod
1. MCC-ethanol	1.2	0.03 – 0.06	1.2	MF
2. MCC-ethanol	3.2	0.10 – 0.23	2.8	SF
1. MCC-acetone	1.5	0.03 – 0.06	1.5	MF
2. MCC-acetone	3.0	0.11 – 0.24	3.0	SF/MF
1. MCC (dry)	2.7	0.03 – 0.08	2.7	MF
2. MCC (dry)	3.5	0.09 – 0.25	2.5	SF
1. NFC-water	2.7	0.03 – 0.12	2.7	MF
2. NFC-water	4.0	0.12 – 0.31	-	Porod
1. NFC-ethanol	1.6	0.03 – 0.11	1.6	MF
2. NFC-ethanol	4.0	0.22 – 0.31	-	Porod
1. NFC (dry)	3.5	0.03 – 0.09	2.5	SF
2. NFC (dry)	4.0	0.23 – 0.31	-	Porod
1. NFC-gel (wet)	2.9	0.03 – 0.11	2.9	MF
2. NFC-gel (wet)	3.8	0.13 – 0.25	2.2	SF
1. NFC-gel (dried)	1.9	0.03 – 0.14	1.9	MF
2. NFC-gel (dried)	4.0	0.20 – 0.40	-	Porod
1. NFC-gel (re-wet)	1.6	0.03 – 0.07	1.6	MF
2. NFC-gel (re-wet)	4.0	0.25 – 0.30	-	Porod

The radius of gyration could be determined for the MCC-water, MCC-ethanol, MCC-acetone, NFC-ethanol, and re-wet NFC-gel samples, because their intensity curves obeyed Guinier law (Eq. 5) at low q-values. The extrapolated curves for the MCC-samples can be seen in Fig. 3. The fitting results for the radius of gyration were 4.0 nm for MCC-water, 4.0 nm for MCC-ethanol, 3.2 nm for NFC-ethanol, and 4.5 nm for both MCC-acetone and re-wet NFC-gel. Also fittings according to the Guinier law for rods (Eq. 10) were made for these samples, and the radii of gyration of the cross-sections were 1.2 nm for MCC-water, 1.5 nm for MCC-ethanol, 1.6 nm for NFC-ethanol, and 2.2 nm for both MCC-acetone and re-wet NFC-gel.

The invariant, and thus the average chord length (using Eq. 8), could be determined for these samples. The result for the average chord length was 2.4 ± 0.1 nm for MCC-water, 1.6 ± 0.1 nm for MCC-ethanol, 1.3 ± 0.1 nm for MCC-acetone, 1.2 ± 0.1 nm for NFC-ethanol and 2.5 ± 0.1 nm for re-wet NFC-gel.

The specific surface of cellulose was determined from Eq. 4 and the results are shown in Table 2. The volume fraction of cellulose was needed for this, and it was determined from the weight fractions, which were measured for all the wet samples. The calculated values for the volume fractions of cellulose were small and included large relative errors. Even a small change in this value causes a large difference in S, and thus the values of S have large error margins (even 50%) in the case of the wet samples. Using the measured volume fractions and the determined invariants for the wet MCC-samples, S values were calculated also using Eq. 7 and these results differed strongly from the previous values for all three samples, which can be explained by the error margins. However, the specific surface of the dry MCC and NFC powders and the dried NFC-gel could be determined reliably using equation 4, because for these samples the measured optical thickness was straightforwardly that of cellulose. The results were 4.9 m^2/g for MCC, 6.4 m^2/g for NFC powder and 7.0 m^2/g for dried NFC-gel.

Table 2. The specific surface of cellulose (S), the average chord length (l_c), the average chord length of the cellulose phase (l_1), and that of the solvent phase (l_2), and the diameter of gyration of the cross-section ($2R_{gc}$) of the studied MCC and NFC samples. S was calculated according to Eq. 4 and l_c according to Eq. 8.

Sample	S [m^2/g]	l_c [nm]	l_1 [nm]	l_2 [nm]	$2R_{gc}$ [nm]
MCC-water	240	2.4	3.0	11	2.4
MCC-ethanol	430	1.6	1.8	20	3.0
MCC-acetone	110	1.3	1.7	5.4	4.4
MCC-dry	5.0	-	-	-	-
NFC-water	660	-	-	-	-
NFC-ethanol	180	1.2	1.8	3.4	3.2
NFC-dry	6.4	-	-	-	-
NFC-gel wet	420	-	-	-	-
NFC-gel dried	7.0	-	-	-	-
NFC-gel re-wet	370	2.5	3.3	8.6	4.4

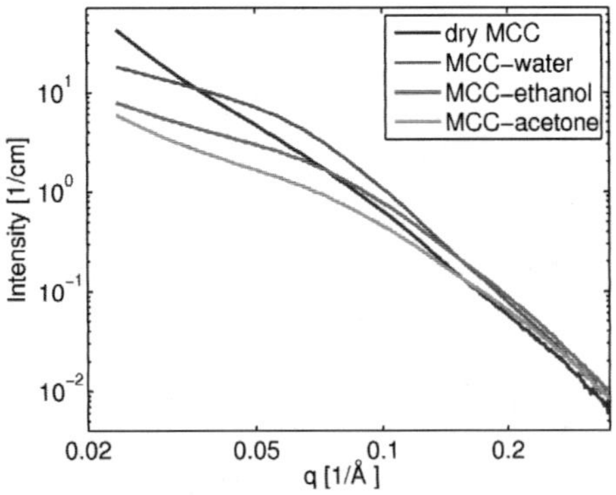

Figure 1. Integrated intensities of dry MCC, MCC-water, MCC-ethanol, and MCC-acetone. Both axis are in logarithmic scale and the intensities are given in absolute units [cm^{-1}].

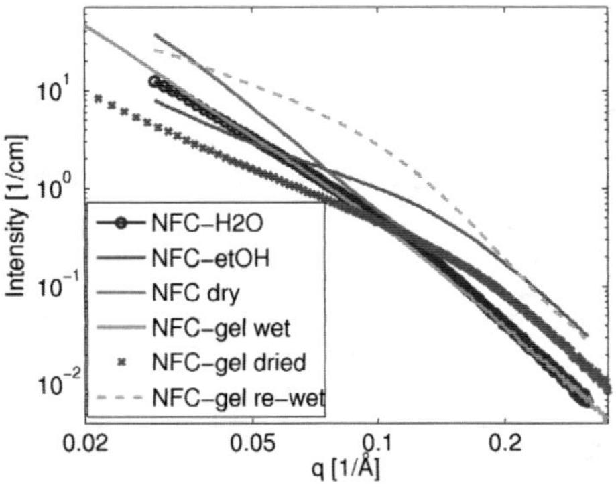

Figure 2. Integrated intensities of NFC-water, NFC-ethanol, dry NFC and wet, dried and re-wet NFC-gel. Both axis are in logarithmic scale and the intensities are given in absolute units [cm^{-1}].

3.2. WAXS results of the NFC samples

According to the WAXS results, the crystallinity of the NFC powder and the dried NFC-gel was the same, 56 ± 3 %. The widths in the cross-sectional directions of the cellulose crystallite

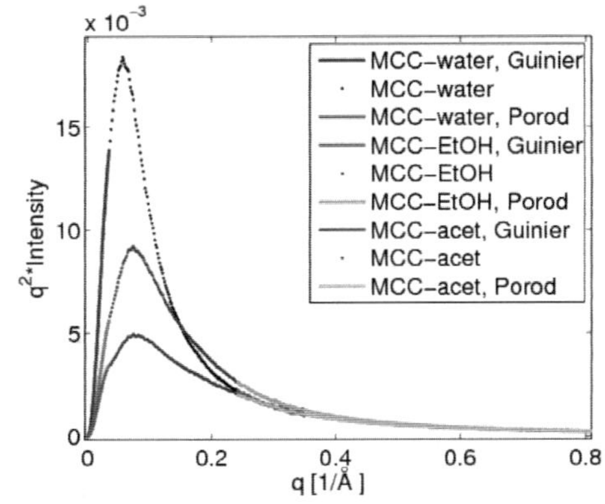

Figure 3. Extrapolations in the Guinier and Porod regions for the MCC-water, MCC-ethanol and MCC-acetone samples.

were 4.5 ± 0.4 nm (*1-10*), 4.9 ± 0.4 nm (*110*) and 5.3 ± 0.2 nm (*200*) for dried NFC-gel, and 4.1 ± 0.4 nm (*1-10*), 5.1 ± 0.4 nm (*110*) and 5.4 ± 0.2 nm (*200*) for dry NFC powder. So also the width of the crystallites was about the same in these samples. These results were very close to the corresponding values of the MCC samples made of wood pulps [4].

4. Discussion

According to the WAXS results, the crystallinity of NFC was smaller, 56 %, than that of cotton based MCC, 65 % [4]. Also the cross-sectional size of cellulose crystallites was smaller in NFC compared to MCC, for which values of 5.2 nm (*1-10*), 7.4 nm (*110*) and 8.8 nm (*200*) were obtained [4]. This can be explained by the fact that MCC was made of cotton linter, which has large crystallites and high crystallinity already in native state [4]. However, the specific surface of cellulose was only slightly larger for dry NFC powder than for dry MCC powder.

The fact, that the specific surface of cellulose S was larger in NFC-water than in MCC-water, indicated that water has been able to affect more the cellulose structure in NFC than in MCC. S was always smaller in the samples immersed in ethanol or acetone than in the samples immersed in water, except in the case of MCC-ethanol. So it was assumed that the volume fractions of ethanol and cellulose were not determined reliably for that sample, and thus the conclusion was made according to the results of all the other samples, that water could diffuse inside the cellulose microfibril bundles more effectively than ethanol and acetone, and the strongest effect occurred in the case of NFC.

Wet NFC-gel and NFC-water did not obey Guinier law like all the other wet samples (Fig. 1 and 2), which may be related to the swelling of the cellulose structures. Water has the ability to alter cellulose structure by interactions on the surfaces of the fibrils causing swelling of the surface chains [23, 11]. The lack of Guinier region at small q values may indicate that in these samples the cellulose microfibril structures are less compact and more heterogeneous than in the other wet samples. However, it is also possible that for wet NFC-gel and NFC-water the Guinier region could be found at smaller q values than measured in this study.

The different average chord length values in MCC-water, MCC-ethanol and MCC-acetone indicated that there were differences between the structures of MCC in various solvents. The diameter of gyration of the cross-section, $2R_{gc}$, was largest in MCC-acetone. The $2R_{gc}$ value was smallest in MCC-water, and it was exactly same as the average chord length value of MCC-water (2.4 nm), whereas for all the other immersed samples $2R_{gc}$ values were larger than the average chord length values. In MCC-ethanol $2R_{gc}$ was about two times the average chord length value, and in MCC-acetone it was about three times that value. Comparing the WAXS results and the diameter of gyration values, water has the largest effect on the cellulose microfibrils among the studied solvents.

Journal of Physics: Conference Series **247** (2010) 012030 doi:10.1088/1742-6596/247/1/012030

Jungnikl et al. gained similar results for the average chord length of their cellulose samples as we gained in this study for MCC and NFC [21]. We gained the average chord length values of 1.3 nm for MCC-acetone, 1.6 nm for MCC-ethanol, 1.2 nm for NFC-ethanol, 2.4 nm for MCC-water, and 2.5 nm for re-wet NFC-gel. Jungnikl et al. studied cellulose samples made of spruce wood, which was extracted with sodium chlorite ($NaClO_2$) for delignification and with sodium hydroxide (NaOH) for extraction of hemicellulose. For native spruce wood and the sample delignified with $NaClO_2$, they got as an average chord length 2.1 nm, for NaOH treated samples $2.7-3.3$ nm, and for the sample treated with H_2O_2 acetic acid 3.9 nm. All the mentioned values were gained from the samples measured in water. [21] The l_c values of the re-wet NFC-gel and MCC-water samples were between that of the native spruce and the NaOH treated cellulose samples, and the NFC/MCC-ethanol and MCC-acetone samples resembled clearly none of the wet samples measured by Jungnikl et al.

However, the l_c values of NFC-ethanol and MCC-acetone agreed very well with the never-dried and re-wet lyocell fibres, which Vickers et al. [25] studied. They determined for the average chord length of the never-dried and re-wet samples 1.3 and 1.2 nm, respectively. For the dry fibres they gained much higher values, from 7.5 nm to 117 nm. For the average chord lengths of water and cellulose phases in the never-dried fibres they gained 5.9 nm and 1.7 nm, respectively, and 2.9 nm and 2.0 nm in the re-wet fibres. [25] The crystalline structure of lyocell is cellulose-II [26], whereas that of cotton-MCC and the studied NFC was cellulose-I, but the characteristic length scales can be compared even if the crystalline structure of cellulose is not the same.

Ishii et al. [27] studied the effect of the sequential immersion in water, acetone, and DMAc on the pore structure of cellulose (softwood pulp and Whatman CF11) and its dissolution in lithium chloride/N,N-dimethylacetamide (LiCl/DMAc). They observed a difference between the DMAc treated and the untreated celluloses only when acetone was used as a medium. According to the specific inner surface and the average chord length values, the amount of small pores was increased due to the DMAc treatment in acetone exchanged cellulose. When water was used as a medium, no difference between the results of untreated and DMAc-treated cellulose was detected, which they explained by the fact that the sorption of the DMAc-treated cellulose in water spoils the effect of the solvent exchange. According to the interaction parameter, which they calculated, acetone is not adsorbed onto cellulose. [27]

Wet NFC-gel, NFC-water, and dry MCC and NFC obeyed power laws with the exponents -2.9, -2.7, -2.7 and -3.5, respectively, at $q = 0.03 - 0.09$ Å^{-1}. Ishii et al. [28] studied the power law behaviour of untreated, solvent-exchanged, and milled softwood pulp samples using SAXS. Solvent-exchanging included the sequential immersions in water, in acetone and finally in DMAc. At $q = 0.016 - 0.034$ Å^{-1}, the SAXS-curves of the untreated, solvent-exchanged and milled cellulose samples obeyed power laws with exponents -3.5, -2.9 and -3.7, respectively. So the solvent-exchanged sample was the only mass fractal, whereas two other samples were surface fractals. [28] Dry NFC powder behaved like untreated and milled softwood pulp of Ishii et al., and wet NFC-gel, NFC-water, and dry MCC behaved like the solvent-exchanged softwood pulp of Ishii et al. But all the other studied MCC and NFC samples had the exponents in the first mentioned q-region between -1.2 and -1.9.

Strømme et al. have studied the fractal dimensions of different cellulose powders including MCC (Avicel) by multilayer BET adsorption of water and nitrogen. They obtained by water adsorption a fractal dimension of 1.43 for MCC (at the length scale of about $0.1 - 10$ nm), whereas by nitrogen adsorption the fractal dimension of MCC was 2.41 (at length scale up to $100 - 1000$ nm). Their conclusion was that the water adsorption altered the structure of highly crystalline cellulose by water interactions on the surface of the fibrils without actual swelling of the material. [24]

5. Conclusions

The effects of water, ethanol, and acetone on the nanostructure of microcrystalline and nanofibrillated cellulose were studied using SAXS and WAXS. The specific surface of cellulose increased strongly when both MCC and NFC were immersed in water, ethanol, or acetone (only MCC) compared to dry cellulose, which indicated, that the fibre bundles opened in all the studied solvents. The largest effect was detected with water on NFC. The scattering intensities of the immersed samples differed from each other and the same structural parameters could not be determined for all the samples, which makes the direct comparison of the results difficult, but, on the other hand, gives information on the differences between the samples. However, important conclusions on the morphology of cellulose microfibrils can be made: Differences between the radii of gyration, crystallite diameters (determined by WAXS) and average chord lengths indicate that the cellulose microfibrils have heterogenous structure when immersed in the solvents. On the basis of the different average chord length values, the solvents affected the structure of MCC and NFC, and the effect depended on the used solvent. The average chord length was smaller for MCC-acetone, MCC-ethanol, and NFC-ethanol than for MCC-water and re-wet NFC-gel. These differences could arise from the better ability of water to diffuse inside the cellulose microfibril bundles than ethanol and acetone. This same conclusion was made also according to the values of the specific surface of cellulose.

Acknowledgments

We would like to thank J. Rettenmaier & Söhne GmbH for providing the nanocellulose samples and Hans-Peter Hentze for support and comments on the manuscript. K.L. thanks the National Graduate School of Materials Physics for the financial support.

References

[1] Fleming K, Gray D G and Matthews S 2001 *Chem. Eur. J.* **7** 1831
[2] Elazzouzi-Hafraoui S, Nishiyama Y, Putaux J L, Heux L, Dubreuil F and Rochas C 2008 *Biomacromolecules* **9** 57
[3] Azizi Samir M A S, Alloin F and Dufresne A 2005 *Biomacromolecules* **6** 612
[4] Leppänen K, Andersson S, Torkkeli M, Knaapila M, Kotelnikova N and Serimaa R 2009 *Cellulose* **16** 999
[5] Pääkkö M, Ankerfors M, Kosonen H, Nykänen A, Ahola S, Österberg M, Ruokolainen J, Laine J, Larsson P T, Ikkala O and Lindström T 2007 *Biomacromolecules* **8** 1934
[6] Azizi Samir M A S, Alloin F, Gorecki W, Sanchez J Y and Dufresne A 2004 *J. Phys. Chem.* B **108** 10845
[7] Boluk Y 2005 *Cellulose* **12** 577
[8] Bochek A M and Kalyuzhnaya L M 2002 *Russian Journal of Applied Chemistry* **75** 989
[9] Awadel-Karim S, Leclerc D, Grondey H, Yatich S K, Nazhad M M and Pazner L 1998 *Holzforschung* **52** 67
[10] Vareli G D, Panagiotis G D and Akrida-Demertzi K 1997 *Z. Lebensm. Unters Forsch.* A **205** 204
[11] Kotelnikova N E, Petropavlovsky G A, Shevelev V A, Vasil'eva G G and Volkova L A 1976 *Cellul. Chem. Technol.* **4** 391
[12] Petropavlovsky G A and Kotelnikova N E 1985 *Acta Polymerica* **36** 118
[13] Battista J A 1975 *Microcrystal polymer science* (New York, McGraw-Hill Book) 208
[14] Mantanis G I, Young R A and Rowell R M 1995 *Cellulose* **2** 1
[15] Schlosser H 2008 *Wochenblatt für Papierfabrikation* **6** 252
[16] Vainio U, Maximova N, Hortling B, Laine J, Stenius P, Simola L K, Gravitis J and Serimaa R 2004 *Langmuir* **20** 9736
[17] Schmidt P W 1991 *J. Appl. Cryst.* **24** 414
[18] Svergun D I and Koch M H J 2003 *Rep. Prog. Phys.* **66** 1735
[19] Glatter O and Kratky O 1982 *Small Angle X-ray Scattering* (London, Academic Press Inc.) 155
[20] Spalla O, Lyonnard S and Testard F 2003 *J. Appl. Cryst.* **36** 338
[21] Jungnikl K, Paris O, Fratzl P and Burgert I 2008 *Cellulose* **15** 407
[22] Nishiyama Y 2009 *J. Wood Sci.* **55** 241
[23] Papirer E, Brendle E, Balard H and Vergelati C 2000 *J. Adhesion Sci. Technol.* **14** 321
[24] Strømme M, Mihranyan A, Ek R and Niklasson A 2003 *J. Phys. Chem.* B **107** 14378
[25] Vickers M E, Briggs N P, Ibbett R N, Payne J J and Smith S B 2001 *Polymer* **42** 8241

[26] Wei M, Yang G, Tian Y, Shao H and Hu X 2009 *Holzforschung* **63** 23
[27] Ishii D, Kanazawa Y, Tatsumi D and Matsumoto T 2007 *J. Appl. Pol. Sci.* **103** 3976
[28] Ishii D, Tatsumi D and Matsumoto T 2003 *Biomacromolecules* **4** 1238

Journal of Physics: Conference Series **247** (2010) 012031 doi:10.1088/1742-6596/247/1/012031

Mineral crystal alignment in mineralized fracture callus determined by 3D small-angle X-ray scattering

Yifei Liu[1], Inderchand Manjubala[1], Paul Roschger[2], Hanna Schell[3], Georg N Duda[3], and Peter Fratzl[1]

[1] Department of Biomaterials, Max Planck Institute of Colloids and Interfaces, 14424 Potsdam, Germany

[2] 4th Medical Department, Ludwig Boltzmann Institute of Osteology at Hanusch Hospital of WGKK and AUVA Trauma Centre Meidling, 1140 Vienna, Austria

[3] Julius Wolff Institut and Center for Musculoskeletal Surgery, Charité- University Medicine Berlin, Augustenburger Platz 1, 13353 Berlin, Germany

E-mail: fratzl@mpikg.mpg.de

Abstract. Callus tissue formed during bone fracture healing is a mixture of different tissue types as revealed by histological analysis. But the structural characteristics of mineral crystals within the healing callus are not well known. Since two-dimensional (2D) scanning small-angle X-ray scattering (sSAXS) patterns showed that the size and orientation of callus crystals vary both spatially and temporally [1] and 2D electron microscopic analysis implies an anisotropic property of the callus morphology, the mineral crystals within the callus are also expected to vary in size and orientation in 3D. Three-dimensional small-angle X-ray scattering (3D SAXS), which combines 2D SAXS patterns collected at different angles of sample tilting, has been previously applied to investigate bone minerals in horse radius [2] and oim/oim mouse femur/tibia [3]. We implement a similar 3D SAXS method but with a different way of data analysis to gather information on the mineral alignment in fracture callus. With the proposed accurate yet fast assessment of 3D SAXS information, it was shown that the plate shaped mineral particles in the healing callus were aligned in groups with their predominant orientations occurring as a fiber texture.

1. Introduction

As a functionally important tissue in all vertebrates, bone supports the body weight, enables functional movement and protects internal organs [4]. For these purposes, compact bone is known to have a fairly complicated hierarchical structure down to the nano level which contributes to its outstanding mechanical performances [5-8]. Although the structural features of bone are relatively well understood, the structural and mechanical properties of the fracture callus formed during bone fracture healing are not yet clear at a similar scale. It is observed by histology that during the formation, maturation and resorption of a callus, there are various types of tissue involved [9, 10]. Yet due to the spatially anisotropic and temporally dynamic property of callus composition, it is difficult to obtain accurate information of the callus characteristics in a three-dimensional manner [11-13].

From material composition point of view, mature bone mainly consists of two phases: organic phase (collagen) and inorganic phase (mineral). Both components as well as their interactions determine the mechanical behavior of bone [14]. One frequently used technique for the investigation of bone mineral crystals is small angle X-ray scattering (SAXS) [15, 16]. The advantage of this technique is that sample preparation is relatively easy and delicate local texture of the specimen can be acquired quantitatively. But the drawback is that only a cross section of the overall scattering pattern from the illuminated sample volume is able to be detected. This would lead to a considerable loss of information or could even result in misleading interpretation from the data, if the mineral crystals are expected to have a highly anisotropic behavior like in the fracture callus.

To overcome this drawback, three-dimensional (scanning and rotational) small angle X-ray scattering (3D SAXS), which was previously applied to investigate bone minerals in horse radius [2] and oim/oim mouse femur/tibia [3], is used here to reveal the spatial distribution of mineral crystal size and orientation in mineralized callus tissue of a sheep osteotomy model. Although the 3D-data accumulation is similarly realized through a controlled sample rotation as in those earlier studies, in the current work we propose a different routine of data analysis. The simple but effective representation of 3D SAXS data and the reconstruction of 3D SAXS patterns enable a direct visualization of mineral alignment in the investigated sample volume, which provides insight into the 3D structural properties of the callus material as well as their relation with its mechanical performance.

2. Materials and Methods

2.1. Sample Preparation

The sample investigated in this study comes from an earlier study [17] on fracture healing in a large animal model (female Merino sheep) of a 3 mm osteotomy stabilized with a monolateral external rigid fixator. Longitudinal section of bone/callus sample was harvested at 3 weeks time point of healing, fixed, dehydrated with ethanol, and embedded in polymethylmethacrylate (PMMA) without decalcifying. Using a low speed diamond saw (Buehler Isomet, Buehler GmbH, Duesseldorf, Germany), a 200 μm section was cut from the embedded block for 3D SAXS evaluation. Experimental regions of interest (ROIs) were chosen within the periosteal callus on the medial side of the proximal fragment (figure 1). The positions of ROIs were defined to have comparison with the previously reported structural properties of the callus [1].

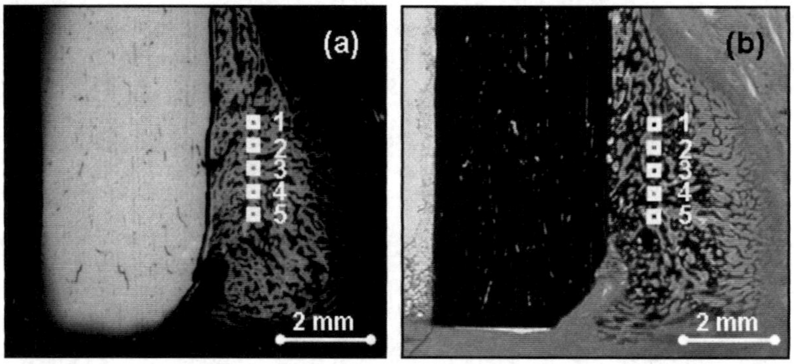

Figure 1. Regions of interest were chosen in the periosteal callus formed at 3 weeks of healing, on the medial side of the proximal fragment. Altogether five positions (ROI 1 to 5, marked in yellow) with a same transversal distance from the adjacent cortex were experimentally evaluated by 3D SAXS. Backscattered electron image (figure a) and Safranin Orange/von Kossa staining (figure b) show an abundant but anisotropic bone formation within the callus.

2.2. SAXS Measurements

3D small-angle X-ray scattering was done with a laboratory SAXS instrument (Nanostar, Bruker AXS, Karlsruhe, Germany). X-ray beam (*Cu-Kα*, λ = 0.1542 nm) is generated with a rotating Cu anode generator (M06XCE– SRA, Mac Science, Japan) operated at 4 kW (40 kV, 100 mA) and is collimated by a double pinhole system (first pinhole diameter 100 µm, second pinhole diameter 300 µm). Sample is mounted on a translation stage with three degrees of freedom (translation axes *x* and *y* within the sample plane and rotation axis *z* perpendicular to the beam). 2D scanning (scanning small-angle X-ray scattering, sSAXS) over the sample is realized via automatic movement of the sample stage in *x* and *y* directions with a resolution of 200 µm (real space resolution) which is equivalent to the thickness of sample slice as well as the beam diameter at the sample position (focal plane). 3D SAXS is done by additionally applying a sample rotation along *z* axis. The scattering pattern is collected with an accumulation time of 3600 seconds on a two-dimensional CCD detector of 1024×1024 pixels with pixel size of 105.26 µm (reciprocal space resolution) placed at a distance of 600 mm behind the specimen. All the chambers are evacuated to avoid air scattering. A schematic overview of the whole experimental setup is shown in figure 2.

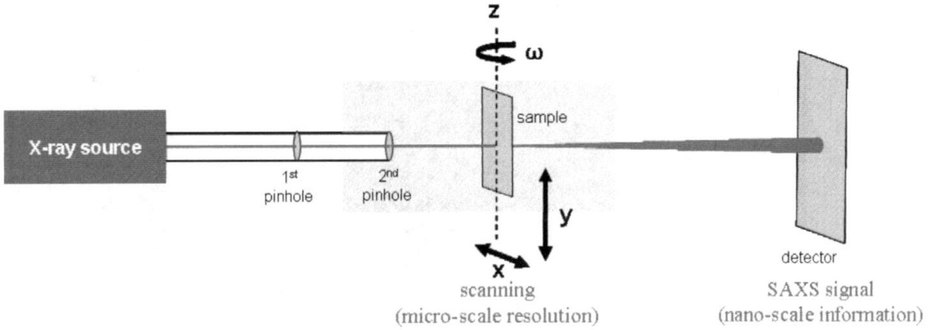

Figure 2. Schematic view of the experimental set-up, which is typically the same for 3D SAXS. Similar to 2D sSAXS, movement of the sample is possible in *x* and *y* directions. But additionally, the sample is rotated around the vertical axis *z* by an angle ω to gain texture information in the third dimension.

In this study, the sample was rotated around the vertical axis *z* in steps of 15° from -45° to 45°, leading to seven rotation angles. For each rotation angle ω ($-45° \leq \omega \leq 45°$), a 2D X-ray radiograph (absorption image) of the whole sample was generated by using a pin-diode. Regions of interest chosen to be located along a vertical line were marked by two parallel sharp knife edges attached onto the sample, and the corresponding x-y motor positions were defined for SAXS measurements according to the seven radiographs. 2D SAXS patterns were by this means always collected at the same five positions on the specimen as illustrated in figure 1 for all the rotation angles, which enables the reconstruction of the five 3D SAXS patterns in reciprocal space.

The calibration of the beam center and scattering distance was done at each angle ω by measuring a silver behenate (AgBh) standard sample. 2D SAXS patterns were all corrected for transmission rate derived from radiography, background scattering (empty beam) and detector noise reduction (dark current).

2.3. Data Analysis

2.3.1. *1D normalization.* Particularly due to the change of effective sample thickness (along the beam direction) during sample rotation, a normalization factor (NF) has to be calculated from the series of 2D SAXS patterns acquired at the same sample position to correct the change in SAXS intensity caused by the change in illuminated sample volume. This was realized by integrating a total SAXS intensity of the 2D SAXS pattern along a vertical line across the beam center (v axis in figure 3b, or z axis in figure 3d, both in reciprocal space). Practically a radial integration was applied to the circulator sectors of -0.5° < χ < 0.5° and -179.5° < χ < 180.5° within a fixed radial range (50 pixels < r < 420 pixels, 0.34 nm^{-1} < q < 2.83 nm^{-1}, 0.5° < 2θ < 4.2°) at all seven ω angle positions. χ is the azimuthal angle as defined in figure 3d. q is the modulus of the scattering vector defined as $q = \frac{4\pi}{\lambda}\sin\theta$, where 2θ is the scattering angle and λ is the X-ray wavelength. Inner radius r_1 was set to be 50 pixels for avoiding scattering influence from the beam stop, and outer radius r_2 was set to be 420 pixels for keeping the integrated region inside the detector range. The integrated intensities within circulator sectors of -0.5° < χ < 0.5° and -179.5° < χ < 180.5° were then averaged to obtain a constant K(ω) depending only on the rotation angle of the specimen. The ratio NF(ω) = K(ω) / K(0) was taken as a normalization factor for each scattering pattern obtained at any given rotation angle ω. Table 1 gives a summary of the normalization factors for all the five regions of interest at all the seven rotation angles.

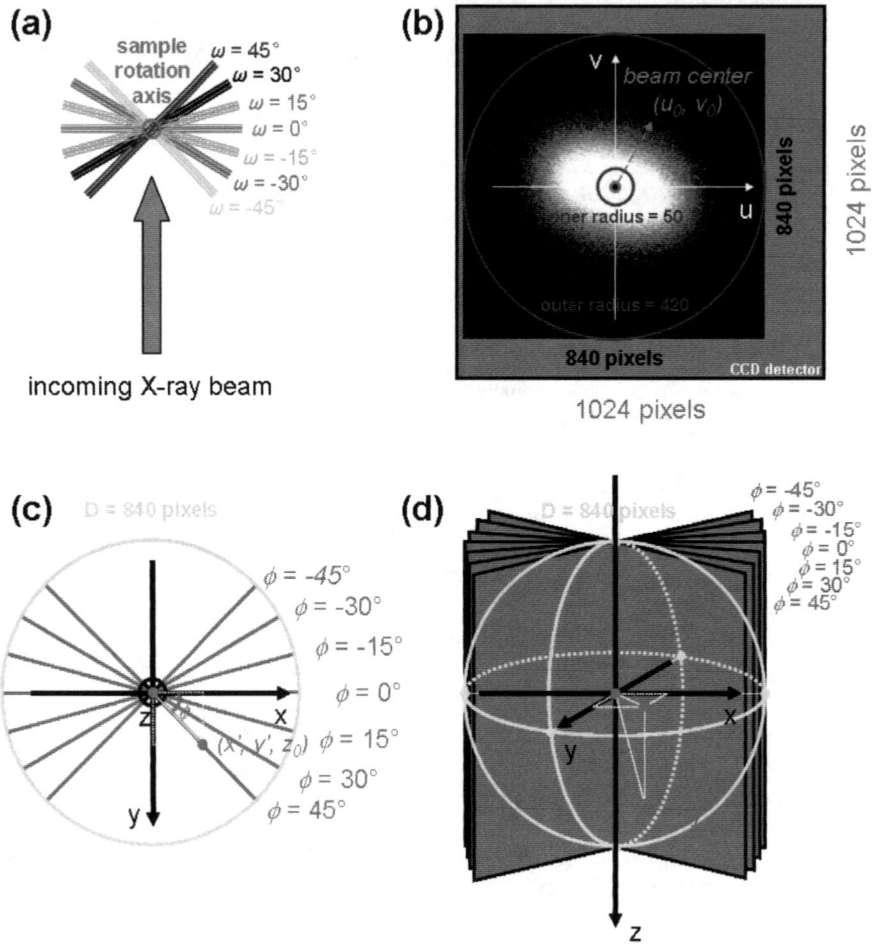

Figure 3. Sample rotation in real space (figure a) is characterized by the rotation angle ω. Measured SAXS pattern (figure b) representing reciprocal space is characterized by the angular position ϕ (figure c) of the detector with respect to the sample, and additionally by the azimuthal angle χ which lies in the detector plane (figure d).

Table 1. Normalization factors NF(ω) of the five regions of interest on the sample at different rotation angles (-45° ≤ ω ≤ 45° with a step of 15°).

	NF (ROI 1)	NF (ROI 2)	NF (ROI 3)	NF (ROI 4)	NF (ROI 5)
ω = 45°	1.4375	1.3716	2.0033	1.2581	1.2163
ω = 30°	1.0244	1.6131	1.3503	1.1230	0.9953
ω = 15°	1.0504	1.1827	1.3110	1.1637	0.9722
ω = 0°	1.0000	1.0000	1.0000	1.0000	1.0000
ω = -15°	0.9307	1.2313	1.3722	0.7827	0.8379
ω = -30°	0.7882	1.7248	1.2627	1.7890	0.7852
ω = -45°	1.0236	1.8978	1.6243	1.9608	0.9563

Using this normalization procedure ensures that the SAXS intensity does not change with rotation in the direction of the rotation axis. In this way, we correct for the changes in effective sample volume when the section is rotated. The limitation of this procedure is, however, that the projection of the X-ray beam on the specimen also changes with rotation. Hence, the procedure is only exact if the nanostructure (as studied by SAXS) stays approximately the same within the different probing volumes when the rotation angle ω varies. Figure 4 describes the intensity profiles (azimuthally integrated) of the first ROI before and after normalization.

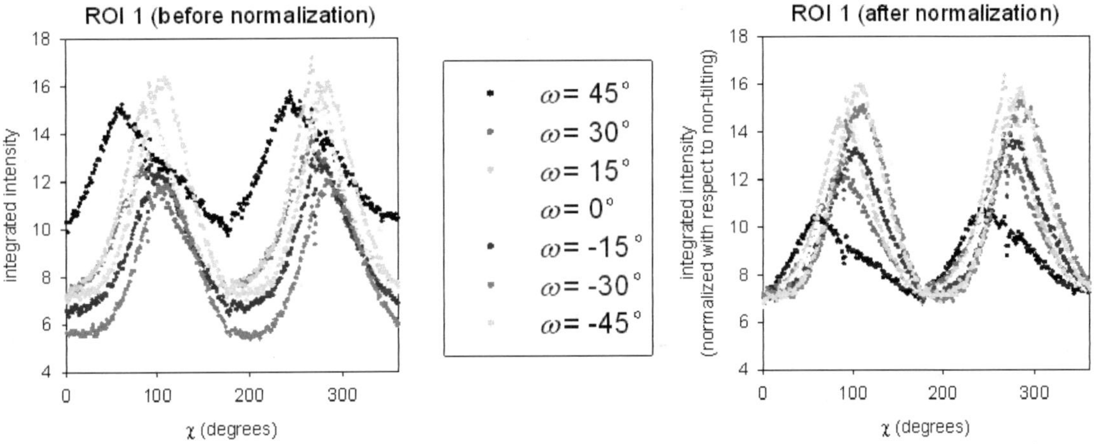

Figure 4. Intensity profiles (azimuthally integrated) of the first region of interest before (left) and after (right) normalization using the factors in Table 1. The measured intensity was initially the highest at $\omega = \pm 45°$ because rotation induced an increase of the illuminated sample volume. This is corrected by the normalization.

2.3.2. *2D representation.* For small-angle X-ray scattering the Ewald sphere can be considered to be flat. Therefore scattering patterns at different rotation angle ω correspond to cross sections of the 3D SAXS "ellipsoids" in reciprocal space at different angular position ϕ. Considering that ω is the angle between the sample plane and the detector plane (figure 3), we have $\phi = -\omega$.

For each of the five ROIs at the sample, 1D $I(\chi)$ profiles were deduced for different ω positions (ϕ positions) by azimuthal integration of all the 2D SAXS patterns within the fixed radial range from r_1 to r_2 around the direction of the incoming X-ray beam. $I(\chi, \phi)$ was then plotted in a 2D manner to illustrate the orientation of the scattering pattern in 3D. Normalization factors (Table 1) were used to scale the whole $I(\chi)$ spectrum for normalizing $I(\chi, \omega)$ at $\chi = 0°$ and $\chi = 180°$. $I(\chi)$ values at unavailable ω positions (ϕ positions) were linearly interpolated.

2.3.3. *3D visualization.* Assume that n 2D measurements $g_1(u,v)$, $g_2(u,v)$,…, $g_n(u,v)$ with detector coordinates u and v (figure 3b) were acquired at angular positions ϕ_1, ϕ_2,…, ϕ_n respectively, and the angular positions satisfy $-45° \leq \phi_1 < \phi_2 < \ldots < \phi_n \leq 45°$ (figure 3c and 3d). Then the value of each voxel in the reconstructed volume, $f(x,y,z)$, can be calculated as (figure 3c and 3d):

$$f(x,y,z)=\begin{cases}\dfrac{(\phi_{i+1}-\hat{\phi})g_i(l,z')+(\hat{\phi}-\phi_i)g_{i+1}(l,z')}{\phi_{i+1}-\phi_i} & \phi_i \leq \hat{\phi} < \phi_{i+1}\\ \dfrac{(\phi_{i+1}+180^\circ-\hat{\phi})g_i(l,z')+(\hat{\phi}-\phi_i-180^\circ)g_{i+1}(l,z')}{\phi_{i+1}-\phi_i} & \phi_i+180^\circ \leq \hat{\phi} < \phi_{i+1}+180^\circ\\ 0 & \text{else}\end{cases} \quad (1),$$

where $$\hat{\phi}=\begin{cases}\arctan\dfrac{x-x_0}{y-y_0} & x-x_0\geq 0, y\neq y_0\\ \arctan\dfrac{x-x_0}{y-y_0}+180^\circ & x-x_0<0, y\neq y_0,\\ 0 & y=y_0\end{cases}$$

$$l=\sqrt{(x-x_0)^2+(y-y_0)^2},$$

and $z'=z-z_0+u_0$.

In these equations, x_0, y_0 and z_0 define the center position (origin) of the reconstructed volume and i is the number of 2D measurement which satisfies $1 \leq i \leq n-1$. The 3D spherical volume with the dimension of d = 840 pixels was reconstructed by equation (1) from the series of 2D measurements done at different sample rotation angle ω. Normalization factors for normalizing $I(\chi, \omega)$ in the previous section were also used here to scale the intensity of the raw SAXS patterns before coordinate conversion.

To interpret the directionality of the reconstructed 3D volume with 2D plots, integral values for all the ray directions with different azimuth angles ϕ ($\phi_1 \leq \phi \leq \phi_n$) and inclination angles χ ($0° \leq \chi < 360°$) from the origin of the volume were calculated (figure 3d), with 1° interval for both ϕ and χ. In order to avoid the anisotropic shadow from the beam stop and to balance the integral length for all the directions, the rays within the lengths of inner radius r_1 (50 pixels) and outer radius r_2 (420 pixels) from the origin (beam center) were integrated. r_1 and r_2 were chosen such to match the effective q range (0.34 nm^{-1} < q < 2.83 nm^{-1}) defined before. The integrations along such ray paths were calculated with Siddon ray tracing algorithm [18].

3. Results and Discussion

Figure 5 illustrates the 2D orientation figure $I(\chi, \phi)$ of all the five investigated sample volumes (200 µm × 200 µm × 200 µm each) at the five ROIs.

In all the five ROIs, apart from the existence of two periodic maximums appearing along the χ axis with a spacing of 180° due to the complete symmetry of the 3D SAXS ellipsoid, it turns out that there are also more than one global maximum along the ϕ axis. The intensity maps always show two separate groups of intensity concentration in this direction, indicating an overlaying of two ellipsoids in 3D reciprocal space. This agrees with results from previous sSAXS measurements which reveal in some cases a 2D SAXS pattern of spark instead of oval [1].

The dash lines (marked in black) in $I(\chi, \phi)$ figures correspond to stacks of mineral crystals with different amount (relative value of normalized intensity) and orientation (peak position of normalized intensity) within the illuminated sample volume. It can be seen that each dash line is made up of a series of neighboring maximums (with respect to the other intensities at the same ϕ position) reflecting a gradual change of the preferred mineral orientation. Knowing that the directionality of SAXS pattern in reciprocal space is perpendicular to the orientation of mineral crystals in real space, it can be further concluded that the 3D SAXS pattern has a tendency of being an oblate spheroid.

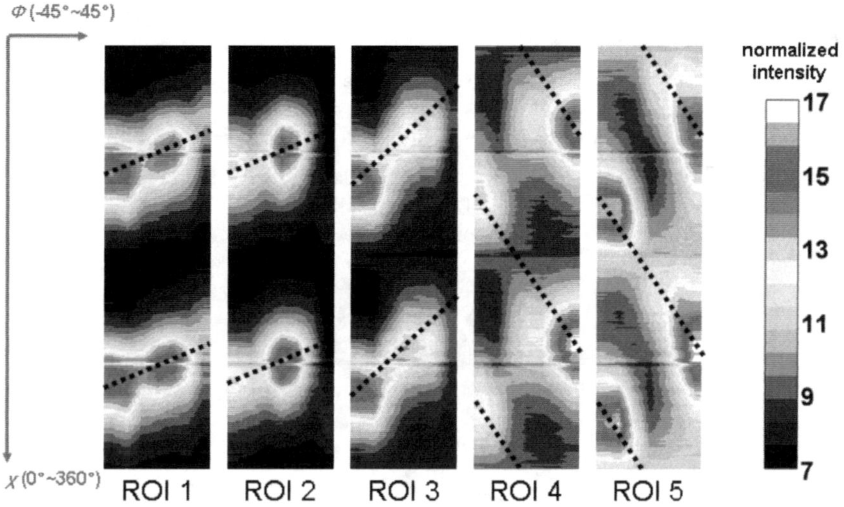

Figure 5. 2D orientation figure $I(\chi, \phi)$ of the five ROIs. Normalized intensities are shown with colour-coding.

Figure 6 demonstrates the integral values for different ϕ and χ angles of one ROI with 2D orientation figure $I(\chi, \phi)$, calculated from the reconstructed 3D volume using the Siddon ray tracing algorithm. Compared to the $I(\chi, \phi)$ plot got from direct linear interpolation (figure 5, ROI 1), the obtained $I(\chi, \phi)$ orientation figure here (figure 6) has higher intensity caused by finer integration scheme. And in the latter figure, the calculation procedure from 1D to 3D and back to 2D has led to some artifacts at $\phi = 0°$, $\chi = 0°$, and $\chi = \pm 180°$. Nevertheless the two results are basically in good agreement with each other. The global maximum of the directionality occurs at the position of $\phi = -30.0° \pm 7.5°$ and $\chi = 285° \pm 2°$. This means, if without the need of 3D visualization, a direct plotting of the 2D orientation figure $I(\chi, \phi)$ should be precise enough and maybe even better (free from artifacts) to depict the mineral alignment in 3D.

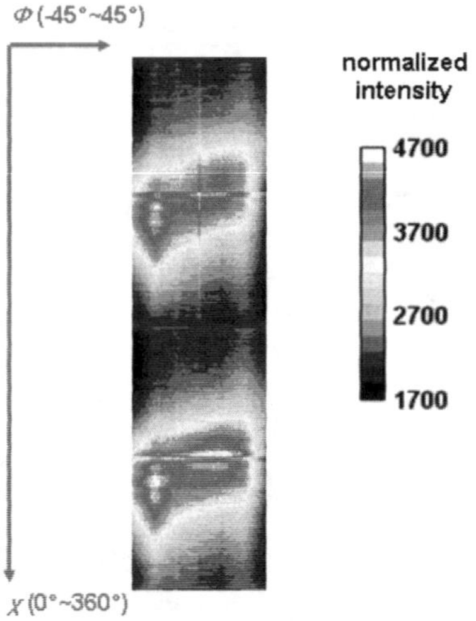

Figure 6. 2D orientation figure $I(\chi, \phi)$ of the first ROI, calculated from the reconstructed 3D volume with Siddon ray tracing algorithm. Normalized intensities are shown with colour-coding.

Based on one of the proposed models of mineralized fibrils [19], SAXS-pattern direction (orientation in reciprocal space, red arrows in figure 7) of the mineral crystal stacks should form a normal plane of the fiber direction (orientation in real space, perpendicular to the plane of figure 7). The general idea is that the X-ray beam (due to the size of its cross-sections) covers many mineralized collagen fibrils (with a typical diameter in the order of 100 nm) where the mineral crystals (plates viewed edge-on in figure 7) are oriented differently in different groups of fibrils. This leads to a SAXS pattern in the shape of an oblate ellipsoid. The direction of the ellipsoid axis can be read from the patterns in figure 5 and corresponds to the perpendicular to the broken line. Since the ellipsoid is not perfect in the sense that the intensity varies around the rotation axis (that is, along the broken line in figure 5), we conclude that there is a certain texture in the orientation distribution of the crystals. Maps such as the ones described in figure 5 can be used to quantify their orientation distribution.

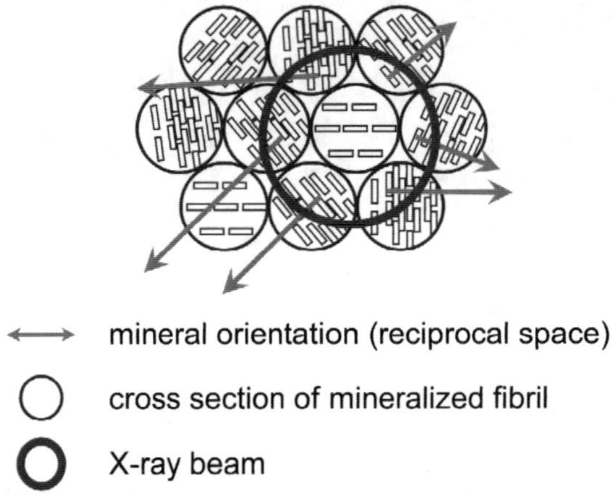

Figure 7. Illustration of circular dispersion of crystals in normal bone illuminated by a 200 μm beam (not drawn to scale), adapted from a previously proposed model of mineralized fibrils [19]. Fiber orientation in real space (not depicted) is perpendicular to the plane of the paper, and the red arrows show mineral orientations in reciprocal space for mineral crystals within the beam area indicated by solid blue ring.

4. Conclusion

In this study, a simple and effective method to visualize and interpret the 3D directionality of SAXS results obtained from sheep fracture callus is presented. Interpretation of the 3D directionality via 2D plotting gives accurate information of the mineral orientation within bony structure at specific positions of the sample. 3D reconstruction of the full SAXS patterns in reciprocal space enables quantitative determination and visualization of the global directionality of the overall SAXS pattern. As a main result of the current study concerning the structural properties of callus minerals, it was shown that the mineral particles were aligned in stacks with their predominant orientations lying in a single plane perpendicular to the fiber direction.

Acknowledgements

This study was supported by a grant of the German Research Foundation (DFG), under the collaborative research centre scheme, SFB 760, Project B1. The authors would like to appreciate G. Dinst and A. Martins for sample preparation, and I. Zenke for SAXS assistance.

Journal of Physics: Conference Series **247** (2010) 012031 doi:10.1088/1742-6596/247/1/012031

References

[1] Liu Y, Manjubala I, Roschger P, Epari D R, Schell H, Lienau J, Bail H J, Duda G N and Fratzl P 2009 The evolution of size and distribution of apatite mineral crystals during bone fracture healing in sheep *Bone* **44** S271-S2

[2] Fratzl P, Schreiber S and Boyde A 1996 Characterization of bone mineral crystals in horse radius by small-angle X-ray scattering *Calcif. Tissue Int.* **58** 341-6

[3] Fratzl P, Paris O, Klaushofer K and Landis W J 1996 Bone mineralization in an osteogenesis imperfecta mouse model studied by small-angle X-ray scattering *J. Clin. Invest.* **97** 396-402

[4] Carter D R and Beaupre G S 2001 *Skeletal function and form: mechanobiology of skeletal development, aging, and regeneration* (New York: Cambridge University Press)

[5] Lakes R 1993 Materials with structural hierarchy *Nature* **361** 511-5

[6] Rho J Y, Kuhn-Spearing L and Zioupos P 1998 Mechanical properties and the hierarchical structure of bone *Med. Eng. Phys.* **20** 92-102

[7] Weiner S and Wagner H D 1998 The material bone: structure-mechanical function relations *Annu. Rev. Mater. Sci.* **28** 271-98

[8] Fratzl P and Weinkamer R 2007 Nature's hierarchical materials *Prog. Mater. Sci.* **52** 1263-334

[9] Hall B 2005 *Bones and cartilage: developmental and evolutionary skeletal biology* (London: Elsevier Academic Press)

[10] Shapiro F 2008 Bone development and its relation to fracture repair. The role of mesenchymal osteoblasts and surface osteoblasts *Eur. Cells Mater.* **15** 53-76

[11] Hildebrand T, Laib A, Muller R, Dequeker J and Ruegsegger P 1999 Direct three-dimensional morphometric analysis of human cancellous bone: microstructural data from spine, femur, iliac crest, and calcaneus *J. Bone Miner. Res.* **14** 1167-74

[12] Cann C E 1988 Quantitative CT for determination of bone mineral density: a review *Radiology* **166** 509-22

[13] Gerstenfeld L C, Alkhiary Y M, Krall E A, Nicholls F H, Stapleton S N, Fitch J L, Bauer M, Kayal R, Graves D T, Jepsen K J and Einhorn T A 2006 Three-dimensional reconstruction of fracture callus morphogenesis *J. Histochem. Cytochem.* **54** 1215-28

[14] Fratzl P, Gupta H S, Paschalis E P and Roschger P 2004 Structure and mechanical quality of the collagen-mineral nano-composite in bone *J. Mater. Chem.* **14** 2115-23

[15] Fratzl P, Fratzlzelman N, Klaushofer K, Vogl G and Koller K 1991 Nucleation and growth of mineral crystals in bone studied by small-angle X-ray scattering *Calcif. Tissue Int.* **48** 407-13

[16] Fratzl P, Groschner M, Vogl G, Plenk H, Eschberger J, Fratzlzelman N, Koller K and Klaushofer K 1992 Mineral crystals in calcified tissues: a comparative study by SAXS *J. Bone Miner. Res.* **7** 329-34

[17] Schell H, Lienau J, Epari D R, Seebeck P, Exner C, Muchow S, Bragulla H, Haas N P and Duda G N 2006 Osteoclastic activity begins early and increases over the course of bone healing *Bone* **38** 547-54

[18] Siddon R L 1985 Fast calculation of the exact radiological path for a three-dimensional CT array *Med. Phys.* **12** 252-5

[19] Rubin M A, Jasiuk I, Taylor J, Rubin J, Ganey T and Apkarian R P 2003 TEM analysis of the nanostructure of normal and osteoporotic human trabecular bone *Bone* **33** 270-82

Journal of Physics: Conference Series **247** (2010) 012032 doi:10.1088/1742-6596/247/1/012032

GISAXS in the study of supramolecular and hybrid liquid crystals

G. Ungar[1], F. Liu[1], X. B. Zeng[1], B. Glettner[2], M. Prehm[2], R. Kieffer[2], C. Tschierske[2]

[1]Department of Engineering Materials, University of Sheffield, Mappin Street, Sheffield S1 3JD, U.K.
[2]Martin-Luther-University Halle-Wittenberg, Kurt-Mothes-Str. 2, D-06120 Halle/Saale, Germany.

Abstract

The use of grazing incidence small and intermediate angle X-ray scattering in the study of structure and alignment of thermotropic liquid crystals is illustrated on selected examples. These include columnar LC phases of a star-shaped mesogen, several honeycomb phases of T-shaped and X-shaped bolaamphiphilic LCs, and gold nanoparticles coated with mesogenic ligands. Sharp Bragg reflections from systems with 2-d and 3-d periodicities are obtained through annealing. Due to nearly perfect surface alignment in most cases, indexing of complex diffraction patterns is facilitated. Honeycomb cells with deformed hexagonal cross-sections, as well as kagome lattice, are shown. The tilt of the reciprocal lattice is shown to help establish the correct structure in the case of the stretched hexagonal honeycombs and the rhombohedral arrays of ordered strings of gold nanoparticles.

Grazing incidence small angle x-ray scattering has become an indispensable technique in the study of surfaces and interfaces ([1,2,3] and references therein), systems which include semiconductor multilayers, metallic nanoparticles, quantum dots, mesoporous films [4,5,6,7,8], lipids and surfactants [9,10,11,12], and block copolymers [13,14,15,16,17]. There are only a few studies of liquid crystalline systems using GISAXS though: spin coated and LB films of phthalocyanines [18], liquid crystalline side chain block copolymers [19], discotic supramolecules in thin films [20,21], dendrimers in thin films [22], self-assembly of block copolymer-based supramolecules in thin films [23] and surface micelles of surfactant [24].

Here we will illustrate the usefulness of grazing incidence small and intermediate angle diffraction in investigations of surface aligned periodic thermotropic LC structures, as well as LC-coated metal nanoparticles, on examples from our recent practice.

Columnar liquid crystals

In the study of columnar LCs a surface alignment technique for recording 2d X-ray diffraction patterns has been introduced by G. Pelzl [25], whereby a LC droplet on a glass surface is irradiated with a beam close to parallel to the slide. The alignment occurs at the LC-air interface and due to its convex curvature the transmitted beam is likely to hit areas under Bragg condition. Even better alignment is obtained in thin films either spin-coated from solution or melt spread on surfaces such as Si (111). This way true grazing incidence diffraction experiments can be carried out. The most useful q range for liquid crystals is ca. 0.3 to 5 nm^{-1}, i.e. small to intermediate scattering angles. Here both ranges are referred to as GISAXS. In the GISAXS experiments described below typical x-ray wavelength used is 1.0 Å with beam size 200 μm x 50 μm on the sample and sample to detector distance of ~75 cm.

Classic columnar LCs are formed by discotic molecules [26]. However, other molecular architectures can also give rise to columnar phases through self-assembly. Here we give the example of a star-shaped compound (compound **1**) [27]

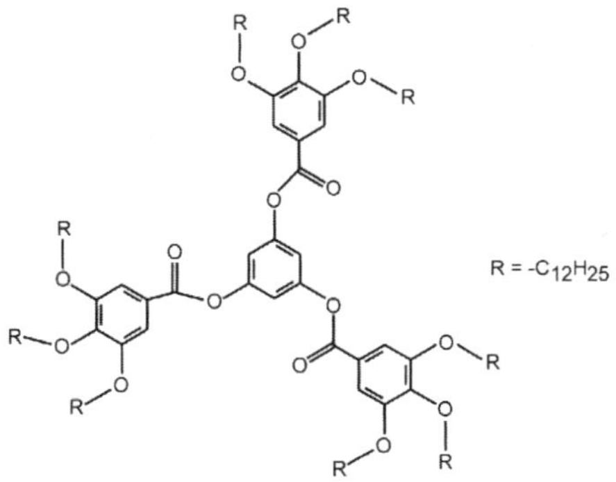

Compound **1**

kindly donated by Prof. M. Lehmann of TU Chemnitz, Germany, which forms a hexagonal columnar phase with a lattice parameter 3.5 nm. Figure 1 shows the GISAXS pattern of compound **1** spin-coated on silicon. The pattern shows a hexagonal array of sharp Bragg spots. The columnar LC has planar orientation, with high in-plane alignment of the columns (normal to *z*) but random orientation in the surface *x,y* plane. Thus the reciprocal lattice is 2-dimensional and, apart from the meridional reflections, all others arise when the columns are parallel to the incident beam and the reciprocal x^*y^* plane crosses the Ewald sphere. For all columns the [10] direction coincides with the *z*-axis. There is virtually no azimuthal spread of the higher order reflections, and the value of the order parameter is therefore exactly –½. The diffuse wide-angle scattering halo centered around q = 14 nm^{-1} confirms the liquid crystal nature of the phase.

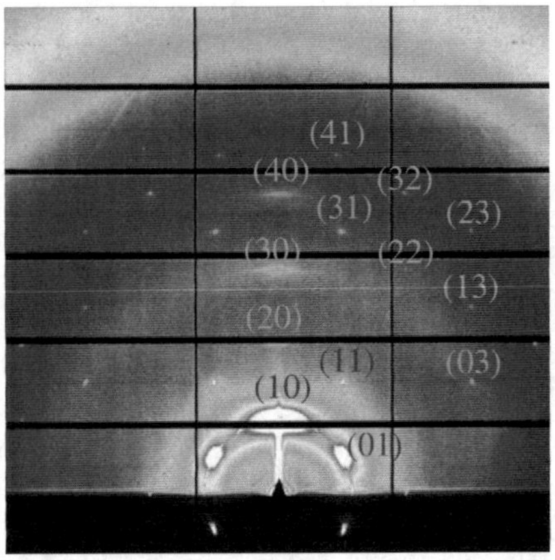

Figure 1. GISAXS-GIWAXS pattern of the columnar LC phase of the star-shaped compound **1** on Si (111) surface. Due to the large detector size (Pilatus-2M, Dectris) *q*-range 0.2 - 14 nm^{-1} is covered with high resolution. Linear intensity scale. Recorded at beamline I16, Diamond Light Source.

Liquid crystal honeycombs

Journal of Physics: Conference Series **247** (2010) 012032 doi:10.1088/1742-6596/247/1/012032

Among the more recently introduced molecular architectures forming liquid crystal phases are polyphilic LC compounds that have a polar H-bonding group at each end, and one or two flexible chains attached laterally to the aromatic rod [28,29]. These molecules combine three or four types of incompatible units which results in a wide range of novel and sometimes highly complex self-assembled patterns with 2-d or 3-d mesoscopic order [30]. The most common class of such structures are honeycomb columnar phases, where the rod-like mesogens form the walls and the side chains fill the cells. The polygonal cross-section of the cells ranges from triangular, via square and pentagonal to hexagonal and extended hexagonal.

Figure 2 shows a GISAXS pattern of compound **2**, which forms a honeycomb phase of hexagonal symmetry between 86°C and 116°C.

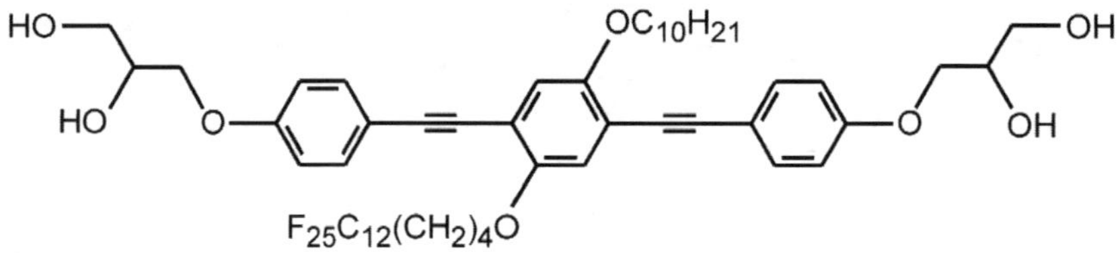

Compound **2**

Here the incidence angle α_i is comparatively high (2°) since higher diffraction orders were targeted, taking advantage of the Bragg-Brentano geometry. Intensities from powder diffraction were used for the reconstruction of the electron density map [31]. This is shown in Figure 3b, along with a snapshot of molecular dynamics simulation based on the experimental unit cell parameters (Figure 3c).

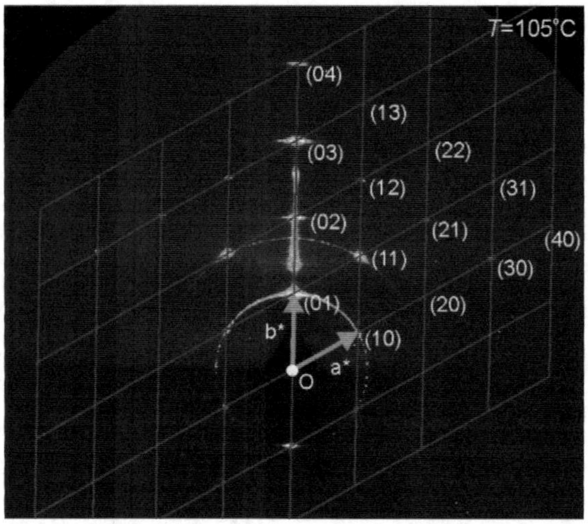

Figure 2. GISAXS pattern of compound **2** on Si surface. The reciprocal net is superimposed. Recorded at BM28 (XMaS line) at ESRF, using the MAR CCD detector.

In spite of the hexagonal reciprocal lattice, the structure of the LC phase is not a simple hexagonal honeycomb, but rather one based on the kagome lattice or trihexagonal tiling (Figure 3a). Thus there is one hexagonal and two triangular channels per unit cell, the former containing the alkyl, and the latter the partially perfluorinated alkyl chains. In this way these two incompatible chains are fully separated.

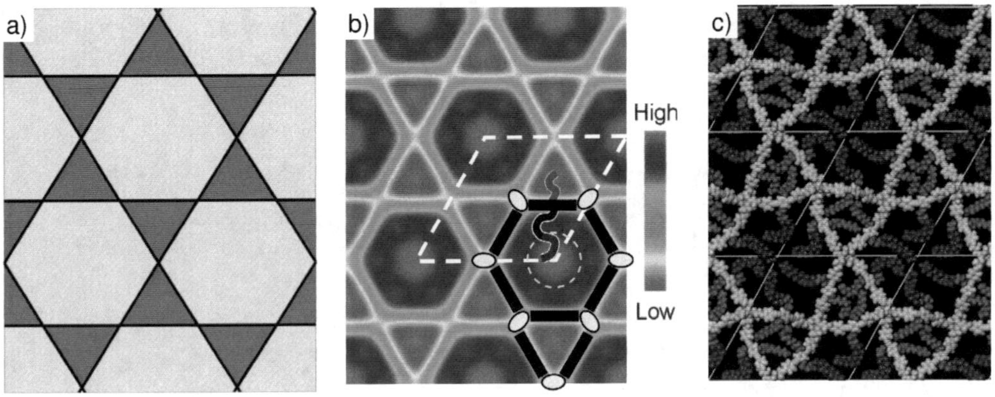

Figure 3. a) Kagome lattice. b) Electron density map calculated from powder diffraction pattern [31]; the schematics of rod-like aromatic cores of eight molecules are overlaid (black lines) with their terminal glycerol groups (white ellipses); one of the cores has its two lateral chains drawn. c) Snapshot of a molecular dynamics simulation (box equal to experimental unit cell with c = 0.45 nm, with periodic boundary conditions. Colour coding: red = alkyl chains, purple = fluoroalkyl chains, green aromatic cores and glycerol units. View down the column axis [31].

The following example illustrates the way GISAXS may be used to test a particular structural model. Compounds **3** and **4** both form rectangular columnar phases, plane group *c*2*mm*. Based on electron density maps reconstructed from powder diffraction intensities [32], on consideration of molecular dimensions and on comparisons with the phase structures of homologous compounds [33], it was suggested that both compounds form honeycombs with stretched hexagons as basic tiling units. In compound 3 the circumference of the hexagon is made up of 10 molecules, four sides consisting of two molecules end-to-end and the remaining two sides of only one molecule each – see Figure 5a [32]. In compound **4** the hexagons contain only 8 molecules in circumference: two molecules in two of the sides and one in each of the four remaining sides (Figure 5b) [33]. The difference between these two structures is due to the smaller volume of the side chain in **4** compared to that in **3**. Homologous compounds with still smaller side chains display honeycombs with regular hexagons, or else with pentagons, squares or triangles as elementary tiles [30,34].

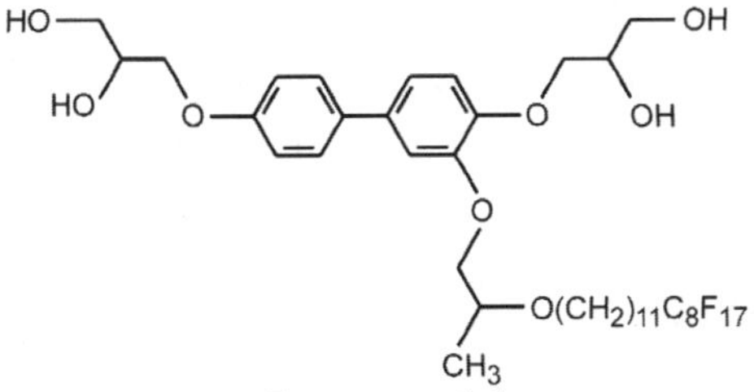

Compound **3**

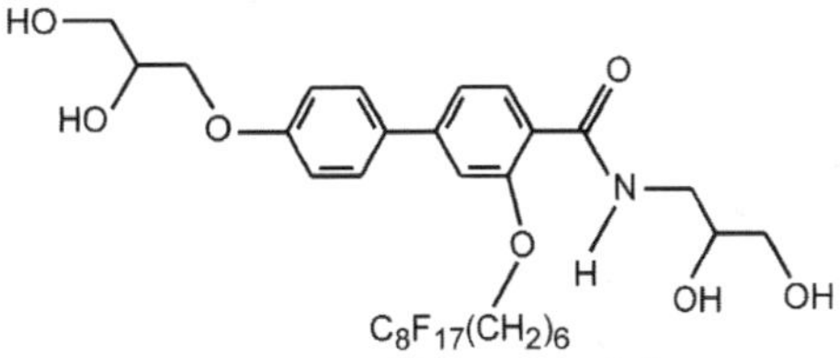

Compound **4**

Figure 4 shows the GISAXS patterns of compounds **3** and **4** on Si surface. The rectangular reciprocal x^*y^* nets again show that the honeycombs lie with their channels parallel with the substrate surface. However, while in Figure 4a one of the unit cell axes is parallel to the surface, in Figure 4b the lattice is tilted with the lattice plane [11] parallel to the surface instead. Figure 4b is therefore effectively a fibre pattern with [110] the fibre axis. This observation is perfectly in line with the structural model proposed for these two phases, as seen in Figure 5. The orientation of the honeycombs is such that the closely spaced edges rest on the substrate surface, causing minimum distortion to the bottom layer of the honeycomb.

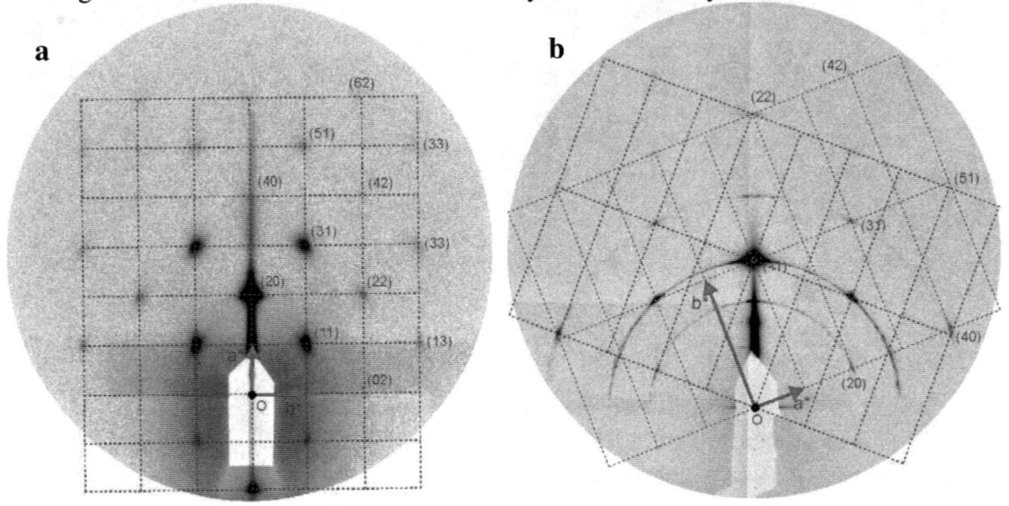

Compound **3** Compound **4**

Figure 4. GISAXS patterns of rectangular columnar *c2mm* phases in a) compound **3** at 60°C and b) compound **4** at 115 °C. The tilt of the reciprocal lattice in b) is explained in Figure 5b.

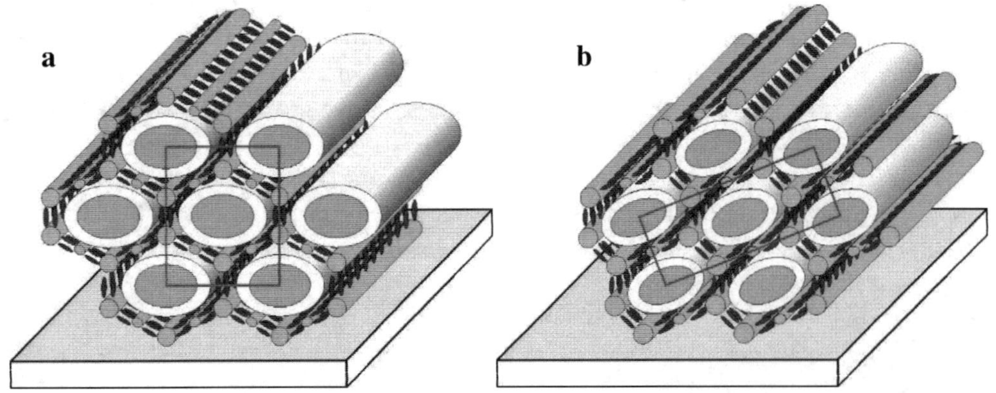

Compound **3** Compound **4**

Figure 5. Schematic structures of the stretched hexagonal honeycomb phases resting on the substrate surface. a) Compound **3** (hexagons with 10 molecules in circumference), b)compound **4** (hexagons with 8 molecules in circumference). Compare with Figure 4.

LC-coated gold nanoparticles

While metal nanoparticles deposited on flat substrates normally give more or less diffuse low-angle scattering at grazing incidence [35,36], more recently ordered monolayer and multilayer superlattices of nanoparticle systems have been achieved by various film preparation techniques, and GISAXS has been instrumental in examining them [37,38,39,40,41,42]. In a different approach, nanoparticles that are coated with mesogenic ligands and annealed have been found to order in the bulk so as to give sharp Bragg reflections of the superlattice [43,44]. As with the examples given above, surface alignment may again be an effective aid in determining the structure of such superlattices. Two examples of such LC-capped 2 nm gold nanoparticles are shown in Figure 6a, the difference between being only in the length of the thioalkyl co-ligand [45]. The material exhibit Schlieren texture between crossed polarizers (Figure 6b), indicating that the mesogens maintain long-range nematic-like orientational order. The GISAXS pattern of a spin-coated and annealed film on Si (111) substrate is shown in Figure 6c for sample **AuL4C12** that has the longer co-ligand chains. Indexing revealed a rhombohedral superlattice of Au nanoparticles, with a = 8.0 nm and c = 3.5 nm. The nanoparticles form strings, the three stings passing through a unit cell being displaced by $c/3$ with respect to each other – see Figure 6d. The strings are tilted relative to the surface normal as this allows the densely packed (111) plane of the superlattice to lie on the Si substrate.

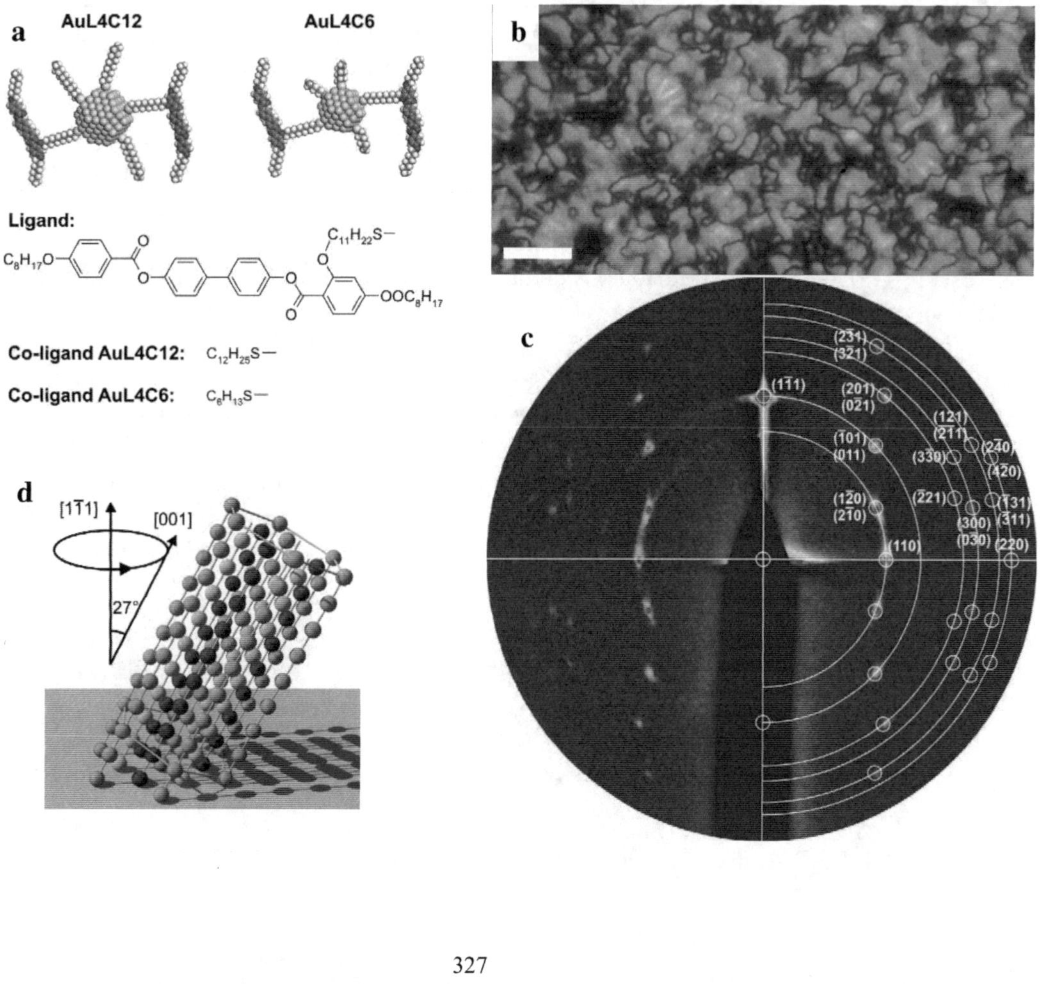

Figure 6. a) Schematic structure of the gold nanoparticles coated with a rod-like mesogen laterally attached via an alkylthiol spacer; dodecylthiol (compound **AuL4C12**) or hexylthiol (compound **AuL4C6**) were used as co-ligands, or surface diluents. b) Micrograph of **AuL4C12** between crossed polarizers at room temperature showing threaded nematic texture; bar = 10 μm. c) GISAXS pattern of a thin layer of **AuL4C12**; the white circles overlaid on the right-hand half of the diffraction pattern indicate the positions of diffraction peaks as calculated from the best-fit unit cell parameters. d) Nanoparticle array in the $R\bar{3}m$ phase resting on the densely packed $(1\bar{1}1)$ plane, as in the thin films recorded by GISAXS; the c-axis is tilted by 27° to the substrate normal. After ref. 44 (with permission of VCH-Wiley).

In comparison, the system **AuL4C6**, having shorter co-ligands, forms a superlattice with nanoparticles virtually touching each other within a string, while the distance between the hexagonally arranged strings is nearly the same as in **AuL4C12**. Here the strings stand normal to the substrate, as there are no well developed crystallographic planes with $l > 0$.

Conclusions

As illustrated by the above examples, GISAXS is a very useful technique for studying alignment and structure of complex self-assembled liquid crystals and supramolecular systems, as well as their composites with nanoparticles. When liquid crystalline alignment is required to create e.g. molecular wires on nanopaterned surfaces, angle resolved GISAXS patterns as a function of the rotation around the surface normal can also yield important information [46]. Another area of application of GISAXS is the study of liquid crystal infiltrated into nanoporous ceramics, such as nanoporous silica or anodic alumina. Depth profiling by varying the incidence angle can also be applied in order to study surface phase transitions. A related subject is the study of alignment layers acting as substrate for controlling the orientation and pretilt in LC devices. With increasing availability of dedicated synchrotron and off-line facilities, it is expected that in forthcoming years the application of GISAXS in research related to thermotropic LCs and supramolecular self-assemblies will undergo considerable expansion.

This work was supported by EPSRC and DFG through the NSF EUROCORES programme SONS-2, Project SCALES. It was also supported by the European Union Collaborative Project NANOGOLD (FP7-NMP-2008-SMALL-2), as well as by the WCU program through the National Research Foundation of Korea funded by the Ministry of Education, Science and Technology (R31-10013). We thank Drs. D. Mannix, S. Brown and P. Thompson for technical help at the BM28 XMaS beamline, ESRF, as well as Drs. S. Collins and A. Bombardi at I16, Diamond Light Source.

[1] Renaud G, Lazzari R and Leroy F 2009, *Surface Science Reports* **64** 255

[2] Smilgies D-M, Busch P, Papadakis C M and Posselt D 2002, *Synchrotron Radiation News* **15** 35

[3] Mueller-Buschbaum P 2006, *Progr. Colloid Polym. Sci.* **132**, 23.

[4] Grosso D, Babonneau F, Albouy P A, Amenitsch H, Balkenende A R, Brunet-Bruneau A and Rivory J 2002, *Chemistry Of Materials* **14** 931

[5] Doshi D A, Gibaud A, Goletto V, Lu M C, Gerung H, Ocko B, Han S M and Brinker C J 2003, *J. Am. Chem. Soc.* **125** 11646

[6] Du P, Li M, Douki K, Li X, Garcia C B W, Jain A, Smilgies D-M, Fetters L J, Gruner S M, Wiesner U and Ober C K 2004, *Advanced Materials* **16** 953

[7] Nagarajan S, Li M, Pai R A, Bosworth J K, Busch P, Smilgies D-M, Ober C K, Russell T P and Watkins J J 2008, *Adv. Mater.* **20** 246

[8] Crossland E, Kamperman M, Nedelcu M, Ducati C, Wiesner U, Toombes G, Hillmyer M, Ludwigs S, Steiner U, Smilgies D-M and Snaith H 2009, *Nano Lett.* **9** 2807

[9] Als-Nielsen J, Jacquemain D, Kjaer K, Leveiller F, Lahav M and Leiserowitz L 1994, *Phys. Rep.* **246** 251

[10] Qian S,Wang W, Yang L and Huang H W 2008, *Proc. Natl. Acad. Sci. USA* **105** 17379

[11] Yang L and Fukuto M 2005, *Phys. Rev. E* **72** 010901

[12] Mills T T, Toombes G E S, Tristram-Nagle S, Smilgies D-M, Feigenson G W and Nagle J F 2008, *Biophys. J.* **95** 669.

[13] Lee B, Park I, Yoon J, Park S, Kim J, Kim K W, Chang T and Ree M 2005, *Macromolecules* **38** 4311

[14] Khanna V, Cochran E W, Hexemer A, Stein G E, Fredrickson G H, Kramer E J, Li X, Wang J and Hahn S F 2006, *Macromolecules* **39** 9346

[15] Busch P, Posselt D, Smilgies D-M, Rauscher M and Papadakis C M 2007, *Macromolecules* **40** 630

[16] Lee C, Kim S H and Russell T P 2009, *Macrom. rapid comm.* **30** 1674

[17] Park S, Lee D H, Xu J, Kim B, Hong S W, Jeong U, Xu T and Russell T P 2009, *Science* **323** 1030

[18] Bryant G C, Cook M J, Ruggiero C, Ryan T G, Thorne A J, Haslam S D and Richardson R M 1994, *Thin Solid Films* **243** 316

[19] Busch P, Krishnan S, Paik M, Toombes G E S, Smilgies D-M, Gruner S M and Ober C K 2007, *Macromolecules* **40** 81

[20] Kim H S, Choi S M, Lee J H, Busch P, Koza S J, Verploegen E A and Pate B D 2008, *Adv. Mater.* **20** 1105

[21] Kim H S, Choi S M, Pate B D and Park P G 2009, *ChemPhysChem* **10** 2642

[22] Leshchiner I D, Agina E V, Boiko N I, Richardson R and Shibaev V P 2008, *Russian Chemical Bulletin* **57** 2101

[23] Tung S-H, Kalarickal N C, Mays J W and Xu T 2008, *Macromolecules* **41** 6453

[24] Fontaine P, Goldmann M, Muller P, Faure M C, Konovalov O and Krafft M P 2005, *J. Amer. Chem. Soc.* **127** 512

[25] Pelzl G, Diele S, Grande S, Jakli A, Lischka C, Kresse H, Schmalfuss H, Wirth I and Weissflog W 1999, *Liquid Crystals* **26** 401

[26] Chandrasekhar S, Sadashiva B K and Suresh K A 1977, *Pramana* **9** 471

[27] Liu B, Zhang X Y and Yang X Z 1992, *Chin. Chem. Lett.* **3** 664

[28] Chen B, Zeng X B, Baumeister U, Ungar G and Tschierske C 2005, *Science* **307** 96

[29] Chen B, Baumeister U, Pelzl G, Das M K, Zeng X B, Ungar G and Tschierske C 2005, *J. Am. Chem. Soc.* **127** 16578

[30] Tschierske C 2007, *Chem. Soc. Rev.* **36** 1930

[31] Glettner B, Liu F, Zeng X B, Prehm M, Baumeister U, Bates M A, Walker M, Boesecke P, Ungar G and Tschierske C 2008, *Angew. Chem. Int. Ed.* **47** 9063

[32] Prehm M, Liu F, Baumeister U, Zeng X B, Ungar G and Tschierske C 2007, *Angew. Chem. Int. Ed.* **46** 7972

[33] Cheng X H, Prehm M, Das M K, Kain J, Baumeister U, Diele S, Leine D, Blume A and Tschierske C 2003, *J. Am. Chem. Soc.* **125** 10977

[34] Liu F, Chen B, Baumeister U, Zeng X B, Ungar G and Tschierske C 2007,*J. Am. Chem. Soc.* **129** 9578

[35] Naudon A, Babonneau D, Thiaudière D and Lequien S 2000, *Physica B* **283** 69

[36] Babonneau D, Videnović I R, Garnier M G and Oelhafen P 2001, *Phys. Rev. B* **63** 195401.

[37] Saunders A E, Ghezelbash A, Smilgies D-M, Sigman M B and Korgel B A 2006, *Nano Letters* **6** 2959

[38] Aleksandrovic V, Greshnykh D, Randjelovic I, Froemsdorf A, Kornowski A, Roth S V, Klinke C and Weller H 2008, *ACS Nano* **2** 1124

[39] Dunphy D, Fan H, Li X, Wang J and Brinker C J 2008, *Langmuir* **24** 10575

[40] Smith D K, Goodfellow B, Smilgies D-M and Korgel B A 2009, *J. Am. Chem. Soc.* **131** 3281

[41] Hanrath T, Choi J J and Smilgies D-M 2009, *ACS Nano* **3** 2975

[42] Bodnarchuk M I, Kovalenko M V, Pichler S, Fritz-Popovski G, Hesser G and Heiss W 2010, *ACS Nano* **4** 423

[43] Donnio B, García-Vázquez P, Gallani J L, Guillon D and Terazzi E 2007, *Adv. Mater.* **19** 3534

[44] Zeng X B, Liu F, Fowler A G, Ungar G, Cseh L, Mehl G H and Macdonald J E 2009, *Adv. Mater.* **21** 1746

[45] Cseh L, and Mehl G H 2007, *J. Mater. Chem.* **17** 311

[46] Bramble J P, Tate D J, Revill D J, Sheikh K H, Henderson J R, Liu F, Zeng X B, Ungar G, Bushby R J and Evans S D 2010, *Adv. Funct. Mat.* **20** 914

Journal of Physics: Conference Series **247** (2010) 012033 doi:10.1088/1742-6596/247/1/012033

Two - dimensional spatially ordered arrays of cobalt nanowires: polarized SANS study

A.P. Chumakov[1], S.V. Grigoriev [1], N.A. Grigoryeva[2], K.S. Napolskii[3], I.V. Roslyakov[3], A.A. Eliseev[3], A.I. Okorokov [1], H. Eckerlebe[4]

[1]Petersburg Nuclear Physics Institute, Gatchina, 188300 St.Petersburg, Russia

[2]Faculty of Physics, St-Petersburg State University, 198504 St-Petersburg, Russia

[3]Department of Materials Science, Moscow State University, 119992 Moscow, Russia

[4]GKSS Forschungszentrum, 21502 Geesthacht, Germany

E-mail: chumakov@lns.pnpi.spb.ru

Abstract. Structural and magnetic properties of two-dimensional spatially ordered system of ferromagnetic cobalt nanowires embedded into Al_2O_3 matrix have been studied using polarized small-angle neutron scattering. The small-angle scattering pattern exhibits many diffraction peaks (up to the third reflection order), which corresponds to the scattering from highly correlated hexagonal structure of pores and magnetic nanowires. A comprehensive analysis of contributions to the scattering intensity, including nonmagnetic (nuclear) contribution, magnetic contribution depending on the magnetic field, and nuclear-magnetic interference indicating the correlation between the magnetic and nuclear structures was carried out. A detailed pattern of the remagnetization of an ordered array of the magnetic nanowires has been obtained. It has been illustrated that polarized neutron scattering provides unique information inaccessible by the standard magnetometry techniques.

1. Introduction

One of the most important challenges in physics and materials science today is the preparation of ordered nanostructure arrays with the controlled properties and dimensions. Anodic aluminium oxide (AAO) films formed by two-step anodization technique [1] or nanoimprint technology [2] are well-known to possess uniform pore structure with hexagonal arrangement of cylindrical channels, which makes them an extremely attractive matrix or reactor for synthesis of nanowires. The unique microstructure of AAO templates opens a pathway for growing ordered arrays of nanowires with ultimately high length at very low diameters (which could be considered as one-dimensional) [3]. These composites could be used as a perfect model objects for deeper understanding of magnetization reversal processes in spatially-ordered systems on nanolevel.

A number of works report controllable growth and investigation of magnetic nanowires, however most of them involve only magnetometry studies, which are known to provide integral characteristic of bulky system and give no information on organization of magnetic material [4]. The small-angle polarized-neutron scattering method is much more rarely used, although this technique is sensitive both to organization [5] and the magnetic properties of the nanostructured media [6,7].

The present study is focused on the SANS investigation of the spatially ordered arrays of Co nanowires grown in AAO templates.

2. Synthesis and Experimental Details

Preparation of spatially ordered arrays of cobalt nanowires was carried out by metal electrocrystallization inside the channels of anodic alumina films with a thickness of 110 μm. In order to obtain better ordering of the channels, the anodic aluminium oxide membranes were prepared by the two-step anodization technique [1, 2, 8]. The anode oxidation of aluminium is performed in a two-electrode electrochemical cell in a 0.3 M $(COOH)_2$ solution at a DC voltage of 40 V for 50 and 42,5 h in the first and second anodization stages, respectively. The procedure of fabrication of electrodes on AAO films is described in details elsewhere [7, 9].
Electrodeposition of cobalt was carried out in three-electrode cell from 0.3M $CoSO_4$ + 0.05M $CoCl_2$ + 0.3M H_3BO_3 solution at deposition potential E_d = -0.8 V versus Ag/AgCl reference electrode using Autolab PGSTAT 302 potentiostat. The Pt wire serves as a counter electrode. The duration of the Co electrodeposition was 1.5 h, which corresponds, according to scanning electron microscopy, to the formation of nanowires with a length of 34.5 μm and a diameter of about 40 nm. Figure 1 presents the SEM pictures of a side view of the AAO film with arrayed Co nanowires on a bottom part (a) and a view of the Co wires obtained after solution of the alumina membrane (b).

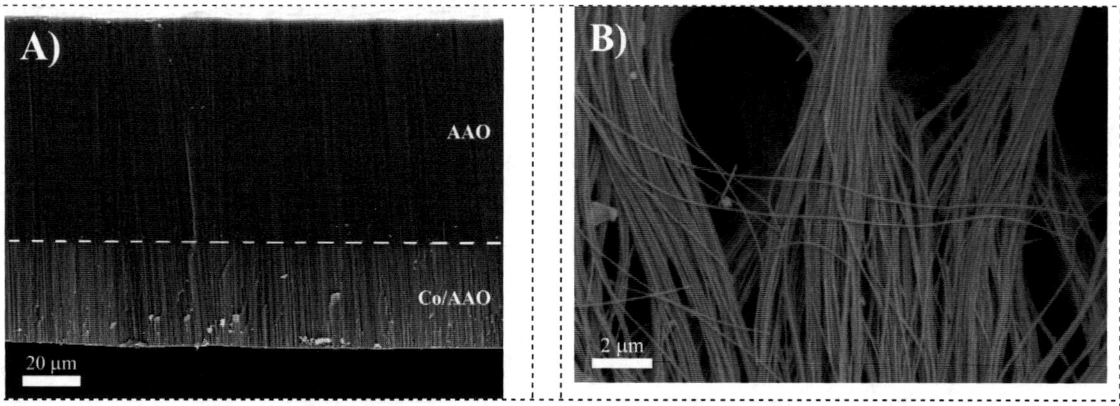

Figure 1. The SEM data: a side view of the anodic alumina film with arrayed Co nanowires (a) and a view of the Co wires after solution of the alumina membrane (b).

Polarized SANS measurements were carried out at the SANS-2 beamline of Geesthacht Neutron Facility (GeNF). Polarized neutron beam with mean wavelength of $\lambda = 0.7$ nm, wavelength spread of $\Delta\lambda/\lambda = 10\%$ and initial polarization $P_0 = 0.95$ was used at. A film with an area of 0.3 cm^2 was oriented perpendicularly to the neutron beam and was uniformly irradiated over the entire area. Such an orientation of the sample corresponds to parallel orientation of the pores and the long axis of the magnetic nanowires to the incident neutron beam. The chosen geometry of the experiment allows the observation of diffraction reflections from the ordered system of pores or magnetic wires in a small-angle scattering range. The sample-to-detector distance of 14 m was used with appropriate collimations to cover scattering vectors Q from 0.025 to 0.17 nm^{-1}. An external magnetic field (up to 800 mT) was applied in the horizontal plane perpendicular to the incident beam (and perpendicular to the long axis of the nanowires). We determine the total (nuclear and magnetic) scattering ($I(Q) = I^+(Q) + I^-(Q)$) and the polarization dependent part of the scattering ($\Delta I(Q) = I^+(Q) - I^-(Q)$), where $I^+(Q)$ and $I^-(Q)$ are the intensities for neutrons polarized parallel (+) and anti-parallel (–) to the magnetic field. The field-dependent scattering intensity is extracted as $I_H(Q) = I(Q,H) - I(Q,0)$. The polarized SANS data were taken at room temperature.

3. Results

A magnetization curve with the magnetic field in the plane of the sample and perpendicular to the nanowires, that corresponds to the polarized SANS experiment, were recorded using SQUID magnetometer Cryogenics S-700 at the temperature 290 K. Figure 2 present the magnetization curve saturated above H = 800 mT. A small hysteresis loop is observed in the range from -300 mT to 300 mT. There is hardly noticeable change in a slope of the magnetization curve in the range of small field within ± 100 mT as compare to the range of larger fields from 100 mT to 300 mT. This feature may be related to the change in a mechanism of the sample's magnetization.

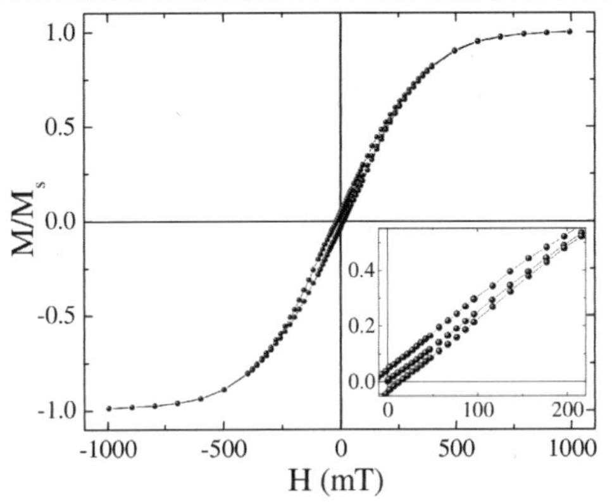

Figure 2. Magnetization curve for Co nanowires embedded into alumina membrane in the field applied perpendicular to the long axis of the wires.

A typical neutron diffraction pattern measured from Co nanowire array grown inside AAO membrane is shown in Figure 3 a. The first and the second order reflections have been detected, suggesting the presence of strong correlations in the positions of the pores (or nanowires) over significant distances. The hexagonally arranged set of reflections demonstrates the 6-fold symmetry of the structure. For the considered geometry the interplanar spacing d_{hk} can be calculated as $d_{hk} = \sqrt{3}a/(2\sqrt{h^2+hk+k^2})$., where a – is a unit cell parameter of 2D superlattice.

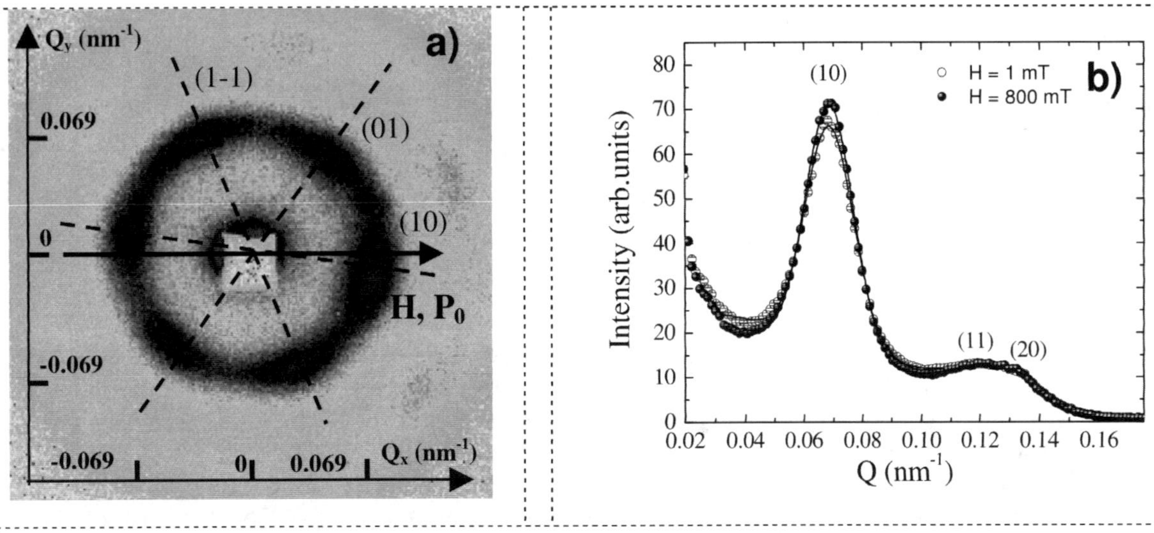

Figure 3. Typical 2D neutron diffraction pattern from Co/Al_2O_3 nanocomposite with spatially ordered

structure (a). Momentum-transfer dependence of the total (nuclear and magnetic) neutron scattering intensity at H = 1 and 800 mT (b).

The widening of reflections to 10-20 degrees in an azimuthal direction, shown in Figure 3 a, indicates that samples have a block structure, with slight disorientation of domains in the plane of the film. Nevertheless, the diffraction patterns, typical for a quasi-single crystal, were recorded using a neutron beam of 5 mm diameter, suggesting correlated orientation of the domains on the irradiated area (~ 0.3 cm^2). Similar results had been observed previously for pure anodic aluminium oxide membranes by SANS techniques [5].

Figure 3 b illustrates the momentum-transfer dependence of the total (nuclear and magnetic) neutron scattering *I*(Q) for the nonmagnetized (H = 1 mT) and magnetized (H = 800 mT) sample. Several diffraction maxima are clearly observed, indicating the highly ordered structure of the Co/Al_2O_3 nanocomposite. The momentum transfer dependence *I*(Q) is satisfactorily reproduced by the sum of the Gaussian components with the centers located at $Q_{10} = 0.069 \pm 0.004$ nm^{-1}, $Q_{11} = 0.119 \pm 0.004$ nm^{-1}, $Q_{20} = 0.138 \pm 0.004$ nm^{-1}, and fixed half-width of $w = 0.0090 \pm 0.0005$ nm^{-1}; coupled with a diffuse small-angle scattering, represented by an exponential decay. The observed diffraction maxima are classified in the hexagonal lattice with the parameter $a = 106 \pm 2$ nm. It should be noted that applying the external magnetic field induces noticeable increase of the diffraction peak intensity in the order of 5 % (see Fig. 3 b).

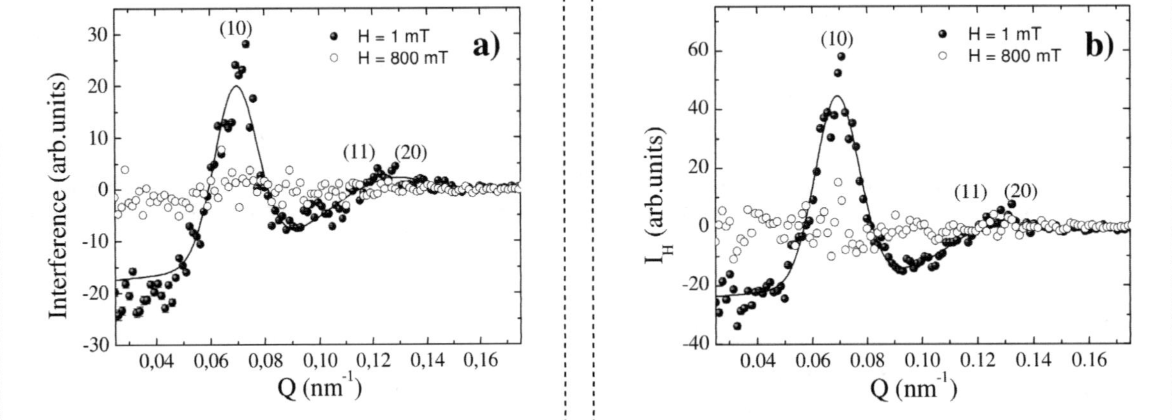

Figure 4. Q-dependence of the nuclear–magnetic interference $\Delta I(Q)$ at H = 1 and 800 mT (a) and magnetic contribution to the scattering $I_H(Q)$ (b).

Figure 4 shows the nuclear-magnetic interference $\Delta I(Q)$ and the field-dependent part of neutron scattering $I_H(Q)$ as a function of scattering vector. The peak positions for the magnetic contributions match the positions of nuclear peaks (at zero magnetic field in Figure 3 b). This confirms the formation of the spatially-ordered magnetic structure, e.g. the ordered array of nanowires with periodicity of 106 nm.

As seen from Figure 4 a, the interference contribution ΔI(Q) is equal to zero in the entire momentum-transfer range for the unmagnetized sample (H = 1 mT). In high magnetic fields, the polarization-dependent scattering consists of two contributions: interference maxima at Q values coinciding with the positions of diffraction reflections in scattering on the ordered structure of nanowires and interference in diffuse small-angle scattering. These two contributions have opposite signs in a good agreement with theory [6]. The interference scattering can be approximated by the sum of the Gaussians with the centers at zero (diffuse scattering) and at $Q = Q_{hk}$ (coherent scattering). The magnetic field dependence of the interference contribution integrated over the azimuthal angle at $|Q_{10}|$

is shown in Fig. 5 a. Note that the interference contribution is proportional to the magnetization, which exhibits no hysteresis within error bars in initial magnetization or in subsequent remagnetization of the system of magnetic nanowires in the field perpendicular to the long axis of wires. It is interesting to note that the increase of intensity is not gradual but has a bump at H = 150 mT and a deep at H = 200 mT.

The magnetic-field-dependent scattering component $I_H(Q)$ shown in Figure 4 b represents the difference between the magnetic cross sections of the sample in two significantly different states, a state close to saturation magnetization (in a field of H = 800 mT) and a completely demagnetized state. The main component of the magnetic cross section at H = 800 mT is the system of the magnetic reflections described by the sum of Gaussians whose positions correspond to the maxima of the nuclear scattering cross section. On the contrary, the main component of the cross section in zero magnetic field is scattering on domains; i.e., diffuse scattering, which is described by a Gaussian at Q = 0 with no contribution to diffraction reflections. The approximation of the experimental data in the frame of this model is shown in Fig. 4 b. The magnetic field dependence of the most intense magnetic reflection integrated over the azimuthal angle at Q = $|Q_{10}|$ is shown in Fig. 5 b. Firstly, the intensity of the magnetic reflection does not change within error bars with the magnetic field in the range from 0 up to 200 mT. Secondly, the increase of intensity with the field starts at H = 200 mT. Thirdly, the remagnetization process proceeds with hysteresis in the field range above 200 mT.

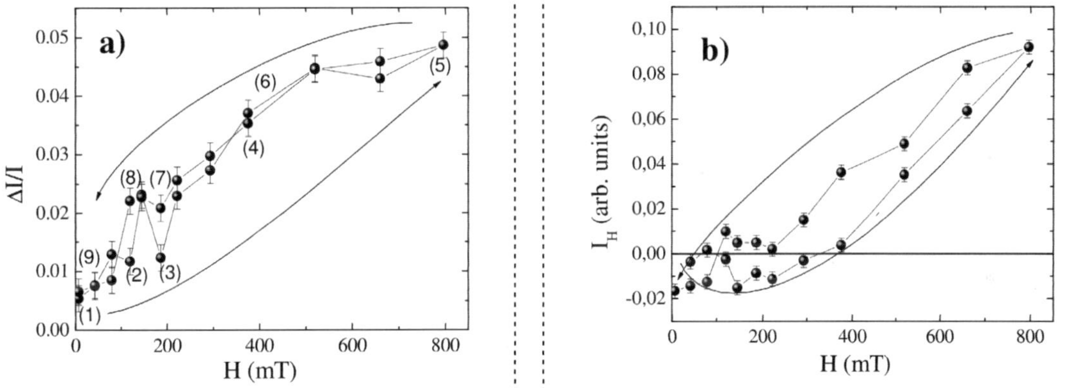

Figure 5. Magnetic field dependence (a) of the interference contribution and (b) the magnetic coherent contribution obtained from the approximation of the experimental data.

The additional features of the intensity of the neutron scattering upon remagnetization in the range of ± 200 mT are well seen in Fig.5. They correspond to the slight change in the slope of the magnetization curve that can be distinguished in Fig.2. The model describing the remagnetization process of the magnetic nanowires array in the perpendicular field is presented in Fig 6. At zero field every nanowire is divided on domains with randomly oriented magnetization, position (1). The orientation of the magnetic moments depends probably on magnetic easy direction of nanowire and the anisotropy and the size of magnetic crystalline particles constituting nanowires. We suppose that domain walls can be bounded to the boundary between crystallites within a nanowire. There is also no reason to believe in correlation between the magnetization in neighboring wires and as a result we expect no interference or pure magnetic contributions to the neutron reflections. When the perpendicular external magnetic field is applied then an orientation of the magnetic moments changes toward the field axis, position (2). The domain walls can be removed in some of nanowires, so that they become single-domain particles. The component of the magnetization along the field axis appears what is reflected in appearance of the nuclear magnetic interference. Further increase in quantity of such single domain nanowires, so that now every wire is a single magnetic domain, position (3),

results in additional dipole interaction between nanowires favoring up-to-down orientations of the magnetic moments in the neighboring wires. This additional dipole interaction competes with the magnetic field and this competition leads to the state where different wires are not correlated. This situation is reflectes in an additional minimum in nuclear-magnetic interference and in pure magnetic contributions to the reflexes. Further increase of the magnetic field starts reorient magnetic moments in directions toward its axis, position (4). The fully saturated magnetization in the nanowires is reached at large magnetic fields, position (5). The subsequent decrease of the magnetic field is shown in positions (6) – (9). The magnetic moments inside the nanowires are differently oriented in positions (6 and 7) as compared to positions (4 and 3), respectively. This configurations of magnetic moments result in hysteresis of magnetic intensity shown in Figure 6 and absence of any hysteresis in the nuclear magnetic interference shown in Figure 5 a. The single-domain nanowires become multidomain again at H = 200 mT, when the dipole interaction competes with the field, position (8). The system becomes fully disordered massive of multidomain nanowires at H = 0, position (9). It should be noted that the described model is a suggestion only and has not been tested by rigorous modeling of the experimental data.
The similar magnetization process was observed in the massive of nickel nanowires [6] but the field range where a disordered state occurs, is limited by H = 100 mT. This difference can be related to the difference in values of the magnetic moments of Co and Ni. It may also well be that crystalline structures of nanowires defined by the crystallization process during synthesis are different for cobalt and nickel.

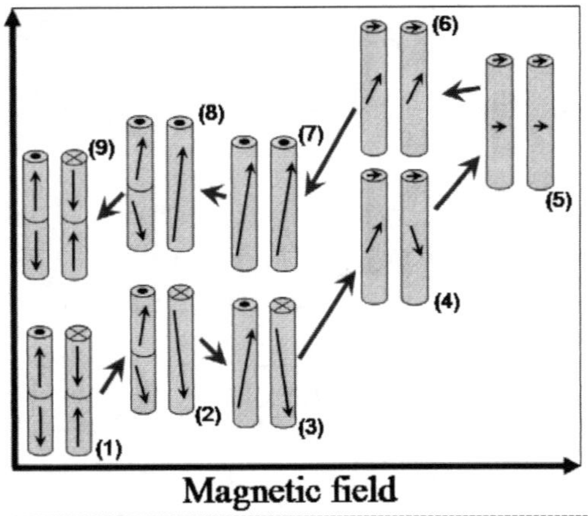

Figure 6. The scheme of the remagnetization model for in magnetic nanowires array. Arrows show the direction of the magnetization while the length of the arrow corresponds rather to the area with homogeneous magnetization but not to the magnetization value.

4. Concluding remarks

In conclusion, we have investigated the process of the remagnetization of the massive of Co nanowires in the magnetic field applied perpendicularly to the long wire's axis. The measurements reveal the presence of hysteresis in field dependence of the magnetic scattering, connected to several processes well reflected in the intensity of the pure magnetic and nuclear-magnetic contributions upon remagnetization. In the ground state (at no field applied) the moments of nanowires are aligned parallel to the long axis, but are randomly distributed between +1 and –1 states and probably in multi-domain state. The magnetic field applied results in a deviation of the average magnetic moment of a wire from its long axes, however, shape and/or crystalline anisotropy, dipole interaction compete with the magnetic field so that the magnetization starts at H larger than 250 mT. Further increase of the field aligns moments in perpendicular direction shown in the rapid increase of scattering intensity of the magnetic reflexes. Thus, the inflection of experimental curve at the $H \approx 250$ mT should be attributed to the change of the magnetization mechanism in the Co nanowires array from incoherent to the coherent rotation. We have represented the model for magnetic moment behavior in cobalt nanowires embedded into anodic alumina matrix. One can also conclude that polarized-neutron scattering provides information inaccessible to the standard magnetometric methods.

Acknowledgments

This work was performed within the framework of a Federal Agency of Science and Innovations (projects 02.513.11.3485). The Russian authors thank RFBR for partial support (Grant № 10-02-00634). This work was supported by the Presidium of the Russian Academy of Sciences under the Program for Basic Research. The Russian authors are grateful to the GKSS Research Centre (Germany) for hospitality.

References

[1] H. Masuda and K. Fukuda, 1995 Science **268**, 1466.
[2] S. Shingubara, J. Nanopart. 2003 Res. **5**, 17.
[3] G. Sauer, G. Brehm, S. Schneider, et al., 2002 J. Appl. Phys. **2**, 3243.
[4] Y. Piao, H. Lim, J. Y. Chang, et al., 2005 Electrochim. Acta **50**, 2997.
[5] S. V. Grigoriev, N. A. Grigorieva, A. V. Syromyatnikov, et al., 2007 Pis'ma Zh. Éksp. Teor. Fiz. **85**, 549 [2007 JETP Lett. **85**, 449].
[6] S.V. Grigoriev, N.A. Grigorieva, A.V. Syromyatnikov, et al., 2007 Pis'ma Zh. Éksp. Teor. Fiz. **85**, 738 [2007 JETP Lett. **85**, 605].
[7] K.S. Napolskii, A.P. Chumakov, S.V. Grigoriev, N.A. Grigoryeva, H. Eckerlebe, A.A. Eliseev, A.V. Lukashin, Yu.D. Tretyakov. // Physica B, 2009, v. **404**, pp. 2568–2571.
[8] O. Jessensky, F. Muller, and U. Gusele, 1998 Appl. Phys. Lett. **72**, 1173.
[9] K.S. Napolskii, P.J. Barczuk, S.Yu. Vassiliev, A.G. Veresov, G.A. Tsirlina and P.J. Kulesza. // Electrochimica Acta, 2007, v. **52**, pp. 7910–7919.

Journal of Physics: Conference Series **247** (2010) 012034 doi:10.1088/1742-6596/247/1/012034

Mapping the microstructure of a friction-stir welded (FSW) Al-Li-Cu alloy

F De Geuser[1], F. Bley[1], A. Denquin[2], A. Deschamps[1]

[1] SIMaP, 1130 Rue de la Piscine BP 75, 38402 St Martin d'Hères Cedex, France
[2] ONERA, 29 avenue de la division Leclerc, B.P. 92, 92230 Châtillon Cedex, France

E-mail: frederic.de-geuser@simap.grenoble-inp.fr

Abstract. The friction-stir welding (FSW) process induces both heat and deformation which lead to inhomogeneous precipitation microstructures in structural hardening alloys. A map of this microstructure can be obtained by 2D scanning SAXS in alloys where (a) a single phase is precipitating and (b) the precipitates are roughly isotropic. In Al-Li-Cu alloys, these conditions are not fullfilled. Very anisotropic precipitates are forming (T_1 and θ') with aspect ratios in the range 10 to 100. The inhomogeneity in the texture of the material (due to the deformation and recrystallisation) is a strong obstacle to the interpretation of the SAXS signal. This paper is an attempt to apply systematic simple interpretation models to characterise the precipitation microstructure across the weld area. It shows that in certain condition, it is possible to apply simple Guinier-type plot to extract the morphologies (length and thickness) of the particles.

1. Introduction

The third generation of Al-Li alloys are subject to many studies due to the urge for weight reduction in aerospace industry. Further weight savings can be obtained by the use of friction-stir welding (FSW) instead of riveting. The FSW process is a solid state welding technique that induces both heat and deformation. Depending on the way they have been affected by the weld, four zones can be defined in the material:

- the unaffected base material (BM)
- the heat affected zone (HAZ)
- the thermo-mechanically affected zone (TMAZ)
- the weld nugget (WN) at the centre of the weld, where recrystallisation has occurred

This leads to a complex precipitate microstructure which creates a gradient of mechanical and corrosion properties. Microstructure mapping of FSW welds has been successfully obtained by small angle X-ray scattering (SAXS) before [1], but in the case of Al-Li-Cu alloys, there are two obstacles to the interpretation of the SAXS profiles. First, at least two phases are likely to form (T_1, Al_2CuLi forming platelets in the 111 planes of aluminium and θ', Al_2Cu, forming platelets in the 001 planes of aluminium). Other phases might form as well (e.g. δ', Al_3Li) but they will not be considered in this study. Then, the strong anisotropy of the T_1 and θ' precipitates might cause misinterpretation of the data, as the texture of the material is affected by the FSW process. This preliminary study aims at assessing the use of SAXS as a tool for mapping the microstructure of Al-Li-Cu alloys in the different zones of the weld.

Table 1. Composition of the Al-Li-Cu alloys (weight %).

Alloy	Cu	Li	Mg	Ag	Zr
2098	3.2-3.8	0.8-1.3	0.25-0.8	0.25-0.6	0.04-0.18
2050	3.2-3.9	0.7-1.3	0.2-0.6	0.2-0.7	0.06-0.14

2. Experimental procedure

2.1. Materials

Two types of alloys have been used in this study. The first is a 2098 alloy that has been studied in another work [2] and the other is a 2050. These two alloys have very similar composition, as can be seen in table 1, and were provided by Alcan. Friction stir welding was performed by Onera for the 2098 and by Eads-IW in the framework of the Coralis project (see acknowledgments) for the 2050. For 2098, FSW was performed in T8 condition. One of the 2098 alloy was then heat treated at 100°C for 1000h to artificially age the alloy. For the 2050 alloy, FSW was performed in T3 condition and then aged to T8 condition.

Two types of precipitates are expected to form in the alloy [3]:

- T_1 (Al_2CuLi): platelets laying in the (111) planes of the aluminium matrix.
- θ' (Al_2Cu): platelets laying in the (001) planes of the aluminium matrix.

Other phases such as δ' (Al_3Li) or S (Al_2CuMg) might form as well but are considered in the present study as negligible (this is justified by TEM experiments, not shown here). Other Cu-rich objects are likely to precipitate: Guinier-Preston (GP) zones and/or θ''. As their habit planes are the aluminium (002), they are difficult to separate from the θ' precipitates. As this distinction is not crucial, we will consider all these objects as θ'-like particles.

2.2. SAXS experiments and interpretation

The small-angle scattering experiments have been performed at the BM02-D2AM beamline at the European synchrotron radiation facility (ESRF). The samples were cut perpendicularly to the weld line and were mechanically polished down to a thickness of about 100μm. The energy was set to 8.3keV and the beam size was about 200μm x 200μm. The sample to detector distance was 33.6cm (as determined by silver behenate diffraction).

We have performed 1D scans across the FSW welds, recording a SAXS image every 200μm. For the 2050 sample, we have performed a 2D scan with a 500μm step to map the whole weld region.

Typical SAXS images collected on the CCD detector can be seen on figure 1 and 2. These images have been extracted from the 1D scans across the weld line. The images present streaking that is typical from anisotropic precipitates in a textured alloy. Except in the base material, the grain texture of the alloy is unknown and varies a lot across the weld, from the recrystallised weld nugget to the thermo-mechanically affected zone.

The anisotropic nature of the SAXS intensity could in principle be accounted for by using intensity sectors instead of azimuthally averaged intensity, or even by fitting directly the image by a 2D intensity model. This is not practical in this particular case for two reasons. First, the actual texture of the material, and more specifically the out of plane texture, is unknown (and varies across the welding plane), so that modelling of the data is difficult. Second, 2D and 1D scans lead to a lot of SAXS images and analysing manually selected intensity sectors would be extremely time-consuming. We use azimuthally averaged images to process the SAXS images, and extract data for platelets particles, bearing in mind the following considerations:

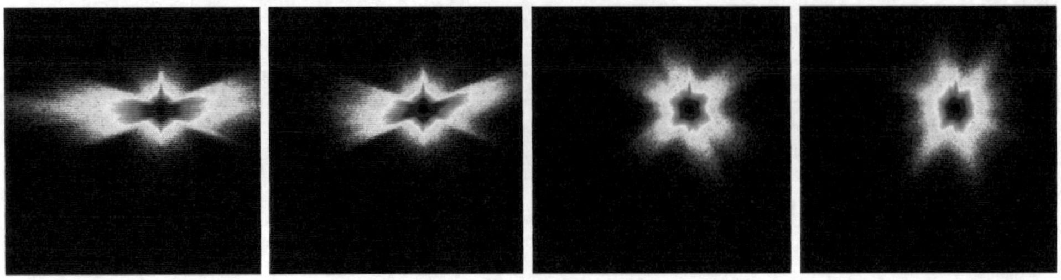

(a) (b) (c) (d)

Figure 1. SAXS images obtained on the 2098 alloy welded in T8 condition: (a) BM, (b) HAZ, (c) TMAZ and (d) WN.

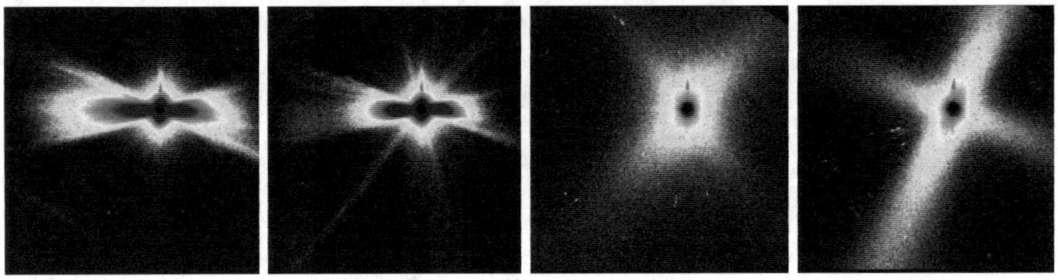

(a) (b) (c) (d)

Figure 2. SAXS images obtained on the 2098 alloy welded in T8 condition after 1000h at 100°C: (a) BM, (b) HAZ, (c) TMAZ and (d) WN.

- The grain texture of the material is not random and not homogeneous.
- Two types of precipitates (T_1 and θ') with different electronic contrast are present in the material in such a way that they are indistinguishable.

If the distribution of orientations around the average texture of the material is broad enough, the averaged SAXS intensity should be characteristic of the morphologies of the precipitates and this should be true no matter the orientation of the sample.

Figure 3(a) shows such an azimuthally averaged SAXS image (double log plot). The high q region follows a classic q^{-4} Porod law, whereas the intermediate q region follows a q^{-2} behaviour, typical from a flat specimen. This behaviour (true in every zone of the weld) justifies the use of the azimuthally averaged intensity. We can then assume that the intensity is of the form (cf. [4], pp 17-51):

$$I(q) = A\frac{2\pi}{q^2}(\Delta\rho)^2 t^2 \exp\left(-q^2\frac{t^2}{12}\right) \tag{1}$$

where A is the area of the particles, $\Delta\rho$, their electronic contrast with the aluminium matrix and t is their thickness. A Guinier-type plot Log(Iq^2) vs. q^2 can then be used to extract the thickness t of the precipitates. Figure 3(b) is an example of such a plot. It clearly shows a linear part that can be used to fit a Guinier law to extract the thickness.

In principle, the relation (1) and its associated Guinier plot can also be used to extract the length of the particles via their cross-section area A. However, although the q=0 intensity of this Guinier plot is proportional to A, it also reflects the texture of the material. This non-uniform texture will not alter the slope of the Guinier plot, but is likely to modify the q=0 intensity.

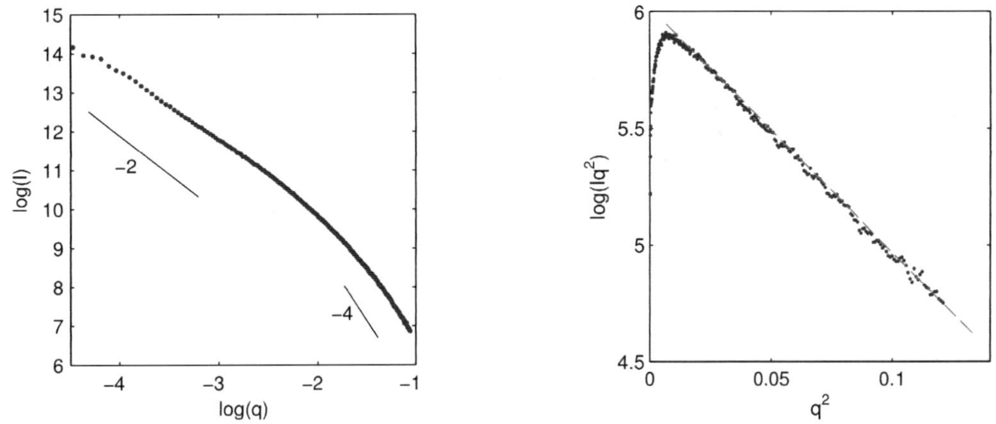

(a) (b)

Figure 3. azimuthally averaged SAXS image extracted from a 1D scan across the weld line for the 2098 alloy welded in T8 condition. (a) Double-log plot. The q^{-2} region before the classic q^{-4} Porod law is typical from flat precipitates. (b) Guinier-type plot Log(Iq^2) vs. q^2. The linear part gives the thickness according to eq. (1).

To obtain the length of the particles, we have performed a classic Guinier plot Log(I) vs. q^2 in the very low q range. Although this is far from being a perfect solution, the initial slope of the Guinier plot should still give the correct tendency for the evolution of the particles length, even if the absolute scale is incorrect.

3. Results and discussion

3.1. Temporal stability of the welded 2098 alloy

Although the 2098 alloy is considered to be stable in T8 condition, the temporal stability of the alloy will be modified by the FSW process. To analyse this modification, an artificially accelerated ageing treatment has been applied (1000h at 100°C) to the 2098 alloy. Figure 4 shows the morphologies obtained from the Guinier plots for the 2098 alloys welded in T8 condition with and without artificial ageing.

Figure 4(a) seems to indicate that little variation in morphology is observed across the weld. The thickness is constant and the length is slightly lower at the centre of the weld. This is slightly misleading in the sense that two important piece of information are missing: the volume fraction and the type of precipitates (T_1 or θ'). Although we know that mainly T_1 is present in the base material [2], it is very likely that θ'-type precipitates are forming in the weld zone. Denquin *et al* [2] have observed only sparse particles in the weld, suggesting a dissolution of the T_1 precipitates. Our results suggest that, if this is the case, the remaining T_1 or the θ'-type precipitates which may have formed are of similar size.

Figure 4(b) shows the morphologies obtained for the 2098 alloy artificially aged 1000h at 100°C. In contrast to what is observed in the non-aged alloys, the 4 FSW zones (BM, HAZ, TMAZ and WN) can be distinguished, and are labeled on the figure. The increase in thickness in the HAZ zone can be interpreted as a thickening of T_1 precipitates and is consistent with figure 2(b) which shows similar patterns than figure 2(a), but in a smaller q-range. In the TMAZ and WN zones, the SAXS patterns clearly change (cf. figures 2(c) and 2(d)) and the morphologies measured in figure 4(b) are most likely due to θ'-type precipitates.

From the 1D scans obtained on the 2098 alloy welded in T8 condition with and without

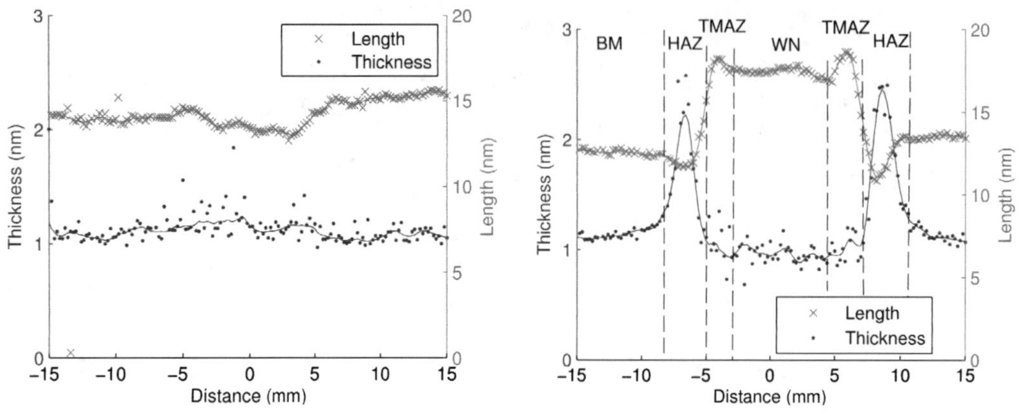

(a) (b)

Figure 4. Precipitates morphologies obtained by SAXS. (a) 2098 alloy T8 FSW and (b) 2098 alloy FSW artificially aged.

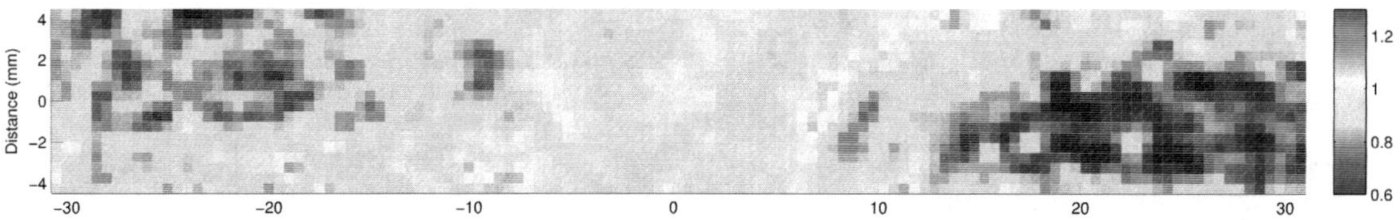

Figure 5. Precipitates thickness map (in nm) obtained on the 2050 welded in T3 condition and aged to T8 condition.

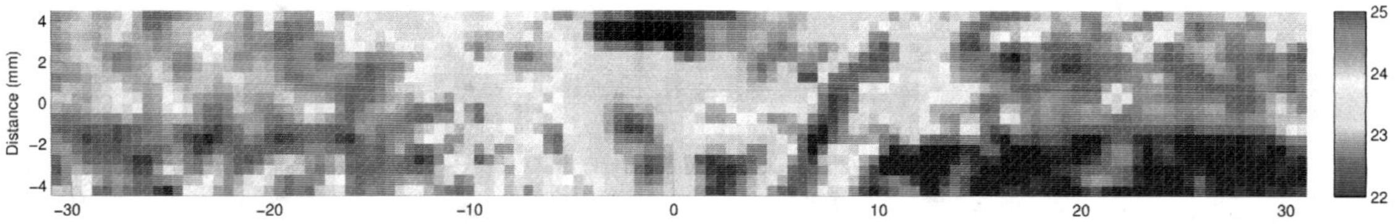

Figure 6. Precipitates length map (in nm) obtained on the 2050 welded in T3 condition and aged to T8 condition.

ageing, it can be concluded that during an artificial ageing, θ'-type precipitates are forming in the TMAZ and WN zones with the solute content issued from the dissolution of the initial T_1 platelets. In the HAZ, the results may be interpreted as a thickening of remaining T_1 platelets.

3.2. 2D microstructure mapping on the 2050 alloy welded in the T8 condition

To further probe the anisotropy of the microstructure, it might be necessary to use 2D scanning SAXS mapping of the FSW weld. This has been attempted on a 2050 Al-Li-Cu alloy welded in the T8 condition. The data has been processed in the same way than for the 1D scans. The results are plotted in figure 5 for the precipitates thicknesses and figure 6 for the lengths.

There is little evolution in terms of precipitates thickness, as in the case of the 2098 alloy welded in T8 condition. In contrast, however, the precipitates are slightly shorter in the weld

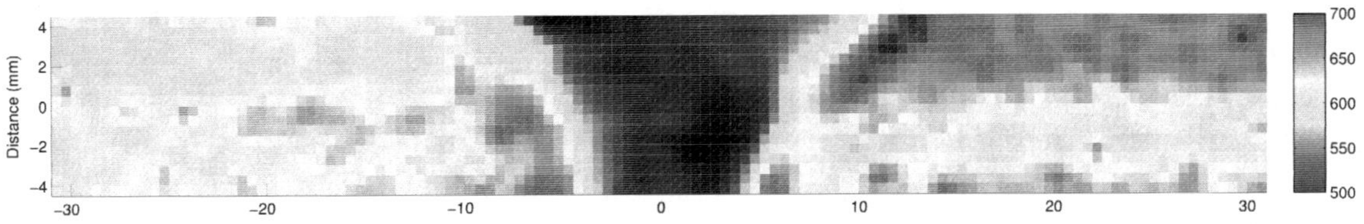

Figure 7. I(q=0) map obtained on the 2050 welded in T3 condition and aged to T8 condition.

nugget than in the base material.

As already discussed, the inhomogeneities in the texture of the material do not permit the use of the integrated intensity as a tool for probing the volume fraction, as the intensity in the plane of the detector is very dependent on the actual texture of the irradiated material. However, if we consider that the Fourier transform of a flat precipitate is a streak in the direction normal to the habit plane of the precipitate, and we consider the fact that this streak will give intensity at large q only when it is oriented not too far from the plane of the detector, it can be expected that the intensity at very low angle will be less influenced by the texture and will better reflect the amount of scattering matter.

With this in mind, we have chosen to use the extrapolated intensity at q=0 as an estimation of the relative volume fraction in the material. Figure 7 shows the evolution map of I(q=0) in the weld area. If I(q=0) is indeed a good indication of the behaviour of the volume fraction, figure 7 seems to show a much lower volume fraction in the weld nugget than in the base material. There is also the indication of an increase of the volume fraction in a zone that is likely to be the TMAZ. These observations are consistent with the fact that T_1 phase nucleation is very sensitive to the presence of dislocations [5]. As the T8 precipitation heat treatment has been applied after the FSW welding process, the low amount of dislocations in the recrystallised nugget is less favourable to T_1 nucleation whereas the TMAZ zone has undergone a further deformation during the weld, with a likely increase in dislocation density, which is in favour of a higher T_1 number density.

4. Conclusion

This preliminary study has shown that scanning SAXS can be used to map the microstructure of a FSW aluminium alloy, even in the case of anisotropic precipitates. However, care must be taken in the interpretation of the results in terms of precipitates volume fraction because of the variation of the material texture due to the deformation induced by the welding process and the recrystallisation that occurs in the weld nugget. The presence of different phases, which is the norm rather than the exception in aluminium alloys, should also be considered carefully, as it is impossible to ascertain which phase is present at which data point. To overcome these limitations, scanning SAXS should then be used in combination with other characterisation techniques such as transmission electron microscopy (TEM) for determining the phase forming in each zones, electron back scattered diffraction (EBSD) to map the texture in the weld area and/or differential scanning calorimetry (DSC) to help with the calibration of the precipitates volume fraction.

Acknowledgments

Part of this study has been financially supported by the French ANR (National Research Agency) under contract number ANR-08-MAPR-0020-05 (project CORALIS). The authors also wish to acknowledge the support of the BM02-D2AM CRG beamline staff at ESRF.

References

[1] Deschamps A, Denquin A, Genevois C and Doisneau-Cottignies B 2005 *Acta Mater.* **53** 2447–2458
[2] Denquin A, Allehaux D, Lapasset G and Ostersehlte H 2008 *Aluminium Alloys – their physical and mechanical properties* vol 2 ed J Hirsch, B Skrotzki G G (Wiley-VCH) pp 1939–1944
[3] Gable B M, Zhu A W, Csontos A A and Starke E A 2001 *J. Light Met.* **1** 1–14
[4] Glatter O and Kratky O (eds) 1982 *Small-Angle X-ray Scattering* (Academic Press)
[5] Ringer S, Muddle B and Polmear I 1995 *Metall. Mater. Trans.* A **26** 1659–1671

Journal of Physics: Conference Series **247** (2010) 012035 doi:10.1088/1742-6596/247/1/012035

SANS investigation of neutron-irradiated Fe-Cr alloys

C Heintze[1], A Ulbricht[1], F Bergner[1] and H Eckerlebe[2]

[1] Forschungszentrum Dresden-Rossendorf, PF 51 01 19, 01314 Dresden, Germany
[2] GKSS-Forschungszentrum Geesthacht GmbH, Max-Planck-Straße 1, 21502 Geesthacht, Germany

E-mail: a.ulbricht@fzd.de

Abstract. The effect of neutron irradiation on the formation of defect/solute clusters in binary iron alloys is important for the understanding of the damage mechanisms in structural materials applied in the field of nuclear technology. The irradiation behaviour of four binary Fe-Cr alloys of commercial purity with chromium contents between 2.5 and 12.5at% and neutron exposure of 0.6 and 1.5 dpa at 300 °C has been studied by SANS. The analysis of the experimental results indicates that there are at least two populations of irradiation-induced features. The appearance of α'-phase particles was shown for Fe-12.5at%Cr and to a less content in Fe-9at%Cr. We have observed a strong and complex effect of Cr and weak or insignificant effects of neutron exposure.

1. Introduction

Ferritic-martensitic chromium steels are candidate materials for future applications in both Generation-IV fission and fusion technology [1, 2]. Experimental investigation of Fe-Cr alloys is important for the understanding of the effect of chromium on the irradiation behaviour of more complex alloys. In the context of structural materials for application in Generation-IV systems, the European Commission has launched a research project named 'Generation IV and Transmutation Materials' (GETMAT) to qualify commercially available materials [3]. One of the work packages is devoted to multi-scale modeling and modeling-oriented experiments for Fe-Cr alloys. The experimental techniques selected for the investigation of irradiation-induced microstructures on the nanometer scale include Atom probe tomography, TEM, Positron annihilation and SANS. The present paper is devoted to the results of SANS experiments performed for Fe-Cr alloys of four different compositions exposed to neutron irradiation. Special emphasis is put towards nuclear scattering.

Experimental evidence and existing knowledge on the effect of the addition of Cr on the irradiation response has recently been reviewed by Malerba et al. [4]. SANS has been applied successfully to the investigation of irradiation-induced microstructures in ferritic-martensitic Cr-steels [5, 6]. However, the effects of chromium and neutron exposure are complex. Therefore, investigation of binary Fe-Cr model alloys can gain additional insight.

In section 2, the materials and irradiation conditions are introduced and experimental details are given. Nuclear scattering curves, size distributions and A-ratios measured for four different Cr-contents and two neutron exposures are summarized in section 3. The nature of the irradiation-induced defects is discussed in section 4.

2. Experiments

2.1. Material and Irradiation

The material used in this work was delivered by SCK•CEN, Mol, Belgium [7]. It was obtained by furnace melting of industrial purity Fe and Cr. After casting, the obtained ingots were cold worked under protective atmosphere to fabricate plates of 9 mm in thickness, treated at 1050°C for 1 h in high vacuum for austenisation and stabilization and tempered at 730°C for 4 h followed by air cooling. Compositions are summarized in Table 1 [7]. Optical microscopy indicates ferritic/martensitic microstructures with ferrite fractions decreasing at increasing Cr content. For the characterization by means of TEM we refer to [7].

Neutron irradiation was performed in the framework of the MIRE experiment in the reactor BR2, Mol, Belgium [7]. The irradiation temperature was (300 ± 5) °C. Neutron flux was 9 x 10^{13} cm^{-2} s^{-1} ($E > 1$ MeV). The total neutron exposure was 0.6 dpa and 1.5 dpa. The irradiated samples were delivered along with the unirradiated references as coupons of dimensions 7 x 7 x 1 mm^3.

Table 1. Composition of investigated Fe-Cr-alloys (values in wt%, rest Fe)

	Mn	Si	P	S	Cr	Ni	C	N	O
Fe-2.5at%Cr	0.009	0.02	0.013	0.0023	2.4	0.044	0.008	0.0117	0.035
Fe-5at%Cr	0.02	0.04	0.011	0.006	4.6	0.06	0.02	0.0127	0.065
Fe-9at%Cr	0.03	0.09	0.012	0.0007	8.4	0.07	0.02	0.0148	0.066
Fe-12.5at%Cr	0.03	0.11	0.05	0.006	11.6	0.09	0.027	0.0237	0.03

2.2. SANS measurements

SANS measurements were performed at the SANS-2 facility of GKSS, Geesthacht, Germany. The samples were placed in a saturation magnetic field of 1.7 T perpendicular to the incident neutron beam direction. A wavelength, λ, of 0.58 nm and three sample-detector distances of 1, 4 and 16 m with corresponding beam collimation lengths were used to cover scattering vectors of magnitude, Q, from 0.1 nm^{-1} to 3 nm^{-1} ($Q = 4\pi\sin\theta/\lambda$, where 2θ is the scattering angle). Scattered neutrons were recorded with an area ^{3}He-detector (50 x 50 cm^2) using 128 x 128 pixels. The measured data were corrected for sample transmission, detector response and background. Macroscopic differential scattering cross-sections, $d\Sigma/d\Omega$, were obtained by calibration against the incoherent scattering of vanadium. The SANS data reduction was carried out using the SANDRA [8] software package. The azimuthal dependence of scattering was used to separate magnetic and nuclear scattering.

In order to calculate the size distribution of scatterers, a dilute two-phase matrix-inclusion microstructure composed of homogeneous spherical scatterers randomly dispersed in a otherwise homogeneous Fe-Cr matrix is assumed except for Fe-12.5at%Cr, where the assumption of a dilute system has to be dropped. For a distribution of homogeneous spherical scatterers of radius, R, in the probed volume, the coherent nuclear scattering cross section, $(d\Sigma/d\Omega)_{nuc}$, can be expressed as:

$$\left(\frac{d\Sigma}{d\Omega}\right)_{nuc}(Q,R) = c_R \Delta\eta^2_{nuc} V(R) |F(Q,R)|^2 \tag{1}$$

with the form factor,

$$|F(Q,R)|^2 = \frac{9(\sin QR - QR\cos QR)^2}{(QR)^6}, \tag{2}$$

the volume of the sphere, $V = 4\pi R^3/3$, and the volume fraction per size increment, c_R. For scatterers forming a bcc lattice coherent with the matrix, the nuclear scattering contrast, $\Delta\eta^2_{nuc}$, is given by:

$$\Delta\eta^2_{nuc} = \left[(n\bar{b}_{nuc})_C - (n\bar{b}_{nuc})_M\right]^2 \tag{3}$$

where $n = 2/a_0^3$ is the atom number density and a_0 is the lattice parameter of the bcc lattice. Subscripts C and M refer to cluster (scatterer) and matrix, respectively, and $\bar{b}_{\mathrm{nuc}}$ is the average nuclear scattering length.

The indirect transformation method [9] was applied to obtain the size distribution of scatterers without assuming a certain type of distribution. For more details see [10]. The lower detection limit of SANS is about 0.5 nm in terms of radius. It is important to emphasize that, as long as the nuclear contrast is not known in advance, only the product, $c_R(R)\Delta\eta^2_{\mathrm{nuc}}$, can be derived, that means, the size distribution can be obtained in relative units only.

The *A*-ratio originally defined as ratio of the scattering cross-sections, d*Σ*/d*Ω*, perpendicular and parallel to the magnetic field direction [11] was calculated as a function of the scattering vector *Q* according to equation (4), where the subscript "mag" indicates the magnetic scattering cross section.

$$A = \frac{(\mathrm{d}\Sigma / \mathrm{d}\Omega)_{\mathrm{mag}}}{(\mathrm{d}\Sigma / \mathrm{d}\Omega)_{\mathrm{nuc}}} + 1 \qquad (4)$$

3. Results

The nuclear scattering curves measured for the unirradiated conditions of the Fe-Cr alloys are shown in figure 1(a). A pronounced effect of Cr on the nuclear scattering was observed especially at low and high values of *Q*. A Q^{-4}-dependence is found to represent the measurements well in the intermediate *Q*-range. The nuclear scattering curves measured for the irradiated conditions of Fe-2.5at%Cr and -12.5at%Cr alloys are shown in figure 1(b).It is assumed in the following that we can neglect an evolution of inhomogeneities already existing in the non-irradiated sample. Basically the corresponding scattering curves of the non-irradiated and the irradiated condition overlap below 0.2 nm^{-1}.

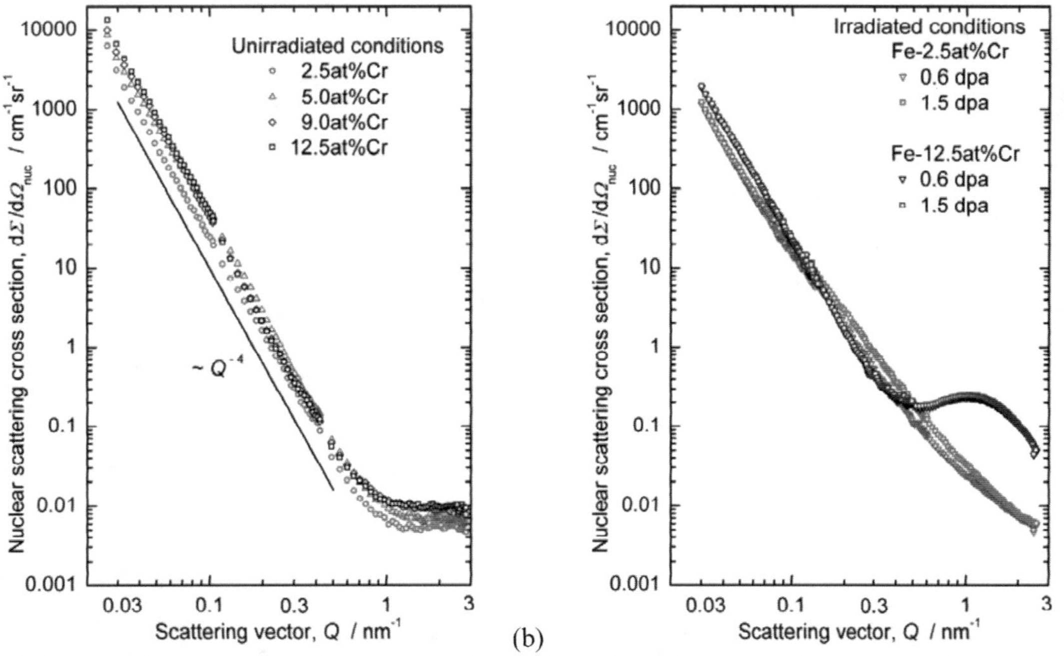

(a) (b)

Figure 1. Measured nuclear scattering cross sections (a) for the unirradiated reference samples and (b) for the irradiated conditions of .Fe-2.5at%Cr and Fe-12.5at%Cr.

The complete set of measured nuclear difference scattering cross sections with the respective unirradiated conditions taken as reference are shown in figures 2(a) to 2(d) as a function of the scattering vector, Q, for Cr contents of 2.5, 5, 9, and 12.5 at%, respectively, and for neutron exposures of 0.6 and 1.5 dpa. The difference scattering data below Q values of 0.3 nm^{-1} are smaller as the measurement errors of the irradiated and unirradiated conditions. Therefore, we used only data with Q above 0.3 nm^{-1} for analysis. The fitted curves are those used in order to reconstruct the volume-related size distributions, $c_R \Delta \eta^2_{\mathrm{nuc}}$, according to equation (1). These size distributions are summarized in figure 3 using the same arbitrary units. The decrease of the scattering cross section observed for Fe-12.5at%Cr at decreasing Q, see figure 2(d), is caused by interference effects typical for concentrated alloys. Therefore, the range of fit was restricted to Q-values beyond the maximum, where the scattering curve is not disturbed by interference effects, as suggested in [12] as first approximation.

For Fe-2.5at%Cr and Fe-9at%Cr, figures 2(a) and 2(c), the scattering curves indicated the contribution of two individual populations of defects. Therefore, the high-Q range of the measured difference scattering curve was fitted in the first step, and the remaining difference was fitted in the second step. We have observed a pronounced bimodality of the reconstructed size distributions (figures 3(a) and (c)). For Fe-5at%Cr the results are not in contradiction with a bimodal size distribution but the components cannot be separated reliably. For Fe-12.5at%Cr a possible bimodality is obscured by the interference effect outlined above.

The A-ratios calculated from the difference cross sections are displayed in figure 4 as a function of the scattering vector, Q, for the different Cr-contents. A pronounced Q-dependence of the A-ratio was observed for Fe-2.5at%Cr and Fe-5at%Cr. The A-ratio of the other alloys is independent of Q.

Journal of Physics: Conference Series **247** (2010) 012035 doi:10.1088/1742-6596/247/1/012035

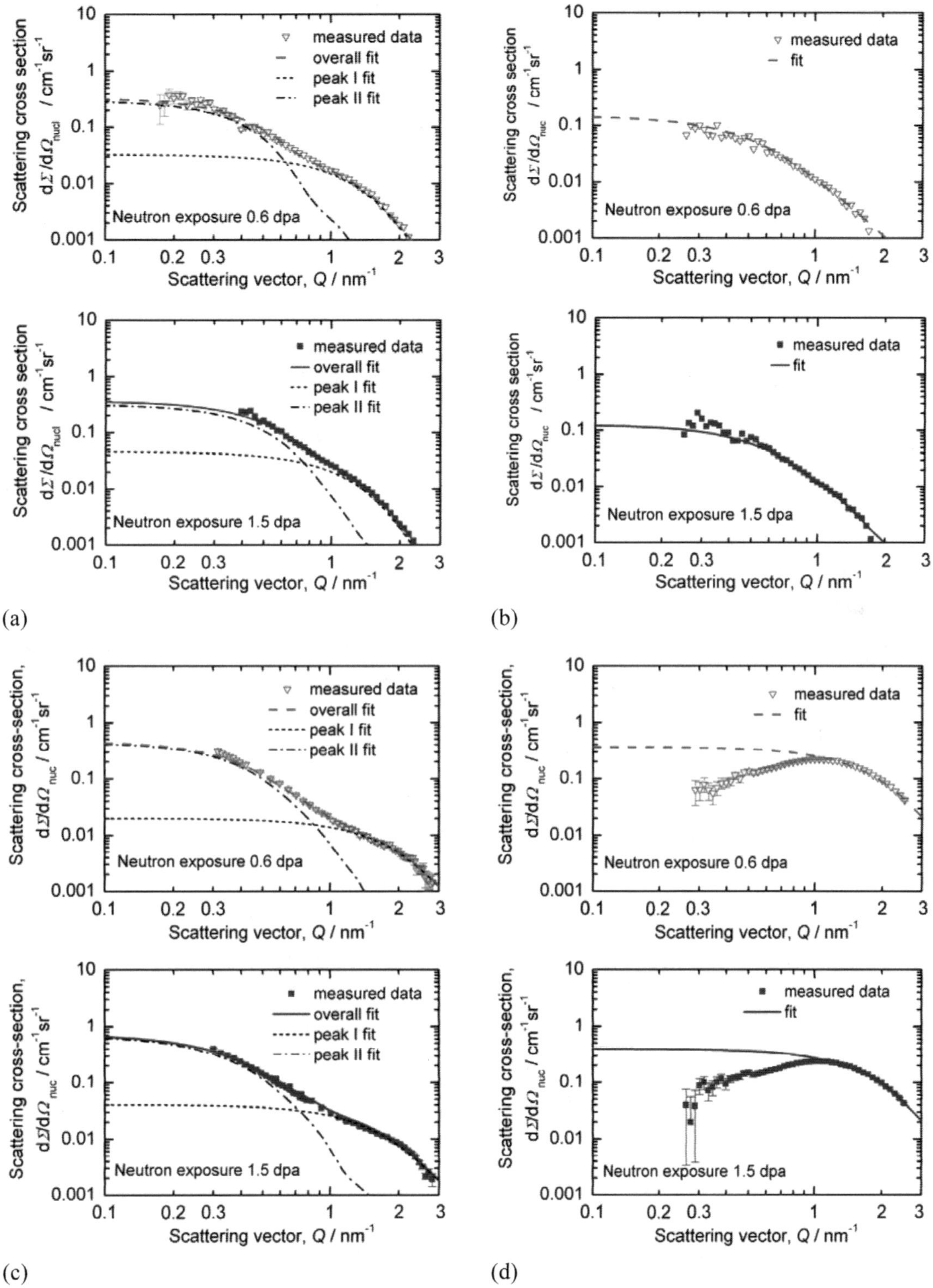

Figure 2. Measured nuclear difference scattering curves with the respective unirradiated conditions as reference for (a) Fe-2.5at%Cr, (b) Fe-5at%Cr, (c) Fe-9at%Cr, and (d) Fe-12.5at%Cr. Lines represent fits needed in order to reconstruct size distributions.

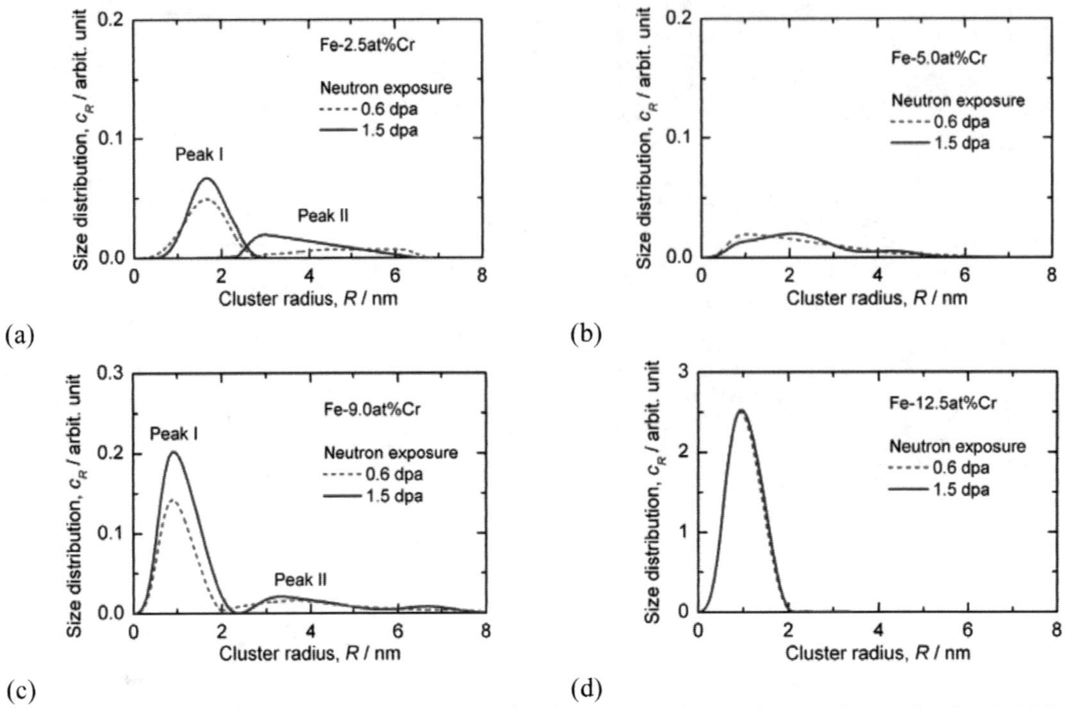

Figure 3. Size distributions of irradiation-induced scatterers in (a) Fe-2.5at%Cr, (b) Fe-5at%Cr, (c) Fe-9at%Cr and (d) Fe-12.5at%Cr.

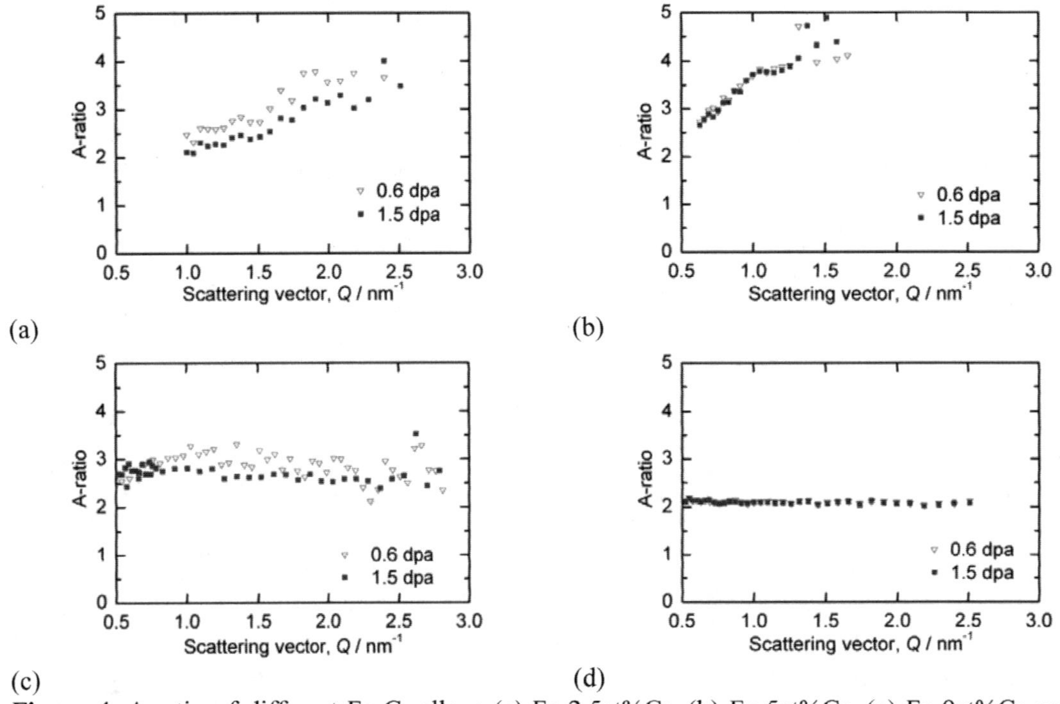

Figure 4. A-ratio of different Fe-Cr alloys (a) Fe-2.5at%Cr, (b) Fe-5at%Cr, (c) Fe-9at%Cr and (d) Fe-12.5at%Cr.

4. Discussion

4.1. Unirradiated conditions
For the unirradiated conditions, the level of incoherent scattering was determined from Porod plots [13, 14] giving more exact results than the direct estimation from the high-Q range of figure 1. The results are plotted in figure 5 as a function of Cr. The indication line was obtained taking into account an ideal mixture of Fe and Cr isotopes.

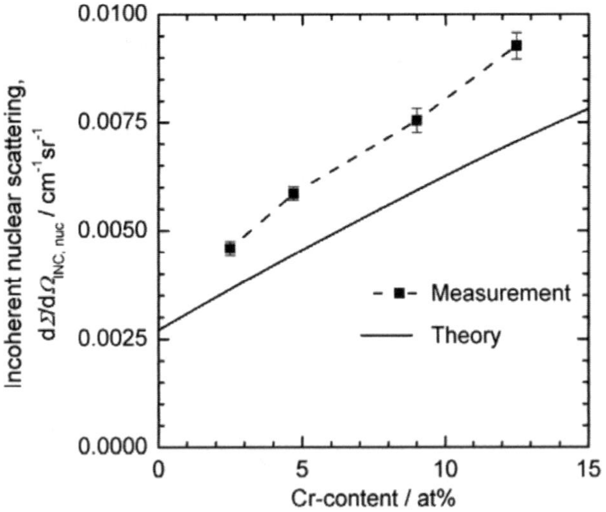

Figure 5. Incoherent nuclear scattering measured for the unirradiated references plotted as function of the chromium content. The indication line characterizes the values expected for an ideal Fe-Cr system.

In general, the measured values are of the same order of magnitude as the indication line and follow the same trend. The deviation of the measured values from the indication line is assumed to reflect the degree of purity of the alloys, which is also an increasing function of the Cr-content (see table 1).

4.2. Effect of neutron irradiation
We have found bimodal size distributions of scatterers for Fe-2.5at%Cr and Fe-9at%Cr in both irradiation conditions. For Fe-5at%Cr and Fe-12.5at%Cr bimodal size distributions are possible but could not be resolved because of small intensity (5at% Cr) and interparticle interference (12.5at% Cr). The radii of the scatterers cover the ranges from 0.5 to 2.5 nm for the first component and from 2.5 to 6 nm for the second component. The two components of the size distribution reflect two different populations of irradiation-induced defects denoted population I and II.

In general, the A-ratio provides integral information on the composition of the scatterers. The Q-dependence of A-ratio observed for Fe-2.5at%Cr and Fe-5at%Cr indicates that the average compositions of the two defect populations are different. A-ratios in excess of 3 are impossible in the binary Fe-Cr system (including vacancies and interstitials) but are characteristic of Cr-rich carbides [5]. Therefore, we assume that population I are irradiation-induced Cr-C-enriched clusters. The amount of impurity carbon (table 1) was checked to be sufficient to explain the observed effect. The size range of population II agrees with the size range of dislocation loops (mean diameters between 6 and 16 nm) detected by means of TEM for the same alloys [7]. However, SANS is not sensitive to undecorated dislocation loops [15]. A possible explanation could be the decoration of loops with Cr- and/or C-atoms reported in the literature [16].

The value of the A-ratio of 2.05 independently of Q for Fe-12.5at%Cr provides evidence for the dominant type of scatterers to be α'-phase particles. In [17] the solubility limit of Cr in Fe was estimated on the basis of the SANS data allowing the nuclear contrast and the volume fraction of α' to be calculated. The estimations of the solubility limit and the volume fraction are 8.5at% and 4.3vol%, respectively. For Fe-9at%Cr the *A*-ratio between 2.05 and 3 indicates both α' and Cr-C-enriched clusters to contribute. In order to estimate volume fractions of Cr-C-enriched clusters the detailed composition is needed. An Atom probe investigation is in preparation for the same materials.

The dependence on neutron exposure of both size (see figure 3) and *A*-ratio (see figure 4) is weak or insignificant. A SANS investigation of the same set of alloys irradiated to 0.06 dpa is in progress and will gain additional information.

5. Conclusion

The SANS investigation of neutron irradiated Fe-Cr alloys indicates the presence of two types of irradiation-induced defect/solute clusters of radii in the ranges from 0.5 nm to 2.5 nm and 2.5 to about 6 nm. α'-phase particles are the dominant type of scatterers for Fe-12.5at%Cr and contributes to scattering for Fe-9at%Cr. A candidate for a second population of scatterers are Cr-C-enriched clusters or decorated loops. We have observed a strong and complex effect of Cr and weak or insignificant effects of neutron exposure.

Acknowledgement

This work was partly supported by the European Commission within the collaborative project GETMAT under Grant Agreement No. 212175. The authors would like to thank M Matijasevic and A Almazouzi (SCK-CEN) for providing the material along with basic documentation and D. Terentyev for helpful discussions.

References

[1] Klueh R L and Nelson A T 2007 *J. Nucl Mater.* **371** 37
[2] Cook I 2006 *Nature* **5** 77
[3] Fazio C, Alamo A, Almazouzi A, De Grandis S, Gomez-Briceno D, Henry J, Malerba L and Rieth M 2009 *J. Nucl. Mater.* **392** 316
[4] Malerba L, Caro A and Wallenius J 2008 J. Nucl. Mater. **382** 112
[5] Mathon M-H, de Carlan Y, Geoffroy G, Averty X, Alamo A and de Novion C-H 2003 *J. Nucl. Mater.* **312** 236
[6] Coppola R, Kampmann R, Magnani M and Staron P 1998 *Acta mater.* **46** 5447
[7] Matijasevic M and Almazouzi A 2008 *J. Nucl Mater.* **377** 147
[8] Biemann P, Haese-Seiller M and Staron P 2000 *Physica B* **276-278** 156
[9] Glatter O 1980 *J. Appl. Cryst.* **13** 7
[10] Ulbricht A, Heintze C, Berger F, Eckerlebe H 2009 Poster presented at TMS 138th Annual Meeting, San Francisco, paper submitted to *J. Nucl. Mater.* manuscript No. JNM-D-09-00100
[11] Solt G, Frisius F, Waeber W B and Tipping P 1993 *Effects of Radiation on Materials: 16th Int. Symp.* eds A S Kumar, D S Gelles, R K Nanstadt, E A Little (Philadelphia: ASTM STP 1175) p 444
[12] Staron P 1997 PhD Thesis, University Hamburg, also published as Report GKSS 97/e/53, GKSS Forschungszentrum Geesthacht GmbH
[13] Porod G 1951 *Kolloidzeitschrift und Zeitschrift für Polymere* **124** 83
[14] Porod G 1952 *Kolloidzeitschrift und Zeitschrift für Polymere* **125** 108
[15] Bergner F, Ulbricht A, Hernandez-Mayoral M and Pranzas P K 2008 *J. Nucl. Mater.* **374** 334
[16] Wakai E, Hishinuma A, Kato Y, Yano H, Takaki S and Abiko K 1995 *J. Phys. IV France* **05** C7-277
[17] Bergner F, Ulbricht A and Heintze C 2009 *Scripta Mater.* **61** 1060

Journal of Physics: Conference Series **247** (2010) 012036 doi:10.1088/1742-6596/247/1/012036

Microstructure of hydrogenated Mg_2Ni studied by SANS

K Mori[1], M Sugiyama[1], K Iwase[2], S Kawabe[1], Y Onodera[1], K Itoh[1], T Otomo[3], and T Fukunaga[1]

[1] Research Reactor Institute, Kyoto University, 2-1010 Asashiro-Nishi, Kumatori, Sennan, Osaka 590-0494, Japan

[2] Frontier Research Center of Applied Nuclear Science, Ibaraki University, 162-1 Shirakata-Shirane, Tokai, Ibaraki 319-1106, Japan

[3] Institute of Materials Structure Science, High Energy Accelerator Research Organization, 1-1 Oho, Tukuba, Ibaraki 305-0801, Japan

E-mail: kmori@rri.kyoto-u.ac.jp

Abstract. X-ray powder diffraction (XRD) and small-angle neutron scattering (SANS) experiments were carried out for the hydrogenated and duterated Mg_2Ni, respectively. According to the results of XRD experiments, both of them coexisted with unhydrogenated (or undeuterated) Mg_2Ni in the hydrogen absorbing cycle. Furthermore, in the SANS experiments, a slope of SANS curve, $I(Q)$, was roughly evaluated by using the following power law: $I(Q) \propto Q^{-m}$, where Q is the magnitude of the scattering vector, and m can be equated with a fractal dimensionality, D_S (= $6 - m$). In conclusion, the hydrogenated and duterated Mg_2Ni showed $D_S \sim 3$ and ~ 2, respectively. The significant difference between D_S's can be understood by considering the scattering length densities, ρ, of Mg_2Ni, Mg_2NiH_4, and Mg_2NiD_4.

1. Introduction

An intermetallic compound of Mg_2Ni, which has the $P6_222$ symmetry in the hexagonal system, is well known as one of the hydrogen storage alloys, alongside $LaNi_5$ and Ti–V alloys [1–5]. In the temperature range 573 – 623 K, Mg_2Ni transforms into Mg_2NiH_4 due to the hydrogen absorption. The hydrogenation temperature is relatively high, although its hydrogen-to-metal ratio (H/M) reaches 1.33. Since such a superior hydrogen storage capacity, all sorts of investigations – sample preparation, hydrogenation, and structural properties – have been conducted so far [6–12].

To understand the hydrogenation mechanism of Mg_2Ni, it is important to clarify the structural features of Mg_2NiH_4 in the hydrogen absorbing cycle. According to the structural studies of Mg_2NiH_4 by the X-ray and neutron powder diffractions (XRD and NPD), the crystal structure can be described by the model based on the *Fm–3m* symmetry in the cubic system around 600 K or the *C2/m* in the monoclinic one at room temperature (RT) [2,14]. However, the XRD and NPD no longer detect a microstructure such as an interface.

The aim of this study is to clarify the detailed interface structure of Mg_2NiH_4 in the hydrogen absorbing cycle by the small-angle neutron scattering (SANS). The SANS is applicable for this purpose, because neutron is more sensitive than X-ray for hydrogen. And also, we can easily control the scattering length density, ρ, of the Mg_2Ni hydride by using both of H_2 and D_2 (i.e., Mg_2NiH_4 and Mg_2NiD_4), as we mention later.

2. Experimental Procedure

We prepared three polycrystalline samples: Mg_2Ni, Mg_2NiH_4, and Mg_2NiD_4. First, Mg_2Ni ingots (KOJUNDO Chemical Laboratory Co. Ltd) were grinded to be powder (less than 212 μm). The Mg_2Ni powder was put into a silica tube, and loaded on a hydrogenation system equipped with a furnace. After heating at 573 K for 1.5 h in a vacuum, it was reacted under 1.3 MPa of H_2 (or D_2) gas at 573 K for 3 h, followed by cooling to RT. The XRD patterns were measured by a X-ray diffractometer using Cu Kα radiation.

SANS experiments were carried out on the SANS-U at Institute for Solid State Physics (ISSP), the University of Tokyo, installed at the cold neutron beam port C1-2 in the JRR-3 reactor of Japan Atomic Energy Agency (JAEA) [13]. The wavelength of the incident neutron beam is 7 Å. The prepared three samples were put into a quartz cell; the sample width is 1 mm long. For each sample, SANS spectrum in the Q-range 0.01 – 0.15 $Å^{-1}$ was collected at RT for 10 min, where Q is the magnitude of the scattering vector ($Q = 4\pi \sin\theta/\lambda$).

3. Results and discussion

The XRD patterns of Mg_2Ni, Mg_2NiH_4, and Mg_2NiD_4 at RT are shown in Fig. 1. Prior to hydrogenation (or deuteration), it was confirmed that our Mg_2Ni sample slightly includes $MgNi_2$ as an impurity phase. Even though the preparation of Mg_2Ni single-phase is difficult, the $MgNi_2$ does not react with H_2 and D_2 [1,2]. After hydrogenation (or deuteration), Bragg reflections of Mg_2NiH_4 (or Mg_2NiD_4) appeared on the XRD patterns, in addition to those of the unhydrogenated (or undeuterated) Mg_2Ni, as shown in Fig. 1(b) and (c). Note that Mg_2NiH_4 and Mg_2NiD_4 have the crystal structure based on the *C2/m* symmetry in the monoclinic system, and the preferred orientation considerably affects the intensities of several Bragg reflections.

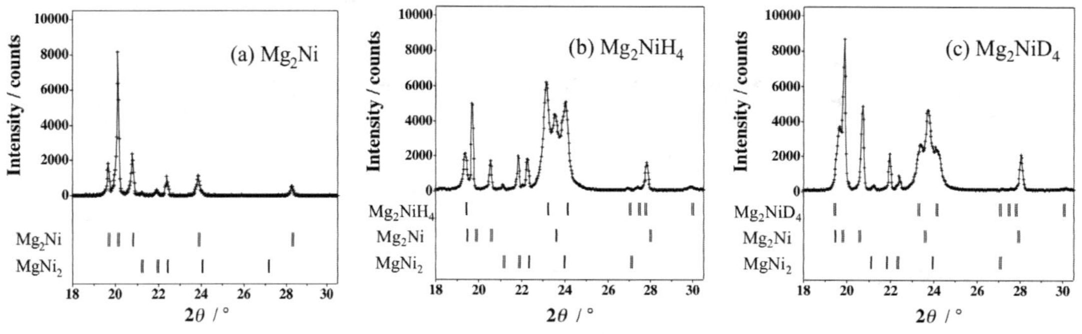

Figure 1. XRD patterns of (a) Mg_2Ni, (b) Mg_2NiH_4, and (c) Mg_2NiD_4 at RT. The vertical marks below the XRD patterns indicate the position of Bragg reflections for Mg_2Ni, $MgNi_2$, and Mg_2NiH_4 (or Mg_2NiD_4), taking into account our preliminary crystal structure analyses.

Fig. 2 demonstrates SANS intensities, $I(Q)$, for Mg_2Ni, Mg_2NiH_4, and Mg_2NiD_4. It is noteworthy that the background intensities of Mg_2NiH_4 are much higher than those of Mg_2Ni and Mg_2NiD_4. This is because the incoherent scattering cross section of H is too large (σ_H = 80.26 barn) [15]. In preliminary SANS analyses, the SANS curves deviated from that of the Porod law, $I(Q) \propto Q^{-4}$. Then, in order to quantify a slope of the SANS curve, we assumed the following power law: $I(Q) = C_0 + C_1Q^{-m}$, where C_0 and C_1 are coefficients, and m is an exponent. As shown in Fig. 2, excellent fits between the observed and calculated SANS data were obtained in the Q-range 0.01 – 0.05 $Å^{-1}$. The obtained m of Mg_2Ni was close to 4. Likewise, the m of Mg_2NiH_4 was almost 3, whereas that of Mg_2NiD_4 was approximately 4.

Before considering the interface structure of Mg_2NiH_4, it is beneficial to evaluate the ρ for all of phases: Mg_2Ni, Mg_2NiH_4, and Mg_2NiD_4. This is because the SANS intensities strongly depend on the differences between ρ's ($\Delta\rho$). The ρ is given by: $\rho = \sum b_j / V_{cell}$, where b_j is a scattering length of the *j*th atoms in a unit cell with a volume, V_{cell}. In this calculation, the lattice parameters of Mg_2Ni were fixed as follows: a = 5.2 Å and c = 13.2 Å in the hexagonal system, and those of Mg_2NiH_4 and Mg_2NiD_4: a = 6.5 Å, b = 6.4 Å, c = 6.6 Å and β = 93.2 ° in the monoclinic one, taking into account our preliminary crystal structure analyses. And also, for each element, the b was provided from *International Tables* [15]: b[H] = –3.7390(11) fm, b[D] = 6.671(4) fm, b[Mg] = 5.375(4) fm, and b[Ni] = 10.3(1) fm.

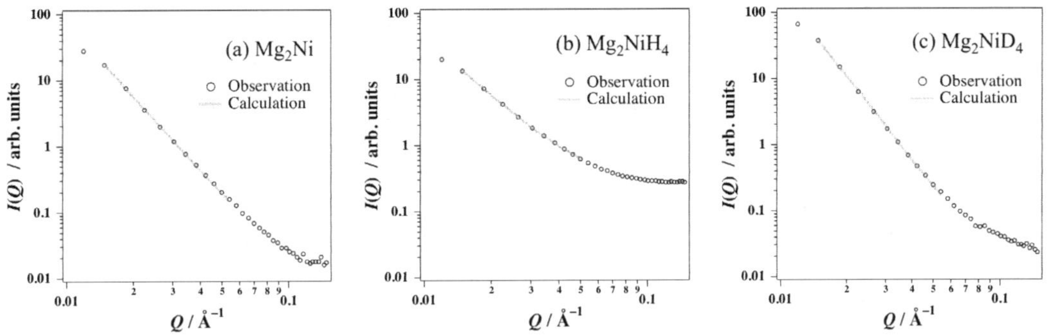

Figure 2. SANS intensities, $I(Q)$, of (a) Mg_2Ni, (b) Mg_2NiH_4, and (c) Mg_2NiD_4 at RT. Broken lines ($0.015 \leq Q \leq 0.05$ Å^{-1}) are the calculated curves on the basis of the following power law: $I(Q) = C_0 + C_1Q^{-m}$, where C_0 and C_1 are coefficients, and m is an exponent.

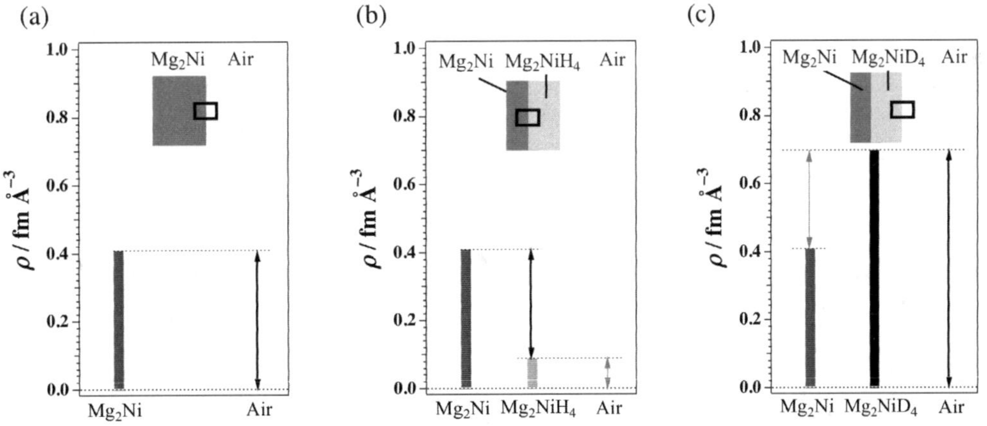

Figure 3. Scattering length densities, ρ, for (a) Mg_2Ni, (b) Mg_2NiH_4 (+ unhydrogenated Mg_2Ni), and (c) Mg_2NiD_4 (+ undeuterated Mg_2Ni), respectively. Suppose that ρ for "Air" is zero. The small rectangles on the schematic illustrations indicate the interface detected by SANS.

Fig. 3 shows the ρ's for (a) Mg_2Ni, (b) Mg_2NiH_4 (+ unhydrogenated Mg_2Ni), and (c) Mg_2NiD_4 (+ undeuterated Mg_2Ni). Suppose that ρ for "Air" is zero. As for Mg_2Ni, it was safely concluded that the obtained m (~ 4) indicates a two-dimensional surface of Mg_2Ni, since m can be equated with fractal dimensionality, D_S (= 6 – m), that is, D_S ~ 2. On the other hand, for Mg_2NiH_4 and Mg_2NiD_4, we should pay attention to what we observe. In the case of Mg_2NiH_4, the $\Delta\rho$ between the unhydrogenated Mg_2Ni

and Mg_2NiH_4 (Mg_2Ni I Mg_2NiH_4) is larger than that between Mg_2NiH_4 and Air (Mg_2NiH_4 I Air), as shown in Fig. 3(b). Therefore, it is most likely that we observe the interface of "Mg_2Ni I Mg_2NiH_4" with a complicated three-dimensional network (since, $D_S \sim 3$). Conversely, for Mg_2NiD_4, a two-dimensional surface of Mg_2NiD_4 ($D_S \sim 2$) would be observed, taking into account the large $\Delta\rho$ between Mg_2NiD_4 and Air (Mg_2NiD_4 I Air).

From the results of XRD and SANS experiments, it seems that Mg_2NiH_4 encroaches on Mg_2Ni particles in the hydrogen absorbing cycle. Especially, in the SANS experiments, the combination of the hydrogenation (H_2) and deuteration (D_2) is considerably beneficial for changing the interface we want to observe. However, the quantitative analysis and the interpretation for the m and D_S would be still inexperienced; further investigations are needed.

4. Conclusions

XRD and SANS experiments were carried out for the hydrogenated and duterated Mg_2Ni, respectively. In the XRD experiments, we confirmed that Mg_2NiH_4 (or Mg_2NiD_4) coexists with the unhydrogenated (or undeuterated) Mg_2Ni. In addition, the surface of Mg_2NiD_4 with $D_S \sim 2$ and the interface between the unhydrogenated Mg_2Ni and Mg_2NiH_4 with $D_S \sim 3$ were observed in the SANS experiments. These results imply that Mg_2NiH_4 encroaches on Mg_2Ni particles in the hydrogen absorbing cycle.

Acknowledgements

We wish to acknowledge the SANS-U group of the ISSP, the university of Tokyo, for the help in SANS experiments. This work is supported by a Grant-in-Aid for Yong Scientists (B) from the Ministry of Education, Science and Culture, Japan.

References

[1] Reilly J J and Wiswall Jr. R H 1968 *Inorg. Chem.* **7** 2254
[2] Schefer J, Fischer P, Hälg W, Stucki F, Schlapbach L, Didisheim J J, Yvon K and Andresen A F 1980 *J. Less-Common Met.* **74** 65
[3] Genossar J and Rudman P S 1981 *J. Phys. Chem. Solids* **42** 611
[4] Akiba E and Iba H 1998 *Intermetallics* **6** 461
[5] Iwase K, Kamiyama T, Nakamura Y, Mori K, Yonemura M, Harjo S, Ishigaki T and Akiba E 2006 *Mater. Transations* **47** 271
[6] Huot J, Akiba E and Takada T 1995 *J. Alloys Comp.* **231** 815
[7] Orimo S, Seto H, Ikeda K, Nagao M and Fujii H 1996 *Physica B* **226** 370
[8] Chen J, Dou S X and Liu H K1996 *J. Alloys Comp.* **244** 184
[9] Orimo S, Ikeda K, Fujii H and Yamamoto K 1997 *J. Alloys Comp.* **260** 143
[10] Ikeda K, Orimo S, Züttel A, Schlapbach L and Fujii H 1998 *J. Alloys Comp.* **280** 279
[11] Orimo S, Züttel A, Ikeda K, Saruki S, Fukunaga T, Fujii H and Schlapbach L 1999 *J. Alloys Comp.* **293-295** 437
[12] Itoh K, Otomo T, Sugiyama M, Mori K and Fukunaga T 2009 *J. Phys.: Conference Series* **144** 012107
[13] Okabe S, Nagao M, Karino T, Watanabe S, Adachi T, Shimizu H and Shibayama M 2005 *J. Appl. Crystallogr.* **38** 1035
[14] Noréus D and Werner P –E 1984 *J. Less-Common Met.* **97** 215
[15] Wilson A J C and Prince E 1999 *International Tables for Crystallography* **Vol. C** (Kluwer Academic Publishers)

Nano-quasicrystal formation in $Zr_{75}Cu_{20}Pt_5$ Glass Ribbons during Annealing Examined by in-situ SWAXS

H. Okuda[1,4], Y. Kashitani[1], R. Arao, S.Ochiai[1], J.Saida[2], S.Sasaki[3] and H. Masunaga[3]

[1]Department of Materials Science and Engineering, Kyoto University, Sakyo-ku Kyoto 606-8501 Japan,
[2]Center for interdisciplinary Sciences, Tohoku University, Aoba Sendai 980-8578, Japan,
[3]JASRI Spring8, Kohto, Sayo 679-5198 Japan.

E-mail: okuda@materials.mbox.media.kyoto-u.ac.jp

Abstract. Microstructure evolutions of nano-quasicrystals precipitated in ZrCuPt metallic glass ribbons during annealing has been examined by in-situ small-angle scattering and wide angle diffraction (SWAXS). The time-resolved measurements were made at a photon energy close to the Zr K absorption edge to obtain reasonable balance between transmission and resolution. During annealing the sample in vacuum, increase in SAXS intensity and diffraction peak at medium angle region was observed. The size obtained from SAXS and diffraction almost agreed, suggesting that the precipitation process in the glass ribbon is described by a growth process of single-crystalline nano-quasicrystals. Development of precipitated microstructure was further discussed by comparing SAXS profiles and diffraction profiles.

1. Introduction

Precipitation process in metallic glasses is an interesting subject to understand metallurgical background of the stability of glass state of the bulk metallic glass from kinetic viewpoint of phase transformation. Zr-Cu base alloys, in particular, are well known as an alloy system that exhibits excellent stability, for example, like ZrCuAlNi quaternary alloys[1-4]. Among them, Zr-based alloys containing small amount of noble metals such as Pt have been extensively examined by Saida et al[5-9], since such alloys quite often gives very small nanocrystals as primary precipitates at the first crystallization stage, which eventually transform into metastable microstructure containing crystalline phases without a primary quasicrystalline phase in the late stage of heat treatment. One of the reasons that such precipitation processes of quasicrystals are interesting is that it may reflect inherent local environment of glass to some extent. Namely, if the origin of the stability of Zr-Cu based glasses are explained by their local icosahedral order in the glass or supercooled liquid state, precipitation of quasicrystals observed by small addition of noble metal elements may reflect such local environment. If it is the case, kinetics of nucleation and growth may be different from conventional one observed in precipitation processes of age-hardenable metallic alloys. From this viewpoint, microstructures obtained from SAXS, which gives information on growth kinetics of the quasicrystalline precipitates

[4] To whom any correspondence should be addressed.

as concentration heterogeneities, and that obtained by diffraction spot of quasicrystals which gives information on kinetics of precipitates as viewed from lattice order, are important factors to discuss the characteristic microstructure. Analyzing kinetics of precipitation, which is a standard procedure in metallurgy of crystalline alloys for aluminium and iron alloys, is not easy for these glasses, since materials far from equilibrium are generally sensible to the slight deviation of initial conditions of preparation. Another difficulty for Zr-Cu base alloys is that they easily form oxides with a nanometer scale, which also give a strong small-angle scattering.
In the present work, simultaneous SWAXS measurements were made in-situ in vacuum to avoid such problems.

2. Experimental

The sample ribbons with a molar composition of $Zr_{70}Cu_{25}Pt_5$ used in the present experiment were prepared by single-role melt-spinning by a Cu roll rotating at 40 m/s at the surface. The ribbon with 0.02 mm in thickness and 1 mm in width was placed in the furnace placed at the sample chamber evacuated by a turbo molecular pump to avoid oxidation during annealing. In-situ measurements were made at beam-line 40B2 of Spring8, Hyogo, Japan. SAXS and the diffraction profiles were measured with Rigaku RAXIS VII simultaneously, since the photon energy of the present measurements, 17.99 keV, and large area of Image Plate, 300x300 mm^2, were enough to record SWAXS on the same plate with sufficient resolution. By recording them on the same image plate, it is easy to distinguish SAXS

Figure 1. Experimental set-up of the present *in-situ* measurements. The sample chamber, center, was evacuated by a turbo molecular pump during *in-situ* measurements. The upstream and downstream pipes with lower vacuum were separated from the sample chamber.

by quasicrystals from SAXS by metastable crystallites or oxide crystallites. This point is very important in the present work, since mixed signals of SAXS from glass or quasicrystals and that from crystalline particles or oxides often lead to erroneous conclusions. The sample was heated with electric furnace installed in the chamber, and SWAXS profiles were recorded every 3 minutes with 150 s of exposure. TEM works by Saida et al.[5,6] shows that quasicrystals are spherical in the present condition.

3. Results and Discussions

Figure 2 gives the SWAXS profiles obtained for the isothermal annealing at 610 K, which were radially averaged from the two-dimensional intensity pattern. For the melt-spun (as Quenched) state, the SAXS intensity profile shows a monotonic decrease with the magnitude of scattering vector, $q=4\pi \sin(\theta)/\lambda$, where θ is the scattering angle and λ is the wavelength of the incident X-rays. At larger q, the intensities increase with q for all the annealing time, which should be mostly the tail of halo peak at much larger q. The SAXS intensities with approximately $q<1$ nm^{-1} increase with annealing time, suggesting that size and volume of the precipitates increase with time. In the larger q region at q~6

nm^{-1}, a Bragg peak was observed, which grows with annealing time. An enlarged figure of the peak is also shown in Fig.2. It is seen that the peak also grows with annealing time, corresponding to the increase in the SAXS intensity.

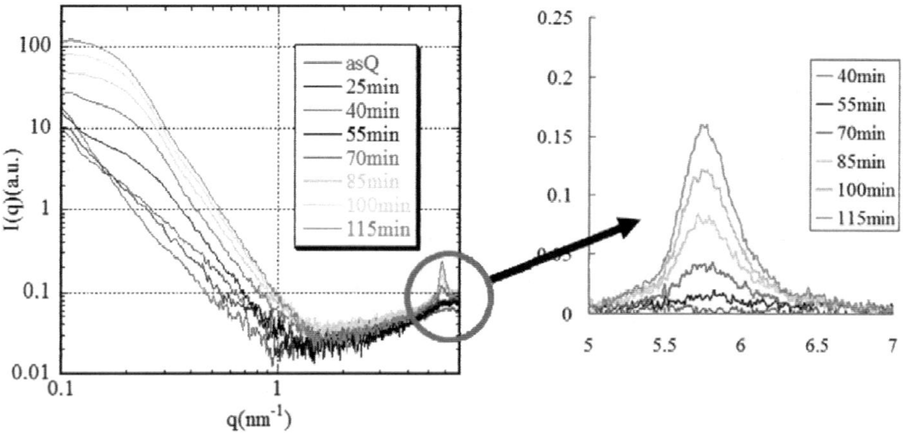

Figure 2. SWAXS intensity of ZrCuPt ribbon annealed at 610 K

Comparing the SAXS part in Fig.2 with SAXS profiles commonly observed for annealed metallic alloys where precipitation hardening is used for designing performance of the materials[10,11], we may notice that interparticle interference peak is not observed in the present sample, which is rather commonly observed for the alloys such as Al-Zn, Al-Zn-Mg or Al-Li[10-12], even for Al-4at%Zn with expected volume fraction of about 3%. This means that the number density of the precipitated quasicrystal is small even in the later stage of annealing where well-defined SAXS profile is observed. The diffraction peak observed at large q also grow with time, suggesting that the growth of quasi-crystalline precipitates. To obtain more quantitative parameters, temporal evolution of the peak intensity for SAXS and diffraction was plotted in Figure 3. When the temporal evolution of SAXS peak intensity and the diffraction peak intensity agree, it suggest that the microstructure of the sample is well approximated by two-phase model, a precipitate having well-defined interface, with a constant concentration contrast between inside and outside of the precipitate and a uniform single-(quasi)crystalline structure in the precipitates. From Fig. 3, the temporal power law for the SAXS was about 5.3 for the intermediate time range as shown in the figure, whereas that for the diffraction is about 3.7. The fact that the temporal powers for the SAXS and the Bragg do not agree each other suggests that the microstructure of the precipitate is not described by a simple two-phase model with a

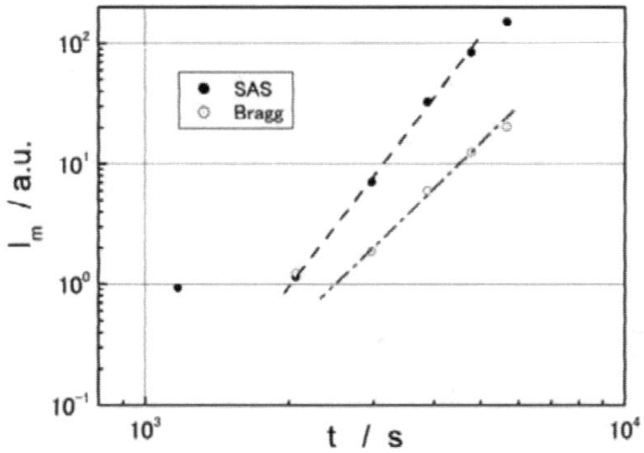

Figure 3. Temporal evolution of the peak intensity for SAS and the diffraction peak at $q=5.8\ nm^{-1}$

fixed structure. To add with, the power is much larger than the ones known for precipitation[13-15]. These results suggest that the precipitation process in the present glass system is different from the precipitation kinetics commonly known for crystalline alloys. Since no interparticle interference is observed in the present SAXS profile, the peak intensity for SAXS is related to $\Delta\rho^2\langle v^2\rangle N$, with $\Delta\rho$ the difference of electron density between the precipitates and the matrix, <v> the volume of the precipitates and N the number density of precipitates. When a two-phase model applies for the precipitate microstructures, increase in I_0 is explained by the increase in the volume, number density, or the contrast.

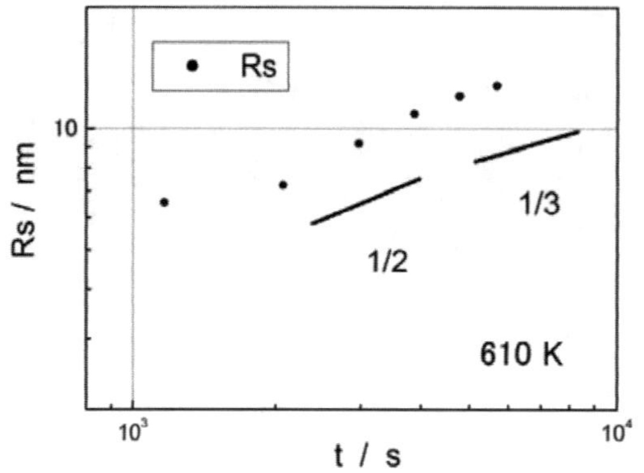

Figure 4. Change of radius, Rs, during annealing.

Figure 4 shows the temporal evolution of spherical radius, Rs, obtained by the Guinier approximation of SAXS profiles. In the initial stage with low scattering intensity, Rs is affected by the SAXS intensity originated from quenched-in fluctuation observed even for as melt-spun sample shown in Fig. 2. In the later stage, however, it is noted that the slope is larger than one half, suggesting that the temporal power law for the growth kinetics is larger than the one expected by the diffusion limited growth[16]. This point is an interesting feature of the quasicrystal growth found in the present experiment, since in most of precipitation processes in crystalline alloys, the temporal power law is one half, or sometimes smaller owing to competitive growth/coarsening or slowing down due to strain effect.

The growth kinetics, i.e., the temporal evolution of the size of quasicrystal is important information as discussed above. In the present measurements, two other parameters on the size of precipitates can be calculated. One is the radius calculated from the Porod regions[17-19], and the other is the one from

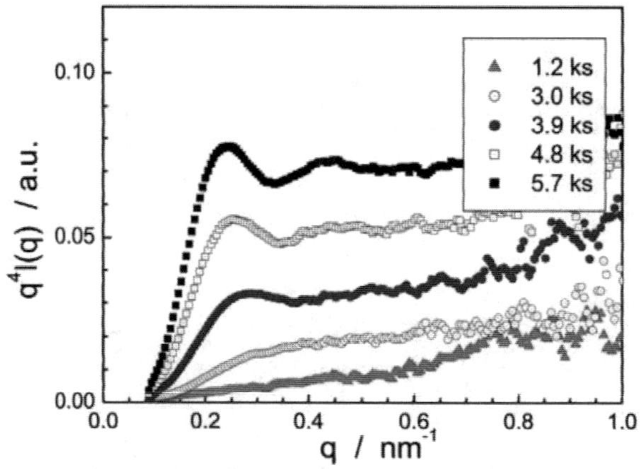

Figure 5. Change of $q^4I(q)$ as a function of q.

the Bragg peak[20]. Assuming that the precipitates are spherical, SAXS intensity from monodisperse size distribution with size R gives a well-defined oscillation with a period that converges to qR= π as q goes to infinity when $q^4I(q)$ is plotted against q. With a broader size distribution, the oscillation of the function, $q^4I(q)$ becomes smaller in amplitude, and the degree of such oscillation is used for the calculation of size distribution from measured SAXS intensity[19]. In other words, such function is a measure of the sharpness of size distribution[21] and also can be used to assess the radius R from the oscillation. Figure 5 gives the function $q^4I(q)$ plotted for several annealing time. It is seen that well-defined first peak is observed in the late stage of annealing. The radius obtained for the first peak of the plot is listed in Table 1. Another size parameter is obtained from the Scherrer analysis of the diffraction peak at q~5.8 nm^{-1}. To obtain the radius of quasicrystallites from diffraction peak, a Gaussian fit of the peak with nonlinear least square fitting was made, and then the size was evaluated by a Guinier approximation. Since the glass matrix is expected in the supercooled fluid state during annealing, we may neglect the effect of strain broadening in the present analysis. The size parameters are summarized in Table 1.

Table 1. Size parameters obtained from the Guinier approximation (Rs), at Porod regions (R_p), and the Scherrer analysis (R_{Bragg})

time / ks	R_s /nm	R_p/ nm	R_{Bragg}/ nm
1.2	(6.6)	NA	NA
2.1	7.2	NA	NA
3.0	9.2	NA	3.3
3.9	10.9	9.5	6.8
4.8	12.0	10.8	9.0
5.7	12.8	11.2	9.4

In the present analysis, one important point to consider on the diffraction profiles is that these profiles are nearly ideal powder patterns. This is confirmed by the fact that the two-dimensional intensity profiles on the Image Plate give smooth ring patterns for the Bragg peak. Then, we need to treat the diffraction profile as a smeared pattern, since the diffraction profile of a quasicrystallite is smeared by the contribution from small quasicrystals whose Bragg spot is located in the vicinity of that of the quasicrystallite. Figure 6 gives a schematic illustration of a smearing effect at a Bragg position. If only one nanocrystal or a nanoQC contributes to the Bragg spot, then it gives a single crystal pattern. When the effect of coherent strain is negligible, a Bragg profile from a single crystal simply reflects the shape factor of the crystal which is essentially identical to that for SAS, multiplied by the structure factor of the nanocrystal or of the nanoQC. Therefore, an order spot profile of spherical $L1_2$ precipitates

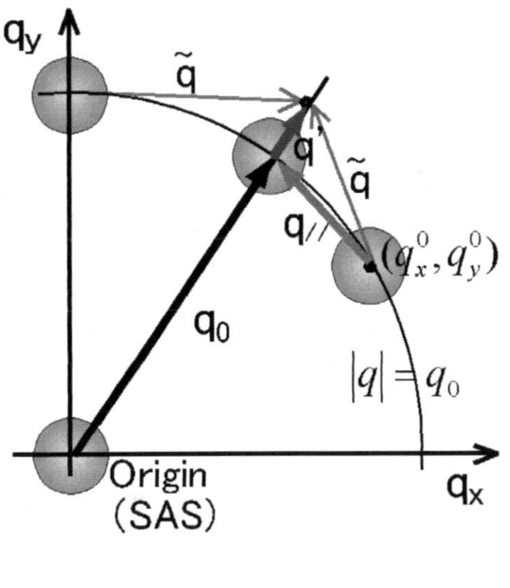

Figure 6. Schematic illustration of the effect of powder pattern on the Bragg profile. The profile in the radial direction is explained by the sum of the contribution from nanoQCs that form the same Debye ring.

in a fcc matrix can be treated in the same manner as that for SAS profile except the interparticle interference function[22]. On the other hand, if the nanoQCs are randomly oriented, a Bragg peak profile across the Debye ring as shown in Fig. 6 is a sum of intensities from nanoQCs contributing to the relevant Debye ring. The Bragg peak profile is then given as

$$I(q') = \int_{|q|=q_0} F^2(\vec{q}'+\vec{q}_{//})P(\vec{q}_{//})dq^2 \qquad (1)$$

where $q_{//}$ is the relative distance between the measuring Bragg point and that for other contributing ones shown in Fig.6 and P(q) is an orientation distribution function. Since the Debye ring was uniform and smooth in the present experiment, P(q) can be regarded as a constant. Then eq. (1) is an integration over a sphere of the Debye ring. This equation is equivalent to a smearing calculation for infinite two-dimensional source size as a formality, if the curvature of the Debye ring is negligible compared with the FWHM of the shape factor, i.e., SAS. This is an acceptable approximation in the present experiment, where SAS intensity appears at $q<<1$ nm^{-1} and the Bragg spot appears at q=5.8 nm^{-1}. When the area of integration in eq. (1) is replaced from a sphere with $|q|=q_0$ by a plane with $q_z=q_0$, a simple moment calculation gives that the radius obtained from diffraction by Guinier approximation has a statistical meaning of ;

$$R_{Bragg}{}^2 = \langle R^6 \rangle / \langle R^4 \rangle$$

and the Porod's law of q^{-4} is changed to q^{-2}. The final slope of q^{-2} for the order spot profile of well-defined $L1_2$ precipitates having powder (polycrystalline) microstructure has been observed in annealed NiAlCr ternary alloys[23]. Considering that Rs is given by

$$R_s{}^2 = \langle R^8 \rangle / \langle R^6 \rangle$$

and R_p by

$$R_p = \langle R^3 \rangle / \langle R^2 \rangle$$

the average size should satisfy the relationship

$$R_S > R_{Bragg} > R_p$$

regardless of the shape of the size distribution function. Table 1 shows that this relationship does not hold even for the late stage where well-defined q^{-4} power law is observed for SAXS. The present results suggest that there is a slightly disordered lattice shell outside the quasicrystal precipitates, which may also modify the growth law of the precipitated from a simple diffusion-limited growth. From SAXS results, on the other hand, two phase model is applicable, suggesting that this slightly disordered shell should have the identical composition as that of the inner core, since the oscillation periodicity in Fig. 5 should be otherwise destroyed.

4. Conclusions

Time-resolved measurements of quasicrystal formation from metallic glass ribbons were made for $Zr_{70}Cu_{25}Pt_5$ ternary alloy. The temporal evolutions of the intensity and the radius were examined about composition heterogeneity and the quasicrystalline structure by using SAS and a diffraction spot. Present analysis revealed that the growth kinetic is different from ordinary diffusion-limited growth, and it may stem from the microstructure characteristics of the precipitates, i.e., less lattice-ordered structure of the shell region. Further examination of the structure by using contrast variation with anomalous scattering at the Zr K edge is now in progress.

references

[1] Inoue A, Zhang T and Masumoto T, *Mater. Trans.* 31(1990)17.
[2] Inoue A, *Acta Mater.* 48(2000)279.

[3] Perker A. And Johnson W J, *Appl. Phys. Lett.*, 63(1993)2342.
[4] Okuda,H, Murase I, Kurosaki R, Nakagawa E, Ochiai S, Yokoyama Y, Saida J, and Inoue K. *Intermetalics,* 14(2006)1038.
[5] Saida J, Matsushita J and Inoue A, *J. Alloys Compd.*, **342**(2000) 18.
[6] Saida J. Sanada T, Sato S , Imafuku M and Inoue A, *Appl. Phys. Lett.*, **91** (2007) 11901.
[7] Murty B S, Ping D H, Ohnuma M and Hono K, *Acta Mater.* **49**(2001)3453.
[8] Murty S, Ping D H Hono K and Inoue A, *Scr. Mater.* **43**(2000)103.
[9] Okuda H, Fukumoto T, Saida J Ochiai S, Sasaki S and Masunaga H, *J. Appl. Cryst.* **41**(2008)675.
[10] Bauer R and Gerold V, *Acta Metall.*, **10**(1962) 637.
[11] Cocco G, Fagherazzi G and Schiffini L, *J. Appl. Cryst.*, **10**(1977)325.
[12] Beddoe R E, Gerstenberg K W and Gerold V, *Decomposition of Alloys, the early stages*, Pergamon Press, Oxford,1984, p110.
[13] Gunton J D, San Miguel M and Sahni P S, *Phase Transition and Critical Phenomena*, vol.8 ed. Domb C and Lebowitz J L, Academic Press, NY,1983 p267.
[14] Hennion M, Ronzaud D and Guyot P, *ActaMetall* **30** (1982)599.
[15] Binder K, Kalos, K and Lebowitz J L, *Adv. Coll. Interf. Sci.*, **10** (1979)173.
[16] Doherty,R. D. *Physical Metallurgy*, 2nd ed., ed. Cahn R. and Haasen, P. Chapt. 14, p966.
[17] Porod G, *Kolloid -Z.*, **124**(1951)83
[18] Feign L A and Svergun D., *Structure Analysis by Small-Angle X-ray and Neutron Scattering*, Plenum, NY,(1987) p230.
[19] Lechter J H and Schmidt P W, *J. Appl. Phys.* **37**(1966),649.
[20] Culity B D, *Elements of X-ray diffraction*, Addison & Wesley,Reading (1978) p102.
[21] Okuda H, Osamura K and Amemiya Y, *Trans. JIM* **34**(1993) 758.
[22] Okuda H, Tanaka M, Osamura K, and Amemiya Y. *J. Appl. Crystallogr.* **30**(1997)592.
[23] Bley, F, Livet F., Okuda H. and Simon J.P *J. Appl. Crystallogr.* **30**(1997)914.

Journal of Physics: Conference Series **247** (2010) 012038 doi:10.1088/1742-6596/247/1/012038

In situ small-angle neutron scattering study of the early stages of precipitation in Fe-25at% Co-9at% Mo and Fe-1at% Cu at 500 °C

P Staron[1], E Eidenberger[2], M Schober[2], M Sharp[1], H Leitner[3], A Schreyer[1], H Clemens[2]

[1]Institute of Materials Research, GKSS Research Centre, Max Planck-Str. 1, 21502 Geesthacht, Germany
[2]Department of Physical Metallurgy and Materials Testing, University of Leoben, Franz Josef-Str. 18, 8700 Leoben, Austria
[3]Christian Doppler Laboratory 'Early Stages of Precipitation', University of Leoben, Franz Josef-Str. 18, 8700 Leoben, Austria

E-mail: Peter.Staron@gkss.de

Abstract. The early stages of precipitate formation in Fe-25at% Co-9at% Mo at 500 °C was investigated by *in situ* small-angle neutron scattering (SANS). The ratio of magnetic to nuclear SANS intensity, which depends on the precipitate composition, and the precipitate sizes were analyzed. The results suggest that in the Fe-Co-Mo alloy an initial stage of precipitate formation with strong compositional changes is completed after 150 s, when the full magnetic signal is established. For comparison, the well-known nucleation and growth reaction in Fe-1at% Cu was studied. Differences in the growth rate of the mean size of the developing structures could be distinguished in the two investigated alloys. While in the Fe-Co-Mo alloy the precipitate structures grow with $t^{0.15}$ to $t^{0.21}$, the mean precipitate radius in the Fe-Cu alloy grows with $t^{0.40}$ in the observed time interval.

1. Introduction

The ternary system Fe-Co-Mo has been investigated since the 1930s, where Köster and Tonn first mentioned a possible strengthening mechanism due to precipitation of intermetallic phases from a strongly supersaturated matrix [1]. Today, it is known that hardness values over 68 HRC can be achieved in this material. These alloys are promising technical materials because they also possess a high thermal microstructural stability in the hardened state [2].

The precipitates that are responsible for this hardening consist of Fe, Co, and Mo and have sizes of a few nanometers. According to literature, formation of the so-called μ-phase of the composition $(Fe,Co)_7Mo_6$ is accountable for the hardening in ternary Fe-Co-Mo alloys. This phase can be considered as a solid solution of the two intermetallic phases Fe_7Mo_6 and Co_7Mo_6 found in the binary systems Fe-Mo and Co-Mo [3, 4]. Both binary phases exhibit an ordered rhombohedral structure with almost identical lattice parameters [4]. While results on mechanical properties of Fe-Co-Mo alloys like hardness, toughness, or tensile strength can be found in the literature [1, 2], little is known about the formation mechanism of the hardening phases in these alloys (i.e. μ-phase or possible metastable precursor phases).

In order to contribute to the clarification of the precipitate formation mechanism of the μ-phase, non-isothermal experiments were conducted by Leitner et al. on a Fe-25at% Co-9at% Mo alloy, employing differential scanning calorimetry and 3D atom probe microscopy (3DAP) [5]. It was shown that the formation of the μ-phase takes place via clustering of Mo in the matrix. The peak hardness of the alloy was reached after solution annealing, followed by quenching to room temperature, and subsequent continuous heating up to about 615 °C. However, the interpretation of the 3DAP results obtained by Leitner et al. [5], especially with regard to precipitate size and volume fraction, was difficult because of the very small analyzed volumes of $7 \times 7 \times 80$ nm^3. The small cross-section of the analyzed volume leads to cutting off larger particles, influencing e.g. the mean particle size.

In contrast to microscopic methods, small-angle neutron scattering (SANS) yields information from a relatively large volume and thus gives mean values over a very large number of precipitates, which makes it very sensitive to small changes in the size and volume fraction of precipitates. In the present study, SANS was used for the first time to observe the early stages of precipitate formation in Fe-25at% Co-9at% Mo at 500 °C in an *in situ* experiment. An Fe-1at% Cu alloy was studied for comparison because it is a simple binary system, in which Cu precipitates are formed by nucleation and growth.

2. Experimental procedure

The exact compositions of the studied samples were Fe-25.1at% Co-9.3at% Mo-0.4at% Si-0.1at% C and Fe-1.0at% Cu. A new *in situ* furnace for ferromagnetic samples was used for the first time for this study, allowing a heating rate of 5 K/s. The samples with a diameter of 25 mm and a thickness of 1.8–2.0 mm were heated in the furnace that was placed between the pole shoes of an electromagnet so that the ferromagnetic samples could be magnetized to saturation during the measurements. The furnace had two ceramic heating elements with direct contact to the sample. The heating elements were in a vacuum chamber with water-cooled outer walls and 1 mm thick Al windows for the neutron beam. A heat shield was not used at the moderate temperature of 500 °C for avoiding additional background. Heating up to a nominal temperature of 500 °C could be performed within about 100 s (figure 1). Temperature calibration was done in a separate run for each material with two thermocouples placed in holes drilled into the illuminated sample volume. The mean temperature within the illuminated volume was 493 °C for the Fe-Co-Mo sample and 495 °C for the Fe-Cu sample. The difference between the two materials is expected to result from different thermal conductivities.

The SANS measurements were carried out at the SANS-2 instrument of GKSS Research Centre, Geesthacht, Germany. Selector-monochromated neutrons with a mean wavelength of 0.58 nm and a wavelength spread of 10% were used. The distance between sample and detector was 1 m. The used area detector had a size of 500 mm × 500 mm and a pixel size of 2.2 mm × 2.2 mm. The data acquisition time at the beginning of the reaction was 30 s. After 30 minutes it was increased to 60 s, after additional 60 minutes it was increased to 120 s. An empty image without a sample in the beam was subtracted from all detector images. Scattering curves were corrected for sample transmission.

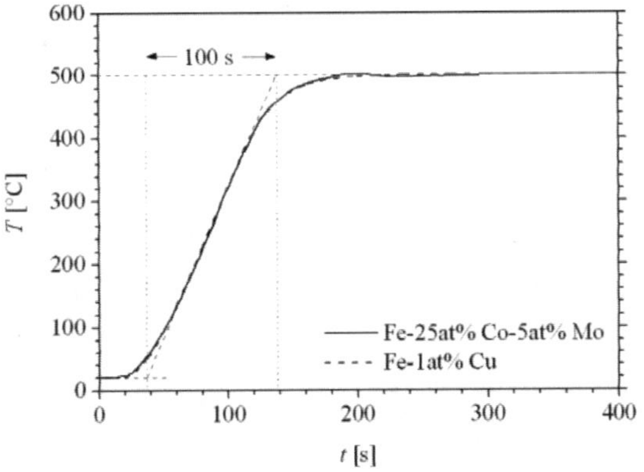

Figure 1. Temperatures measured directly at the two samples in two different runs.

Absolute cross sections were obtained by calibration with a Vanadium sample.

Details about the small-angle scattering technique can be found e.g. in [6, 7, 8]. When matrix and precipitates have different magnetizations there is a magnetic scattering contrast $\Delta\eta_m$ in addition to the nuclear scattering contrast $\Delta\eta_n$. η is the scattering length density, which is the mean coherent scattering length per volume. While nuclear scattering is isotropic, magnetic scattering depends on the angle α between scattering vector and magnetization. The total contrast can be written as the sum

$$(\Delta\eta)^2 = (\Delta\eta_n)^2 + \sin^2\alpha\,(\Delta\eta_m)^2 . \tag{1}$$

Scattering curves of the nuclear and magnetic cross sections were calculated as azimuthal mean values within sectors parallel and perpendicular to the external field. The sector size was increased as much as possible to improve the statistics without significant smearing. The smearing effect is large for $\alpha = 0°$ where large intensity gradients are present in azimuthal direction. Finally, 20° wide sectors were used.

3. Results

3.1. Scattering curves

Selected nuclear and magnetic scattering curves for both materials are shown in figures 2 and 3. Already after 66 s, a strong intensity increase is observed in the Fe-Co-Mo alloy (figure 2). A maximum is observed in these scattering curves, resulting from a high precipitate number density. In general, Fe-Co-Mo shows stronger scattering intensity than Fe-Cu. However, in the very early stages, both intensities are quite low; therefore, integrated intensities will be discussed in the following.

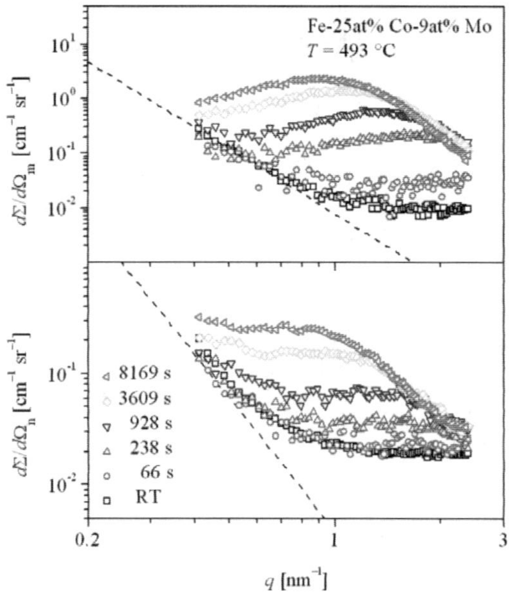

Figure 2. Magnetic (top) and nuclear (bottom) scattering curves of Fe-25at% Co-9at% Mo at RT and after various aging times at 493 °C. The dashed line marks the q^{-4}-slope of contributions from larger inhomogeneities.

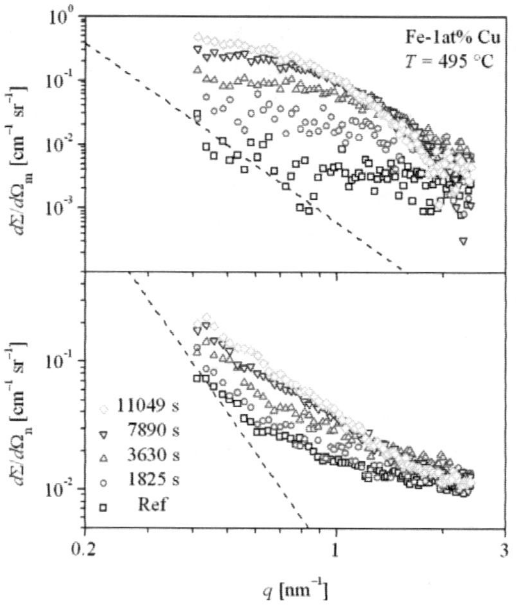

Figure 3. Magnetic (top) and nuclear (bottom) scattering curves of Fe-1at% Cu during incubation period (reference) and after various aging times at 495 °C. The dashed line marks the q^{-4}-slope of contributions from larger inhomogeneities.

3.2. Integrated intensity

The integrated small-angle scattering intensity is defined as [6]

$$Q = 4\pi \int_0^\infty \frac{d\Sigma}{d\Omega}(q) q^2 dq = (2\pi)^3 f(1-f)(\Delta\eta)^2 \tag{2}$$

and is a measure for the product of particle volume fraction f and square of scattering contrast $\Delta\eta$. The baseline of Q, given by its value before precipitation starts, was subtracted for both alloys since only the contribution of newly formed precipitates was of interest. It is thus assumed that the scattering contributions to the baseline do not change significantly during decomposition.

The ratio R of integrated intensities calculated from magnetic and nuclear scattering curves,

$$R = \frac{Q_m}{Q_n} = \frac{(\Delta\eta_m)^2}{(\Delta\eta_n)^2}, \tag{3}$$

reflects the ratio of scattering length density differences squared, $(\Delta\eta_m)^2$ and $(\Delta\eta_n)^2$, as long as both scattering contributions originate from the same structure with the same volume fraction f. In general, $\Delta\eta_m$ is given by the mean magnetic moments per atom in matrix and precipitate. It does not change with the chemical composition of the precipitates as long as they are not ferromagnetic. $\Delta\eta_n$, on the other hand, changes with chemical composition. Thus, R is an indicator for chemical changes in the precipitate composition leading to changes in $\Delta\eta_n$ and also $\Delta\eta_m$. The hardening precipitates in the Fe-Co-Mo alloy are expected to be non-ferromagnetic with an approximate composition in later decomposition stages of 35at% Fe-41at% Mo-18at% Co-6at% Si, which was observed in a recent 3DAP study in a sample that was heated up to 615 °C [9]. In the case of ferromagnetic precipitates, the quantitative interpretation of R is more complicated because changes in the mean magnetic moment per atom also lead to changes in R.

Integrated magnetic and nuclear intensities as well as the ratio R is shown for both alloys in figures 4 and 5. Figure 6 shows a comparison of Q_m and measured hardness (HV10) for both alloys.

3.3. Mean precipitate size

The interference maximum in the scattering curves of the Fe-Co-Mo alloy results from the combination of precipitate volume fraction, precipitate size and morphology, and concentration gradients between precipitate and matrix phase. While the precipitates are growing, the maximum

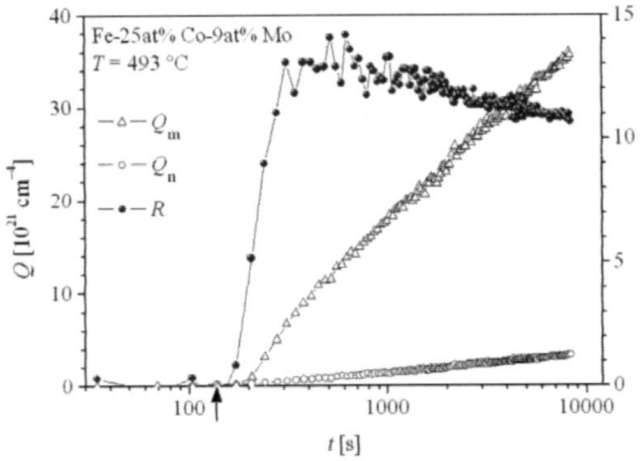

Figure 4. Integrated SANS intensities Q_m and Q_n after baseline subtraction and ratio $R = Q_m/Q_n$ for Fe-25at% Co-9at% Mo. The arrow marks the time when 493 °C are almost reached.

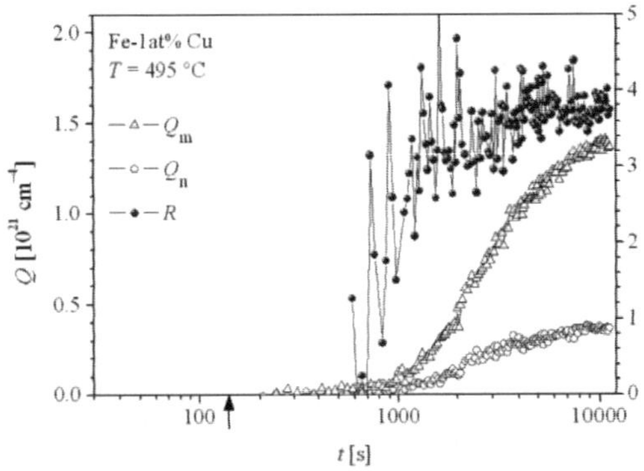

Figure 5. Integrated SANS intensities Q_m and Q_n after baseline subtraction and ratio $R = Q_m/Q_n$ for Fe-1at% Cu. The arrow marks the time when 495 °C are almost reached.

shifts to smaller q. The inverse of q_{max} can be used as a measure that is proportional to the precipitate size [10]. Even if there were no particles with a sharp interface, but rather an interconnected structure of two phases [9], the inverse of q_{max} still describes a characteristic length of the system. The maximum of each scattering curve was fitted by a Gaussian curve and q_{max}^{-1} plotted as a function of ageing time (figure 7).

In the case of Fe-Cu, the Guinier approximation was used for determining the Guinier radius R_G [7]. In a plot of $\ln(d\Sigma/d\Omega)$ over q^2, there is a linear part for $q \cdot R_G \leq 2$. As in the calculation of Q, the scattering curve of the reference state before precipitation starts was subtracted for the Guinier plot. From the slope s of the linear part, R_G is calculated as $\sqrt{-3s}$. The Guinier radius is a mean size of the scattering particles. In the case of spherical particles, the sphere radius is $R_S = \sqrt{5/3}\,R_G$. R_S is plotted as a function of ageing time (figure 7).

4. Discussion

In the Fe-Co-Mo system, Q was found to increase from the very beginning of heating (figure 4). After a temperature of 493 °C is reached, the increase of R from 0 to 13 takes 150 s. During this period, Q_m has a higher slope than in later stages, while Q_n does not show this change in slope. This means that the magnetic contrast is building up during the first 150 s. As was observed with the atom probe [5], precipitate formation starts with a simultaneous enrichment of Mo and depletion of Fe and Co. With the exchange of these elements, the magnetic moment per atom decreases in the precipitate region and $\Delta\eta_m$ increases. After 150 s $\Delta\eta_m$ reaches a plateau. Thus, from this data it is concluded that there are strong compositional changes in the newly formed precipitates during the first 150 s. In the following, R decreases slowly due to smaller changes that can occur in the matrix as well as in the precipitate composition while Q still increases.

The initial behavior of the Fe-Cu system is different: An incubation period of about 800 s is observed after the aging temperature of 495 °C is reached, during which the intensity is close to zero (figure 5). This behavior is typical for the nucleation mechanism where a nucleus of critical size that is stable and can grow must be formed by statistical diffusion processes. One must keep in mind,

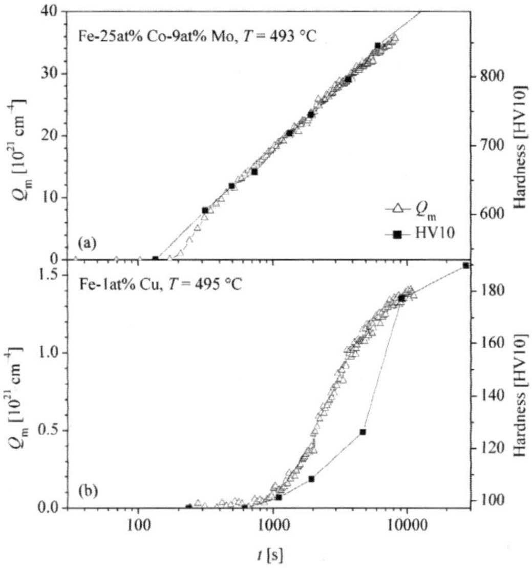

Figure 6. Comparison of magnetic integrated intensity Q_m and hardness for a) Fe-25at% Co-9at% Mo and b) Fe-1at% Cu.

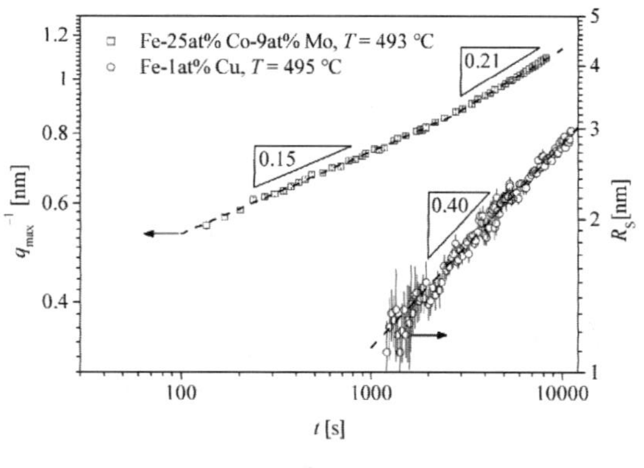

Figure 7. Reciprocal of the position of the maximum in the scattering curves for Fe-25at% Co-9at% Mo and sphere radius determined from the Guinier approximation for Fe-1at% Cu as a function of ageing time.

however, that there is a detection limit for small numbers of very small particles; the true incubation time could thus be smaller than 800 s. At the end of the incubation time R increases and reaches a value of about 3.2 after 1300 s. It continues to increase slower up to about 3.8 after 10000 s (figure 5). The initial relatively fast increase of R fits to the picture of classical nucleation, in which overcritical clusters already have the final equilibrium concentration; in the Fe-Cu case from the equilibrium phase diagram it is usually expected to be almost pure Cu. However, the increase in R afterwards hints to a further change in precipitate composition. In addition, the apparent maximum value of R of about 3.8 is significantly smaller than expected for pure non-ferromagnetic Cu precipitates in Fe ($R \approx 12$). Recent studies have suggested that the Cu precipitates can reach their equilibrium composition only slowly [11], and can contain significant amounts of Fe. It is, however, not yet fully clear if this is an explanation for the observed R-behavior. In general, due to the definition of the scattering length density η changes in R can be due to changes either in the composition or in the mean volume per atom, but also due to changes in the mean magnetic moment per atom in matrix and precipitate. As a consequence, the small precipitates in Fe-Cu also would have to be tested for a significant magnetic moment.

Comparing the results for both materials, they seem to behave similar as far as the initial compositional changes in the precipitates are concerned. Also in the Fe-Co-Mo system, the compositional changes producing precipitates are very fast, resembling classical nucleation. Nevertheless, the actual decomposition mechanism in this highly supersaturated system is expected to deviate significantly from classical nucleation. Such deviations in highly supersaturated systems were already observed in Ni-Al alloys [12, 13]. For Ni-Al, the decomposition could be described with a clusterdynamic decomposition model taking into account not only high initial supersaturations, but also small nucleation energies. It was suggested that the character of the decomposition reaction changed from classical nucleation and growth to a nucleation-free diffusion-controlled reaction with vanishing nucleation energy at increasing supersaturations.

In the literature, the formation of precipitates in Fe-Co-Mo alloys was sometimes described to proceed via spinodal decomposition [14]. From the present results it is concluded that, should spinodal decomposition play a role in the system studied here at 500 °C, it is most likely during the first 150 s where the strong increase of R indicates strong compositional changes. After 150 s the precipitates should be close to their final composition and only smaller compositional changes seem to occur as indicated by the small changes in R. The substantiation of this conclusion will be subject of further work involving TEM and 3DAP investigations.

A comparison of integrated intensity Q_m and hardness (HV10) of both materials also shows that there are significant differences in the two precipitation reactions (figure 6). While integrated intensity and hardness show similar behavior for each material, the shape of the corresponding curves for both materials is different. In both materials maximum hardness as well as maximum precipitate volume fraction was not reached within the observed period. The observed differences as well as precipitate volume fractions will be addressed in a forthcoming study.

Another difference in the decomposition reaction is observed in the mean precipitate size (figure 7). In the Fe-Co-Mo alloy q_{max}^{-1} evolves with $t^{0.15}$ up to 2700 s, followed by a faster increase with $t^{0.21}$. In the Fe-Cu alloy R_S grows with $t^{0.40}$ in the observed time interval, which is somewhere between the expected behaviour for particle growth ($t^{1/2}$) and coarsening ($t^{1/3}$). Since the maximum volume fraction was not reached in the observed time interval, both alloys were well before the coarsening regime. The very small growth exponent of the Fe-Co-Mo alloy and its increase could indicate quite early decomposition stages. Although similar exponents had also been observed in the context of spinodal decomposition [15, 16], they do not seem to be evidence for spinodal decomposition in this case for the reason discussed above. On the other hand, the much larger growth exponent close to 0.5 of the Fe-Cu alloy could indicate the growth phase of the classical nucleation and growth type of decomposition [17]. A full analysis of the data will become possible with numerical decomposition models.

5. Conclusions and Outlook

In the present study the early stages of decomposition in Fe-Co-Mo at 493 °C were investigated for the first time using *in situ* small-angle neutron scattering. From a comparison of nuclear and magnetic small-angle scattering intensity an initial stage in the formation of hardening precipitates was identified, which is completed after about 150 s. In this stage strong compositional changes occur, while further compositional changes afterwards are much smaller. From this result it was concluded that, if spinodal decomposition took place in the decomposition reaction, it is most likely in this initial stage. After the initial stage, growth exponents of 0.15 and 0.21 were observed for the inverse of the position of the maximum in the scattering curve.

Fe-1at% Cu was used as a reference system, in which the nucleation and growth of Cu precipitates at 495 °C after an incubation time of 800 s could be observed. A growth exponent of the mean precipitate radius of 0.40 was observed, indicating a strong growth of stable nuclei.

Additional *in situ* SANS experiments at different temperatures will help to understand the early decomposition stages in the investigated materials. Small-angle X-ray scattering experiments at a synchrotron source are also planned [18], which can give much better time resolution through the high intensity of the source. Moreover, 3DAP and TEM studies of both materials are currently being performed. The combination of these experimental techniques with suitable decomposition models is expected to lead to a comprehensive description of the precipitation reactions in the investigated materials. Especially the question if and when spinodal decomposition takes place in the system Fe-Co-Mo has to be addressed.

Acknowledgement
G. Kozik from GKSS is gratefully acknowledged for designing and providing the *in situ* furnace and for technical assistance during the SANS measurements.

References

[1] Köster W, Tonn W 1932 Das System Eisen-Kobalt-Molybdän. *Arch Eisenhüttenwes* **5** 627–630
[2] Geller J A 1983 *Instrumentalniye Staly* (Moscow: Metallurgia)
[3] Houdremont E 1956 *Handbuch der Sonderstahlkunde*, vol 2 (Düsseldorf: Verlag Stahleisen)
[4] Hansen M 1958 *Constitution of binary alloys* (New York: McGraw-Hill)
[5] Leitner H, Schober M, Clemens H, Caliskanoglu D, Danoix F 2008 Precipitation behavior of an Fe-Co-Mo-alloy during non-isothermal ageing *Int J Mat Res* **99** 367–374
[6] Kostorz G 1979 Small-angle scattering and its applications to materials science *Treatise on Materials Science and Technology – Neutron Scattering* vol 15 ed G Kostorz, H Herman (New York: Academic Press) 227–289
[7] Guinier A, Fournet G 1955 *Small-angle scattering of X-rays* (New York: Wiley, London: Chapman&Hall)
[8] Pranzas K 2008 *Neutrons and synchrotron radiation in engineering materials science* ed W Reimers, A R Pyzalla, A Schreyer, H Clemens (Weinheim: Wiley-VCH)
[9] Eidenberger E, Stergar E, Leitner H, Scheu C, Staron P, Clemens H 2008 Precipitates in a Fe-Co-Mo alloy characterized by complementary methods *Berg- und Hüttenmännische Monatshefte (BHM)* **153** 247–252
[10] Fratzl P, Lebowitz J L, Penrose O, Amar J 1991 Scaling function, self-similarity, and the morphology of phase-separating systems *Phys Rev* B **44** 4794–4811
[11] Nagano T, Enomoto M 2006 Simulation of the growth of copper critical nucleus in dilute bcc Fe-Cu alloys *Scripta Mater* **55** 223–226
[12] Staron P, Kampmann R 2000 Early-stage decomposition kinetics in Ni-Al alloys: I. Small and wide-angle neutron scattering investigation on Ni-13 at% Al and clusterdynamic modeling. *Acta Mater* **48** 701–712
[13] Staron P, Kampmann R 2000 Early-stage decomposition kinetics in Ni-Al alloys: II. clusterdynamic modeling of former experimental results *Acta Mater* **48** 713–720

[14] Doi M, Tanabe H, Miyazaki T 1987 Development of ultra high strength steels based on spinodal decomposition *J Mater Sci* **22** 1328–1334

[15] Ujihara T, Osamura K 2000 Kinetic analysis of spinodal decomposition process in Fe-Cr alloys by small-angle neutron scattering *Acta Mater* **48** 1629–1637

[16] Binder K, Stauffer D 1974 Theory for the slowing down of the relaxation and spinodal decomposition of binary mixtures *Phys Rev Lett* **33** 1006–1009

[17] Wagner R, Kampmann R 1991 *Homogeneous second phase precipitation Materials Science and Technology: A Comprehensive Treatment* vol 5 ed R W Cahn, P Haasen, E J Kramer (Weinheim: Wiley-VCH) 213–303

[18] Zickler G A, Eidenberger E, Leitner H, Stergar E, Clemens H, Staron P, Lippmann T, Schreyer A 2008 In-situ small-angle X-ray scattering study of the precipitation behavior in a Fe-25at.% Co-9at.% Mo alloy *Mater Charact* **59** 1809–1813

Microstructure of creep-exposed single crystal nickel base superalloy CSMX4

P Strunz[1,2], J Zrník[3,4], A Epishin[5], T Link[6], S Balog[7]

[1] Nuclear Physics Institute, CZ-25068 Řež near Prague, Czech Republic
[2] Research Centre Řež, CZ-25068 Řež near Prague, Czech Republic
[3] Technical University of Košice, Deptartment of Materials Science, Slovakia
[4] COMTES FHT Inc., CZ-33441 Dobřany, Czech Republic
[5] Federal Institute for Materials Research and Testing, Unter den Eichen 87, 12205 Berlin, Germany
[6] Technical University Berlin,10623 Berlin, Germany
[7] Laboratory for Neutron Scattering, PSI & ETH Zürich, CH-5232 Villigen, Switzerland
E-mail: strunz@ujf.cas.cz

Abstract. Main changes which appear in microstructure of Ni base superalloys during creep exposure is the directional coarsening (rafting) of originally cubic γ' precipitates embedded in γ matrix. The progress in γ' degradation can be assessed with help of the analysis of geometrical parameters of the γ' microstructure. Determination of the morphological changes of γ' in creep-exposed single-crystal Ni-base superalloy CMSX4 using SANS was carried out. A new sample design (conic specimen) was tested in order to get continually changing stress levels in one sample and corresponding variety of microstructural change. Evolution of rafts with stress magnitude is clearly observable from the parameters obtained using SANS (the mean thickness of the γ channel in different crystallographic directions of the single-crystal, the evolution of the γ' precipitate size and distance, the evolution of the specific interface). The use of conic specimen and spatial scan facilitate determination of microstructure evolution in dependence on the exposure parameters.

1. Introduction

Turbine blades in gas turbines, fabricated of Ni base superalloys, operate under the creep and creep-fatigue conditions. One of the main changes which appear in microstructure during creep exposure is the directional coarsening (rafting) of originally cubic γ' precipitates embedded in γ matrix. For residual lifetime estimation, it is desirable to have a non-destructive method which could reliably evaluate the progress in γ' degradation in dependence on exposure time and relate these changes to the magnitude of stress the blade is experiencing. The available literature indicates that necessary information can be obtained from the analysis of geometrical parameters of the γ' microstructure and γ/γ' misfit [1, 2]. Using the bulk sensitive small-angle neutron scattering (SANS), morphological changes of γ' phase resulting from the operation condition can be detected. SANS method [3] has proved to contribute successfully to the morphology assessment of γ' in nickel base superalloys [4-8].

The aim of this investigation was to evaluate the morphological changes of γ' in creep-exposed single-crystal Ni-base superalloy CMSX4 and relate them to the various applied stress levels. The secondary aim of the experiment was to test a possibility to investigate the microstructural parameters using a new special sample design: a conic specimen. As a result of the diameter change along the longitudinal axis, a continuous change of applied stress magnitude acting during the creep test exists which causes different microstructural change at different positions along the specimen.

2. Experimental

Currently, the most widely used material for high pressure propulsion turbines is single-crystal Ni-base superalloy CMSX-4 (the representative of the 2nd generation of single-crystal nickel base superalloys) with increased content of Re which strengthens the γ' precipitates and retards their coarsening [9]. The alloy underwent the following heat treatment: solutioning at 1581K/18h/AC (gas-fan quenched) + 1413K/6h/AC + 1145K/24h/AC, which led to large volume fraction (≈ 70%) of cubic γ'-precipitates.

Then, the conic specimen (Fig. 1, left) of CMSX-4 was exposed to a maximum stress 135 MPa for 100h under uniaxial tensile creep load along [001] direction (the temperature during the creep was held constant at 1100°C). The conic shape of the specimen enabled to receive various stress levels along the specimen axis (Fig. 1, right). The resulting stress was calculated from the known load and sample geometry. Afterwards, a slice was cut form the specimen along the vertical axis (i.e. [001]) in order to produce a sample of an appropriate thickness (≈2.5 mm) for the ex-situ SANS experiment. Thus, the spatial scan of the sample could depict microstructure for exposures between 35 and 135 MPa.

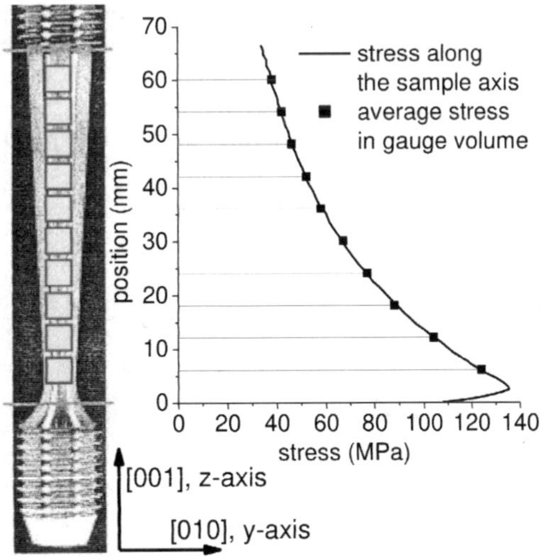

Figure 1. The tested specimen (left) exposed at 1100°C for 100h. The continuous change of applied stress was ensured by preparing the creep specimen in the form of a cone. Ten points scanned during the SANS experiment are marked. Stress dependence (σ_{min}=35MPa, σ_{max}=135MPa) on the position along the sample axis and average stress in the measured gauge volumes (right).

SANS measurement was carried out at the SANS II instrument of SINQ user lab of Paul Scherrer Institute, Villigen, Switzerland [10], which is equipped with a two-dimensional (2D) position sensitive detector. The samples were first oriented (ω-scan, tilt) in such a way that the [100] crystallographic direction was parallel to the beam and the [001] crystallographic direction was vertical.

The spatial adjustment into the beam was done using linear table. The sample (Fig. 1) was scanned in vertical direction (i.e. along [001]) at room temperature. The used small slit (height 5.2 mm, width 4.9 mm) ensured that the morphology gradient within the gauge volume (≈50 mm^3) is not excessively large.

The scattering data were collected at several geometries (the sample-to-detector distance varied from 1 m to 6 m, and the neutron wavelength λ from 4.55 Å to 19.6 Å). The covered range of the scattering vector **Q** was 1.2×10^{-3} Å^{-1} to 0.3 Å^{-1} (i.e. 1.2×10^{-2} nm^{-1} $< Q <$ 3 nm^{-1}), where the magnitude $Q = |\mathbf{Q}| = |\mathbf{k}-\mathbf{k}_0|$, $\mathbf{k}_0$ and $\mathbf{k}$ being the wave vectors of incident and scattered neutrons, respectively, and $|\mathbf{k}| = |\mathbf{k}_0| = 2\pi/\lambda$ due to the elastic scattering. The measured raw data were corrected for background scattering, sample volume and normalized to the macroscopic differential cross section [3]. The data measured at large Q-magnitudes indicated the presence of a second population of γ' precipitates of cuboidal shape with size about 10 nm. These were formed in the γ-phase channels on cooling from the high temperature. The most interesting- from the point of view of rafting - is, however, the measurement performed at very low Q-magnitudes as it brings out information on the rafting process of the original large cubic precipitates.

3. Results

The two of the obtained 2D scattering curves (for σ= 38 and 124 MPa) are displayed in Figs. 2a and 2b. They were taken at the opposite ends of the SANS sample (the most top and the most bottom scanned areas shown in Fig. 1). All the other scattering curves taken at the points in between these two extremes exhibit a continuous change of the γ' shape. Therefore, only two extreme cases are displayed. Nevertheless, all then scanned points were equally evaluated and provide thus dependencies of microstructural parameters on the stress level.

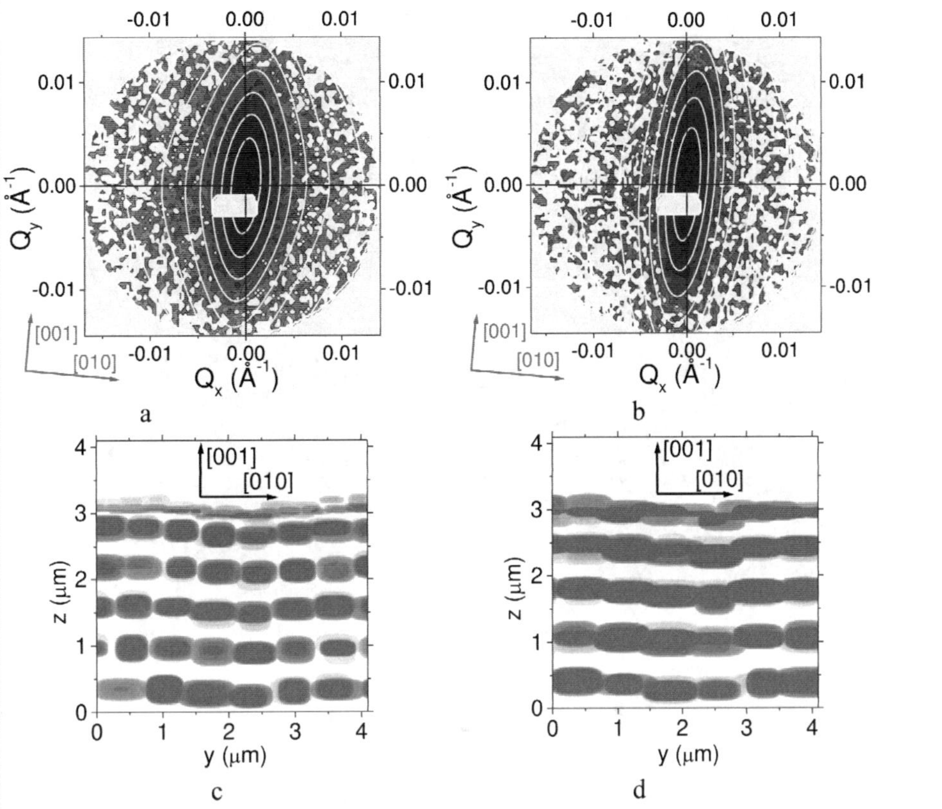

Figure 2. Scattering curves at two of the scanned points: (a) minimum and (b) maximum stress, respectively. Orientation: [100] parallel to the beam, [001] nearly vertical. Measured data are shown as the colour-scale maps whereas the optimum 2D fits are depicted by the white equiintensity lines. (c), (d): sections through the optimum 3D models (gray: γ' rafts; white: matrix) for the (a) and (b) fits, respectively.

The remaining part of the specimen, not used for SANS sample, was investigated by SEM in order to analyze microstructural change of γ'. SEM micrographs (presented in Fig. 3) indicate still partially non-rafted microstructure in the part of the bar exposed to the lowest stress, while the highest stress causes already advanced rafting.

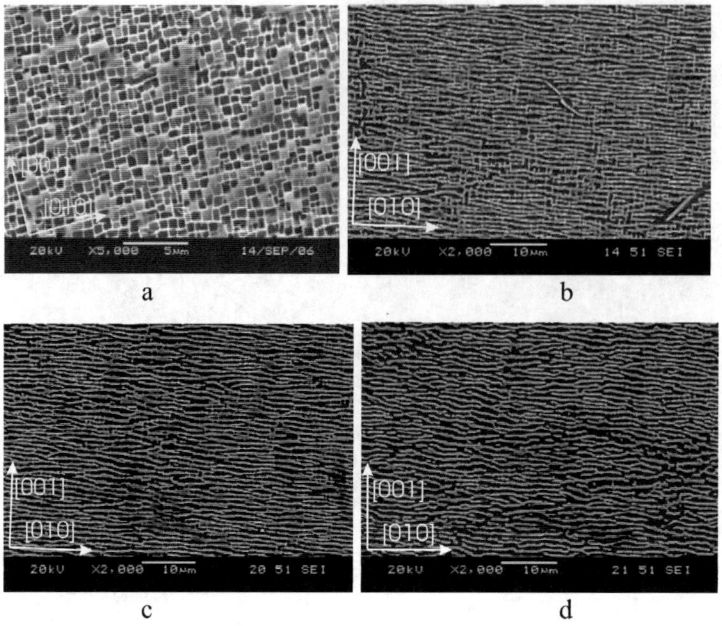

Figure 3. SEM micrographs indicating γ'-morphology changes: (a) original microstructure before exposure, (b) cut from the large-diameter part of the bar (the lowest stress), (c) central part of the specimen and (d) cut from the small-diameter part of the bar (the highest stress).

Qualitatively, the 2D scattering curves confirm the results of the electron microscopy performed on the specimen from the same bar.

4. Microstructural parameters evaluation

SANS results were used to evaluate the change of microstructural parameters of the γ' phase along the conic specimen. Collected scattering data were processed with a numerical modelling technique [11] based on the simulation of a scattering profile generated from a three-dimensional (3D) microstructural model of a particle system. Thus, the modelled scattering curve also covers the interparticle-interference effect. Moreover, multiple scattering corrections [12] are included as well in this procedure. The calculated 2D profile is then matched with the experimental curve in order to find microstructural parameters.

A model of (partially) agglomerated γ'-precipitates was employed for the modelling of rafting. Due to the large size of the precipitates in the order of μm, the measured data does not provide direct information on their size as well as on the distance between them. However, the information of the γ channel thickness between precipitates can be obtained as it is significantly smaller. Then, the relative change of size and distance of precipitates itself can be calculated from the model providing that the volume fraction is known. As the thermal history of the sample is identical at all the locations, an assumption was made that the volume fraction of large precipitates is the same for all the scanned points. The constant volume fraction 55% was used for the modelling. If correct, then the absolute distance and size of precipitates in the respective directions could be obtained. As we do not know the actual volume fraction, however, we can take the reported size and distance parameters evolution only as a relative measure of the real size and distance.

Fig. 2c and 2d display two models corresponding to the optimum fits to the SANS data for the two scanned points at opposite ends of the conic specimen.

The detailed evaluation of the data has brought a series of morphological parameters in dependence on the applied stress. In Fig. 4, the evolution of various parameters with the stress level is depicted. As can be seen from the evolution of the channel thickness (Fig. 4a), various extent of rafting of γ' precipitates embedded in the γ phase matrix was observed at variously exposed positions inside the sample. The tendency corresponds to the expected evolution: the larger stress, the more advanced rafting process. However, the used conditions for exposing the CMSX-4 single crystal do not allow to observe initial stages of rafting process, as can be seen from already large split of the thickness in x- and y- direction on one side and in z-direction on the other, even at the lowest applied stress level.

The relative change of the size and distance of precipitates can be observed in Fig. 4b. As the precipitate size in x- and y-directions is larger than the average distance between the precipitates in these directions (see overlapping precipitates in Fig. 2d), a clear occurrence of rafting can be deduced also from these parameters.

The last graph, Fig. 4c, displays the evolution of the specific interface between γ and γ' phase. The total specific interface as well as its projections to x, y and z-directions decrease with the stress level, as expected. The more pronounced is the decrease of the projection into x- and y-directions than into z-direction, as the former is caused by the disappearance of γ channels in [001] direction during the rafting process. Comparison between magnitudes in the respective directions confirms that the rafting advanced even for the lowest stress. The determined total specific interface between to be 37000 and 50000 cm^2/cm^3 can be used only as indication of the real absolute value as it depends on the input volume fraction as described above.

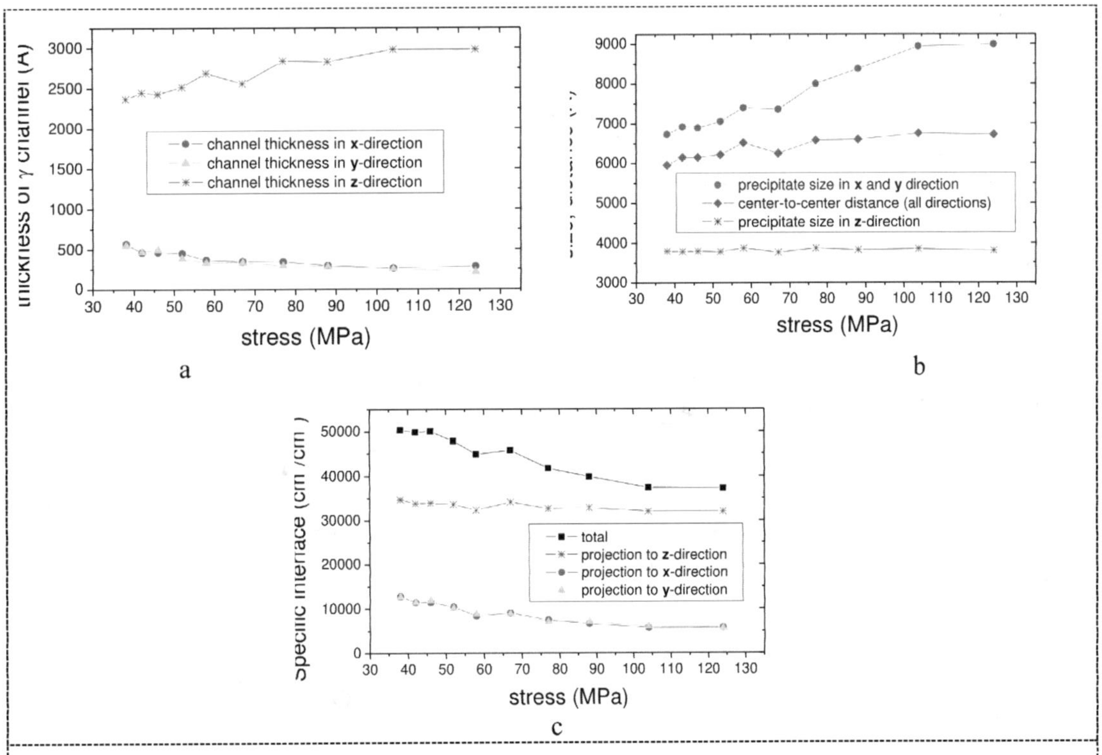

Figure 4. Resulting evolution of parameters. (a) the mean thickness of the γ' channel in the respective directions (x–direction is [100], y–direction is [010] and z–direction is [001]), (b) the relative evolution of the average precipitate size and distance, (c) the relative evolution of the specific interface - total and projection to the respective directions.

5. Conclusions

A different extent of γ'-precipitates rafting was observed for variously exposed positions inside the conic specimen. Evolution of rafts with stress level is clearly observable from the parameters obtained from SANS curves. The tendency corresponds to the expected evolution: the larger stress, the more advanced rafting process. Qualitatively, the 2D scattering curves confirmed the results of the electron microscopy performed on the specimen from the same bar.

The use of conic specimen and spatial scan largely facilitate determination of microstructure evolution in dependence on exposure parameters. The detailed evaluation of the SANS data brought a series of morphological parameters in dependence on the applied stress, particularly the mean thickness of the γ channel in the respective crystallographic directions ([100], [010] and [001]) of the single-crystal CMSX-4 specimen, the evolution of the γ' precipitate size and distance as well as the evolution of the specific interface.

The results confirmed that SANS method can be a proper non-destructive method for evaluation of structural changes caused by creep exposure in Ni base superalloys.

Acknowledgments
This research project was supported by EC under the FP6 (Key Action: Strengthening the European Research Infrastructures), contract No. RII3-CT-2003-505925 (NMI3). P. Strunz gratefully acknowledges support by MSM2672244501 and IRP AV0Z10480505 projects.

References

[1] Mughrabi H, Biermann H, Ungar T 1992 *Superalloys 1992*, (Warrendale, PA: TMS) p 599
[2] Link T, Epishin A, Brickner U, Portella P D 2000 *Acta Mater.* **8** 1981
[3] Kostorz G 1979 *Neutron Scattering*, ed. G Kostorz (New York: Academic Press) p 227-289
[4] Zrník J, Strunz P, Horňák P, Vrchovinský V, Wiedenmann A 2002 *Applied Physics A* **74** 1155
[5] Zickler G A, Schnitzer R, Radis R, Hochfellner R, Schweins R, Stockinger M and Leitner H 2009 Mat. Sci. Eng. A **523** 295-303
[6] Gilles R 2005 *Zeitschrift für Metallkunde* **96** 325-334
[7] Ratel N, Deme B, Bastie P and Caron P 2008 *Scripta Materialia* **59** 1167-1170
[8] Aizawa K, Tomimitsu H, Tamaki H and Yoshinari A 2000 *Journal of Applied Crystallography* **33** 847-850
[9] Durand-Charre M 1997 *The Microstructure of Superalloys* (Amsterdam, The Netherlands: Gordon and Breach Science Publishers) p 5
[10] Strunz P, Mortensen K and Janssen S 2004 *Physica B* **350** e783
[11] Strunz P, Gilles R, Mukherji D and Wiedenmann A 2003 *J. Appl. Cryst.* **36** 854-859
[12] Schelten J and Schmatz W 1980 *J. Appl. Cryst.* **13** 385-390

SCFT simulation and SANS study on spatial distribution of solvents in microphase separation induced by a differentiating non-solvent in a semi-dilute solution of an ultra-high-molecular-weight block copolymer

K Ando[1,2], T Yamanaka[1,2], S Okamoto[*1,2], N Sakamoto[3], D Yamaguchi[4], S Koizumi[4], H Hasegawa[5] and N Koshikawa[6]

[1] Department of Materials Science and Engineering, Nagoya Institute of Technology, Gokiso-cho, Showa-ku, Nagoya 466-8555, JAPAN
[2] ISS Advanced materials research and Application Center, Nagoya Institute of Technology, Gokiso-cho, Showa-ku, Nagoya 466-8555, JAPAN
[3] Analysis and Simulation Center, AsahiKASEI Corp., 2-1 Samejima, Fuji, 416-8501, JAPAN
[4] Advanced Science Research Center, Japan Atomic Energy Agency, Tokai, Ibaraki 319-1195, JAPAN
[5] Department of Polymer Chemistry, Graduate School of Engineering, Kyoto University, Kyoto-daigaku Katsura A3-014, Nishikyo-ku, Kyoto 615-8530, JAPAN
[6] Space Environment Utilization Center, Tsukuba Space Center, Japan Aerospace Exploration Agency, 2-1-1 Sengen, Tsukuba-shi, Ibaraki 305-8505, JAPAN

E-mail: okamoto.shigeru@nitech.ac.jp

Abstract. Lyotropic order-disorder transition (ODT) in a semi-dilute solution of an ultra-high-molecular-weight block copolymer using a neutral solvent and a differentiating non-solvent was investigated. The role of the differentiating non-solvent in the ODT was revealed by the spatial distribution of the solvents simulated theoretically by the self-consistent field (SCF) theory approach and demonstrated experimentally by small-angle neutron scattering (SANS). The solutions of polystyrene-*block*-poly (*tert*-butyl methacrylate) (PS-*b*-P*t*BuMA) with an ultra-high-molecular-weight (UHMW) in a mixture of tetrahydrofuran (THF)/ water as a good and neutral solvent and a differentiating non-solvent, respectively, were used. The contrast variation method by SANS experimentally demonstrated that THF and water collectively move into the P*t*BuMA phase and that the volume fraction ratios of water to THF in both phases are almost equivalent to each other. This means that the solvent mixture of THF and water behaves as a single and strongly selective solvent. The collective movement of the solvents was well reproduced by the simulation. Interestingly, it was also suggested that THF gather in the vicinity of styrene-(*tert*-butyl methacrylate) interface to shield the unfavorable strong interaction between the PS and the P*t*BuMA phases and that water gather in the center of the P*t*BuMA phase to decrease the unfavorable contact with the PS phase.

* To whom any correspondence should be addressed.

1. Introduction

The most interesting feature of block copolymers (BCPs) is that they spontaneously form various microdomain structures with periodicity on the order of the molecular size, such as spheres, cylinders, lamellae and gyroids by the change of volume fraction of the constituent blocks. Many researches have been extensively conducted on microdomain structures in bulk [1-3] or solutions of neutral solvents [4-8]. Selective solvents provide richer phase behavior with thermotropic and lyotropic order-disorder transition (ODT) and order-order transition (OOT) and also expand the reachable morphologies [9-12]. These characteristics enable us to apply BCPs to functional materials. BCPs with ultra-high molecular weight (UHMW), for example, have the potentials to form promising photonic crystals since the periodicities are on the order of visible light [13-15]. However, it is difficult to obtain highly ordered structures of UHMW BCPs in bulk or concentrated solutions because of their high viscosity. Lodge et al. reported a strongly selective solvent enhances the segregation power between constituent block chains [11, 12]. Thus we prepared semi-dilute solutions of BCPs using strongly selective solvent mixtures. This low concentration sufficiently decreased the viscosity while the segregation power was kept high enough to induce microphase separation. As a result, we obtained highly ordered photonic copolymers and succeeded in fabrication of non-linear optical materials [16]. Fuse et al. first reported a microphase separation induced by solvent mixture of a good and selective solvent in a low concentration solution [17]. We also found that the addition of "differentiating non-solvent" to a semi-dilute BCP solution of a neutral solvent induces microphase separation. Here we define "a differentiating non-solvent" as a solvent that is miscible with a neutral solvent and immiscible with either constituent blocks but that has slight affinity to one of the blocks. Interestingly, the solvent mixture showed high selectivity to one component. To our knowledge, a systematic investigation on ODTs and OOTs induced by a strongly selective solvent in a semi-dilute solution has not been conducted. In this paper, we employ a semi-dilute solution of polystyrene-*b*-poly(*t*-butyl methacrylate) using tetrahydrofuran and water as a neutral and a differentiating non-solvent, respectively. Water has slight affinity for *tert*-butyl methacrylate because of the polarity of the carbonyl groups. This system shows lyotropic ODT, OOTs and macrophase separation, which will be described elsewhere. Here we investigate the role of the differentiating non-solvent in the mechanism of the microphase separation. Especially, we focus on the spatial distribution of the differentiating non-solvent and that of the neutral or common solvent under the influence of the differentiating non-solvent, which will be revealed by using computer simulation on the basis of the self-consistent field (SCF) theory and the contrast variation method by small angle neutron scattering (SANS) with the combination of deuterated and protonated solvents.

2. Experimental

2.1. Materials.

A polystyrene-*block*-poly (*tert*-butyl methacrylate) (PS-*b*-P*t*BuMA) diblock copolymer with an UHMW was synthesized by the anionic living polymerization procedures under high vacuum. Prior to polymerization, styrene and *tert*-butyl methacrylate monomers were purified from commercial product by *tert*-butyl magnesium chloride and triethyl aluminium, respectively. Tetrahydofuran (THF), the polymerization solvent, was also purified under vacuum by sodium. Polystyrene was polymerized first using *sec*-BuLi as the initiator in THF at -50 °C for 2 hours. Then *tert*-butyl methacrylate was added to form PS-*b*-P*t*BuMA diblock copolymer and allowed to proceed for 24 h, followed by the termination with methanol. The molecular weight and the polydispersity of PS-*b*-P*t*BuMA were characterized by gel permeation chromatography (GPC). The weight-averaged molecular weight, M_w, and PDI were estimated to be 8.2×10^5 g/mol and 1.2, respectively. The volume fraction of polystyrene, f_{PS} was evaluated to be 0.43 by ^{1}H-NMR. Hence the degrees of polymerization of PS and P*t*BuMA were 2910 and 2970, respectively.

2.2. Small-angle neutron scattering (SANS).

Small-angle neutron scattering measurements with a SANS-J-II spectrometer were performed at JRR-3 (Japan Research Reactor-3) in the JAEA (Japan Atomic Energy Agency, Tokai, Japan). The wavelength (λ) was 0.665 nm. The scattering intensity was detected by a 5 inch size cross-wired position-sensitive photomutiplier tube (R3239, provided by Hamamatsu Photonics Co. Ltd.) with a thin sheet of ZnS/6LiF scintillator (0.2 mm thickness) [18]. The cameralength between a sample and the detector was 3.5 m. The scattering vector is defined by $q = (4\pi\sin\theta)/\lambda$, where 2θ is the scattering angle. All the samples were filled in a quartz cell with the sample thickness of 2mm thickness. Exposure time was 1800 sec. The observed two-dimensional scattering intensities were azimuthally averaged and corrected for transmission, sample thickness, electric background, empty cell contributions and incoherent scattering. They were scaled to absolute intensity by using Al as a secondary standard. BCP solutions were prepared using tetrahydrofuran (THF) as a neutral solvent and water as a differentiating non-solvent. Four combinations of deuterated/protonated THF (d-THF and h-THF) and deuterated/protonated water (D_2O and H_2O) were prepared with the ratios of d-THF/h-THF/ D_2O/ H_2O = 1/0/1/0, 1/0/0/1, 0/1/1/0 and 0/1/0/1 by volume, for the contrast variation measurements. For the other experiments, deuterated THF and D_2O were used. Polymer concentration was kept constant at 9.7 vol % for all the solutions.

2.3.1 Theory.

We used computer simulation software, SUSHI, based on the self-consistent field (SCF) theory in the OCTA system [19] to calculate equilibrium concentration distribution ϕ of K-type segments taking the chain conformation into account using the path integral formalism. To calculate the statistical distribution of the chain conformation, they focused on a single chain in a system composed of mutually interacting many chains, and approximate this tagged chain by a single ideal chain in a mean field. For that reason, this mean field is called self-consistent field. Here we briefly summarise the theory described by Honda et al. [19]. The self-consistent field is a potentials field determined by the interaction between segments and the incompressible conditions and is given by

$$V_K(r) = W_K(r) + \gamma(r), \tag{1}$$

where $\gamma(r)$ is a potential decided by incompressiblity condition and $W_K(r)$ is the mean field resulting from the interaction between segments, and is given by

$$W_K(r) = \sum_{K'} \chi_{KK'} \phi_{K'}(r). \tag{2}$$

The subchain is independent statistically and the statistical weight can be described by

$$\frac{\partial}{\partial i} Q_K(i',\vec{r}';i,\vec{r}) = \left[\frac{b^2}{6}\nabla^2 - \beta V_K(\vec{r})\right] Q_K(i',\vec{r}';i,\vec{r}). \tag{3}$$

By solving equation 3, path integral is determined and the concentration distribution ϕ of K-type segments can be calculated by

$$\phi_K(r) = M_K \frac{\int_0^N ds \int dr_0 \int dr_N Q_K(0,r_0;s,r) Q_K(s,r;N,r_N)}{\int dr_0 \int dr_N Q_K(0,r_0;N,r_N)}. \tag{4}$$

the self-consistent field $V_K(\boldsymbol{r})$ depends on the segment concentration $\phi_K(\boldsymbol{r})$. Then, $\phi_K(\boldsymbol{r})$ is calculated from the path integral $Q_K(i, \boldsymbol{r})$ through equation 4. Finally, the path integral $Q_K(i, \boldsymbol{r})$ is obtained by solving equation 3 that contain the self-consistent field $V_K(\boldsymbol{r})$. Therefore, $V_K(\boldsymbol{r})$, $\phi_K(\boldsymbol{r})$ and $Q_K(i, \boldsymbol{r})$ define themselves in a recursive manner or self-consistently. In our case, the volume fraction of the components $\phi_\alpha(\boldsymbol{r})$ (α = PS, P*t*BuMA, THF, water) are functions for which the total free energy attains a minimum.

2.3.2 Simulation condition.

We performed one-dimensional simulation on the basis of SCF theory as described above. The polymer concentration, ϕ_P, was fixed at 9.7vol %. The order-disorder transition and the order-order

Journal of Physics: Conference Series **247** (2010) 012040 doi:10.1088/1742-6596/247/1/012040

Table 1. Interaction parameter χ between A and B.

	PS	P*t*BuMA	THF	water
PS	0	1.46	0	4.07
P*t*BuMA		0	0	0.01
THF			0	0
Water				0

transition between lamellar and gyroid and cylinder occurred with increase of ϕ_w at this polymer concentration, which will be described in detail elsewhere. Below ϕ_w =3.6 vol %, the system was in the disordered state by using UV-vis spectrometer. SAXS measurements revealed lamellar structure were formed at ϕ_w = 3.6 vol % and cylinder structure were formed at ϕ_w =6.4 - 9.1 vol %. The spectroscopic measurements also revealed that gyroid structure were formed ϕ_w = 5 and 5.4 vol %. Rigorously speaking, a comparison between the free energies of all the different possible morpholgies is needed to determine the most stable one. However, we neglected these comparisons for the simplicity of discussion in this study: We concentrated on the strong effect of differentiating non-solvent on the drastic selectivity increase of the solvent mixture, the resulting phase separation phenomenon and the solvent distribution in the phases. We employed the system size optimization method that minimizes the free energy density of the system by optimizing the length of the one-dimensional simulation space on which reflection boundary condition is satisfied [20]: The system size, X, is automatically optimized around X=256. The degrees of polymerization was set N_{PS}:N_{PtBuMA}:N_{THF}:N_{Water} = 29:39:1:1. Note that N_{PS}:N_{PtBuMA} = 29 : 39 is employed in order that the ratio of the values reflect the volume fraction determined by ^{1}H-NMR, i.e., f_{PS} = 0.43 ≈ 29 / (29 + 39), even though the actual degrees of polymerization of the synthesized copolymer is 2910 and 2970, respectively. The Flory-Huggins interaction parameters, χs, between the components are shown in table 1. For simplicity, the value of χs between THF and the other was set at 0, because THF mixes freely with water and is very good solvent for PS and P*t*BuMA. The other values of χs were determined to match the calculated phase diagram with the experimental phase diagram.

3. Result and Discussion

3.1 Result of simulation.

In figure 1, we show the calculated composition profiles of the four components, PS, P*t*BuMA, THF and water, at ϕ_P =9.7 vol % and ϕ_w =1.8 vol %. The left axis indicates the concentrations of PS, P*t*BuMA and water (designated as ϕ_{PS}, ϕ_{PtBuMA} and ϕ_w, respectively) and the right axis indicates the concentration of THF (designated as ϕ_{THF}) and the bottom axis indicates the spatial axis, x. The red,

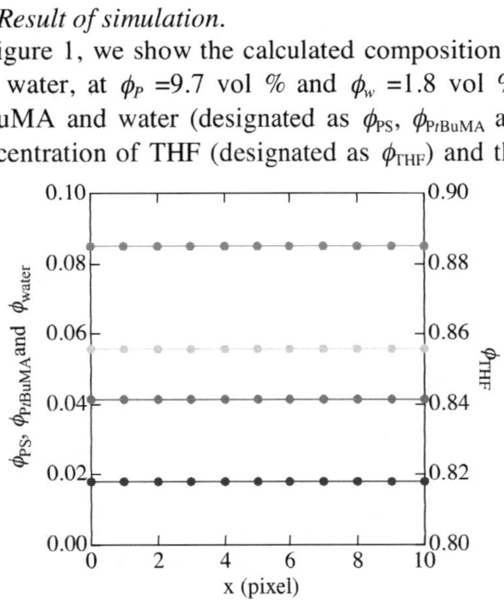

Figure 1. Spatial distribution of all the components at ϕ_w = 1.8 vol %.

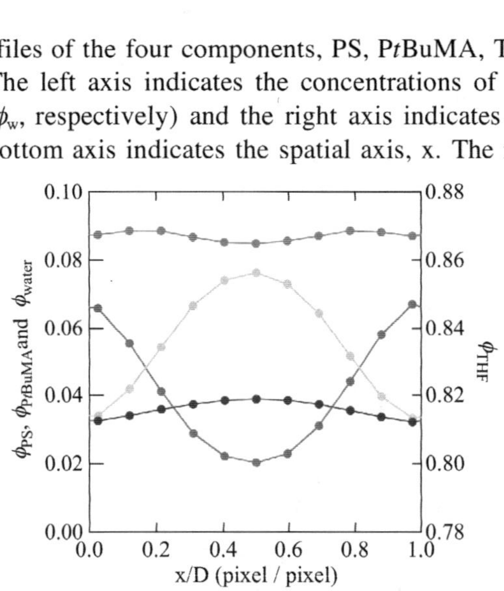

Figure 2. Spatial distribution of all the components at ϕ_w = 3.6 vol %.

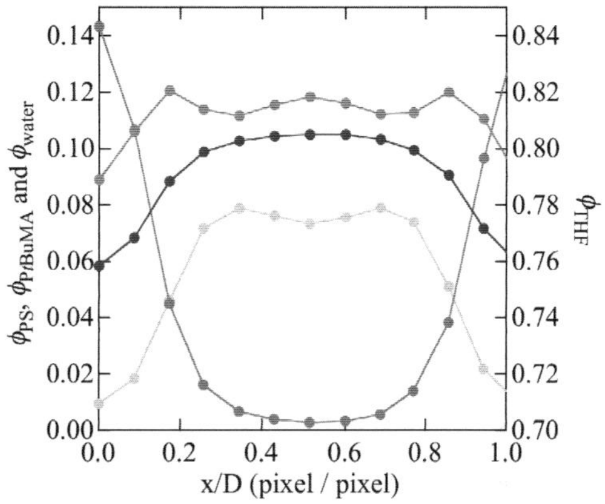

Figure 3. Spatial distribution of all the components at ϕ_w = 9.1 vol %.

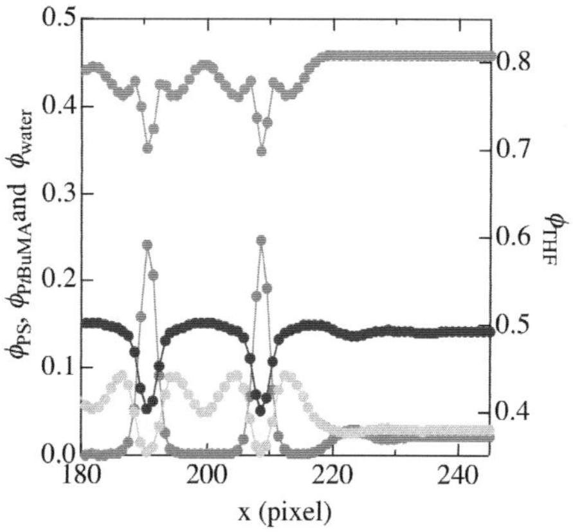

Figure 4. Spatial distribution of all the components at ϕ_w = 13 vol %.

green, purple and blue colors indicate the components of St, *t*BuMA, THF and water, respectively. As can be seen, the distributions of all the components are uniform over the whole system, i.e., the system is in the disordered state, which is in good agreement with the experimental data observed by UV-bis spectrometer. Next, the composition profiles was calculated for ϕ_w = 3.6 vol % as exhibited in figure 2. For the simplicity, the profiles were shown over one period with the x-axis normalized by the domain periodicity, D. Here the value of D was 10.8 pixels. All the components periodically fluctuate around their average compositions, i.e., the system went into the ordered state by the increase of the solvent selectivity prompted by the addition of enough amount of water. Here it is worthy noting that the neutral solvent, THF, gathers in the vicinity of the phase boundary between PS and P*t*BuMA. This shield the unfavorable contact between the two components at the interface, as first reported by Lodge et al. [21]. It should be also noted that the water molecules locate in the middle of the P*t*BuMA, i.e., away from the interface, which also decreases the unfavorable contact [21] .We defined PS and P*t*BuMA phases as the regions where the concentrations are higher than their average values, from which the volume fractions of the phases were measured and will be discussed later in the next section. Figure 3 shows the calculated composition profile for ϕ_w =9.1 vol %. Here the profiles were plotted over one period with the x-axis normalized by the value of D = 13.1 pixels. The fluctuations of all the components were enhanced. Surprisingly, the spatial fluctuation of the neutral solvent, THF, was also enhanced to the same extent as the others, i.e., THF behaved as a selective solvent in the presence of water. The major amount of THF and water locate in the P*t*BuMA phase. This is qualitatively rationalized as follows; if THF uniformly distributes over the whole system while water is selectively introduced into the P*t*BuMA phase, the macrophase separation between THF and water should take place. This segregation brings about a huge entropy loss because the majority of the components (90.7 vol.% by volume or over 99 mol% by the numbers) consists of the small molecules, THF and water. Thus the segregation is energetically unfavorable. Figure 4 indicate the calculated composition profile for ϕ_w =13 vol %. The amount of water added to the system exceeded the upper limit of solubility in P*t*BuMA phase, so that the macrophase separation between solvents and BCP occurred. Indeed, two types of the composition fluctuation were observed: in the region at $0 \leq X \leq 210$, all the components periodically fluctuate around their average composition (microphase separation), while at $X \geq 210$, the space were mostly filled with solvents. Note that the typical profiles of the domain structures over two periods in the polymer-rich phase and those of an appropriate size in the solvent-rich phase were shown for the simplicity.

3.2 Small-angle neutron scattering (SANS).

As Takenaka et al. and Endo et al. reported, the contrast variation method by the SANS is a useful

Table 2. The scattering length density of polymers and solvents.

(cm^{-2})	ρ_{PS}	ρ_{PtBuMA}	$\rho_{h\text{-}THF}$	$\rho_{d\text{-}THF}$	ρ_{H2O}	ρ_{D2O}
$\rho_\beta \times 10^{10\,a}$	1.34	0.537	0.183	6.39	-0.560	6.34

[a] β = PS, P*t*BuMA, h-THF, d-THF, H_2O and D_2O

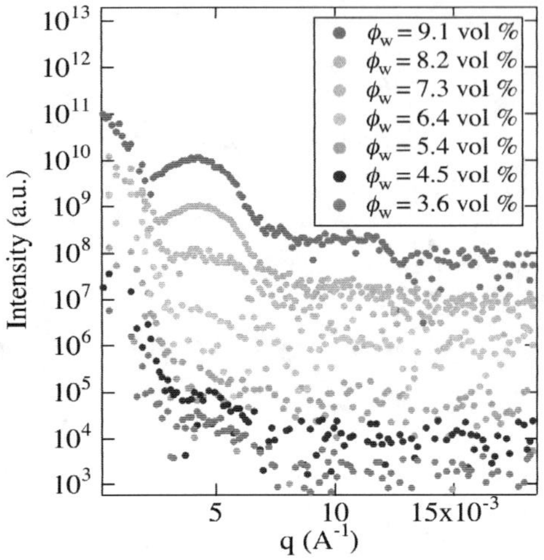

Figure 5. 1-D scattering profiles for the solution of deuterated THF and D_2O at the polymer concentration of 9.7 vol % with increase of ϕ_w.

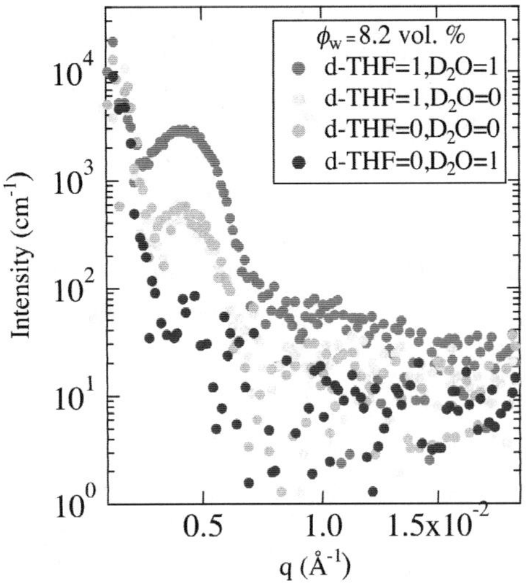

Figure 6. 1-D scattering profiles at ϕ_w = 8.2 vol %.

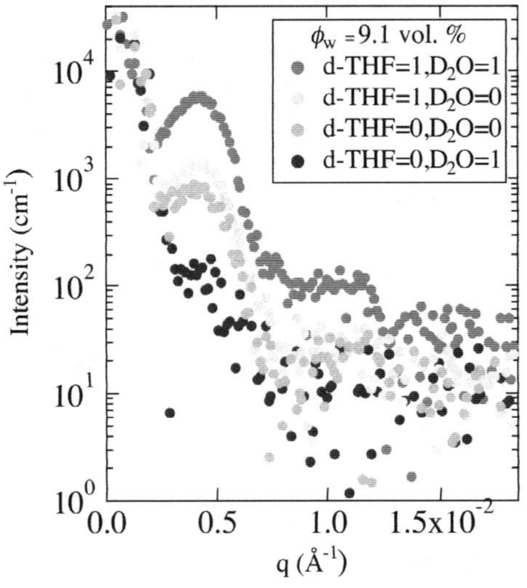

Figure 7. 1-D scattering profiles at ϕ_w = 9.1 vol %.

measurement for evaluation of the distribution of a specific component in complex system [22, 23]. Lodge et al., for example, experimentally demonstrated the distribution of solvent to the interface between different phases in BCP solutions near order-disorder transitions by the SANS contrast variation method as described above [21]. Here, we performed SANS contrast variation measurements in order to investigate the distribution of the BCP and the solvent molecules in the semi-dilute solution. Figure 5 shows 1-D SANS from the BCP dissolved in deuterated THF and deuterated water at the constant polymer concentration of 9.7 vol % with varying the water concentration at ϕ_w = 3.6 - 9.1

vol %. The peak position gradually shifted toward the lower q-region with increasing the amount of water. The scattering length densities of the polymer chains and deuterated and protonated solvents are shown in table 2. The value of the scattering length density of PS ($\rho_{h\text{-}PS} = 1.34 \times 10^{10}\,\mathrm{cm}^{-2}$) is higher than that of P*t*BuMA ($\rho_{h\text{-}PtBuMA} = 0.537 \times 10^{10}\,\mathrm{cm}^{-2}$). At $\phi_w = 3.6$ and 4.5 vol %, the system was in the weak segregation regime and thus the scattering peak was very weak. Subsequently peaks disappeared at $\phi_w = 5.4$ and 6.4 vol % because water selectively went into the P*t*BuMA phase as was described by the computer simulation in the previous section, thus the scattering length density of the P*t*BuMA phase increased and eventually the scattering length density matched between the two phases at these water concentrations. Above $\phi_w = 7.3$ vol %, the scattering length density difference reversed and became larger and the system went into the stronger segregation state, so that the scattering intensity became strong. The red profile at $\phi_w = 9.1$ vol % in figure 5 showed the multiple scattering maxima, indicating that well-ordered microdomain structures were formed under these conditions. At $\phi_w = 8.2$ vol % and 9.1 vol % where the relatively strong intensity was observed, a contrast variation experiment by SANS were performed using deuterated and protonated THF and water. This method allows us to determine the concentrations of the components in each phase by changing scattering contrast with combinations of the protonated and deuterated solvents. Figures 6 and 7 show the scattering intensity profile of the solutions at $\phi_w = 8.2$ vol % and $\phi_w = 9.1$ vol %. The scattering profiles of the red, yellow, green and blue colors indicate that the solvent mixture consists of deuterated THF and D_2O (designated as "d-THF = 1 and D_2O = 1"), deuterated THF and H_2O (d-THF = 1 and D_2O = 0), protonated THF and D_2O (d-THF = 0 and D_2O = 1), and protonated THF and H_2O (d-THF = 0 and D_2O = 0), respectively. The red scattering profiles for "d-THF = 1 and D_2O = 1" in figures 6 and 7 showed the multiple scattering maxima, which also indicate well-ordered microdomain structures were formed. The scattering intensity, $I(q)$, is proportional to the square of scattering length density difference, $\Delta\rho$, between the PS and the P*t*BuMA phases, the lattice factor, $Z(q)$, and the form factor, $P(q)$, respectively.

$$I(q) \approx \Delta\rho^2 Z(q) P(q) \tag{5}$$

The SANS profiles in figures 6 and 7 have the primary intensity maxima at ca. 4.32×10^{-3} and ca. 4.19×10^{-3} Å^{-1}, respectively. The peak position and the shape of the profiles remained unchanged while the peak intensity altered with the contrast variation. Therefore, we consider that the product, $Z(q)P(q)$, was constant throughout all the contrast. The equation 5 is simplified as follows;

$$\begin{aligned} I(q) &\approx \Delta\rho^2 \\ &= \Big[\phi_{PS\_PS} \times \rho_{PS} + \phi_{w\_PS} \times \rho_{water} + \phi_{THF\_PS} \times \rho_{THF} \\ &\quad - \left(\phi_{PtBuMA\_PtBuMA} \times \rho_{PtBuMA} + \phi_{w\_PtBuMA} \times \rho_{water} + \phi_{THF\_PtBuMA} \times \rho_{THF}\right)\Big]^2 \end{aligned} \tag{6}$$

where ϕ_{i_PS} (i = PS, water and THF) and ϕ_{φ_PtBuMA} (j = P*t*BuMA, water and THF) are volume fractions of i-th and j-th species in the PS and the P*t*BuMA phases, respectively. The values of ρ_{water} and ρ_{THF} indicate the scattering length densities of water and THF: $\rho_{water} = \rho_{H2O}$ or ρ_{D2O} and $\rho_{THF} = \rho_{h\text{-}THF}$ or $\rho_{h\text{-}THF}$. Note that we should consider the four combinations of deuterated/protonated water (D_2O and H_2O)and deuterated/protonated THF (d- THF and h-THF) for equation 6. The scattering intensity from the solution with the ratio of d-THF/h-THF/ D_2O/ H_2O = 1/0/1/0 is designated as $I_{dd}(q)$. In the same way, $I_{dh}(q)$, $I_{hd}(q)$, $I_{hh}(q)$ are used. The ratios of the scattering intensities $I_{dd}(q)$ / $I_{dd}(q)$, $I_{dh}(q)$ / $I_{dd}(q)$, $I_{hd}(q)$ / $I_{dd}(q)$, $I_{hh}(q)$ / $I_{dd}(q)$ were calculated. Table 3 shows comparison of the ratios of the experimental and calculated scattering intensities. All the experimental ratios were consistent with the calculated ratios except for I_{hd}/ I_{dd} at $\phi_w = 8.2$ vol %. We consider this slight error occurred due to the statistical precision because $I_{hd}(q)$ was very weak. Table 4 shows the composition of each component in each phase determined from the values in table 3. The values of ϕ_{water_PtBuMA} and ϕ_{THF_PtBuMA} at $\phi_w = 8.2$ vol % are 9.80 and 84.8 vol %, respectively, and greater than the average values (8.2 and 82.1 vol %, respectively), while ϕ_{water_PS} and ϕ_{THF_PS} are smaller than the averages. Thus ϕ_{PS_PS} is greater than the

Table 3. Ratio of experimental and calculated peak intensities at ϕ_w = 8.2 and ϕ_w = 9.1 vol %.

	ϕ_w = 8.2 vol %		ϕ_w = 9.1 vol %	
	experimental	calculated	experimental	calculated
I_{dd}/I_{dd}	1	1	1	1
I_{dh}/I_{dd}	0.18	0.18	0.23	0.23
I_{hd}/I_{dd}	0	0.02	0.02	0.02
I_{hh}/I_{dd}	0.18	0.18	0.14	0.14

Table 4. Composition of each component in each phase.

	ϕ_w = 8.2 vol %		ϕ_w = 9.1 vol %	
	PS phase	P*t*BuMA phase	PS phase	P*t*BuMA phase
PS	16.3	0	19.0	0
P*t*BuMA	0	5.40	0	4.40
THF	78.0	84.8	75.2	84.6
water	5.70	9.8	5.80	11.0

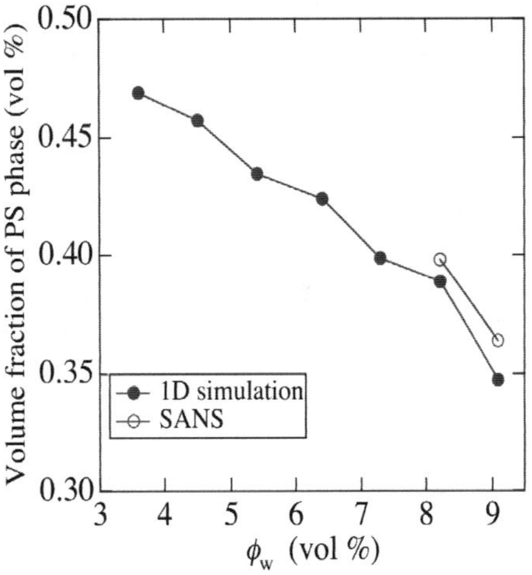

Figure 8. The change of volume fraction of PS phase as a function of ϕ_w.

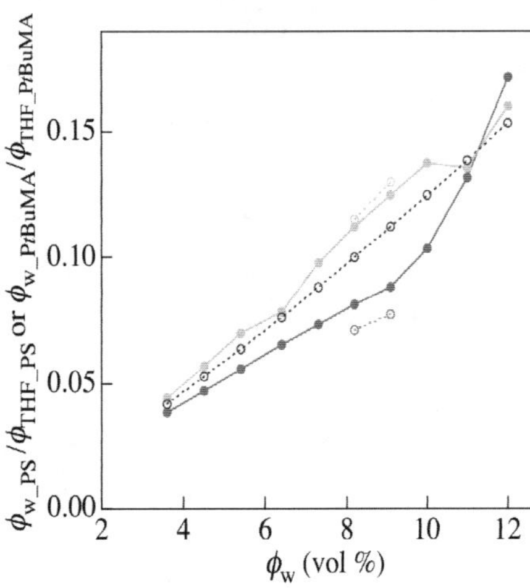

Figure 9. The volume fraction ratios between water and THF in both phases as a function of ϕ_w.

average (9.7 vol %) and ϕ_{PtBuMA_PtBuMA} is smaller than that (9.7 vol %). Note that both of ϕ_{PS_PS} and ϕ_{PtBuMA_PtBuMA} should be 9.7 vol % if the solvents uniformly distribute over the whole system. These tendencies are enhanced at ϕ_w = 9.1 vol % as shown in table 4. The volume fraction of the PS phase at ϕ_w = 8.2 and 9.1 vol % are shown by the open circles as a function of ϕ_w in figure 8, where the calculated values obtained by SUSHI are also plotted by the closed circles. Both of the experimental and the theoretical values decreased with the increase of ϕ_w. In other words, the PS phase was de-swelled while the P*t*BuMA phase was swelled. This means that the solvents moved from the PS phase into the P*t*BuMA phase progressively as ϕ_w increased. Here it is interesting to investigate the ratio of the water content to the THF content in each phase. The values of ϕ_{water_PS} / ϕ_{THF_PS} and ϕ_{water_PtBuMA} / ϕ_{THF_PtBuMA} are shown by the red and the green open circles, respectively, as a function of ϕ_w in figure 9. The theoretical values obtained by SUSHI are also plotted by the red and the green closed circles. The experimental values are in good agreement with the theoretical ones and approximately close to those averaged over the whole system shown by the blue open circles. These results of figures 8 and 9 means that the ratio of the water content to the THF one was kept almost constant while both of water and THF moves from the PS phase into the P*t*BuMA phase by the addition of water. Therefore we consider that water and THF distribute together as a single solvent.

4. Conclusion

The mechanism of the lyotropic microphase separation induced by the differentiating non-solvent in the semi-dilute solution of an UHMW PS-b-P*t*BuMA was studied by simulation on the basis of SCFT and SANS. The composition of each component in each phase was evaluated by the SANS contrast variation method. It was found that at higher ϕ_w, major amount of water and THF selectively located in the P*t*BuMA phase than in the PS phase. The two solvents, THF and water, collectively moved from the PS phase to the P*t*BuMA phase by the addition of water in order to avoid unfavourable entropy loss by the separation of the solvents. In other words, de-swelling of PS and swelling of P*t*BuMA occurred. Additionally, the ratios of the amounts of the solvents in both of the phases are close to each other. This means that the solvent mixture behaved as a strongly selective single solvent and enhanced the segregation power between the PS and the P*t*BuMA phases. These tendencies were well reproduced by the simulation based on the SCF theory. Two- and three-dimensional simulations of this semi-dilute solution system are still under investigation in order to reproduce the lyotropic OOT and will be reported in future.

Acknowledgement

A part of this work was performed as the "International space station applied research partnership program of the Japan Aerospace Exploration Agency (JAXA) and Nagoya Institute of technology." This work is supported in part by the Grant-in-aid from the Japanese Ministry of Education, Culture, Sports, Science and Technology (17550189, 21015011, and 21550208).

This work was partially supported by Polymer Photonic Crystal Research working group of JAXA Space Biology and Microgravity Science Committee.

The synchrotron radiation experiments were performed at the BL40B2 in the SPring-8 with the approval of the Japan Synchrotron Radiation Research Institute (JASRI) (Proposal No. 2009A1419 and 2009B1343).

The SANS experiments were conducted at SANS-J with the approval of the Japan Atomic Energy Agency (JAEA) (Proposal No. 2008B-A12 and 2009A-A70.)

References

[1] Hamley I W 2004 *Developments in BLOCK COPOLYMER Science and Technology*
[2] Hamley I W 1998 *The Physics of Block Copolymers, Oxford University Press: New York*
[3] Bates F S and Fredrickson G H 1990 *Annu. Rev. Phys. Chem.* **41** 525
[4] Balsara N P, Perahia D, Safinya C R,Tirrell M and Lodge T P 1992 *Macromolecules* **25** 3896-3901
[5] Shibayama M, Hashimoto T, Hasegawa H and Kawai H 1983 *Macromolecules* **16** 1427-33
[6] Lodge T P, Pan C, Jin X, Liu Z, Zhao J, Maurer W W and Bates F S 1995 *J. Polym. Sci., Polym Phys. Ed.* **33** 2289
[7] Sakamoto N, Hashimoto T, Han C D, Kim D and Vaidya N Y 1996 *Macromolecules* **30** 5321-30
[8] Huang C I, ChapmanB R, Lodge T P and Balsara N P 1998 *Macromolecules* **31** 9384-86
[9] Shibayama M, Hashimoto T and Kawai H 1983 *Macromolecules* **16** 16-28
[10] Hashimoto T, Shibayama M, Kawai H, Watanabe H and Kotaka T 1983 *Macromolecules* **16** 361-71
[11] Hanley K J, Lodge T P and Huang C I 2000 *Macromolecules* **33** 5918-31
[12] Lodge T P, Pudil B, Hanley K J 2002 *Macromolecules* **35** 4707-17
[13] Urbas A M, Thomas E L, Kriegs H, Fytas G, Penciu RS and Economoou L N 2003 *Physical Review Letters* **90** 108302-1-108302-4
[14] Yoon J, Lee W and Thomas E L 2006 *Nano Lett.* **6** 2211-14
[15] Yoon J, Mathers R T, Coates G W and Thomas E L 2006 *Macromolecules* **39** 1913-19
[16] Tsuchiya K, Nagayasu S, Okamoto S, Hayakawa T, Hihara T, Yamamoto K, Takumi I, Hara S,

Hasegawa H, Akasaka S and Koshikawa N 2008 *Optics Express* **16** 5362-71
[17] Fuse C, Okabe S, Sugihara S, Aoshima S and Shibayama M 2004 *Macromolecules* **37** 7791-98
[18] Koizumi S, Iwase H, Suzuki J, Oku T, Motokawa R, Sasao H, Tanaka H, Yamaguchi D, Shimizu H M and Hashimoto T 2007 *J. Appl. Cryst.* **40** s474-s479
[19] Honda T, Kodama H Roan J R, Morita H, Urashita S, Hasegawa R, Yokomizo K, Kawakatsu T and Doi M 2004 *SUSHI Users Manual*
[20] Honda Tand Kawakatasu T 2006 *Macromolecules* **39** 2340-2349
[21] Lodge T P, Hamersky M W, Hanley K J and Huang C I 1997 *Macromolecules* **30** 6139-49
[22] Takenaka M, Nishitsuji S, Amino N, Ishikawa Y, Yamaguhi D and Koizumi S 2009 *Macromolecules* **42** 308-11
[23] Endo H, Mihailescu M, Monkenbusch M, Allgaier J, Gompper G, Richter D, Jakobs B, Sottmann T, Strey R and Grillo I 2000 *J. Chem. Phys.* **115** 580

Journal of Physics: Conference Series **247** (2010) 012041 doi:10.1088/1742-6596/247/1/012041

Small angle X-ray scattering study of poly(N-isopropyl acrylamide) based cryogels near the volume-phase transition temperature

Mohand Chalal[1,2,3] Françoise Ehrburger-Dolle[3,7], Isabelle Morfin[3], Maria-Rosa Aguilar de Armas[4,5], Maria-Luisa López[4,5], Françoise Bley[6]

[1]Laboratoire d'Electronique Quantique, Faculté de Physique, USTHB Alger, 16111 Alger, Algeria
[2]Département de Physique, Faculté des Sciences, UMBB, 1 Avenue de l'Indépendance, 35000 Boumerdès, Algeria
[3]Laboratoire de Spectrométrie Physique, UMR 5588 CNRS/UJF, 38402 Saint Martin d'Hères, France
[4]Instituto de Ciencia y Tecnología de Polímeros, CSIC and CIBER-BBN, 28006 Madrid, Spain
[5]Networking Biomedical Research Centre in Bioengineering, Biomaterials and Nanomedicine, CIBER-BBN, Spain
[6]Science et Ingénierie des Matériaux et Procédés, UMR 5266 CNRS/INPG/UJF, 38402 Saint Martin d'Hères, France
[7]Corresponding author

E-mail: francoise.ehrburger-dolle@ujf-grenoble.fr

Abstract. The structural modifications induced by changes in temperature are investigated by Small-Angle X-ray Scattering (SAXS) over a broad range of q-values (3.5×10^{-2} - 12 nm^{-1}) in cryogels based on *N*-isopropylacrylamide (NIPA) and/or 2-Hydroxyethyl methacrylate-L-Lactide-Dextran (HEMA-LLA-D) macromer. Various copolymeric cryogels of these two monomers are prepared by cryopolymerization yielding macroporous gels (*cryogels*). For the plain pNIPA cryogel, the SAXS curves obtained at each temperature are well fitted by a sum of four equations describing respectively the scattering resulting from the gel surface (power law), from the solid-like (Guinier equation) and liquid-like (Ornstein-Zernike equation) heterogeneities and from the chain-chain correlation yielding a broad peak (pseudo-Voigt equation) in the high-q domain. The temperature dependence of the parameters obtained from the fit is analyzed and discussed. It is shown that the existence of an isoscattering (or isosbestic) point observed in pNIPA gels and in some copolymers is related to features observed by Differential Scanning Calorimetry and swelling ratio measurements.

1. Introduction

Intelligent polymeric materiels which exhibit response to external stimuli such as temperature have received considerable interest in recent years because of their applications in the field of biomaterials, biomedicine and biotechnology [1]. Poly(*N*-isopropylacrylamide) (pNIPA) gel is a typical example of a temperature sensitive gel as it exhibits a volume phase transition at a critical temperature (T_c) of about 34 °C in aqueous media [2]. Below T_c, pNIPA hydrogels are swollen, hydrated, and hydrophilic. Above T_c, the gels shrink due to the distortion of the hydrophilic/hydrophobic balance in the network

structure. The rate of response of pNIPA hydrogels is low due to the formation of a dense "skin layer" of the shrunken gel, which prevents the mass transport of water out of the shrinking gel [3]. One possibility to circumvent this problem inherent to macroscopic gels (or macrogels) consists in the synthesis of a macroporous gel [4]. The macropores serve as channels that facilitate convective transport of liquid released during the shrinkage of the gel constituting the thin macropore walls. An efficient way to obtain macroporous gels is the cryotropic gelation technique [5], implying polymerization at a sub-zero temperature, yielding *cryogels*. These highly porous polymeric materials exhibit a broad variety of porosities and morphologies [6-10] allowing the preparation of macroporous gels with properties tailored for a given application, particularly in the field of biochemistry or bioengineering (*e.g.*, tissue engineering scaffolds [11]).
For many applications, the mechanical properties of pNIPA must be improved. Increasing the degree of cross-linking which increases the elastic modulus of the gel decreases the degree of swelling and accordingly, the amplitude of the volume change at T_c. The most often proposed solution to overcome this drawback is the preparation of copolymer hydrogels. Copolymerization of NIPA with acrylic acid (AA) yields gels that have a higher swelling ratio and a higher critical temperature [12]. The introduction of dextran based derivatives in pNIPA copolymers was shown to improve the mechanical properties, to increase the swelling ratio and also to confer to the copolymer hydrogel a better reversibility [13]. The present investigation is devoted to a series of NIPA based copolymers synthesized for drug release purposes. Previously synthesized L-lactide-dextran (LLA-D) based macromers were associated with NIPA in order to combine the thermosensitive characteristics of NIPA with the high hydrophilicity of dextran (D). The presence of hydrolyzable LLA combined with 2-hydroxyethyl methacrylate (HEMA) allows the degradation of the cross-links in these copolymers and eventually permits the solid gel to transform into a liquid solution.

As information about the internal structure of the gels and its variation as a function of temperature is essential for using gels as functional materials, this field has been widely investigated since the last 30 years [14]. It appears that Small-Angle Scattering techniques [15] constitute experimental methods that are well adapted to investigate the multiscale structure of polymer gels. The usual q-domain for Small Angle X-ray Scattering (SAXS) or Small Angle Neutron Scattering SANS (10^{-2} to 1 nm^{-1}) allowing structural information between roughly 100 and 1 nm in the real space can be extended up to a few μm by Small Angle Light Scattering (SALS) for transparent gels.

The first extensive study of pNIPA gels (macrogels) by SANS near the volume phase transition was published by Shibayama *et al.* in 1992 [16]. Since that time, several papers reporting SAXS or SANS studies of thermally induced structural modifications in pNIPA gels [17-21] have been published. Because of their fast kinetic response, microgels [22, 23] or core-shell colloidal particles [24, 25] were also extensively investigated. The first paper devoted to the comparison of SAXS curves obtained for macrogels and cryogels appeared recently [26]. It was shown that combining SAXS and Wide-Angle X-ray Scattering (WAXS) by extending the q-domain generally investigated (0.05 - 1 nm^{-1}) up to larger q-values (12 nm^{-1}) brings new information about the changes in the molecular arrangement below and above the transition temperature.

The aim of the present paper is to show the structural modifications occurring during the stepwise increase of temperature not only at the mesoscale but also, for the first time, at the molecular scale in plain pNIPA and p(HEMA-LLA-D) cryogels and in NIPA-*co*-HEMA-LLA-D copolymers with different NIPA/(HEMA-LLA-D) ratios. The macroscopic structure (macropore size distribution and wall thickness) of these cryogels and its change with temperature have been investigated previously by bi-photonic fluorescence microscopy [27]. The present paper is devoted to the study by SAXS of the internal structure of these walls, at the meso and nanoscale at different temperatures between 18 and 37°C. The observed features will be related to that occurring at the macroscopic scale obtained by Differential Scanning Calorimetry (DSC) and swelling ratio (*SR*) measurements.

2. Materials and experimental methods

2.1. Cryogel sample preparation

Macroporous hydrogels were prepared by free radical cryopolymerization of the synthetic macromer, HEMA-LLA-D, and NIPA. Copolymers (NIPA-*co*-HEMA-LLA-D) with different NIPA/(HEMA-LLA-D) ratios (80/20, 60/40 and 40/60 w/w) were prepared. In the following, these copolymer samples will be labeled as coNIPA(80/20), coNIPA(60/40) and coNIPA(40/60) respectively. Synthesis of the macromer was described elsewhere [11]. The polymerization reactions were carried out in tubular-shape glass moulds. NIPA monomer and HEMA-LLA-D macromer were dissolved in water to reach a final concentration of 6% wt/v. The cross-linker, *N,N'*-methylenebisacrylamide (MBAA), was dissolved in this mixture (6.6 wt% of total amount monomer/macromer) and nitrogen was passed-through the solution for 15 min. *N,N,N',N'*-Tetramethylethylenediamine (TEMED) (1 wt %) used for reaction activation was added and the solution was cooled in an ice bath for 5 min. The radical initiator, ammonium persulfate (APS) (1 wt %), was added and the reaction mixture was stirred during 1 min. 1 ml of the reaction mixture was injected into the glass mold. The solution was frozen at -20°C for 1 hour and then at -12°C during 16 hours. The samples were subsequently thawed at room temperature. In all cases, the yield of the polymerization reactions exceeded 85%. The cryogel matrix contained in each glass mold was washed by passing distilled water to remove any possible unreacted monomer and other ingredients and subsequently dried in air until reaching a constant weight. From a chemical point of view, the copolymers consist of chains of dextran (D) with HEMA-LLA side chains connecting pNIPA chains (figure 1). The synthesis of plain pNIPA cross-linked with MBAA and that of the pHEMA-LLA-D cryogels was described earlier [11, 26].

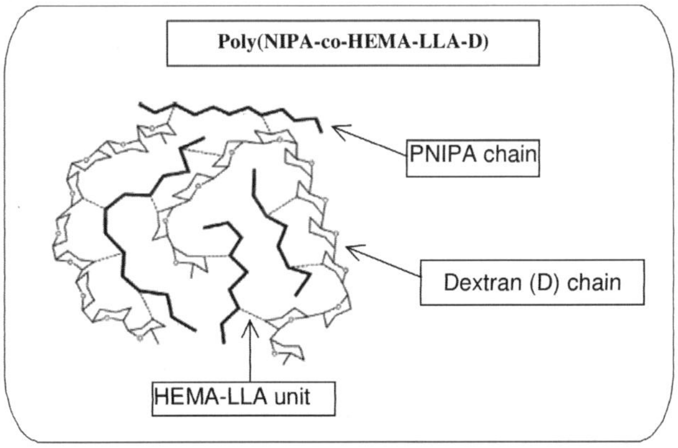

Figure 1. Scheme of the chemical structure of the NIPA based copolymers.

2.2. SAXS measurements

SAXS experiments were performed at the European Synchrotron Radiation Facility (ESRF)[a] Grenoble, France, on the French CRG beamline D2AM. The incident energy was set to 16 keV ($\lambda = 0.77$ Å). The detector, an indirect illumination CCD camera (Princeton Instruments), with pixel size equal to 50 μm was located at distances of 2 and 0.3 m from the sample. These configurations provided data for wave vectors q ($q = \left(\frac{4\pi}{\lambda}\right)\sin\left(\frac{\theta}{2}\right)$) ranging between 3.5×10^{-2} and 12 nm^{-1}. The beam-stop was a small pillar (2 mm diameter lead wire).

The swollen samples were placed in a stainless steel holder closed by two mica windows 1 mm apart. The sample holder was inserted in a small furnace allowing temperature to vary between 10 and 40°C with an accuracy better than 0.2 °C. Heating was provided by a Thermocoax® wire coiled

[a] http://www.esrf.eu/

around the furnace concentrically to a cold water circuit. Temperature control was achieved by means of an Eurotherm® controller and sample temperature was measured by a Pt resistance located close to the cryogel sample. SAXS measurements were performed at different temperatures between 18 and 37°C. The duration of each temperature plateau was 16 minutes. After about 8 minutes, the shape of the curves did not change anymore as equilibrium was reached. Temperature control and data acquisition were automated.

The CCD images were processed by means of the software *bm2img* available on the beamline[b]. All intensity curves were corrected by taking into account the flat field, dark current and distortion. Scattering of the same thickness of pure water was subtracted from the total intensity.

3. Results and discussion

3.1. Evolution of the SAXS curves with the NIPA/(HEMA-LLA-D) ratio at room temperature

Figure 2a shows that the SAXS curves obtained for the pNIPA, the p(HEMA-LLA-D) and the coNIPA cryogels are very different from each other. The dashed lines plotted in figure 2b were obtained from the sum $xI_{pNIPA}+(100-x)I_{p(HEMA-LLA-D)}$ with x=40 or 60, assuming a mixture of pNIPA and p(HEMA-LLA-D). It appears that these curves do not fit the experimental ones. It follows that the copolymer cryogels do not consist of two independent components. They have their own structure that depends on the concentration of pNIPA and p(HEMA-LLA-D).

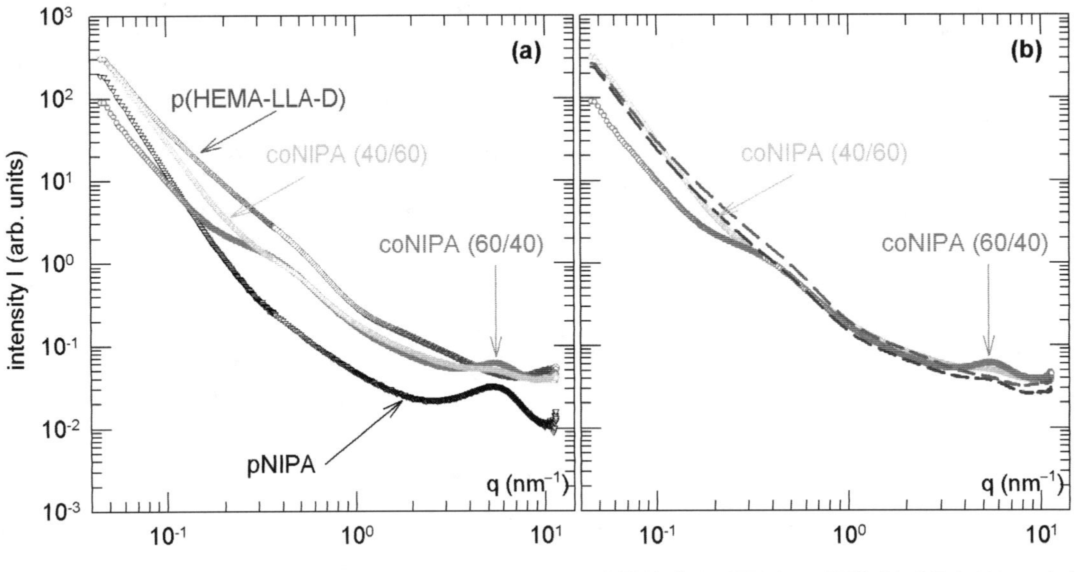

Figure 2. SAXS curves obtained at room temperature (24°C) for pNIPA, p(HEMA-LLA-D) and their copolymers (a) and comparison with the curves calculated (dashed lines) by assuming a mixture of pNIPA and p(HEMA-LLA-D) in the proportions 60/40 and 40/60 (b).

3.2. Evolution of the SAXS curves with temperature

Figure 3 collects the SAXS curves obtained for the cryogels at different temperatures: the plain pNIPA (a), the 60/40 copolymer p(NIPA-co-HEMA-LLA-D) (b) and the p(HEMA-LLA-D) (c). The pNIPA graph reveals, as expected, a strong temperature dependence.

[b] http://www.esrf.fr/exp_facilities/BM2/BM2.html

Journal of Physics: Conference Series **247** (2010) 012041 doi:10.1088/1742-6596/247/1/012041

For pNIPA an isoscattering point is observed at 3.633 nm^{-1} up to 30°C. This point is no longer observed at 32°C and above, *i.e.*, above the volume phase transition temperature (T_c=31.4°C, see next paragraph). The SAXS curves obtained for pNIPA have been extensively analyzed elsewhere [28]. A summary of the results will be given in the next paragraph.

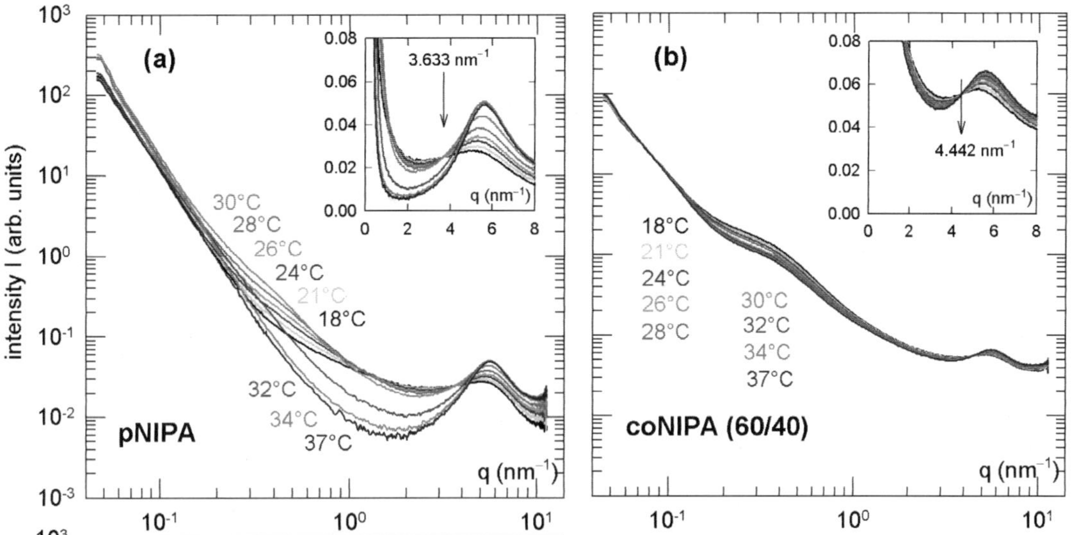

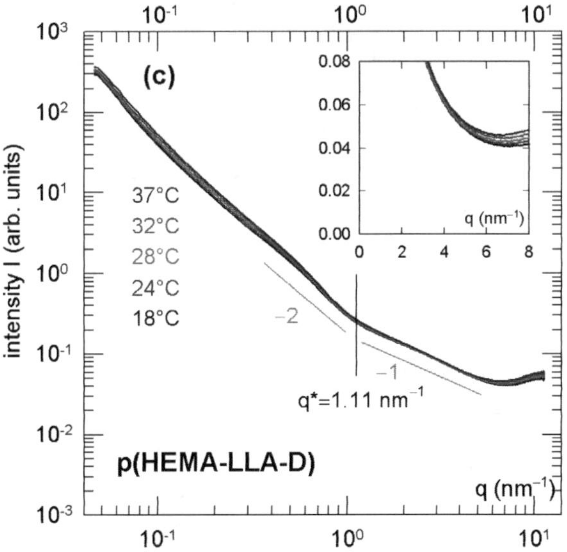

Figure 3. Evolution of the SAXS curves with temperature:
(a) pNIPA;
(b) coNIPA (60/40)
(c) p(HEMA-LLA-D).
The insert shows the curves plotted in linear coordinates displaying an isoscattering point located at 3.633±0.012 nm^{-1} for pNIPA (curves a) and at 4.442 nm^{-1} for coNIPA (60/40) (curves b). No isoscattering point for p(HEMA-LLA-D) (curves c).

As shown in figure 3c, the temperature increase has a much weaker effect on the SAXS curves obtained for the p(HEMA-LLA-D) cryogel than for the pNIPA one. In the low-*q* domain and above 5 nm^{-1}, the intensity slightly increases with T but no isoscattering point can be observed. A precise analysis of these curves is by far more difficult than that of pNIPA and would probably require measurements with neutrons on deuterated samples. Meanwhile, it may be noticed that the curves display a change of slope around q^*=1.11 nm^{-1}. Below q*, the absolute value of the slope is slightly larger than 2. Above q^*, it is close to 1. Such a feature could be attributed to the crossover between the scattering by coil ($q< q^*$) and a rod ($q> q^*$) allowing the determination of the persistence length L_p of a polymer chain by means of the following relation $L_p = 1.91/q^*$ [29] yielding L_p=1.72 nm. This value agrees with the persistence length for dextran (1.5 - 1.8 nm depending on the molecular weight) reported by White and Deen [30].

As for the 60/40 copolymer cryogel, the intensity decreases with temperature in the 0.1 - 1 nm^{-1} *q*-domain while the contrary is observed for pNIPA (figure 3a) for T ≤ 30°C. Also the curves display an isoscattering point over the whole range of temperature but located a larger *q*-value (4.442±0.012 nm^{-1}). The origin and the physical meaning of the isoscattering points will be discussed in a next paragraph.

3.3. Analysis of the SAXS curves obtained for pNIPA at different temperatures

3.3.1. Fit of the experimental data. In the q domain investigated by a large majority of researchers (q<1 nm^{-1}), small-angle scattering of neutrons (SANS) or X-rays (SAXS) by neutral gels is assumed to result from the sum of two contributions [16]:

- the scattering from density fluctuations in thermal blobs which exist also in semi-dilute polymer solutions (liquidlike or solutionlike fluctuations) and which is described by the Ornstein-Zernike (OZ) equation

$$I_{OZ}(q) = \frac{I_{OZ}(0)}{1+q^2\xi^2} \tag{1}$$

where ξ is the size of the thermal blobs (polymer rich domains) and $I_{OZ}(0)$ is the value of the O-Z intensity when $q \to 0$

- the scattering from solidlike density fluctuation due to the gel cross-linking often described by a Guinier equation

$$I_G(q) = I_G(0)\exp\left[-\frac{R_G^2 q^2}{3}\right] \tag{2}$$

where R_G is the radius of gyration related to the diameter a by $a = 2R_G\sqrt{5/3}$ for spherical objects and $I_G(0)$ is the value of the Guinier intensity when $q \to 0$.

In the present case, experiments were extended in the WAXS domain in order to investigate the chain-chain correlation peak recently revealed in cryogels and gels [26]. Without the support of a relevant model, such peaks can be fitted with a pseudo-Voigt equation [31] writing as

$$I_V(q) = a\left[\frac{c}{1+(q-q_1)^2\xi_1^2} + (1-c)\exp\left(-\frac{(q-q_1)^2\xi_1^2}{2}\right)\right] \tag{3}$$

where a corresponds to the maximum of the peak located at q_1 (assumed to correspond to $d = 2\pi/q_1$ in the real space), c is the fraction of the lorentzian contribution and ξ_1 is related to the size of the organized domains.

Finally, because the gel objects investigated are the walls (12 μm thick at room temperature [27]) of the macropores, a low-*q* intensity turn-up resulting from the scattering of the gel surface is expected and actually observed. Thus, the low-*q* part of the intensity curve $I(q)$ will be described by the following power law

$$I_P(q) = I_{P_0} q^{-p} \tag{4}$$

where the exponent p is related to the surface fractal dimension D_s by $D_s = 6 - p$ and I_{P_0} is the prefactor.

It follows that the equation $I_{fit}(q)$ used for the fit of the experimental curve $I(q)$ consists in the sum of four contributions

$$I_{fit}(q) = I_P(q) + I_G(q) + I_{OZ}(q) + I_V(q) \tag{5}$$

involving 10 parameters in a non-linear regression. The regression procedure was achieved by means of the Marquardt-Levenberg algorithm provided by SigmaPlot 10.0. Because the data in $I(q)$ extend over more than 4 decades, a weight w=1/ $[I(q)]^2$ was used. The upper q limit for the regression was fixed at 6.2 nm^{-1} in order to avoid the contribution of expected scattering features occurring at higher q-values. The large number of parameters involved in the fit requires a careful analysis of the residuals. The results of this analysis [28] indicate that the fitting equation obtained by the non-linear regression can be considered as describing satisfactorily the experimental data. The comparison between the fitting curve and the experimental points plotted in figure 4 leads to the same conclusion.

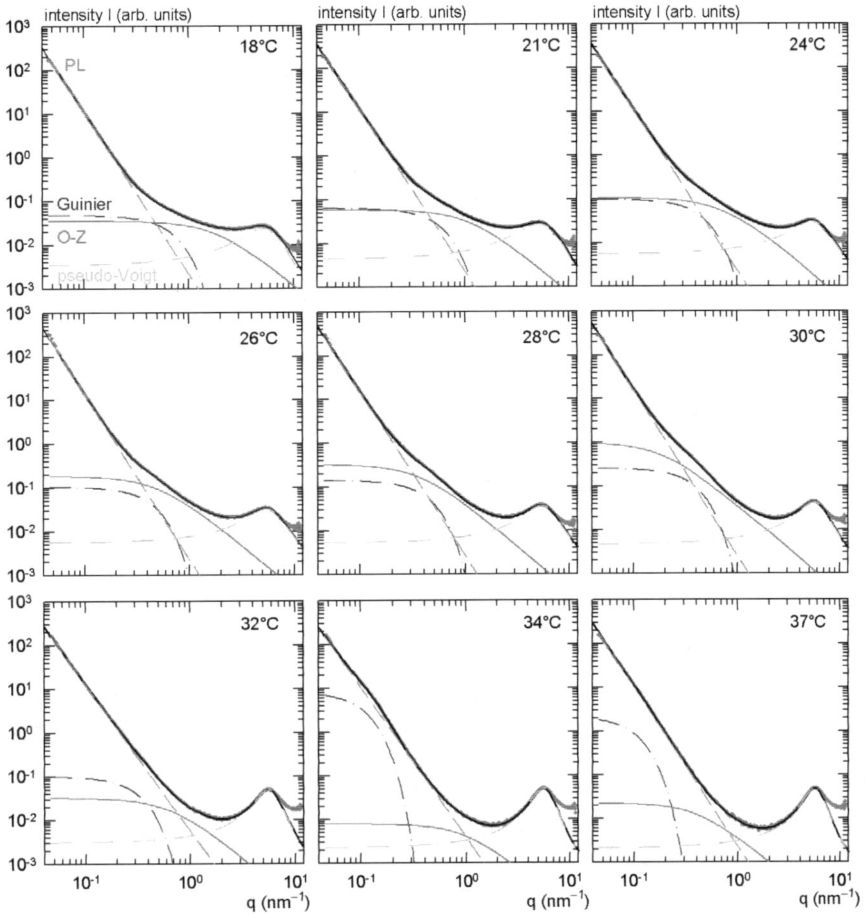

Figure 4. Logarithmic plots of the SAXS curves obtained for the pNIPA cryogel at different temperatures (filled gray circles) and the fitting curves (black continuous line) with their four components (power law, Guinier, Ornstein-Zernike and pseudo-Voigt).

The evolution with temperature of the OZ and pseudo-Voigt calculated curves is shown on figures 5a and 5b. Below 32°C (figure 5a), the intensity of the chain-chain correlation peak increases. Above 32°C (figure 5b), the OZ contribution drops significantly while the peak remains nearly unchanged and similar to that obtained for the dry cryogel as already observed previously in other cryogels [26]. Figure 5c shows the sum of the OZ and pseudo-Voigt curves calculated from the fit. These curves display (see insert) the same isoscattering point as the experimental curves plotted in figure 1. This result strengthens the validity of the fitting procedure and proves that the isoscattering point is related to the structure of the thermal blobs. Actually, such a result is also expected from the curves plotted in figure 4 which show that the power law and the Guinier scattering do not contribute anymore to the intensity in the isoscattering point region.

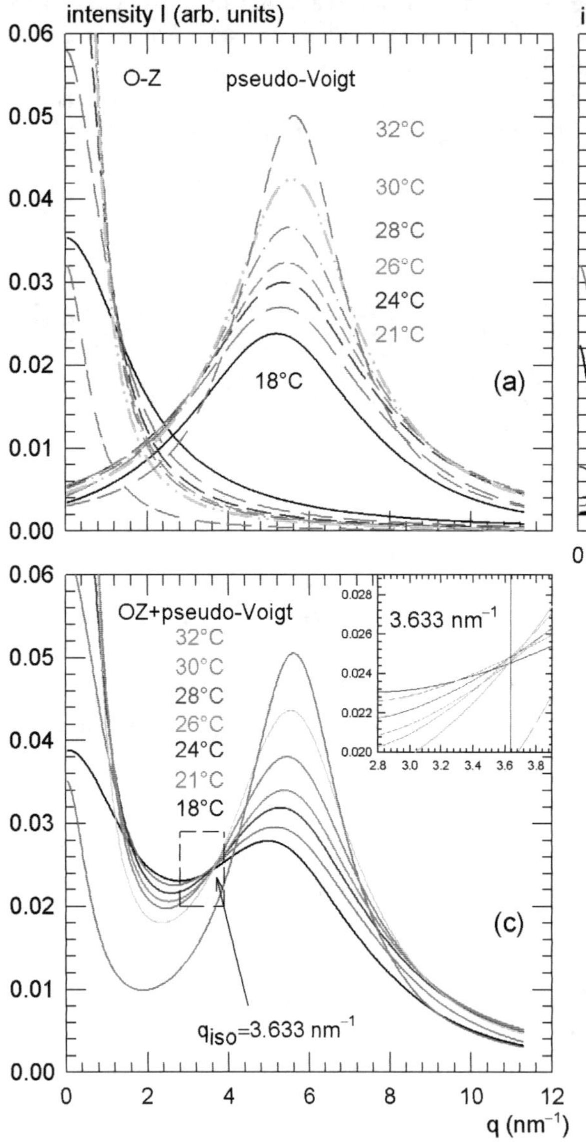

Figure 5. Evolution of the Ornstein-Zernike (OZ) and the pseudo-Voigt curves calculated by using the parameters obtained from the fit:
(a) between 18 and 32°C
(b) at 32, 34 and 37°C and for the dry cryogel
(c) sum of the O-Z and Voigt curves displaying the same isoscattering point (insert) as the experimental curves.

3.3.2. *Evolution of the fitting parameters with temperature.* Figure 6a shows the variation of the OZ parameters ξ and $I_{OZ}(0)$ defined in equation (1). By means of swelling and calorimetry experiments, Li and Tanaka [32] have shown that pNIPA gels belong to the Ising universality class. Thus, ξ and $I_{OZ}(0)$ are expected to diverge at a critical temperature T_c according to the following equations, near T_c:

$$\xi = A|T_c - T|^{-\nu} \tag{6}$$

$$I_{OZ}(0) = B|T_c - T|^{-\gamma} \tag{7}$$

In these conditions, one expects ν=0.631 and γ=1.238. These exponents are slightly larger than the ones involved in the mean field model (0.5 and 1 for ν and γ respectively). In any case, equations (6) and (7) imply that ξ scales as $[I_{OZ}(0)]^{\nu/\gamma}$ with ν/γ close to 0.5. This remark allows the determination of the temperature domain in which the fit to equations (6) and (7) is relevant.

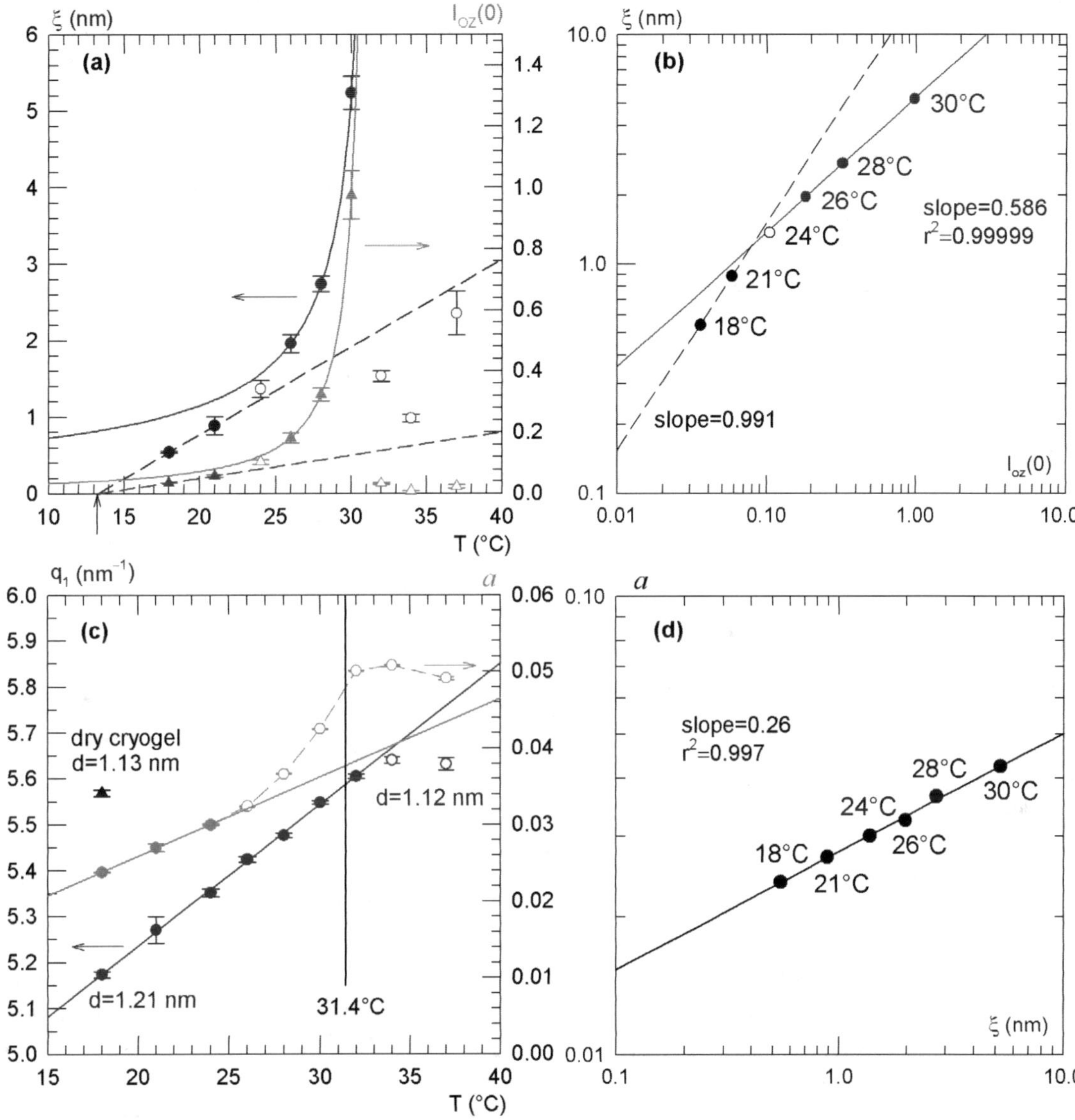

Figure 6. **(a)** Variation of the correlation length ξ (left axis) and the intensity at q=0, $I_{OZ}(0)$ (right axis) with temperature. The solid lines result from the fit with equations (6) and (7). The dashed lines drawn through the points obtained at 18 and 21°C extrapolate to 13.3°C (arrow). **(b)** logarithmic plot of the variation of the parameter ξ with $I_{OZ}(0)$. The solid line results from the linear regression of the data measured at 26, 28 and 30°C. The slope of the dashed line joining the points at 18 and 21°C is close to 1; **(c)** evolution of the position q_1 (left axis) and the intensity a (right axis) of the high-q peak resulting from the fit with the pseudo-Voigt equation (equation 3). The solid lines result from a linear regression of the data (closed symbols); **(d)** Variation of the pseudo-Voigt intensity a (equation 3) with the blob size ξ (equation 1).

As shown in figure 6b, this scaling law is verified only for three temperatures (26, 28 and 30°C, may be four if the point at 24°C is taken into account) yielding an exponent equal to 0.586 which is quite small. Nevertheless, a non-linear regression with equations (6) and (7) performed over the three point yields (figure 6a) (i), the same T_c value (31.35 and 31.36 in equations (6) and (7), respectively) and (ii), critical exponents ν=0.712 and γ=1.22. Including the point at 24°C for the non-linear regression yields a significant difference between the values of T_c obtained from the fit by equation (6) (32.30°C) and equation (7) (31.90°C), and unrealistic exponents (ν=1.015 and γ=1.523). Further, the critical temperature (31.4°C) deduced from Differential Scanning Calorimetry (DSC) measurements (see paragraph 3.3) and resulting from the "melting" of the water cages around the isopropyl groups [33] agrees perfectly with that determined by the non-linear regression performed between 26 and 30°C. Thus, despite the fact that the number of data points is very low, the result confirms the validity of an Ising model for describing the transition in the present pNIPA cryogel as it was for the gels investigated by Shibayama *et al.* [16], yet the value of ν is slightly larger than the theoretical one (0.631). It could be observed in figure 6a that the lines drawn through the points corresponding to 18 and 21°C for ξ and $I_{OZ}(0)$ extrapolate to the same temperature (13.3°C). Also, the slope of this line in the logarithmic plot of ξ *versus* $I_{OZ}(0)$ (Figure 6b) is very close to 1. These features do not prove but suggests a linear increase of ξ and $I_{OZ}(0)$ in the low temperature domain and incites one to further investigate this question in the future.

Equation (1) indicates that the intensity $I_{OZ}(q)$ scale as q^{-2} for $q\xi$>1, in the asymptotic regime. In fact, this asymptotic regime is not visible because of the high-q extra scattering resulting from an inter-chain correlation revealed by the broad peak fitted with the pseudo-Voigt equation (3). The variation of the peak position q_1 and the peak height a as a function of temperature are plotted on figure 6c. This figure shows that the peak height a increases linearly with temperature up to 24°C and remains nearly constant above 32°C. Between 24 and 32°C, the rate of increase becomes larger. It appears (figure 6d) that a scales as the blob size ξ with an exponent equal to 0.26. The physical meaning of the value of this exponent is not obvious. Meanwhile, the scaling law suggests that the increase of a corresponds to the increase of the number of scatterers related to the increase of the size ξ of the thermal blobs. Conversely, the peak position q_1 displays a weak monotonous linear increase (between 5.18 and 5.64 nm^{-1}) with temperature up to 34°C. The values of q_1 are very similar to that obtained for the different cryogels investigated earlier [26] below and above the critical temperature but slightly larger than the value (5.7 nm^{-1}) reported for a weakly cross-linked pNIPA gel at 40°C [20]. The increase of q_1 (figure 8) corresponds, in the real space, to a decrease of the distance d between the dehydrated (*i.e.*, without "water cages" around the isopropyl terminal groups in the side-chain) polymer chains. d can be estimated by means of the Bragg equation $d = 2\pi / q_1$ from 1.21 to 1.12 nm. As discussed earlier [26], such lengths are smaller or close to twice that of the side chain. It follows that the peak may actually originate from a local packing by means of hydrophobic association of isopropyl groups of two neighbor chains and/or hydrogen bonds between amide groups [33]. Thus, it may be assumed that the scattering object from which this peak originates are the collapsed hydrophobic clusters (or nanopockets) revealed by Ahmed *et al.* [34] by means of UV Resonance Raman experiments, which are present in the thermal blobs.

The weak variation of q_1 (or d) with temperature could be attributed to tiny changes in the conformation of the pNIPA chains resulting from the change of hydrogen bonding of water at the C=O or N–H sites. The strong increase of a (concordant with the increase of ξ) between 26 and 30°C could be related to the abrupt spectral change in the amide (AmI) bands observed between 30 and 40°C in the pNIPA nanoparticles investigated in [34], for which the transition temperature is 33°C. Figures 5, 6 and 8 show that collapsed clusters are present at lower temperatures, between 18 and 24°C. This feature is consistent with the ones reported in [34]. It is also consistent with the conclusions drawn about the peak intensity in a gel as compared to the cryogel at room temperature [26]. Meanwhile, it is difficult to explain why the values of ξ obtained at 18°C (0.543 nm) and 21°C (0.886 nm) are smaller than the corresponding value of d (1.21 and 1.19, respectively) deduced from the peak position q_1. For the dry sample, q_1 (=5.56 nm^{-1}) is slightly smaller than in the fully collapsed cryogel at 37°C

(q_1=5.63 nm^{-1}). Accordingly, the inter-chain distance would be slightly larger in the dry sample (1.13 or 1.03 nm) than in the collapsed aqueous cryogel (1.11 or 1.02 nm). The analysis of the UV Raman resonance spectra of the dry pNIPA led Ahmed *et al.* [34] to suggest the existence of interchain hydrogen bonding between pendent amides. Conversely, in the collapsed hydrogel, amides seem to be hydrogen bonded to water and not directly to each others. At this point, and because of the very small differences in the values of *d*, it is difficult to go further into this discussion.

The evolution with temperature of the other parameters extracted from the fit is thoroughly analyzed elsewhere [28].

3.3.3. *Origin of the isoscattering point in pNIPA.* The analysis of the OZ and pseudo-Voigt equation, that of the corresponding parameters ($I_{OZ}(0)$, ξ and a), and their correlation render the origin of the isoscattering point more straightforward. In the q-range investigated (q>2 nm^{-1}), the variation of the scattered intensity $I(q,T)$ is equal to $I_{OZ}(q,T)+I_V(q,T)$. Both functions contain temperature dependent parameters for which the temperature dependence is described by equations that remain the same over a given range of temperature. It follows that $I(q,T)$ is expressed by a linear combination of two functions, in which case, an isoscattering point q_{iso},I_{iso} is expected [35]. The isoscattering point does no longer exist above T_c because the temperature dependence of the parameters in $I_{OZ}(q,T)$ changes. In the present case, q_{iso} equals 3.633 nm^{-1}. The physical meaning of this particular q-value is still unclear. Additionally, it must be noted that this isoscattering point does not indicate that the system transforms from one state to an other one as reported in many situations [31, 36-39]. Here, the isoscattering point results from the fact that the number of nanopockets giving rise to the high-q peak is directly related to the size of the thermal blob (diverging at T_c=31.4°C). In the domain of temperature ranging between 32 and 37°C, i.e., above the temperature of the transition occurring at the molecular level and characterized by the divergence of the O-Z parameters, it is still necessary to include an O-Z equation to fit the experimental curves (figure 4). Meanwhile, the intensity in quite small and the values of ξ vary between 1 and 2.5 nm (figure 6a). This feature makes sense because the system is still a gel. However, the gel network is different from what it was below T_c and the isoscattering point vanishes accordingly.

3.3.4. *Analysis of the isoscattering points in the NIPA based copolymer cryogels.* Figure 7 shows linear plots of the SAXS curves measured at different temperatures for the cryogels investigated. As already seen in figure 3, there is no isoscattering point for p(HEMA-LLA-D). For coNIPA-40/60 containing the lowest amount of NIPA, there is a q-value below which the curves are merging into a single curve, as a result of the significant contribution of the temperature independent scattering of HEMA-LLA-D. Strictly speaking, this point is not an isoscattering point. Meanwhile, it may be concluded that all coNIPA cryogels display an isoscattering point as does the pNIPA cryogel. It follows that as for pNIPA, the SAXS curves can be described by a linear combination of equations. The shift in q and in intensity could result from the temperature independent contribution coming from HEMA-LLA-D which increases the overall level of the scattered intensity as its concentration increases in the copolymer. It follows also that the effect of the increase of temperature is the same in the coNIPA gels as in the pNIPA gel: the increase of ξ and $I_{OZ}(0)$ is related to that of the peak intensity a up to the critical temperature T_c=31.4°C. Examination of the curves obtained for coNIPA-80/20 (figure 7b, insert) containing a large amount of NIPA indicates that, as for pNIPA, the isoscattering point vanishes above 30°C. Thus this sample should, at least, partly collapse above a critical temperature located between 30 and 32°C. On the contrary, for coNIPA-60/40 (figure 7a), the isoscattering point exists over the whole range of temperature investigated. It follows that the critical temperature, if any, is located above 37°C. The same comment applies probably to coNIPA-40/60 (figure 7b)

The pertinence of the conclusions deduced from the isoscattering points will be analyzed by means of two other independent methods: DSC and swelling ratio (SR) measurements presented in the next paragraph.

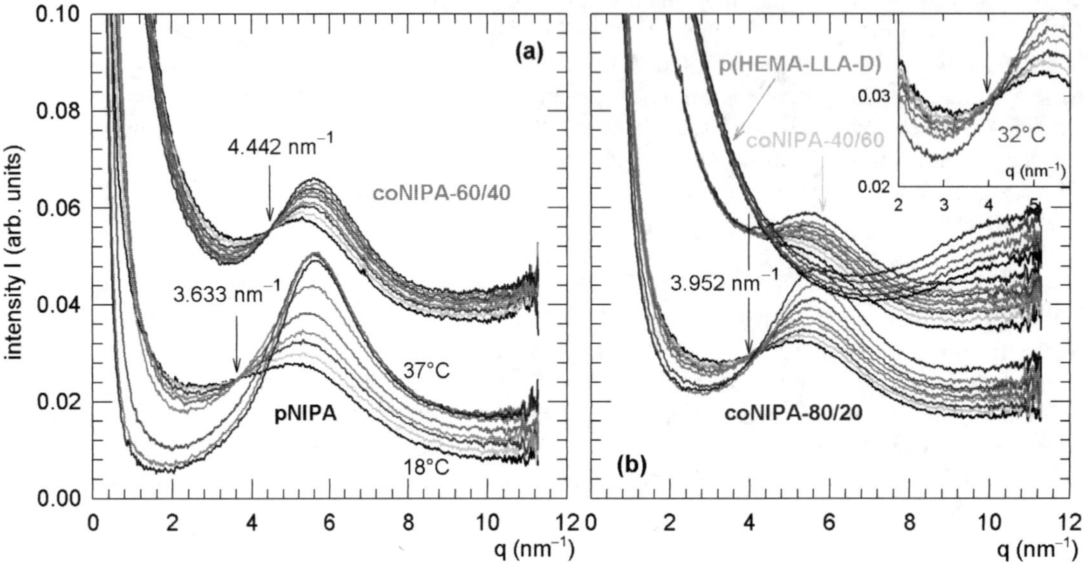

Figure 7. SAXS curves measured at 18, 21, 24, 26, 28, 30, 32, 34 and 37°C for pNIPA and coNIPA-60/40 (**a**) and coNIPA-80/20, coNIPA-40/60 and p(HEMA-LLA-D) (**b**). For p(HEMA-LLA-D) the temperatures corresponding to the curves are 18, 24, 28 32 and 37°C (from bottom to top at high q).

4. Relation between information obtained by SAXS and that obtained by other methods

4.1. Differential Scanning Calorimetry (DSC)

Differential Scanning Calorimetry measurements were achieved by means of MICRO-DSC III (Setaram). The heating rate was 0.1 °C/min for all samples except for pNIPA (0.5 °C/min). The cryogel samples were placed in a closed cell. The reference cell was filled with water in order to enhance the signal. The DSC curves obtained for all samples are collected on figure 8a. The curves are shifted along the ordinate axis by addition/subtraction of a constant in order to bring together the baselines on the high temperature side. This figure shows major differences between the DSC curves.

The DSC curves obtained for pNIPA and for coNIPA-80/20 are plotted in figure 8b. Both curves exhibit an endothermic peak characterized by the same extrapolated temperature (31.4°C). This endothermic peak corresponds to the dissociation of the hydrophobic interaction [40]. In other words, it results from the "melting" of the water cages around the isopropyl groups [33]. As already mentioned, this temperature agrees perfectly with that deduced from the divergence of the Ornstein-Zernike parameters ξ and $I_{OZ}(0)$ in subsection 3.3.2. Because the temperatures are the same, it may be concluded that the mechanism is the same in both cryogels: coNIPA-80/20 should collapse, at least partly, at 31.4°C. This results agrees with the fact that the isoscattering point no longer exists at 32°C and above (figure 7b). In contrast with the DSC curve obtained for pNIPA, that obtained for the copolymer displays a supplementary endothermic peak around 22°C. Because the DSC device did not permit to explore lower temperatures, it was not possible to refine this endothermic effect also observed for coNIPA-60/40 and coNIPA-40/60 and may be, to a lesser extent, for p(HEMA-LLA-D) (figures 8c and d). This peak is quite broad and can hardly be attributed to the same type of transition as the peak located at 31.4°C. A possible origin could be desorption of water molecules bounded to the HEMA-LLA-D macromer.

For the copolymer containing less NIPA (coNIPA-60/40 and coNIPA-40/60) the DSC curves (figure 8c) reveal the absence of the endothermic peak at 31.4°C *i.e.,* no melting and therefore, no collapse of the gel. This feature also agrees with the existence of the isoscattering point over the whole

Journal of Physics: Conference Series **247** (2010) 012041 doi:10.1088/1742-6596/247/1/012041

range of temperatures investigated for coNIPA-60/40 (figure 7a). Shibayama *et al.* [40] reported a shift of the endothermic peak toward a higher temperature for NIPA based copolymers containing acrylic acid or dimethylacrylamide. The intensity of the endothermic peak decreased with the decrease of the NIPA fraction and eventually, the peak vanished. The situation seems different for the copolymers investigated here. Thus, it is likely that the absence of the NIPA endothermic peak below 40°C and the evidence of an isoscattering point up to 37°C does not suggest an elevation of the collapse temperature above these limits. This question needs however to be further investigated. Also, the role played by the comonomer and, particularly dextran, needs to be addressed in the future.

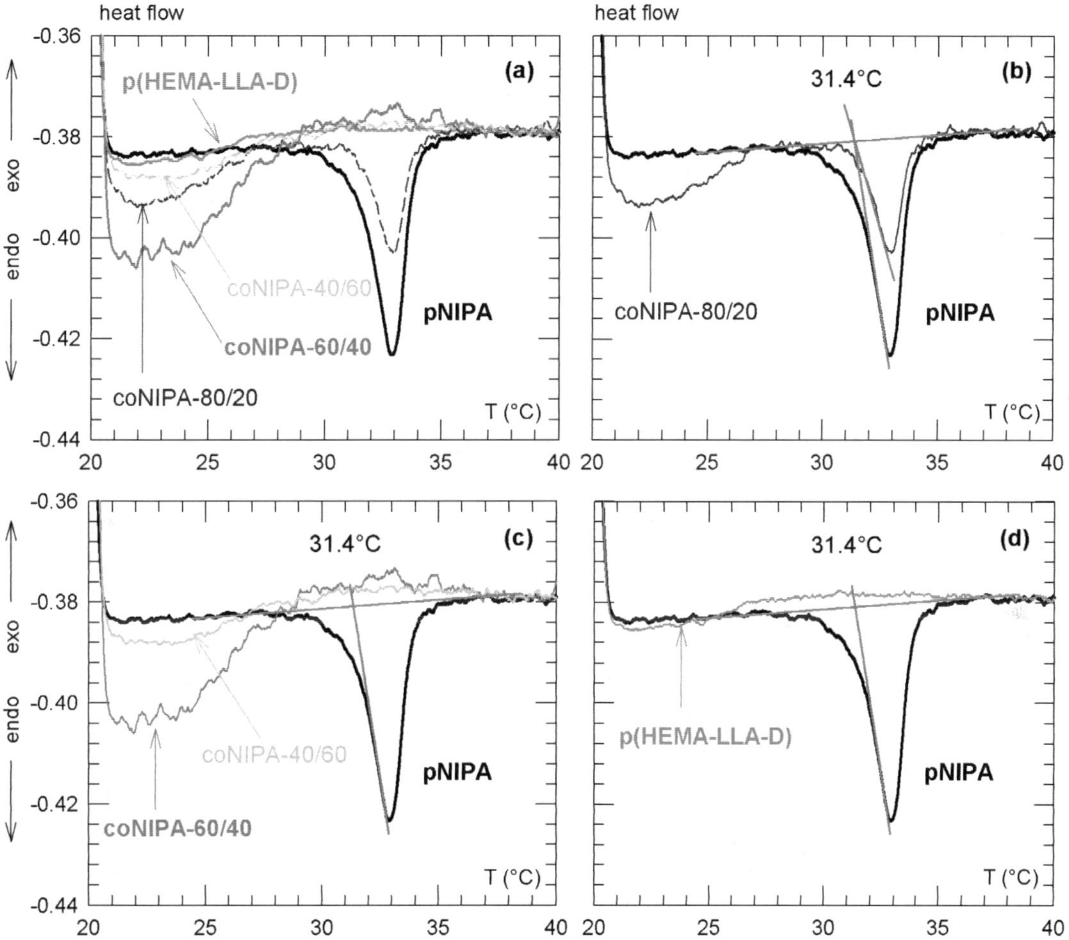

Figure 8. DSC curves measured for the different cryogels at 0.1 °C/min (for pNIPA, the heating rate was 0.5°C/min). The curves are shifted by addition of the constant needed to bring together the baselines on the high temperature side. The heat flow, expressed in mW is not normalized.

4.2. Swelling ratio and macroporous texture

The macroscopic evidence of the physico-chemical phase transition is revealed by significant decrease of the sample volume (the volume phase transition) related to the decrease of the degree of swelling or *swelling ratio* defined by $SR = (M_{swollen} - M_{dry})/M_{dry}$ where $M_{swollen}$ is the weight of the sample swollen in water and M_{dry} is the weight of the dry cryogel. The variation of *SR* with temperature is plotted on figure 9a in linear coordinates. In figure 9b, a logarithmic ordinate axis is used in order to compare the shape of the curves.

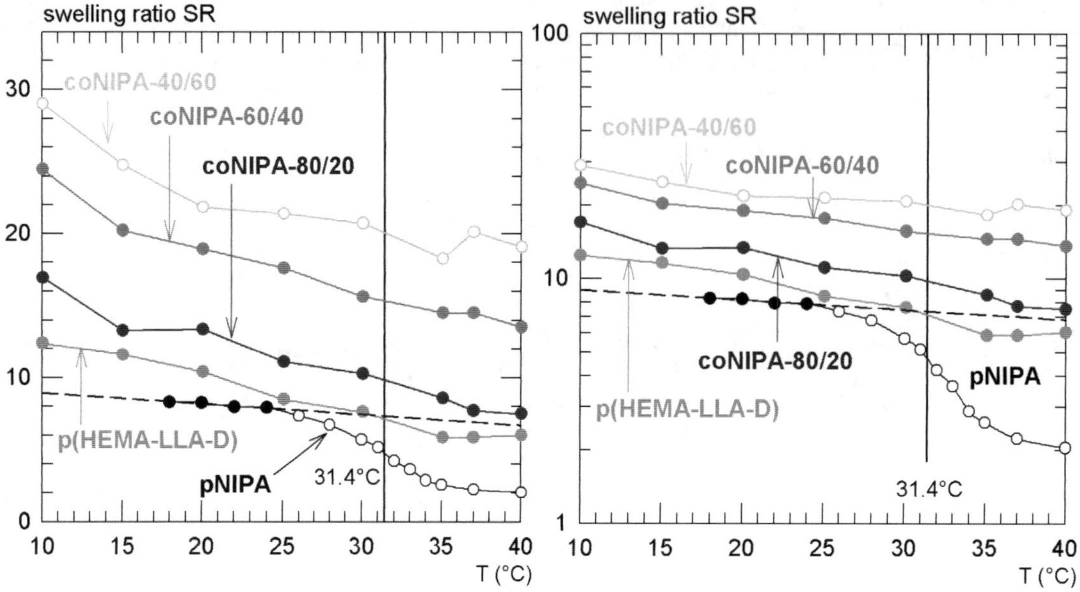

Figure 9. Evolution with temperature of the swelling ratio SR for the different cryogel samples plotted with a linear **(a)** and a logarithmic **(b)** ordinate axis.

For the pNIPA cryogel, the swelling ratio begins to decrease significantly at 26°C. This temperature corresponds to the beginning of the phase transition domain reported in section 3.3.3. It appears that the curve *SR(T)* displays an inflexion point at the critical temperature T_c=31.4°C [28]. Clearly, the *SR* curves obtained for the copolymers and for p(HEMA-LLA-D) do not show the same feature as pNIPA. Meanwhile all copolymer cryogels show a diminution of their swelling ratio as the temperature increases. Figure 9b suggests that the relative diminution of *SR* is the largest for coNIPA-80/20. The decrease of *SR* seems to become smoother as the amount of NIPA in the copolymer decreases from 80 to 40. For coNIPA-80/20, the volume transition expected from the SAXS and the DSC measurements does not appear clearly in the *SR(T)* curves. Meanwhile, this cryogel shows a slightly larger decrease above T_c which is not observed for the other copolymer cryogels and which could agree with the SAXS and DSC features.

Figure 9 also reveals a noticeable feature: the significant increase of *SR* as the amount of NIPA decreases, *i.e.*, as the amount of the HEMA-LLA-D macromer increases. This feature could be explained by the fact that dextran is more hydrophilic than NIPA. However, the degree of swelling of p(HEMA-LLA-D) is only slightly larger than that of pNIPA. Thus it may be concluded that the particular chemical structure of the copolymer gels renders the system more hydrophilic than each of its comonomer alone. The temperature stability of the swollen samples could proceed from a similar mechanism.

The origin of the nearly continuous decrease of *SR* in the copolymer cryogels and in p(HEMA-LLA-D) is still difficult to explain by structural changes occurring at the meso/nanoscale. However, it must be noted that the *SR* curves shown in figure 9 characterize macroporous gels; thus they take into account not only the water in the swollen gel but also the water filling the macroporosity.

The macroscopic structure (macropore size distribution and wall thickness) of these cryogels have been investigated by bi-photonic fluorescence microscopy [27]. It was shown that the macropore size distribution goes progressively from a relatively narrow distribution peaking at 37 μm in the pNIPA cryogel to a very broad distribution extending up to 200 μm in the p(HEMA-LLA-D) cryogel when the

NIPA/(HEMA-LLA-D) decreases from 80/20 to 60/40 and to 40/60. Also the wall thickness in the copolymer cryogels was smaller (6±2 μm) than in the pNIPA cryogel (12±2 μm at 23°C) and nearly independent on temperature. Thus, as a first approximation, the macroporous texture of the different cryogels agrees with their swelling ratio.

For pNIPA, the wall thickness decreases from 12±2 μm at 23°C to 10±2 μm at 28°C and to about 4 μm at 34°C. The mean pore size varies between 37.2, 43.7 and 33.6 nm accordingly. It must be noted that the larger mean pore size at 28°C than at 24°C results from a significant decrease of the contribution of the smaller pores yielding a non-symmetrical distribution. The diminution of the number of small pores also participates to the deswelling of the macroporous gels. In order to check this features on the copolymer cryogels, bi-photonic fluorescence microscopy measurements at intermediate temperatures will be performed in the near future.

5. Conclusions

The aim of the work described in this paper was to show the structural modifications occurring during the stepwise increase of temperature not only at the mesoscale but also, for the first time, at the molecular scale in plain pNIPA and p(HEMA-LLA-D) cryogels and in NIPA-*co*-HEMA-LLA-D cryogels with different NIPA/(HEMA-LLA-D) ratios. In order to analyze and explain the results obtained for the copolymer cryogels it was necessary first to thoughtfully analyze that obtained for plain pNIPA. To this end, the experimental curves were fitted by means of a sum of four equations in order to take into account the scattering of each component of the system.

The analysis of the parameters yielding structural information at the meso and nano-scale were compared with the macroscopic volume change and the thermodynamic effects over the same range of temperature. It is shown that the critical temperature T_c=31.4°C determined from the divergence of the Ornstein Zernike equation is the same as that of the endothermic effect measured by DSC resulting from the "*melting*" of the water cages surrounding the side chains (dissociation of the hydrophobic solvatation). The evolution of the OZ parameters ξ and $I_{OZ}(0)$ with temperature confirms the validity of an Ising model for describing the transition in this pNIPA cryogel. The scaling law between ξ and the intensity a of the chain-chain correlation peak shown here for the first time explains the origin of the isoscattering point observed up to T_c.

The SAXS curves obtained for the copolymer cryogels do not result from a simple combination of curves obtained for each polymer. Thus, fitting these curves was not possible. Meanwhile, each bunch of isothermal SAXS curves shows also an isoscattering point which allows the comparison with plain pNIPA. Particularly, the analysis of the isoscattering points indicates the absence of a gel collapse which agrees with DSC and swelling ratio measurements.

Finally, the SAXS experiments presented here bring a very new series of information about the structural changes induced by temperature in pNIPA gels and, therefrom, in a new series of NIPA based copolymer cryogels. It is also worth mentioning that the information deduced from the analysis of the SAXS data agrees quite well with the ones obtained by other independent methods.

Acknowledgements
Preparation of the cryogels investigated was performed in the context of the FP6-Network of Excellence: EXPERTISSUES, "Novel Therapeutic Strategies of Tissue Engineering of Bone and Cartilage Using Second Generation Biomimetic Scaffolds". Partial financial support by CICYT-MAT2007-63355 is acknowledged. The authors are grateful to the ESRF, Grenoble, for access to the French CRG beamline D2AM (proposal No.0201702) and they acknowledge the help of its technical staff, Jean-François Berar, Nathalie Boudet, Bernard Caillot, Stephan Arnaud. The authors wish to thank Nimet Bölgen, Erhan Piskin and Julio San Román for their collaboration in the synthesis of the cryogels, Isabelle Grillo for her help in the DSC measurements and Jean-Louis Chemin for the design of the furnace used for the SAXS measurements. They also thank Roger Casalegno and Omar Ziane for fruitful discussions.

References

[1] Galaev I and Mattiasson B 2007 *Smart Polymers: Applications in Biotechnology and Biomedicine* (CRC Press, Taylor & Francis Group)
[2] Hirokawa T and Tanaka T 1984 *J. Chem. Phys.* **81** 6379–80
[3] Matsuo E S and Tanaka T 1988 *J.Chem. Phys.* **89** 1695–703
[4] Okay O 2007 Macroporous Hydrogels from Smart Polymers *Smart Polymers: Applications in Biotechnology and Biomedicine* ed Galaev I and Mattiasson B (CRC Press, Taylor & Francis Group) pp 269-93
[5] Lozinsky V I 2008 *Russ. Chem. Bull.* **57** 1015-32
[6] Lozinski V I, Plieva F M, Galaev I Y and Mattiasson B 2002 *Bioseparation* **10** 163-88
[7] Lozinsky V I, Galaev I Y, Plieva F M, Savina I N, Jungvid H and Mattiasson B 2003 *Trends in Biotechnology* **21** 445-51
[8] Plieva F M, Karlsson M, Aguilar M R, Gomez D, Mikhalovsky S and Galaev I Y 2005 *Soft Matter* **1** 303-9
[9] Plieva F, Huiting X, Galaev I Y, Bergenståhl B and Mattiasson B. J. 2006 *Mater. Chem* **16** 4065-73
[10] Plieva F M, Galaev I Y and Mattiasson B 2007 *J. Sep. Sci.* **30** 1657–71
[11] Bölgen N, Plieva F, Galaev I Y, Mattiasson B and Piskin E 2007 *J. Biomater. Sci. Polymer Edn.* **18** 1165-79
[12] Hirose H and Shibayama M 1998 *Macromolecules* **31** 5336-42
[13] Dong J, Chen L, Ding Y and Han W 2005 *Macromol. Chem. Phys.* **206** 1973-80
[14] Arndt K F, Krahl F, Richter S and Steiner G 2009 Swelling-Related Processes in Hydrogels *Hydrogel Sensor and Actuators: Engineering and Technology* (*Springer Series on Chemical Sensors and Biosensors* vol. 6) ed Gerlach G and Arndt K F (Springer-Verlag: Berlin-Heidelberg) pp 69-136
[15] Shibayama M 2006 *Bull. Chem. Soc. Jpn.* **79** 1799-819
[16] Shibayama M, Tanaka T and Han C C 1992 *J. Chem. Phys.* **97** 6829-41
[17] Liao G, Xie Y, Ludwig K F, Bansil R and Gallagher P 1999 *Phys. Rev. E* **60** 4473-81
[18] Norisuye T, Kida Y, Masui N, Tran-Cong-Miyata Q, Maekawa Y, Yoshida M and Shibayama M 2003 *Macromolecules* **36** 6202-12
[19] Koizumi S, Monkenbusch M, Richter D, Schwahn D and Farago B 2004 *J. Chem. Phys.* **121** 12721-31
[20] Kosik K, Wilk E, Geissler E and Laszlo K 2007 *Macromolecules* **40** 2141-47
[21] Hirokawa Y, Okamoto T, Kimishima K, Jinnai H, Koizumi S, Aizawa K and Hashimoto T 2008 *Macromolecules* **41** 8210-19
[22] Kratz K, Hellweg T and Eimer W 2001 *Polymer* **42** 6631-39
[23] Fernandez-Barbero A, Fernandez-Nieves A, Grillo I and Lopez-Cabarcos E 2002 *Phys. Rev. E.* **66** 051803
[24] Seelenmeyer S, Deike I, Rosenfeldt S, Norhausen Ch, Dingenouts N, Ballauff M, Narayanan T and Linder P 2001 *J. Chem. Phys.* **114** 10471-78
[25] Karg M, Wellert S, Pastoriza-Santos I, Lapp A, Liz-Marzan L M and Hellweg T 2008 *Phys. Chem Chem. Phys.* **10** 6708-16
[26] Perez P, Plieva F M, Gallardo A, San Roman J, Aguilar M R, Morfin I, Ehrburger-Dolle F, Bley F, Mikhalovsky S, Galaev I Y and Mattiasson B 2008 *Biomacromolecules* **9** 66–74
[27] Chalal M, Ehrburger-Dolle F, Morfin I, Vial J C, Aguilar de Armas M R, San Roman J, Bölgen N, Piskin E, Ziane O and Casalegno R 2009 *Macromolecules* **42** 2749-55
[28] Chalal M, Ehrburger-Dolle F, Morfin I, Bley F, Aguilar de Armas M R, Lopez Donnaire M L, San Roman J, Bölgen N, Piskin E, Ziane O and Casalegno R 2010 *Macromolecules* **43** 2009-2017
[29] Buhler E and Boué F 2004 *Macromolecules* **37** 1600-10
[30] White J A and Deen W M 2002 *Biophys. J.* **82** 2081-89

Journal of Physics: Conference Series **247** (2010) 012041 doi:10.1088/1742-6596/247/1/012041

[31] Vargas Pereira F, Merlo A A, Bley F, Morfin I, Ritter O M, Pesce da Silveira N and Ehrburger-Dolle F 2008 *Liquid Crystals* **35** 299-313
[32] Li Y and Tanaka J 1989 *J. Chem. Phys.* **90** 5161-66
[33] Cho E C, Lee J and Cho K 2003 *Macromolecules* **36** 9929-34
[34] Ahmed Z, Gooding E A, Pimenov K V, Wang L and Asher S A 2009 *J. Phys. Chem. B* **113** 4248-56
[35] Panda D and Datta A 2008 *Appl. Spectrosc.* **62** 341-44
[36] Dewettinck K, Foubert I, Basiura M and Goderis B 2004 *Cryst. Growth Des.* **4** 1295-302
[37] Nicolai T, Pouzot M, Durand D, Weijers M and Visschers R W 2006 *Europhys. Lett.* **73** 299-305
[38] Dubois E, Cabuil V, Boué F and Perzynski R 1999 *J. Chem. Phys.* **111** 7147-60
[39] Gibaud T, Stradner A, Oberdisse J, Lindner P, Pedersen J S, Oliveira C L P and Schurtenberger P 26 Jan 2009 *arXiv:0901.4077v1 [cond-mat.soft]*
[40] Shibayama M, Mizutani S, and Nomura S 1996 *Macromolecules* 2019-24

Modelling Small Angle Neutron Scattering Data from Electrospun Fibres

Saeed D Mohan, Fred, J Davis, Robert H Olley and Geoffrey R Mitchell

Polymer Science Centre, University of Reading, Reading RG6 6AF UK

E-mail: s.d.mohan@reading.ac.uk

Abstract. Electrospinning is a technique employed to produce nanoscale to microscale sized fibres by the application of a high voltage to a spinneret containing a polymer solution. Here we examine how small angle neutron scattering data can be modelled to analyse the polymer chain conformation. We prepared 1:1 blends of deuterated and hydrogenated atactic-polystyrene fibres from solutions in N, N-Dimethylformamide and Methyl Ethyl Ketone. The fibres themselves often contain pores or voiding within the internal structure on the length scales that can interfere with scattering experiments. A model to fit the scattering data in order to obtain values for the radius of gyration of the polymer molecules within the fibres has been developed, that includes in the scattering from the voids. Using this model we find that the radius of gyration is 20% larger than in the bulk state and the chains are slightly extended parallel to the fibre axis.

1. Introduction

Electrospinning is a method that can be used to produce nanoscale to microscale sized polymer fibres. This is done by the application of a high voltage to a spinneret containing a polymer solution. At the tip of the spinneret the polymer droplet deforms into a Taylor cone due to the electric field applied [1]. A polymer jet is extruded from the Taylor cone to a grounded collector. During transit the majority of solvent evaporates of the material leaving behind solid fibres which are deposited in a randomly orientated fibre mat on to the grounded collector. There are many factors that can influence fibre production that have to be taken into account. For example fibre diameters can be controlled by variations in spinneret tip to collector distance or a variation in the voltage applied, thereby altering the field strength [2]. There are also the solvent effects [3] on the spinnability along with solution viscosity [4] and the molecular weight of the polymer [5]. A polymer with a sufficiently high molecular weight is required in order for chain entanglements to occur which is crucial to the formation of fibres in the electrospinning process [4].

An understanding of the chain geometry during this electrospinning process can be developed through analysis of the chain conformations in the fibres by small angle neutron scattering. However, there are factors that have to be taken into account during data analysis. Electrospun polymer fibres have been known to contain a porous surface and an internally voided structure [6, 7]. The surface porosity is thought to be due to water vapour condensing on the fibre surface and spinodal decomposition occurring [6]. The internal void structure is a result of the fibre forming an outer skin that traps solvent within the internal structure. The solvent will subsequently evaporate off slowly leaving behind the voided structure [7]. These are important factors in data analysis of electrospun fibres as both can be on a length scale recorded using small angle neutron scattering experiments. In this paper we discuss how we can resolve the issues caused by the voided structure of the fibres in order to analyze chain conformations in isotopically labeled fibre samples.

2. Materials and Methods

2.1. Polymers
Hydrogenated atactic-polystyrene and deuterated atactic-polystyrene with molecular weight characteristics displayed in Table 1 were used for the small angle neutron scattering experiments.

Table 1: Molecular weight characteristics of polymers used in small angle neutron scattering.

Polymer	Label	*Molecular Weight (g/mol)*	Mw/Mn
Hydrogenated Polystyrene	HPS-1	241,000	4.8
Hydrogenated Polystyrene	HPS-2	136,000	4.4
Deuterated Polystyrene	DPS-1	205,000	3.8
Deuterated Polystyrene	DPS-2	128,000	3.8

2.2. Electrospinning
Solutions for electrospinning were prepared in 1:1 blends of HPS-1/DPS-1 and HPS-2/DPS-2 by volume at concentrations of 0.2g/ml, 0.3g/ml and 0.4g/ml. Two different solvents were used in the electrospinning process, N, N – Dimethylformamide (DMF) and Methyl Ethyl Ketone (MEK). The polymer was weighed out in the correct proportions to the solvent and then left to dissolve on a shaker (Bühler) for 2 days.

The electrospinning equipment was set up in a horizontal configuration. Solutions were placed in a 5ml Luer lock syringe with a 22G (0.41mm inner diameter) needle attached. The syringe containing the solution was placed inside a syringe pump and set to a flow rate of 0.13ml/min. Samples were prepared at needle tip to collector distances of 20cm and 30cm and applied voltages of 10kV, 12.5kV and 15kV. Randomly orientated fibre mats were collected onto a grounded aluminium electrode and left to dry for a minimum of 24 hours to ensure that any remaining solvent had evaporated. The mass of samples varied between 100mg – 300mg, which was sufficient to enable a good signal noise ratio to be obtained in ~30 minutes during neutron scattering experiments.

Several sets of aligned samples have also been collected by electrospinning onto a grounded rotating collector with a tangential velocity of the collector surface ~5ms^{-1}. These samples were prepared at the same applied voltages and needle tip to collector distances as described above. Trial experiments determined that this rotational speed would not mechanically extend the fibres but provide sufficient levels of fibre alignment on the collector necessary for a study of the chain conformations parallel and perpendicular to the fibre axis.

2.3. Scanning Electron Microscopy
Scanning electron microscopy (SEM) was performed on using an FEI Quanta 600 SEM equipped with a field emission gun and a Cambridge Instrument SEM 360. All samples were coated with gold prior to examination. Fibre diameters were measured from images obtained using the Scandium software. A grid was overlaid over the image and measurements made on several fibres in each grid square to ensure unbiased sampling.

2.4. Small Angle Neutron Scattering
Small Angle Neutron Scattering (SANS) was performed using the LOQ instrument at the STFC ISIS facility, UK. A wavelength, λ, range of 2.2Å – 10Å was used giving a scattering vector range |Q|(Å^{-1}) range of 0.006Å^{-1} – 0.27Å^{-1} where:

$$|\underline{Q}| = 4\pi \sin\theta / \lambda \tag{1}$$

and 2θ is the scattering angle.

Fibres were easily removed from the aluminium electrode and placed into an open ended cylindrical shaped sample holder. The sample holder was covered with aluminium foil to hold the sample in place and the whole assembly was placed inside the sample holder rack on the LOQ instrument.

Scattering experiments were run for ~30 minutes on each sample. All scattering experiments were carried out at 25°C. Data was collected onto the two dimensional ORDELA area detector. The data was converted to an absolute scale by measurements of known polystyrene blends. An empty sample holder was prepared in a similar fashion to that described above and the measured scattering was subtracted as a background. The incoherent background was incorporated into the data fitting method. Data were radially averaged and reduced to a one dimensional differential scattering cross section as a function of the scattering vector. Data analysis was performed using the non-linear least squares data fitting program FISH [8].

3. Results and Discussion

3.1. SEM of fibres

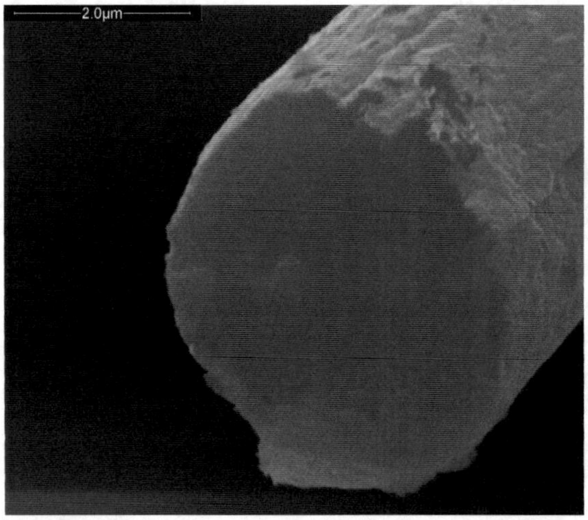

Figure 1a: Cross section of a fibre prepared from a solution of 0.2g/ml HPS-1/DPS-1 in DMF at a collector distance of 20cm with an applied voltage of 15kV

Figure 1b: Cross section of a fibre prepared from a solution of 0.2g/ml HPS-1/DPS-1 in MEK at a collector distance of 20cm with an applied voltage of 15kV

Images of fibres prepared from DMF and MEK obtained using SEM are shown in Figure 1. Fibres that have been prepared from DMF (Figure 1a) have a cylindrical shape, whereas those prepared from MEK (Figure 1b) show a more ribbon like structure. This ribbon like structure is thought to be due to the collapse of the fibre during the electrospinning process itself [2].

Journal of Physics: Conference Series **247** (2010) 012042 doi:10.1088/1742-6596/247/1/012042

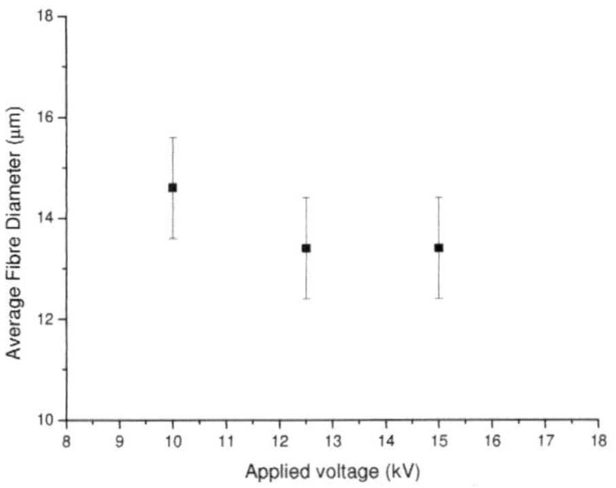

Figure 2a: Variation in average fibre diameter as a function of applied voltage for 0.3g/ml solution of HPS-1/DPS-1

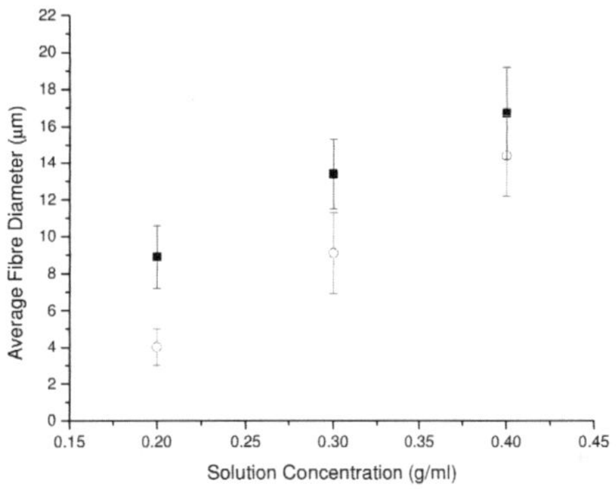

Figure 2b: Variation in average fibre diameter with solution concentration they were prepared from. Samples were electrospun at 10kV with a collector distance of 30cm □ HPS-1/DPS-1 and □ HPS-2/DPS-2.

Samples were prepared under different spinning conditions with variations in the collector distance, applied voltage and solution concentration. By increasing the applied voltage in the electrospinning process, fibres are produced with smaller diameters (Figure 2a), as there is a greater acceleration to the grounded collector, causing more stretching [2]. Figure 2b shows that the fibre diameter increases with solution concentration. Figure 2b also shows that the higher the molecular weight polymer produces larger diameter fibres, due to an increase in viscosity [4]. SEM analysis of fibre cross sections reveal an internal structure with voids present as shown in Figure 3. The voids vary in shape and size from fibre to fibre, in sizes from several hundred nanometres to tens of nanometres. The latter are small enough to be detected in SANS.

Figure 3: Cross section of a fibre prepared from a 0.4g/ml solution of HPS-2/DPS-2 in DMF at collector distance of 30cm and an applied voltage of 10kV.

3.2. SANS of Fibres

In order to analyse the chain conformation in these polymer fibres, a model that takes the voids into account has been developed so that a value for the radius of gyration can be extracted. Firstly, fibres were prepared using only deuterated polymers from both solvents so that the contrast arises only from the voids themselves and other inhomogenieties present in the fibres. Fitting a model, $I(Q) = kQ^{-N}$, where k is scaling factor, to the scattering data from the purely deuterated fibres prepared from both solvents gave a value for N of 4, as shown in Figure 4a and 4b for fibres sun from DMF and MEK respectively. A value of N = 4 is consistent with scattering arising from the sharp interfaces of the voids [9]. Both sets of deuterated fibres prepared from different solvents display a scaling of Q^{-4} although the level of scattering from the fibres prepared from MEK is much lower than those prepared from DMF. Fibres prepared from DMF display higher levels of voiding in the internal structure than those prepared from MEK explaining this higher level of scattering. Attempts to include a contribution from the surface porosity did not lead to an improved fit and we concluded that the majority of the scattering arises from the internal structure rather than any surface features.

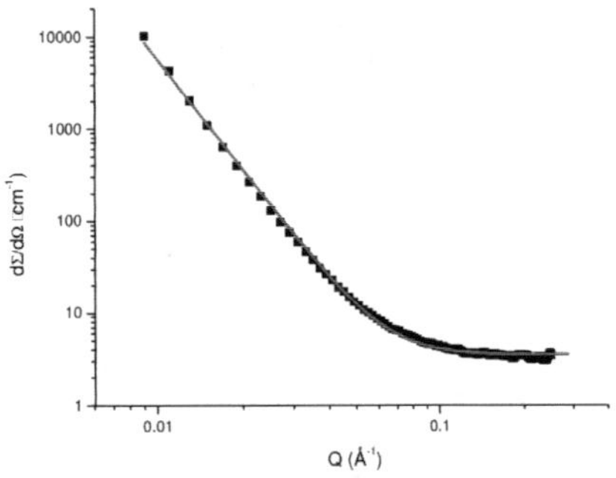

Figure 4a: Log-log plot of scattering data and data fit for the Q^{-N} model to a purely deuterated polystyrene fibre spun from DMF. Fibres were prepared from 0.3g/ml solution of DPS-1 in DMF at an applied voltage of 10kV and collector distance of 20cm. ▪ is the scattered data, solid line is the model fit.

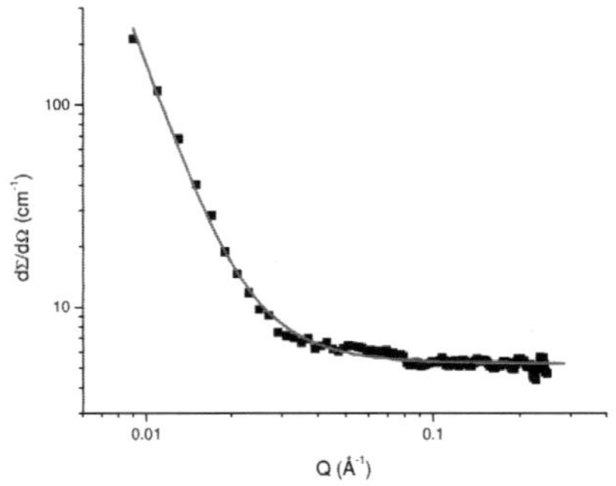

Figure 4b: Log-log plot of scattering data and data fit for the Q^{-N} model to a purely deuterated polystyrene fibre spun from MEK. Fibres were prepared from 0.3g/ml solution of DPS-1 in MEK at an applied voltage of 10kV and collector distance of 20cm. ▪ is the scattered data, solid line is the model fit.

Further analysis was performed on the deuterated fibres by an application of the Shultz polydisperse sphere model. The distribution function for the model is defined in Equation 2 [8]:

$$f(R) = \left[\frac{z+1}{\overline{R}}\right]^{z+1} R^{z} \exp\left[-\left(\frac{z+1}{\overline{R}}R\right)\right]\frac{1}{\Gamma(z+1)} \quad (2)$$

Where $\overline{R}$ is the mean sphere radius, z is the parameter related to the width of the distribution.

We have taken the voids to be spherical in nature as observed in the SEM images and we have applied the model to the scattering from the deuterated fibres. Figure 5 shows a polydisperse sphere model fit to the deuterated fibres. A mean void size of approximately 21nm ± 2nm for fibres prepared from DMF and fibres prepared from MEK show a higher mean void size at 25nm ± 2nm. These values are in broad agreement with those obtained from the image analysis of the SEM micrographs. Void sizes measured in the SEM analysis vary between 20nm to several hundred nanometres in fibres prepared from DMF, with a mean void size of 19.5nm. Fibres prepared from MEK show a smaller variation in void sizes ranging from 20nm-140nm with a mean size of 35nm.

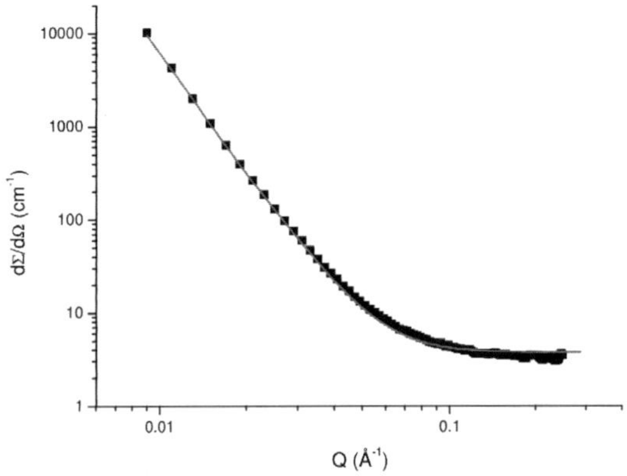

Figure 5: Log-log plot of scattering data and data fit for the Shultz Polydisperse Sphere model to a purely deuterated polystyrene fibre spun from DMF. ▪ is the scattered data, solid line is the model fit.

The previous section shows that I (Q) =kQ^{-4} provides a good fit to the void scattering. We use a Shultz Polydisperse Gaussian coil model combined with this Q^{-4} model to fit the complete scattering from fibres prepared using isotopically labelled mixtures. This model is defined in equation 3 [8]:

$$I(Q) = \frac{A}{\left((1+U)y^2\right)}\left\{y - 1 + (1+Uy)^{-1/U}\right\} + BQ^{-4} \quad (3)$$

Where A = 2x(1-x)V($\Delta\rho$)2 x is the concentration of the labelled polymer, V is the volume, $\Delta\rho$ is the difference in the scattering length, y = (QRg)2/(1+2U) and U = 1 – M_W/M_N, Rg is the radius of gyration and B is a scaling constant. In this model, values are fixed by the molecular weight characteristics of the polymer apart from Rg and B which are allowed to vary. Rg was initially set to that of the bulk state material and B started at 0.

This model was successfully applied to the electrospun fibres prepared from isotopically labelled mixtures of polystyrene. This is demonstrated for fibres prepared from a 0.2g/ml solution prepared at 10kV with needle tip to collector distance of 20cm for HPS-1/DPS-1 in DMF (Figure 6a) and MEK (Figure 6b).

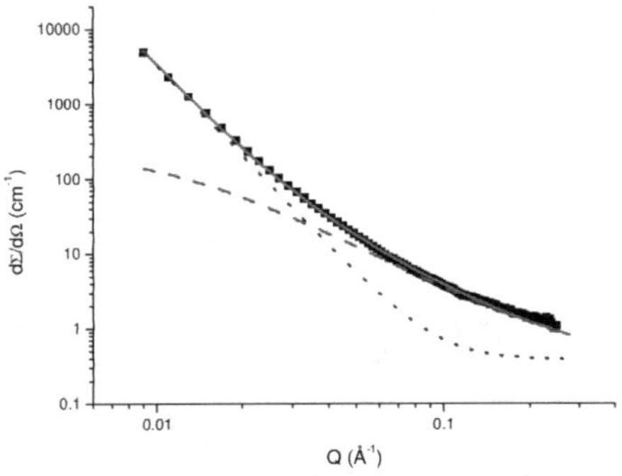

Figure 6a: Log-log plot of scattering data and data fits for a fibre spun from 0.2g/ml HPS-1/DPS-1in DMF at a voltage of 10kV and collector distance 20cm. ▪ is the scattered data, solid line is the combined model fit, dashed line is the polydisperse Gaussian coil fit and the dotted line is the Q^{-N} fit.

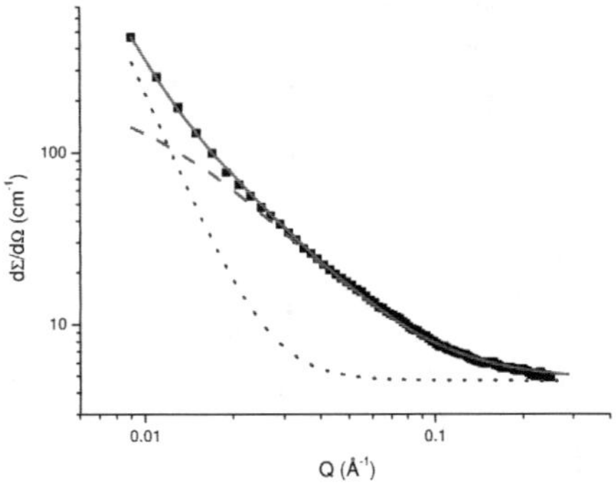

Figure 6b: Log-log plot of scattering data and data fits for a fibre spun from 0.2g/ml HPS-1/DPS-1in MEK at a voltage of 10kV and collector distance 20cm. ▪ is the scattered data, solid line is the combined model fit, dashed line is the polydisperse Gaussian coil fit and the dotted line is the Q^{-N} fit.

Figure 6a and 6b shows that the contribution of the scattering from the voids dominates at low Q. The scattering from the voids is less in fibres that have been prepared from MEK. This is possibly due to the collapsed structure of the fibre reducing the trapped solvent present. Values of the radius of gyration extracted using this approach for the samples referred to in Figure 6 are 157Å ± 10Å for the fibre prepared from DMF and 156Å ± 10Å for the fibre prepared from MEK. These values are ~ 20% above that found in the bulk state material (130Å) as is expected from a good solvent which suggests that little rearrangement of the chains has taken place during what can be considered to be flash drying during the electrospinning process.

The data recorded for samples with a macroscopic alignment of the electrospun fibres were subjected to a similar analysis but using sections taken from the 2-d data between 60° and 120° and -30° to 30° where 90° corresponds to the fibre axis. Application of the model defined in Equation 3 was used to measure the radius of gyration for the isotopically labelled aligned samples for each data set at 60° to 120° (parallel to fibre axis) and -30° to 30° (perpendicular to fibre axis). In the aligned samples there is a larger radius of gyration value parallel to the fibre axis when compared to the perpendicular component showing an extension of the chain conformation. The largest difference observed was an extension of ~ 10% parallel to the fibre axis. The fibres are not perfectly aligned with respect to the main axis of rotation of the collector. There is an angular distribution of fibres, therefore the scattering observed is a convolution of the angular distribution of the fibres from the main axis of rotation for the collector and the scattering for an individual fibre. The angular variation of the fibres varies between samples and work is currently underway to include this deconvolution process in the analysis procedure.

4. Conclusions

SANS from electrospun fibres indicates that there is a high level of voiding within samples prepared from DMF, as confirmed by SEM analysis. A model has been developed and successfully applied to extract the values for the radius of gyration of the polymer chains within the fibres. This analysis shows that the radius of gyration in the fibres is ~ 20% greater than that of the bulk state as expected for a good solvent system. Similar values were recorded for samples prepared using MEK as the solvent. This analysis was extended to analyse the anisotropy of the chain conformations, by comparing the dimensions parallel and perpendicular to the fibre axis. This revealed anisotropy in the radius of gyration which was up to 10% greater in the direct parallel to the fibre axis.

References

[1] Taylor G 1964 *Proc. R. Soc. London Ser. A.* **280** 383 – 397
[2] Ramakrishna S, Fujihara K, Teo WE, Lim TC and Ma Z 2005 *An Introduction to Electrospinning and Nanofibres* (World Scientific Publishing Co.)

[3] Jarusuwannapoom T, Hongrojjanawiwat W, Jitjaicham S, Wannatong L, Nithitanakul M, Pattamaprom C, Koombhongse P, Rangkupan R and Supaphol P 2005 *European Polymer Journal* **41** 409-421
[4] Wang C, Hsu CH, and Lin JH 2006 *Macromolecules* **39** 7662-7672
[5] Eda G, Liu J and Shivkumar S 2007 *Materials Letters* **61** 1451–1455
[6] Medeiros ES, Mattoso LHC, Offeman RD, Wood DF and Orts WJ 2008 *Canadian Journal Of Chemistry* **86** 590-599
[7] Demir MM 2010 *eXPRESS Polymer Letters* **4.1** 2–8
[8] Heenan RK RAL REPORT 89-129
[9] Pethrick RA and Dawkins JV 1999 *Modern Techniques for Polymer Characterisation* (John Wiley)

Network Microstructure of PVA Cast Gels Observed by SAXS Measurements

Emiko Otsuka[1], Masaaki Sugiyama[2] and Atsushi Suzuki[1]

[1] Graduate School of Environment and Information Sciences, Yokohama National University, 79-7 Tokiwadai, Hodogaya-ku, Yokohama, 240-8501 Japan.
[2] Research Reactor Institute, Kyoto University, 2 Asashiro-Nishi, Kumatori-cho, Sennan-gun, Osaka 590-0494 Japan.

E-mail: eotsuka@ynu.ac.jp

Abstract. We report the relationship between macroscopic properties (swelling ratio, breaking stress and strain) and the network microstructure of PVA cast gel. Swelling in water solvent and drying in air were alternately repeated, where water was exchanged by new water before each swelling. The effects of the number of water exchange, *N* on the macro- and microscopic properties were examined in detail. As a result, the swelling ratio of PVA cast gel decreased, and both of breaking stress and strain increased with increasing *N*. The measurements using X-ray diffraction (XRD) indicated that the average distance between the microcrystallites decreased due to the formation of the additional microcrystallites during the repeated processes of water exchange with drying. On the other hand, the measurements using Fourier Transform Infrared (FT-IR) Spectroscopy suggested that the microcrystallites and hydrogen bonds between polymers increased during the repeated processes.

1. Introduction

Poly(vinyl alcohol) (PVA) gels are useful biomaterials due to its low toxicity and high biocompatibility, and many studies on their practical applications have been reported by researchers in a variety of fields [1,2]. PVA can be physically crosslinked by the freeze-thawing method in water [3-5] or in a dimethyl sulfoxide (DMSO)/water mixed solvent [6-9], which introduces microcrystallites as crosslinks. On the other hand, PVA can be chemically crosslinked by crosslinkers, such as glutaraldehyde (GA) [10,11] or multifunctional reactive compounds [12]. Chemical gels can be also obtained by crosslinking PVA in solid state or in solution by irradiation of γ-ray [13] or electron beam [14,15]. Mechanical properties of physically and chemically crosslinked PVA gel have been extensively studied [16-18].

Recently, a simple method to obtain a physically crosslinked PVA gel (called "PVA cast gel") was successfully developed [19]. The materials used in this technique were water and PVA powders, and the method just involved drying at room temperature; no special chemicals were used. The formation of hydrogen bonds and microcrystallites was identified during the drying process (gelation process) using X-ray diffraction (XRD) technique, Fourier transform infrared (FT-IR) spectroscopy, and measurements of the swelling ratio under repeated water exchanges [20]. The dried sample could swell and was insoluble in water even after sufficient water exchange, which indicated that PVA cast gel was a physically crosslinked polymer network, in other words, a swollen hydrogel.

In this paper, we report the swelling and mechanical properties in pure water of a thin plate of physically crosslinked PVA gels prepared by a cast-drying method. The swelling ratio was measured whenever the dried PVA cast gel reached the equilibrium swollen-state in water during the repeated water exchange with drying. For each swollen gel during the repeated processes, the breaking stress, σ_r, and strain, ε_r, were obtained by a uniaxial loading test. The network structure was analyzed using XRD and FT-IR measurements. The macroscopic properties are discussed in terms of the formation of microcrystallites and hydrogen bonds by the exchange of solvent with drying.

2. Experimental

2.1. Sample preparation

PVA powder was kindly supplied by Kuraray Co., Ltd. (Cat. No. PVA117), and was used without further purification. The average degree of polymerization was 1,700, and the average degree of hydrolysis was between 98 and 99 mol%. An aqueous PVA solution was obtained by dissolving 15.0 wt% of the PVA powder into deionized and distilled water at 90 °C for more than 2 hrs. The PVA solution (15.0 g) was poured into a plastic dish of polyethylene with an inner diameter of 85 mm and left in air at room temperature (ca. 25 °C). The solution became very viscous within several days and solidified (thin disk) within a week or more. The sample was prepared by a single process of drying of the PVA aqueous solution at room temperature. The single gelation method named as "cast-drying method".

2.2. Measurement of the swelling ratio

Figure 1 shows the schematic diagram of the process of repeated water exchange with drying. The as-prepared sample (PVA cast gel film) was put into 1,000 mL of pure water at room temperature and kept for 24 hrs to reach the equilibrium state. After the weight of the swollen sample, W_t, was measured in air, the sample was dried in air at 30 °C for 24 hrs, and hereafter the dehydrated weight, W_d, was measured. This process of the swelling in water and the drying at room temperature were alternately repeated, where water was exchanged by new water before each swelling. The swelling ratio, which is defined here as W_t/W_d, was calculated. This process of water exchange with drying was repeated 9 times (the number of the repeated processes, $N = 9$).

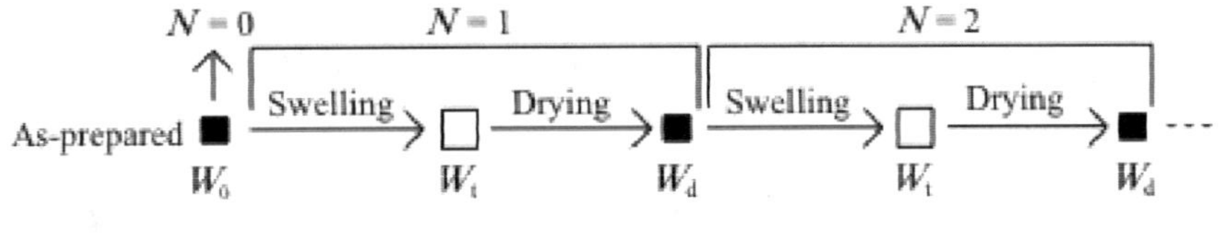

Figure 1. Illustration of the treatment for the measurement of swelling ratio of PVA cast gel during repeated water exchange with drying.

2.3. Measurement of the stress-strain curve

For each swollen gel during the repeated processes, σ_r and ε_r were obtained by a uniaxial loading test. In this test, the gel was shaped using a dumbbell cutter in accordance with JIS K-6251-8. The ends of the dumbbell-shaped gel were clamped in rigid, acrylic grips. The gel was stretched in pure water at room temperature with a constant rate of elongational stretch between grips. The forces generated during deformation were measured using a double leaf spring with four strain gauges. Using a food dye two markers were applied to the gel by hand in the gauge section. Images of the gel and markers were obtained during the tests using a video camera, which were captured and saved in a computer. The distance between the markers was obtained by analyzing the images after the test. The typical stretch rate in the gauge section was 0.63 mm/sec. From the measured values of force and elongation, the nominal stress, σ, and strain, ε, were calculated.

2.4. SAXS, SWAXS and ATR FT-IR measurements

To determine the network structures, SAXS measurements were performed at room temperature using a small angle X-ray scattering (SAXS) equipment with the wavelength of 0.1488 nm at BL10C in the Photon Factory, High Energy Accelerator Research Organization (KEK), and using additionally a small and wide angle X-ray scattering (SWAXS) equipment with the wavelength of 0.1542 nm operated at 40 kV and 50 mA (SAXSess, Anton Paar, PANalytical).

To study the change in molecular structures (e.g., the formation of hydrogen bonds) of the PVA cast gel during the water exchange with drying, the measurements of FT-IR with an Attenuated Total Reflection (ATR) method were carried out at room temperature. The ATR FT-IR spectra were recorded on an FT-IR spectrophotometer (Jasco FTIR610) equipped with an ATR attachment with a horizontal ZnSe crystal (Jasco PRO400S). The resolution of the spectra was 4 cm^{-1}, and the scans were repeated 100 times. The IR absorbance spectra were normalized based on CH_2 bending vibration that appears at ca. 1,500 ~ 1,400 cm^{-1}. The components of absorption band were deduced from the second-derivative of spectra and were resolved applying the curve-fitting method. All spectra were found to involve CH_2 bending vibration consisting of five peaks, including the peak at 1,427 cm^{-1} [21]. The total absorbance area of five peaks was assumed equivalent under the present experimental conditions, and the spectra were normalized by the total area.

3. Results and Discussion

3.1. Swelling and mechanical properties of PVA cast gels during water exchange with drying

Figure 2 shows W_t/W_d of PVA cast gel as a function of N. The W_t/W_d decreased rapidly from N = 1 to 3, and hereafter continued to decrease gradually.

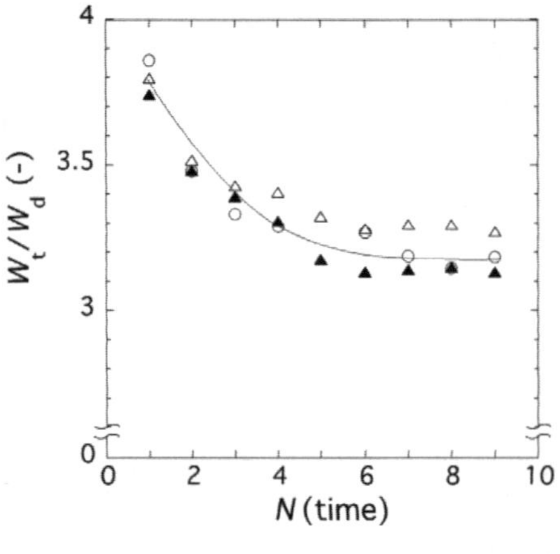

Figure 2. Swelling ratio of PVA cast gels during water exchange with drying. The results of three samples from two different PVA cast gel films, which were prepared by the same method at different times, are presented. Open circles; the gel obtained from one PVA cast gel film used in the uniaxial loading test. Open and closed triangles; the gels obtained from the other PVA cast gel.

Figure 3 (a) shows the nominal stress-strain (σ-ε) curve of each swollen gel. The slope in the low ε-range, which corresponds to Young's modulus, E, increased apparently with increasing N (Fig. 3(b)).

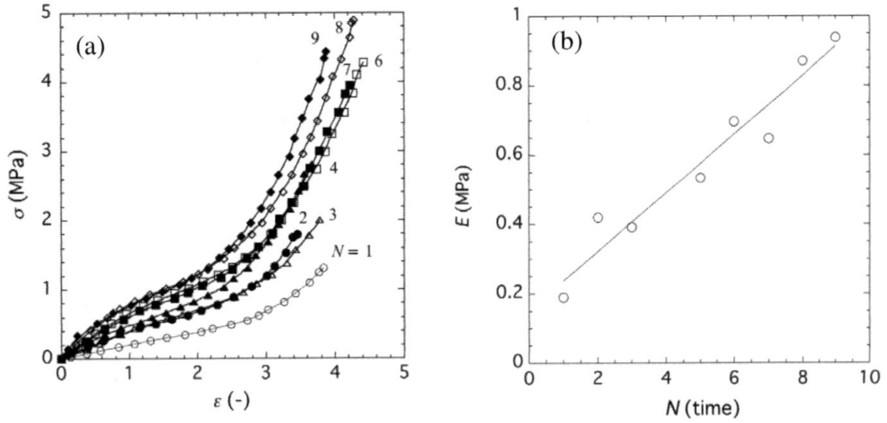

Figure 3. (a) Stress-strain (σ-ε) curves during the simple elongation test of PVA cast gels during water exchange with drying, where the gels were ruptured at the maximum ε, and (b) Young's modulus of the samples, determined from the σ-ε curve in (a).

Figure 4 shows the relationship between W_t/W_d and σ_r or ε_r of swollen gels, which includes the data in Fig. 2. Taking the results of Fig. 2 into account, we found that with increasing N, W_t/W_d decreased and both of σ_r and ε_r increased. These results should be attributed to the changes in the network microstructures by the repeated water exchange with drying. It has been well recognized that W_t/W_d decreases and E increases with increasing the crosslinks [22,23]. Therefore, one possible origin of the changes in the network microstructures might be related with the increment of the crosslinks by the repeated treatment.

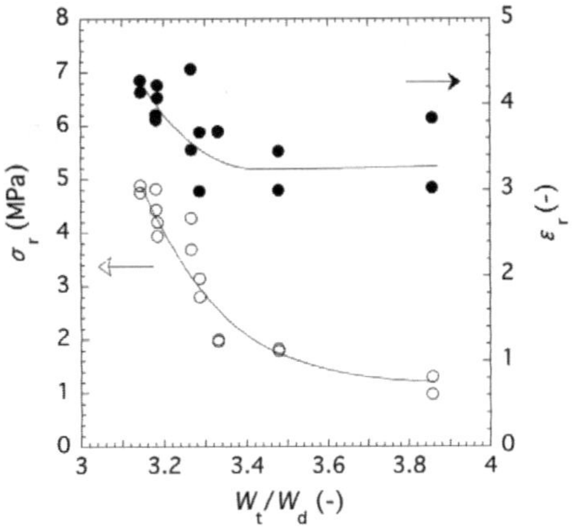

Figure 4. Relationship between the breaking stress and strain and W_t/W_d of PVA cast gels during water exchange with drying.

3.2. Network structure of PVA cast gel

Figure 5 shows SAXS spectra of three as-prepared PVA cast films with W_t/W_d = ca. 1.05 at $N = 0$, which were prepared by the same method at different times. Each spectrum was fitted by the Lorenz function, $I(q) = I_0 + (2A/\pi)[(w/\{4(q - q_c)^2 + w^2\})]$, where I_0, A, $w/2$, and q_c are the background, the area, the half-width, and the peak-top position, respectively. Although the fitting parameter, A was largely different from each other, the parameters, w and q_c were almost the same with errors less than 5 and 0.7 %, respectively. A broad peak centered at q_c = ca. 0.64 nm^{-1} was characteristic to the PVA cast film, which corresponded to the average distance between the microcrystallites, L = ca. 9.8 nm.

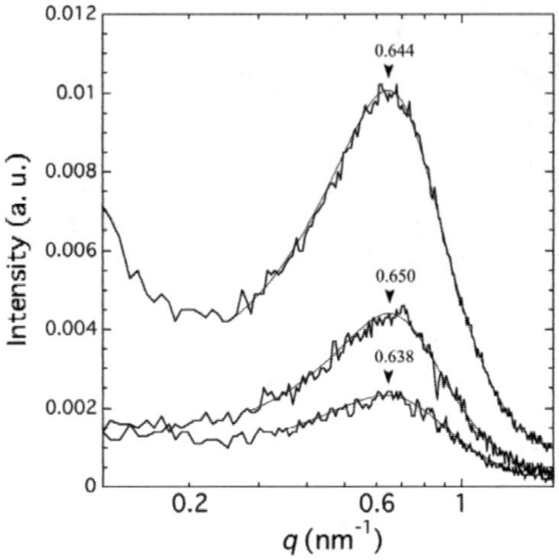

Figure 5. SAXS spectra of PVA cast gel film ($N = 0$). Three different samples prepared by the same method at different times were used.

Figure 6(a) shows SWAXS spectra of the PVA cast gel film in a swelling swollen process. Intersection of two extrapolations of lines before and after the broad peak in dried gel with W_t/W_d = 1.05 was q = 0.65 nm^{-1} (L = 9.7 nm). The peak position agreed well with q_c obtained by the SAXS measurements (Fig. 5). With an increase in W_t/W_d (from 1.05 to 2.6) by slowly adding water, the peak position shifted to a lower wavelength, and finally disappeared above W_t/W_d = 3.3 since the scattering from water became dominant in the spectrum. As is shown in Fig. 6(b), L increased at first, and gradually increased when W_t/W_d exceeded 1.8. Therefore, we can approximately consider L of gel with $W_t/W_d = 2$ to be L of the swollen gel.

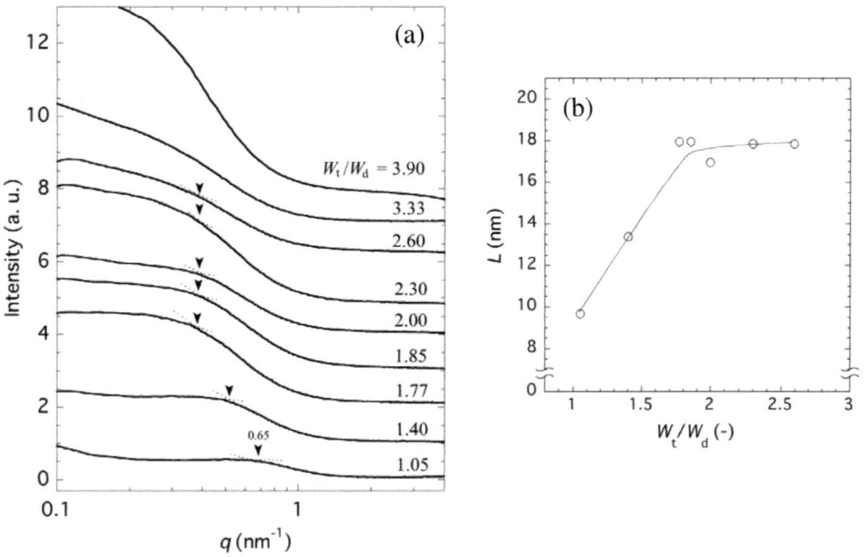

Figure 6. (a) SWAXS spectra of as-prepared PVA cast gels with increasing W_t/W_d. The vertical axis of each spectrum is shifted for clarity. (b) L of PVA cast gels during increasing W_t/W_d.

Figure 7(a) shows SWAXS spectra of the PVA cast gels in the swollen state with $W_t/W_d = 2$. With increasing N, the peak position shifted slightly but evidently to a higher wavelength, indicating that L decreased by the repeated water exchange with drying (Fig. 7(b)). This observation was consistent with the decrement of the swelling ratio, as is shown in Fig. 2.

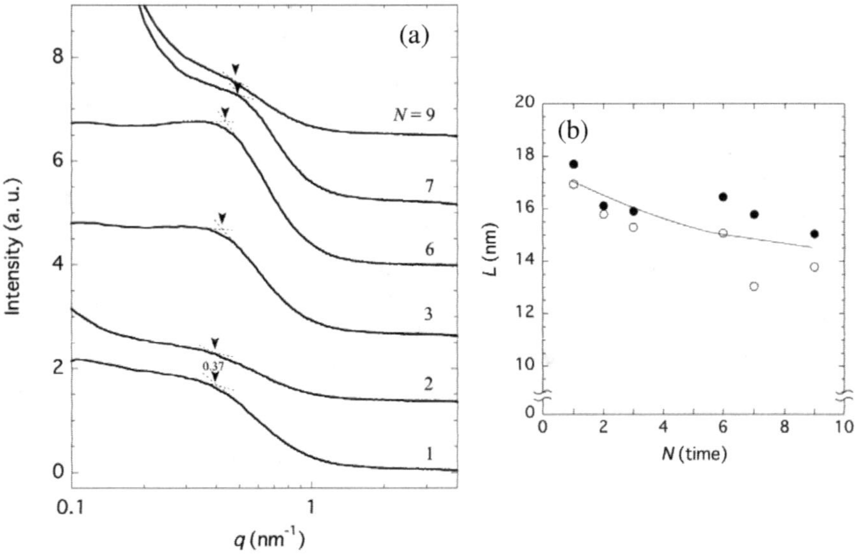

Figure 7. (a) SWAXS spectra of PVA cast gels in $W_t/W_d = 2$ during water exchange with drying. The vertical axis of each spectrum is shifted for clarity. (b) L as a function of N. Two sets of data were obtained using different samples prepared by the same method at different times.

Figure 8(a) shows ATR FT-IR spectra of the dried gels with increasing *N*. The large peak at around 3,300 cm^{-1} was due to the O-H stretching of the non-hydrogen bonded and hydrogen bonded O-H groups [24]. Its peak height was almost constant during repeated water exchange with drying. On the other hand, the C-O stretching at 1,143 cm^{-1} due to the formation of microcrystallites [25] slightly changed with increasing *N*. Figure 8(b) shows the peak area at 1,143 cm^{-1}, which was obtained by deconvoluting the spectra to remove the effects of the other peaks. The peak area at 1,143 cm^{-1} increased from $N = 0$ to 1 and became almost constant after $N = 1$, resulting from the formation of additional microcrystallites by the repeated water exchanges with drying.
From the present results of SAXS, SWAXS and ATR FT-IR spectra, we found that the repeated processes of the water exchange with drying resulted in the formation of additional crosslinks due to the hydrogen bonds and the microcrystallites. The microstructures of polymer network characterized by the microcrystallites have important roles in determining the macroscopic swelling and mechanical properties.

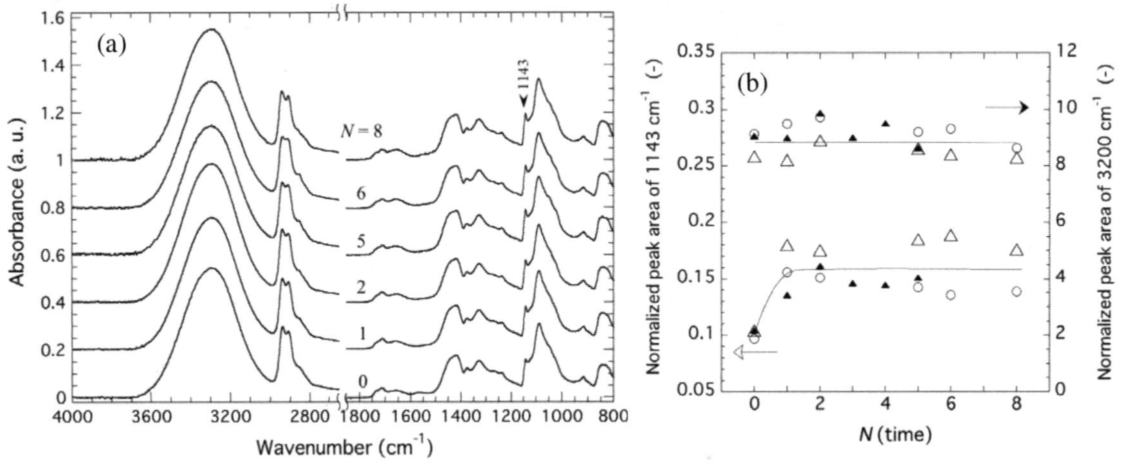

Figure 8. (a) ATR FT-IR spectra of PVA cast gels in dried state during water exchange with drying. The vertical axis of each spectrum is shifted for clarity. (b) The peak area at 1,143 cm^{-1} of the spectra, where each symbol indicates the same meaning described in Fig. 2.

4. Conclusion

We have studied the effects of water exchange with drying on the equilibrium swelling ratio and the mechanical properties of PVA cast gel prepared by a cast-drying method. From the measurements of XRD and FT-IR techniques, the formation of additional microcrystallites due to hydrogen bonds between -OH groups of PVA were detected during the repeated water exchange with drying. These microscopic analyses were consistent with the macroscopic observations that the swelling ratio decreased and the breaking stress and strain increased at first and became constant thereafter. After sufficient water exchanges with drying, the swelling ratio should approach constant, therefore the network structure of PVA cast gel might become stable.

References

[1] Finch CA (ed) 1997 *Polyvinyl Alcohol-Developments, 2nd Edition* (New Jersey-Wiley)
[2] Hassan C M and Peppas N A 2000 *Adv. Polym. Sci.* **153** 37
[3] Peppas N A 1975 *Makromol. Chem.* **176** 3443
[4] Nambu M 1982 *Japanese Patent Kokai* **57**/130543
[5] Nambu M 1984 *Japanese Patent Kokai* **59**/56446
[6] Takeshita H, Kanaya T, Nishida K and Kaji K 2001 *Macromolecules* **34** 7894
[7] Yamaura K, Hamaguchi M, Rokudai N, Tamura S and Matsuzawa S 1994 *J. Appl. Polym. Sci.* **51** 2041
[8] Tanigami T, Murase K, Yamaura K and Matsuzawa S 1994 *Polymer* **35** 2573
[9] Takahashi N, Kanaya T, Nishida K and Kaji K 2003 *Polymer* **44** 4075
[10] An Y, Koyama T, Hanabusa K, Shirai H, Ikeda J, Yoneno H and Itoh T 1995 *Polymer* **36** 2297
[11] Praptowidodo V S 2005 *J. Mol. Struct.* **739** 207
[12] Mansur H S, Oréfice R L and Mansur A A P 2004 *Polymer* **45** 7193
[13] Wang B, Kodama M, Mukataka S and Kokufuta E 1998 Polymer Gels and Networks **6** 71
[14] Peppas N A and Merrill E W 1977 *J. Appl. Polym. Sci.* **21** 1763
[15] Yoshii F, Zhanshan Y, Isobe K, Shinozaki K and Makuuchi K 1999 *Radiat. Phys. Chem.* **55** 133
[16] Takigawa T, Kashihara H, Urayama K and Masuda T 1992 *Polymer* **33** 2334
[17] Urayama K, Ogasawara S and Takigawa T 2006 *Polymer* **47** 6868
[18] Meissner B and Matejka L 2008 *Polymer* **49** 2560
[19] Otsuka E and Suzuki A 2009 *J. Appl. Polym. Sci.* **114** 10
[20] Otsuka E and Suzuki A 2009 *Prog. Colloid. Polym. Sci.* **136** 121
[21] Kumar G N H, Rao J L, Gopal N O, Narasimhulu K V, Chakradhar R P S and Rajilu A V 2004 *Polymer* **45** 5407
[22] Flory P J 1953 *Principles of Polymer Chemistry* (New York-Cornell University Press)
[23] Yokoyama F, Masada I, Shimamura K, Ikawa K and Monobe K 1986 *Coll. Polym. Sci.* **264** 595
[24] Miya M, Iwamoto R and Mima S 1984 *J. Polym. Sci. Polym. Phys. Ed.* **22** 1149
[25] Lee J, Lee K J and Jang J 2008 *Polym. Test.* **27** 360

Journal of Physics: Conference Series **247** (2010) 012044 doi:10.1088/1742-6596/247/1/012044

In-situ observation of dynamic water behavior in polymer electrolyte fuel cell by combined method of Small-Angle Neutron Scattering and Neutron Radiography

A Putra[1], H Iwase[2], D Yamaguchi[1], S Koizumi[1,5], Y Maekawa[3], M Matsubayashi[4], T Hashimoto[1]

[1]Advanced Science Research Center, Japan Atomic Energy Agency, Tokai, Ibaraki 319-1195, JAPAN
[2]Neutron Science Laboratory, Institute Solid State Physics, University of Tokyo, Tokai, Ibaraki, 319-1106, JAPAN
[3]Quantum Beam Science Directorate, Japan Atomic Energy Agency, Takasaki, Gunma 370-1292, JAPAN
[4]Quantum Beam Science Directorate, Japan Atomic Energy Agency, Tokai, Ibaraki 319-1195, JAPAN

E-mail: koizumi.satoshi@jaea.go.jp

Abstract. In-situ observation was conducted on an operating polymer electrolyte fuel cell with a combined method of small-angle neutron scattering (SANS) and neutron radiography (NR). The combined measurement system has been recently developed to visualize water in a wide length scale from nano- to millimeter and successfully detected a spatial distribution of the water generated in individual cell elements; NR macroscopically detected the water in a gas diffusion layer and a flow-field, whereas SANS microscopically did in a membrane electrode assembly. In particular SANS was found to be a strong tool to make a rather precise analysis on the water content inside of ion conducting channels of polymer electrolyte membrane.

1. Introduction

A Fuel cell has attracted a lot of interests as a next-generation power supply. Particularly a polymer electrolyte fuel cell (PEFC) is promising replacement for conventional power sources because it exhibits efficient energy conversion without the chemical pollution associated with fuel combustion. The cell has an operating temperature near 80 °C, removing the need for thermal shielding and allowing for instant start-up. Additionally, the direct conversion of chemical energy to electrical energy side steps the inefficient intermediate conversion to heat and can thus result in efficiencies as high as 90%. As a result the PEFC has promising application in transportation and consumer electronic and considerable amount of research funding is being directed towards improving the efficiency, reliability, and cost effectiveness of a PEFC system [1].

Efficient water management has become increasingly important in the effort to optimize fuel cell energy production [2,3]. During an operation of PEFC, water is generated in individual cell elements (a polymer electrolyte membrane (PEM), catalysts, gas diffusion layers (GDL), and flow fields) as a result of electrochemical reaction, transportation and exclusion of water. Because the proton conductivity of the PEM is highly dependent upon its internal water content, the cell must balance water production and removal to create equilibrium where the PEM is highly saturated without flooding either of the electrodes. A deficiency of water in the flow channel can potentially dry out the membrane, reducing conductivity and efficiency. However, flooding the channels with water impedes gas flow and also reduces efficiency. Therefore, it is crucial to determine simultaneously and in-situ the distribution and diffusion of the water appeared in the individual cell elements belonging to a given single fuel cell under operation.

In order to evaluate the effectiveness of water management schemes, a method capable of looking inside an operational fuel cell and visualize movement of water dynamic could be very useful. So far, many of in situ observations on the water in an operating PEFC have been attempted by several methods [4-15], however scientist are still developing the best way to visualize the water generated inside an operating PEFC in a wide length scale from nanometer (microscopic scale) to millimeter (macroscopic scale). In fact it is an ambitious attempt to cover the whole length scale from nano to millimetre at once. Nevertheless, any difficulties can be overcome by taxing one's ingenuity. Here in order to attain our purpose, utilization of neutron beam as a probe for water seems a promising solution. For example, in the microscopic scale, Small-angle X-ray scattering (SAXS) [4] and small-angle neutron scattering (SANS) [5, 12] have successfully observed a scattering maximum, the so-called "ionomer peak" [16], due to the ion-cluster swollen by water [4, 5, 16, 17] in the PEM. While the measurement of ionomer peak of PEM under the operating condition of PEFC is hardly achieved by SAXS since X-ray has small magnitudes in the scattering lengths, *b*, for the "light" elements such as C, H, O and the scattering from those elements is easily overwhelmed by that from the other metallic elements. Contrarily, neutron is endowed with potential for *in-situ* observation of the materials consisting of light elements by following points; (i) neutron has no charge and interacts not with electrons but with nuclei of the atoms through nuclear forces (ii) the energy level of neutron (an order of 10meV) is much lower than that of X-ray (an order of keV) when they are compared with at the similar wavelength of Å order. Because of the character (i), neutron shows generally high penetration and complementary scattering properties as compared with those of X-ray. That is, neutron can pass through some kind of metal, e.g., aluminium, nearly 100% (when its thickness is around 1cm), while neutron is fairly scattered by the materials composed of light elements, like water. Another merit of neutron given by character (ii) is that during the irradiation neutron makes nearly no damages on the specimen, such as degradation through bond breaking and following generation of radicals. Furthermore, owing to the complexity of nucleus-neutron interaction *b* can be greatly changed even between isotopes of the same element and in particular the difference in *b* between hydrogen (H) and deuterium (D) corresponds to an outstanding case. With all those merits of neutron are taken into account, in this study we conducted an *in-situ* observation of operating PEFC by using SANS technique. As will be described in detail in the experimental section, the accessible length scales by SANS are from nanometer to several micrometers, and thus, crucial information regarding to the water distribution within PEM and catalyst layers is well-detectable by this technique. Meanwhile, the water distribution in the flow field as well as in GDL is related to an issue of macroscopic scale and beyond the scope of SANS. However, neutron has already established the other solution to observe those macroscopic scales larger than several tens of microns, that is, a real-space imaging technique, neutron radiography (NR). Here it should be stressed again that the character (i) effectively enhances the value of NR in observation of operating PEFC. As neutron can completely pass the outer metallic parts and is highly sensitive to hydrogen, NR is one of the best methods to visualize the water distribution inside of PEFEC in the macroscopic length scale [7-11, 13-15].

Naturally, SANS and NR, have been developed independently to investigate static or dynamic structural properties of materials in micro- and macro-scales, respectively. From now on, it certainly gives great advantage to make both of two measurements be conducted in the same instrument and to acquire the data from micro- to macro-scales at once. To realize this idea, a neutron imaging system is installed between the sample position and detector chamber of the SANS spectrometer [18]. Though this contrivable method its self is already describe elsewhere[18], here we introduce typical SANS data obtained from operating PEFC giving the significance of comparison with the simultaneous NR data.

2. Experimental section

2.1. Combined SANS and NR set-up

Figure 1 shows a schematic illustration for the concept of combined measurement of SANS with NR in one instrument. As shown in figure 1a the basic instrument is the focusing and polarized neutron ultra-small-angle scattering spectrometer (SANS-J-II) at research reactor (JRR-3) at Japan Atomic Energy Agency (JAEA), Tokai, Japan [19]. The total length of the spectrometer from the 1st aperture, which is located at just behind the velocity selector, to detector is 20m, and the sample position is located at the centre between the 1st aperture and detector. Thus, both of the distances from 1st aperture to sample and from sample to detector are equal to 10m and such long distances make an advantage of high spatial resolution for both SANS and NR measurements. The other detailed specifications of SANS-J-II spectrometer are described elsewhere [18, 19] and here we mention the typical collimations employed in measurements. The incident neutron was usually monochromatized by using a velocity selector to the wavelength of λ = 6.5 Å and $\Delta\lambda/\lambda$ = 0.13. With utilizing the conventional pinhole collimation (i.e., D_1 and D_2, which denote the sizes of 1st and 2nd apertures, respectively, are set to D_1 = 20x20 mm^2 and D_2 = 8 mmϕ) and the ^{3}He main position sensitive detector (PSD), whose diameter and spatial resolution were 0.58 m and 5 mm, respectively, and the camera length was changed from 10m to 2.5m, we can cover the q-range from q = 3x10^{-3} to 0.2 Å^{-1}, where q is magnitude of wave vector given by $q=(4\pi/\lambda)\sin(\theta)$ with 2θ being the scattering angle. Furthermore, a high-resolution position sensitive scintillation detector (5 inch in size with 0.5 mm spatial resolution) was utilized for focusing ultra-small-angle neutron scattering (focusing USANS), to access the range of q = 3x10^{-4} ~ 3x10^{-3} Å^{-1}. In all, by employing both pinhole SANS and focusing USANS, a wide range of q from 3x10^{-4} to 0.2 Å^{-1}, corresponding to from several micrometers to nanometers in real space, can be measured on SANS-J-II instrument.

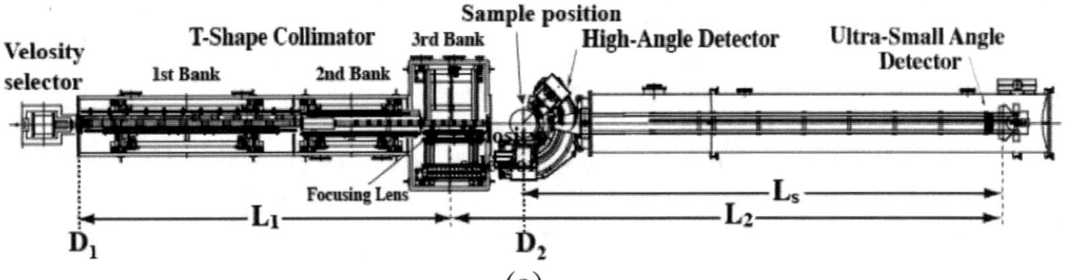

(a)

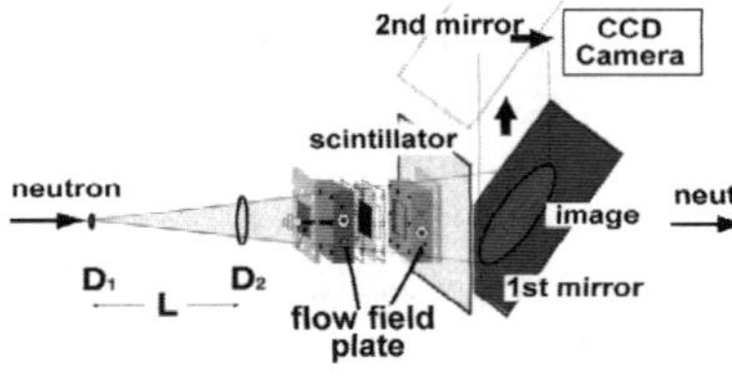

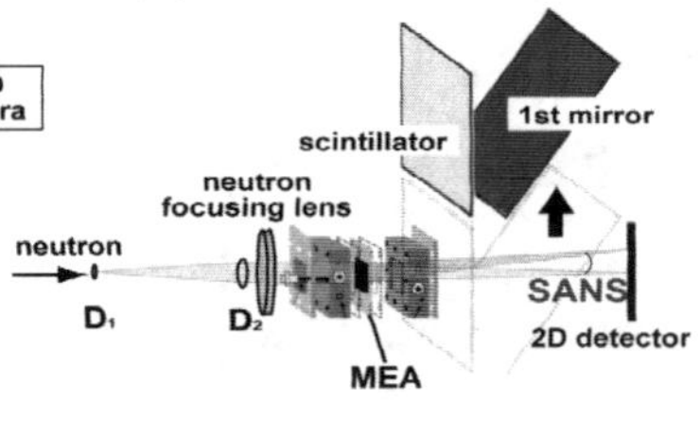

(b) (c)

Figure 1. Schematic diagram of a focusing and polarized neutron ultra-small-angle scattering spectrometer (SANS-J-II) (a), and illustration of a combined method of neutron radiography (b) and small-angle neutron scattering (c) installed at the SANS-J-II spectrometer at research reactor (JRR-3) Japan Atomic Energy Agency (JAEA), Tokai, Japan.

To combine SANS with NR, we installed a neutron imaging system just behind the sample position of SANS-J-II (cf. figure 1). Figure 1b and 1c schematically show our switching mode of NR and SANS measurements, respectively. The imaging system was composed of a sheet of $ZnS/^6LiF$ scintillator (0.2 mm thickness), two optical mirrors, optical lens (MicroNikkor, 55 mm, Nikon Co.), and a full frame transfer type CCD camera (C4880, Hamamatsu Photonics Co. Ltd.), respectively. The CCD camera was placed far from a direct neutron beam to protect from radiation damage. The scintillator and first optical mirror were mounted on a vertical elevating bench. For NR measurements, we selected D_1 = 20x20 and D_2 = 50x50 mm^2, and L = 10000 mm, giving L/D_1 = 500 which was matched with the spatial resolution of the imaging system of 184 μm. When we observed SANS from PEFC, a scintillator sheet and the first optical mirror (defined as a NR set), were removed out from a beam line, and SANS were acquired by a 2D detector (figure. 1c).

2.2. Fuel Cell set-up

In this study we have made the following modifications on the standard single fuel cell provided by Japan Automobile Research Institute (JARI)[20] to optimize it for the measurements of SANS and NR. On a JARI standard single fuel cell, an active area of MEA was originally 50 x 50 mm^2 and a flow channel for H_2 and air gases was a single parallel serpentine, the width, depth and pitch of which were 1mm, each. Figure 2 shows a front view of JARI cell on which the shaded square corresponds to the 50 x 50 mm^2 of active area of MEA as well as the area irradiated by neutron when NR measurement was conducted. While the circles with numbers correspond to the spot size of 10mmϕ and indicate the irradiated area of each SANS measurement performed at different nine positions. As will be discussed later, the position dependence of SANS profile should be related to the distribution of water at microscopic scale. With each standard dimension of the JARI cell being kept, we replaced some of cell elements to enhance the transmission of neutron. Graphite flow field plates and cupper current collectors, which cause strong small-angle scattering for neutron, were replaced with those made of aluminum and further coated with a gold thin layer (1μm thickness) to achieve chemical stability and low electric resistance. Clamping plates were also made of thinner aluminum. Rubber sheet heaters were placed not to cover a MEA area of 50x50 mm^2. Transmission of the modified cell with those replacements described above was dramatically improved from 0.01 to 0.86 for a cold neutron λ = 6.5 Å. As will be discussed later a current-voltage curve was compared between the original standard JARI cell and the modified one and confirmed that it was well reproduced through the modification.

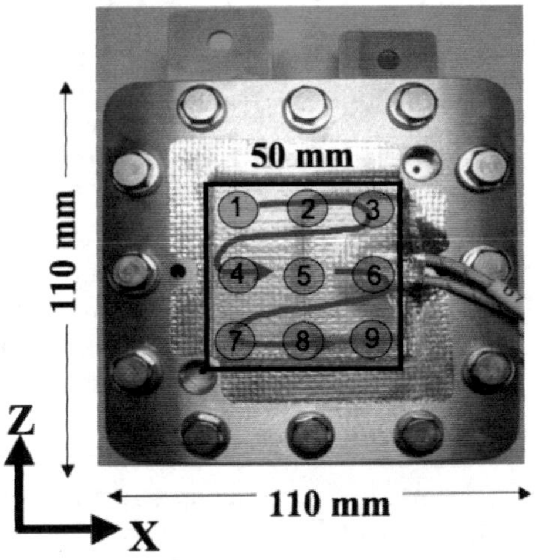

Figure 2. Single fuel cell designed by Japan Automobile Research Institute, JARI is modified as suitable for SANS and NR measurements. The numbered spot indicates the position examined by SANS, whereas the area (blue) indicates the area examined by NR.

A membrane electrode assembly (MEA) element consisted of a Nafion 212 (DuPont) sandwiched with carbon-supported catalysts of Pt (0.5 mg/cm^2) and Pt-Ru (0.5 mg/cm^2) for the cathode and anode, respectively was prepared by Eiwa Co. Ltd. Japan. Carbon papers were used as the GDL (190 μm thickness) for both anode and cathode. Hydrogen gas (99.99 %) was supplied from a hydrogen generator (OPGU-2200, HORIBA STEC Co., Ltd. Japan) and air was supplied from a common gas bomb. The other description related to PEFC operating system is elsewhere [18].

2.3. In-situ observation of dynamic water by combined SANS and NR method

In this study, the fuel cell temperature was controlled at 80 °C, and dew-points of the anode and the cathode at 80°C, respectively. The flow rates of hydrogen for the anode and these of air for the cathode were set at 180 ml/min and 800 ml/min, respectively. The current density was increased step-wisely from 0 to 640 mA/cm^2 with duration of 5 min at each current density. Before starting PEFC in operation, PEFC was purged with dry nitrogen for 15 min, and it was confirmed by NR that the no water was left inside the PEFC.

NR observation of operating PEFC was done for every increased current density from 0 to 640 mA/cm^2. Exposure time was 210 s for each observation. After subtracting a dark image, the all images were normalized by these of the dried cell obtained before operation. Other specifications for the measurement are described in sec. 2.1.

SANS measurements were also performed on the same operating PEFC at each current density from 0 to 640 mA/cm^2 alternately with NR ones. Furthermore, at each current density SANS profiles were measured for nine times at the different positions as described in sec. 2.2 (cf. figure. 2) to get the information about the distribution of water in microscopic level. The measuring time spent for each profiles was 80 s. Other detailed conditions for the SANS measurement are described in sec. 2.1, or elsewhere[18]. The obtained data were analyzed by SANS-J-II data analysis and IGOR-Pro version 5.02 program. A scattering profile from each measurement was subtracted by the scattering profile of operating PEFC without MEA.

3. Results and discussion

In figure 3 the polarization curves obtained from a standard (JARI) cell and from a modified one are shown. On the same figure, the distance between ion channels, L_{ion}, which is determined by SANS measurement and thus obtained from a center part of modified cell (i.e., position 5 in figure 2), is also plotted as a function of the current density. This L_{ion}-current density curve facilitates understanding of relationship between the electrochemical performance of PEFC and the water distribution at microscopic scale in PEM. Note that a more detailed explanation of L_{ion} will be described later when we illustrate a general feature of SANS profile obtained for operating PEFC system (cf. figure 5). According to the polarization curves in figure 3, performance of the modified cell is slightly better than that of the standard cell, i.e., at the intermediate range of current density (300-400 mA/cm^2), the cell voltage (V_{cell}) of modified cell is slightly higher (by ca. 0.03-0.05 V) than that of standard cell. However, both of polarization curves of standard and modified cells exhibit basically the same behavior and can be divided into following three stage; (1) A rapid initial fall in V_{cell} occurred at the initial increment of current density from 0 mA/cm^2. At the same time L_{ion} did also abruptly decrease by ca. 1.5 Å (from 39.5 to 38 Å), however, for the moment we cannot judge if there is a significant correlation between those decrement in V_{cell} and L_{ion}. (2) After the first rapid fall, the V_{cell} behaved rather stable and gradually reduced from ca. 0.8 to 0.65V during the increment of current density from 50 to 450 mA/cm^2. In the same stage, L_{ion} also exhibited an interesting behavior; at the smaller range of current density (from 50 to 150 mA/cm^2) L_{ion} was stable at 38 Å. While in the range of current density between 150 and 450 mA/cm^2, L_{ion} certainly increased from 38 to 44 Å. It seems to exist an explainable correlation between the behaviors of V_{cell} and L_{ion} in this stage. A gradual decrement in V_{cell} with increasing current density can be attributed to the sight increments of activation over voltage as well as resistance over voltage but hardly to the increment of concentration over voltage. The ground for excluding the concentration over voltage here is the observed increment in L_{ion} at the same region. In fact, L_{ion} tends upward when the ion channels are swollen by water (cf. the symbols of dry and swollen MEA, □ and ■, in figure 3). Therefore a constant increment of L_{ion} in the range of current density from 150 to 450 mA/cm^2 well reflects the increment of water content in ion channels. Such an increment is, on one hand, resulted from the enhancement in reaction rate of water production, and on the other hand, leading to an enhancement in transportation rate of protons, instead of causing a concentration over voltage. (3) In contrast, when the current density became over 500 mA/cm^2, the behaviors of both the V_{cell} and L_{ion} were completely changed from those observed in stage 2. Namely, V_{cell} rapidly fell down with increasing the current density and eventually PEFC stopped to operate at the current density of 640 mA/cm^2. While L_{ion} kept nearly constant at around 44 Å, suggesting a saturation of swelling of ion channel with water or some equilibrium swelling of ion channel attained by the balance between the inflow and outflow of water. In any case, the rapid decrement of V_{cell} observed on this stage tends to be attributed mainly to the concentration over voltage and to be connected with a curious correlation of L_{ion} behavior. Flooding of water occurred inside of the cell is, for instance, a well-known cause to impede smooth supply of reactant stuff, and one of the probable scenario to bring about both of a drop of V_{cell} and a saturation of L_{ion} simultaneously. However, to make sure such an interpretation, getting further information about water distribution by NR seems indispensable.

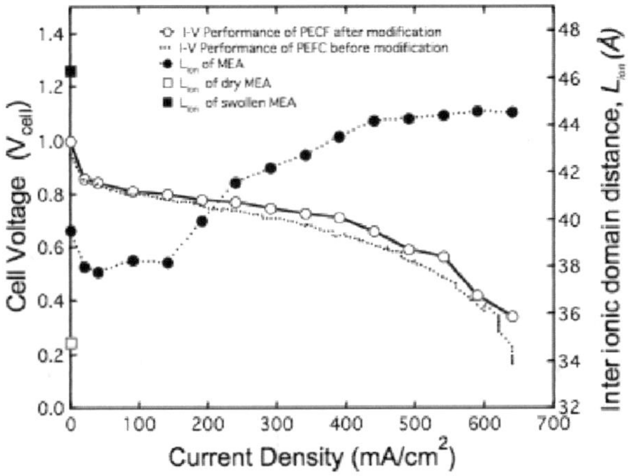

Figure 3. Cell voltage of PEFC before and after replacement of cell element for the combined method of SANS and NR, and inter ionic domain distance as function of current density at position 5.

Figure 4 shows NR images obtained for the modified cell operated at current densities from 0 to 640 mA/cm^2. Note that the inlets of anode (H_2) and cathode (air) gasses were at lower and upper right corners of the images and, their outlets were at upper and lower left corner of the images, respectively. After subtracting a dark image, the all images are normalized by these of the dried cell obtained before operation. When the current density was lower than 190 mA/cm^2, all of the observable area through the NR measurement is almost transparent, indicating that macroscopically water does not appear in the flow field plates. On the other hand, when the current density was ranging from 240 to 390 mA/cm^2, water started to appear as a shadow in the image at the upper right portion, near to the air inlet, the. The dark shadow was clearly observed along the single parallel serpentine. When the current density further increased and came to 440 mA/cm^2, the water started to accumulate widely from upper to middle parts of the NR image, and at 640 mA/cm^2, almost all area in the flow field-plates was fully filled with water. In particular, darker and more shadowed area was observed mainly on the portions of right and left edges in NR image where 90° turn regions of serpentine route exists, indicating a lot of water droplets generated in the flow fields. In contrast, the entire area irradiated by neutron becomes slightly dark, which originate from the accumulated water distributed in MEA or GDL. After acquisition of the above knowledge on the macroscopic distribution of water, the above interpretation as to the behaviors of V_{cell} and L_{ion} in figure 3 becomes more convincing. Namely, a rapid decrement in V_{cell} or a constant L_{ion} observed in stage 3 is most likely caused by the flooding of water occurred in the major part of flow field and GDL, as is clearly captured by the NR images of bottom row in figure

4. Such a comprehensive understanding concerning the water distribution should be a great merit of combined measurements of SANS and NR and will be further discussed later (cf. figure 6 and 7). However, before making a comparison of SANS and NR data, it is necessary to illustrate the overall SANS profile obtained for the modified cell under operating condition.

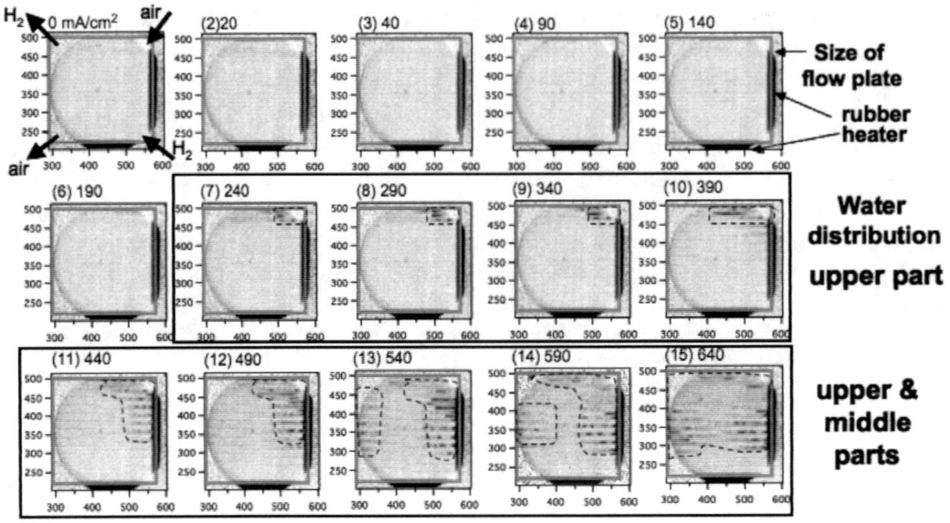

Figure 4. Neutron radiography images (50 x 50 mm^2) obtained for PEFC with single parallel serpentine flow field at operated current densities of 0 to 640 mA/cm^2. Anodic inlet and outlet for H_2, and cathodic inlet and outlet for air supplies are shown by arrows.

Figure 5 shows SANS profiles obtained at position 5 of modified cell (cf. figure 2) and with increasing current density from 0 to 640 mA/cm^2, where the profiles taken at various current density from (1) 0 mA/cm^2 to (2) 640 mA/cm^2 are superposed. According to the difference in power-law behavior as well as in the change of profile with increasing current density, the scattering profiles can be divided into two regions; (i) One possesses a power-law index of ca. -4 and is appeared in the q region of 0.003 ~ 0.02 Å^{-1}, which is attributed to the carbon-supported catalysts packed in MEA. Within this region, the scattering intensity was slightly decreased with increasing current density (cf. a thick gray arrow indicating downward at q= 0.005 Å^{-1}). (ii) The other possesses a power-law index of -2 and dominate the q region of 0.035 ~ 0.2 Å^{-1}, which is reflecting the structure of Nafion membrane in MEA. The scattering peak appeared at around q = 0.15 Å^{-1} corresponds to "ionomer peak" and this peak intensity was increased with increasing current density due to the swelling of ion-cluster by water [16, 18]. Assuming that the Nafion membrane, which is under the PEFC operation and swollen by water, is separated into the following two phase; (i) crystalline domains, which consists of polytetrafluoroethylene (PTFE) backbone and strongly expels water, and (ii) ion conducting channel, which is induced by and composed of the water surrounded by the sulfonic acid side chains of Nafion. Then, the scattering intensity in the second region (Nafion membrane) can be formulated as follow:

$$I(q) = I_{Cry}q^{-\alpha} + I_{ion}\exp[-C(q-q_m)^2] \qquad (C>0) \tag{1}$$

where $I_{Cry}q^{-\alpha}$ is scattering intensity of crystalline domain phase with slope, $\alpha = 2$, and $I_{ion}\exp[C(q\text{-}q_m)^2]$ is scattering peak due to the spatial distribution of the ion conducting channel of Nafion approximated by a Gaussian function.

From the scattering intensity of ion conducting channel, I_{ion}, peak-height is related to the magnitude in scattering length difference between water in the ion channel and matrix phase of the Nafion membrane. High peak indicated large contrast and therefore large amount of water is expected to be included in the ion channel. The I_{ion} can be calculated as follow:

$$I_{Ion} = \Delta b_{ion}^2\ \varphi(1-\varphi) \tag{2}$$

where φ is the volume fraction occupied by ion channel in the membrane, and Δb_{ion} is the difference in the scattering length density between the ion channel (here approximately consists of only water) and the matrix phase of Nafion membrane as given by following eq. 3.

$$\Delta b_{ion} = \phi\ b_{D_2O} + (1-\phi)\ b_{H_2O} - b_{Nafion} \tag{3}$$

The coherent scattering length of j-th molecule, b_j ($j = D_2O$, H_2O and *Nafion*), is calculated by $b_j = [N_A\rho_j\sum n_i b_i]/M_{w,j}$, where N_A is Avogadros's number, ρ_j is the mass density (g/cm^3) of j-th molecule, n_i is number of atomic species i in j-th molecule, b_i is the bound scattering length of atomic species i, and $M_{w,j}$ is the molar mass (g/mol) of j-th molecule. ϕ is the fraction of deuterated water (D_2O) in the ion channel when D_2O and H_2O coexist in the system. In this study $\phi = 0$ due to only H_2O produced in the experiment. The $b_{H_2O} = 5.6 \times 10^{-7}$ Å^{-2}, and $b_{Nafion} = 4.11 \times 10^{-6}$ Å^{-2}. Meanwhile, C, which controls the peak-width, is related to inhomogeneity of the interdomain distance of ion channel. Wider peak indicated large inhomogeneity in interdomain distance. And, q_m, peak-position, where the maximum intensity of the peak is located, is related to L_{ion}, i.e., the magnitude of interdomain distance between ion channels itself, as has already been introduced in figure 3. L_{ion} is determined from q_m by $L_{ion}=2\pi/q_m$. Shift of the peak to the lower q indicates increase in L_{ion} and as a consequence suggesting the accumulation of water in ion channel. In summary, as the current density increases we found the following features on SANS profile: the scattering intensity at $q=$ 0.005 Å^{-1} slightly decreases; whereas the scattering peak at around $q = 0.15$ Å^{-1} systematically increases, and its q-value shifts to a lower value. These findings are attributed to appearance of the water in MEA in nanometer scale.

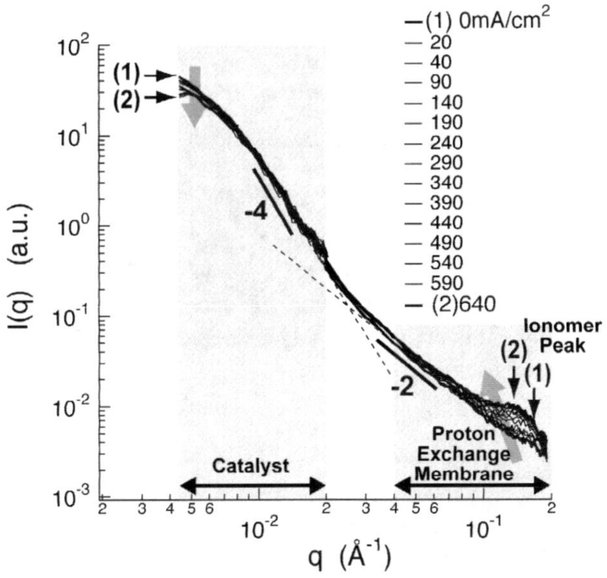

Figure 5. SANS profiles obtained for an operating PEFC at steady state at position 5 with increasing a current density from 0 to 640 mA/cm^2.

Next we will show the advantages of the combined NR and SANS method to visualize water in operating PEFC. As shown in figure 4 that at the current densities of 0 to 190 mA/cm^2, the NR images in all observable area is almost transparent, indicating that water does not appear macroscopically in the flow field plates. However, when these range of current densities was applied, actually the water already appeared in the MEA, which was detected by SANS as shown in figure 6. This figure shows L_{ion} determined by SANS measurement at each position from 1 to 9 (cf. those positions for SANS measurement have been assigned in figure 2) as a function of current density. As we discern in the figure 6, even no current density was applied in the operating PEFC, L_{ion} at any positions is bigger than that of in dried MEA (35 Å). This indicates that the water still remained in the ion channel after purging process of residual water from inside of PEFC with nitrogen gas. At low current densities (smaller than 190 mA/cm^2), the accumulation of water at upper positions of MEA (position 1, 2, and 3) was higher than that of other positions as indicated by higher L_{ion} values. Further, by increasing the current density up to 390 mA/cm^2, distribution of water in the ion channel increased rapidly indicated by drastic increase in $L_{\rm ion}$. Especially, at the position 1, 2 and 3, water appeared not only microscopically in MEA, but also macroscopically in the flow field as recorded by NR (cf figure 4). At higher current density (above 390 mA/cm^2), there was no change in $L_{\rm ion}$ at the position 1, 2 and 3 indicating that saturation as to swelling of ion channel with water was reached in those positions. Excess of water seemed to flow over to the flow-field plates and dropped to the lower flow field until reached the water outlet of the operating PEFC. At the same time, at the lower positions of SANS measurements, the distribution of water in ion channel increased rapidly to reach the maximum of $L_{\rm ion}$, and the excess of water also distributed to the flow field. In such a way, by applying combined SANS and NR method, we can make a seamless visualization of water inside of PEFC in wide range scale from nano- to millimeter order.

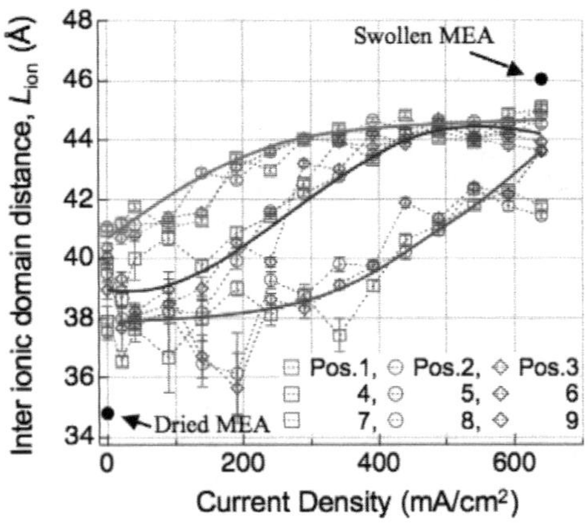

Figure 6. Inter-ionic domain distance, L_{ion} of Nafion in MEA from 9 position measurements determined by SANS as a function of the current density.

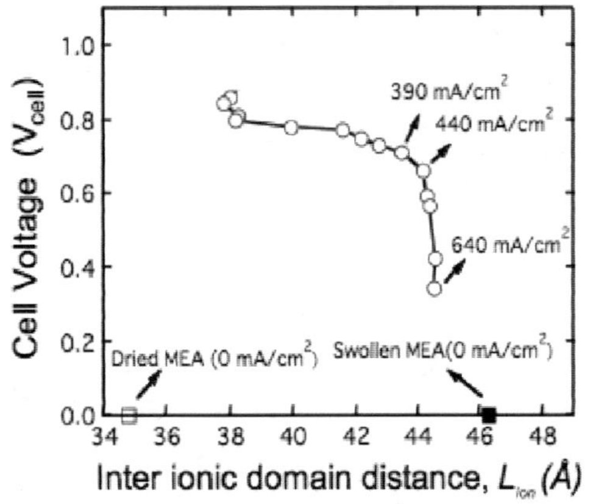

Figure 7. Relationship between electrical properties and water appearance in the MEA. The inter ionic domain distance, L_{ion} is taken from the position 5 as representative of whole measurements.

Moreover as has been partly argued in figure 3, due to the *in-situ* observation by SANS, we can estimate the relationship between microscopic water distribution during the operating of PEFC and electrical properties of the MEA. In figure 7, $V_{\rm cell}$ is plotted as a function of $L_{\rm ion}$, of which data is estimated from the SANS profile taken at position 5, as an attempt to visualize the relationship between these two parameters more directly. Note that the curve of figure 7 is hereafter denoted as "$L_{\rm ion}$- $V_{\rm cell}$" curve for simplicity. In this figure an increase in $L_{\rm ion}$ from 38 Å to about 44 Å, indicating an accumulation process of water in ion conducting channel, is clearly recognized as nearly horizontal part of $L_{\rm ion}$- $V_{\rm cell}$ curve. In this stage, where current densities being higher than 190 mA/cm^2 and up to 440 mA/cm^2, the performance of the MEA did not show any significance changed which was showed by small decrease in the cell voltage (0.8 to 0.71 V). However, at current densities higher than 440 mA/cm^2, an abrupt decrease in V_{cell}, indicating that the performance of the PEFC lowered, is observed in L_{ion}- V_{cell} curve at the position of $L_{ion} \sim 44$ Å where inter ionic distance was still smaller than that of swollen MEA (46 Å). In summing up those observed information, it may be speculated that the cause of lowering of the performance in this PEFC system was not the limitation of proton conductivity of Nafion membrane but the limitation of reaction rate at catalyst layer or the limitation of the rate of discharging water in flow field as well as GDL.

4. Conclusion

In-situ observation was successfully conducted on the operating PEFC system with applying the combined measurement system of SANS and NR, which has been recently developed. Due to a complementary combination in accessible length scale between SANS and NR, nearly seamless observation was possible as to water widely distributed from nano- to millimetre scale. In particular, SANS was found to be a powerful tool to estimate the amount of water in swollen state of PEM as well as that in catalyst layer. Through this study it was well confirmed that microscopic distribution of water (obtained by SANS) as well as macroscopic one (obtained by NR) are strongly correlated to the electrochemical performance of PEFC system.

Acknowledgement

The authors would like to thank Prof. Tsutsumi (Ibaraki University and FC Development Co., Ltd., Japan) for helpful supports on operation of a fuel cell. This work was supported by research and development of polymer electrolyte fuel cell technology project of New Energy and Industrial Technology Development Organization (NEDO), Japan.

References

[1] Vielstich W, Lamm A and Gasteiger H 2003 *Handbook of Fuel Cells-Fundamentals, Technology, Application*, Wiley, New York
[2] Nguyen T V and White R E 1993 *J. Electrochem. Soc.* **140** 2178
[3] Yi J S and Nguyen T V 1998 *J. Electrochem. Soc.* **145** 1149
[4] Okawara A, Horie H, Miyake H, Nishimura H, Saijo K and Hashimoto T 1992 *Polymer* **33** 579
[5] Mosdale R, Gebel G and Pineri M 1999 *J. Mem. Sci.* **118** 269
[6] Bellows R J, Lin M Y, Arif M, Thompson A K and Jacobson D 1999 *J. Electrochem. Soc.* **146** 1099
[7] Satija R, Jacobson D L, Arif M and Werner S A 2004 *J. Pow. Sour.* **129** 238
[8] Pekula N, Heller K, Chuang P A, Turhan A, Mench M M, Brenizer J S and Unlu K 2005 *Nucl. Instr. and Meth. A* **542** 134
[9] Ludlow D L, Calebrese C M, Yu S H, Dannehy C S, Jacobson D L, Hussey D S, Arif M, Jensen M K and Eisman G A 2006 *J. Power Sources* **162** 271
[10] Kramer D, Zang J, Shimoi R, Lehmann E, Wokaun A, Shinohara K, and Scherer G G 2006 *Electrochim. Acta* **51** 2603
[11] Zhang J, Kramer D, Shimoi R, Ono Y, Lehmann E, Wokaun A, Shinohara K and Schere G G 2006 *Electrochim. Acta* **51** 2715
[12] Feindel K W, Bergens S H and Wasylishen E 2006 *J. Am. Chem. Soc.* **128** 14192
[13] Xu F, Diat O, Gebel G and Morin A 2007 *J. Electrochem. Soc.* **154** B1389
[14] Kim T, Sim C M and Kim M H, 2008 *Appl. Rad. Isotopes* **66** 593
[15] Ueda T, Takenaka N, Asano H, Tanimoto K, Mochiki K, kawabata Y and Matsubayashi M 2008 *J. Power Energy Sys.* **2** 997
[16] Gierke T D, Munn G E and Wilson E I 1981 *J. Polym Sci., Polym. Phys.* **19** 1687
[17] Perrin J C, Lyonnard S and Volino F 2007 *J. Phys. Chem. C* **111** 3393
[18] Iwase H, Koizumi S, Likura H, matsubayashi M, Yamaguchi D, Maekawa Y and Hasimoto T 2009 *Nucl. Instr. and Meth A* **605** 95

[19] Koizumi S, Iwse H, Suzuki J, Oku T, Motokawa R, Sasao H, Tanaka H, Yamaguchi D, Shimizu H M and Hasimoto T 2007 *J. Appl. Cryst.* **40** s474
[20] Hashimasa Y, Numata T, Moriya K and Watanabe S 2006 *J. Power Sources* **155** 182

SAXS studies on added-salt species dependence of conformation of fully-charged poly(*L*-glutamate) in aqueous solutions

Shigeru Shimizu[1], Yoshio Muroga[2], Shozo Iida[3], Masanobu Miyahara[1], Masato Ishibashi[1], Hiroki Ikake[1], Kimio Kurita[1]

[1] College of Science and Technology, Nihon University, Tokyo 101-8308, JAPAN
[2] Graduate School of Engineering, Nagoya University, Nagoya 464-8603, JAPAN
[3] Graduate School of Science, Nagoya University, Nagoya 464-8603, JAPAN

E-mail: shimizu@chem.cst.nihon-u.ac.jp

Abstract. The effect of counter-ion species on the local conformation of fully-charged poly(*L*-glutamate) in added-salt aqueous solutions were studied by small-angle X-ray scattering(SAXS). In the case of presence of monovalent counter-ion, the scattering behaviour could be explained by taking into consideration the excluded volume effect and intermolecular-interface interaction on the scattering function of wormlike chain. On the other hand, since SAXS profile of the solution in the presence of divalent counter-ion well fitted to the scattering function of dendrimer, it suggested that the conformation of polymer chain in such solution would be more compact conformation than that of presence of monovalent ion.

1. Introduction

Some of polypeptide chains exhibit a transition between helical state and random-coiled state, and the transition is usefully employed as a model for the phenomena of folding or denaturation of proteins. Poly(*L*- or *D*-glutamic acid) (PLGA or PDGA) shows such a conformational transition when the pH of the solution is changed. Many studies have been directed to the analysis of the structure in each conformational state: By intrinsic viscosity and light scattering, Holtzer et al. [1] have shown that the conformation of fully-charged poly(sodium *L*-glutamate) PLGA-Na is represented by random coil with unperturbed effective bond length of ca. 8Å. Muroga et al.[2] have estimated the averaged helical-sequence length of PDGA-Na in the helical- and helix/coil transition-states by small-angle X-ray scattering(SAXS).

In general, counter-ions as well as bi-ions[3] affect the conformation of charged PLGA. The bond formed between carboxylate-ion and counter-ion bears both features of ionic- and covalent-natures and their fraction would depend on the ionisation energy of the counter-ion. Accordingly, the conformation of PLGA-salt should also depend on the species of counter-ions, but the counter-ion species dependence of the conformation has not been fully clarified yet.

In the present work, our attention is directed to the local conformation of fully-charged PLGA in random-coiled state in added-salt aqueous solution, and the counter-ion species dependence of the conformation is studied with respect to Na^{+}, Li^{+}, $(CH_3)_4N^{+}$(TMA^{+}), Mg^{++}, Ca^{++} and Ba^{++}, where these ions are known to have different ionisation energy. The conformation was clarified by analyzing SAXS curve in a high scattering-vector range, where the data would reflect a part of a polymer chain and the analytical result was expected to be insensitive to the molecular-weight distribution of the sample and free from possible excluded volume effect.

2. Experimental Section

2.1. Samples and Preparations

PLGA-Na of $M_w = 1.78 \times 10^4$(Sample-I) and 5.48×10^4(Sample-II) (Sigma-Aldlich Co.) were converted to the acid form PLGA by passing them through a column filled with a mixed-bed of ion-exchange resins, Amberlite IR-120B and IR-400. The potentiometric titration of PLGA was carried out in 0.05M NaOH aqueous solution with 0.2M NaCl under the atmosphere of N_2 gas at room temperature with Accumet Model-15 pH meter(Fisher Scientific Co.). The degree of neutralisation of carboxylic-group α for the sample employed in SAXS was adjusted with respective hydroxide on the basis of the titration data. The concentration C_s of added salts, NaCl, LiCl, $(CH_3)_4NCl$(TMACl), $MgCl_2$, $CaCl_2$ and $BaCl_2$, was 0.02 or 0.2 mol/L and the polymer concentration C_p ranged from 2.5×10^{-3} to 1.0×10^{-2} g/mL (Table1).

2.2. SAXS measurements

SAXS measurements were performed at the Photon Factory of High Energy Accelerator Research Organization, Tsukuba, Japan. The wavelength of incident X-ray λ was 1.488Å. The scattered X-ray was detected by a one-dimensional position sensitive proportional detector(PSPC) with 512 channels. The absolute value of scattering vector $|\boldsymbol{q}|(= q = (4\pi/\lambda)\sin(\theta/2)$, where θ is a scattering angle) per channel was evaluated with diffraction from a reference sample, chicken tendon collagen. The sample cell, made of stainless steel, had quartz windows of 20μm thickness whose intervals were 1mm. The excess scattered intensity $I(q)$ of the sample solution over the solvent was obtained after transmission corrections for solutions and solvents. The details of the apparatus and the measurement are described elsewhere[4].

3. Results and Discussion

Figure 1(a) shows the scattering profile for PLGA-Na with $C_p = 5.0 \times 10^{-3}$ g/mL (○) and 1.0×10^{-2} g/mL (×) in 0.2M NaCl aqueous solution, as an example, plotted in the form of $I(q)$ vs q. As is shown in Figure 1(b), if $I(q)$ is reduced to $I(q)/C_p$, the two scattering profiles satisfactorily coincided on a single curve over q ranging from 0.015 to 0.35 Å. Therefore, it was safely assumed that scattering data for all samples with C_p of 2.5×10^{-3} to 1.0×10^{-2} g/mL would reflect only intramolecular-interferences, at least, in that q range.

In general, $I(q)$ of a rodlike polymer chain is given by[5]:

$$I(q) \propto I_{\text{thin}}(q) I_{\text{cs}}(q) \tag{1}$$

where $I_{\text{thin}}(q)$ is scattering intensity of a hypothetical chain with no cross-section and $I_{\text{cs}}(q)$ is the cross-sectional factor. $I_{\text{thin}}(q)$ is proportional to $1/q$ for a needle-like rod and $I_{\text{cs}}(q)$ is proportional to $\exp(-\frac{1}{2} < R_{\text{cs}}^2 > q^2)$ for a chain with mean-squared radius of a cross-section $< R_{\text{cs}}^2 >$. In a q range of:

$$\frac{1}{L_p^2} < q^2 < \frac{1}{< R_{\text{cs}}^2 >}, \tag{2}$$

where L_p is persistence length, the local structure of a polymer chain could be regarded as a rigid-rod. Thus, when $I(q)$ is plotted in the form of $\ln(I(q)q)$ vs q^2, one can draw a straight line in a q range of Eq.(2), the slope of which , $-slope$, is related to $< R_{\text{cs}}^2 >$ through the relation of $< R_{\text{cs}}^2 >= 2 \cdot (-slope)$.

In Figure 2 is shown an example for the plot of $\ln(I(q)q)$ vs q^2 for PLGA-salts in the presence of (a) Na^+ and (b) Mg^{++}. $< R_{\text{cs}}^2 >^{1/2}$ for all samples thus evaluated is listed in Table 1. For

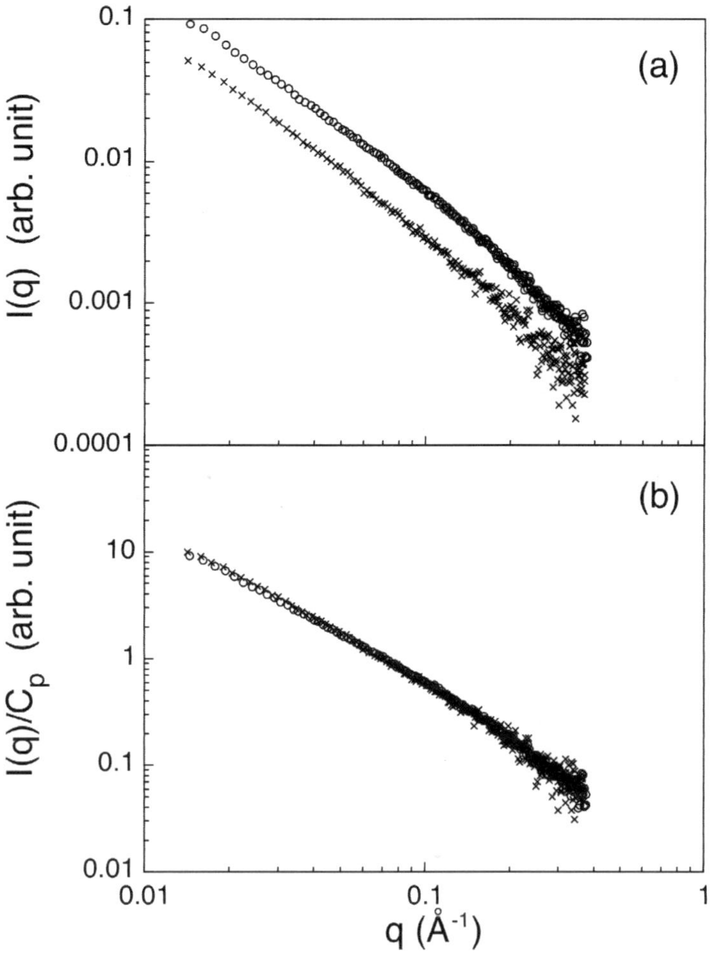

Figure 1. SAXS profile of Sample-I of $\alpha = 0.85$ in 0.2M NaCl. (a): Plot of $I(q)$ vs. q for (○) $C_{\rm p} = 5.0 \times 10^{-3}$ g/mL and (×) $C_{\rm p} = 1.0 \times 10^{-2}$ g/mL. (b): Plot of $I(q)/C_{\rm p}$ vs q.

all counter-ion species, $< R_{\rm cs}^2 >^{1/2}$ is 4.0 ± 0.5 Å, which is slightly smaller than that expected for PLGA-Na if its side chain takes a fully-stretched form, 5.2 – 5.6 Å[6].

The Kratky plot $I(q)q^2$ vs q is shown for all samples in Figure 3. From the appearance of their profiles, they seem to be classified into three categories: Category (I) for Li^+(a), Na^+(b), TMA^+(c), and Mg^{++}(d), where the profile is represented by a smoothly-increasing curve when q is increased; Category (II) for Ca^{++}(e) , where it is represented by a horizontal curve in a higher q range; Category (III) for Ba^{++}(f), where it is represented by a curve with a bump in a small q range. The features in Category (I), (II) and (III) are observed in the Kratky plot of a wormlike chain, Debye function for Gaussian chain and gel-like structure having high segment-density, respectively.

The profiles (a)-(d) in Figure 3 were compared with scattering function of wormlike chain

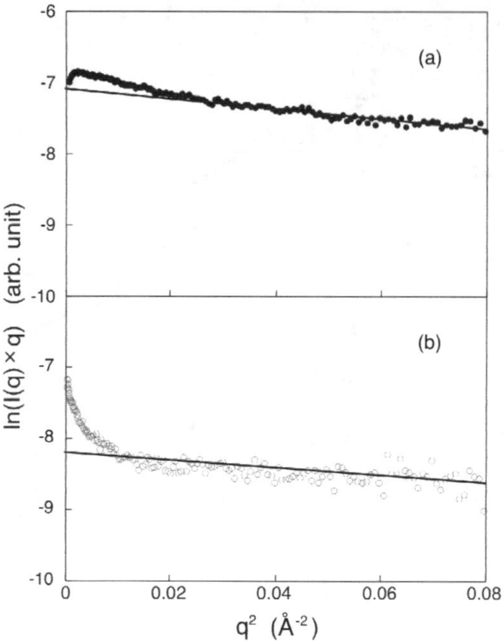

Figure 2. Plot of $\ln(I(q)q)$ vs. q^2 for Sample-I. ● : Sample-I in NaCl, $\alpha = 0.85$, $C_s = 0.02$M, $C_p = 1.0 \times 10^{-2}$ g/mL. ○ : Sample-II in $MgCl_2$, α=0.77, $C_s = 0.02$M, $C_p = 2.5 \times 10^{-3}$ g/mL.

Table 1. Root mean-square radius of cross-section $< R_{cs}^2 >^{1/2}$ for PLGA in random-coil region.

Sample	Added salt	α	C_s (mol/L)	C_p (g/mL)	$< R_{cs}^2 >^{1/2}$(Å)
Sample-I	LiCl	0.84	0.2	0.010	4.0±0.5
Sample-I	NaCl	0.85	0.2	0.010	4.0±0.5
Sample-I	NaCl	0.85	0.2	0.010	4.0±0.5
Sample-II	NaCl	0.85	0.02	0.005	4.0±0.5
Sample-I	TMACl	0.84	0.2	0.010	4.0±0.5
Sample-II	$MgCl_2$	0.77	0.02	0.0025	4.0±0.5
Sample-II	$CaCl_2$	0.86	0.02	0.0050	4.0±0.5
Sample-II	$CaCl_2$	0.90	0.02	0.0050	4.0±0.5
Sample-II	$CaCl_2$	0.85	0.02	0.0025	4.0±0.5
Sample-II	$BaCl_2$	0.83	0.02	0.0025	4.0±0.5
Sample-II	$BaCl_2$	0.86	0.02	0.0050	4.0±0.5
Sample-II	$BaCl_2$	0.90	0.02	0.0050	4.0±0.5

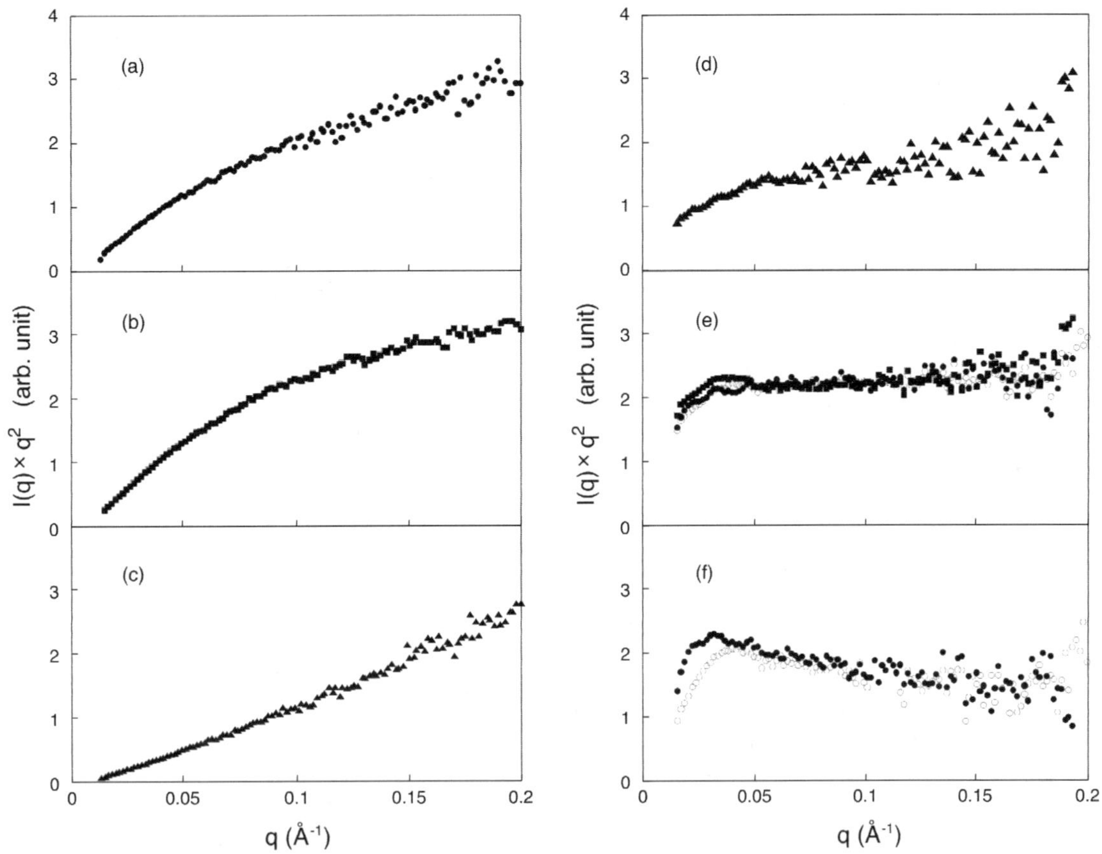

Figure 3. Kratky plot of PLGA aqueous solution in the presence of several added salt species without cross-section correction. (a) ●: Sample-I in LiCl, α=0.84, C_s = 0.2M, C_p = 0.01 g/mL, (b) ■: Sample-I in NaCl, α=0.85, C_s = 0.2M, C_p = 0.01 g/mL, (c) ▲: Sample-I in TMACl, α=0.85, C_s = 0.2M, C_p = 0.01 g/mL, (d) ▲: Sample-II in $MgCl_2$, α=0.77, C_s = 0.02M, $C_p = 2.5 \times 10^{-3}$ g/mL, (e) ○: Sample-II in $CaCl_2$, α=0.84, C_s = 0.02M, $C_p = 2.5 \times 10^{-3}$ g/mL, (f) ●: Sample-II in $BaCl_2$, α=0.83, C_s = 0.02M, $C_p = 2.5 \times 10^{-3}$ g/mL, (e)○ : Sample-II in $CaCl_2$, α=0.84, C_s = 0.02M, $C_p = 2.5 \times 10^{-3}$ g/mL.

$P(q)$[7][8]:

$$P(q) = 2u^{-2}\Big[\exp(-u) - 1 + u\Big] + \frac{4}{15L_r} + \frac{7}{15L_r u} - \left(\frac{11}{15L_r} + \frac{7}{15L_r u}\right) \times \exp(-u) \tag{3}$$

where u and L_r are defined by the contour length of a polymer chain L and the persistence length L_p as

$$u = \frac{L L_p q^2}{3} \tag{4}$$

$$L_r = \frac{L}{2L_p} \tag{5}$$

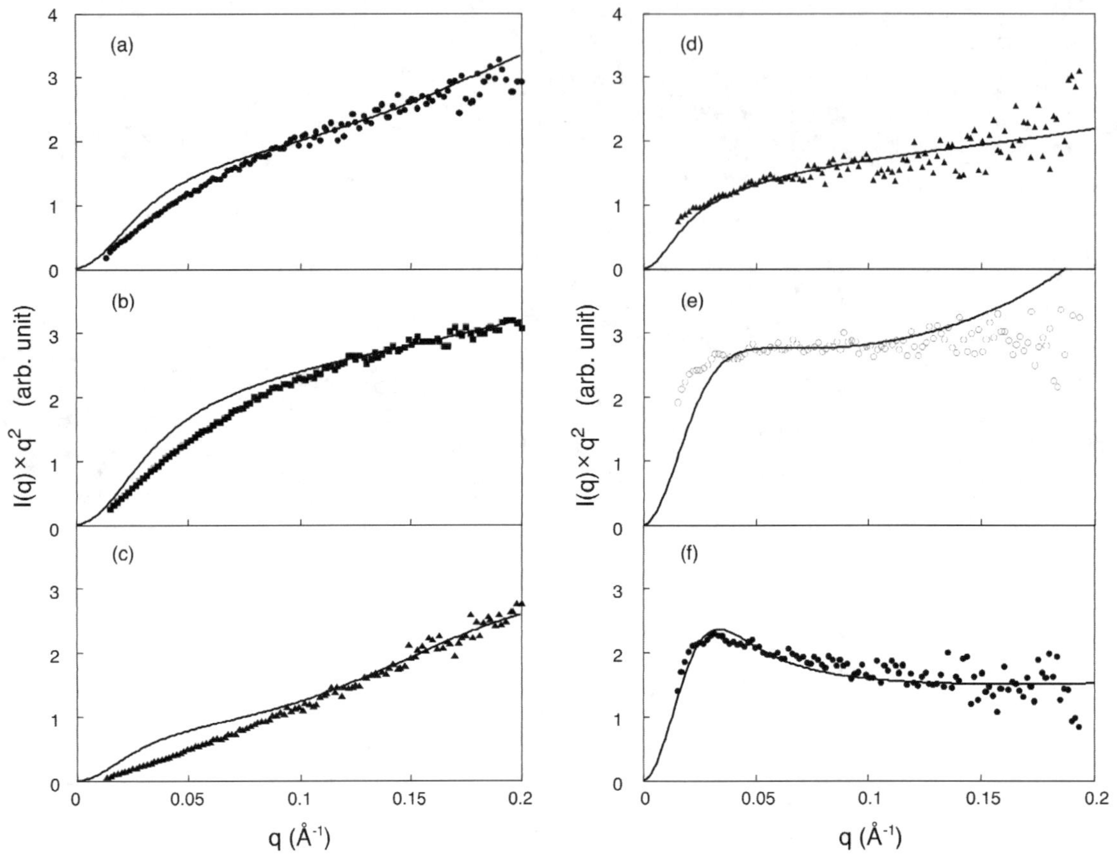

Figure 4. Comparison between $I(q)q^2$ and theoretical scattering curves(solid curve). Observed curves: (a) ●: Sample-I in LiCl, α=0.84, C_s =0.2M, C_p = 0.01 g/mL, (b) ■: Sample-I in NaCl, α=0.85, C_s = 0.2M, C_p = 0.01 g/mL, (c) ▲: Sample-I in TMACl, α=0.85, C_s = 0.2M, C_p = 0.01 g/mL, (d) ▲: Sample-II in $MgCl_2$, α=0.77, C_s = 0.02M, C_p = 2.5 × 10^{-3} g/mL, (e) ○: Sample-II in $CaCl_2$, α=0.84, C_s = 0.02M, C_p = 2.5 × 10^{-3} g/mL, (f) ●: Sample-II in $BaCl_2$, α=0.83, C_s = 0.02M, C_p = 2.5 × 10^{-3} g/mL.

Eq.(3) is valid under the conditions,

$$L_r > 10 \quad \text{and} \quad (2L_p q)^2 < 10 \tag{6}$$

A solid curve in (a)-(d) of Figure 4 is a theoretical curve computed with Eq.(3) . It is seen that the observed data in a higher q range is well mimicked by the theoretical curve, although the data in a lower q range deviates downward from the curve. When a slight effect of excluded-volume effect[7] and intermolecular-interferences[9][10] are taken into account, however, these deviations were significantly improved, as is shown by a solid curve in Figure 4. The persistence length L_p thus evaluated is 5Å for PLGA-Li, 6Å for PLGA-Na, 6Å for PLGA-Mg and 8Å for PLGA-TMA, respectively (Table 2). Such a difference in L_p among PLGA-salts having different counter-ions might suggest that bulkiness of counter-ion condensed around the polymer chain seriously affects the stiffness of the chain. In Figure 4(e), the observed data are well compared with the theoretical scattering function for wormlike chain, with L_p of 6 Å.

Table 2. The persistence length L_p estimated by the scattering function of wormlike chain with the excluded volume effect and the intermolecular interaction.

Solvent	C_s (mol/L)	L_p(Å)
LiCl aq.	0.2	5.0
NaCl aq.	0.2	6.0
TMACl aq.	0.2	8.0
$MgCl_2$ aq.	0.02	6.0
$CaCl_2$ aq.	0.02	6.0

The Kratky plot of PLGA-Ba was analysed with $P(q)$ for regular-network polymer, a so-called dendrimer[11]:

$$S(q) = N_b\left[S^S_{sb}(q) + S^f_{sb}(q) + S^a_{sb}(q)\right] + N_b(N_b - 1)S_{ib}(q) \tag{7}$$

where $S^S_{sb}(q)$ is one intrabranch self correlation part, $S^f_{sb}(q)$ one intrabranch cross-correlation part between blocks that originate from same, $S^a_{sb}(q)$ one intrabranch cross-correlation part form different stems, $S_{ib}(q)$ one intrabranch correlations part, and these correlation parts are described in terms of segment length b, N generations of monomeric blocks, $f + 1$ functionality parameter, N_b number of branches and n number of monomers forming Gaussian links with segments. The solid curves in Figure 4(f) are theoretical curves calculated by Eq.(7). The curve shows a good fit to the data when the parameters of as $b = 10$, $N = 4$, $N_b = 2$, $f = 2$, $n = 10$ are chosen. This result would suggest that a gel-like network structure is developed in PLGA-Ba.

In general, the bond formed between carboxylate-ion and counter-ion bears both ionic- and covalent-bonding natures and their relative weights would depend on the ionisation energy of the counter-ion. That is, the ionic-feature of the bond would be more accentuated as the ionisation energy of counter-ion becomes lower. In the case of divalent counter-ions, the increase of the ionic-feature of the bond would lead to the inter- or intra-molecular bridging of the divalent-ion to two carboxylate-ions and, as a result, to the formation of the network chain. The present result that the conformation of PLGA-salts would become more compact and gel-like in the order of $Ba^{++} > Ca^{++} > Mg^{++}$ is compatible with the fact that the first ionisation energy is in the order of Ba^{++}(5.2 eV) < Ca^{++}(6.1 eV) < Mg^{++}(7.6 eV).

4. Conclusions

The local conformation so fully charged PLGA in the presence of counter-ion was studied by small-angle X-ray scattering. In the presence of Li^+, Na^+, TMA^+, Mg^{++} and Ca^{++}, their SAXS curves were well explained the scattering function of the wormlike chain by taking into account the excluded volume effect and intermolecular interferences. In the presence of Ba^{++}, however, the SAXS curve was only explained by the scattering function of dendrimer.

5. Acknowledgments

This work was partly supported by grant of Institute of Quantum Science, Nihon University.

References

[1] Hawkins R B and Holtzer A 1972 *Macromolecules* **5** 294
[2] Muroga Y, Tagawa H, Hiragi Y, Ueki T, Kataoka M, Izumi Y and Amemiya Y 1988 *Macromolecules* **21** 2756
[3] Shimizu S, Muroga Y, Hyono T and Kurita K 2007 *J. Appl. Cryst.* **40** s553

[4] Ueki T, Hiragi Y, Izumi Y, Tagawa H, Kataoka M, Muroga Y, Matsushita T and Amemiya Y 1982/1983 *Photon Factory Activity Rep.* VI70
[5] Glatter O and Kratky O (eds) 1982 *Small Angle X-ray Scattering* (New York: Academic Press)
[6] Muroga Y, Suzuki K, Kawaguchi Y and Nagasawa M 1972 *Biopolymers* **11** 137
[7] Sharp P and Bloomfield V A 1968 *Biopolymers* **6** 1201
[8] Schmid C W, Rinehart F P and Hearst J E 1971 *Biopolymers* **10** 883
[9] Shimode M, Miura M, Urakawa H, Yamanaka S and Kajiwara K 1996 *SEN'I GAKKAISHI* **52** 301–309
[10] Coviello T, Maeda H, Yuguchi Y, Urakawa H, Kajiwara K, Dentini M and Crescenzi V 1998 *Macromolecules* **31** 1602
[11] Hammouda B 1992 *J. Polym. Sci.: Part B: Polym. Phys.* **30** 1387

Journal of Physics: Conference Series **247** (2010) 012046 doi:10.1088/1742-6596/247/1/012046

Nanoparticle aggregation behaviour in polymer nanocomposites: bulk vs. thin films

Him Cheng Wong and Joao T Cabral[1]

Department of Chemical Engineering, Imperial College London, SW7 2AZ.

Email: j.cabral@imperial.ac.uk

Abstract. We report on the stability and cluster formation in C_{60} fullerene-polystyrene mixtures, using a combination of small angle neutron scattering, optical and atomic force microscopy. Bulk nanocomposites are found to be stable for C_{60} loadings up to 1-2% mass fraction in PS of M_w = 270 kg/mol and approximately 160°C. At larger fullerene concentration, the coherent scattering intensity grows with time, within 10s of min, and is well described by an asymptotic relationship with rate compatible with Arrhenius temperature dependence. The nanocomposite structure can be described by fractal scattering with a broad range of dimensions D_f ~1.85-2.4 and polydisperse fractal sizes, up to hundreds of nm, depending on annealing temperature and time. Confinement in thin films of tens to hundreds of nm is shown to result in well-defined surface topographies, ranging from isolated clusters to percolated spinodal-like undulations with tuneable wavelength and amplitude. These results provide insight into the control of dispersion and self-assembly of nanoparticles in polymer matrices and thus their exploitation in functional materials, including organic photovoltaics and thin film coatings.

1. Introduction

The introduction of trace amounts of nanoparticles (NPs) in polymers (P) can significantly enhance and tune their mechanical, electrical and thermal properties. The performance of these so-called polymer nanocomposites (PNCs) is however largely governed by the dispersion of the NP, which remains "the greatest stumbling block to the large-scale production and commercialisation of nanocomposites" [1]. While well dispersed NPs can enhance the material's mechanical properties and processibility [2], NP segregation can also be advantageous, for example in controlling thin film stability [3-5] and morphology in advanced applications including photovoltaics [6]. Composites based on conducting polymers and fullerene (C_{60}) derivatives are particularly attractive for applications in flexible electronics, light-emitting displays, and photovoltaics since the discovery of photoinduced electron transfer between conducting polymers (donor) and C_{60} (acceptor) [7]. Clustering of C_{60} in selected polymer matrices [8, 9] and solvents [10, 11] has been well documented, as well as in the pure solid state, where C_{60} clusters with aggregation number of up to 55 were found by mass spectrometry [12]. Systematic studies are however needed to understand fullerene association in polymer matrices in detail, and thus the effect of processing, in order to explore its potential as performance enhancing fillers.

---

[1] To whom any correspondence should be addressed.

Particle aggregation studies can be carried out both in direct (or real) and reciprocal spaces, with well defined advantages and limitations. Real space studies involve detailed analysis of microscopy images and are often limited by image overlapping of 3-dimensional structures, lack of contrast, and non-representative local sampling. As a complementary technique to microscopy, small angle scattering (SAS) measures the spatial correlations in a mesoscopic range, typically from 1-1000 nm, depending on the radiation and angle of scatter. SAS is one of the best established techniques to elucidate the structural features of NP in nanocomposites such as its radii of gyration (R_g), shape, surface-to-volume-ratio, aggregation nature and size distribution. Sorensen et al. [13] and Rieker et al. [14] have adopted small angle X-ray scattering to characterise the fractal nature of carbon black aggregates, while structural information of C_{60} NP in benzene solution has been studied extensively using static and dynamic light scattering [10]. A recent study by Narayanan et al. [15] combined transmission electron microscopy and ultra small angle X-ray scattering to study the nanostructural features of fumes silica NPs in poly(vinyl acetate), fully characterising the size distribution and fractal nature of the silica NP.

In this paper, we investigate the aggregation /association behaviour of C_{60} NP in bulk polystyrene (PS) and compare it with its thin film counterpart, using reciprocal and real techniques, respectively. In the first part, we probe the structural information of C_{60} NP in a bulk hydrogenous polystyrene matrix using SANS. We compute the time evolution of the scattering intensity as a function of wavenumber q to study the NP aggregation as a function of particle loading and annealing temperature. The growth rate of the scattering intensity is then fitted to an empirical kinetic equation in order to determine the growth kinetics of the aggregation process. We then obtain fractal exponents for nanocomposites with representative particle concentration and annealing temperatures and times. Finally, a combination of optical microscopy (OM) and atomic force microscopy (AFM) is employed to elucidate the association of NPs in thin films of entangled PS-C_{60} annealed above glass transition temperature (T_g). Confinement in thin films (10s-100s of nm) breaks the symmetry of bulk association and leads to clustering of fullerenes in a spinodal pattern under specific conditions, providing a route to tuneable, functional particle-polymer structures [16].

2. Experimental section

2.1. Materials used and sample preparation

Bulk and thin film samples of PS-C_{60} nanocomposites were prepared using rapid precipitation and spin coating method, respectively, following procedures which were reported elsewhere [16, 17]. In short, bulk composites were prepared by separate dissolution of polymer and nanoparticle in toluene dilute solution, ultrasonication, mixing and further sonication and finally co-precipitation in a 5-fold excess of methanol (non-solvent). After drying under full vacuum at 100°C for 3 days, the resulting nanocomposite fibre was hot-pressed at 150°C for 6 min, below the agglomeration threshold limit [17], into 0.33 mm thick, 16 mm diameter disks; three individual films were then stacked and wrapped in aluminium foil, resulting in approximately 1 mm thickness (and 60% neutron transmission). Neutron contrast (neutron scattering length density) of C_{60}, hydrogenous PS and deuterated PS is respectively 5.7, 1.41 and 6.42 $\mu Å^{-2}$. To maximize contrast, we opt to study C_{60} association in hydrogenous polymer matrix, which requires careful incoherent background subtraction.

Thin films were spun cast onto silicon wafers (with a 2 nm silicon oxide layer) from ultrasonicated dilute toluene solutions of 0.1-2.5 % mass fraction to obtain film thicknesses between 10 to 150 nm. Spun cast thin film samples were dried at room temperature for at least one day prior to annealing.

Fullerenes were purchased from MER Corp. (99+% purity) and were used as received. Polystyrene was obtained from BP Chemicals and re-precipitated into a five-fold excess of non-solvent methanol and characterised by Gel Permeation Chromatography (GPC), yielding molecular mass M_w = 270 kg/mol and polydispersity index PDI = 2.3, thus radius of gyration R_g~14 nm.

2.2. Small angle neutron scattering

SANS experiment were carried out at two small-angle neutron scattering diffractometers: (a) LOQ (ISIS, Oxfordshire), a fixed-geometry instrument at the pulsed source with the sample-detector distance of 4 m which translates to a wide wavenumber range (0.009 $Å^{-1} < q < 0.249$ $Å^{-1}$) due to the polychromatic nature of the incident beam (wavelength 2 Å to 10 Å); (b) PAXE at the (Laboratoire Léon Brillouin, Saclay) configured to λ = 15 Å and sample detector distance of 5.11 m, yielding a wavenumber window between 0.0033 $Å^{-1}$ and 0.048 $Å^{-1}$. SANS data was calibrated using a primary standard (isotopic polymer blend) and the direct beam. The corresponding length scale window accessible by combination of the two instruments is approximately 2 nm < d < 30 nm.

In order to study C_{60} aggregation kinetics, 'fresh' (i.e. as pressed) nanocomposites are kept below T_g. where the structure is frozen, and then rapidly 'jump' to the experimental temperature (above T_g) while the synchronised time-resolved SANS measurement is triggered. In this work, we have employed a custom-made neutron scattering 'kinetic' cell, depicted in Figure 1. The cell consists of a brass pre-heating block, an experimental block with quartz windows and 45° exit cone, and a thin brass sample carrier which moves the sample quickly (within 2 s) between the two blocks.

In addition to time-resolved measurements, bulk PS-C_{60} samples were also measured quiescently: (a) 'unannealed' (as melt pressed) samples; (b) samples which were annealed ex-situ in vacuum oven at 160 ± 10°C for 12 h ('ex-situ annealed' samples). The latter samples are measured ex-situ while the unannealed samples were measured while annealing in-situ. Typical acquisition times were 20 min for quiescent specimens and 10 min for time-resolved measurements.

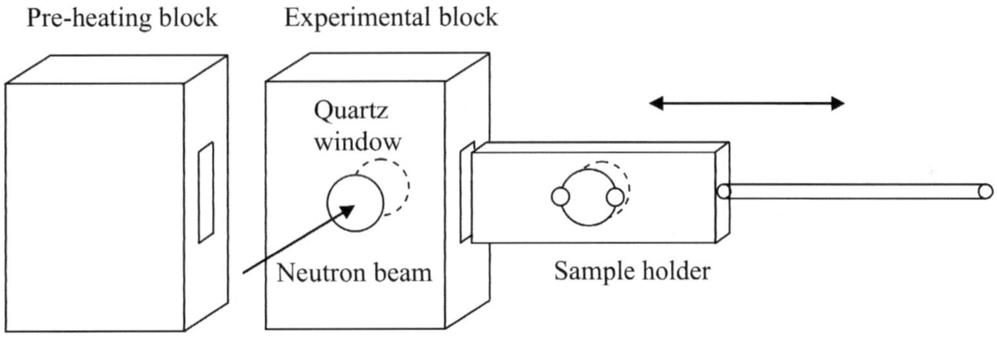

Figure 1. Schematic of the 'Temperature Jump' neutron kinetics scattering cell, depicting the pre-heating and experimental ovens and motorised sample holder. [18]

2.3. Optical and atomic force microscopy

The morphology of polymer nanocomposites thin films was studied in-situ using a custom made hot stage mounted on a reflection optical microscope (Olympus BX41M) equipped with a CCD camera (Adimec 1000m) controlled by LabVIEW. The surface topography was further characterized by tapping-mode AFM (Innova, Veeco) using super-sharp TESP-SS probes. The data was analysed by fast-Fourier Transform and radial averaging (ImageJ, NIH, USA) and NI Vision assistant to obtain particle sizes and in-plane structure periodicity.

3. Results and discussion

A systematic C_{60} aggregation study in PS matrix as a function of C_{60} loading and annealing temperature using SANS was conducted and representative results are shown in Figure 2. After annealing at 180°C for a short period of time (14 min), excess forward scattering can be observed for samples containing more than 2% mass fraction of C_{60}. Scattering of 1% PNC sample and neat PS is identical, indicating good C_{60} dispersion below 2%.

In the second series of experiment, we study the effect of annealing temperature on the dispersion of C_{60} NPs at a constant 2% mass fraction. Figure 2(b) demonstrates that annealing temperature plays an important role in governing the NP aggregation process and indicates a sharp onset of cluster-formation between 160°C and 180°C. In short, the experiments indicate a NP aggregation threshold concentration at 1-2% and temperature above 160°C.

We have previously conducted a complementary aggregation study using differential scanning calorimetry (DSC) [17, 19]. C_{60} is shown to increase the T_g of PS by up to 4°C at approximately 4%.Upon annealing above the threshold temperature (T^*), the T_g of the nanocomposites which exceed the threshold concentration (ϕ_c) will recover towards a lower plateau value. We associate this T_g recovery to NP agglomeration, thereby decreasing surface area in contact with polymer matrix. It is observed that the onset of T_g deviation between 'fresh' and 'annealed' samples occurs between 1-2% mass fraction (ϕ_c) (Figure 3(a)) and above 160°C (T^*) (Figure 3(b)), in agreement with previous [17] and current SANS results which show significant excess forward scattering beyond the same threshold. The good agreement between our DSC and SANS results indicates that the T_g is a suitable gauge for the extent of dispersion of NP composites [17].

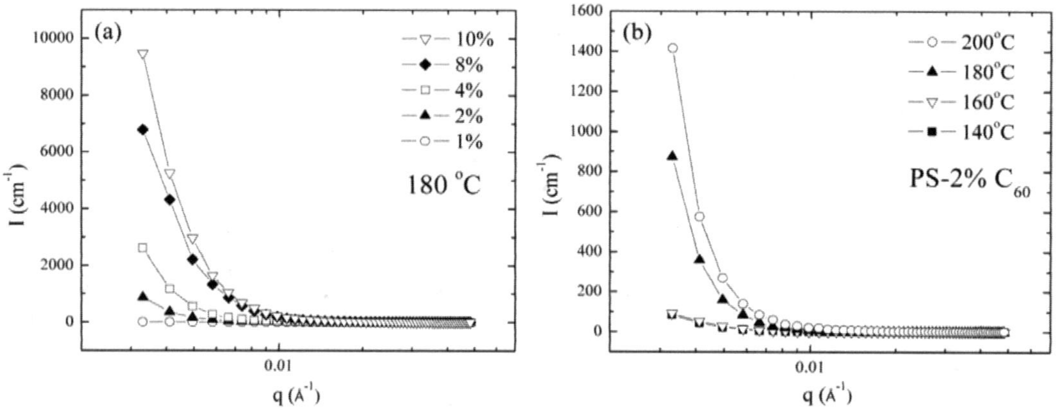

Figure 2. (a) SANS spectra of bulk polystyrene (PS) nanocomposites containing C_{60} mass fractions ranging from 1% to 10%, annealed at 180°C for constant time 14 min (b) effect of annealing temperature on PS-2% C_{60} at constant time 14 min.

3.1. NP aggregation kinetics in bulk

The time dependence of the scattering intensity of a PS-2% C_{60} composite annealed at 180°C as a function of wavenumber is shown in Figure 4(a). This representation is a temporal cross section of the 2% specimen in Figure 2(a). The intensity rises rapidly with time and, in the q-range examined, the growth kinetics can be approximated to an asymptotic behaviour where $I_0 = I(t{=}0)$ and $I_\infty = I(t \rightarrow \infty)$ are the scattering intensity before and after the annealing period; K(q) is the growth rate [s^{-1}]:

$$I(q,t) = I_0 + (I_\infty - I_0)(1 - e^{-K(q)t}) \qquad (1)$$

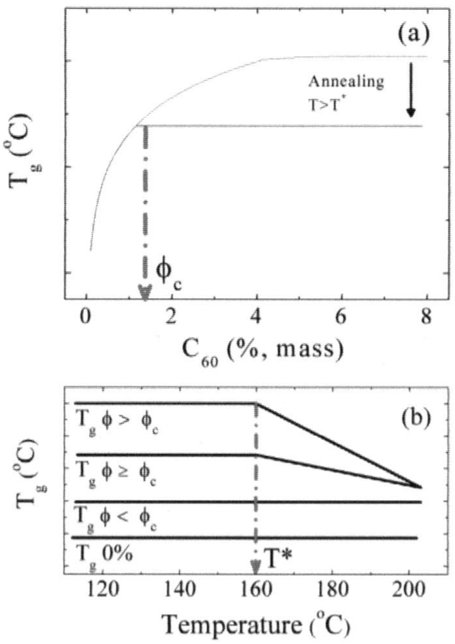

Figure 3. Schematics illustrating the effect of thermal annealing on the glass transition temperature (T_g) of the nanocomposites as a function of (a) C_{60} loading and (b) annealing temperature. DSC agrees remarkably well SANS in determining agglomeration threshold ϕ_c and T^*.

K(q) varies slowly with q across the wavenumber range studied, with up to 30% changes from the highest and lowest q, as shown in Figure 4(a). Next, we concentrate on the intensity at the lowest q accessible for comparison purposes between composites of different C_{60} loading. Figure 4(b) plots the time dependence of the scattered intensity from nanocomposites with different C_{60} concentrations annealed at 180°C. As expected, scattering intensity of 1% sample shows weak time dependence in comparison with the 2% specimen.

Above the threshold aggregation concentration, at constant annealing temperature of 180°C the NP growth at 4% loading is faster relative to 2%, as evidenced by the slight increase in growth constant (K ≈ 0.07 s^{-1} from K ≈ 0.06 s^{-1}) and shorter annealing time (20 min from 50 min) to reach 95% of the 'plateau' value. Further work is necessary to elucidate how the NP aggregation kinetics changes with annealing temperature in samples exceeding the threshold concentration. The scattering intensity with for 8% and 10% does not change with time, as shown in Figure 4(b), indicating that the specimens are agglomerated at the outset and the 'saturation' state is reached even before annealing.

At 2% mass fraction, which is the upper limit of the aggregation threshold, the kinetics of the C_{60} NP aggregation is also investigated as a function of annealing temperature (Figure 4(c)). The time evolution of the scattering intensity for each annealing temperature is well described with the asymptotic relation above. The growth constant K for each annealing temperature is obtained from the fit and plotted against inverse temperature, as shown in the inset of Figure 4(c). The growth constant is predictably decreasing with annealing temperature and can be described by an Arrhenius temperature dependence, as expected at high temperatures (well above T_g). Each annealing temperature appears to have its own 'plateau' value, which would imply that not only the clustering kinetics but also the asymptotic morphology depends on temperature. Further work is necessary to extend the time window and intermediate temperature range, as the matrix viscosity varies rapidly with T according to WLF relationship.

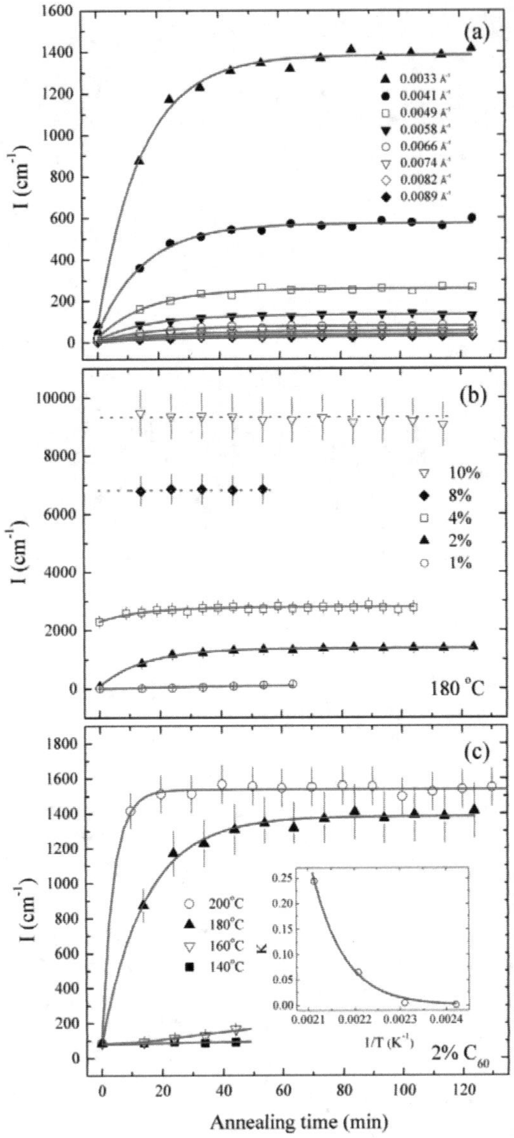

Figure 4. (a) Growth kinetics of C_{60} aggregates in bulk sample containing 2% mass fraction of C_{60} under in-situ thermal annealing at 180°C, expressed in terms of different q values. The q value that we adopt in subsequent kinetics study (b and c) is 0.0033 $Å^{-1}$. Bulk C_{60} aggregation kinetics as a function of (b) C_{60} mass fraction under isothermal annealing at T = 180°C and (c) annealing temperature for sample containing 2% mass fraction of C_{60}. Inset in (c) shows the Arrhenius temperature dependence of the K constant. Red solid lines are best fits according to Eq. (1) while red dash lines are linear guide-to-the-eye.

At short times, the intensity growth with time can be described by both an asymptotic relationship (Eq. (1)) and a logarithmic function [17]. Indeed, the Maclaurin second order polynomial approximation of $(1-e^{-x})$ and $\ln(1+x)$ is identical. However, at longer times, Eq. 1 appears to describe the intensity growth data more accurately and is hence adopted in this work. Further it has a clearer physical meaning as it predicts an asymptotic structure at long times, as large clusters are formed, instead of continuous growth.

3.2. NP fractal dimension

(Nano)particle aggregates may form 'statistically self-similar' structures within a range of lengthscales and are thus treated as fractals [20]. Numerous soft matter systems such as linear or branched polymers, colloids and nanoparticles exhibit fractal behaviour. The scattering intensity I (q) can be related to the scattering vector q via the following power-law equation:

$$I(q) \sim q^{-D_f} \qquad (2)$$

where D_f is called fractal or Hausdorff dimension which can be extracted as the negative slope of the scattering data plotted in log-log scale. Several growth mechanisms that lead to fractal aggregates have been proposed based on computer simulations. Witten and Sandler [21, 22] first proposed the diffusion limited aggregation (DLA) growth model which involves an immobilized growth site and one diffusing particle which randomly walks from a distant lattice point until it collides with the stationary growth site. The repetition of this process generates remarkable self-similar fractal structures. DLA is a fast aggregation process which is only limited by the diffusion time of the particle and the resulting D_f of the structure in two and three dimensions is ~ 1.7 [21] and ~ 2.5 [23], respectively. Another model of fractal formation involves multiple aggregations of relatively monodisperse particles. This so-called cluster-cluster (DLCA) aggregation proposed by Jullien and co-workers and Meakin [24, 25] is in contrast to the exclusive small - large cluster growth of DLA. Two particles next to each other will aggregate and continue to diffuse, becoming cluster sites themselves to form larger fractal structures, with D_f ~ 1.8 – 2.05 in three dimensions [26].

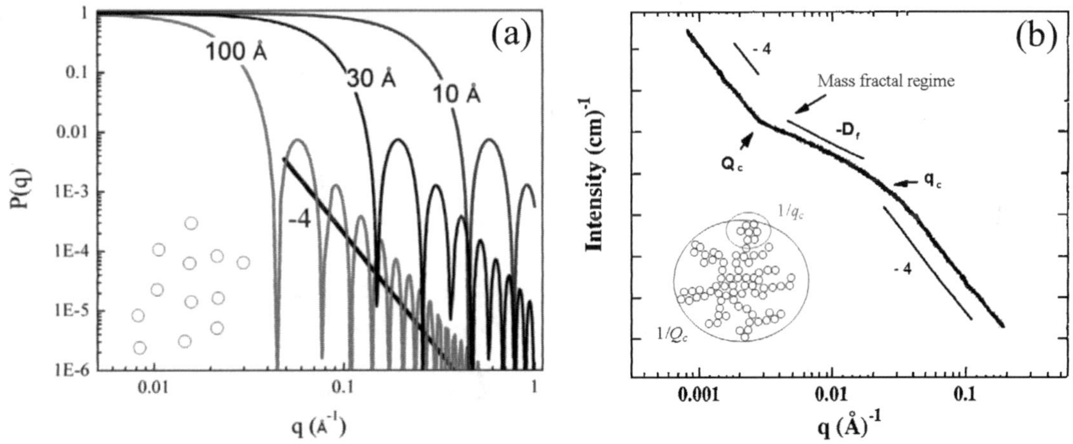

Figure 5. (a) Computed form factor of a sphere with radius r with respect to wavenumber q. Note the Porod regime towards high q value has -4 q dependence. (b) Typical scattering pattern from fractal system (See text below).

There are generally 3 power-law regimes in typical scattering data from fractals (Figure 5(b)). At low- q, the upturn in scattering intensity is attributed to scattering from the fractal agglomerates which gives a power-law exponent of -4 [13]. Whereas scattering from the surface of the individual particle happen at the high- q regime which is also known as the Porod power law regime (slope of -4 is again observed) (Figure 5(a)). In between these two boundaries, at the 'intermediate' q range, lies the fractal regime which has a smaller D_f that describes the type of fractals in the system. The individual primary aggregate size is inversely proportional to the q value (via 1/d) at the crossover point q_c from the fractal regime to Porod regime at high q whereas another crossover value Q_c indicates departure from fractal regime to agglomerate regime at low q-regime, corresponding to the size of an agglomerate.

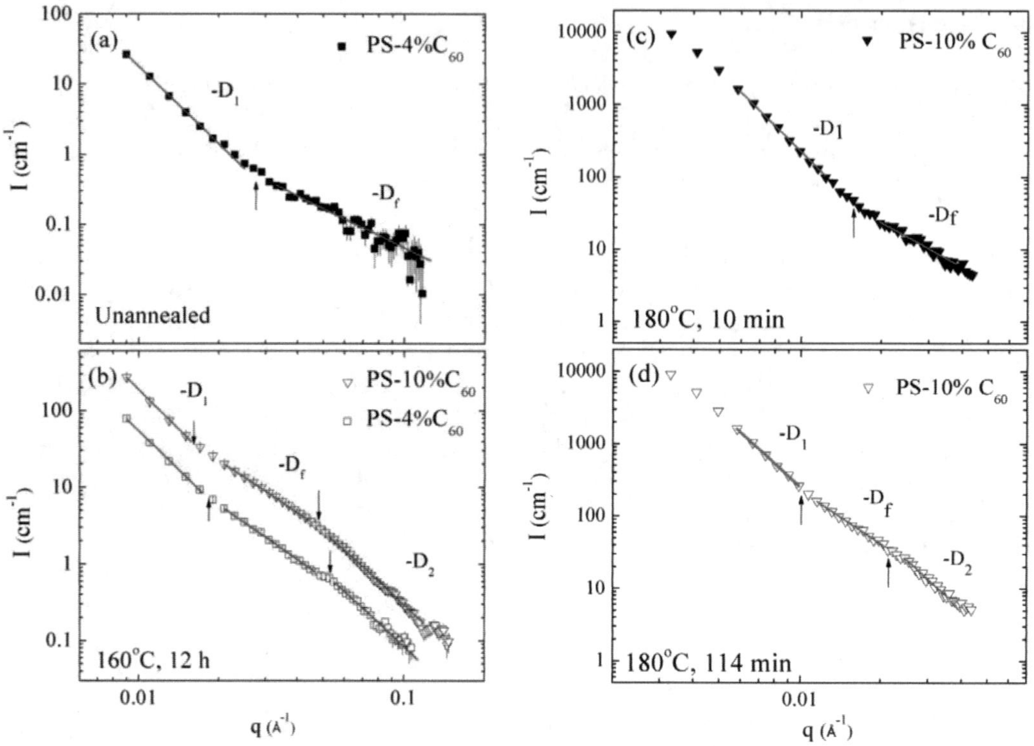

Figure 6. Selected log-log plots of SANS spectra of bulk PS-C_{60} nanocomposites. (a) PS-4% C_{60} composite before annealing and (b) PS-4% and 10% annealed ex-situ at 160°C for 12 h, measured at LOQ. PS-10% C_{60} annealed in-situ at 180°C measured after (c) 10 min and (d) 114 min (PAXE, LLB). Red solid lines are fits to a power-law equation, giving the mass fractal exponent D_f; arrows indicate crossover points Q_c and q_c. The exponents and the crossover points Q_c and q_c are tabulated in Table 1.

Upon subtraction of the incoherent background, normalised by the mass of the PS alone, the coherent scattering from bulk nanocomposites with loading exceeding 2% C_{60} are well described by a power law, with exponents D_f of 1.85 to 2.4 (Figure 6). Samples with lower C_{60} loading and typical acquisition times of 20 min do not provide enough statistics to yield permit a reasonable estimate of D_f. Results for 4 and 10% C_{60} loading are shown in Figure 6 with various thermal treatments. We first consider a relatively low annealing temperature of 160°C in Figures 6(a), (b): the unannealed PS-4% composite (Figure 6(a)) does not show a q_c; upon 12 h annealing (Figure 6(b)), both 4% and 10% specimens show a distinct q_c within the measured q-window. The latter shows lower Q_c and q_c crossover points, as expected for larger fractal structures and primary aggregates. Separately, we consider the effect of higher temperature annealing at 180°C on PS-10% C_{60} composites: Figure 6(c) and (d) correspond to annealed times of 10 min and 114 min at 180°C. With time, the structure factor shifts towards low q, corresponding to a reduction in Q_c and the emergence of a q_c within the observed window. However, the Q_c and q_c values in (d) are still lower those on (b) indicating that the extent of agglomeration is greater after 2 h at 180 than 12h at 160°C. In addition, the comparable Q_c values for fresh and annealed PS-10% C_{60} samples at 160°C (12 h) and 180°C (2 h) suggests the composites cluster from the outset and their structure varies slowly thereafter. The summary of q_c, Q_c and D_f are presented in Table 1.

The primary cluster size ($1/q_c$) is estimated from the high- q fractal cutoff q_c as ranging from 2-5 nm (commensurate with C_{60} size of 1 nm). The apparent fractal sizes estimated from $1/Q_c$ are of the order of 10s of nm. However, agglomerates larger than 250 nm have been observed by TEM on bulk

PS-C_{60} nanocomposites [27] and micron-sized clusters are visible in thin films (10-100nm) upon thermal annealing (discussed below). Such dimensions are is outside the spatial range of the present SANS study. The discrepancy of fractal sizes suggests that these 'secondary' aggregates (size ~ $1/Q_c$) further associate to form micron-sized agglomerates in a fractal manner, resulting in polydisperse particle size distributions. Based on DSC measurements, we have indeed proposed earlier that aggregates become polydisperse beyond 1% mass fraction of C_{60}. Future work should cover such large lengthscales by USANS and USAXS and size distribution analysis accessible based on relevant fitting models. Small-angle light scattering is unfeasible due to low light transmission above 0.1% C_{60}.

Table 1. Exponents from power law fits and crossover points of SANS data shown in Figure 6.

Sample	D_1	D_f	D_2	Q_c [Å^{-1}]	q_c [Å^{-1}]
Unannealed PS-4% C_{60} LOQ	3.66	1.85	-	0.03	-
PS-4% C_{60} LOQ (160±10°C, 12 h)	3.39	2.36	3.22	0.018	0.053
PS-10% C_{60} LOQ (160±10°C, 12 h)	3.42	2.2	3.21	0.016	0.049
PS-10% C_{60} PAXE (180°C, 10 min)	3.69	2.04	-	0.014	-
PS-10% C_{60} PAXE (180°C, 114 min)	3.44	2.4	3.34	0.01	0.021

The fractal dimensions obtained agree well with that excepted for mass fractal aggregates. However, the broad range of fractal dimensions from 1.85 to 2.4 means their growth mechanisms could follow both the diffusion limited aggregation (DLA) and diffusion limited cluster-cluster aggregation (DLCA) growth model. The slight higher fractal dimension of annealed 4% and 10% sample (D_f = 2.2 to 2.4) could be rationalized with increase in polydispersity at high C_{60} loading [17] or branching of the mass-fractal aggregates [15]. The latter explanation is in agreement with the DLA growth model as the fast and Brownian diffusion [28] of NP is more likely to be colliding with a branched arm than the center of the cluster site. The lower D_f of 2.04 for the 10% sample in the early stage of annealing could suggest that clusters are less branched.

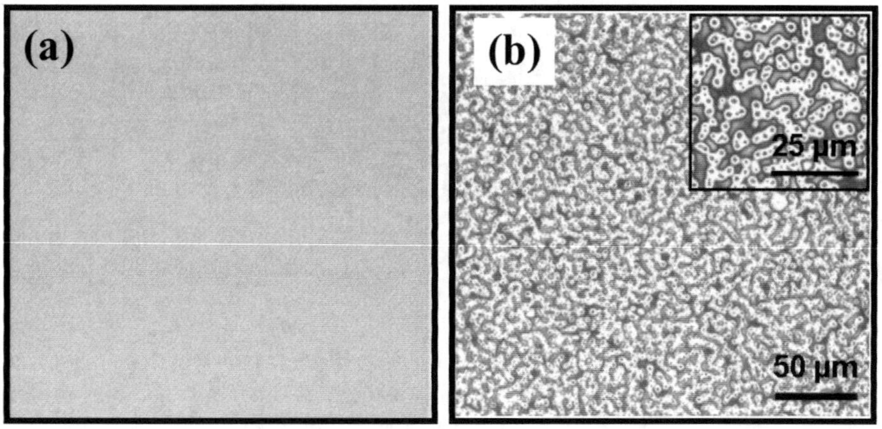

Figure 7. Optical micrograph of (a) neat PS and (b) PS thin film containing 5% mass fraction of C_{60} (thickness [h] = 150 nm) after annealing for 30 min at 180°C. Upon annealing, the neat film remains uniform while the filled counterpart shows a spinodal clustering morphology, reported in [16].

Nanoparticle aggregation under 2D thin film confinement

Spatial confinement in thin films is expected to break the fractal growth symmetry in the bulk and possibly result in the coupling of lengthscales. Further, the glass transition temperature of polymers is known to be altered due to the confinement in nanopores and thin polymer films, and depends strongly on surface interactions. [29, 30] Following our investigation of NP in bulk nanocomposites, NP clustering in the polymer nanocomposite thin film counterpart is investigated in real space using optical and atomic force microscopy.

Figure 7 compares the lateral morphology of a high M_w (270 kg/mol) PS thin film (thickness [h] = 150 nm) and its nanocomposite (5% mass fraction of C_{60}) counterpart upon annealing at 180°C for 30 min. While the neat film remains uniform and stable within the experimental time scale of up to 10 hours, C_{60} NP in the nanocomposites thin films aggregates in a spinodal pattern within minutes (see inset of Figure 7(b)). Decreasing C_{60} loading (2wt%) results in nucleation and growth of clusters; these may percolate at long annealing times(>days) while above a threshold concentration exceeding 2%, forming interconnected spinodal-like morphologies, as depicted in Figure 7(b).

AFM topography scans of annealed nanocomposites thin films (Figure 8) shows that the films do not dewet from the substrate. Instead, C_{60} aggregates grow and are covered by a PS top layer (with lowest surface energy) forming peak-like structures, resulting an undulating surface topography that develops with time, with amplitude reaching between 100 - 200 nm. Selective dissolution of PS in the sample using tetrahydrofuran (THF) reveals that the micron-sized C_{60} aggregates (insoluble in THF) are anchored on the substrate, retaining their spinodal morphology (data not shown here).

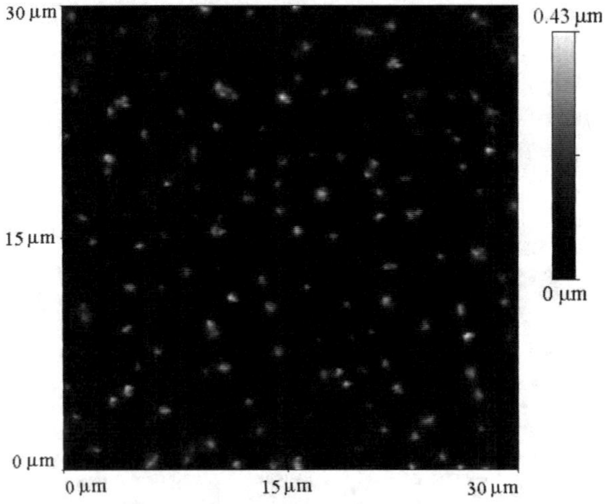

Figure 8. AFM topography scan of a representative PS-C_{60} nanocomposite thin film (h = 135nm) annealed at 180°C for 30min.

Detailed quantitative analysis of this so-called spinodal clustering morphology has been reported elsewhere [16] and is discussed in terms of a dominant lengthscale λ^* of the spinodal structure and size of individual clusters. The coarsening kinetics of λ^* follows $\lambda^* \sim t^{1/3}$ for relatively thick films (170 nm) and, at longer times, the in-plane structure eventually pins. In addition, the time dependence of the cluster size and number also follows a similar asymptotic behaviour, in comparison to the bulk. The effect of film thickness, matrix molecular weight [16] and annealing temperature [31] are beyond the scope of this paper and will be reported elsewhere, elucidating the underlying mechanism of this spinodal clustering process. The effect of thin film confinement, the presence of a substrate wall and the non-equilibrium solidification process by spin coating might break the symmetry of cluster growth,

Journal of Physics: Conference Series **247** (2010) 012046 doi:10.1088/1742-6596/247/1/012046

leading to the observed clustering behaviour. Lateral diffusion of the NP in polymer nanocomposites thin films is expected to be considerably faster than the vertical diffusion process (i.e., surface attraction/wetting) [32].We interpret this spinodal association behaviour as an interplay between particle-particle clustering and particle-substrate attraction (or 'entropic-push') which is largely governed by a polymer matrix of a given M_w and film thickness. [16].

4. Concluding remarks

The stability and structure formation of model polymer-nanoparticle mixtures, such as the present PS-C_{60}, in both bulk and thin films, has significant impact on the function and performance of the nanocomposite system. This work establishes precise conditions for miscibility and characterises the kinetics of the aggregation process in bulk and in confinement. We establish that C_{60} is miscible up to 1-2 % in PS-270k and approximately 160°C based on small angle neutron scattering and calorimetry experiments. In the bulk, coherent scattering intensity grows asymptotically with kinetics described by an empirical rate K, exhibiting Arrhenius dependence. Fractals with a broad range of dimensions D_f ~1.85-2.4 are found, with polydisperse fractal sizes, up to 100s of nm. Confinement in thin films, with thickness ranging from 10-100s of nm, results in symmetry breaking and the emergence of a surface topography with regular spinodal morphology. The wavelength and amplitude of these periodic, interconnected structures can be precisely tuned by the processing parameters temperature, time and concentration. Understanding nanoparticle association and collective self-assembly processes has thus great potential in growing technological applications, particularly in organic photovoltaics and thin film coatings.

References

[1] Balazs A, Emrick T and Russell T P 2006 *Science* **314** 1107.
[2] Mackay M E, Dao T, Tuteja A, Ho D L, Van Horn B, Kim H C and Hawker C J 2003 *Nat. Mater.* **2** 762.
[3] Barnes K A, Karim A, Douglas J F, Nakatani A I, Gruell H and Amis E J 2000 *Macromolecules* **33** 4177.
[4] Holmes M A, Mackay M E and Giunta R K 2007 *J. Nanopart. Res.* **9** 753.
[5] Yaklin M A, Duxbury P M and Mackay M E 2008 *Soft Matter* **4** 2441.
[6] Hoppe H and Sacriciftci N S 2006 *J. Mater. Chem.* **16** 45.
[7] Sariciftci N S, Smilowitz L, Heeger A J, Wudl F 1992 *Science* **258** 1474.
[8] Weng D, Lee H K, Levon K, Mao J, Scrivens W A, Stephens E B and Tour J M 1999 *Eur. Polym. J.* **35** 867.
[9] Mackay M E, Tuteja A, Duxbury P M, Hawker C J, Van Horn B, Guan Z, Chen G and Krishnan R S 2006 *Science* **311** 1740.
[10] Qicong Y, Marecek, J and Chu B 1994 *J. Chem. Phys.***101** 2665.
[11] Zolotukhin I V, Yanchenko L I and Belonogov E K 1998 *JETP Letters* **67** 720.
[12] Martin T P, Naher U, Schaber H and Zimmermann U 1993 *Phys. Rev. Lett* **70** 3079.
[13] Sorensen C M, Oh C, Schmidt P W, Rieker T P 1998 *Phys. Rev. E.* **58** 4666.
[14] Rieker T P, Misono S, Ehrburger-Dolle F 1999 *Langmuir* **15** 914.
[15] Narayanan R A, Thiyagarajan P, Zhu A J, Ash B J, Shofner M L, Schadler L S, Kumar S K and Sternstein S S 2007 *Polymer* **48** 5734.
[16] Wong H C and Cabral J T 2010 *Phys. Rev. Lett.* In press.
[17] Wong H C, Sanz A, Douglas J F and Cabral J T 2010 *J. Mol. Liq.* **153** 79.
[18] Cabral J T 2002 PhD Thesis, Imperial College, London
[19] Sanz A, Wong H C, Douglas J F and Cabral J T 2009 *Submitted.*
[20] Mandelbrot B 1977 *Fractals, Form, Chance and Dimension* (Freeman, San Francisco)

Journal of Physics: Conference Series **247** (2010) 012046 doi:10.1088/1742-6596/247/1/012046

[21] Witten T A and Sander L M 1981 *Phys. Rev. Lett.* **47** 1400.
[22] Witten T A and Sander L M 1983 *Phys. Rev. B.* **27** 5686.
[23] Sander L M, Cheng Z M and Richter R 1983 *Phys. Rev. B.***28** 6394.
[24] Meakin P 1983 *Phys. Rev. Lett.* **51** 1119.
[25] Kolb M, Botet R and Jullien R 1983 *Phys. Rev. Lett.* **51** 1123.
[26] Meakin P 1984 *Kinetics of aggregation and gelation* (New-Holland, New York)
[27] Tuteja A, Duxbury P M and Mackay M E 2007 *Macromolecules*, **40** 9427.
[28] Tuteja A, Mackay M E, Narayanan S, Asokan S and Wong M S 2007 *Nano Lett.***7** 1276.
[29] Alcoutlabi M and McKenna G B 2005 *J. Phys.: Condens. Matter* **17** R461.
[30] Keddie J L, Jones R A L and Cory R A 1994 *Europhys. Lett.* **27** 59.
[31] Wong H C and Cabral J T 2009 *In prepraration*.
[32] Narayanan S, Lee D R, Guico R S, Sinha S K and Wang J 2005 *Phys. Rev. Lett.* **94** 145504.

Combined time-resolved SAXS and X-ray Spectroscopy methods

Wim Bras, Sergey Nikitenko, Giuseppe Portale, Andy Beale, Ad van der Eerden, Dirk Detollenaere

Netherlands Organisation for Scientific Research (NWO)

DUBBLE@ESRF, BP220, F38043 Grenoble Cedex, France

Utrecht University

Inorganic Chemistry and Catalysis, Utrecht University, Sorbonnelaan 16, 3584 CA Utrecht, The Netherlands

Wim.Bras@esrf.fr

Abstract. Recently developed equipment suitable for quasi simultaneous data collection of SAXS/WAXS and X-ray spectroscopy is discussed. The main applications for this technique are foreseen to be time-resolved studies in inorganic materials relevant for catalysis research and ceramics. The equipment is described and experimental limitations are discussed. Experiments on nanoparticle synthesis and the formation of inorganic materials for catalysis are used as examples of the usefulness of the developed equipment.

1. Introduction

At synchrotron radiation laboratories the combination of several experimental techniques in a single experiment using time-resolved X-ray scattering experiments has become wide spread in the last decade. The synergy between the different data sets that can be obtained is clear in cases where faster time resolution is required, where sample inhomogeneity occurs or where different sample environments are not guaranteed to reproduce the same conditions in independent experiments. The latter can be the case when for instance high temperature furnaces with different X-ray windows are used.

The earliest combinations of SAXS with other techniques can be traced to the use of Differential Scanning Calorimetry(1) and with Wide Angle Scattering (WAXS)(2, 3). The latter is an obvious combination since it is technologically relatively simple to implement and it allows the length scales over which structures can be studied to extend from the conventional SAXS regime to inter atomic distances. Both the combinations of SAXS and WAXS as well as DSC are so wide spread to the extent that most of the recently designed X-ray scattering beam lines are constructed as SAXS/WAXS beam lines. Later combinations with light scattering and birefringence(4) managed to extend the accessible length scale range to the micrometer range. However, the latter set-up is rather complicated and at present has found only limited applications.

Combinations with spectroscopic techniques like Raman scattering, UV-Vis, Fourier Transform Infra Red spectroscopy and dielectric spectroscopy have also been successfully implemented (5-10). These techniques provide chemical information which is not accessible with X-ray scattering techniques. However, these experimental schemes again have not found a wide spread use yet. Especially for Raman scattering where modern fibre optics make it relatively easy to engineer a combined set-up it is puzzling to see that there are relatively few studies. It is possible that the relative high price of this type of equipment plays a role in this but it is more likely that still many potential users are not aware of the possibilities.

Even though there are in some cases clear advantages in combining techniques it is almost inevitable that in such experiments the data quality of at least one of the techniques will be compromised to some extent. For the earlier generation of SAXS/WAXS beam lines where the divergence of the X-ray beam was still relatively high one had to be aware that the over- and under-focus of the X-ray beam on the two detectors could influence the data adversely. For the third generation SR sources, with their much lower divergences, this is much less of an issue since effectively the 'focal depth of field' is larger than the distance separating the SAXS and WAXS detector. However, there are many other pitfalls. The combination of techniques for combinations sake is something that should be avoided.

Also care has to be taken in the interpretation of the different data sets with respect to the sensitivity of the techniques to certain events. A fine example of this is the debate that ranged for some time early this century in the polymer crystallisation literature. Here the issue was that depending on the simultaneous or time-staggered occurrence of a scattering signal in the SAXS data or the earliest moment of detection of diffraction peaks in the WAXS regime, in a sample that was quenched below the crystallisation temperature, one would be able to distinguish between a crystallisation model that could be described by a spinodal decomposition like mechanism or by a more accepted nucleation and growth mechanism(11-14). Obviously it is much more difficult to prove that something is not there then to prove that a diffraction peak is present. However, since in the SAXS data one used the invariant, and therefore could utilise all the pixels of the SAXS detector, in the WAXS data the relevant diffraction peaks were only mapped upon a limited number of detector pixels. This led to a huge difference in data quality between the two techniques. Besides that, effects like Scherrer broadening due to small crystallite size and thermal broadening due to the elevated temperatures that were required also played a role. Effectively one required a much better WAXS detector than SAXS detector to be able to obtain the same data quality (15). In hindsight one can safely say that the outcome of many experiments was influenced more by the detector quality than by the crystallisation behaviour of the polymers under investigation. In this context it is maybe also worthwhile to observe that a relative large number of crystallisation studies on inorganic materials find the presence of an amorphous precursor before the nano-particles become nano-crystallites. This can be the case but it also is feasible that one deals with disordered nanocrystals which generate diffraction peaks below the detection threshold (16). In the specific case of the crystallisation of semicrystalline polymers it appears that the most wide spread opinion at present is that there indeed is a mesophase observed before the crystallisation starts but there still exists difference of opinion about what exactly this mesophase is. However, from an experimental point of view it is impossible to access the real onset of crystallisation via WAXS since this technique is intrinsically not sensitive enough to measure non-perfect nanocrystals at elevated temperatures (15). The SAXS/WAXS data clearly has to be complemented with data from other experimental techniques and modelling before the ultimate answer will be found.

In the domain of X-ray based techniques the cases where scattering and diffraction might not provide all the answers one can in the first place try to utilise 'total scattering' methods (17, 18) which can provide information about the order inside the nanoparticles but in other cases one can revert to the spectroscopic techniques that have been mentioned above. Another technique that can generate data relevant for the earliest stages of a crystallisation process or that can shine light on the structure of small disordered domains is X-Ray Absorption Spectroscopy (XAS). The technique probes the fine structure that can be observed in the X-ray absorption spectra near an absorption edge. XAS can be subdivided in two regions. The first region, close (5-50 eV) to the absorption edge is called NEXAFS or XANES, both acronyms implying X-ray Absorption Near Edge Structure. This is the region where information about valence and the coordination of the probe atom can be found. Beyond this energy region (i.e. > 50 eV) one generally speaks of X-ray Absorption Fine Structure (EXAFS). The information that can be obtained from this regime concerns atomic pair distribution function. The experimental requirements for both techniques are very similar and it is beyond the scope of this manuscript to detail the differences (19). XAS has in the past been successfully combined with WAXS (20, 21) but the combination with SAXS, which adds an extra length scale, is relatively new (22, 23).

XAS is a spectroscopic technique where the transmission of a sample has to be measured as function of photon energy which is varied over the absorption edge of a probe-atom (24) This in practice means that there is a conflicting requirement with carrying out X-ray scattering methods simultaneously since this requires in general constant photon energy. This situation can be remedied by performing the experiments quasi simultaneous in which the XAS data is obtained during an energy scan and the scattering data in a brief interlude between the XAS scans. See figure 1.

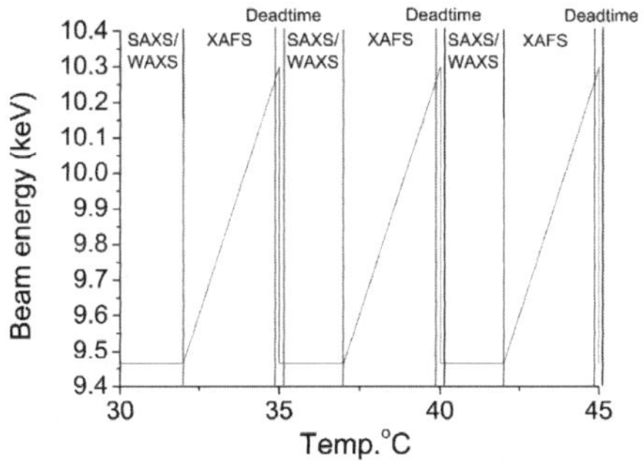

Figure 1 Data acquisition scheme for quasi- simultaneous SAXS/WAXS/XAS experiments. The horizontal scale represents the sample parameter that is varied, in this example the temperature.

The time-resolution that can be achieved in this way is dominated by the time required to carry out an energy scan since the data acquisition time of X-ray scattering can be limited to 1-5 seconds. This can be a function of the mechanics of the monochromator or of the counting statistics required to obtain an acceptable data quality. With a conventional X-ray monochromator the mechanical limitations will impose a lower limit of about 30-60 seconds/frame. Faster speeds can be achieved with special monochromators(25). This scanning technique is conventionally named QuXAFS and should not be mistaken with the energy dispersive variety of XAS where a polychromatic beam is used and the time resolution can be reduced to milliseconds (26, 27) at the expense of a reduced energy resolution and therefore lower XAS data quality.

In materials science there is a multitude of experimental problems where millisecond time-resolution is not required but where the time scale of structural developments can be tuned by choosing parameters like temperature or chemical concentration. One can think about for instance thermally induced crystallization processes or catalytic reactions. It is for these type of experiments that we have developed the combined SAXS/WAXS/XAS equipment (23). Here we will discuss some results and limitations of the technique.

2. Results

2.1. Instrumentation

A technical drawing of the combined SAXS/WAXS/XAFS set-up is shown in figure 2. For a full technical description one is referred to our earlier work (23).

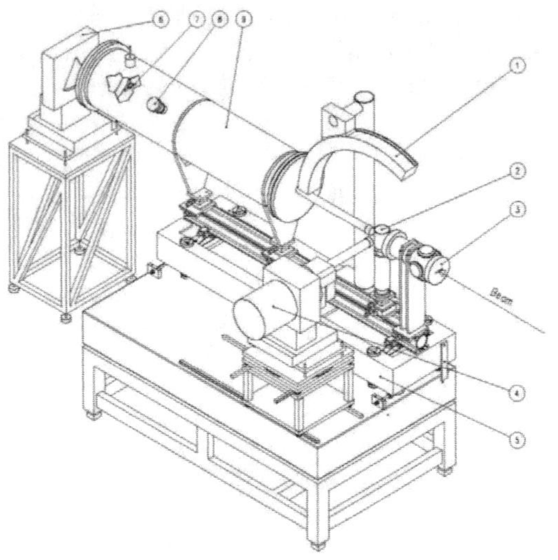

Figure 2 Technical drawing of the combined SAXS/WAXS/EXAFS set up on beam line BM26A at the European Synchrotron Radiation Source. 1- position sensitive INEL detector, 2 – sample position, 3 – intensity monitor for incoming beam, 4 – fluorescence EXAFS detector, 5 – height adjustable optical bench, 6 – position sensitive SAXS detector, 7 – Beam stop and transmitted intensity monitor, 8 – evacuation port, 9 – SAXS flight tube. (reprinted from (23))

The DUBBLE BM26A beam line receives dipole radiation of an ESRF 0.4 Tesla bending magnet. The lower energy limit is determined by the absorption of some Be windows in the beam line and is about 4 keV which means that Ti is the element from which one can still comfortably obtain XAS data. The higher energy range is limited by the mechanics of the monochromator and is around 42 keV, i.e. the Ce K-edge.

At present not sufficient experiments have been carried out to provide a detailed analysis of what time resolution is achievable for which elemental absorption edge. However, the rate limiting step is the collection of the XAS data. In the case that one only wants to collect XANES data, i.e. over a relative small energy range, a time resolution of 30-60 seconds is required for the energy scan and data collection. When the energy has to be scanned over the larger range required for EXAFS this will increase considerably.

For photon energies up to 20 keV it is still feasible to resolve SAXS diffraction distances pertaining to real space structures of around 40 nm. For lower energies this can be increased to about 80 nm.

2.2. Nanoparticle synthesis

An example where both SAXS as well as XANES/EXAFS data were required to shed light onto a formation process and chemical state of samples can be found in the formation of Co-nanostructured particles. Co nanoparticles have been formed by mixing Cobalt bis(2-ethylhexyl)sulfosuccinate, $Co(AOT)_2$, reverse micelles dispersion in n-heptane together with 1 wt% in ethanol sodium borohydride solution, according to the reaction

$$Co(AOT)_2 + 2NaBH_4 \rightarrow Co + 2NaAOT + B_2H_6 + H_2$$

.In this case the high fluidity of this solution of a reversed micellar solution was used not only for the formation of the 2-6 nm sized nanoparticles but also for directing the shape of nanoparticle superstructures. The size is generally reckoned to be a function of the amount of reducing agent added to the solution but the shape can be influenced by the water content since one can tune the microstructure from the micelles from cylindrical to spherical by varying the concentration. These nanoparticles can then either remain in solution with a surrounding protective layer or, in case of existing attractive potentials, can start to form elongated nanostructures.

The nanoparticle size is well suited for SAXS investigations but due to the small size it is very hard to obtain any diffraction information. The Scherrer broadening from which the diffraction peaks suffer, assuming that the order is already high enough to create diffraction, will be so much that it will be difficult to detect above the background due to the scattering of other components of the sample. The internal structure of these nanoparticles 'building blocks' as well as the nanostructure created when they do have an attractive potential, can be studied with X-ray absorption techniques. In figure 3 the SAXS data for three samples with identical composition but obtained by different treatment routes are shown: A) Co nanoparticles separated from the reacting mixture immediately after complete Co(II) reduction; B) Co nanoparticles formed after complete Co(II) reduction and aged in the same reacting mixture for 1 day; C) Co nanoparticles of sample A aged at room temperature in air for about one day. From these data it could be derived that self-assembled elongated microstructure (q^{-1} slope) was formed using preparation methods A and C (28). The average Co nanoclusters size was 1nm. For sample B, aging of Co nanoparticles in reaction mixture produces increase of the size of the Co nanoparticles and a mixture of clusters of 10nm and 21nm average radii was found via SAXS data modeling (28).

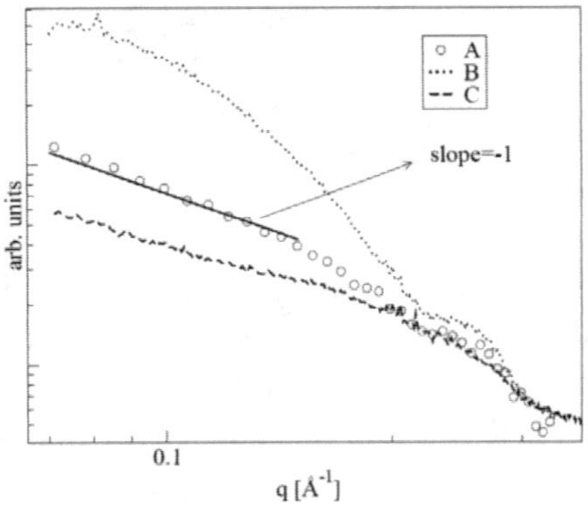

Figure 3 SAXS data of AOT reverse micellar solutions with Co-nanoparticles of three samples with identical composition but with different preparation methods. (reprinted with permission from (28))

The corresponding EXAFS spectra are shown in figure 4. From these spectra the parameters regarding the Co-Co interaction can be derived. We again observe rather different results depending on the sample processing pathway. For sample 'B' the lowest coordination number and shortest Co-Co distance of all three samples is found. In contrast the disorder, i.e. deviation from the bulk crystalline lattice is highest.

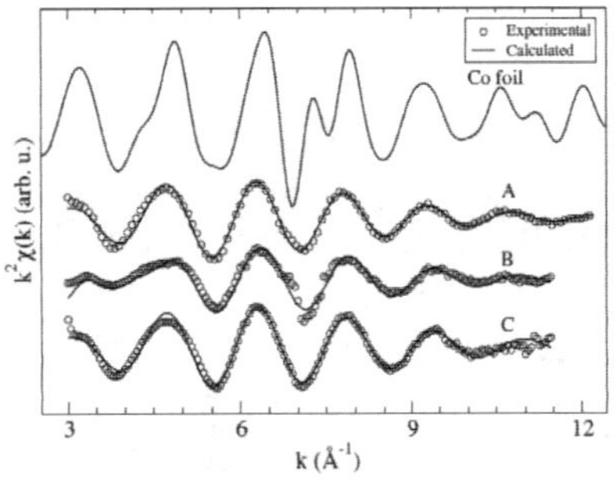

Figure 4 k-weighted EXAFS spectra of the same samples corresponding to the SAXS patterns shown in figure 3. (reprinted with permission from (28)

On the basis of the combined SAXS and EXAFS results one can derive a model for the micro- and nanostructure found in this sample. The schematic drawing for samples prepared according to the 'A' and 'C' recipes is shown in figure 4. The disordered Co nanoparticles are enclosed in cylindrical vesicles and form an elongated loose necklace. The average cluster diameter was found close to 1-2nm. Using the 'B' procedure, the microstructure is mainly unchanged, but a significant fraction of nanoparticles has a larger diameter, close to 4nm.

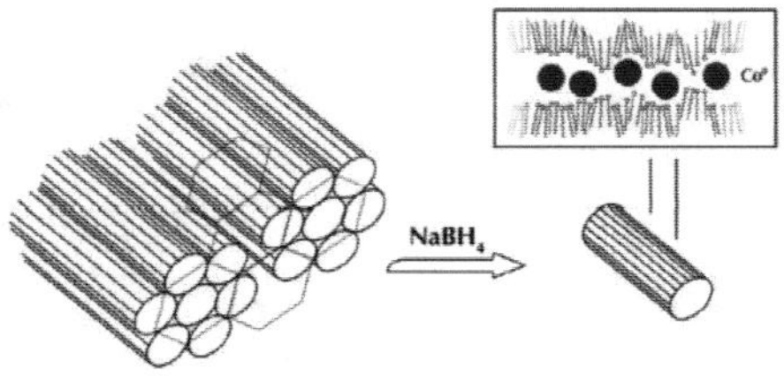

Figure 5 Schematic drawing of the nano- and microstructure of reversed micellar solutions with Co-nanoparticles obtained with the sample preparation method 'B' (as described in the text) (reprinted with permission from (28)

2.3. Inorganic catalytic systems

A second example in which the combination of SAXS/WAXS and EXAFS played a role in elucidating the growth process can be found in the study of the formation of inorganic microporous crystalline materials which can for instance be used as catalyst carriers. Here structure formation has to be studied at a very broad range of length scales. The very smallest scale is at the molecular level where the local coordination can be studied by spectroscopy. The nano scale (primary building elements) can be studied with both spectroscopy as well as diffraction and the longer range order, i. e. the crystallite formation is typically a length scale that falls in the remit of SAXS.

The specific material studied in this example is zinc-substituted microporous aluminophosphate used for the creation of molecular sieves. The starting material was a gel which was heated during the experiment and thus inducing structure formation. The initial state is amorphous, as derived from the absence of diffraction peaks in the WAXS spectrum. See figure 6a. However, at elevated temperatures the materials start to crystallize as evidenced by the occurrence of diffraction peaks. The XAFS (both XANES and EXAFS) data allow us to follow the changes in Zn^{2+} coordination number during the crystallisation process and this changes considerably during the crystallisation process when the Zn is incorporated in the crystalline fraction. See figure 6b. The interesting point to note is that we find clear changes in both the EXAFS as well as the diffraction curves but a SAXS pattern that is only slightly changing which indicates the presence of domain sizes of about 12 nm. See figure 6c. This is rather peculiar since one would expect to see at least some changes due to the fact that crystallisation is rarely such a benign event that no changes occur. A possible cause for this could be that the Zn^{2+} ions responsible for 'inducing' the formation of a particular microporous phase only comprise 2.5 wt% and therefore most of the material in the reaction gel remains amorphous/uncrystallised(29).

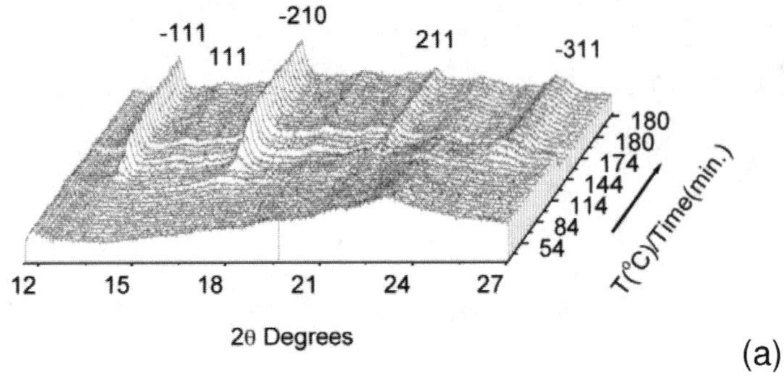

(a)

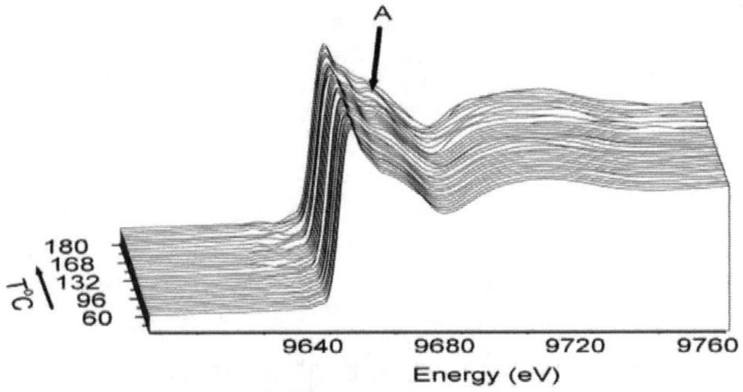

(b)

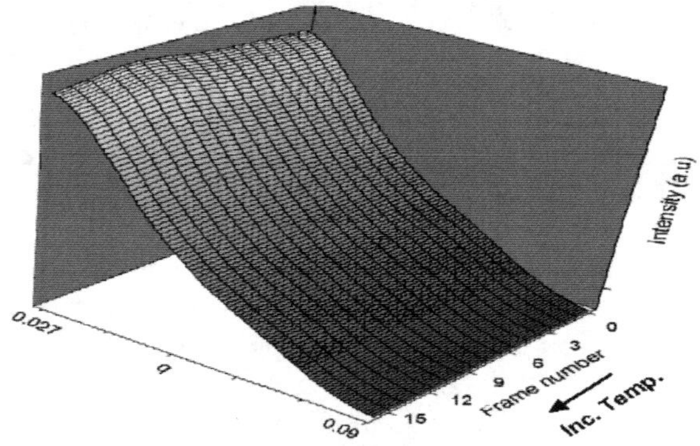

(c)

Figure 6a Time-resolved WAXS patterns from a crystallising gel. 6b- corresponding X-ray spectroscopy patterns, 6c corresponding SAXS patterns. The figures do have different orientations for clarity. Reprinted with permission from (22)

However, some clarity in the situation can be obtained when we consider one of the parameters that is most sensitive to changes, i.e. the invariant.

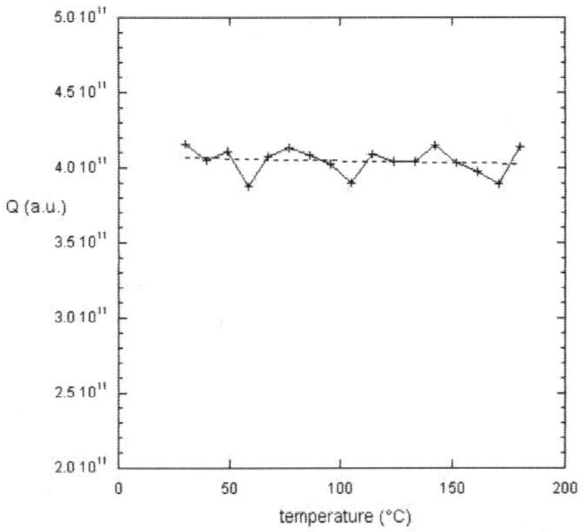

Figure 7 Time development of the invariant, Q, for the scattering experiment described above and in figure 6. (reprinted from (23))

For a two phase system the invariant can be given by: $Q = \int I(\vec{q})dV = \int_{q=0}^{q=\infty} I(q)q^2dq = \langle n_e \rangle^2 \phi_1\phi_2$ in which $\langle n_e \rangle^2$ is the electron density contrast between the two phases and $\phi_{i,j}$ the volume fractions of the two phases. In figure 7 the time development of the invariant is sketched and one finds that within the experimental error margin Q remains unchanged. Since we do know from the SAXS data that the domain size also remains unchanged the only logical conclusion, apart from a non functioning detector, as was mentioned previously, is that in the phase separated system one of the phases crystallizes whilst the other remains amorphous. This process takes place without material transport between the two phases and with no appreciable changes in density of the crystallizing phase. As was mentioned in the introduction it is often rather difficult to prove the absence of any change and therefore the proper functioning of the detector was checked.

3. Limitations

The system that we have developed is unfortunately not universally applicable. In this section we will discuss some of the limitations.

The energy range over which this type of technique combination can be applied is limited. Obviously one would like to perform the diffraction and scattering experiments close to the energy imposed by the XAS experiment. For WAXS or powder diffraction this is in general possible provided that there is a position sensitive detector available that can operate at this energy and has the required spatial resolution. For gas filled detectors, like the Inel (30), the photon detection efficiency is rather low when the photon energy exceeds 15 keV. The low count rate capability of this type of detector is in general not a major problem thanks to the limited time resolution for which the combined set-up has been developed. However, if this is combined with a low efficiency then this becomes a major problem and different detectors have to be used. This can be CCD based systems but these will suffer from a smaller accessible q-range. Also at higher energies the diffraction pattern will be compressed and one might loose too much resolution in which case one has to revert to a diffractometer style of data acquisition which will increase the time required to obtain a data frame.

Similar arguments can be given for the SAXS detector with the obvious caveat that a real photon counting detector would be the best choice which unfortunately rules out a large number of CCD detectors. A second effect of an increase in photon energy is a loss in low angle resolution. The parasitic scatter cone of a beam line is determined by the slit settings and the beam size. Since this is unlikely to change much as function of energy but the scattering patterns get compressed this means that the real low angle part of the scattering pattern will disappear in the parasitic scattering cone. This effect can partially be overcome by using a micro focus beam but this has as a down side that the possibility that radiation damage can occur in the small exposed volume increases.

An alternative solution would be to accept that the energies at which the different data sets have to be taken are further removed from each other. This solution is less favourable due to the longer energy scanning time required. This will most likely also require an adjustment, not only in the monochromator settings, but also in those of the other optical components of the beam line. In theory this can be achieved by storing these positions in a look-up table but this is only a theoretical option and in practice either the XAS or the SAXS data quality will suffer a loss in quality although with the newer generation of monochromators these limitations have become less important.

A limitation that exists in the case of the EXAFS experiments is that the doping with the probe atom should be high enough in order to generate sufficient statistics. The maximum length of the energy scan required to obtain the EXAFS results is dictated by the time scale on which structural developments take place. If this is too short then obviously it will be impossible to obtain relevant data. An example where the application of the technique failed was in the case of temperature induced cordierite glass ($Mg_2Al_4Si_5O_{18}$) crystallisation which was doped a crystallisation enhancer in the form of 0.34 mol% Cr_2O_3 (31). This system would be an ideal candidate for investigations with three techniques. Already an extensive study has been made with a combined SAXS and WAXS experiment. However, the Cr doping was found too low to be able the elucidate the role of this atom in the initial stages of crystallisation when the required time-resolution was on the order of 1 minute/frame. Increasing the X-ray intensity to counteract this requires also detectors that can handle the increased count rate but also the sample has to be able to withstand this flux for an extended period. With modern beam lines, even for inorganic samples, this is not always the case (32).

4. Conclusions

We have constructed a combined SAXS/WAXS/EXAFS set up which can for certain classes of samples generate data sets in which the fact that the data is obtained simultaneously allows the elucidation of the different steps in growth processes ranging from the molecular scale to long range structural features. Under favourable conditions time resolutions of 1 minute/frame can be achieved which is compatible with numerous experiments in for instance inorganic synthesis of micro-structured materials.

The applied technique is useable for samples where the EXAFS edges are not too far removed from the energy range where SAXS data can be obtained. Also limitations exist when there are low concentrations of the probe atom in the sample. However, in the cases that the combination of these three technique can be applied it generates a wealth of information which is difficult to surpass by other means.

5. Acknowledgements

The data and figures on the AOT Co-nanoparticle synthesis were graciously provided by Dr. A. Longo. The DUBBLE staff are acknowledged for technical support. NWO/FWO is thanked for granting the beam time on BM26A. Piet Lassing is thanked for the technical drawings.

Figures 3, 4 and 5 have been reprinted with permission from A. Longo et al. Journal of Applied Physics 105, 114308, Copyright 2009, American Institute of Physics. Figure 6 has been reprinted with permission from A. Beale et al. Journal American Chemical Society 128, 2006, 12386

6. References

1. Russell, T. P., and J. T. Koberstein. 1985. Simultaneous Differential Scanning Calorimetry And Small-Angle X-Ray-Scattering. Journal Of Polymer Science Part B-Polymer Physics 23:1109-1115.
2. Bark, M., H. G. Zachmann, R. Alamo, and L. Mandelkern. 1992. Investigations Of The Crystallization Of Polyethylene By Means Of Simultaneous Small-Angle And Wide-Angle X-Ray-Scattering. Makromolekulare Chemie-Macromolecular Chemistry And Physics 193:2363-2377.
3. Bras, W., G. E. Derbyshire, A. J. Ryan, G. R. Mant, A. Felton, R. A. Lewis, C. J. Hall, and G. N. Greaves. 1993. Simultaneous Time Resolved Saxs and Waxs Experiments Using Synchrotron Radiation. Nuclear Instruments & Methods in Physics Research Section a-Accelerators Spectrometers Detectors and Associated Equipment 326:587-591.
4. Fernandez-Ballester, L., T. Gough, F. Meneau, W. Bras, F. Ania, J. C. Francisco, and J. A. Kornfield. 2008. Simultaneous birefringence, small- and wide-angle X-ray scattering to detect precursors and characterize morphology development during flow-induced crystallization of polymers. Journal Of Synchrotron Radiation 15:185-190.
5. Bryant, G. K., H. F. Gleeson, A. J. Ryan, J. P. A. Fairclough, D. Bogg, J. G. P. Goossens, and W. Bras. 1998. Raman spectroscopy combined with small angle x-ray scattering and wide angle x-ray scattering as a tool for the study of phase transitions in polymers. Review of Scientific Instruments 69:2114-2117.
6. Rastogi, S., J. G. P. Goossens, and P. J. Lemstra. 1998. An in situ SAXS/WAXS/Raman spectroscopy study on the phase behavior of syndiotactic polystyrene (sPS)/solvent systems: Compound formation and solvent (dis)ordering. Macromolecules 31:2983-2998.
7. Bras, W., G. E. Derbyshire, D. Bogg, J. Cooke, M. J. Elwell, B. U. Komanschek, S. Naylor, and A. J. Ryan. 1995. Simultaneous Studies of Reaction-Kinetics and Structure Development in Polymer Processing. Science 267:996-999.
8. Wurm, A., R. Soliman, J. G. P. Goossens, W. Bras, and C. Schick. 2005. Evidence of pre-crystalline-order in super-cooled polymer melts revealed from simultaneous dielectric spectroscopy and SAXS. Journal of Non-Crystalline Solids 351:2773-2779.
9. Ezquerra, T. A., I. Sics, A. Nogales, Z. Denchev, and F. J. Balta-Calleja. 2002. Simultaneous crystalline-amorphous phase evolution during crystallization of polymer systems. Europhysics Letters 59:417-422.
10. Mesu, J. G., A. M. Beale, F. M. F. de Groot, and B. M. Weckhuysen. 2006. Probing the influence of X-rays on aqueous copper solutions using time-resolved in situ combined video/X-ray absorption near-edge/ultraviolet-visible spectroscopy. Journal Of Physical Chemistry B 110:17671-17677.
11. Terrill, N. J., P. A. Fairclough, E. Towns-Andrews, B. U. Komanschek, R. J. Young, and A. J. Ryan. 1998. Density fluctuations: the nucleation event in isotactic polypropylene crystallization. Polymer 39:2381-2385.
12. Chu, B., and B. S. Hsiao. 2001. Small-angle X-ray scattering of polymers. Chemical Reviews 101:1727-1761.

13. Wang, Z. G., B. S. Hsiao, E. B. Sirota, and S. Srinivas. 2000. A simultaneous small- and wide-angle X-ray scattering study of the early stages of melt crystallization in polyethylene. Polymer 41:8825-8832.
14. Strobl, G. 2000. From the melt via mesomorphic and granular crystalline layers to lamellar crystallites: A major route followed in polymer crystallization? Eur. Phys. J. E 3:165-183.
15. Bras, W., I. P. Dolbnya, D. Detollenaere, R. van Tol, M. Malfois, G. N. Greaves, A. J. Ryan, and E. Heeley. 2003. Recent experiments on a combined small-angle/wide-angle X-ray scattering beam line at the ESRF. Journal of Applied Crystallography 36:791-794.
16. Gilbert, B. 2008. Finite size effects on the real-space pair distribution function of nanoparticles. Journal of Applied Crystallography 41:554-562.
17. Chupas, P. J., K. W. Chapman, G. Jennings, P. L. Lee, and C. P. Grey. 2007. Watching nanoparticles grow: The mechanism and kinetics for the formation of TiO2-supported platinum nanoparticles. Journal of the American Chemical Society 129:13822-+.
18. Michel, F. M., L. Ehm, S. M. Antao, P. L. Lee, P. J. Chupas, G. Liu, D. R. Strongin, M. A. A. Schoonen, B. L. Phillips, and J. B. Parise. 2007. The structure of ferrihydrite, a nanocrystalline material. Science 316:1726-1729.
19. Koningsberger, D. C., and P. R., editors. 1988. X-Ray Absorption. Wiley.
20. Dent, A. J., M. Oversluizen, G. N. Greaves, M. A. Roberts, G. Sankar, C. R. A. Catlow, and J. M. Thomas. 1995. A Furnace Design for Use in Combined X-Ray-Absorption and Diffraction up to a Temperature of 1200-Degrees-C - Study of Cordierite Ceramic Formation Using Fluorescence Qexafs/Xrd. Physica B 209:253-255.
21. Sankar, G., P. A. Wright, S. Natarajan, J. M. Thomas, G. N. Greaves, A. J. Dent, B. R. Dobson, C. A. Ramsdale, and R. H. Jones. 1993. Combined Quexafs-Xrd - a New Technique in High-Temperature Materials Chemistry - an Illustrative in-Situ Study of the Zinc Oxide-Enhanced Solid-State Production of Cordierite from a Precursor Zeolite. Journal of Physical Chemistry 97:9550-9554.
22. Beale, A. M., A. M. J. van der Eerden, S. D. M. Jacques, O. Leynaud, M. G. O'Brien, F. Meneau, S. Nikitenko, W. Bras, and B. M. Weckhuysen. 2006. A combined SAXS/WAXS/XAFS setup capable of observing concurrent changes across the nano-to-micrometer size range in inorganic solid crystallization processes. Journal Of The American Chemical Society 128:12386-12387.
23. Nikitenko, S., A. M. Beale, A. M. J. van der Eerden, S. D. M. Jacques, O. Leynaud, M. G. O'Brien, D. Detollenaere, R. Kaptein, B. M. Weckhuysen, and W. Bras. 2008. Implementation of a combined SAXS/WAXS/QEXAFS set-up for time-resolved in situ experiments. Journal Of Synchrotron Radiation 15:632-640.
24. D. C. Koningsberger, D. C., and R. Prins, editors. 1988. X-Ray Absorption. Wiley.
25. Frahm, R., M. Richwin, and D. Lutzenkirchen-Hecht. 2005. Recent advances and new applications of time-resolved X-ray absorption spectroscopy. Physica Scripta T115:974-976.
26. Dent, A. J. 2002. Development of time-resolved XAFS instrumentation for quick EXAFS and energy-dispersive EXAFS measurements on catalyst systems. Topics in Catalysis 18:27-35.
27. Newton, M. A., B. Jyoti, A. J. Dent, S. G. Fiddy, and J. Evans. 2004. Synchronous, time resolved, diffuse reflectance FT-IR, energy dispersive EXAFS (EDE) and mass spectrometric investigation of the behaviour of Rh catalysts during NO reduction by CO. Chemical Communications:2382-2383.
28. Longo, A., F. Giordano, F. Giannici, A. Martorana, G. Portale, A. Ruggirello, and V. T. Liveri. 2009. Combined small-angle x-ray scattering/extended x-ray absorption fine structure study of coated Co nanoclusters in bis(2-ethylhexyl)sulfosuccinate. Journal of Applied Physics 105.
29. O'Brien, M. G., A. M. Beale, C. R. A. Catlow, and B. M. Weckhuysen. 2006. Unique organic-inorganic interactions leading to a structure-directed microporous aluminophosphate

crystallization as observed with in situ Raman spectroscopy. Journal of the American Chemical Society 128:11744-11745.
30. Evain, M., P. Deniard, A. Jouanneaux, and R. Brec. 1993. POTENTIAL OF THE INEL X-RAY POSITION-SENSITIVE DETECTOR - A GENERAL STUDY OF THE DEBYE-SCHERRER SETTING. Journal of Applied Crystallography 26:563-569.
31. Bras, W., S. M. Clark, G. N. Greaves, M. Kunz, W. van Beek, and V. Radmilovic. 2009. Nanocrystal Growth in Cordierite Glass Ceramics Studied with X-ray Scattering. Crystal Growth & Design 9:1297-1305.
32. Mesu, J. G., A. M. J. van der Eerden, F. M. F. de Groot, and B. M. Weckhuysen. 2005. Synchrotron radiation effects on catalytic systems as probed with a combined in-situ UV-Vis/XAFS spectroscopic setup. Journal Of Physical Chemistry B 109:4042-4047.

Journal of Physics: Conference Series 247 (2010) 012048 doi:10.1088/1742-6596/247/1/012048

XPCS study of dynamic correlation in polyurethane gel–carbonyl iron composite under magnetic field

Helena Grigoriew[1], Lutz Wiegart[2], Anna Boczkowska[3] Monika Mirkowska[4]
[1] Institute of Nuclear Chemistry and Technology, Warsaw 03195, Poland
[2] ESRF, Grenoble, 38043 Cedex 9, France
[3] Faculty of Materials Science Warsaw University of Technology, Poland
[4] Institute of Electronic Materials Technology, Warsaw, Poland

E-mail: haga311@hotmail.com

Abstract. An X-ray photon correlation spectroscopy (XPCS) study of composite-type material consisting of polyurethane gel and carbonyl iron micrometric spheres was performed under magnetic fields of 0, 300 and 600 mT. The onion-like spheres structure was destroyed during the composite processing. The following conclusions were obtained from the study: -The polyurethane matrix is preferred as a source for the observed dynamic effects. –Below 300mT the material dynamics in direction of the outer magnetic field are very clear. –For 600 mT the dependence of the dynamics on magnetic field direction disappears, but the correlation rate is much higher. These findings may be caused by a disturbance of the polymer mesostructure by larger strain leading to its cross-linking.

1. Introduction

The polyurethane gel-carbonyl iron composite (PU) belongs to magnetorheological elastomers which reveal a change of their mechanical properties under an external magnetic field. These elastomers consist of magnetically permeable particles (such as iron or other ferromagnetic particles) of sizes ranging from a few to several hundreds of micrometers, added to a viscoelastic polymeric material prior to cross-linking. The origin of the magnetorheological property of such materials is still a major challenge for condensed matter physics although the interest in these materials has increased rapidly during last years due to their potential application in many fields of industry.

According to our knowledge, no XPCS study of magnetorheological composite dynamics was published so far. The nearest materials studied by XPCS under a magnetic field are ferrofluids, consisting of small, nanometric ferromagnetic particles in a pure viscous liquid as a carrier. It occurred that due to the magnetic dipole-dipole interaction between particles under a magnetic field, the particles joined in dimers, trimers or chains oriented along the field direction. XPCS indicated in these cases that the correlation effects and diffusion should be slowed down perpendicular to the field direction and enhanced parallel to it, caused by an unfolding of individual particles and / or the formation of particles assemblies [1-3].

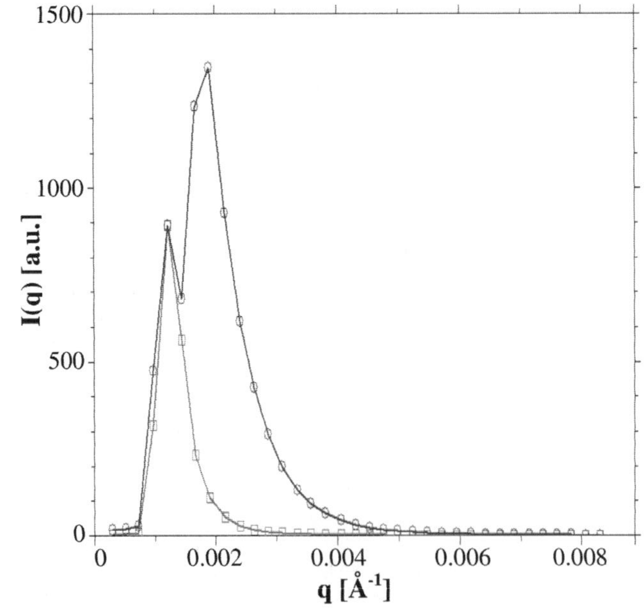

Fig.1 SANS curve of loose carbonyl iron particles (blue) and PU composite (red)

The gels and polymers dynamic relaxation influenced by various parameters, as temperature, aging etc. were also studied using the method [4-7], and some features of their dynamic relaxation are close to those observed in the present work.

A possible model of the dynamic response of the studied composite PU to an outer magnetic field should take into account the suppression of the field-induced particle motion along the field lines by the viscoelastic polymer matrix. Then the strain is generated in both components of the composite. Slow dynamic relaxation of the strain in supercooled liquids was studied also with the attempt XPCS [8-10] to describe the mechanism of their dynamic response.

The great size of carbonyl iron (CI) particle in relation to the X-ray scattering scale should be a reason that X-rays will register mainly single particles, not particle groups. On our initial SANS image of the loose CI spheres an intense broad diffraction ring, centered at about 420 Å was detected which is typical for the CI sphere onion-like structure with such inter-shell distance. However, SANS images of PU show no such diffraction ring, which seems to be an evidence of the complete destruction of the shells structure during the processing (Fig.1) [11]. The destroyed shell structure rather should not be able to react on the stress induced by magnetic field in a specific way, as it is e.g. for compact hard sphere metallic structure, but it rather will be a sum of not coordinated very weak events generated in the shell remains.

2. Experimental

Materials. The investigated material was produced at the Faculty of Materials Science, Warsaw University of Technology. It consists of a polyurethane gel matrix and embedded carbonyl iron (CI) spheres of 7 μm size. The polyurethane gel is formed of isocyanate and polyol compounds with a proportion of hard and soft segments equal to 30/70. To obtain magnetorheological composite (PU), the polyurethane gel was processed together with 11.5 vol% of CI for 24 hours under a magnetic field of 300 mT at a temperature of 25°C. As it was shown using SEM, short CI chains were formed inside the material during the processing.[12].

The XPCS experiment was carried out at the ID10C branch of the Troika Beamline at the European Synchrotron Radiation Facility (ESRF). The sample in form of a slice of PU of about 40 μm thickness was held in the sample cell located between electromagnet poles filled with helium to avoid air scattering effects on the long way of the X-ray. The magnetic field lines were oriented perpendicular to the X-beam and parallel to the sample surface and the embedded inside CI chains, respectively (Fig.2).

The angle (q) and time (t) resolved measurements of the speckle pattern, I(q, t), were recorded using a partially coherent X-beam of 1.75 Å wavelength (to suppress any X-ray fluorescence signal from Fe of 1.74 Å), by a direct illumination CCD area detector (Princeton Instruments) of pixel size 22.5 microns located 2.3 m downstream of the sample. It occurred that the relaxation times (t_o) ranged from $10x10^2$ sec to $100x10^2$ sec. The speckle patterns were recorded with about 1 sec acquisition time and 6 sec lag-time. The long-time series of the sample measurements, I(q,t), obtained from the speckle patterns, were performed under magnetic fields of: 0, 300 and 600 mT. An on-line correlation software was applied to calculate the one-time averaged intensity autocorrelation functions, $g^{(2)}$ (q,t) according to the formula:

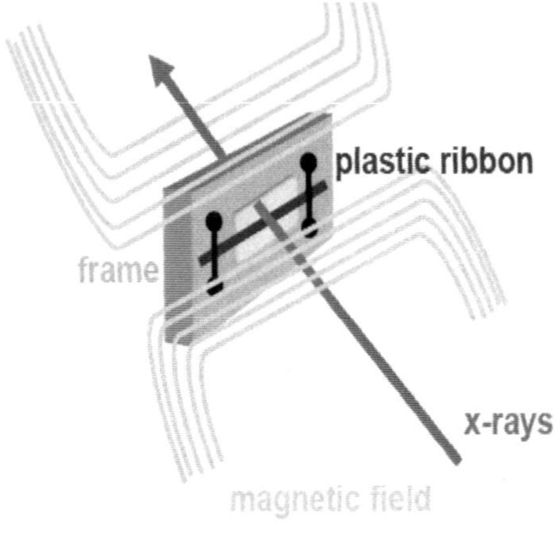

Fig.2 Sketch of the experimental arrangement

$$g^{(2)}(q,t)=\langle I(q,t')I(q,t'+t)\rangle/\langle I(q,t')^2\rangle$$

where: $q = 4\pi \sin\theta / \lambda$, t – lag time, t' –time, and the brackets denote a time average.
Software masks were used to determine the dependence of the correlation functions $g^{(2)}$ (q,t) on the direction of the magnetic field, i.e. for the scattering parallel (par) and perpendicular (per) to the field lines.

3. Results and Discussion

The $g^{(2)}$(q,t) curves gained at 0 mT, i.e .reference curves, show an extremely slow dynamics with correlation times reaching about 10^4 sec. Under a field of 300 mT the dynamics is faster (correlation time ~ 2×10^3 sec) and differences dependent on the directions are visible. Under a field of 600 mT the correlation time becomes even faster, on the average 1000 sec, while the anisotropy disappeared.

All correlation functions $g^{(2)}$ (q,t) exhibit a simple exponential character, as observed for some polymers for temperature below T_c, when besides of viscous also an solid factor of relaxation exists, but without an "stretched" form arising from an additional relaxation barrier [8]. In this work, the simply exponential character of the curves could be expected, because this polyurethane structure is of the gel type, i.e. it is not as strong ordered as it can be for this kind of polymers.

The correlation rates, Γ (1/sec), were found by fitting exponential functions, according to the formula: $y = a\,e^{-2\Gamma t}$, to the subsequent $g^{(2)}$ (t,q) curves in their regions of correlation. Fig. 3 shows Γ (1/sec) curves for subsequent magnetic field values (0, 300 and 600 mT) in sets of 2 curves at each direction (par and per). Within each set the higher relaxation rate is observed in the direction parallel to the magnetic field (par), the lower one in the direction perpendicular to the magnetic lines (per).

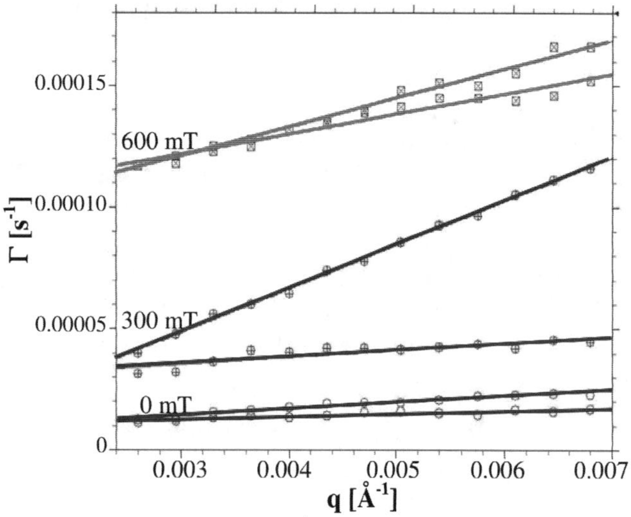

Fig.3. Correlation rates $\Gamma[s^{-1},q]$. For each pair of curves, the higher curve is from measurements made in direction of magnetic field lines, and lower one in direction perpendicular to them.

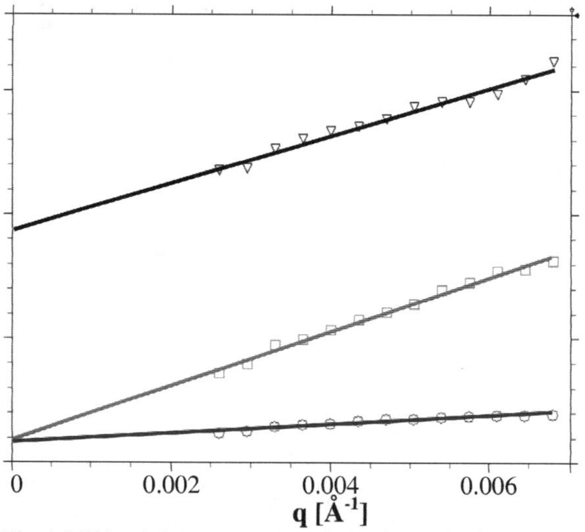

Fig.4. Γ(1/sec) for measurements performed without the mask (isotropic average) with straight lines extrapolated to q= 0

Table I. Diffusion coefficients [Å/sec]

Magnetic field	0 mT	300 mT	600 mT
Parallel	0.003	0.018	0.011
perpendicular	0.001	0.003	0.009

The next parameter determined from the measurements is the diffusion coefficient, D [Å/sec]. In the case of linear Γ, as obtained for all Γ(q) curves, the relation between relaxation rate and diffusion coefficient is given by the formula: Γ=D x q. The values of the diffusion coefficient, D [Å/sec], were found from the slopes of straight lines fitted to Γ (1/sec) (Table I).

The discussion of the obtained results dependent on magnetic field value (Fig.3 and Table I): is as follows:

For the reference sample at 0 mT very moderate values of Γ and D, exhibiting only a very weak anisotropy, are observed. The formation of short chains of CI particles as documented by SEM, aligned in the parallel (par) direction, do not result in a distinct anisotropic dynamics, manifested in the rate of correlation or diffusion. It is an argument for insignificant role of the CI particles in dynamic behavior of the composite.

For the intermediate value of the magnetic field (300 mT), the same parameters are strongly dependent on the magnetic field direction and of larger values. The obtained sample response to the outer magnetic field is similar to the response of other polymers and gels to an absorbed additional internal energy, for example to an increase in temperature, where at higher temperature the Γ slope becomes more steep [9].

Contrary, at 600mT the dependence of Γ and D on the magnetic field direction completely disappears. This could be caused by reaching a kind of saturation, but the unexpectedly large correlation rates Γ, suggest a much more fundamental change in the system. To get a clear picture of these changes, without direction-dependent parameters, we considered the Γ(q) curves obtained from XPCS measurements in isotropic mode (without mask). Figure 4 shows the corresponding data for all used magnetic fields together with a linear fit extrapolating to q = 0.

A dramatic increase of the isotropic Γ(0) values of about ten times for 600 mT as compared with to 300 mT is observed, whereas the slopes of the both Γ (q) curves are similar. As it is known, for $q \rightarrow 0$, $\Gamma \rightarrow E / \eta$, where E is the elasticity yielding a frequency-independent storage modulus, and η is the shear viscosity of the material [9]. A change in the ratio of elasticity and shear viscosity can presumably be associated with a modification of the gel mesostructure.

Gels are materials with a tendency to change their mesostructure to a more dense and more cross-linked phase even without any outer influence, which is known as an aging process [13]. As it was observed in many cases that the mesostructure disturbance is followed by its quick rebuilding through cross-linking and domains growth inside a mesostructure [13].

The polymer strain, caused by the urge of the CI particle to shift along the field line, during an increase of the magnetic field to 600 mT, would reach a high enough value to cause a disturbance of the gel mesostructure. The next step upon these changes is a rebuilding and reinforcement of the mesostructure according to one of the gelation mechanisms [13]. This reinforcement might cause here the decrease of susceptibility of the dynamic features in one direction, and domain growth results in an increase of elasticity E, observed as larger correlation rates, Γ(0).

On the other hand, the magnetorheological effect in PU was measured at the Warsaw University of Technology as the material's response to an outer magnetic field in form of the displacement of a bend sample deflection.

Figure 5[12] shows the displacement effect for a PU sample under the outer magnetic field in the range 0 – 900 mT, for various contents of CI. For 11.5vol% CI (the sample used for the XPCS study) an exponential-type displacement is observable in the range of 0 – 600 mT, but for higher magnetic fields no further displacement was obtained [12].These findings also agree with the hypotheses of reinforcement of the PU mesostructure by a magnetic field of more than 600 mT.

Journal of Physics: Conference Series **247** (2010) 012048 doi:10.1088/1742-6596/247/1/012048

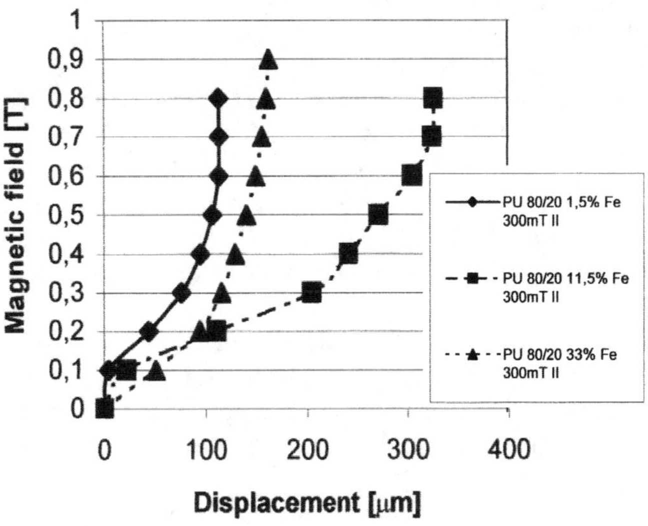

Fig.5 Magnetorheological effect of the studied PU, as the displacement of bend sample deflection upon magnetic field (squares) [12]

4.Concluding remarks

The results of the study seem to prefer polyurethane gel matrix as the source of the main dynamical effects observed by XPCS, arising from a stress transferred from the embedded magnetic particles to the matrix under an external magnetic field.

The magnetic field strongly affects the material dynamics. For its moderate value of 300 mT the dependence of the material dynamics on the direction of the external magnetic field is very clear with the parallel direction showing larger relaxation rates, $\Gamma(q)$.

For higher magnetic field (600 mT) the material's dynamic response to the magnetic field changes dramatically. The dependence of the obtained parameters on the magnetic field direction disappears, but the correlation rate increases by about ten times. This dynamical behavior as a function of the magnetic field strength can be caused by a disturbance of the polymer mesostructure at higher fields, leading to its cross-linking and reinforcement. Further studies are needed to confirm the hypothesis about a rebuilding of the mesostructure.

Acknowledgements. We thank Dr. Torok Gyula for technical assistance in SANS measurements, and M.Sc. Stefan Awietjan for experimental help.

5. References

1 J. Lal, D. Abernathy, L. Auvray, O. Diat, G. Gruebel, Eur. Phys. J. E-4 (2001), 263
2 P. Boesecke, O. Diat, J. Appl. Cryst. 30,(1997), 867
3 A. Robert, E. Wandersman, E. Dubois, V. Dupuis, R. Perzynski, Europhys. Lett. 75 (5), (2006) 764
4 H. Mendil, P. Baroni, L. Noirez, Eur. Phys. J. E19 (2006) 77
5 S. G. J. Mochrie, A. M. Mayes, A. R. Sandy, M. Sutton, S. Brauer, Phys. Rev. Lett. 78 (1997) 1275
6 K. N. Pham, S. U. Egelhaaf, P. N. Pusey, W. C. K. Poon, Phys. Rev. E 69 (2004) 0115031
7 M. Okamoto, T. Norisuye, M. Shibayama, Macromolecules 34 (2001) 8496
8 K. Trachenko, Phys. Rev B 75 (2007) 212201
9 Y. Chushkin, C. Caronna, A. Madsen, EPL 83 (2008) 36001, doi: 10.1209/0295-5075/83/36001
10 V. Viasnoff, F. Lequeux, Phys. Rev Lett. 89 (2002) 065701
11 Unpublished data of SANS experiment at Budapest Neutron Centre, Hungary
12 A. Boczkowska, S. Awietjan, K. Babski, R. Wróblewski, M. Leonowicz, Proc of SPIE, vol. 6170 61700R-1 to -11
13 C. J. Brinker, G. W. Scherer "Sol-Gel Science: the Physics and Chemistry of Sol-Gel Processing.", Academic Press, Boston, 1990

Journal of Physics: Conference Series **247** (2010) 012049 doi:10.1088/1742-6596/247/1/012049

Crystallisation Pathways of Polymorphic Triacylglycerols Induced by Mechanical Energy

Y.L. Lee[a], R.I. Ristic[a*], L.L.DeMatos[a], and C.M. Martin[b]

[a)]Chemical and Process Engineering Department, University of Sheffield, Mappin street, Sheffield S1 3JD, UK

[b)] CCLRC Daresbury Laboratory, Warrington, Cheshire, WA4 4AD, UK

The aim of these studies is to establish sound scientific principles to guide nucleation rate and the selection of a desired polymorph via the application of mechanical energy - ultrasound (US) irradiation. When delivered to a metastable liquid, before the offset of nucleation and under constant temperature and supercooling conditions, the wave nature of this simple form of energy should be critical for defining different crystallisation pathways of polymorphic materials including polymorph selection. To test this hypothesis, we crystallized a melt-grown trilaurin (LLL), a typical polymorphic triacylglycerols (TGA's), with and without US by using *in-situ* simultaneous synchrotron radiation time-resolved small-angle X-ray scattering (SAXS) and wide angle X-ray scattering (WAXS), SAXS/WAXS. Without US application, both polymorphic forms β' and β crystallized. With US treatment of the super cooled melt, the following effects were observed: (a) a marked decrease of induction times (b) an increased nucleation rate, and (c) selective crystallization of only β-form when crystallised at 25 and 30°C with input powers of 20 and 100 W and a sonication time of 2 s. Combining the existing knowledge on the dynamic nucleation of collapsing cavities and a qualitatively developed (P-T) phase diagram for the TGA's, it was possible to describe, for the first time, the behaviour of the most important parameters and the events that characterize the crystallization of these systems. It was shown that the interplay of sonication and the temperature of supercooled melts are critical to the selection of a stable β form.

*** Corresponding Author. Tel: +44 (0)114 2227516 Email: r.i.ristic@sheffield.ac.uk.**

1. Introduction

There has been an increasing interest in the application of ultrasonic waves to promote crystallisation from different phases [1, 2]. The application of power ultrasound during crystallisation (sono-crystallisation) may influence nucleation rate, morphology of particles, rate of crystallization as well as polymorphism [3-5]. The control of the above properties is of capital importance to the crystallisation of a wide range of materials used in processing industries dealing with foods, cosmetics, speciality chemicals, pharmaceuticals, etc. These characteristics can directly influence the efficiency of downstream processes such as separation, filterability, powder flow, caking, compaction, tabletting, and dissolution. In particular, the identification and control of polymorphism is essential to the process development of foods, speciality chemicals, and of special importance to the formulation and patenting of pharmaceutical materials.

In food industries, the following processes are closely related to polymorphic control: the processing of end-products made of fat crystals (such as chocolate, margarine and shortening), cream whipping and the separation of specific lipid materials from natural resources like oils, vegetable and animal fats, etc. In addition, it is well known that the polymorphism of these materials is quite susceptible to external factors governing crystallisation such as temperature, pressure and mechanical waves. This makes crystallisation of fats even more interesting to study since, in the course of crystallisation, the functional properties of fats can be manipulated using these parameters in a controllable manner, and hence industrial processes that lead to products of a desired quality. Thus, knowledge of the polymorphic state of polycrystalline fat is essential for the proper manufacture of fat containing products. In industry, the polymorphic control is usually practiced by tempering or seeding [6]. The former is mostly applied to the chocolate or ice cream industries, whereas the latter refers to the use of certain additives such as seed crystals or templates. One of the challenges in the above processes is mastering polymorphic control by using ultrasonication.

A number of papers concerning sonocrystallization of fats have been published [7-9]. Quite recently, sonocrystallization has been examined in confectionery fats, indicating the possibility that tempering can be achieved by sonocrystallisation [4]. This work has been followed by the studies of the polymorphic behaviour of tripalmitin (PPP) and cocoa butter in the presence of ultrasound energy [5]. The main experimental findings can be summarized as follows: (a) the nucleation rate of PPP was enhanced and induction time was shortened by ultrasound application; (b) the polymorphic form V of cocoa butter was directly crystallized when ultrasound, under the optimal conditions of temperature and the duration of sonication, was applied. However, a key drawback of this study was that X-ray analysis of the crystallizing material has been done ex-situ using conventional X-ray facilities of limited power and resolution. Therefore, these results did not reflect the real-time crystallization process and as such, were not sufficiently informative. It is therefore quite obvious that such results inevitably lead to a speculative and somewhat uncertain interpretation of the crystallization process.

In this paper, the crystallisation behaviour of a typical triacylglycerol, trilaurin (LLL), is examined *in-situ* using synchrotron radiation time-resolved small angle X-ray scattering (SAXS) and wide-angle X-ray scattering (WAXS) techniques. The results were considered within a developed framework based on a strong-coupling between a qualitative (T-P) diagram for LLL and the high pressure modified free energy for homogeneous nucleation. The potential for polymorph control has been investigated by using interplay between sonication power and supercooling of the melt.

2. Experimental Section

2.1 Materials

Trilaurin (LLL) was used as a sample and purchased from Sigma Aldrich (UK). It had a purity of more than 98% and it was used without further purification.

2.2 Methods

All experiments were conducted at the Station 6.2, at Daresbury SRS. This station enabled the performance of *in-situ* time-resolved small-angle X-ray scattering and wide-angle X-ray scattering (SAXS/WAXS) combined with a bespoke sonocrystallisation system, assembled from commercial and custom-made components (Fig 1).

The ambient of the analysing sample was manipulated and maintained using a custom-built crystallisation jacketed-cell with connections to two circulating thermostatic water-baths set at different temperatures. The origin of the water flowing in/out of the jacketed-cell was alternated between either water-baths depending of one's need to either heat the sample (for melting) or cool down to the temperature of crystallisation. The windows of the cell are made from polyimide material (Kapton) of 25-μm thickness. The thickness of the sample that is exposed to the X-ray radiation was 1mm. The temperature of a sample was measured by a thermocouple, with an accuracy of ±0.5°C.

Journal of Physics: Conference Series **247** (2010) 012049 doi:10.1088/1742-6596/247/1/012049

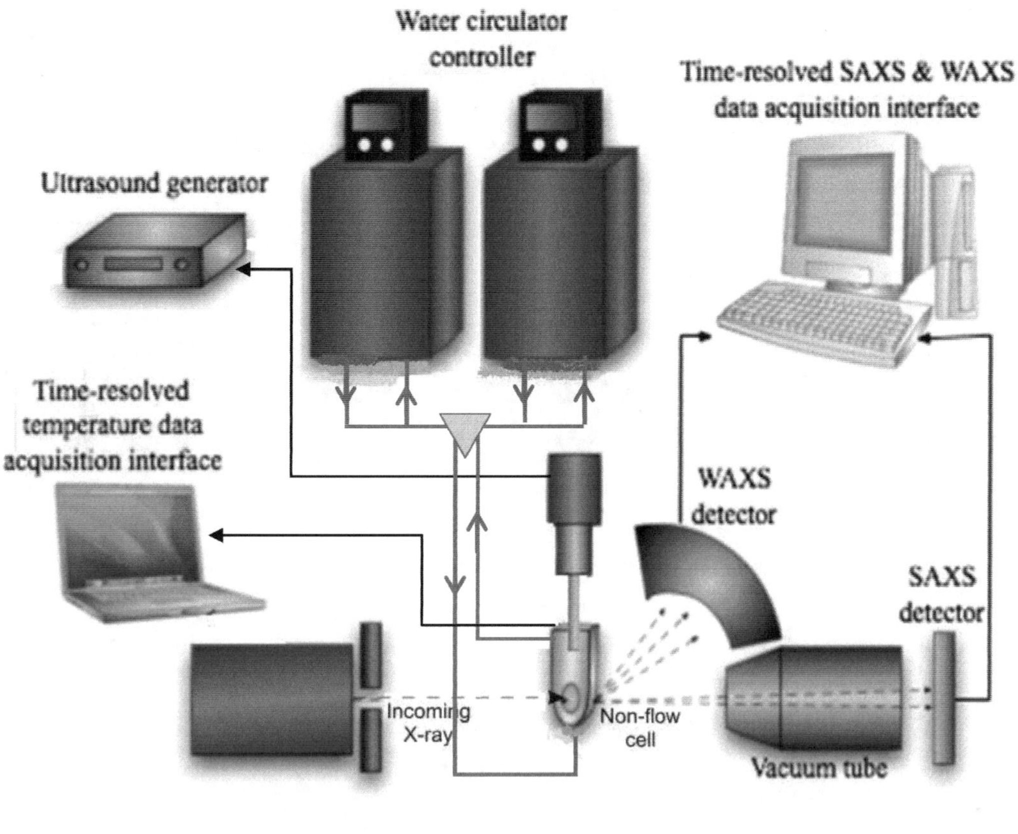

Figure 1. Schematic representation of combined sonocrystallisation and SAXS-WAXS set-up used at Daresbury SRS station 6.2.

A sample of 30ml was put in the cell and melted by heating to 60 °C. Then, the sample was cooled down below its melting temperature to a desired crystallisation temperature by switching from hot to cool circulating water. A selected supercooling of the melt was achieved approximately within 5 to 10 min. From this moment, $t = 0$, simultaneous collection of WAXS and SAXS spectra began. These spectra were monitored and collected until the melt had entirely crystallised and evolved into the stable β-phase. Exactly the same procedure and supercoolings were used for sonication experiments. Sonication was initiated with an electronically controlled ultrasound generator, Model P100/3-20 (Sonic Systems Ltd, UK), once the melt achieved a desired crystallisation temperature. The duration of the sonication of the melt was of 2 seconds which was applied at various sonication powers.

Spectra were taken at 10s interval. The induction times and polymorphic forms of the crystallised samples were determined by time-resolved simultaneous SAXS/WAXS measurements. Both these spectra are used to identify a particular polymorph, where SAXS defines the lamellar structure with long-spacing values and WAXS defines a corresponding short interplanar distance of a subcell [9]. SAXS, rather than WAXS, was used to measure induction times of the experiment due to its higher detecting system sensitivity. The camera length of SAXS was 3.5 m. For WAXS, this length was 0.3 m. Both SAXS and WAXS data analysis were calibrated according to the standard method using wet rat tail collagen and silicon. The system was operating at a wavelength of 0. 14 nm.

3. Results

3.1 Polymorphic forms of LLL

There are three known polymorphic forms of LLL crystals, α, β' and β with α-form being the least stable and β-form the most stable. The main difference between these polymorphs is found in the angle of tilt from the base plane in the crystal lattice (Fig. 2a). Figure 2(b) shows the crystal physical appearance of β' and β-forms of LLL under an optical microscope. The melting points and the angles of tilt for the three polymorphs of LLL are presented in Table 1.

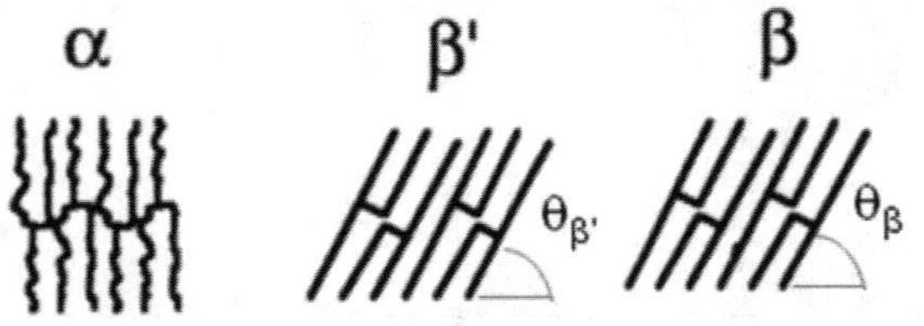

(a)

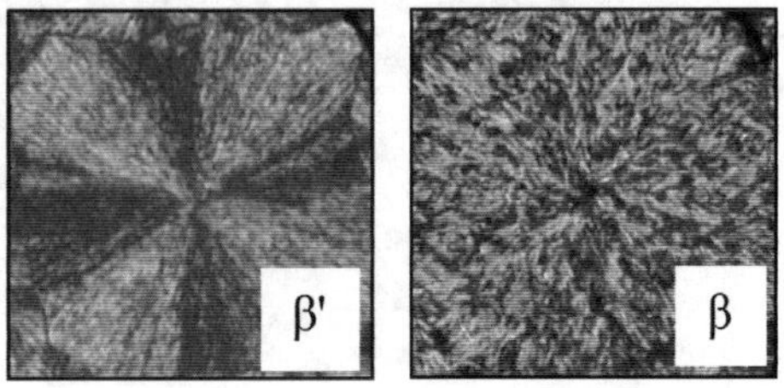

(b)

Figure 2. (a) Polymorphic lamellar structure of LLL. (b) Optical microscope images of LLL β' and β-forms.

Table 1. Melting points and the calculated angle of tilt of the lamellae structure in LLL [10].

	Polymorphic Forms		
	α-form	β'-form	β-form
Melting point, T_m (°C)	15.0	35.0	46.5
Angle of tilt, θ (°)	90°	66°	63°

3.2 Studies of Induction Times and Polymorphic Transformation

Figure 3(a) shows typical time-resolved synchrotron radiation X-ray diffraction SAXS and WAXS spectra from the crystallising LLL melt at constant crystallisation temperature, T_{cryst} = 25°C, without sonication. The spectra were taken at 10s interval. However, for clarity of presentation and to avoid a highly dense layout, not all spectra are reproduced in the Figures. The time is regarded as zero, $t = 0$ s, at the moment when a desired crystallisation temperature, in this case T_{cryst} = 25°C, is achieved. The driving force (supercooling) for melt crystallisation is the difference between the melting point of a substance and the crystallisation temperature. In this case, the degree of metastability for each polymorphic form is:

$$\Delta T_{cryst}^{\beta'} = T_m^{\beta'} - T_{cryst} \quad (1)$$

and

$$\Delta T_{cryst}^{\beta} = T_m^{\beta} - T_{cryst} \quad (2)$$

where $T_m^{\beta'}$ and T_m^{β} are the melting temperatures of β' and β respectively.

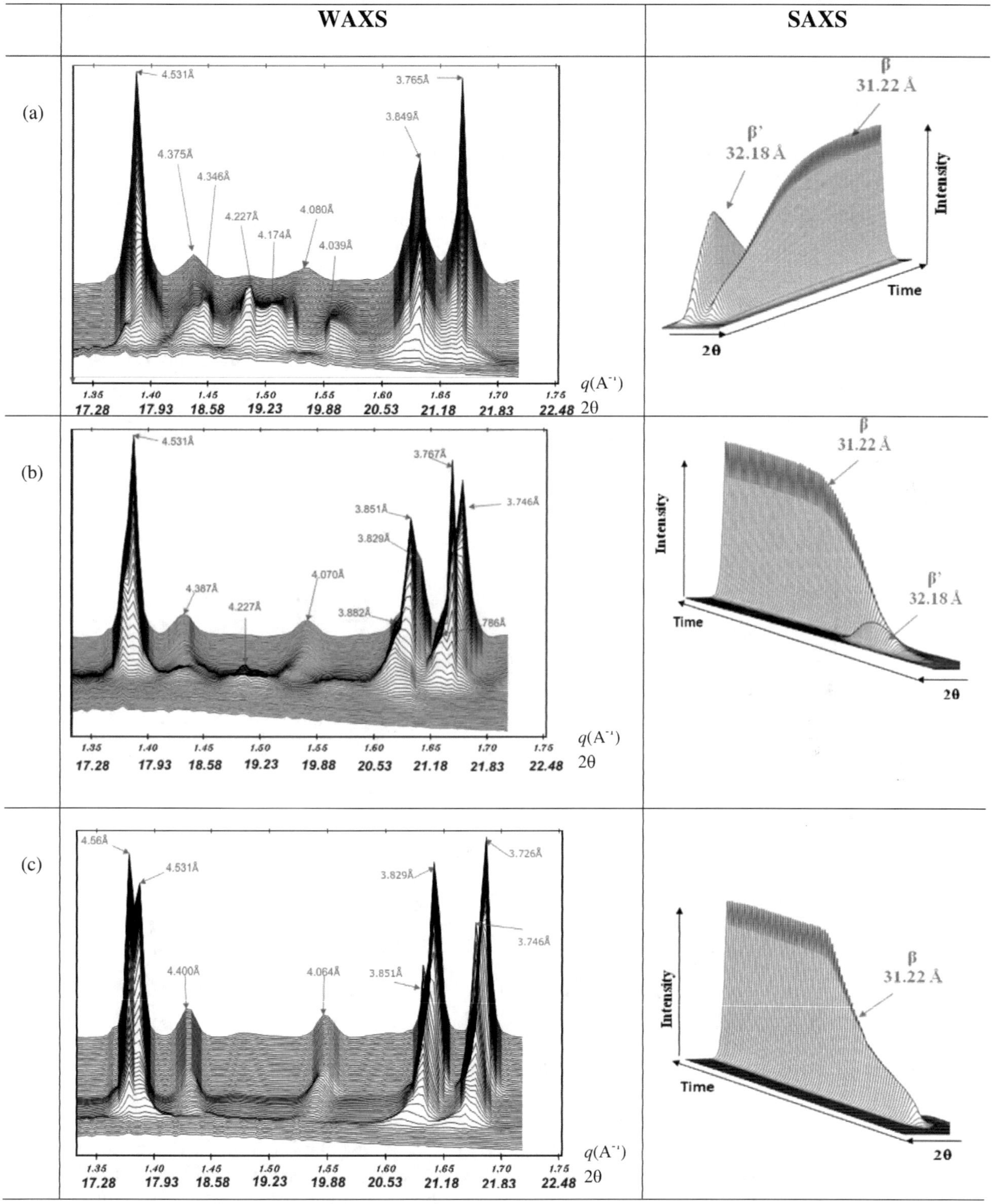

Figure 3. Time-resolved SAXS/WAXS data of crystallising melt of LLL at T_{cryst} = 25: (a) without ultrasound; (b) US input power = 5W, 2s sonication time and; (c) US input power = 100W, 2s sonication time.

The diffracted peaks observed in SAXS Figure 3(a) confirmed that both β'- and β-form crystallised, with β'-form crystallised first. Induction time was determined as the length of time measured from the moment the temperature of the system has cooled down to the crystallisation temperature, T_{cryst} = 25°C to the moment at which the first appreciable peak of the SAXS spectrum was detected. The induction time determined in this way was for β'-form 280 s and for β-form 540 s. This agrees with the Ostwald rule of stages which implies that the preferred polymorph to crystallise first is the less stable one.

Figure 3(b) shows the time-resolved WAXS and SAXS spectra for crystallisation of LLL obtained at an input power of 5 W while the other external factors were kept the same as those in the non-sonicated case. Upon comparison between the SAXS spectrum of the sonicated and that of non-sonicated crystallising melts, it becomes clear that both polymorphic forms crystallised, with the metastable β'-form arising first. However, unlike in non-sonicated case, the induction time in the US irradiated melt is 250s for β' and 420 s for β. Both values are appreciably smaller than those for nonsonicated melt. This suggests that even a small input power of the ultrasound may have a quite strong effect on increasing the nucleation rate, *J*, for both phases ($J \sim 1/\tau$). The metastable form did not last for a long and disappeared quickly when β-form begins to crystallise.

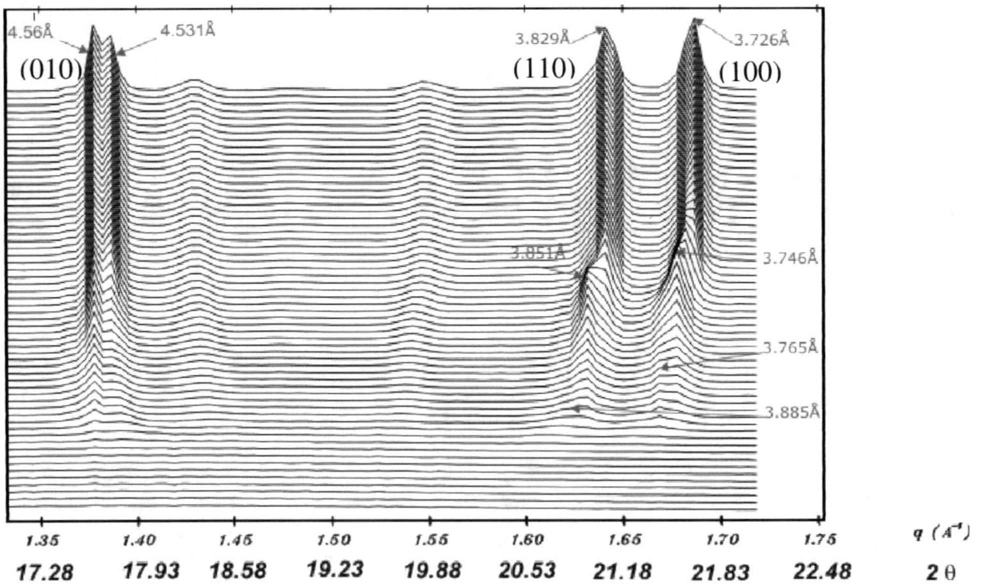

Figure 3 (d). The same WAXS spectra of 3 (c) plotted in a different projection. Peaks (110) and (100) are seen to shift to higher angles of diffraction, 2θ.

When a considerably higher input power of 100 W was applied under the same other external conditions, the corresponding SAXS/WAXS spectra showed two markedly new effects caused by the applied sonication power (Fig 3(c) and 3(d)). The first one is manifested through the prevention of the nucleation of the metastable β' (selection of β), while the second one through an apparent shift in the reflections (110) and (100) of the β WAXS spectra towards slightly larger 2θ. It is important to notice that these shifts were only observed in the experiments where ultrasound energy was applied and only the stable β-form crystallised. A similar effect, with a selection of only stable β form, was observed at the input power of 20 W (WAXS/SAXS spectra not shown here), but with a relatively longer induction time of 140 s compared to that of 90 s obtained for the input power of 100 W.

Finally, we find that crystallisation temperature may have a significant effect on the induction times of the two polymorphic phases. Using the same external conditions as those for T_{cryst} = 25°C, but now with a higher crystallisation temperature, T_{cryst} = 30°C, we find that the general features of the effect remain unchanged, but the inductions times become somewhat bigger compared to those obtained at T_{cryst} = 25°C (Table 2).

Table 2 summarizes induction times and the corresponding external conditions of supercooling, crystallisation temperature, input power of the ultrasound and its duration, under which the experiments were conducted.

Table 2. Induction times τ (s) for each polymorphic form of LLL under different crystallisation temperatures, T_{cryst} (°C), and supercooling, ΔT_{cryst} (°C), with and without ultrasound.

			LLL Polymorph	
Crystallisation temperature, T_{cryst}	Induction time		β'-form	β-form
T_{cryst} = 25°C		Supercooling, ΔT_{cryst}	10°C	21.5°C
	τ	US = 0s	280s	540s
		US = 2s, 5W	250s	420s
		US = 2s, 20W		140s
		US = 2s, 100W		90s
T_{cryst} = 30°C		Supercooling, ΔT_{cryst}	5°C	16.5°C
	τ	US = 0s	340s	580s
		US = 2s, 5W	300s	460s
		US = 2s, 100W		240s

The overall observations presented in this section lead to the following findings: (1) the induction time of either metastable or stable polymorphic form of LLL seems to be quite susceptible to the input power of applied sonication; our experiments showed that even the applied power of 5 W for 2 s is sufficient to reduce induction times substantially compared to those without sonication, (2) the selection of the stable polymorphic form β is achievable at higher levels of the input power at the same other external conditions of crystallisation, (3) the induction time of the selected stable form reduces with input power and temperature and (4) WAXS spectra of ultrasonically selected β shows at the initial stage of the growth of this form the peak shifts of the (110) and (100) reflections.

4. Discussion

In the following we shall try to give an account for the key findings, (1) to (4), presented in the previous section.

4.1 Consideration of Induction Time

4.1.1 Qualitative Temperature-Pressure Diagram

The crystallisation of LLL under the influence of ultrasound, in general, is shown to reduce the induction times of the crystallisation of the polymorphs. To analyse this finding we make use of a qualitative temperature-pressure (T-P) diagram. Since this diagram is still not available for any TGAs including LLL, we shall define it qualitatively by using the Clapeyron equation and relevant experimental data concerning this material. This equation applies to equilibrium between any two phases and is given by

$$\left(\frac{dT}{dP}\right)_{equil} = T\frac{\Delta V}{\Delta H} \qquad (3)$$

where $\left(\frac{dT}{dP}\right)_{equil}$ is the change in the melting temperature T with pressure P, ΔV is the corresponding molar volume change due to melting and ΔH is the enthalpy of melting. For most substances including fats, $\left(\frac{dT}{dP}\right)_{equil}$ is positive. It is assumed that this derivative is constant for a wide range of the (T-P) phase diagram. The thermodynamic properties characterising the right side of eq. 3 are experimentally known apart from ΔV. The latter one can be calculated as $\Delta V = M\left(\frac{\rho_s - \rho_l}{\rho_s \cdot \rho_l}\right)$, where M molar mass, ρ_s is the density of solid phase and ρ_l is the density of an isotropic liquid phase (melt).

The density of any phase of a fat is determined from an experimental dependence that describes the change in the density of a particular phase as a function of the number of carbon atoms present in the acyl-chains in the fat molecule (Figure 4). These dependencies were obtained by applying a least-squares fitting procedure to the available experimental data [10-13].

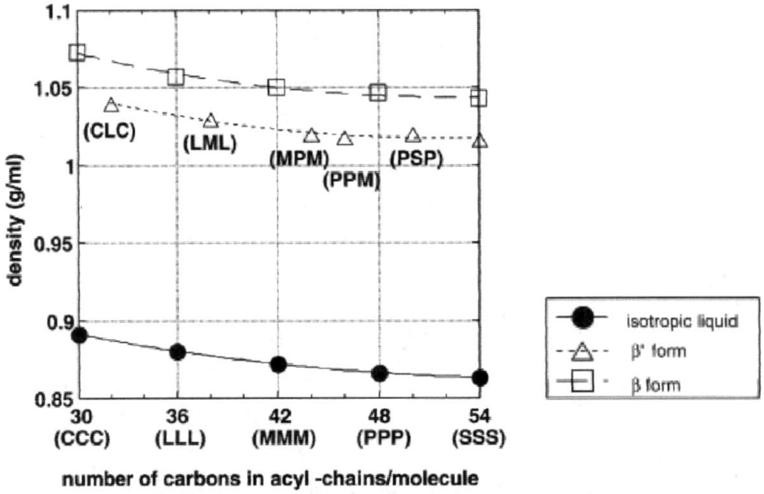

Figure 4. The density of fats as a function of the number of carbon atoms present in their corresponding acyl-chains. Solid and broken lines are the least-square fitted curves for the three physical states shown.

Table 3 summarises the values of the densities found from Fig 4 for each form of LLL, the calculated corresponding values of ΔV and experimentally known changes of the enthalpy of melting for the two polymorphs.

Table 3. Thermodynamic properties of polymorphic forms of LLL.

	Polymorph LLL	
	β'	β
ΔH (kcal/mol)	19.4	19.5
Molar volume change, $\Delta V \times 10^{-6} \left(\mathrm{m^3/mol}\right)$	106.3	123.1
Density of solid phase, $\rho_s \left(\mathrm{g/cm^3}\right)$	1.03	1.06
Density of liquid (melt) $\rho_l \left(\mathrm{g/cm^3}\right)$	0.88	

To get a qualitative idea about (T-P) phase diagram for both β' and β, we express Eq. (3) in its explicit form for each polymorph and find the ratio

$$\frac{\left(\frac{dT}{dP}\right)^{\beta'}_{equil}}{\left(\frac{dT}{dP}\right)^{\beta}_{equil}} = \frac{T_m^{\beta'}(\Delta V)^{\beta'} / (\Delta H)^{\beta'}}{T_m^{\beta}(\Delta V)^{\beta} / (\Delta H)^{\beta}} = \frac{T_m^{\beta'}\left(\rho_s^{\beta'} - \rho_l\right) / (\Delta H)^{\beta'} \cdot \rho_s^{\beta'}}{T_m^{\beta}\left(\rho_s^{\beta} - \rho_l\right) / (\Delta H)^{\beta} \cdot \rho_s^{\beta}} \qquad (5)$$

When the corresponding experimental values $T_m^{\beta'}, \rho_s^{\beta'}, \rho_l, (\Delta H)^{\beta'}, T_m^{\beta}, \rho_s^{\beta}$ and $(\Delta H)^{\beta}$ are substituted into this equation, the calculated ratio of the gradients is 1.30. The fact that this value is greater than 1 suggests that an equilibrium straight line of the *β'* form should be steeper than that corresponding to β. This suggestion leads us to a possible qualitative (T-P) diagram (Figure 5). Although this diagram was worked out by using thermodynamic parameters for LLL, it can be easily demonstrated that its general appearance would remain the same for any other fat from the family of TGA presented in Fig 4. Therefore, Fig 5 should reflect the equilibrium between the solid and liquid phases for both polymorphs of a fat over a significantly large range of temperatures and pressures. Due to the difference in gradients between the two polymorphs, the straight equilibrium lines are expected to cross one another at some point C, defined by a particular pair of temperature and pressure.

When ultrasound is applied to a liquid (melt), two events occur simultaneously; (i) the mechanical waves of ultrasound induce periodic change in the liquid [14] and (ii) the liquid absorbs energy from the applied ultrasound [15].

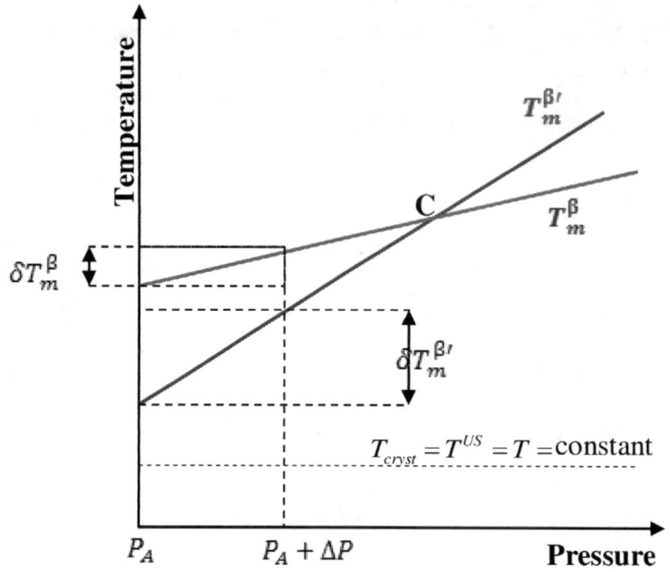

Figure 5. Qualitative Temperature vs. Pressure diagram for β' and β-forms of fats.

If the applied ultrasound wave has a pressure amplitude that is higher than that of the ambient pressure of the liquid, the transient cavitation bubbles will be formed then collapsed followed by the development of extremely high pressure in their vicinity [7]. An increase in the local pressure consequently causes a rise in temperature. The effect of local changes in pressure and temperature in the crystallisation of LLL is discussed in details in the following two sections.

4.1.2 Change in Local Pressure

The mechanism for the collapse of a cavitation bubble is still not fully understood. It is less clear how a very high pressure localised in space and time develops in the vicinity of the collapsing bubble. Initially, it has been speculated that a bubble overexpands in the negative part of an applied acoustic wave, caves in sharply in the positive half to conceivably 1% of its initial radius, and then rebounds, sending out a pressure pulse (shock wave) [16-18]. These speculations have been recently confirmed experimentally to a considerable extent [19]. However, the measurements of the real pressure and temperature in a collapsing bubble and in the consequently developed shock-wave still remain unresolved. Therefore, to get some idea about the order of magnitude of the parameters that characterise these transient phenomena, we have to rely on the data that result from different models [20-22]. Estimated calculations for a moderately supercolled liquid show that at a point approximately10^{-4} cm from the bubble wall (of diameter ~ 10^{-5} cm the liquid is subjected to average pressure of the order of 10 kbar over a period of 10^{-9} s [17]. It has been shown that at positive pressure pulses of this order or higher, nucleation in supercooled liquids is likely to be initiated [20-23]. This kind of nucleation has often been considered as a dynamic nucleation, because it is caused by the relative motion of the crystallising liquid [24]. In the following section, we shall try to give an account for this phenomenon in sonicated LLL melt using a combination of the classical theory of homogeneous nucleation of a supercooled liquid [25] and the melting point shift [26] whose magnitude is related to the pressure change via the Clapeyron equation.

According to the classical theory for homogeneous nucleation of an undercooled liquid [23, 25], the rate of nucleation, J, is given by:

$$J = \left(\frac{nkT}{h}\right)\exp\left(\frac{-\Delta G_a}{kT}\right)\exp(-\Delta W) \tag{6}$$

where n is the number of molecules/cm^3, k is the Boltzmann constant, T is the absolute temperature, h is the Planck's constant, ΔG_a is the activation energy for a molecule to cross the liquid crystalline face, ΔW is the height of the free-energy nucleation barrier for a cluster to overcome it and grow irreversibly as a crystal. This barrier is given by:

$$\Delta W = \frac{16\pi}{3}\frac{\gamma^3}{kT\rho^2(\Delta H^2)\left(\frac{\Delta T}{T_m}\right)^2} \tag{7}$$

in which γ is the free energy between the liquid and crystalline interface, ρ is the density of the solid phase, ΔH is the molar enthalpy of melting, T_m is the melting temperature $\Delta T = T_m - T = T_m - T_{cryst} = \Delta T_{cryst}$ is the supercooling. In addition to this, we assume that the structure of a nucleation cluster of either phase is the same as that of the corresponding crystal phase.

As we saw earlier, application of ultrasound to a metastable liquid results in high pressure pulses generated in the vicinity of collapsing bubbles in the undercooled melt. As a consequence, a sudden onset of nucleation is likely to occur in these small high pressure volumes of the liquid because of the displacement of the melting point [20]. More specifically, an increase of pressure towards its own maximum in such a small volume of the melt, may well result in the local increase of enthalpy, solid-liquid interface energy, density and melting point. If ΔP is the difference between the pressure in the vicinity of a collapsing bubble and the ambient (atmospheric) pressure, then the new enthalpy at the new melting point is given by $\Delta H^{US} = \Delta H(1+\delta)$, where $\delta = \frac{\Delta P \Delta V}{\Delta H}$ in which ΔV is the difference in specific volumes between the undercooled melt and the crystalline phase. A further consequence of the ultrasound perturbation is reflected on the changes of other thermodynamic quantities; the internal density of a nucleus and the corresponding interfacial free energy between the nucleus and the liquid turn into new values, ρ^{US} and γ^{US}, respectively. These changes are given by $\frac{\gamma^{US}}{\rho^{US 2/3}} = \frac{(1+\delta)\gamma}{\rho^{2/3}}$. Similarly, a displaced melting point due to sonication (T_m^{US}), due to sonication is given by $T_m^{US} = T_m(1+\delta)$. Hence, it is easy to show that $\frac{\Delta H^{US}}{T_m^{US}} = \frac{\Delta H}{T_m}$. Therefore, the new height of the free-energy nucleation barrier in the vicinity of the collapsing cavity becomes

$$\Delta W^{US} = \frac{16\pi}{3}\frac{\gamma^3(1+\delta)^3}{kT^{US}\rho^2(\Delta H^2)\left(\frac{\Delta T^{US}}{T_m}\right)^2} \tag{8}$$

where T^{US} is the local change in the temperature of the melt and $\Delta T^{US} = T_m^{US} - T^{US}$ is the corresponding undercooling. Both T^{US} and ΔT^{US} can be obtained from an isothermal approximation in a qualitative *(T-P)* diagram. For the isothermal process: $T^{US} = T = T_{cryst}$. Using the Clapeyron equation and the straight melting lines (Figure 5), we get $\Delta T^{US} = \Delta T_{cryst} + \delta T_m$, where ΔT_{cryst} is the initial supercooling and δT_m results from ΔP.

Equation (8) can be now expressed as

$$\Delta W^{us} = \frac{16\pi}{3} \frac{\gamma^3 (1+\delta)^3}{k T_{cryst} \rho^2 (\Delta H^2) \left(\frac{\Delta T_{cryst}}{T_m} + \delta \right)^2} \tag{9}$$

This expression gives an estimate of the free-energy barrier for nucleation in a very small volume of very high-pressure around the collapsing bubble created by applied sonication. For further analysis it is appropriate to consider a ratio of this equation and equation (7) which gives a free-energy barrier for nucleation from non-sonicated melt:

$$\frac{\Delta W^{us}}{\Delta W} = \left(\frac{\Delta T_{cryst}}{T_m} \right)^2 \frac{(1+\delta)^3}{\left(\frac{\Delta T_{cryst}}{T_m} + \delta \right)^2} = \left(\frac{\Delta T_{cryst}}{T_m} \right)^2 \frac{\left(1 + \frac{\Delta P \Delta H}{\Delta H} \right)^3}{\left(\frac{\Delta T_{cryst}}{T_m} + \frac{\Delta P \Delta H}{\Delta H} \right)^2} \tag{10}$$

Replacing the corresponding experimental values from tables 1, 2, 3 and taking $\Delta P \approx 10\,\text{kbar}$, this ratio can be then estimated for either of the two polymorphs of LLL. A simple calculation shows that for both β' and β, $\frac{\Delta W^{US}}{\Delta W} << 1$. This implies that the height of the free-energy barrier for a cluster to overcome, before it stands a chance of surviving and growing irreversibly as a crystal, is much smaller in the melt around the collapsing bubble than that in the non-sonicated melt. Referring back to Equation (6), reduction in the free-energy barrier for nucleation around the collapsing cavity can trigger nucleation in one or more spots, resulting in an increase of the nucleation rate, *J*. Since $J \propto 1/\tau$, it implies reduction in induction times [27]. This agrees with the results obtained and presented in Table 2 where the induction times are reduced remarkably upon application of ultrasound for crystallisation of LLL. However, this model does not explain why in some cases, only the stable beta form was crystallised. The proposed model indicates that cavitation is a necessary but not sufficient condition for dynamic nucleation; additional factors need to be taken into account.

At this stage we argue that factors such as the duration of the final stages of collapsing cavities, the lifetimes of the subcritical nuclei and stability of subcritical nuclei may play a critical role in the nucleation of a particular polymorph. In a fairly supercooled melt similar to those used in our experiments, the duration of the final stage of a collapsing bubble, at which a high pressure pulse is developed in its vicinity, is of the order of nanoseconds [9, 22]. This suggests that in order to generate a particular polymorph, this time should be of the order or longer than an average lifetime for a cluster (subcritical nuclei). The lifetime scales of the less stable β' clusters are probably too short compared to the duration of the collapsing cavities and therefore, β'clusters were dissolving before achieving their critical size. Comparing to the more stable β clusters which are also subjected to the same local change of pressure, it is reasonable to assume that an average lifetime of this type of clusters is comparable or bigger than the duration of the final stage of the collapsing bubble. Apparently, in this case, application of sonication acts as a catalyst which promotes the nucleation and hence irreversible precipitation of the β form.

At the end, we give a brief comment on how locally induced ΔP might influence ΔG_a, the activation energy for a molecule to cross the liquid interface. It is likely that a locally increased pressure will result in an increase of the viscosity of the supercooled melt. For relatively small molecules such as LLL, this effect should be negligible while for relatively large molecules, an increase in pressure could significantly reduce the nucleation rate [28]. However, since the induction times observed in this work (Table 2) are pronouncedly reduced due to sonication, we believe that ΔG_a has not been influenced noticeably by ΔP.

4.2 Effect of Temperature

A comparison of induction times at the crystallisation temperature of 25°C with their counterparts obtained at T_{cryst} = 30 °C (Table 2), shows a quite clear increase of induction times with temperature regardless of the melt was sonicated or not. These findings are in a good agreement with equations (6) to (9). In order words, under the same conditions of sonication, the closer the crystallisation temperature to a melting point of either polymolymorph, the lower the corresponding supercoolings and hence, the higher the free-energy barriers for the nucleation of both β' and β form. The longer induction times observed at higher temperature imply that the thermal energy may have a significant destructive effect on the nucleation processes of both polymorphic forms.

It is important to mention that, in principle, the stability of an externally imposed crystallisation temperature can be affected by inevitably generated heat that stems from the application of sonication and the crystallisation itself.

The heat generated by sonication is caused by the dissipation of acoustic energy and the heat formation in the highly compressed gas in the cavitation bubbles. The former one is a consequence of the interaction between the mechanical wave and the liquid, which is manifested through viscous losses, heat conduction losses, and the losses related to internal molecular processes [29]. The latter one is transferred into the liquid shell between the collapsing bubbles and the liquid by radiation and conduction.

Finally, the last contribution to the overall increase of temperature may come from the precipitation of the solid phases which is accompanied by the release of the heat of the heat of crystallisation.

The observed temperature increase due to the above factors had an appreciable value of approximately 2°C for a relatively short period of time during and after the sonication and then would cease to exist for the rest period of crystallisation. The contribution of this effect is rather difficult to separate. We believe that the overall heat corresponding to the observed temperature increase, apart from decelerating growth of the already nucleated melt, should not have any significant effect on the observed differences in the induction times of the nucleation processes because all experiments were performed at same external conditions of the input sonication power and its duration and above all, the temperature increase was almost the same.

4.3 Shifts in β-form WAXS Peaks

The peak shifts in WAXS spectra of different TGA's have been observed earlier at different – nonsonicated conditions [30-32]. These authors associate the peak shifts occurrence in the course of crystallisation to a possible existence of either new metastable phases or remains of β' whose growth precedes the growth of the stable β phase. However, a great deal of controversy still remains regarding the existence of possible crystalline phases and the mechanisms for phase transformations. Apparently, the main reason for this lies in a quite limited knowledge of a three-dimensional atomic-level crystal structure of the family of these materials [11]. This has been partly because of problems in growing TGA single crystals to a sufficient size and perfection for X-ray diffraction studies. A second complication relates to the fact that metastable polymorphs including β' have a strong tendency to transform to the more stable β form. As a consequence of these two factors, a quantitative characterisation concerning crystal lattice distortion of these possible transient polymorphs is not available at present. Therefore, there is a significant amount of uncertainty of whether peak shifts are caused by new polymorphs or by a pronounced imperfection of a stable β-form at its initial stage of growth.

At this stage we argue that the peak shifts of the (110) and (100) reflections of the WAXS spectra of the β-form of LLL (Fig 3 (c) and (d)) may be influenced by a marked presence of the strain developed in the particles of this phase just after they have been nucleated. This phenomenon is well known and is typical for very small particles in inorganic and organic systems [33-34]. This strain is usually caused by liquid inclusions captured by morphologically unstable facets of the growing particles [35]. A shifted reflection may be a result of the X-ray diffraction from an ensemble of strongly structurally defected particles and the presence of the traces of amorphous liquid inclusions in them. At the later stage of growth, it is quite likely that liquid inclusions will solidify and that the previously induced strain will relax on the growing facets by forming dislocations in the particles. These processes are time dependent and accompanied by a pronounced increase in crystal perfection, which finally leads to the peak positions of undistorted β particles. An additional reason that leads us to the proposed scenario lies in the fact that the position of the SAXS spectra (Fig 3 (c)) remains unchanged throughout the crystallisation process. Due to the lack of atomic crystal structure and consequently the anisotropy of elastic constants within the material, the question of why these particular reflections are most affected remains open.

5. Conclusions

The effects and mechanisms by which ultrasound energy may influence the crystallisation of polymorphic LLL have been examined by synchrotron radiation time resolved SAXS/WAXS measurements. A striking reduction of the induction times (and an increase in the corresponding nucleation rates) has been observed in the sonicated supercooled melts. These results are explained in the light of two major events that may be critical for the observed effects: (1) the ultrasound-stimulated nucleation occurred in the high pressure vicinity of the collapsing bubbles because of the melting point shift and the consequent increase of supercooling and (2) a pronounced decrease of the free energy barrier for the nucleation compared to that without sonication. At the fixed conditions of sonication power and its duration, it was shown that at the higher crystallisation temperature the longer induction times were needed for the onset of nucleation events. At the initial stage of precipitation, the selected crystallisation temperatures were slightly and to the same extent increasing, by 2°C on average, due to the heat generated by the absorption of the applied ultrasound energy and the crystallisation heat. In spite of fact that a total heat generated in this way is inevitable and almost the same regardless of the crystallisation temperature, the corresponding increase of this temperature should not have any fundamental influence on the nucleation behaviour of the two polymorphs versus the externally imposed temperature. It has also been shown that is possible to select the stable polymorph phase β of LLL. The selection of this phase requires the interplay between sonication power, crystallisation temperature and supercooling. At the end, an account for the peak shifts of the (110) and (100) reflections of the β -WAXS spectra has suggested and related to the presence of anysotropically distributed unrelaxed strain caused by the capture of the liquid inclusions at the initial stage of growth.

We believe that apart from making some progress in clarifying the presented complexity, the above findings suggest new research challenges to be followed. An ultimate goal is to develop a theoretical model, and more importantly, one with a certain degree of predictability of a desired polymorph. At present, none of these two is available. To achieve this we need a better understanding of the following issues: (i) (T-P) phase diagram for β' and β forms, (ii) stability (lifetime) of the nuclei of these polymorphic forms as a function of temperature, (iii) the mechanism and lifetime of collapsing cavities, a basic mechanism of the dynamic nucleation in the vicinity of a collapsing bubble, (iv) crystal structure on an atomic level of most of TGA and (v) structural defects and thermal behaviour of these materials. The work presented here should be a generic one and as such applicable not only to LLL, but also to any TGA.

Acknowledgment. We gratefully acknowledge the financial support of this work by the U.K. EPSRC through grant EP/C512731/1. We acknowledge the support of the director and staff of the Daresbury Laboratory through the provision of facilities for SAXS/WAXS experiments.

References

1. Mason, T.J., Paniwnyk, L., Lorimer, J.P., *Ultrason. Sonochem.,* 1996, **3,** *S253-S260*
2. Mason, T.J. in: *Ultrasound in Food Processing;* Eds. M.J. Povey, T.J. Mason, Blackie Academic & Professional, London, 1998, pp: 105-126
3. Oshaka, K., Trinh E.H., *Appl. Phys. Lett.,* 1998, **73,** pp: 129- 131.
4. Baxter, J.F., Moris, G.H., Gaim-Marsoner, G., *European Patent Application,* 95306833.5, 1995.
5. Higaki, K., Ueno, S., Koyano,T., Sato,K., *J. Am. Oil Chem. Soc., 2001,* **78**, pp: 513-518.
6. Sato, K., *Chem. Eng. Sci.,* **2001,** 56/57, pp: 2255-2265
7. Hem, S.L. *Ultrasonics,* 1967, **5**, pp: 202-2007
8. Enamoto, T., Sung, T.H., Nakagawa, Z.E., Lee, S.C. *J. Mat. Sci.*1992**, 27,** pp**:** 5239-5243
9. Ohsaka, K. Trinh, E. H. *Appl. Phys. Lett.* 1998, **73**, pp:129-131
10. Small, D.M., *Handbook of Lipid Research: The Physical Chemistry of Lipids: From Alkanes to Phospholipids*. 1986, Plenum Press, New York and London. pp: 97-101, 123-125, 347-352, 368-369

11. Sato, K., Goto, M., Yano, J. Honda, K., Kodali, D.R., Small, D. *J. Lipid Res.*, 2001. **42**(3) pp: 338-345.
12. Van Langevelde, A., van Malseen, K., Driessen, R., Goubitz, K., Hollander, F., Peschar, R., Zwart, P., Schenk, H. *Acta Crystallographica Section B*, 2000. **56,** pp: 1103-1111.
13. Jensen, L.H. and A.J. Mabis, *Refinement of the structure of [beta]-tricaprin.* Acta Crystallographica, 1966. **21**(5): p. 770-781.
14. Wu, C.C. and P.H. Roberts, *Shock-wave propagation in a sonoluminescing gas bubble.* Physical Review Letters, 1993. **70** (22): pp:3424.
15. Mason, W.P., *Physical Acoustics, Vol. 2A*. 1965, New York: Academic Press.
16. Apfel, R.E., *Acoustic cavitation prediction.* The Journal of the Acoustical Society of America, 1981. **69**(6) pp: 1624-1633.
17. Prosperetti, A., *Frontiers in Physical Acoustics*, D. Sette, Editor. 1986: North-Holland, New York.
18. Hickling, R. and M.S. Plesset, *Collapse and Rebound of a Spherical Bubble in Water.* Physics of Fluids, 1964. **7**(1): pp: 7-14.
19. Claus-Dieter, O., Kurz T., Geisler, R., Lindau, O. and Lauterborn, W. *Phil. Trans. R. Soc. Lond. A*. 1999. **357** pp: 269-293.
20. Hickling, R. *Nature*, 1965. **206** pp: 915-917.
21. Hickling, R. *Nature*, 1965. **207** pp: 742-742.
22. Hickling, R. *Physical Review Letters*, 1994. **73**(21) pp: 2853.
23. Lee, C.P. and T.G. Wang, *J. Appl. Phys.* 1992. **71**(11) pp: 5721-5723.
24. Chalmers, B., *Principles of Solidification.* 1982, Malabar, FL: R.E. Krieger Publishing.
25. Turnabull, D. and Fisher, J.C., *J.Chem. Phys.* 1949. **17** pp:71-73.
26. Chalmers, B., *Liquids: Structures, properties, Solid Interactions*, T.J. Hughel, Editor. 1965, Elsevier: New York. pp: 308.
27. Mullin, J.W., *Crystallisation,3rd Ed.* 2003. Butterworth Heinmann Ltd, Oxford.
28. Bridgman, B.W., *The Physics of High Pressure*. 1971, New York: Macmillan.
29. Kinsler, L.E., et al., *Fundamentals of Acoustics, 4th Ed.* 2000, New York: Wiley & Sons.
30. Sato, K., Arishima, T., Wang, Z.H., Ojima, K, Sagi, N. and Mori, H., *JAOCS.* 1989. **66** pp:664-674.
31. Marangoni, A. *Inform, American Oil Chemists Society,* May 2001. **12** pp: 479-481.
32. Kellens, M., Meeuseen, W., Gerke, R. and Reynaers, H. *Chemistry and Physics of Lipids*, 1991. **58**(1-2) pp. 131-144.
33. Ristic, R.I., Sherwood, J.N. and Wojciechowski, K., 1988. *J. Crystal Growth* **91** pp.163
34. Prasad, K.V.R., Ristic, R.I., Sheen, D.B. and Sherwood, J.N. *Int. J. Pharmaceutics* 2001, **215** pp: 29-44.
35. Chernov, A.A., Coriell, S.R. and Murray, B.T. *Journal of Crystal Growth*, 1993. **132** pp: 405-413.

Journal of Physics: Conference Series **247** (2010) 012050 doi:10.1088/1742-6596/247/1/012050

Complex fluids under microflow probed by SAXS: rapid microfabrication and analysis

Hazel P. Martin[a], Nicholas J. Brooks[b], John M. Seddon[b], Nick J. Terrill[c], Paul F. Luckham[a], Adam J. Kowalski[d], João T. Cabral[a]

[a] Department of Chemical Engineering, Imperial College London, Exhibition Road, London SW7 2AZ, United Kingdom

[b] Department of Chemistry, Imperial College London, Exhibition Road, London SW7 2AZ, United Kingdom

[c] Diamond Light Source Ltd., Diamond House, Harwell Science and Innovation Campus, Chilton, Didcot, Oxfordshire, OX11 0DE, United Kingdom

[d] Unilever Research & Development, Port Sunlight Laboratory, Bebington, Wirral,CH63 3JW, United Kingdom

E-mail: `j.cabral@imperial.ac.uk`

Abstract. We report a combined microfluidic and online synchrotron small-angle X-ray scattering (SAXS) study of complex surfactant mixtures under flow. We investigate the influence of a series of flow constrictions, generating well-defined, periodic extensional flow fields, on the microstructure of two model surfactant mixtures containing SDS and CTAC. Specifically, the lamella spacing, orientation and structural order are reported and correlated with the imposed flow field: geometry, flow velocity and residence time. The design, fabrication and operation of a microfluidic system using rapid prototyping is described in detail. We show that polydimethyl siloxane (PDMS), ubiquitous in microfabrication, provides a suitable matrix for SAXS microdevices provided that: (i) PDMS thickness are kept to a minimum while retaining structural integrity ($\sim$1000μm) and (ii) scattering from the structure of interest is sufficiently decoupled from the amorphous background scattering. The combination SAXS-microfluidics provides unprecedented opportunities to elucidate the non-equilibrium structure formation and relaxation of complex fluids, demonstrated here for concentrated surfactant mixtures.

1. Introduction

Microfluidic technologies have evolved rapidly in the last two decades and find a multitude of applications, particularly in the fields of chemistry and biology. Microdevices [1–3] miniaturize and expedite discovery, synthesis of new chemicals and drugs, [4–6] facilitate process automation and provide novel opportunities for analytical tools such as separation techniques and large mapping and screening for protein crystallisation. [7,8] Although far less exploited, microfluidics provides unprecedented opportunities in physical sciences. [9,10] High degree of flow field control and uniformity are achieved through exploiting the change of fundamental physics upon reducing processes to the microscale. [11] Particularly, viscous dissipation dominates mass transport whilst inertial effects may be generally neglected. With very little or no inertia, flow control is increased with disappearance of numerous flow instabilities making prominent other phenomena and their effects. [12,13] Further, non-Newtonian rheology of complex fluids in confinement opens so far unexplored fluid mechanics, for example when fluid characteristic timescales and flow dynamics

Journal of Physics: Conference Series **247** (2010) 012050 doi:10.1088/1742-6596/247/1/012050

become comparable. [14] In this study, we take advantage of this particular advantage offered by microfluidics to impose highly defined and controlled flow fields to investigate flow contributions to ordering surfactant systems under processing. Such mixtures are ubiquitous in personal care and food industries and their non-equilibrium behaviour under shear have been extensively investigated. [15] For surfactant mixtures containing surfactant, co-surfactant and water, an interesting reversible phase transition from lamellar phase to multilamellar phase with shear is observed. [16, 17] Typical shear flows imposed on these systems are generally pure shear or extensional shear. The range of applied shear may be extended with microfluidics and allow the decoupling of different components [9] and the understanding of their impacts on the flow. Microfluidics brings an innovative approach to applying a designed and controlled flow while simultaneously analysing its effects on the microstructures of a system of interest.

In this article, we present a novel approach combining Small-Angle X-ray Scattering (SAXS) measurements online in a microfluidic chip which is illustrated by the results obtained for a complex fluid microflow containing a surfactant, co-surfactant and water. A few other microfluidic SAXS experiments have already been explored with different materials such as polycarbonates, polyimides, [18], which scatter very little X-rays. The importance of the chosen material for fabricating microdevices, which will be used with X-rays is reviewed by Greaves and Manz. [19] The micro-fabrication technique presented in this paper offers many advantages: a rapid prototyping fabrication technique, the use of materials chosen according to the studied flow and their scattering behaviour with X-rays, and the flexibility of measuring at any desired location on the entire chip with minimal beam damage.

2. Experimental

2.1. Microfluidic device fabrication

The microfluidic chips were custom made for SAXS measurements using a modified soft lithography method with polydimethylsiloxane (PDMS) and glass. Both materials were chosen for their compliance with soft lithography, replication fidelity, low cost, chemical compatibility with the studied systems, and relatively low X-ray scattering properties. The soft lithography methodology employed is schematised in Figure 1.

The microchannels are designed using a CAD program and the negative is printed on acetate sheets at high resolution of 20000 dpi (CAD-Art, California) to create a photomask. Photoresist SU-8 2025 (Microchem) is spin coated with desired thickness for the microchannel height on a silicon wafer ([100] single side polish, Compart Technology). The photomask is then used to polymerise the photoresist and form the microchannels on the silicon wafer using a dose of 380 mJcm^{-2} UV light. The unpolymerised photoresist is washed away. A thin film of approximately 1 to 2 mm of PDMS monomer mixed with curing agent is cast on the silicon wafer. Separately, a block of 1cm thick PDMS is made in the shape of the window that will cover the majority of microchannels. The block is placed in a 40kHz oxygen plasma chamber (Diener) at a power of 60W for 5 minutes twice with turning the slab in between so to glassify all its surface. Once the surface of the PDMS block transformed into glass, the block is positioned on the silicon wafer on top of the PDMS layer where the microchannels are located. A second thicker layer of PDMS and curing agent is poured around the glassified PDMS block to provide mechanical strength. Once the PDMS cured, the central block is lifted to reveal a window on top of the microchannels where the SAXS measurements will be performed. The microfluidic chip is cut out and separated from the silicon wafer. Holes are punched where the silicon tubings will be fixed within the thicker PDMS walls and the PDMS stamp is irreversibly sealed to a glass coverslip. A schematic of ready to use microfluidic chip is shown in Figure 2.

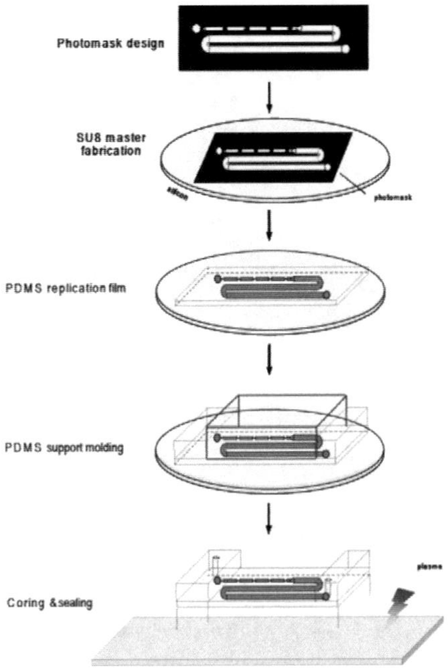

Figure 1. Soft lithography methodology employed for SAXS microfluidic chip fabrication: Photomask design and print; SU8 master fabrication; microdevice replication in PDMS thin film; PDMS structural reinforcement; coring for inlet/outlet ports and irreversible oxygen plasma sealing.

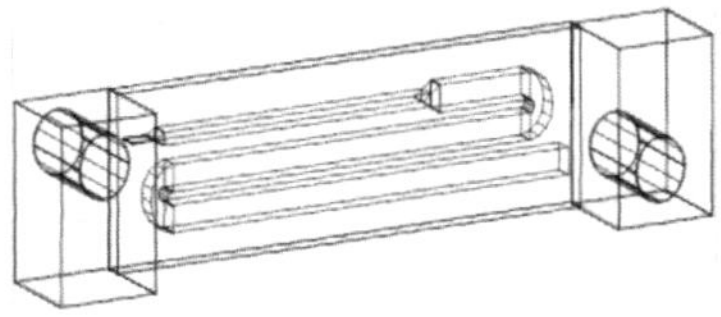

Figure 2. Wireframe design of an illustrative tubular micro-SAXS chip.

The channel sizes were approximately 70 microns in height, 16 cm in length and with widths varying from 100 to 2000 μm. Figure 3 depicts a profilometry of the microchannels used for the SAXS measurements. Three-dimensional surface maps were obtained using vertical scanning interferometry on a Wyko NT9100 Optical Profiling System (cf. Figure 3).

2.2. Surfactant systems: CTAC and SDS

Two model surfactant systems were investigated. One system consists of a mixture of 17.4 wt% of hexadecyl trimethyl ammonium chloride (CTAC: $C_{16}H_{33}N(CH_3)_3{}^+Cl^-$), 20 wt% of 1-pentanol ($C_5H_{11}OH$) and 62.6 wt% of water. The CTAC/Pentanol/Water system is compared to another surfactant system made up of 6.5 wt% sodium dodecyl sulfate ($C_{12}H_{25}NaO_4S$), 7.9 wt% 1-octanol ($C_8H_{17}OH$) in brine (20g/L NaCl in water). The phase behaviour of SDS/Octanol/Brine was reported earlier by Roux and co-workers. [17,20,21] The thermodynamics of the CTAC system is reported elsewhere. [22] Both systems were chosen for their lamellar equilibrium phase forming multi-lamellar vesicles under shear. In both cases, the surfactant is dissolved into the aqueous phase and then the co-surfactant is added to the solution and vigorously mixed by shaking the sample for approximately one minute.

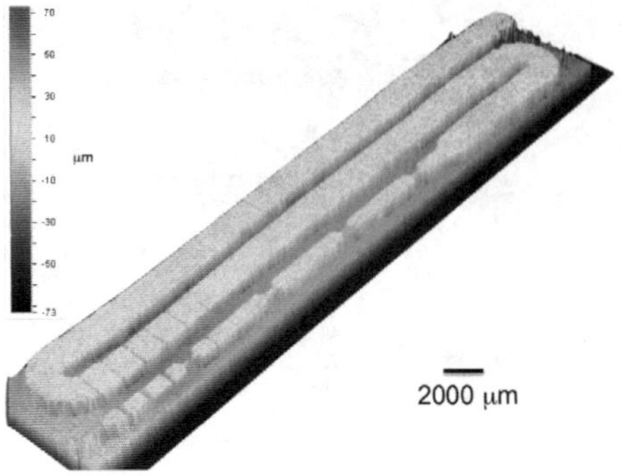

Figure 3. Profilometer scan of a representative contraction/expansion microchip.

2.3. SAXS measurements and microdevice setup

Synchrotron SAXS measurements were carried out at the I22 beamline at Diamond Light Source, Oxfordshire, U.K with an X-ray beam of λ=1.4Å, energy of approximately 8.9keV and beam cross-section of 150x100 μm^2. The microdevice comprises PDMS and glass, chemically compatible with the systems under flow. To minimise background scattering, the glass and PDMS thicknesses were kept to a minimum: a glass microscope coverslip (1x3 $inch^2$) of thickness 150 μm was used as the base material, imparting mechanical strength to the microdevice. X-ray scattering of the PDMS as a function of thickness is depicted in Figure 4 and defined an acceptable range of elastomer thickness, namely that it should be kept below 1.5mm; a cm-thick layer of PDMS reinforces the microdevice outside the scattering region (cf. Figures 1 and 2).

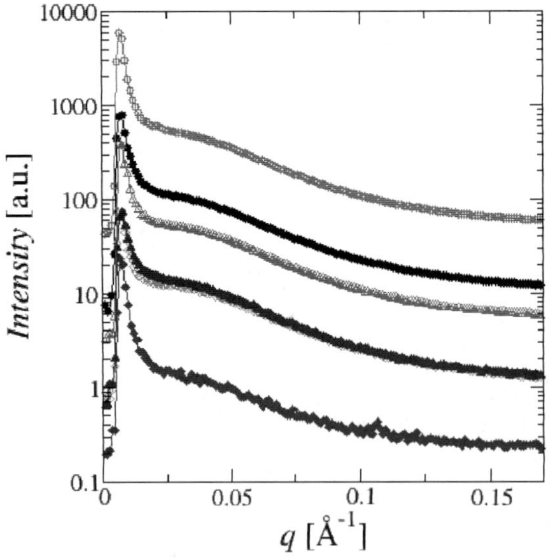

Figure 4. Radially averaged SAXS profile of PDMS microchips with varying PDMS thicknesses ranging from approximately 0.9 mm (top) to 1.8 mm (bottom) and constant glass coverslip thickness of 150 μm; indicating strong attenuation with increasing PDMS thickness. Device thickness is kept around 1 mm near the microchannels and background subtraction is carried out by measuring spectra adjacent, but outside, the channel to minimise uncertainty.

The scattering range of interest is important in the choice of appropriate background materials, as illustrated in Figure 5c where two different systems, CTAC and SDS surfactants systems, were analysed with SAXS in the PDMS/glass microchips. The characteristic scattering halo of PDMS is located at a q-range between 0.01 and 0.1Å^{-1}. In the case of our chosen flow with the CTAC surfactant system, its scattering peak is located at the end of the scattering q-range of the PDMS with q = 0.1065Å^{-1} and is therefore easier to pick up the Bragg peak for the system under flow. Whereas with the SDS surfactant system, its characteristic scattering peak is located at a rather low scattering angle at approx. 0.04Å^{-1} and overlaps with the scattering region of PDMS. Although a careful background subtraction reveals the scattering pattern of the SDS system under flow, results presented in this paper will concentrate on the CTAC system with stronger and more distinguishable SAXS measurements.

The microchip was firmly clamped on a upright stand mounted on a programmable XYZ stage and connected to a Harvard PHD 2200 syringe pump controlled by LabVIEW. The microchip holder was leveled on the XYZ stage and the sample-detector distance (defined as the z-coordinate) remained unchanged throughout the measurements. Locating the exact position of the X-ray beam can be technically challenging when performing SAXS measurements on a microdevice positioned on a XYZ stage. By placing small markers of lead tape beside various reference locations around the network of microchannels and by scanning the chip in both the x and y directions to detect the lead tape markers, the exact coordinates of the microchannels were determined.

3. Results and discussion

3.1. Microdevice background scattering and optimisation

Figure 5a depicts the 1D-SAXS of the two surfactant systems at equilibrium. The scattering peak for the SDS system is located at lower q (q = 0.0409Å^{-1}) than the CTAC scattering peak (q = 0.1065Å^{-1}). Both systems form lamellae at these concentrations and the scattering peaks thus correspond to the interlamellar distances, respectively 153.7$\AA$ for the SDS/Octanol/brine and 59Åfor the CTAC/Pentanol/Water. Upon shear (Figure 5b), the interlamellar distances decrease by typically 0 to 5Å. The scattering patterns of both systems in the microfluidic chip is shown in Figure 5c. The background scattering contribution of the microfluidic chip is also graphically represented. As observed, the scattering peak of CTAC/Pentanol/Water is clearly visible despite the background contribution since its location is at higher q than the PDMS hallow. In contrast, the scattering peak of the SDS/Octanol/Brine mixture under flow is masked due to the microfluidic chip scattering. Therefore, a thin PDMS/glass microfluidic chip appears to be of good choice for SAXS measurements of a CTAC/Pentanol/Water flow, while a more optimal choice of materials, or careful background subtraction, would be necessary to perform online SAXS measurements on SDS/Octanol/Brine system.

3.2. Beam damage and reference locations

Another important aspect to consider for SAXS measurements is the impact and X-ray beam damage on microchips. The X-ray beam cross-section employed was 150x100 μm and measurement acquisition times ranged from 10s to several minutes cumulatively. Since the measured system is under flow, the actual exposure time of the specimen is determined by the flow velocity and is typically 0.01s. Figure 6 shows how at the same beam location on the microfluidic chip, the scattering intensity is not affected by multiple acquisitions of 10s and Figure 7 is an optical micrograph of a microfluidic chip post-SAXS measurements highlighting how the PDMS is minimally marked by the X-ray beam, which turns out to be an advantage in terms of precisely locating on the chip measurements made by the X-rays.

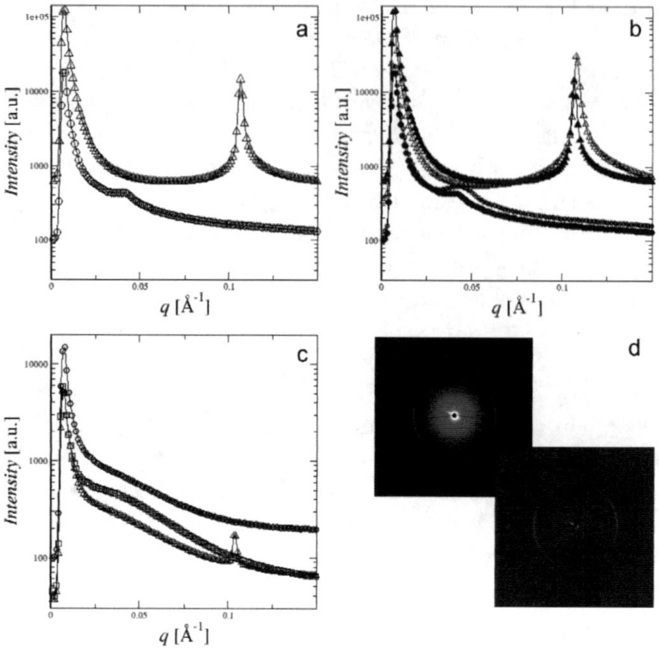

Figure 5. (a): 1D-SAXS graph of quiescent SDS/Octanol/aq. NaCl (○) and quiescent CTAC/Pentanol/Water (△); (b): 1D-SAXS graph of quiescent (•) and sheared (○) SDS/Octanol/aq. NaCl, and quiescent (▲) and sheared (△) CTAC/Pentanol/Water; (c): 1D-SAXS of microflows of SDS/Octanol/aq. NaCl (○) and CTAC/Pentanol/Water (△) and microfluidic chip background only (□). (d) 2D-SAXS pattern for a CTAC microflow specimen before (top) and after (bottom) background subtraction.

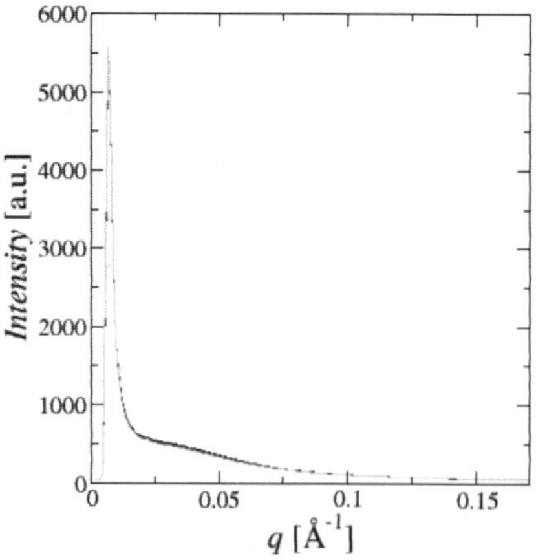

Figure 6. Radially averaged SAXS profile of three sequential measurements (black, red and green consecutively and each of 10s acquisition time) acquired for PDMS microchip at the same location showing minimal SAXS beam damage.

3.3. Surfactant mixtures under periodic microfluidic flow

We present two examples of online microfluidic SAXS measurements. First, the flow profile of the CTAC/Pentanol/Water system across the width of a simple microfluidic channel is illustrated

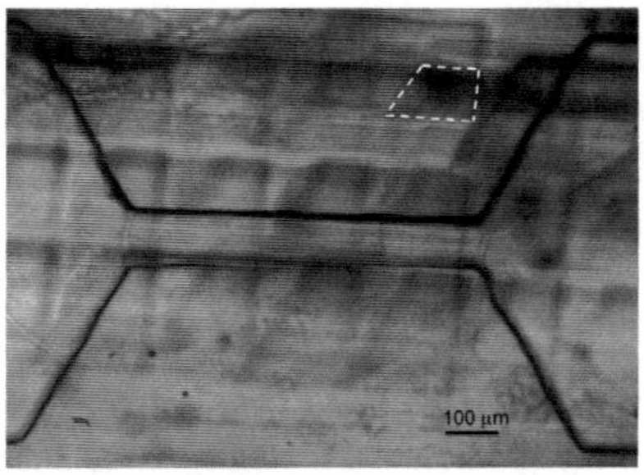

Figure 7. Optical micrograph of a contraction junction in a microfluidic chip after X-ray exposure. Slight beam damage marks are visible (example of a beam mark highlighted with a white dash line) and utilised to assist indexing in X-ray spatial maps.

in Figure 8 at different locations down the microfluidic chip. The flow rate in this example is of 1 ml/h corresponding to an average flow velocity of approximately 7 mm/s. These three flow profiles at sequential locations down a microchannel characterise the microstructural alignment under flow. For the first flow profile at x_a, the lamellar structures in the CTAC system are orientated with the flow near the channel walls and reach an orientation orthogonal to the flow in the middle of the channel. For the following two flow profiles at x_b and x_c, the microstructures relax in time through an unpatterned wide microchannel and become generally aligned with the flow. This orientation flip can be rationalised as follows: the first flow profile was acquired immediately after a microchannel width expansion from 1000 μm to 2000 μm causing a flow deceleration which in turn forces the lamellae to rotate perpendicular to the flow at the centreline. The microstructures then relax back to flow alignment as shown in the following two flow profiles.

Next, we report on the influence of a series of periodic contraction/expansion flows on the CTAC surfactant mixture, shown in Figure 9. The surfactant system was first flowed through a series of four constrictions from widths of 1000 to 100 μm and length of 700 μm with a flow rate of 5 mL/h (Figure 9a). The effects of the extensional flow field are seen in both orientation and size of the structures present. The lamellae sheets change orientation after the constriction to align themselves perpendicularly to the flow at the centreline. The 2D-SAXS patterns clearly show this orientation flip in Figure 9b in accord with the observation of Figure 8. The change in orientation is analysed by plotting the integrated scattering intensity at different azimuthal angles (Figure 9c). For structures aligned with the flow, the integrated scattering intensity is maximum at azimuthal angles of 90 and 270 °, while for those aligned perpendicularly to the flow, peaks are expected at 0 and 180 °. Along the channel width, shifts in the azimuthal angles of the intensity peaks are clearly observed: close to the walls lamellae are aligned with the flow and the maximum intensity is thus found at 90 and 270 ° associated with wall shear; in the flow centreline, the intensity maxima sharpens at azimuthal angles of 0 and 180 ° corresponding to highly oriented lamellae structures perpendicularly aligned to the flow. As for the effects on the interlamellar spacing, Figures 9d and 9e show that the extensional flow field had no immediate effect as the scattering peak location remains constant within measurement uncertainty in q across the width of the channel. Figure 9f plots the relative intensity of the scattering peak across the channel width. The structural order therefore depends on channel position and is evidently associated to the parabolic velocity profile that occurs in pipe flow. Upon a series of constrictions followed by relaxation (not shown here), the rheological properties of the CTAC system and the lamellar spacing are significantly altered, depending on the microchannel geometry. The combination of synchrotron SAXS and microflow processing thus permits continuous monitoring of structural changes and the impact of a well-defined flow fields on non-equilibrium phase behaviour of complex mixtures.

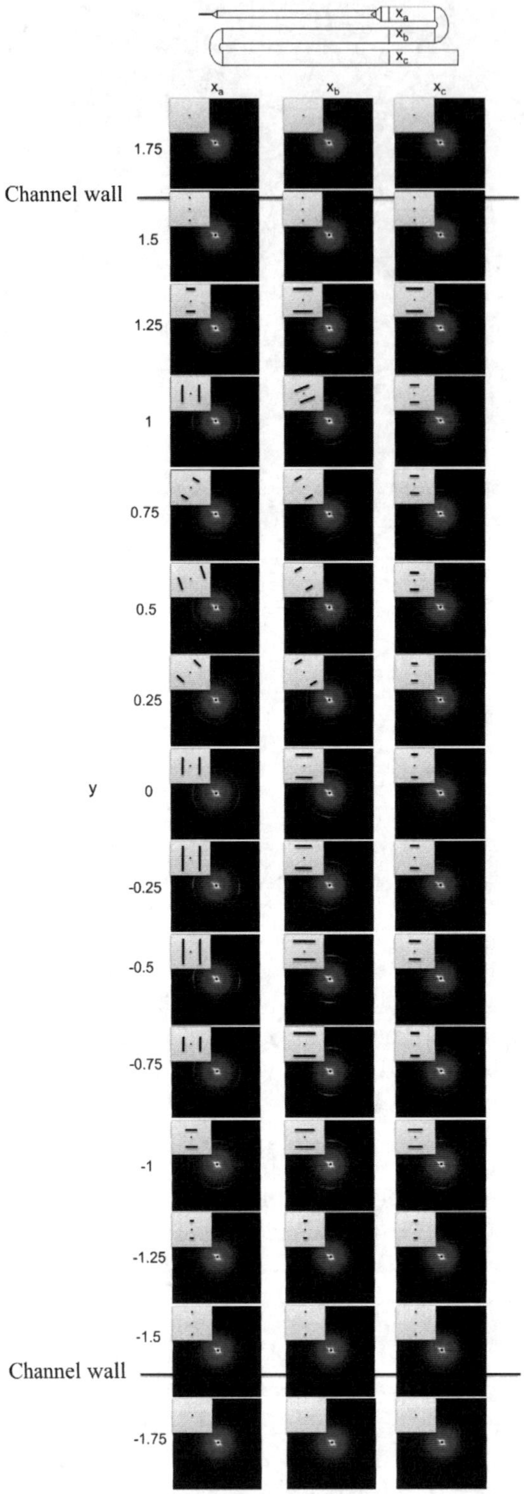

Figure 8. 2D-SAXS profiles of CTAC/Pentanol/Water across a microchannel of width 2000 μm at different locations indicated a, b, and c along the microfluidic channel at a flow rate of 1 ml/h corresponding to residence times of 9, 26 and 73s. Coordinates on the left axis indicate position orthogonal to flow.Top left schematic in each 2D-SAXS profiles relate to the orientation and intensity of the scattering pattern.

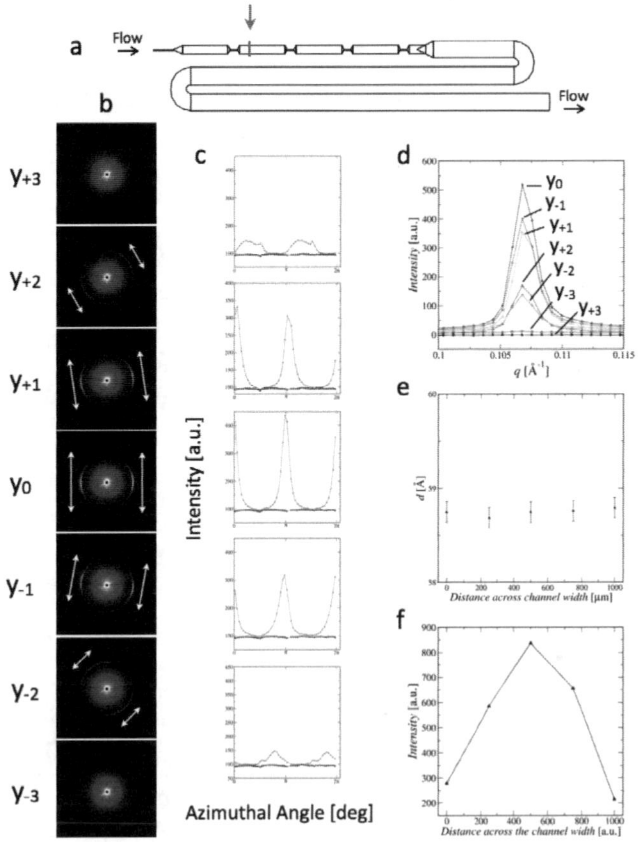

Figure 9. (a) microchannel design for a series of four 100 μm wide constrictions; (b) 2D-SAXS flow profiles of CTAC/Pentanol/Water across microchannel width (1000 μm) located after the first constriction, subject to a flow rate of 5 ml/h. y_0 indicates the centre of the channel and on either sides y_1, y_2, y_{-1}, and y_{-2} annotate the locations where the SAXS measurement was acquired across the width of the channel on, each spaced out by 250 μm. White arrows indicate the orientation of the lamella sheets and are proportional to the scattering intensity; (c) azimuthal scattering intensity at the structural peak; (d) 1D-SAXS patterns corresponding to the locations indicated in panel (b); (e) lamellar spacing as a function of the position across the channel; (f) intensity of the structural peak as a function of the position across the channel.

4. Concluding remarks

Online microfluidic SAXS measurements on model surfactant mixtures were successfully carried out in custom made microchips. These measurements were possible after a careful choice of materials and methodology for microchip fabrication for the systems of interest. A modified rapid prototyping approach enabled fast, effective and inexpensive fabrication of chips. The SAXS results illustrate the insight obtainable for surfactant microstructures under continuous flow: size, orientation and order of lamellar sheets were monitored under straight and contraction/expansion flows. Online microfluidic SAXS measurements elucidate the impact of specific flow fields on surfactant systems, providing guidance for industrial processes and further understanding of fundamental behaviour of such complex systems under external fields.

Acknowledgements The authors acknowledge EPSRC and Unilever for financial support, CASE/CNA05/63. We thank the Diamond Light Source (Oxfordshire) for beamtime and Claire Pizzey (DLS) and Marc Malfois (DLS) for assistance. We are grateful to George Wang for help with profilometry.

References

[1] Reyes D R, Iossifidis D, Auroux P A and Manz A 2002 *Analytical Chemistry* **74** 2623–2636
[2] Auroux P A, Iossifidis D, Reyes D R and Manz A 2002 *Analytical Chemistry* **74** 2637–2652
[3] Vilkner T, Janasek D and Manz A 2004 *Analytical Chemistry* **76** 3373–3385
[4] Cygan Z T, Cabral J T, Beers K L and Amis E J 2005 *Langmuir* **21** 3629–3634
[5] Wu T, Mei Y, Cabral J T, Xu C and Beers K L 2004 *Journal of the American Chemical Society* **126** 9880–9881
[6] Valera F, Cabral J T, Blackmond D G and et al 2010 (in press) *Angewandte Chemie International Edition*
[7] Hansen C L, Skordalakes E, Berger J M and Quake S R 2002 *Proceedings of the National Academy of Sciences of the United States of America* **99** 16531–16536
[8] Li L, Mustafi D, Fu Q, Tereshko V, Chen D L L, Tice J D and Ismagilov R F 2006 *Proceedings of the National Academy of Sciences of the United States of America* **103** 19243–19248
[9] Hudson S D, Phelan F R, Handler M D, Cabral J T, Migler K B and Amis E J 2004 *Applied Physics Letters* **85** 335–337
[10] Cabral J T and Hudson S D 2006 *Lab on a Chip* **6** 427–436
[11] Squires T M and Quake S R 2005 *Reviews of Modern Physics* **77** 977–1026
[12] Stroock A D, Dertinger S K W, Ajdari A, Mezic I, Stone H A and Whitesides G M 2002 *Science* **295** 647–651
[13] Stone H A, Stroock A D and Ajdari A 2004 *Annual Review of Fluid Mechanics* **36** 381–411
[14] Rodd L E, Cooper-White J J, Boger D V and McKinley G H 2007 *Journal of Non-Newtonian Fluid Mechanics* **143** 170–191
[15] Richtering W 2001 *Current Opinion In Colloid & Interface Science* **6** 446–450
[16] Partal P, Kowalski A J, Machin D, Kiratzis N, Berni M G and Lawrence C J 2001 *Langmuir* **17** 1331–1337
[17] Sierro P and Roux D 1997 *Physical Review Letters* **78** 1496–1499
[18] Barrett R, Faucon M, Lopez J, Cristobal G, Destremaut F, Dodge A, Guillot P, Laval P, Masselon C and Salmon J 2006 *Lab On a Chip* **6** 494–499
[19] Greaves E and Manz A 2005 *Lab On a Chip* **5** 382–391
[20] Herve P, Roux D, Bellocq A M, Nallet F and Gulik-Krzywicki T 1993 *Journal de Physique II* **3** 1255–1270
[21] Panizza P, Colin A, Coulon C and Roux D 1998 *Eur. Phys. J. B* **4** 65–74
[22] Martin H P 2010 *PhD Thesis Imperial College London*

New insights of the nucleation and growth process of gold nanoparticles via in situ coupling of SAXS and XANES

Jörg Polte[a], Ralph Kraehnert[b], Martin Radtke[a], Uwe Reinholz[a], Heinrich Riesemeier[a], Andreas F. Thünemann[a], Franziska Emmerling[a,*]

a) BAM Ferderal Institute for Materials Research and Testing, Richard-Willstätter-Str. 11, D-12489, Germany. b) Technische Universität Berlin, Technische Chemie, Straße des 17. Juni 124, D-10623 Berlin, Germany. Department of Chemistry.

franziska.emmerling@bam.de

Abstract. Although metallic nanoparticles play an important role in the area of nanotechnology, a coherent mechanistic explanation for the evolution of the particles during their chemical synthesis has not yet been provided in many cases. To gain a profound understanding of the growth mechanism of colloidal nanoparticles, new approaches using Small Angle X-Ray Scattering (SAXS) combined with X-ray absorption near-edge structure (XANES) are presented. This combination allows for insights into two prominent syntheses routes of gold nanoparticles (GNP): The "slow" reaction using sodium citrate (30-90 min) as a reducing agent and the "fast" reaction employing $NaBH_4$ (within few seconds). In the first case data derived with the coupled XANES and SAXS suggests a four-step particle formation mechanism. For the second system a time resolution in the order of 100-200 ms was achieved by coupling a common laboratory SAXS instrument with a microstructured mixer, which allows data acquisition in a continuous-flow mode. The results indicate a coalescence driven growth process. Based on the capabilities to deduce the size, number and polydispersity of the particles, the results of both methods enable the development of mechanistic schemes explaining the different phases of particle formation and growth, thus providing a basis for improved control over the synthesis processes.

1. Introduction

Gold nanoparticles (GNP) have attracted much attention due to their unique properties and promising applications covering the areas of biotechnology [1], catalysis [2] and optoelectronics [3]. Despite their widespread use, in many cases, a deeper understanding of the underlying formation processes is missing. Hence the size and shape control of nanoparticles often remained somehow "alchemistical". Therefore, a great deal of attention has been focused on understanding the process of nanoparticle formation [4-7]. When such phenomena are fundamentally understood, it becomes possible to tune the size and shape of nanoparticles and consequently improve their performance. This could potentially lead to the development of novel "tuned" materials. However, the realization of such designed nanoparticles is still difficult, since critical stages throughout the formation process are still not sufficiently understood.

In this context time resolved *in situ* experimentation represents the most promising approach by which the key steps of nanoparticle formation can be revealed. Currently, many studies applying various analytical techniques to study this processes have been reported in literature [8-14]. However, these studies mostly focus on data obtained using imaging techniques and optical spectroscopy, which often presents an incomplete picture of the particle synthesis. New insights, necessary for the interpretation of the underlying processes can be obtained by using SAXS, which offers an integral information about size, shape and polydispersity of particles in solution [15]. Additional secondary information derived i.e. from spectroscopic techniques then enables comprehensive studies. Employing such techniques simultaneously and in-situ ensures that data obtained with different analytical tools can be directly related to another. In consequence, corresponding time-resolved measurement provide access to process kinetics and the identification of possible intermediates.

The typical time resolution achieved in SAXS experiment amounts to several seconds using synchrotron radiation, and several minutes with a common X-ray tube. Studies of fast synthesis procedures often require a time resolution in the second or millisecond range, which is typically realized as a "stopped flow experiment" [14]. Unfortunately, for such stopped-flow experiments the use of synchrotron radiation, and in many cases a specially equipped beamlines, is essential [16].

In this contribution we present novel approaches of SAXS for detailed studies of slow and fast nucleation and growth processes of gold nanoparticles in solution phase. The studies provide a profound understanding of the mechanism of particle formation, which is essential for tuning their size and morphology and thus their properties. The possibilities of obtaining complementary information using SAXS in combination with XANES (small angle x-ray-scattering/x-ray near edge structure) are discussed, and illustrated for case studies. Whereas synchrotron SAXS experiments can be performed within several seconds, the time needed for a XANES measurement with sufficient quality was about 2-3 minutes and determined the time resolution of the presented SAXS/XANES coupling [17-18]. This technique was used to investigate the most prominent GNP synthesis - the reduction of $HAuCl_4$ with sodium citrate [19]. For the fast reduction of $HAuCl_4$ with $NaBH_4$ we demonstrate how a continuous-flow setup is able to achieve a sufficient time-resolution (200 ms) with a common laboratory SAXS instrument.

2. Experimental section

2.1. Small angle X-ray scattering (SAXS)

Small angle X-ray scattering is a common method to determine the shape and size of particles in colloidal solutions. The measured scattering signal originates from the contrast between the (growing) nanoparticles and the solvent. The scattering vector q is defined in terms of the scattering angle 2θ and the wavelength λ of the radiation: thus $q = \frac{4\pi}{\lambda}\sin(\theta/2)$. A detailed overview on SAXS data evaluation procedure using model fit functions is given by Pedersen [20]. The curve fitting was carried out using SANS Analysis 3_v3.00 implemented in the software suite IGOR [21]. For the nanoparticle synthesis discussed in the following the experimental scattering curves could be fitted with a model which describes the scattered intensity of hard spheres having a Schultz-Zimm distribution.

The Schultz-Zimm distribution is given by

$$f(r) = (z+1)^{z+1} x^2 \frac{\exp[-(z+1)x]}{R_{Avg}\Gamma(z+1)} \quad (1)$$

where R_{Avg} is the mean radius, $x = r/R_{Avg}$, z is related to the polydispersity ($p = \sigma / R_{Avg}$) by $z = 1/p^2 - 1$m where σ^2 is the variance of the distribution.

The scattering intensity of non aggregated particles can be assumed to be proportional to the form factor of a single particle. Thus the scattering intensity of monodisperse hard spheres with volume V is given by

$$I(q,r) = scale \cdot P(qr) = scale\left[\frac{3V(\Delta\rho)(\sin(qr) - qr\cos(qr))}{(qr)^3}\right]^2 \qquad (2)$$

where $P(qr)$ is the form factor of a single hard sphere and $\Delta\rho$ the scattering length density.

In case of polydisperse spherical particles one has to sum the scattering intensities over all particle sizes weighted by their frequency or to integrate using a size distribution function, respectively. It is common to use the Schulz-Zimm distribution for polydisperse particles [22]. Hence the scattering intensity is given by

$$I(q) = scale\int_0^\infty f(r)\cdot P(qr)dr \qquad (3)$$

An analytical solution of that integral can be found in Kotlarchyk *et al.* [23].

In order to analyze the growth mechanism of nanoparticles the number of particles is an important part of information. This can be obtained by using the general relation of $I(q = 0)$ for a single particle which is independent of its shape and size, i.e. $I_s = (\Delta\rho)^2 V^2$. Thus the scattered intensity $I(q = 0)$ of polydisperse particles can be written as

$$I(q=0) = N\langle V^2\rangle(\Delta\rho)^2 \qquad (4)$$

where N is the number of particles and $\langle V^2\rangle$ the mean value of V^2. $I(q = 0)$ can not be measured directly due to the overlapping of the scattering intensity with the primary beam, and as such, $I(q = 0)$ is accessible via the extrapolation of $I(q)$ for $q \to 0$. Additionally the number of particles can be obtained.

2.2. X-ray absorption near edge spectroscopy (XANES).

XANES measurements were performed at the μSpot beamline of BESSY II [24] . The beam was monochromatized using the double-crystal monochromator (DCM) installed at the beamline. A torroidal mirror in front of the DCM suppresses the harmonics and focuses the beam. This constellation provides a precise energy adjustment with an energy resolution of about $\Delta E/E = 2*10^{-4}$, i.e. 2.4 eV at the energy of the Au L III edge (11.919 keV). The excitation energy was varied from 11889 eV to 11933 eV in steps of 1.2 eV, resulting in 20 recorded XANES-channels. The fluorescence radiation of the sample was detected at an angle of approximately 90° with respect to the X-ray beam. The X-ray fluorescence of the Au L_{III} line was detected with a silicon drift detector (KETEK, model AXAS) at a working distance of 30 mm. The measurement time per XANES scan was determined by the progress of the observed reaction and the needed statistics (relative error less than 1%). All measured XANES spectra were normalized to 0 at the lowest and 1 at the highest used excitation energy. XANES was applied to a gold foil and to a $HAuCl_4$ solution as reference for the final and initial oxidation state. The oxidation state of Au during the process of nanoparticle synthesis was modelled as superposition of the start (Au(III)) and the final state (Au(0)).

2.3. Micro-structured mixer

A micro-structured static caterpillar mixer CPMM R600 SO (IMM, Mainz, Germany) was used at a total liquid flow rate of 50 mL min^{-1} to continuously join and mix equal volumes of the two respective reactant solutions according to the split-and-recombine mixing principle. The principle of the mixing device is described elsewhere [25].

2.4. Preparation method

2.4.1. Citrate reduction

The GNP were synthesized at 75°C according to the procedure described by Turkevich *et al.* [19] in which the chemical reduction of the gold precursor $HAuCl_4$ occurs by dissolved trisodium citrate. The aqueous solutions containing 0.25 mmol/L and 2.5 mmol/L of gold precursor and citrate, respectively. Before mixing the solutions 35 mL aqueous (Millipore) solution of gold precursor (7 mg $HAuCl_4$ x $3H_2O$, Aldrich) and 35 mL aqueous solution of trisodium citrate (51.45 mg $Na_3C_6H_5O_7$ x 2 H_2O, Aldrich) were prepared and pre-heated to the reaction temperature (75°C). The synthesis was carried out under stirring in a flask immersed in a temperature-controlled water bath, adding the citrate solution to the gold solution.

2.4.2. $NaBH_4$ reduction.
The GNP synthesis by reduction of tetrachloroauric acid using sodium borohydride was adopted from a procedure published by Wagner *et al.* [26]. As for the citrate reducing the final gold concentration was 0.25 mmol/L. Briefly, 39.28 mg tetrachloroauric acid were dissolved in 200 mL water (Millipore) and 15.10 mg sodium borohydride were dissolved in 200 mL water. Solutions of $NaBH_4$, were prepared freshly for each experiment.

2.5. SAXS / XANES coupling using an ultrasonic levitator as a sample holder
The combined SAXS and XANES-analysis was carried out at the µSpot beamline at BESSY II (Berlin, Germany) using an ultrasonic levitator as a sample holder (further information see [27]). The focusing scheme of the beamline is designed to provide a divergence < 1 mrad (horizontally and vertically) and a beam diameter of 100 µm at a photon flux of 5×10^9 s^{-1} at a ring current of 100 mA. Data from both measurements were recorded continuously and in parallel, since the small changes of the wavelength required for the XANES measurement can be neglected in the evaluation of the SAXS data. The two dimensional SAXS data were converted into one dimensional diagrams of scattered intensities versus scattering vector q using the computer program FIT2D [28]. For each experimental data point a fresh sample was extracted from the reacting batch solution and injected in the ultrasonic levitator. The small sample droplet was exposed for ca. 130 s to the incoming x-ray beam.

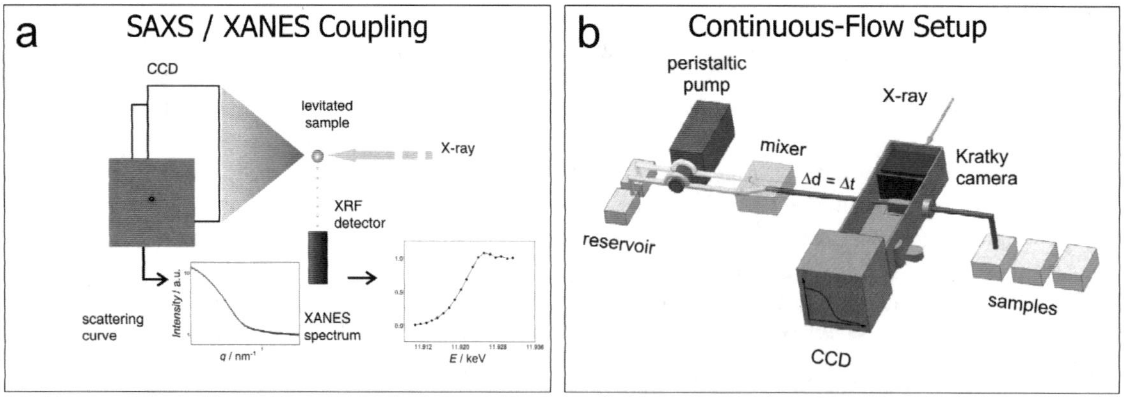

Figure 1 a) Scheme of the experimental setup employed for SAXS and XANES analysis on droplets of the reaction solution during GNP formation. b) Experimental setup for particle synthesis in continuous-flow mode coupling a micro-structured static mixer directly to SAXS-analysis in a flow cell.

2.6. Continuous flow setup (CFS) for ms-time resolution using a conventional X-ray tube
In Figure 1b the CFS using a micro mixer, peristaltic pump and a conventional SAXS instrument (SAXSess, Anton Paar, Graz, Austria) is illustrated. The instrument uses focusing multilayer optics in combination with a slit collimation to provide an intense monochromatic primary beam with low

background. The interconnection of the micro mixer to the SAXS instrument was established via a delay coil (Teflon tubing). The solutions were transported via a peristaltic pump (Ismatec Laboratoriumstechnik GmbH, Wertheim-Mondfeld, Germany) at a flow rate of 50 mL min^{-1} through a flow capillary (quartz, inner diameter of 1 mm and wall thickness of 10 μm) embedded in the SAXS instrument. The delay time between the micro-mixer and the SAXS instrument represents the variable reaction time and was adjusted by varying the length and diameter of the teflon tubing between the two devices which enables the SAXS analysis at different stages of the reaction. The range of the corresponding addressable reaction times ranges from 100 ms (19 cm tube length at 1 mm diameter) to about 136 s (10 m tube length at 3.8 mm diameter). For colloid samples with reaction times exceeding 140 s, liquid samples were collected at the mixer outlet and aged while stirring in a beaker. For the SAXS analysis of such samples the aged batch solution was directly injected into the flow cell. Tubing, flow capillary and glassware were cleaned consecutively with aqua regia, water and isopropanol prior to each experiment. The successful cleaning of the SAXS capillary was confirmed via recording a scattering curve of ultrapure deionized water.

3. Results and discussion

Figure 2 shows the evolution of the SAXS and XANES data recorded at different reaction times for (i) the method coupling applied to levitated sample droplets, and (ii) the continuous-flow setup. In case of the method coupling (i) the XANES spectra indicate an initially slow reduction process but a strong decrease in the Au(III) content after 50 minutes, which coincides with an increase in the intensity of the SAXS curves recorded in parallel. The corresponding data evaluation of SAXS and XANES (Figure 3a-c) allows the differentiation of three phases during the growth process. The first phase of the growth is characterized by coalescence processes, since the volume fraction of particles remains almost constant (about 15%) whereas the number of particles decreases significantly. As a result of this process the mean radius of the particles increases from 2 to 4 nm accompanied by a decrease of the polydispersity from 50% to approx. 20%. The observed coalescence-related increase in particle size corresponds to the XANES data, which indicate a slow reduction of the precursor (see Figure 3c). In a second phase between 25 and 50 min, the particles grow continuously to a mean radius of ca. 5 nm (Figure 3a). The number of particles remains almost constant and the polydispersity decreases to 14% (Figure 3a and b). This is characteristic for the so-called “focusing effect” or “growth by diffusion” model first described by Reiss *et al.* [29-30]. The final phase consists of a rapid consumption of the remaining Au(III) precursor (Figure 3c), which leads to an increase in the particle size and a further decrease of polydispersity to 10%. The final mean radius of the GNP is 7.6 nm. This last phase of an accelerated consumption of Au(III) at a GNP radius of about 5 nm can be attributed to a surface enhanced reduction. The deduced mechanism consisting of four steps is displayed in Figure 4a. After a fast nucleation (step 1), the resulting entities undergo coalescence processes (step 2) forming colloids with a mean radius of about 4 nm. This is followed by a diffusional growth to a mean radius of about 5 nm (step 3) and completed by an autocatalytic growth (step 4).

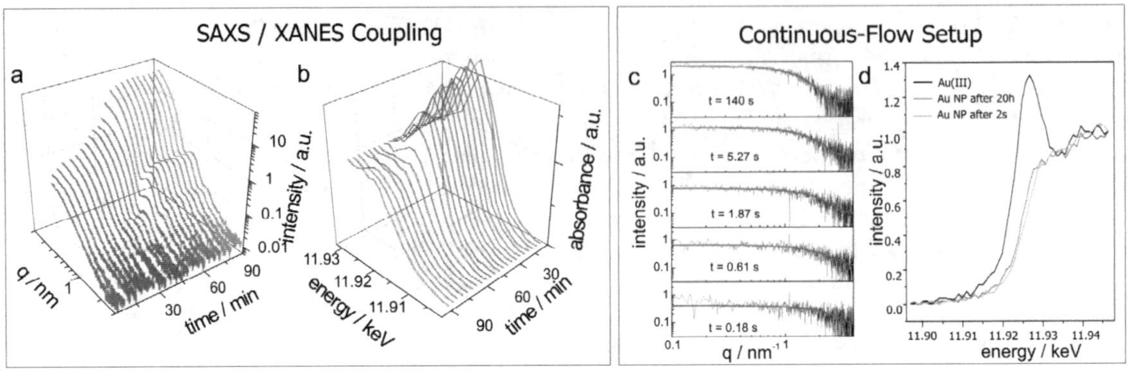

i ii

Figure 2 Data obtained form SAXS and XANES measurements under different experimental conditions. i) SAXS (a) and XANES (b) data as a function of time obtained during the reduction of $AuCl_4^-$ ions with sodium citrate. Red curves belong to the first phase of the experiment; green curves belong to the final phase of the experiment. The SAXS data indicate the formation of spherical nanoparticles with a narrow size distribution. The XANES data show, that the precursor material is reduced throughout the process. ii) SAXS (c) and XANES (d) data obtained during the reduction of $AuCl_4^-$ ions using sodium borohydride. The SAXS curves derived directly after the start of the experiment show the formation and growth of spherical particles. The XANES data illustrate that the reduction is completed after 2 s. Figures reproduced with permission from [17-18].

The rapid formation of GNP by reducing $AuCl_4^-$ with sodium borohydride was followed by SAXS using a continuous flow setup. The SAXS curves obtained directly after mixing the reactants indicate already the presence of particles. XANES measurements under similar conditions reveal the completed reduction of the precursor at this stage of the synthesis (Figure 2 c and d). Even after completed reduction, evaluation of the scattering curves show the growth of spherical nanoparticles starting from a mean radius of about 0.8 nm to 1.5 nm within 10 s, whereas the volume fraction as well as the polydispersity remain constant throughout the growth process. Subsequently, the particles grow within one hour up to a size of about 1.7 nm, showing up to 140 s no significant change in polydispersity(Figure 3d-f). (Note: An increase in polydispersity is experimentally observed after 140 s, which coincides with switching the observation method from in-situ SAXS in continuous flow to SAXS analysis on colloid samples aged under agitation in a separate beaker. The associated change is therefore probably due to the method of sample aging at times exceeding 140 s or the sample transfer, and not the original particle growth mechanism.) Since the volume fraction of particles remains constant throughout the time-scale of the experiment, and reduction of the gold precursor completes within the first 100 ms, the subsequent formation of particles is only due to coalescence processes. Hence, the growth process starts with rapid reduction followed by a nucleation process of gold clusters (step 1) which undergo coalescence processes until the final size is reached (step 2), which is defined by a yet unknown stabilization (see Figure 4 b). This stabilization is most probably of electrostatic origin.

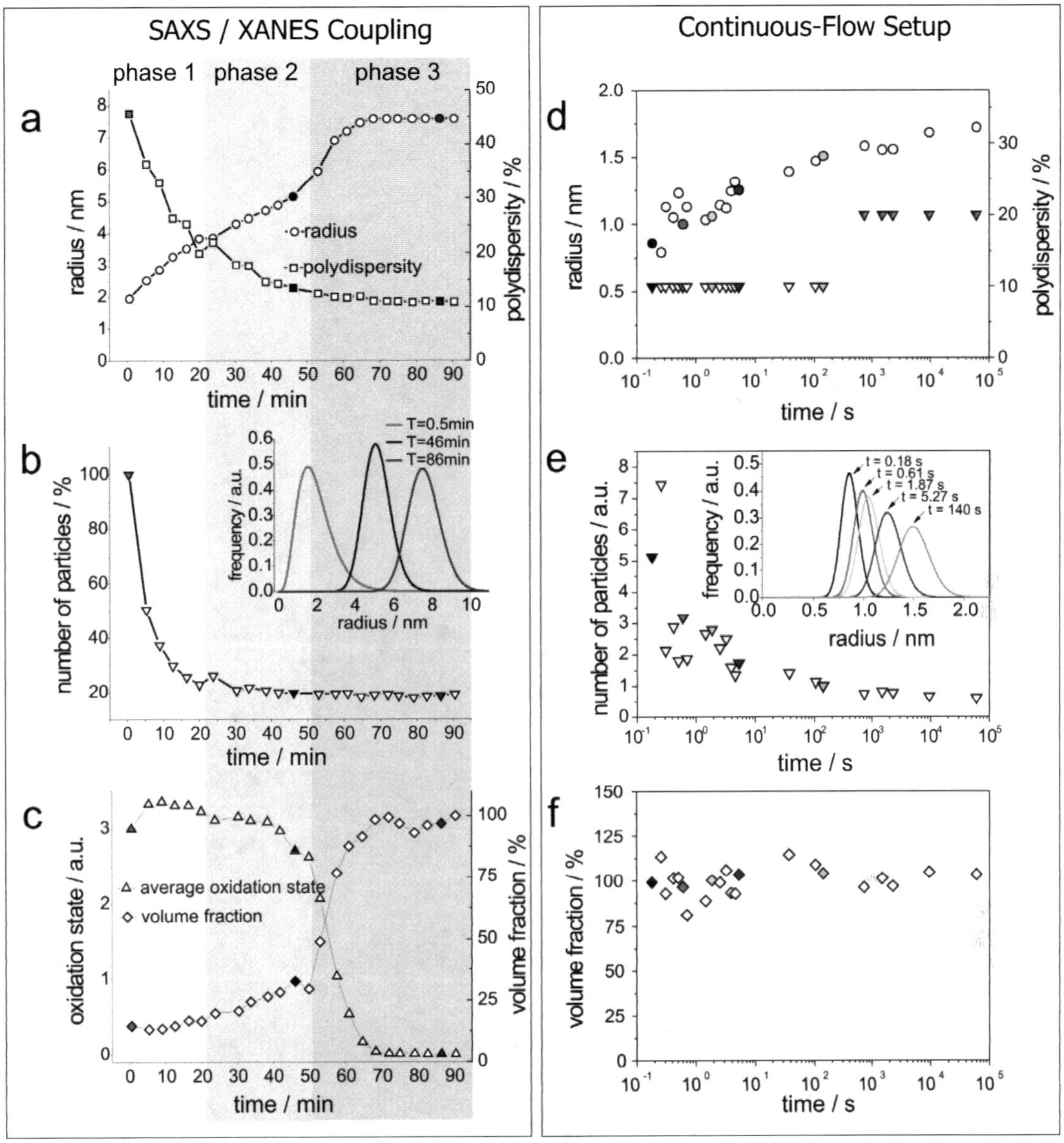

Figure 3 Results of the evaluated SAXS and XANES data plotted vs. reaction time (for the continuous-flow setup in log scale) derived from scattering curves recorded during the synthesis of gold nanoparticles. a) Mean sphere radius and normalized number of particles plotted vs. reaction time derived from SAXS. b) Polydispersity of particles as a function of time and corresponding particle size distribution of the Schulz-Zimm distribution exemplarily shown for three reaction times (inset). c) Average formal oxidation state of gold derived from normalized XANES spectra plotted vs. reaction time; volume fraction of particles calculated from SAXS data. d) Evolution of radius (bullets) and polydispersity (triangles) of gold nanoparticles throughout the reaction process. The values were derived applying a solid sphere model with a Schultz-Zimm distribution whereas the polydispersity was constrained to values of 10% (aging in continuous flow) and 20% (further aging in a stirred beaker). e) The number of gold nanoparticles derived from the fit (normalized to the last measurement). Inset: Schultz-Zimm distribution of particle size at different reaction times. f) Volume fraction which represents the relative total volume of all gold nanoparticles calculated from the evaluated SAXS data. Figures reproduced with permission from [17-18].

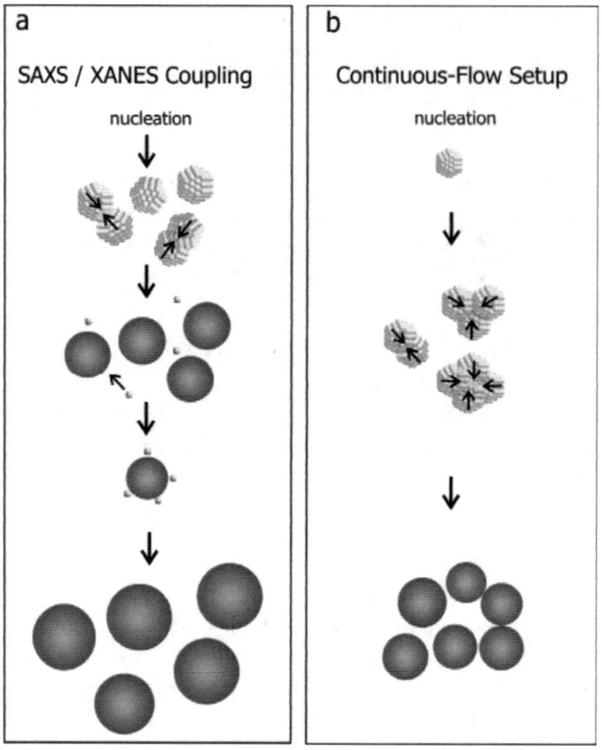

Figure 4 Schematic illustration for the process of gold nanoparticle formation deduced for (a) employing sodium citrate as reducing agent and (b) using $NaBH_4$.

4. Conclusions

The presented case studies for the synthesis of colloidal gold nanoparticles via different chemical routes shows that the developed methods for analyzing nanoparticles with SAXS in a time-resolved manner provide valuable information about the underlying growth mechanism. XANES analysis carried out in parallel provided the required complementary chemical information regarding the rates of precursor consumption. The presented experimental setups are suitable for both, rapid synthesis processes requiring a time resolution in the millisecond range, as well as reaction occurring within minutes to hours. In particular, application of the methods is not limited to the studied specific systems of gold nanoaparticle, but applicable to a wide range of other colloid synthesis reactions.

References

[1] Storhoff J J, Elghanian R, Mucic R C, Mirkin C A and Letsinger R L 1998 One-pot colorimetric differentiation of polynucleotides with single base imperfections using gold nanoparticle probes *Journal of the American Chemical Society* **120** 1959-64
[2] Bond G C and Thompson D T 2006 Status of catalysis by gold following an AURICAT Workshop *Applied Catalysis a-General* **302** 1-4

[3] Treguer M, Rocco F, Lelong G, Le Nestour A, Cardinal T, Maali A and Lounis B 2005 Fluorescent silver oligomeric clusters and colloidal particles *Solid State Sciences* **7** 812-8

[4] Pong B K, Elim H I, Chong J X, Ji W, Trout B L and Lee J Y 2007 New insights on the nanoparticle growth mechanism in the citrate reduction of Gold(III) salt: Formation of the au nanowire intermediate and its nonlinear optical properties *Journal of Physical Chemistry C* **111** 6281-7

[5] Park J, Joo J, Kwon S G, Jang Y and Hyeon T 2007 Synthesis of monodisperse spherical nanocrystals *Angewandte Chemie-International Edition* **46** 4630-60

[6] Angermund K, Buhl M, Dinjus E, Endruschat U, Gassner F, Haubold H G, Hormes J, Kohl G, Mauschick F T, Modrow H, Mortel R, Mynott R, Tesche B, Vad T, Waldofner N and Bonnemann H 2002 Nanoscopic Pt colloids in the "embryonic state" *Angewandte Chemie-International Edition* **41** 4041-4

[7] Watzky M A and Finke R G 1997 Transition metal nanocluster formation kinetic and mechanistic studies. A new mechanism when hydrogen is the reductant: Slow, continuous nucleation and fast autocatalytic surface growth *Journal of the American Chemical Society* **119** 10382-400

[8] Thunemann A F, Kegel J, Polte J and Emmerling F 2008 Superparamagnetic maghemite nanorods: Analysis by coupling field-flow fractionation and small-angle X-ray scattering *Analytical Chemistry* **80** 5905-11

[9] Teranishi T and Miyake M 1998 Size control of palladium nanoparticles and their crystal structures *Chemistry of Materials* **10** 594-600

[10] Daniells S T, Overweg A R, Makkee M and Moulijn J A 2005 The mechanism of low-temperature CO oxidation with Au/Fe2O3 catalysts: a combined Mossbauer, FT-IR, and TAP reactor study *Journal of Catalysis* **230** 52-65

[11] Meneau F, Cristol S, Sankar G, Dolbnya I P, Bras W, Catlow C R A, Thomas J M and Greaves G N 2003 In situ study of the formation of CdS nanoparticles by small-angle X-ray scattering *Journal of Applied Crystallography* **36** 718-21

[12] Meneses C T, Flores W H and Sasaki J M 2007 Direct observation of the formation of nanoparticles by in situ time-resolved X-ray absorption spectroscopy *Chemistry of Materials* **19** 1024-7

[13] Tanaka H, Koizumi S, Hashimoto T, Itoh H, Satoh M, Naka K and Chujo Y 2007 Combined in situ and time-resolved SANS and SAXS studies of chemical reactions at specific sites and self-assembling processes of reaction products: Reduction of palladium ions in self-assembled polyamidoamine dendrimers as a template *Macromolecules* **40** 4327-37

[14] Abecassis B, Testard F, Spalla O and Barboux P 2007 Probing in situ the nucleation and growth of gold nanoparticles by small-angle x-ray scattering *Nano Letters* **7** 1723-7

[15] Fratzl P 2003 Small-angle scattering in materials science - a short review of applications in alloys, ceramics and composite materials *Journal of Applied Crystallography* **36** 397-404

[16] Narayanan T, Diat O and Bosecke P 2001 SAXS and USAXS on the high brilliance beamline at the ESRF *Nuclear Instruments & Methods in Physics Research Section a-Accelerators Spectrometers Detectors and Associated Equipment* **467** 1005-9

[17] Polte J, Erler R, Thunemann A F, Sokolov S, Ahner T T, Rademann K, Emmerling F and Kraehnert R 2010 Nucleation and Growth of Gold Nanoparticles Studied via in situ Small Angle X-ray Scattering at Millisecond Time Resolution *Acs Nano* **4** 1076-

82
[18] Polte J, Ahner T T, Delissen F, Sokolov S, Emmerling F, Thunemann A F and Kraehnert R 2010 Mechanism of gold nanoparticle formation in the classical citrate synthesis method derived from coupled in situ XANES and SAXS evaluation *J Am Chem Soc* **132** 1296-301
[19] Enustun B V and Turkevich J 1963 Coagulation of Colloidal Gold *Journal of the American Chemical Society* **85** 3317-&
[20] Pedersen J S 1997 Analysis of small-angle scattering data from colloids and polymer solutions: modeling and least-squares fitting *Advances in Colloid and Interface Science* **70** 171-210
[21] Kline S R 2006 Reduction and analysis of SANS and USANS data using IGOR Pro *Journal of Applied Crystallography* **39** 895-900
[22] Kotlarchyk M, Stephens R B and Huang J S 1988 Study of Schultz Distribution to Model Polydispersity of Microemulsion Droplets *Journal of Physical Chemistry* **92** 1533-8
[23] Kotlarchyk M and Chen S H 1983 Analysis of Small-Angle Neutron-Scattering Spectra from Polydisperse Interacting Colloids *Journal of Chemical Physics* **79** 2461-9
[24] Paris O, Li C H, Siegel S, Weseloh G, Emmerling F, Riesemeier H, Erko A and Fratzl P 2007 A new experimental station for simultaneous X-ray microbeam scanning for small- and wide-angle scattering and fluorescence at BESSY II *Journal of Applied Crystallography* **40** S466-S70
[25] Lob P, Lowe H and Hessel V 2004 Fluorinations, chlorinations and brominations of organic compounds in micro reactors *Journal of Fluorine Chemistry* **125** 1677-94
[26] Wagner J, Tshikhudo T R and Koehler J M 2008 Microfluidic generation of metal nanoparticles by borohydride reduction *Chemical Engineering Journal* **135** S104-S9
[27] Leiterer J, Delissen F, Emmerling F, Thunemann A F and Panne U 2008 Structure analysis using acoustically levitated droplets *Analytical and Bioanalytical Chemistry* **391** 1221-8
[28] Hammersley A P, Svensson S O, Hanfland M, Fitch A N and Hausermann D 1996 Two-dimensional detector software: From real detector to idealised image or two-theta scan *High Pressure Res.* **14** 235-48
[29] Reiss H 1952 The Statistical Mechanical Theory of Irreversible Condensation .1. *Journal of Chemical Physics* **20** 1216-27
[30] Peng X G, Wickham J and Alivisatos A P 1998 Kinetics of II-VI and III-V colloidal semiconductor nanocrystal growth: "Focusing" of size distributions *Journal of the American Chemical Society* **120** 5343-4

RheoSAXS studies of anisotropic complex fluids under shear

J P de Silva[1], D Petermann[1], B Kasmi[1], M Impéror-Clerc[1] *, P Davidson[1], B Pansu[1], F Meneau[2], J Perez[2], E Paineau[3], I Bihannic[3], L J Michot[3] and C Baravian[4]

[1] Laboratoire de Physique des Solides, UMR 8502 - Université Paris-Sud, F-91405 Orsay, France.
[2] SOLEIL Synchrotron, L'Orme des Merisiers Saint-Aubin, BP 48 91192 Gif-sur-Yvette, France.
[3] Laboratoire Environnement et Minéralurgie, Nancy University, CNRS, UMR 7569, BP 40, 54501 Vandoeuvre Lès Nancy Cedex, France
[4] Laboratoire d'Energétique et de Mécanique Théorique et Appliquée, Nancy University, CNRS, UMR 7563, 2, Avenue de la Forêt de Haye, BP 160, 54504 Vandoeuvre Lès Nancy Cedex, France

Email: imperor@lps.u-psud.fr

Abstract. We discuss the application of *in-situ* rheological small angle X-ray scattering experiments to the study of complex fluids under shear, implemented using custom Couette cylinder rheometers mounted on the SWING beamline of the SOLEIL Synchrotron. We discuss several applications of this technique to the study of phase transitions in nanoparticle doped liquid crystals and shear alignment of clay suspensions. The concurrent capture of rheological and scattering data provides vital information that relates macroscopic properties such as viscosity to the microstructure of the fluid.

1. Introduction

The study of complex fluids and of their rheological properties is an active area of research with a wide range of applications.[1] The bulk macroscopic properties of complex fluids, such as rheological properties such as shear viscosity, for example, will often depend on the structure of the fluid on a microscopic level. Thus it is often convenient to combine various techniques in order to study macroscopic and microscopic properties at the same time. Investigations using small angle scattering techniques (SAS) (light, neutrons, X-rays) can be performed on such fluids directly under flow; this allows the simultaneous recording of rheological quantities (such as flow-curves and viscosity profiles) and structural information. In the case of strongly anisotropic fluids, small angle X-ray or neutron scattering (SAXS/SANS) experiments are particularly well suited to following orientation phenomena of structured fluids under flow. Using new rheological small-angle (Rheo-SAXS) instruments developed at the SWING SAXS beamline of the SOLEIL synchrotron, we present some examples of results demonstrating in detail the behaviour under shear of doped surfactant lamellar phases and colloidal clay suspensions.

* Author to whom correspondence should be addressed

2. Rheo-SAXS setups

The first Rheo-SAS instruments[2-5] were developed in the 1990's and today several such instruments are available around the world, both at neutron and synchrotron facilities. Following the design of these previous instruments, a concentric cylinder Couette shear cell geometry has been chosen for the new instruments installed on the SWING beamline. Two different instruments are available: a custom built Couette cell and a commercial Anton Paar MCR 501 rheometer, both of which are directly installed on the beamline as illustrated in figure 1 below. Between these two instruments a great amount of flexibility exists for the user in terms of the kind of experiments that can be performed.

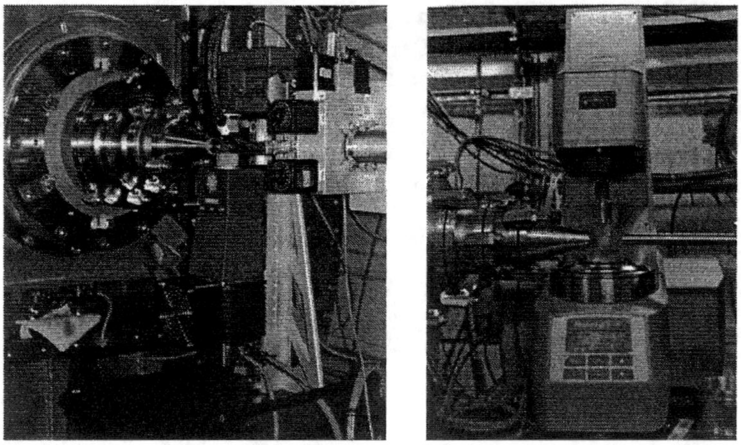

Figure 1. Custom Couette instrument (left) and the commercial MCR-501 rheometer (right) installed on the beamline. The nose of the detector is visible on the left in both images.

The mean stress $<\sigma>$ and shear rate $<\dot{\gamma}>$ in Couette cylinder geometry for a Newtonian fluid, assuming no shear-banding or wall slippage and to a linear approximation, are given by[6]

$$<\sigma>=\frac{(R_e^2+R_i^2)}{4\pi H R_i^2 R_e^2}M,$$

$$<\dot{\gamma}>=\frac{(R_e^2+R_i^2)}{(R_e^2-R_i^2)}\Omega,$$

where R_i and R_e are the interior and exterior Couette cylinder radii respectively, H is the cylinder height, M is the measured torque on the interior cylinder and Ω is the angular velocity. The shear viscosity η can now be calculated through the relation $<\sigma> = \eta <\dot{\gamma}>$. This calculation is valid if the contribution of the fluid located in between the bases of the two Couette cylinders is negligible. We find that, for example, if the bases of the exterior cylinder 'cup' and the interior cylinder are both conical, which may be useful for reducing sample volume, then a correction is applied to the viscosity calculation form of the shear viscosity again for a Newtonian fluid and is now given by

$$\eta=\frac{M}{4\pi\Omega}\left[\frac{HR_i^2R_e^2}{(R_e^2-R_i^2)}+\frac{R_i^4}{4.232(18-H)}\right]^{-1},$$

where all parameters are as previously defined. The custom shear device is fully integrated into SWING's control systems, while the Anton-Paar rheometer is controlled by the accompanying RheoPlus software.

Two types of measurements are made, in 'tangential' or 'radial' geometry. These positions correspond to the X-ray beam either parallel or perpendicular to the flow direction, as illustrated in figure 2. These two positions are realised by displacing the entire rheometer laterally with respect to the incoming X-ray beam. The combination of the scattering patterns of these two configurations make it possible to reconstruct in 3D the orientation distribution of anisotropic nanoparticles with respect to the shear flow directions (velocity direction and velocity gradient direction).

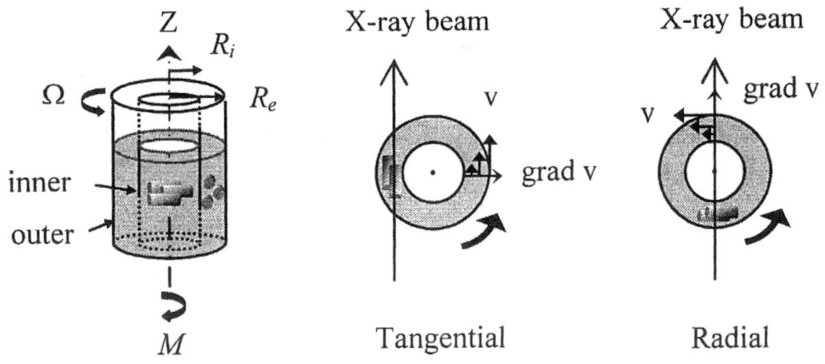

Figure 2. Schematic illustration of the concentric cylinder Couette cell in tangential and radial geometries.

2.1. Description of the instruments

In the case of both instruments (Table 1), identical concentric cylinder Couette cells are fabricated of polycarbonate because it is an inexpensive material and it is relatively easy to obtain small wall thicknesses (less than 0.3 mm in this case), its absorption coefficient for X-rays is small (the attenuation length at 12 keV is 3.8 cm^{-1}), and the scattering contribution in the small q-range is reasonably low. Measurements are then made at an energy value ranging from typically 10 to 12 keV.

The transmission $T_n(r)$ of the cell at some radial position r for any particular medium n is given by

$$T_n(r) = \exp\frac{-2\left(\sqrt{R_e^2 - r^2} - \sqrt{R_i^2 - r^2}\right)}{\mu_n},$$

for $r < R_i$, $\sqrt{R_i^2 - r^2} = 0$ for $R_e > r > R_i$, where μ_n is the absorption coefficient of the medium involved. When multiple media are involved, such as when the cell is filled, the compound transmission of all the included media can simply be taken. In order to account for the width of the X-ray beam in the horizontal plane (around 200 microns), the compound transmission function is convoluted with a Gaussian function where the FWHM is of 200 microns.

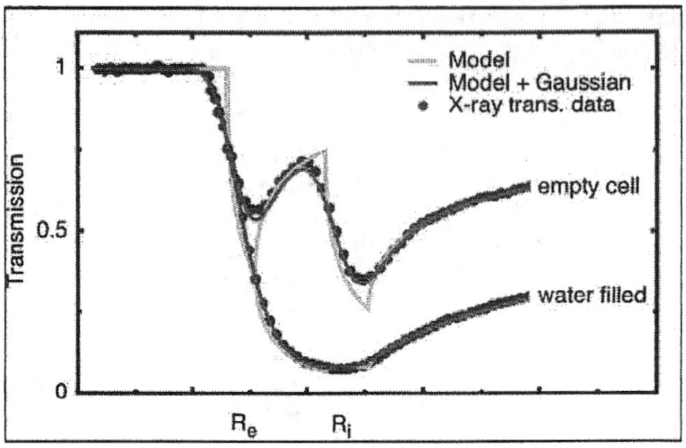

Figure 3. X-ray transmission data and model fits scanning the cell gap (0.5 mm) corresponding to the tangential position.

X-ray transmission data and model simulations are presented in figure 3, in this case scanning across the gap corresponding to the tangential geometry. Good agreement is found between the convoluted transmission model and the X-ray transmission data for the empty and water-filled polycarbonate cell. Absorption coefficients of polycarbonate and water at 12 keV are taken as 3.8 and 2.27 cm^{-1} respectively.

2.1.1. *Custom Couette apparatus.* A complete Couette shear cell apparatus where the lower outer cylinder cup is rotated by a precision stepper motor has been constructed. The speed of rotation is controlled (thus we impose the shear rate). The main characteristics are the possibility to impose large shear rate values, a range from 1 s^{-1} to 5000 s^{-1} is accessible with this setup, without any Taylor-Couette instabilities that develop only when the inner cylinder is rotated. The standard polycarbonate cells have an average radius of 10 mm, a height of 17 mm and gap values ($R_e - R_i$) of 0.5, 0.2 or 0.1 mm with a sample volume of around 1 ml. Smaller cells of average radius 3 mm and height 10 mm permit one to work with very small sample volumes (about 0.2 ml), which is particularly suited to the case of biological samples for example.[7] The measurement of the torque transferred by the sample to the inner cylinder is made using a force sensor at a distance of 20 mm from the rotation axis. The accessible torque range is from 10^{-5} to 2×10^{-2} Nm corresponding to a stress range from 1 to 2000 Pa. If needed, the cell is thermostated by water circulating within the fixed inner cylinder. Constant shear rates and oscillating stresses can be applied to the sample using this apparatus.

	Custom Couette	Commercial rheometer
Min torque (mN.m)	0.01	5×10^{-5}
Max torque (mN.m)	20	200
Min speed (rpm)	0.6	10^{-6}
Max speed (rpm)	960	3×10^{3}
Average radius (mm)	3/10	10
Gap size (mm)	0.2/0.1, 0.2 or 0.5	0.2 or 0.5
Sample volume (ml)	0.2/1	1

Table 1. Technical data for both instruments equipped with similar polycarbonate SAXS cells.

2.1.2. *Commercial rheometer.* A commercial rheometer, the Anton Paar MCR 501, has also been directly installed and integrated into the beamline. Identical polycarbonate cells to those described for the previous apparatus have been fabricated by Anton Paar for this rheometer (Table 1). The two main advantages compared to the custom setup are, firstly, that very low stress values are accessible (minimum torque value of 5×10^{-8} Nm), which is particularly important for dilute colloidal suspensions, and secondly, that a wider range of rheological measurements are possible, including imposed shear rate or stress, oscillotary stress mode, creep-recovery, etc. Because the inner cylinder is rotating rather than the outer one, Taylor-Couette instabilities along the neutral axis at high shear rates may be observed, which is not the case with the previous setup. Temperature control is made via thermostated air circulation around the cell.

2.2. Calibration procedure

The calibration of the cells for each instrument has been performed using Newtonian standards of known viscosity such as PDMS oils and hexadecane. The results compare favourably with calculations based purely on the physical dimensions of the cell. X-ray data conversion to absolute intensity can be performed in the radial geometry by subtracting the scattering of the empty cell from that of the same cell filled with water, as the sample thickness is equal to twice the gap size in this geometry. In the tangential position, transmission scans across the gap allows one to accurately find the gap centre and the tangential sample thickness, taking into account the width of the X-ray beam in the horizontal plane and the absorption coefficients of polycarbonate and sample (see previous description and figure 3). Making the assumption that along the neutral *Z*-axis the scattered intensity is the same in the radial and tangential geometries, then the radial and tangential intensities can be both scaled to absolute units.

3. Examples of anisotropic complex fluids

We will now give two examples (doped lamellar phases and clay suspensions) of some of the first results obtained using these RheoSAXS instruments in order to illustrate the type of structural information that they provide. Both examples concern very anisotropic complex fluids, which undergo orientational changes and/or structural transitions under shear. Experiments were performed at 11 keV and the distance between the sample and the CCD detector was 1.4 m for the lamellar phases and 6 m for the clay suspensions.

3.1. Doped lamellar phases

Concerning the rheological properties of surfactant lamellar phases, a wide range of properties may be observed, including shear-alignment and thinning of the lamellar phases and shear-induced transitions from lamellae to multilamellar vesicles (MLVs) that have been already observed in a wide range of systems using a variety of optical and particle scattering techniques.[1] Our aim is to determine how the addition of inorganic nanoparticles to such a lamellar phase will influence and modify the rheological behaviour.

3.1.1. *Doping with spherical nanoparticles.* Small spherical polyoxometallate (POM) inorganic nanoparticles derived from phosphotungstic acid (diameter 1.1 nm) can be incorporated into the liquid crystalline lamellar phase of Brij30/water, where Brij30 is a non-ionic surfactant mainly composed of $C_{12}E_4$.[8] It is observed in this ternary system that the POMs are preferentially located close to the EO groups of the surfactant membrane rather than remaining dispersed within water between the bilayers, while the inter-lamellar spacing remains unmodified by the presence of the particles.[8]

In pure systems under shear, previous studies of $C_{12}E_4$/water[9] and similar $C_{10}E_3$/water[10] lamellar phases have characterised the shear viscosity and a structural transition from the lamellar state to a multi lamellar vesicle (MLV) or 'onion' phase. On increasing the POMs doping concentration, this

Journal of Physics: Conference Series **247** (2010) 012052 doi:10.1088/1742-6596/247/1/012052

shear-induced transition is offset to higher shear rates (for applied shear rates less than 50 s^{-1}). The SAXS signature of the MLV state can be seen in Figure 4.

For a shear rate of 5 s^{-1}, three resolution-limited Bragg peaks due to the lamellar stacking are recorded. The intensity modulations, with the second order peak weaker than the first one, are related to the spatial distribution of the POMs nanoparticles, that are preferentially located close to the EO groups of the surfactant membrane.[8] In the tangential geometry, the intensity of the Bragg peaks is reinforced in the horizontal direction, indicating a preferred alignment of the lamellar planes under shear parallel to the Couette cell walls. In the small q-range, still in the tangential geometry, the parasitic anisotropic contribution of the polycarbonate cell walls can be seen.

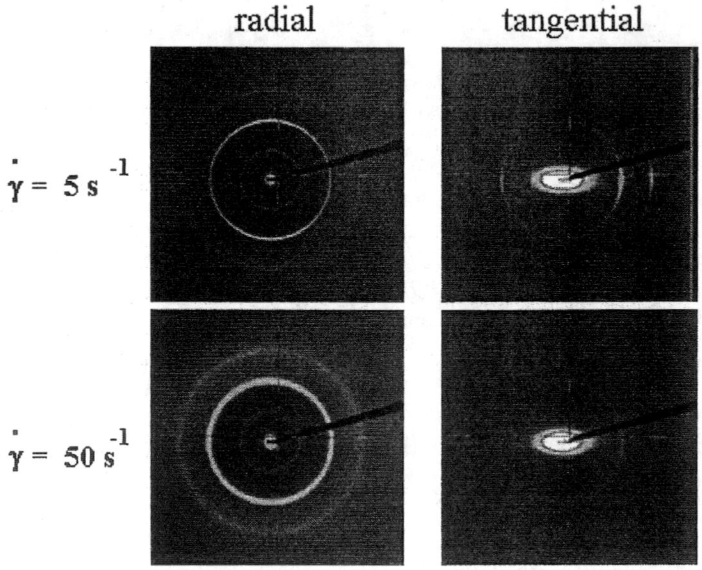

Figure 4. SAXS patterns under shear of the lamellar phase of Brij30 (40 % vol. fraction) doped with POMs nanoparticles (4% vol. fraction in water).

At a shear rate of 50 s^{-1}, the sample is in the MLV state, as can be followed by the viscosity evolution during the RheoSAXS experiment (figure 5). Both tangential and radial patterns give rise to completely isotropic signals. Because the intensity repartition of the peaks is unchanged, one can deduce that the POMs remain encapsulated within the onion membranes, otherwise, the first order Bragg peak will be the most intense, as in the case of the pure surfactant phase. Moreover, a broadening of the Bragg peaks is clearly detected.[12] From the FWHM, an order of magnitude of the coherence length of the lamellar order of about 0.5 micrometers is deduced. Such an effect would not be detected using SANS[11] because it requires sufficiently high resolution for the scattering q-vector. As derived from optical measurements under shear (see figure 5),[12] a typical size for the MLVs is around 10 microns in radius, corresponding to about 1000 lamellar layers. This means that one MLV is formed of several coherent lamellar domains separated by defects rather than a perfect stacking of spherical curved multilayers. The RheoSAXS experiments also allow us to assign the change in the viscosity to the structural transition towards the MLV state.

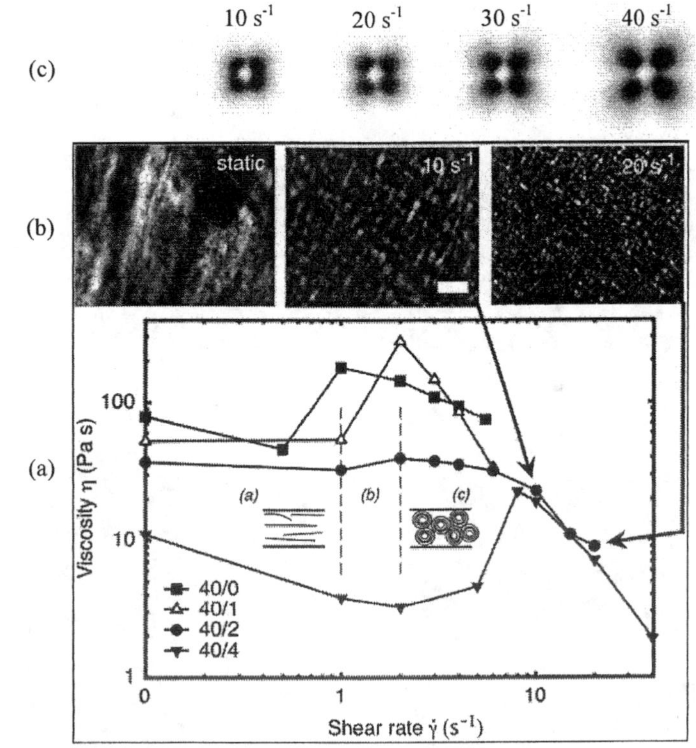

Figure 5. (a) Rheological curves for different doping concentrations (40% surfactant and 0-4% POMs vol. fractions) recorded during the RheoSAXS experiments. The lamellar—MLV transition under shear is confirmed by the optical texture in polarized light (b) and by the characteristic four-lobes SALS patterns (c).[12]

3.1.2. Confinement of anisotropic nanoparticles. Lamellar phases of a non-ionic surfactant doped with anisotropic rod-like goethite (FeOOH iron oxide) particles were studied with shear rates up to 5000 s^{-1}; the data suggest that the goethite particles embedded within the lamellar phase become strongly aligned at much lower shear rates than for a pure goethite solution alone. More precisely we observe that the goethite particles contained within the lamellar phase show strong alignment at a shear rate of 100 s^{-1}, while a pure goethite phase exhibits no alignment at this shear rate; in order to achieve a comparable degree of alignment in the goethite alone, a much higher shear rate of around 5000 s^{-1} is required. This result proves that strong coupling between the host lamellar phase and the goethite nanoparticles is evident.

3.2. Suspensions of disk-shaped clay particles

Suspensions of natural clays exhibit a wide range of rheological properties. In the case of beidellite swelling clays, the colloidal particles are very anisotropic, and can be approximated by a plate-like shape of average diameter 210 nm (polydispersity 40%) and thickness of 0.64 nm.

On increasing the concentration in water, this clay forms a birefringent nematic gel phase.[13] Under shear flow, for shear rates larger than 100 s^{-1}, the 2D SAXS patterns (figure 6) in the tangential geometry exhibit a strong anisotropy that is more and more pronounced upon increasing the shear rate up to 2000 s^{-1}. At the same time, in the radial geometry (figure 6), almost no anisotropy is observed.

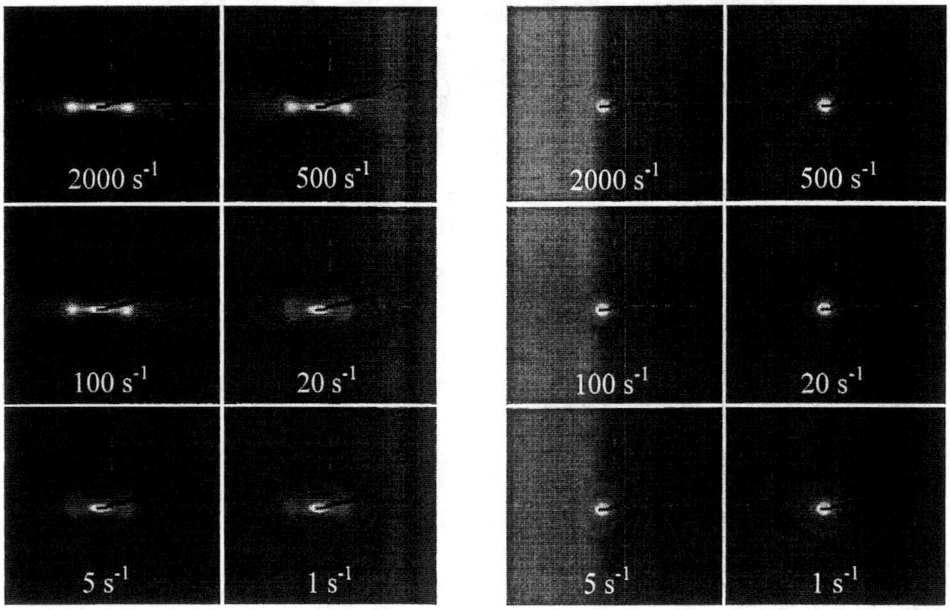

Figure 6. Evolution of the SAXS pattern of a beidellite clay suspension (2.2 % vol. fraction in water) with shear rate in tangential geometry (grad *v*, *Z*) (left) and radial geometry (*v*, *Z*) (right).

By combining the radial and tangential patterns, the complete 3D-orientation distribution for the normal of the plate-like particles can be obtained.[14] As the anisotropy of the particles is very large, one can approximate the scattering of an individual particle (form factor contribution only) by an infinite line parallel to the normal to this particle. Because the scattering signal is confined at high shear rates along the velocity gradient direction (horizontal direction in the tangential geometry), the main orientation of the platelets under flow is the one with the normal parallel to the velocity gradient (orientation (1) in figure 7), as other orientations (2) and (3) are much less favoured.

The anisotropy of the SAXS pattern reveals the shear alignment of the particles. Moreover, even for small particle volume fractions (a few percent), strong modulations of the scattering associated with a structure factor contribution are observed like in Figure 6. This is in agreement with a stabilisation of this colloidal suspension by long-range electrostatic interactions.[13]

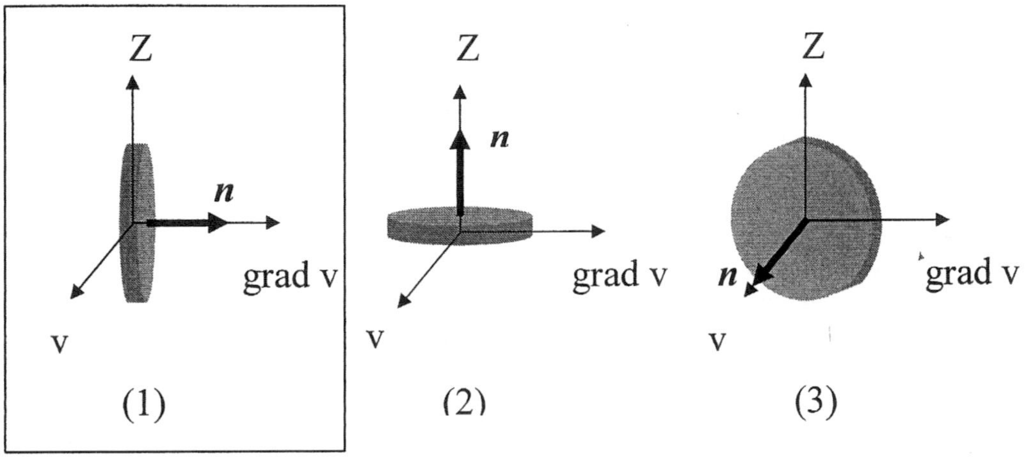

Figure 7. Orientation of the plate-like particles with respect to the shear flow. Orientation (1) is predominant.

4. Conclusions

Rheo-SAXS instruments in Couette shear cell geometry have been successfully implemented on the SWING beamline of the SOLEIL synchrotron. Such apparatus permits the simultaneous recording of rheological properties and structural information for strongly anisotropic complex fluids. This non-invasive technique is therefore an extremely useful aid to properly building the link between the orientational and microscopic structure of these materials and their macroscopic behaviour under flow. Two examples (doped lamellar phases and clay suspensions) of the first results obtained are given and illustrate the type of structural information that these instruments provide.

Acknowledgments

JPdS thanks `Triangle de la Physique' for a postdoctoral fellowship and financial support from the French ANR (N° ANR-07-BLANC-0194) is acknowledged. We are indebted to Anton Paar France for all their assistance concerning the MCR rheometer.

References

[1] Larson R G 1998 *The Structure and Rheology of Complex Fluids* (Oxford University Press)
[2] Safinya C R, Sirota E B, Plano R J, 1991 *Phys. Rev. Lett.* **66** 1986
[3] Lindner P 1992 *Physica B* 180-181 499
[4] Pople J A, Hamley I W, Diakun G P 1998 *Review of scientific instruments* **69** 3015
[5] Camerel F, Gabriel J C P, Batail P, Panine P, Davidson P 2003 *Langmuir* **19** 10028
[6] Salmon J B, 2003 PhD Thesis n°2696, Université de Bordeaux I
[7] Bellomo E G, Davidson P, Impéror-Clerc M, Deming T, 2004, *J. Am. Chem. Soc.* **126** 9101
[8] Poulos A S, Constantin D, Davidson P, Impéror M, Pansu B, Panine P, Nicole L, Sanchez C 2008 *Langmuir* **24** 6285
[9] Müller S, Börschig C, Gronski W, Schmidt C, Roux D 1999 *Langmuir* **15** 7558
[10] Oliviero C, Coppola L, Gianferri R, Nicotera I, Olsson U, 2003 *Colloids and Surfaces A* **228** 85
[11] Nettesheim F, Zipfel J, Olsson U, Renth F, Lindner P, Richtering W, 2003 Langmuir **19** 3603-3618
[12] de Silva J P, Poulos A S, Pansu B, Davidson P, Constantin D, Kasmi B, Petermann D, Asnacios S, Meneau F, Impéror M, *submitted to EPJ E*
[13] Paineau E, Antonova K, Baravian C, Bihannic I, Davidson P, Dozov I, Impéror-Clerc M, Levitz P, Madsen A, Meneau F, Michot L J, 2009 *J. Phys. Chem B* **113** 15858
[14] Bihannic I, Baravian C, Duval J F L, Paineau E, Meneau F, Levitz P, de Silva J P, Davidson P, Michot L

AUTHOR INDEX

AUTHOR INDEX

AUTHOR INDEX